# North American Cornucopia

## TOP 100

### Indigenous Food Plants

# North American Cornucopia

## TOP 100 Indigenous Food Plants

Ernest Small

CRC Press
Taylor & Francis Group
Boca Raton London New York

CRC Press is an imprint of the
Taylor & Francis Group, an **informa** business

CRC Press
Taylor & Francis Group
6000 Broken Sound Parkway NW, Suite 300
Boca Raton, FL 33487-2742



Printed on acid-free paper
Version Date: 20130815

International Standard Book Number-13: 978-1-4665-8592-8 (Hardback)



**Visit the Taylor & Francis Web site at**
**http://www.taylorandfrancis.com**

**and the CRC Press Web site at**
**http://www.crcpress.com**

# Dedication

This book is dedicated to North America's Indigenous Peoples. Most knowledge of the continent's indigenous food plants is based on millennia of observations and experimentation by North America's Native Inhabitants. They have endowed the world with an enduring legacy of wisdom, upon which this volume is ultimately based.

Of course, both sexes have contributed to advancing human usage of food plants, but the lioness has been responsible for the lion's share of progress. It is difficult to overestimate the historical importance of indigenous women in advancing agriculture. As the gatherers of wild foods, as planters and caretakers of crops (as shown in the illustration below), and as preparers of meals, women were responsible for most progress in selecting superior forms of edible plants, inventing

**FIGURE 1** A 2009 United States golden dollar honoring Native American agriculture. The coin shows a woman planting seeds of corn, beans and squash, the three most important traditionally planted crops of various Native North Americans. The combination is known as the "Three Sisters," a mixture that produces excellent nutrition. Beans are rich sources of protein, particularly the essential amino acid lysine, but are usually deficient in the sulfur-containing amino acids methionine and cystine. By contrast, cereal grains like corn contain lower amount of proteins, and these are deficient in lysine but adequate in sulfur-containing amino acids. Together, beans and corn provide a balance of essential protein components, which is especially necessary for survival in diets deficient in meat. Squash is a good source of complex carbohydrates, fiber, vitamins, minerals, and other needed dietary substances. The three crops benefit by being grown closely together in a technique called "companion planting." The corn provides a framework for the beans to climb upon, eliminating the need for poles. Nitrogen-fixing bacteria in nodules on the roots of the bean plants provide nitrogen that fertilizes the soil. The squash spreads on the ground, shading out weeds and creating a microclimate to retain moisture in the soil. The prickly hairs of the squash vines also deter pests.

technologies for increasing crop yield, and transforming often barely palatable foods into marvels of taste. It is no exaggeration to say that indigenous women were the force that ultimately has provided mankind with an abundance of food. Agriculture was the key driver over the last 10,000 years that produced our modern civilization, and so indigenous women in North America (and indeed everywhere) need to be recognized for their indispensable culinary contributions to modern society.

# Contents

# Preface

The subject of food plants is paradoxical. The North American food industry now offers more than 300,000 food items, and an average supermarket often stocks more than 30,000, the great majority derived from plants. This astonishing numerical choice, however, is misleading: less than a dozen plants—such as wheat, rice, corn, and potato—furnish the bulk of almost everything that most humans currently eat. Moreover, fewer than 100 species individually generate a world annual value of at least a billion dollars. Paradoxically, North America, the leading economic center of the world, is the native home of only one of these economic superstars, the sunflower. Indeed, plants of foreign origin, particularly from Eurasia and South America, dominate North American agriculture. Still another paradox is the fact that, in the name of foreign aid, North Americans have dedicated huge resources to the development of new tropical and subtropical crops, largely ignoring the undeveloped botanical treasures at home. The final paradox worth noting concerns the hundreds of guides that have been written about the edible plants of North America: virtually all of these include, indeed highlight introduced, weedy foreign plants (consider "*Stalking the Wild Asparagus*"), leaving a quite misleading impression of the genuine native food resources of the continent.

Of over a thousand indigenous North American species that were consumed in past times, only a few dozen currently have appreciable commercial value. These and many others with promising prospects of commercial development are the subject of this book.

Since the marketplace has judged most indigenous edible plants of North America to be inferior to those imported from foreign lands, why dedicate a book to them? There are several urgent reasons. Current crops are inadequate to meet the food demands of the world's huge, growing population—a billion people now go to bed hungry, and it is questionable whether the more than 10 billion expected by 2100 can be fed. Many North American plants have characteristics that are especially promising for creating new varieties needed to expand food production. Moreover, today's staple food crops are so universally grown that they have become less profitable, and there are excellent prospects of generating new competitive crops from North American natives. Sadly, although indispensable, agriculture is the chief source of degradation of the environment, and the leading threat to the planet's life-supporting ecological systems. North American native plants can contribute to the generation of new crops that are more compatible with the ecology of the world than those now cultivated. Still another important reason for documenting the indigenous food crops of North America is their inherent health benefits; many are nutritionally invaluable and can contribute to reducing the tragic damage to health caused by modern diets. Until recently, the creation of genuinely new crops from wild species, not just new varieties of familiar staples, has been so demanding a task that very few such projects have been able to garner the necessary support and resources. Today, however, advances in biotechnology suggest that transforming wild plants will be much easier in the near future, and so this book's prioritization of the best prospects should be welcome.

North American indigenous food plants are important economically and agriculturally, but the subject is of far broader importance. As detailed in this book, food changed us from apes into humans, and food will determine our future. The eighteenth-century English poet Alexander Pope wrote "The proper study of mankind is man." Food is the most critical need of mankind, and of course everyone is interested in "what's for dinner?" Aside from their commercial and culinary importance, North American indigenous food plants are fascinating subjects of health concerns, enjoyment and conservation of the natural flora, gardening, history, and science. Toward the goal of disseminating the important information in this book to a wide audience, the presentation is user-friendly, concise, and well illustrated in the hope that nonspecialists will find the topics both informative and entertaining.

# Acknowledgments

Brenda Brookes expertly electronically enhanced and arranged most of the illustrations and maps. Her skill and patience in carrying out this very large and challenging task are greatly appreciated.

This book presents over 600 color illustrations of the plants featured in this book. Considerable effort went into choosing the best available illustrations from various sources, and the artists and photographers deserve high praise for their skill and, in most cases, for their kindness in allowing reproduction. Sources of all plant illustrations are given in their captions. These include drawings and paintings scanned for this book from old publications for which copyright has expired; photos downloaded from the Web that are either in the public domain or licensed for reuse; and copyrighted photos for which permission was obtained for reproduction in this book (the photographers are acknowledged in the captions). Information sources employed to prepare the distribution maps are detailed in Chapter 1.

Notations of the form "CC By 3.0" refer to Creative Commons licenses (version 3.0 in this example); similarly "Flickr/CC-attribution" refers to photos licensed for reproduction under the Creative Commons umbrella, with required attribution of the photographer, that appear on the Flickr organization's online photo database. In some cases, photographers use partial names (Sten, Sue), codes (Jeronimo2412, Magnolia 1000), or phrases (ilovebutter, Amy Love Yah, fishermans-daughter); occasionally presentation of specific information has been required as a prerequisite for use of the illustration.

Several illustrations are by artists and photographers associated with the research facilities of my institute, Agriculture & Agri-Food Canada (AAFC), Ottawa. The following AAFC artists contributed the number of drawings or paintings in parenthesis: Barry Flahey (16), Brenda Brookes (12), and Susan Rigby (1). Photographs by AAFC staff included Brenda Brookes (13) and Eric Johnson (1). My friends Sharon and Gerry Channer contributed four photos.

Several of my professional colleagues have been important fountains of knowledge on the plants discussed in this book. Paul Catling co-authored dozens of publications with me on indigenous food plants of North America (these are cited in the appropriate chapters). Jacques Cayouette verified the eastern Canadian distribution of the maps. Stephen Darbyshire clarified many problems and contributed five photos.

This book cites over 3000 publications, most consulted through the courtesy of AAFC library staff. I particularly thank Lise Robillard for her patience and persistence in acquiring many of these.

I thank the staff of Taylor & Francis for competently and efficiently transforming the rather massive and complex manuscript into the final book.

The information presented here has been summarized from a vast literature. It has been said that plagiarism is copying the work of one worker, but scholarship is copying the work of many (or less accurately in the words of Wilson Mizner (1876–1933), "Copy from one, it's plagiarism; copy from two, it's research."). Certainly, a work of this nature owes a great debt to a very wide range of academic and practical writers, too numerous to acknowledge individually. The literature cited in each chapter and in the Appendix to this book identifies most of the major contributions to the accumulated knowledge of North America's edible plants.

# Author

**Dr. Ernest Small** earned a doctorate in plant evolution from the University of California at Los Angeles in 1969 and has since been employed with the Research Branch of Agriculture and Agri-Food Canada, where he presently holds the status of Principal Research Scientist. He is the author of 13 previous books, five of which received or were nominated for major awards. He has also authored over 300 scientific publications on plants. Dr. Small's career has included dozens of appearances as an expert botanical witness in court cases, acting as an adviser to national governments, presenting numerous invited university and professional association lectures, supervising postgraduate students at various universities, participating in international societies and committees, journal editing, and media interviews. Dr. Small is widely known for his work on hemp and marijuana, which has included development of a standard strain that has been the basis of all licensed medicinal marijuana in Canada for more than a decade, supplying over 100,000 patients. He has received several professional honors, including election as a Fellow of the Linnean Society of London; the G.M. Cooley Prize of the American Association of Plant Taxonomists for work on the marijuana plant; the Agcellence Award for distinguished contributions to agriculture; the George Lawson Medal, the most prestigious award of the Canadian Botanical Association, for lifetime contributions to botany; the Queen Elizabeth II Diamond Jubilee Medal for outstanding professional achievements; and the Lane Anderson Award, a $10,000 prize for science popularization, received for *Top 100 Food Plants,* a companion volume to this work. A second companion volume, *Top 100 Exotic Food Plants,* was a finalist for a Botanical and Horticultural Libraries' award.

# Executive Summary

This book is dedicated to the most important 100 indigenous food plants of North America north of Mexico that have achieved commercial success or have substantial market potential. Thirty-four of these are already domesticated, with advanced cultivated varieties (i.e., cultivars); some are major crops (e.g., pecan, squash, sunflower) and some are moderately successful (e.g., blackberries, plums, raspberries), although many are minor (e.g., dwarf cape gooseberry, groundnut, saskatoon). A few have cultivated forms that are little different from the wild species (e.g., hawthorn, mountain mint, winter purslane), and the remainder are completely undomesticated wild plants (e.g., cloudberry, piñon pine, sassafras). Food products from 34 of the 100 inclusions are obtained exclusively or substantially from cultivation (e.g., cranberry, hop, Jerusalem artichoke); crops from 55 are exclusively or substantially wildcrafted (i.e., harvested from wild plants in nature; e.g., bilberry, sugar maple, pawpaw); and 11 crops are obtained both from wild and cultivated plants (e.g., blueberries, cherries, wild rice). Several of the species currently widely harvested for nonedible purposes have potential to be developed as food sources (e.g., American ginseng, jojoba, saw palmetto). Many of the plants are promising as new crops for environments that are considered agriculturally marginal, notably for shallow water (e.g., cattail, Ogeechee lime, reed), semi-deserts (e.g., cactus pear, devil's claw, tepary bean), and very cold climates (e.g., crowberry, salmonberry, scurvy grass). Several of the wild plants are obsolete but were once of immense value to Native Americans, and their amazing productivity suggests that they could be resurrected as modern food plants (e.g., camas, prairie turnip, yampah). Several species have only recently been found to have commercial food value and are the subject of experimental research (e.g., azolla, dwarf glasswort, golden chia), and some could be significant in the immensely profitable food oilseed sector (e.g., blue waxweed, coast tarweed, buffalo gourd). As suggested by this sampling of the 100 plants examined, there are numerous, varied, interesting, and promising indigenous North American crops that warrant research and development.

An introductory chapter discusses just what indigenous plants are, how the top 100 North American food plants were chosen, potential toxicity of some of the species, and the dietary importance of North American food plants toward improving today's unhealthy eating patterns. Also examined are how mankind has evolved in relation to food plants, with particular regard to how this has determined the crops that now dominate North America; the factors that have severely limited historical crop development in North America by comparison with other areas of the world; and the considerations that bear on the future competitive economic development of North American food plants.

Most of the book is made up of 100 chapters, each dedicated to a particular crop (one species or a group of related species). A user-friendly standard format is employed. The initial section entitled "Names" provides information on scientific and English names of the plants. This is followed by a section called "Geography and Ecology of Wild Plants," accompanied by a map showing the North American distribution range. Next is "Plant Portrait," which comprises a basic description of the plant, its history, and its economic and social importance. The following section, "Culinary Portrait," is concerned with food uses. Under a subsection titled "Culinary Vocabulary," information is given on the names of especially important foods prepared from the plants, as well as on a variety of related culinary words, phrases, and terms. The next section is "Prospects," which provides an analysis of the economic future of the crop. This is followed by "Curiosities of Science and Technology," which contains notable and interesting scientific or technological observations and accomplishments that complement the main textual material. Finally, a section called "Key Information Sources" provides selected references to books and articles on the subject of each chapter. A subsection entitled "Specialty Cookbooks" presents references to food preparation using the particular plant in question. There are more than 3000 literature citations in the book, and the text is complemented by over 600 color photos and paintings of extremely high quality.

# Common Name Guide to North American Food Plants Discussed in Detail

**Note:** Many of these plants are also known by other names. See the Index of Common Names (page 717)

# Scientific Name Guide to North American Food Plants Discussed in Detail

***Note:*** Some of these plants are also known by scientific synonyms. See the Index of Scientific Names (page 731), which also provides information on many other species for which limited information is provided.

# Cautions

Toxicity problems are encountered with numerous food plants, not just the indigenous food plants of North America. These include individual allergies and sensitivities to foods, suggesting that unfamiliar foods should be sampled in very small amounts initially; the phenomenon of sensitivity developing only after eating a given food on several occasions; and the possibility that food plants have been contaminated during growth or culinary preparation. Additionally, some food plants can be eaten safely (at least in moderation) by the majority even though some of the constituents are slightly toxic, but can be potentially dangerous for some people, particularly pregnant women, infants, the elderly, and those with medical conditions.

Some of the plants included in this book pose significant risks. The majority of staple food plants have been altered by selection to reduce or eliminate dangerous constituents. By contrast, almost two-thirds of the harvested species discussed in this volume have not been altered by human selection. Most conventional food plants have been sold in commerce for many years, and so have established a record for safety. By contrast, some of the species discussed in this book have been consumed relatively infrequently, and the safety of their consumption in large quantities or over long periods is less certain, and needs to be established.

The kinds of toxicity situations that can be encountered are quite varied. Sometimes, the degree of maturity determines the degree of danger; for example, American persimmons can produce dangerous bezoars—stone-like masses of undigested material—in people with certain stomach conditions, but this is less likely for mature than for immature fruit. Sometimes, parts of a plant are edible whereas other parts are poisonous; for example, the mature fruit of May-apple is edible, while other parts of the plant are very toxic. Sometimes, people have learned to detoxify plants; for example, the danger of consuming pokeweed is traditionally reduced by boiling in several changes of discarded water. Sometimes, some varieties of a species are more edible (less toxic) than others, a situation which appears to apply to Labrador tea. A discussion of the possible dangers of consumption of many of the indigenous food plants of North America is presented in Chapter 1. Information on the toxic potential of most of the plants discussed in this book is also given in the chapters dealing with them, but it should not be assumed that the safety information is complete.

Special dangers are associated with the collection of wild food plants from nature—most notably, the possibilities of mistaking a poisonous plant for an edible one, or of simply accidentally harvesting toxic plants. This book is not intended to be a manual for harvesting food plants from the wild; for an extensive list of competent edible plant field guides for the United States and Canada, see the Appendix to this book.

Most of the plants discussed in this book have traditionally been used in folk medicine, and some are still employed in herbal cures. Discussions are given in this book on obsolete folkloric uses, modern accepted medicinal treatments, and potential applications. In no case should this information be employed for personal diagnosis or medication. The medicinal use of some plants is potentially hazardous and should only be carried out under the supervision of trained, knowledgeable, licensed, health-care professionals.

A chief concern of this book is profit potential of North American indigenous plants from a food perspective, and this has resulted in the inclusion of a number of species that in their present state are rather unpalatable or even dangerous to consume. However, most of the plants are quite pleasant in taste, and many are among the world's most desired foods. The occasional dangers posed by some of them should not detract from the desirability of the majority, which add to the joy of eating and the improvement of health.

# 1 Introduction

## PURPOSE OF THIS BOOK

### General Goal

Plants that we eat are the most indispensable resource that sustains us and civilization. In the early twenty-first century, with the world population surpassing 7 billion, agricultural production is facing dire challenges. The supply of new prime crop lands has virtually been exhausted, and in many areas the best soils are being occupied by urban development or degraded by poor management. Water is increasingly scarce. Problems such as climate change; environmental and biodiversity degradation; pollution of land, air, and water are formidable.

More so than any other large region of the planet, North America, one of the greatest of the world's food-producing regions, still has considerable potential for substantial increases in food output. Curiously, as discussed in this chapter, the continent's crop production is based mostly on plants of foreign origin. However, many North American native food plants have impressive potential for meeting the expanding food needs not just domestically but of the world. Toward this ultimate goal, this book is an analysis of the top 100 indigenous (native) food plants of North America north of Mexico that are most likely to be commercially successful in the future marketplace.

### Specific Goals

The rationale for focusing on North American native food plants, many of which are currently of minor importance, includes the following:

1. It is commonplace in life for international stars to overshadow local talent and suppress their development, and the world's major staple food crops, mostly not native to North America, indeed monopolize attention; however, many indigenous North American food plants have considerable potential for (a) improvement and market expansion, at least to the point of becoming significant minor or alternative crops in some regions and (b) contributing genes for the improvement of established crops. An assessment of their status and prospects is required to prioritize research and development.
2. Native food plants in the United States and Canada are a precious economic heritage that needs to be protected for future generations. Today, gene banks concentrate on preserving seeds of the world's leading crop plants, with comparatively little attention to North American wild plants. An assessment of the most important native food plants of North America should contribute to prioritization of protective measures for them, both in nature and in germplasm collections.
3. Wild plants are tough, often very capable of thriving in agriculturally marginal areas—almost the only kind of situation left in the world for growing more crops. Wild native food plants and cultivated varieties selected from them are adapted to the climate and soils in their native areas of North America, and in many cases are resistant to local insects and diseases. Accordingly, they have features that make them particularly suitable for selecting new crops for North America and indeed for foreign areas with similar climates.
4. Food is an indispensable world resource with an uncertain future. The world's present overconcentration on a small number of major food plants is hazardous: it has led to market surpluses, inefficient subsidization, vast monocultures that are ecologically destructive, and

the dangers of crop failures from hazards such as climate change and epidemics of disease. The North American food crops featured in this book can contribute to crop diversification and the reduction of problems associated with growing a restricted set of elite crops.

5. The major focus of this book is on agricultural aspects of native food crops—their actual or at least potential value as cultivated plants. However, the wild-growing plants also represent an important source of income. All of the species examined grow outside of cultivation in regions of North America, and many are collected by commercial harvesters and wild food enthusiasts. The information presented in this book should aid in the appropriate management of the wild-growing plant food resources of North America.
6. Indigenous American food crops have special cultural significance for indigenous peoples of North America. Moreover, these food plants also have special nutritional significance for Native Americans, since there is strong evidence that North American indigenous groups tend to be genetically adapted to their ancestors' traditional diet that is restricted in calories and easily assimilated carbohydrates (like table sugar), but high in fiber. The obesity and diabetes epidemics that are afflicting indigenous people is due to the modern rich but paradoxically unhealthy diet, and a return to traditional plant foods is often advocated. More generally, North Americans and people everywhere are realizing that increasing consumption of fruits, vegetables, and unsaturated fats and oils is key to avoiding the dietary diseases of modern civilization, and the plants featured in this book have great potential to improve everyone's health.
7. Food plants represent a topic of great importance and interest and have been examined from various practical, academic, and philosophical perspectives. Inevitably, bias has resulted from the particular perspectives that have been adopted. In particular, environmentalists are focused on the deleterious effects of monocrop development and destruction of the landscape, while agribusiness and political interests are usually concerned with profits and economic development. The crops discussed here have or can have commercial value and necessarily are part of the growing interest in finding sustainable solutions to agricultural development. Interest groups also include theoreticians of historical crop evolution, who have advanced a plethora of theories on how and why societies in different part of the world developed agriculture; Native Americans and other commercial wildcrafters, who still collect such plants as wild rice and pine nuts for large-scale commercial sale; the health professional community, concerned about both the benefits and risks associated with many of the plants; rural inhabitants, on whose property many of the most attractive food plants grow; and numerous people who have been attracted to the practice of collecting and eating wild food plants, stimulated by books, clubs, and courses devoted to the subject. This book attempts to present essential facts about North America's food plants, in an objective and balanced fashion, fairly presenting their importance and potential, but at the same time conceding associated problems.
8. This is not a textbook, but is intended to complement the books and literature assigned to students in economic botany, agriculture, and resource management courses. Food plants are the most important of the world's plants. North America is one of the world's critical agricultural areas, and students should certainly become familiar with the continent's leading native food plants.
9. Although this monograph deals with a subject of major scientific, agricultural, economic, and environmental importance, edible plants are of general interest. People like to eat, garden, and interact with nature, and the plants discussed here meet these objectives. Moreover, plants that are or have been important are culturally and historically significant and are associated with fascinating occurrences. The topic of food plants is not only entertaining but also a means of increasing public education. Toward these goals, the presentation is user-friendly, concise, and well illustrated, in the hope that nonspecialists will find the topics interesting.

A distinguished panel of experts prepared a report on the need to develop little-known indigenous crops in tropical regions (National Academy of Sciences 1975). The majority of the plants examined in this volume on North American indigenous plants are little-known temperate region crops, and the need to develop them is parallel to the need for tropical crops. Very little has changed since the 1975 report. A key portion of its very apt introduction follows:

> The strain on world resources posed by rapid population growth, dwindling supplies of nonrenewable resources, and shortages of food puts economic botany in the mainstream of human concern.
>
> Throughout human history man has used some 3000 plant species for food; at least 150 of them have been commercially cultivated to some extent. But over the centuries the tendency has been to concentrate on fewer and fewer. Today, most of the people in the world are fed by about 20 crops—cereals such as wheat, rice, maize, millet, and sorghum; root crops such as potato, sweet potato, and cassava; legumes such as peas, beans, peanuts (groundnuts), and soybeans; and sugar cane, sugar beet, coconuts, and bananas. These plants are the main bulwark between mankind and starvation. It is a very small bastion.
>
> Yet as the prospects for food shortage becomes more acute, people must depend increasingly on plants rather than animals for the protein in their diet. As is well recognized, research is urgently needed to increase the yield of these food plants. However, reliance on a small number of plants carries great risk, for monocultures are extremely vulnerable to catastrophic failure brought about by disease or variations in climate. To help feed, clothe, and house a rapidly increasing world population, it is timely to consider neglected or little-known species.

## FORMAT OF PRESENTATION

This book is formatted in the same fashion as two of my previous volumes on food plants (Small 2009, 2011a), which contain allied information not repeated here. Chapters are presented in alphabetical sequence of common (also known as vernacular and colloquial) names.

Following the common English name, the Latin name of the plant family is given, in turn followed by the most frequent English name of the family. Traditional family names end in *aceae.* For eight of the first plant families recognized by botanists, there are alternative names that do not end in this suffix. In these cases, the family name ending in aceae is given first (current recommended editorial practice), followed by the traditional family name in parenthesis.

The first major section is entitled "Names," which begins by presenting the correct scientific (Latin) name of the species under discussion, sometimes followed by frequently used Latin synonyms. This is followed by bulleted entries giving information on alternative English names, the derivation of names, and related matters.

The next section is "Geography and Ecology of Wild Plants," which is self-explanatory. Distribution maps were prepared for all species, based on consideration of published information, frequently cited in each chapter's bibliography. Consulted online mapping information included Flora of North America (see http://www.efloras.org/flora_page.aspx?flora_id=1); Atlas of United States trees (see http://esp.cr.usgs.gov/data/atlas/little/); USDA PLANTS Database (see http://plants.usda.gov/java/); BONAP's North American Plant Atlas (see http://www.bonap.org/); and E-Flora BC (see http://www.geog.ubc.ca/biodiversity/eflora/). Online map information for Canada is much less complete than for the United States. An extensive list of manuals consulted, which provide information on Canadian distributions, is available in Small et al. (2012).

The next section, called "Plant Portrait," provides a basic botanical and agricultural description of the plant, as well as of its history, and its economic and social importance.

The following section, "Culinary Portrait," is concerned with food uses, particularly industrial and technological aspects. Under a subsection titled "Culinary Vocabulary," information is given (in bulleted format) on the names of especially important foods prepared from the plants, as well as on a variety of related culinary words, phrases, and terms.

The next section "Prospects" attempts to provide a realistic assessment of the probability of future success of the crop.

The following section entitled "Curiosities of Science and Technology," again in bulleted format, contains notable and interesting scientific or technological observations and accomplishments that complement the main text.

The final section called "Key Information Sources" presents selected references to books and articles on the subject of each chapter. These references are intended to direct the interested reader to important reviews, studies, and background material. A subsection entitled "Specialty Cookbooks" presents references to food preparation using the particular plant in question.

## DEFINING INDIGENOUS STATUS

This book highlights food plants that are indigenous to the United States and Canada, so it is necessary to be clear on what this means. For rigorous analyses of the concept of indigenous status, see Ratcliffe (1977) and Peterken (1981). Basically, a species is "indigenous" (or "native") to a given geographical area if it naturally reproduces there and is present in that location as the result of natural processes, without the influence of humans. Contrarily, if it has been transported (deliberately or not) to a location because of human activity and reproduces there without the assistance of humans, it is "introduced" (or "naturalized"). A nonindigenous species that occurs with some frequency in an area because it is often released or escapes, but does not persist indefinitely because of a lack of adaptation to that area, is said to be "spontaneous," "adventive," or "casual." Ships once frequently dumped sand ballast in foreign locations, thus introducing the seeds of foreign species, which showed up regularly near international ports, but rarely persisted for long. The chief difficulty with determining whether a species is indigenous or introduced is the time dimension. Of course, geological and climate changes during the billions of years of Earth's history has meant that all species migrated extensively. At the minimum, indigenous status needs to be assessed starting at least no earlier than the last ice age. However, determining the prerecorded history location of some plants is very difficult and uncertain with respect to the possible influence of humans. In practice, in the Americas, plant species believed to be naturally present in an area since Columbus' 1492 discovery of the New World are almost universally accepted as indigenous to those areas. Of the estimated 18,000 species that grow outside of cultivation in North America north of the Mexican border, about 3000 are thought to be naturalized (Thorne 1993).

Harvesting and consuming edible plants growing outside of cultivation in the United States and Canada is a popular pastime, and there are over 100 books on the subject (see the Appendix to this book). Almost none of these works distinguishes indigenous from introduced plants, and frequently many introductions are featured for their attractive taste. Table 1.1 presents information on some of the introduced species commonly highlighted. Some of these (e.g., asparagus, dandelion, watercress) are in fact cultivated as vegetables, and indeed cultivated varieties are available, which are much tastier than the wild plants encountered in nature.

## HOW THE CHOICE WAS MADE OF THE TOP 100 INDIGENOUS FOOD PLANTS OF NORTH AMERICA

Plant species used (or at least useable) as food crops were considered for inclusion in this book if at least part of their indigenous range is in the United States and/or Canada. (As noted by Harlan [1992], the word "crop" appropriately includes plants that are harvested from both cultivated and wild sources.) Several of the species have extensive ranges, sometimes including other continents; several are utilized primarily outside of North America; some of the species were domesticated primarily (perhaps even entirely) outside of North America; some very important crops have parentage tracing to more than one species, and sometimes a North American species was a relatively

**TABLE 1.1**
**Examples of Introduced (Naturalized, Nonindigenous) Weedy Plants in North America that Are Commonly Highlighted in "Wild Food" Books**

| Common Name | Scientific Name | Food Use | Area of Origin |
|---|---|---|---|
| Amaranth | *Amaranthus retroflexus* L. | Vegetable (young leaves) | Tropical Americas |
| Asparagus | *Asparagus officinalis* L. | Vegetable (young shoots) | Eurasia, Africa |
| Barberry (common) | *Berberis vulgaris* L. | Fruit | Eurasia, Africa |
| Borage | *Borago officinalis* L. | Flavorant (flowers, young leaves) | Eurasia, Africa |
| Burdock | *Arctium lappa* L. | Vegetable (root) | Eurasia |
| Catnip | *Nepeta cataria* L. | Flavorant (leaves; tea) | Eurasia |
| Chickweed | *Stellaria media* (L.) Vill. | Vegetable (leaves) | Eurasia |
| Chicory | *Cichorium intybus* L. | Flavorant (root; coffee substitute) | Eurasia, Africa |
| Clover (red and white) | *Trifolium pratense* L. (red clover), *T. repens* L. (white clover) | Flavorant, vegetable (young leaves, flowers) | Eurasia, Africa |
| Coltsfoot | *Tussilago farfara* L. | Vegetable (flowers and leaves); flavorant (leaves); and tea (do not consume this toxic plant!) | Eurasia, Africa |
| Dandelion | *Taraxacum officinale* F.H. Wigg. aggr. | Vegetable (leaves) | Eurasia |
| Day-lily (orange day-lily, common yellow day-lily) | *Hemerocallis fulva* (L.) L. | Vegetable (young flowers, young tubers) | Asia |
| Day-lily (yellow day-lily) | *Hemerocallis lilioasphodelus* L. (*H. flava* (L.) L.) | As above | Eurasia |
| Garlic mustard | *Allaria petiolata* (M. Bieb.) Cavara & Grande | Flavorant (young leaves) | Eurasia, Africa |
| Horseradish | *Armoracia rusticana* G. Gaertn. et al. | Flavorant (root) | Europe |
| Japanese knotweed | *Reynoutria japonica* Houtt. | Vegetable (young shoots) | Asia |
| Lamb's quarters | *Chenopodium album* L. | Vegetable (leaves) | Europe (range obscure elsewhere) |
| Mallow (common; cheeseweed) | *Malva neglecta* L. | Vegetable (leaves) | Eurasia, Africa |
| Peppermint | *Mentha ×piperita* L. | Flavorant (leaves) | Hybrid of European origin |
| Plantain (common) | *Plantago major* L. | Vegetable (young leaves) | Eurasia |
| Purslane | *Portulaca oleracea* L. | Vegetable (stems and leaves) | Eurasia (some consider it native to North America) |
| Salsify (common) | *Tragopogon porrifolius* L. | Vegetable (root) | Eurasia, Africa |
| Sheep's sorrel | *Rumex acetosella* L. | Flavorant (young leaves) | Eurasia, Africa |
| Spearmint | *Mentha spicata* L. | Flavorant (leaves) | Eurasia |
| Watercress | *Nasturtium officinale* W.T. Aiton | Vegetable (leaves) | Eurasia, Africa |
| Wintercress (upland cress, yellow rocket) | *Barbarea vulgaris* W.T. Aiton | Vegetable (leaves) | Eurasia, Africa |

minor contributor to this parentage; none of these considerations was taken into account, since they do not alter the fact that the crops have at least some "citizenship" in North America north of Mexico. However, only the more prominent of the tropical food species that barely enter the United States across the Mexican border (especially in Florida) were included, as the tropical environment is not representative of most of the region under consideration in this book. For some species, natural presence in the United States or Canada has been disputed (i.e., some have contended that the plants were introduced), and the decision to include them was made on the basis of the majority viewpoint.

The relative merit of food plants can be rated by various criteria, such as superiority of taste, nutritional content, suitability for climates or soils, beneficial or detrimental effects on biodiversity and the environment, and social spin-offs. Plants have numerous kinds of "values," some intangible (such as beauty), some invaluable but difficult to evaluate (notably ecological services, such as stabilizing the local water table), and some that lend themselves to commercial evaluation for multiple products (as noted below, many of the food plants discussed in this book produce nonfood commodities). Philosophically, the viewpoint has been expressed that other species do not exist to serve humans and should be equally respected. All of the above considerations are important, but in the real world money usually talks loudest, and the selection of plants to include in this book is based simply on present or potential profitability.

In evaluating candidate species to include, I surveyed the literature on edible indigenous plants in North America (cited in the Appendix to this book; also examined were previous comprehensive surveys of edible plants, including Morton [1963], Arnason et al. [1981], Turner [1981], Prescott-Allen and Prescott-Allen [1990], Kuhnlein and Turner [1991], Facciola [1998], Wiersema and León [1999], Moerman [2010], and Plants For A Future [http://www.pfaf.org/user/default.aspx]). These sources collectively represent an exhaustive survey of indigenous North American plants used for food (indeed, in some cases of most of the significant food plants of the world). However, the majority of the more than 1500 edible native North American species cited in these works have no market significance. Unfortunately, comprehensive economic statistics are not available for most indigenous North American food plants that are marketed. To determine what species to include, an analysis of indigenous North American food plants was conducted and reported in Table 1.2, and summarized in Table 1.3. As noted in these tables, 50 species (groups of species in a few cases), including plants that are exclusively or mainly cultivated for food (such as sunflower and squash) or exclusively or mainly wildcrafted (collected from wild plants) for food (such as sugar maple, cloudberry, and lingonberry), were identified, which have achieved at least minor commercial importance in the marketplace (i.e., they are regularly encountered for sale as raw food or as processed food products). Next, plants encountered with at least some frequency (sometimes locally) for sale at the "cottage industry" level were identified ("cottage industry" refers to small-scale production and marketing of the plant or products processed from it, generally by one or just a few cooperating individuals, typically with limited investment of capital and frequently to supplement income from other sources). Examples include American persimmon, Ogeechee lime, and Labrador tea (see Table 1.2, for crops evaluated as having cottage industry importance). This accounts for an additional 28 species (Table 1.3). American chestnut was included because, although almost extinct from a deadly fungus blight, there is potential for this wonderful nut to regain its former glory. Several species (American ginseng, coast tarweed, jojoba, and saw palmetto) were included because they are cultivated or harvested extensively for a nonfood product, but have potential to be selected or modified for a food product. Several species were included because the crops are promising enough that experimental cultivation has been carried out for food production (azolla, blue waxweed, buffalo gourd, and roses). Honewort was included because the Japanese form of the species is already a significant food crop in Asia. Camas and prairie turnip were included because of my conclusion that they are the most promising commercial prospects of the numerous abandoned food plants once widely harvested by Native Americans. The remaining 10 species were added on the basis of my personal intuition (or educated guess) that they are promising prospects (or in the case of the marginally palatable Joshua tree, because it at least makes for fascinating reading).

**TABLE 1.2**
**Analysis of Cultivation Status and Commercial Importance of the Species Selected as the Top 100 Indigenous Food Plants**

| Species | Food Class (Primary Food Use, in the Case of Multiple Uses) | Mainly Cultivated for Food or Other Product; or Mainly Wildcrafted[a] | Food Cultivars Selected ("Elementary" Means Primitive Selections, Little Altered from Wild) | Commercial Importance (Potential, Cottage Industry[b], Minor, Major, Huge) |
|---|---|---|---|---|
| Acerola (*Malpighia emarginata*) | Fruit | Cultivated for food | Yes | Major |
| American chestnut (*Castanea dentata*) | Nut | Wildcrafted for food | No | Potential |
| American ginseng (*Panax quinquefolius*) | Vegetable | Cultivated and wildcrafted for medicinals (used incidentally for food) | No | Major |
| American persimmon (*Diospyros virginiana*) | Fruit | Wildcrafted for food | No | Cottage industry |
| Anise hyssop (*Agastache foeniculum*) | Flavorant | Cultivated for food | Yes | Minor |
| Aronia (chokeberry; *Aronia melanocarpa*) | Fruit | Cultivated for food | Yes | Minor |
| Azolla (mosquito fern; *Azolla* species) | Vegetable | Cultivated for forage (used incidentally for food) | No | Potential |
| Bergamot (*Monarda* species) | Flavorant | Cultivated for ornament and food | Yes | Minor |
| Bilberry (*Vaccinium myrtillus*) | Fruit | Wildcrafted for food (rarely cultivated) | No | Minor |
| Blackberries and dewberries (*Rubus* species and hybrids) | Fruit | Cultivated for food | Yes | Major |
| Black walnut (*Juglans nigra*) | Nut | Cultivated for ornament and lumber (secondarily harvested for food); wildcrafted for food | Yes | Minor |
| Blueberries (*Vaccinium* species) | Fruit | Cultivated and wildcrafted for food | Yes | Major |
| Blue honeysuckle (*Lonicera caerulea*) | Fruit | Cultivated for food | Yes | Minor |

*(Continued)*

**TABLE 1.2 (*Continued*)**
**Analysis of Cultivation Status and Commercial Importance of the Species Selected as the Top 100 Indigenous Food Plants**

| Species | Food Class (Primary Food Use, in the Case of Multiple Uses) | Mainly Cultivated for Food or Other Product; or Mainly Wildcrafted[a] | Food Cultivars Selected ("Elementary" Means Primitive Selections, Little Altered from Wild) | Commercial Importance (Potential, Cottage Industry[b], Minor, Major, Huge) |
|---|---|---|---|---|
| Blue waxweed (*Cuphea viscosissima*) | Edible oil | Cultivated experimentally for food | No | Potential |
| Buffaloberries (*Shepherdia* species) | Fruit | Cultivated for ornament, wildcrafted for food | No | Potential |
| Buffalo gourd (*Cucurbita foetidissima*) | Edible oil | Wildcrafted for food (cultivated experimentally) | No | Potential |
| Butternut (*Juglans cinerea*) | Nut | Cultivated for ornament, wildcrafted for food | Yes | Cottage industry |
| Cabbage palmetto (*Sabal palmetto*) | Vegetable | Wildcrafted for food | No | Cottage industry |
| Cactus pear (*Opuntia ficus-indica*) | Fruit | Cultivated for food | Yes | Major |
| California bay (*Umbellularia californica*) | Flavorant | Wildcrafted for food | No | Cottage industry |
| Camas (*Camassia quamash*) | Vegetable | Wildcrafted for food | No | Potential |
| Canada garlic (*Allium canadense*) | Culinary herb | Wildcrafted for food | No | Potential |
| Cattails (*Typha* species) | Miscellaneous | Wildcrafted for food | No | Potential |
| Cherries (*Prunus* species) | Fruit | Cultivated and wildcrafted for food | Yes | Minor |
| Chinkapin (*Castanea pumila*) | Nut | Wildcrafted for food | No | Cottage industry |
| Chive (*Allium schoenoprasum*) | Culinary herb | Cultivated for food | Yes | Major |
| Chufa (*Cyperus esculentus*) | Vegetable | Cultivated for food | No | Minor |
| Cloudberry (*Rubus chamaemorus*) | Fruit | Wildcrafted for food | No | Minor |

| | | | | |
|---|---|---|---|---|
| Coast tarweed (*Madia sativa*) | Edible oil | Cultivated mainly for industrial oil, to a small extent for food | No | Minor |
| Cranberry (*Vaccinium macrocarpon*) | Fruit | Cultivated for food | Yes | Major |
| Cranberry bush (*Viburnum opulus* var. *americanum*) | Fruit | Cultivated for ornament, wildcrafted for food | No | Cottage industry |
| Crowberry (*Empetrum nigrum*) | Fruit | Wildcrafted for food | No | Cottage industry |
| Devil's claw (*Proboscidea* species) | Vegetable | Cultivated for fiber and food | Yes | Cottage industry |
| Duck potato (*Sagittaria latifolia*) | Vegetable | Wildcrafted for food | No | Potential |
| Dwarf cape gooseberry (*Physalis grisea*) | Vegetable | Cultivated for food | Yes | Minor |
| Dwarf glasswort (*Salicornia bigelovii*) | Vegetable | Cultivated for industrial oil and food | No | Minor |
| Elder (*Sambucus nigra*) | Fruit | Cultivated for ornament and food | Yes | Minor |
| Epazote (*Dysphania ambrosiodes*) | Flavorant | Cultivated and wildcrafted for food | No | Minor |
| Evening primrose (*Oenothera biennis*) | Edible oil | Cultivated for food | Yes | Minor |
| Fiddlehead fern (*Matteuccia struthiopteris*) | Vegetable | Wildcrafted (rarely cultivated) for food | No | Minor |
| Golden chia (*Salvia columbariae*) | Pseudocereal | Cultivated for food | No | Minor |
| Golden currant (*Ribes aureum*) | Fruit | Wildcrafted for food | No | Cottage industry |
| Grapes (*Vitis* species) | Fruit | Cultivated for food | Yes | Major |
| Groundnut (apios; *Apios americana*) | Vegetable | Cultivated for food | Yes | Minor |
| Hawthorns (and mayhaws; *Crataegus* species) | Fruit | Cultivated and wildcrafted for food | Elementary | Minor |

(*Continued*)

**TABLE 1.2 (*Continued*)**
**Analysis of Cultivation Status and Commercial Importance of the Species Selected as the Top 100 Indigenous Food Plants**

| Species | Food Class (Primary Food Use, in the Case of Multiple Uses) | Mainly Cultivated for Food or Other Product; or Mainly Wildcrafted[a] | Food Cultivars Selected ("Elementary" Means Primitive Selections, Little Altered from Wild) | Commercial Importance (Potential, Cottage Industry[b], Minor, Major, Huge) |
|---|---|---|---|---|
| Hazelnuts (*Corylus* species) American hazelnut (*C. americana*) Beaked hazelnut (*C. cornuta*) | Nut | Cultivated for food (at least in the form of hybrids) | Yes | Minor |
| Hickories (*Carya* species) | Nut | Wildcrafted for food | Yes | Cottage industry |
| Hog peanut (*Amphicarpaea bracteata*) | Vegetable | Wildcrafted for food | No | Potential |
| Honewort (*Cryptotaenia canadensis*) | Vegetable | Wildcrafted for food | No | Potential |
| Hop (*Humulus lupulus*) | Flavorant | Cultivated for food | Yes | Major |
| Huckleberries (*Vaccinium* and *Gaylussacia* species) | Fruit | Wildcrafted for food | No | Minor |
| Jerusalem artichoke (*Helianthus tuberosus*) | Vegetable | Cultivated for food, fuel, and industrial products | Yes | Minor |
| Jojoba (*Simmondsia chinensis*) | Edible oil (potential) | Cultivated for industrial oil (incidental use as food) | Yes | Potential (for food); a major industrial oil crop |
| Joshua tree (*Yucca brevifolia*) | Vegetable | Wildcrafted for food | No | Potential |
| Juniper (*Juniperus communis*) | Culinary herb | Cultivated for ornament, wildcrafted for food | No | Minor |
| Labrador tea (*Ledum* species) | Culinary herb | Wildcrafted for food | No | Cottage industry |
| Lingonberry (*Vaccinium vitis-idaea*) | Fruit | Wildcrafted for food | No | Minor |
| May-apple (*Podophyllum peltatum*) | Fruit | Cultivated for medicinals, wildcrafted for food | No | Cottage industry |
| Maypop (*Passiflora incarnata*) | Fruit | Wildcrafted for food | No | Potential |

| | | | | |
|---|---|---|---|---|
| Mesquite (*Prosopis* species) | Flavorant | Wildcrafted for food | No | Minor |
| Mexican oregano (*Lippia graveolens*) | Flavorant | Cultivated for food | Yes | Major |
| Mountain mint (*Pycnanthemum pilosum*) | Flavorant | Wildcrafted and cultivated experimentally for food; cultivated for ornament | Elementary | Cottage industry |
| Nettle (*Urtica dioica*) | Vegetable | Wildcrafted for food | No | Cottage industry |
| Nodding onion (*Allium cernuum*) | Vegetable | Wildcrafted for food | No | Cottage industry |
| Northern gooseberry (*Ribes hirtellum*) | Fruit | Wildcrafted for food | No | Cottage industry |
| Ogeechee lime (*Nyssa ogeche*) | Flavorant | Wildcrafted (sometimes cultivated) for food | No | Minor |
| Oregon grape (*Mahonia aquifolium*) | Fruit | Cultivated for ornament, wildcrafted for food | No | Cottage industry |
| Paper birch (*Betula papyrifera*) | Flavorant | Wildcrafted for food | No | Cottage industry |
| Paradise tree (*Simarouba glauca*) | Fruit | Cultivated for food | Yes | Major |
| Pawpaw (*Asimina triloba*) | Fruit | Wildcrafted for food | No | Cottage industry |
| Pecan (*Carya illinoinensis*) | Nut | Cultivated for food | Yes | Major |
| Piñon pine (*Pinus edulis*) | Nut | Wildcrafted for food | No | Minor |
| Plums (*Prunus species*) | Fruit | Cultivated and wildcrafted for food | Yes | Minor |
| Pokeweed (*Phytolacca americana*) | Vegetable | Wildcrafted for food | No | Cottage industry |
| Prairie turnip (*Pediomelum esculentum*) | Vegetable | Wildcrafted for food | No | Potential |
| Raspberries (*Rubus* species and hybrids) | Fruit | Cultivated for food | Yes | Major |
| Red mulberry (*Morus rubra*) | Fruit | Wildcrafted for food | No | Cottage industry |
| Reed (*Phragmites australis*) | Vegetable | Wildcrafted for food | No | Potential |

(*Continued*)

**TABLE 1.2 (*Continued*)**

**Analysis of Cultivation Status and Commercial Importance of the Species Selected as the Top 100 Indigenous Food Plants**

| Species | Food Class (Primary Food Use, in the Case of Multiple Uses) | Mainly Cultivated for Food or Other Product; or Mainly Wildcrafted[a] | Food Cultivars Selected ("Elementary" Means Primitive Selections, Little Altered from Wild) | Commercial Importance (Potential, Cottage Industry[b], Minor, Major, Huge) |
|---|---|---|---|---|
| Roses (*Rosa* species) | Fruit | Cultivated for ornament, wildcrafted for food harvested experimentally for food | No | Potential |
| Saguaro (*Carnegiea gigantea*) | Fruit | Wildcrafted for food | No | Potential |
| Salal (*Gaultheria shallon*) | Fruit | Wildcrafted for food | No | Cottage industry |
| Salmonberry (*Rubus spectabilis*) | Fruit | Wildcrafted for food | No | Cottage industry |
| Saskatoon (*Amelanchier alnifolia*) | Fruit | Cultivated for food | Yes | Minor |
| Sassafras (*Sassafras albidum*) | Flavorant | Wildcrafted for food | No | Minor |
| Saw palmetto (*Serenoa repens*) | Vegetable | Cultivated for ornament and pharmaceuticals, wildcrafted for food | No | Potential |
| Scotch lovage (*Ligusticum scoticum*) | Vegetable | Wildcrafted for food | No | Cottage industry |
| Scurvy grass (*Cochlearia officinalis*) | Vegetable | Wildcrafted for food | No | Potential |
| Sea grape (*Coccoloba uvifera*) | Fruit | Wildcrafted for food | No | Cottage industry |
| Spice bush (*Lindera benzoin*) | Flavorant | Wildcrafted for food | No | Cottage industry |
| Squash (*Cucurbita pepo*) | Vegetable | Cultivated for food | Yes | Major |
| Strawberries (*Fragaria* species) | Fruit | Wildcrafted and cultivated for food | No | Cottage industry |

| | | | | |
|---|---|---|---|---|
| Sugar maple (*Acer saccharum*) | Flavorant | Wildcrafted for food | No | Major |
| Sunflower (*Helianthus annuus*) | Edible oil | Cultivated for food | Yes | Huge |
| Sweet gale (*Myrica gale*) | Culinary herb | Wildcrafted for food | No | Cottage industry |
| Tepary bean (*Phaseolus acutifolius*) | Vegetable | Cultivated for food | Yes | Minor |
| Wild leek (ramp; *Allium tricoccum*) | Vegetable | Wildcrafted for food | No | Minor |
| Wild rice (*Zizania* species) | Cereal | Cultivated and wildcrafted for food | Yes | Major |
| Winter purslane (*Claytonia perfoliata*) | Vegetable | Cultivated and wildcrafted for food | Elementary | Minor |
| Yampah (*Perideridia gairdneri*) | Vegetable | Wildcrafted for food | No | Potential |
| Yerba buena (*Clinopodium douglasii*) | Flavorant | Cultivated for ornament and sometimes for food | No | Potential |

[a] "Wildcrafted" means collected from uncultivated plants growing in nature (or at least in uncultivated landscapes such as roadsides and abandoned fields).

[b] As noted in the text, "cottage industry" refers to small-scale production and marketing of the plant or products processed from it, generally by one or just a few cooperating individuals, typically with limited investment of capital and frequently to supplement income.

**TABLE 1.3**
**Percentages of Scores for Categories of Information Reported in Table 1.2**

| Commercial Importance (Potential, Cottage Industry, Minor, Major, Huge) | Food Cultivars Selected | Source of Food (Wildcrafted or Cultivated) |
|---|---|---|
| Potential: 22 | Modern cultivars: 34 | Wildcrafted only or predominantly: 55 |
| Cottage industry: 28 | Elementary selections, not notably altered from wild form: 3 | Both wildcrafted and cultivated: 11 |
| Minor: 33 | No cultivars: 63 | Cultivated only or predominantly: 34 |
| Major: 16 | | |
| Huge: 1 | | |

## POTENTIAL DANGERS FROM CONSUMPTION OF INDIGENOUS FOOD PLANTS OF NORTH AMERICA

### Natural Toxins

Plants manufacture "primary metabolites"—chemicals critical to key functions of life, such as proteins, simple sugars, carbohydrates, fatty acids, and vitamins. These are often found in substantial quantities in plant species and provide much of the nutrition needed by humans. However, plants also manufacture thousands of "secondary metabolites"—a wide array of chemicals, some representing metabolic waste products, but most of undetermined significance. Some secondary metabolites are key attractive flavoring and odor constituents of spices and beverages; others are used in medicine as therapeutic or health-promoting agents; however, many are toxic to humans, and indeed often protect plants against microorganisms, insects, and animal predators by virtue of bad taste or smell or simply by toxicity. Almost all plant species evolved into their present state before humans came into existence, so it is clear that the toxins are not intended specifically to harm people. Regardless of why toxic chemicals are present in given plants, they are often a danger to the consumption of plant parts that are otherwise tasty and nutritious. Anthropological studies have indicated that some hunter–gatherer societies ate hundreds of plant species, sometimes more than a thousand, so undoubtedly a very wide range of toxic secondary chemicals have been consumed. The human liver is equipped (with enzymes) to detoxify small amounts of numerous natural toxins, but often cannot tolerate a large amount of a given toxin. This is suggestive of adaptation to eating an extensive variety of plants and indicates that a chief danger is consumption of a large amount of material of a plant whose safety is not certain.

Plants often produce concentrated levels of toxins in organs that require protection against herbivores. The meristems of shoots (especially the growing tips of branches or stems) are frequently relatively toxic, because these areas are often tender and especially nutritional, and so are otherwise attractive as food for animals. Underground storage organs, by which plants often survive over winter or during drought, often contain large amounts of carbohydrates (sometimes fat) and may also be protected by high levels of toxins. Seeds usually contain food to give the seedling a good start and once again are often very toxic. The most significant kinds of food plant, cereals, pseudocereals, and pulses (discussed in detail in the following) are harvested for their seeds, so seed toxicity is of fundamental importance.

Fleshy fruits are produced by plants to enlist animals as a means of distributing the contained seeds. Such fruits are typically sour, bitter, and/or quite poisonous when they are "green" (immature), because the plants do not wish to attract animals until the seeds have matured. At maturity, fleshy fruits tend to lack toxins that can poison animals serving to distribute the seeds. Birds often are natural seed distributors, but sometimes berries that birds consume with impunity are very toxic to people. Humans have exploited a greater variety of fruits as food than any other kind of edible plant part. In every region of the world where people have used the native vegetation as sources of food, native fruits are always a great attraction because many are delicious and safe to eat. Not surprisingly, in North America, many strictly wild fruits are among the plants included in this book, and several have been domesticated.

In common language, "nuts" are hard-textured, usually oily, kernel-like edible plant materials housed in a frequently hard shell. Some so-called nuts are quite different, for example peanuts. In technical botanical terminology, nuts are large single hardened achenes, that is, single-seeded dry indehiscent fruits in which the seed coat is not part of the fruit. A sunflower seed is a nut in this sense. When the kernel (which is an embryo) is protected against herbivores by a hard shell, it is unnecessary for it to be additionally protected by toxins, and so kernels of wild nut trees have been a very important source of food for humans. Several wild North American nut trees are among the plants included in this book.

Humans have utilized a wide range of techniques to detoxify wild plants to make them edible (Johns and Kubo 1988). Alternatively, toxic chemicals have usually been eliminated or strongly reduced by selection during the process of domestication. Advanced cultivars are used to produce food in approximately one-third of the species highlighted in this book (e.g., blueberry, grape, hop), and there is little concern about the presence of natural toxins in these.

Almost two-thirds of the harvested species discussed in this book have not been altered by human selection. Food products from many of these have been sold in commerce for many years and so have established a record for safety (e.g., bilberry, huckleberries, sugar maple). However, some of the species have been consumed relatively infrequently, and the safety of their consumption in large quantities or over long periods is less certain, and needs to be established. Some of the species have hazardous constituents (notably American persimmon, epazote, jojoba, may-apple, pokeweed, and sassafras), and their potential toxicity should be appreciated before they are consumed.

Mexican oregano (most "oregano" in the southern United States is Mexican oregano) is an interesting example of a commonly eaten plant that is suspected of having toxic potential: it can reduce reproductive capacity, probably because of the antifertility and carcinogenic constituent lapachenole. However, small amounts of toxins are present in numerous common foods and have not aroused much concern (see Small 2009). As noted above, the human liver is designed to detoxify poisons consumed in small amounts. Accordingly, the fact that many plants do have small amounts of toxic chemicals should not by itself arouse great concern. Nevertheless, common sense dictates that overindulging on poorly known, uncommon food plants should be avoided.

## Allergies

Plants contain a phenomenally varied array of chemicals. Some of these are toxic to all or at least most people, and such toxins were discussed earlier. However, some chemicals produce toxic reactions in only a minority of people. Often "allergies" are defined as negative reactions to proteins, while negative reactions to other food constituents (such as flavor chemicals) are defined as "food chemical intolerance." Both allergies and food chemical intolerance are widespread. The most commonly reported allergenic plant foods are legumes (plants of the pea family), tree nuts, and wheat, and the substances responsible are known to be proteins. Allergies are typically relatively serious negative reactions and eating an unfamiliar food, sometimes even touching it, has the potential of triggering a bad reaction in susceptible individuals. However, food chemical intolerance is often relatively minor, and in such cases might better be termed food chemical sensitivity. On occasion, individuals who regularly contact a given plant become sensitized to it, and people need to be aware of this potential.

Elder is an interesting case of variation in sensitivity. The fruit of American elder is generally considered safe, although uncooked berries seem to cause nausea in some individuals. Elder flowers, which are occasionally fried in batter to make fritters, are reported to produce diarrhea in some people.

Common sense should be exercised when consuming any unfamiliar food, especially when it is not widely marketed. It is wise to limit consumption the first time it is offered, and indeed the second time to be sure one has not become sensitized.

## Microbiological Safety

Many of the food plants discussed in this book are wild plants from which people collect edible plant parts and prepare them in processed products for personal use or for sale as small business ventures. Plant materials need to be collected from areas free of harmful organisms. Aquatic plants in particular should be from water that is not polluted. Foods need to be prepared and stored carefully under appropriate conditions to prevent deterioration by molds and introduction of toxic bacteria. Major food companies and business are almost always very careful to prevent harm to the public because of the huge damage to reputation that results, the fact that deep pockets tend to be sued, and monitoring by government. However, dozens of the least well-known plants in this book are largely left to small businesses, with their trade mostly or entirely restricted to local areas and Internet sales. It is essential that foods prepared for this niche marketplace conform to the highest standards, and buyers need to patronize businesses that they trust.

Fiddlehead fern is an interesting example of a plant whose consumption has been associated with some disturbing cases reported in past decades of diarrhea, nausea, vomiting, abdominal cramps, headaches, and dehydration, particularly among the elderly and infants. Some have speculated that this vegetable has toxic constituents, but this is unproven. More likely, the cases of sickness were due to microbial contamination. Fiddlehead is now a widely marketed niche vegetable, and it is not considered to be dangerous. However, as noted in the chapter on this species, the recommendation is now commonly made for safety reasons that it be boiled for a period longer than is ideal to maximize its taste.

## Chemical Pollution

Much of the world's environment has been polluted by pesticides, industrial contaminants, and a wide variety of discarded compounds, and in some cases plants growing in affected sites can absorb toxic levels of various chemicals. People collecting wild plants for personal use can fail to notice that they are in an area with dangerous levels of soil pollutants, with the result that the foods collected are unhealthy.

## Misidentification

By far, the chief danger for wild food collectors is the erroneous identification of a poisonous plant as one that is edible. Some mushrooms are deadly and have even resulted in the death of experts who simply mistook them for edible species. Similarly, a few species discussed in this book can easily be confused with extremely poisonous species, and in such cases warnings are given. Plants with edible underground or underwater storage organs sometimes grow intermixed with plants with very toxic storage organs, and the two can inadvertently be collected together. This book is not a guide to safe wild food collecting, and other sources (particularly those cited in the Appendix to this book) should be examined for this purpose.

## Medicinal Usage

For some of the species discussed in this book, information is presented on health applications. This includes some modern accepted medicinal treatments, potential applications being explored, and obsolete, folkloric medical uses. In no case should this information be used for medical diagnosis or medication. The medicinal use of some plants is potentially hazardous and should only be carried out under the supervision of trained, knowledgeable, licensed, health-care professionals.

## HUMAN EXPLOITATION OF CROPS FOR FOOD

North American indigenous food plants are best understood in the context of what is known about the historical and biological relationships of humans and food plants. There is an enormous literature on this, and the following condensed account reviews the most important topics.

### Brief History of Agriculture

Humans closely resembling modern people (*Homo sapiens*) appear to have existed for at least the last 100,000 years (Hancock 2004), and perhaps twice as long. For most of this period, the lifestyle was reminiscent of our primate ancestors: people lived in small groups (clans) and were semisedentary (perhaps with long-term base camps), remaining for periods in localized areas where hunting, fishing, and foraging provided sufficient food. For the most part, people were nomadic, wandering in territories of sufficient size to obtain enough food and other materials for survival. In time, some tribes learned to provide measures of care for the species that they relied on, since such protection served to increase their food supply. A common practice of many archaic peoples was to regularly burn land to encourage new plant growth for human food or for better grazing for animal prey. Also to encourage the growth of wild food plants, they would have occasionally irrigated them, competitive plants would be removed or pruned, and wild seeds might even have been planted. When the level of care advanced to the point of taking control of the nurturing of those species up to their harvest, "agriculture" had been invented. Agriculture was invented independently in different areas of the world, at different times dating back at least 12,000 years. Agricultural systems evolved in widely separated regions of the Old and New Worlds, notably between 5000 and 11,000 years ago. The prehistoric transition from subsistence on free-living wild species to agriculture has been termed the "Neolithic Revolution" or "the "Origin of Agriculture." By 2000 years ago, agriculture had become the main source of food in the world.

An enduring mystery is why Stone Age mankind shifted from hunting and gathering to agriculture. Some students of the topic have regarded nature as a kind of paradise which supplied all of mankind's needs, and early farming as backbreaking and time-consuming; conversely, others note the dangers and uncertainty of reliance on foraging and hunting. Agronomists, archaeologists, anthropologists, biologists, demographers, and historians have advanced several theories about what caused humans to raise animals and cultivate plants, and eventually to domesticate them (see Hancock 2004; Weisdorf 2005). Some examples of the numerous hypotheses of why agriculture started follow: (1) Population growth in a region may have diminished or exhausted the supply of wild food, necessitating agriculture. (2) Farming may have arisen in seaside cities, where the occupants dependably obtained most of their food from the ocean, but had the luxury of experimenting with nearby garden plots. (3) Religious ceremonies involving animal sacrifice and establishing of plants may have been the root of initiating the practice of agriculture. Most theories accounting for the invention of agriculture are predicated on the establishment of human settlements of appreciable size, and of human ingenuity having reached a minimum level of sophistication. Indeed, agriculture has been closely associated with the appearance and success of the world's principal civilizations. Just what constitutes "civilization" is arguable, but almost invariably advanced human cultures became highly urbanized, developing impressive cities and architecture. Sometimes fishing or pillage (booty and tribute from conquered peoples) was the basis of success of large settlements, but in most cases where human populations became concentrated into villages of substantial size, it probably became increasingly difficult to harvest sufficient wild food from the local area, and this would have stimulated the development of agriculture. It has been argued that rather than cities stimulating the development of agriculture, agriculture stimulated the development of cities (i.e., the efficient production of food in a small area was responsible for that area developing into a city). There remains considerable disagreement about how and why agriculture originated. Harlan (1992) suggested that the reasons for the development of agriculture may be diverse, and no single model adequately accounts for all or even most occurrences.

Several trends have dominated the historical development of agriculture: (1) Geographically, agriculture arose and developed to a very high level in certain regions of the world and was virtually absent in other regions. (2) Through cultural diffusion and international conquest, agricultural traditions and crops were transferred among countries, with massive suppression of the harvest of most ancient crops, dominance of regional agriculture by a relatively small number of species, and dominance of world agriculture by a few dozen species. (3) The species that dominate agriculture come from a very narrow range of the world's biodiversity: primarily from a very small number (several dozen) of social mammals (species living in groups, usually controlled by a leader, and hence susceptible to human control) and from selected terrestrial, vascular plant families (mostly angiosperms), that is, flowering plants, occasionally gymnosperms (mostly conifers) and spore-producing plants (such as ferns) (Table 1.6).

## Domestication as a Natural, Coevolutionary Phenomenon

Almost all important species currently used in agriculture are domesticated. In common language, "domestication" often refers to wild animals that have been "tamed," that is, habituated to humans so that they cause fewer difficulties when associated in various ways with people. In biology, domestication is the process of selecting individuals of a species that have characteristics that make them useful to people, the selection occurring over generations, so that the desired traits become genetically fixed. Domestication is usually conceived of as a form of "artificial" selection, which is true if one defines artificial as selection related to human activity. However, some (e.g., Darlington 1973) have argued that unconscious, that is, nondeliberate selection is as "natural" as the natural selection that occurs in nature. Domestication is a form of evolution (which can be simply defined as changes in gene frequencies in a population). In some respects, domestication is also a form of symbiosis (an arrangement between species in which both partners benefit), since most of the genes of both humans and the domesticated creatures are maintained and the number of individuals is increased. On the other hand, many consider domestication to be a form of slavery, in which the domesticated creatures have lost important characteristics. (This latter viewpoint is anthropomorphic, that is, framed from a perspective that assumes the affected species have human-like attributes and values.)

Rindos (1984, chapter 3) stated "Domestication clearly cannot be held to be an exclusively human-mediated phenomenon." This is because man is not the only animal that has usurped the freedom of other species, caring for them but at the same time using them as a source of food. Wood wasps and over 40 species of ambrosia beetles cultivate fungi as food sources, inoculating wood with a fungus which they consume. These insects even assemble piles of wood chips, add feces, and remove competitive fungi—activities quite comparable to tilling and fertilizing soil and removing weeds. Some ants and termites also cultivated fungi. Most termites rely on protozoa in their guts to digest cellulose in wood (remarkably, in turn the protozoa rely on symbiotic bacteria embedded on their surface to produce some of the necessary digestive enzymes). However, certain tropical termites lack the ability to digest cellulose and prepare fungus gardens in their nests. "Cultivator ants" are extraordinarily good farmers: they construct a pulpy organic bed of chewed material (including plant debris, flowers, foliage, and excrement) in special chambers, inoculate the beds with fungi, remove foreign fungi, and regulate temperature and humidity through aeration tunnels. All of these symbiotic relationships have coevolved, and both the insects and the fungi having become partly or completely dependent on each other for survival. Similarly, some ant species care for some aphid species, the latter referred to as "ant-cows" because the ants pasture the aphids on vegetation and consume secreted sweet juice from the latter. There are even ants that allow butterfly larvae to feed on the aphids in return for consuming honeydew from the butterflies.

There is support for the relationship between man and crops also having resulted in coevolutionary changes in both humans and domesticates, that is, not only have characters evolved in the plants that better adapt them to survival in the context of their cultivation by mankind but people have also changed in response to the characteristics of the plants. In particular, there has been considerable speculation that the dentition of modern humans has evolved especially to suit a diet of cultivated grains (Rindos 1984). The best evidence that humans have been adaptively genetically changed by domesticating a species is the pattern of adult lactose tolerance. The enzyme lactase that permits infants to digest milk persists into adulthood in human populations indigenous to areas of the world where cows have been raised historically for milk, but not in areas where cows were not raised. There is a similar pattern of adaptation to wheat gluten in the Old World, celiac disease (intolerance to gluten) being less common in areas where wheat has been cultivated for long periods historically. On the whole, however, humans have evolved in response to dependence on domesticates very much less than the domesticates, which have often undergone profound alteration.

## Benefits and Costs of Domestication

Human dependence on domesticates is profound; probably only about 1 billion people could survive on Earth by relying only on wild species for food, and humans are now engaged in a struggle to generate enough food to sustain our increasing numbers. Based on this challenge, one extreme view is that the invention of agriculture and consequent departure from a lifestyle dependent on hunting wild animals and gathering wild plants is "the worst mistake in the history of the human race" (Diamond 1987). Whether life was better for most preagricultural people than it is for us today is debatable, but the subject of this section is to compare lifestyles of the species we have domesticated with their wild ancestors.

Dogs are mankind's first and most familiar domesticate and illustrate quite dramatically the advantages and disadvantages of domestication. Dog breeds are remarkably diversified so much, so that they differ in appearance as much as do many species within a genus, although they belong to just one species. Dogs are also remarkably physically and behaviorally specialized for tasks, such as herding, guarding, and hunting. The advantages to humans are obvious: dogs provide a range of invaluable services, and in comparison the wolf ancestors of dogs were and remain competitors for resources. The advantages to dogs are more or less clear: most (but not all) are treated humanely, and indeed many are pampered. The number of dogs vastly outnumbers the number of wolves (their ancestors), and indeed whereas the future of dogs is assured, wolves are threatened with extinction. Almost all dog breeds are so modified from wolves that they can no longer survive in nature, but even under the care of humans they have many problems. The selection of dog breeds has been carried out with intensive inbreeding, and dog breeds are plagued by numerous genetically based diseases. Selection for appearance such as floppy ears, short legs, elongated body, and flat face causes suffering in many of the breeds. The modern English bulldog is so massive that birth of the puppies requires cesarean section. Of particular interest, a "domestication syndrome" of characteristics is discernible in dogs by comparison with wolves: dogs are mostly smaller (less threatening to humans and requiring less food); brains are often proportionately smaller (this seems to maintain a desirable juvenile behavior in dogs); teeth are proportionately smaller and jaws are usually shorter (dogs usually do not need to hunt for their own food); dogs keep their tails erect (a behavior mostly confined to pack leaders among wolves); coat coloration, adaptive in the wolf as camouflage, has become more varied (to suit the aesthetic tastes of people); most dogs bark much more than wolves (no need to keep silent to avoid predators or alert game); and dogs reproduce readily at all times of the year (wolves are seasonal). Some of these changes are the result of directed selection of characteristics desired by man, and others are the result of decreased selective pressure for features that are no longer essential for survival in the wild.

Highly domesticated plants by comparison with their wild ancestors have benefitted and suffered very much as have dogs. The wild ancestors (if they still exist) of modern varieties of crops are extremely well suited to their natural wild habitats and to coexisting with other species such as herbivores, disease-causing organisms, and pollinators. Most highly domesticated plants are so altered that they can no longer survive in nature, and many can only be propagated by man. The majority of the most valuable domesticates are grown as monocultures and have been selected to thrive in very crowded conditions. Like dog breeds, many specialized forms (cultivars) have been selected for particular attributes and their continued existence as pure types demands very careful measures to maintain their genetic purity. The narrow genetic base of cultivars often makes them very susceptible to insects and microorganisms that the wild ancestors tolerate well. Domesticates have been selected for high productivity, and this often means the photosynthetic organs (leaves) are larger than those of the wild ancestors. Advanced cultivars invariably have been selected for high harvest index, that is, for a large percentage of harvested tissues in relation to overall plant weight, and this produces plants that are relatively poorly adapted to stresses (compare domestic pigs to incredibly athletic and ferocious wild boars). Roots of advanced cultivars are generally suited to relatively shallow soils, where water and fertilizer are provided, but this means that irrigation and fertilization are usually necessary. The particular parts of the plant that are desired (seeds, fruit, leaves, underground storage organs) have been selected for characteristics such as size, shape, and taste characteristics, but this attracts herbivores, requiring protective measures by farmers. Toxic constituents that protect the wild ancestors have usually been eliminated from the edible parts of the plant, and often also from the plant in general, also necessitating the use of protective measures, especially pesticides. Similarly, tough, fibrous tissues have been eliminated or reduced in edible parts, making these more attractive not just to people but also to insects. Protective thorns or prickles, if present, have also been frequently eliminated. The shape of many plants has often been customized for ease of picking; rambling vines have been transformed into shrubs and trees have been miniaturized, making the plants uncompetitive in their natural wild habitats.

Nonfleshy fruits harvested for their seeds (especially the cereals) are among the most important of domesticated plants and are usually characterized by a particular domestication syndrome compared to the wild plants (the domesticate has several of the following features): (1) Loss of fruit "disarticulation" or "dehiscence" (these terms are distinguished in the following). The fruits of wild plants (especially one-seeded fruits) characteristically develop an abscission (breakage) zone at the base of the fruit so that the fruit is released (disarticulated); by contrast, in domesticated plants the desired fruits often remain on the plant for easy collection. The fruits of wild plants with multiple seeds often "dehisce," that is, a pore or slit is developed through which the seeds are dispersed (the seeds are said to have "shattered"); by contrast, in domesticated plants the desired seeds often remain within the fruits, which remain attached to the plant for easy collection. (2) Increase in fruit size. (3) Decrease in hull (wall) thickness. (4) Loss of seed dormancy (efficient harvest requires all plants to germinate, develop, and mature simultaneously). Like many nonfleshy fruits, edible fleshy fruits have also been often selected for increased size and lack of disarticulation.

Figures 1.1 through 1.3 depict the most important stages and modes of plant domestication.

## Domestication and Development of Dominant Staple Crops

Thirty-seven of the 100 North American crops discussed in detail in this book have been domesticated (Table 1.3). The future success of the remaining most promising crops of the continent is mainly dependent on their being domesticated, and consequently it is critical to appreciate the nature of domestication (illustrated in Figures 1.1 through 1.3).

In the earliest stages of the development of agriculture, wild animals were confined and allowed to reproduce and wild plants were cultivated. This reduced or eliminated the need for hunting and foraging, but wild species have numerous drawbacks that limit their productivity and usefulness. With domestication, productivity (on an area basis) and usefulness increased, usually greatly.

Domestication is an expected consequence of long-continued raising of animals and growing of plants in confined areas, for food purposes. In the past, humans inevitably (often unconsciously) favored mutations (or other genetic occurrences such as hybridization) that facilitated and increased the quality, quantity, and efficiency of food produced. Very importantly, however, a degree of domestication also can occur when humans simply selectively harvest wild plants growing in nature. Normally, wild plants subjected to herbivory tend to evolve protective adaptations against being eaten (except in the case of fleshy fruits that are designed to attract herbivores for seed dissemination), not features that make them more attractive to consumers. Humans at the hunter–gatherer stage tended to set up temporary camps and create trails among their camps. Abandoned campsites and trails would tend to be open (unshaded), frequently near lakes or streams, and the soils would be enriched by deposition of organic materials (excrement and unusable remains of harvested animals and plants). Moreover, seeds and roots from gathered plant materials that humans would have selected for their attractiveness would also be deposited in these open, fertilized areas. This amounts to selective planting of desirable food-producing plants in situations where the plants will receive excellent light and soil—a precursor of cultivation. Black (1978) suggested that some fruit and nut trees and shrubs were transplanted beyond their natural ranges by indigenous groups. Inevitably, people would have noticed and eagerly harvested food from the plants that were growing along their trails and former campsites, especially in rubbish, and such plants would have been among the first that would have been considered for deliberate planting. This explanation is variously known as the "rubbish heap" or "dump heap" hypothesis; in archaeology, rubbish heaps are referred to as "kitchen middens." It is interesting that, in parallel, some monkeys have been shown to create "monkey gardens"—concentrations of preferred food plants in an area where they

**FIGURE 1.1** (a) A predomestication ecosystem, in which edible plants have evolved natural relationships with wild animals, including prehuman primates. The plants have not been altered by interaction with humans. (Painting courtesy of B. Flahey.) (b) Incipient domestication, by which interactions with humans alters the distribution and characteristics of edible plants. In this case, a clan is patronizing a plant with superior edibility, possibly scattering its seeds in the local area where it grows well, and thereby increasing the frequency of genes determining desirable characteristics. (Painting courtesy of B. Flahey.)

**FIGURE 1.2** (a) Unintentional introduction of foreign plants to North America, some of which may be edible, some of which may have been introduced in pre-Columbian times and seemingly became established as part of the indigenous flora, some of which may have been domesticated before their introduction, and some of which may have been domesticated after their introduction. As discussed in the text, sometimes distinguishing these possibilities can clarify whether a plant should be considered "indigenous." (Painting courtesy of B. Flahey.) (b) Escape of domesticated edible plants from cultivation and settlement in the wild. As discussed in the text, this can range between merely occasional occurrence and widespread establishment. This has resulted in the creation of some very noxious (albeit somewhat edible) weeds. (Painting courtesy of B. Flahey.)

have discarded seeds (Rindos 1984). Plants that grow vigorously on human-cleared vegetation are known as weeds. It is no accident that many, probably the majority of the world's major domesticated crops are related to or are known to have originated from weeds. The ability to be weedy clearly preadapts plants to being domesticated (for example, as reviewed by Small (2011b), a dozen weedy species of *Medicago* were domesticated in the last several decades, mostly in Australia where there are no indigenous species of the genus). Merely by harvesting the most desirable plants from a geographical area and depositing their seeds in areas frequented by people, a subpopulation has been created that differs genetically from the remainder of the species. The phrase "incipient domestication" is applicable to this situation, but has generally been applied to the very early successes of cultivating crops.

How much does a plant have to be changed from its wild source of origin before it is considered to be "domesticated"? Pickersgill (2007) noted "Domestication is generally considered to be the end-point of a continuum that starts with exploitation of wild plants … and terminates in fixation, through human selection, of morphological and hence genetic differences distinguishing a domesticate from its wild progenitor." However, just how much change is necessary is subjective and therefore so is the concept of a "domesticate" (or "domesticant"). Hawkes (1983, p. 6) wrote "Possibly

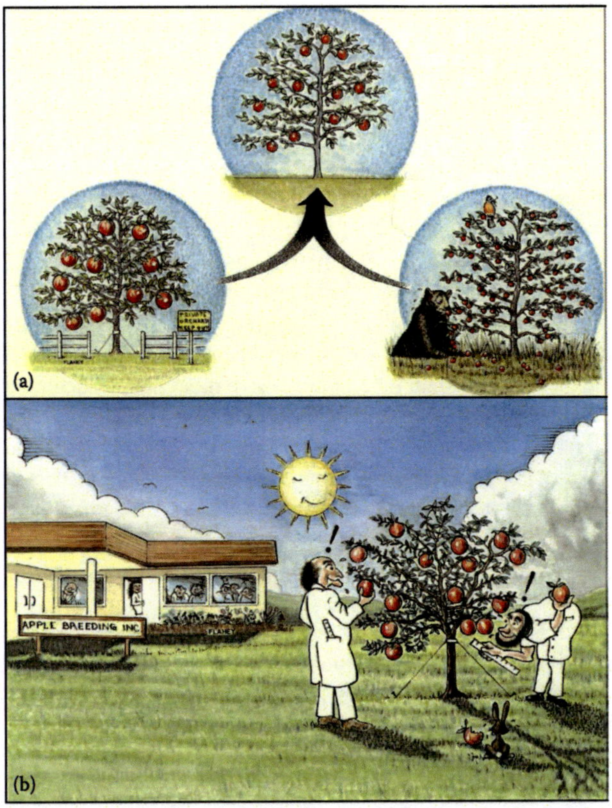


**FIGURE 1.3** (a) Hybridization between domesticated and wild plants. Continuing hybridization can result in gene transfer (introgression), which can weaken or invigorate adaptation and alter the survival of either or both of the wild and domesticated plants. (Painting courtesy of B. Flahey.) (b) Advanced modern breeding. Biotechnology now has the potential of transferring genes among all species (even some that are extinct), of synthesizing artificial genes and perhaps even organisms, and possibly also of creating production systems that are independent of living creatures. (Painting courtesy of B. Flahey.)

only about 200 species have been domesticated as crops." This is certainly a gross underestimate (well over 100 crop species are listed in Table 1.5), but deciding on the number of domesticated food crops is arbitrary, since there is no criterion for how different a cultivated plants needs to be from its wild ancestor to qualify as being domesticated.

Domestication is a continuing process (although man-made hybrids and cultivated forms that have had their chromosome count deliberately doubled sometimes represent "instant domestication"). "Cultivars" are named, distinctive, plant domesticates. These are usually dramatically altered from their wild ancestors, although some cultivars are merely simple mutants, or selections from the wild, and many have been grown in cultivation only for a few generations. Note that in Tables 1.2 and 1.3, 34 of the 100 crops examined in this book are considered to have advanced cultivars and 3 have merely "elementary" cultivars.

## Geography, Ecology, and History of Plant Domestication

Most important crops were domesticated before the dawn of written history (Hawkes 1983), and various lines of evidence need to be considered to trace the place of origin and identify the ancestors (if they are still extant). In many cases, the ancestors are indigenous in one location and the cultivated crops were domesticated elsewhere, sometimes in one primary location and sometimes also in one or more secondary locations. Sometimes domestication took place independently in different places at different times.

De Candolle (1882) and Vavilov (1926), early students of the geography of plant domestication, concluded that most crop plants originated in several distinctly defined areas of the world, predominantly within the tropics and subtropics and usually in mountainous regions. Vavilov noted that in the Old World, crop domestication usually occurred approximately between the Tropic of Capricorn in the southern hemisphere and about 45°N of the equator; and in the New World, crop domestication occurred approximately between the Tropic of Cancer (north latitude 23.5°, through Mexico) and the Tropic of Capricorn (south latitude 23.5°).

Most major crops originated from regions with pronounced seasons, and in such environments plants must store nutrients during the hostile season (when drought or cold prevent growth). Seeds and underground storage organs (roots, tubers) are where most plants accumulate nutrient reserves to survive from one season to the next. Annual plants in seasonal environments produce seeds with stored nutrients at maturity, while many perennial plants store nutrients in underground organs to survive droughts and/or cold periods. Seeds, roots, and tubers are the principal sources of food for most people, and the main crops now used to produce seeds and underground storage organs arose to an appreciable extent in different areas of the world: the cereals are frequently from areas with markedly different wet and dry seasons, and root/tuber crops are often from tropical lowlands with markedly dry periods.

Both plant and animal biodiversity tend to increase from the poles to the equator. It seems logical that in areas with greater numbers of species, more are good candidates for becoming crops. Indeed, there are fewer species in polar regions, and no significant crops. However, although there are many more species in tropical regions, the tropics are not especially important as the site of origin of the world's most important crops.

Bioclimatic factors have clearly been important in determining where important crops arise. "Biomes" are major, regional or global biotic communities, with distinctive dominating plant forms adapted to particular climates and ecological circumstances. Terrestrial biomes include tundra, desert, shrublands, savanna, seacoasts, rainforests, deciduous forests, and evergreen coniferous forests. Domesticated plants have arisen from most biomes, but much more so from some than others. Hawkes (1998) explained the general lack of domesticates from tropical forests, despite their huge numbers of plants and animals, as due to a lack of need for cultivated crops because wild fruits, nuts, and starchy tubers are available throughout the year. Conversely, he attributed the large numbers of domesticates from southern Europe as due to a lack of sufficient wild food plants to provide yearlong nourishment. Hawkes interpreted the large frequency of domesticates from mountainous regions as due to (1) a very wide range of genetic variation adaptive to the very wide habitat variation present, (2) genetic isolation in localized areas that facilitated local fixation of adapted genotypes, and (3) isolated populations of people who could select and perpetuate the most desirable kinds.

Vavilov (1926, 1949–1950) recognized eight principal (primary) centers of apparent crop domestication, none of these in North America north of Mexico; although his work laid the foundation for much of the historical biogeography of crop origins, some of his conclusions are now considered obsolete. Harlan (1992), perhaps the second most influential student (after Vavilov) of the geography of crop domestication, accepted that the origin of many crop plants is indeed associated with some regions, but contended that numerous crops trace their origins to rather diffuse areas. He identified three relatively well-defined "centers" of agricultural origin: the Near East (Jordan, Syria, Turkey, Iraq, and Iran), Mesoamerica (Mexico and Central America), and north China. Harlan also identified three relatively diffuse "noncenters": Africa, Southeast Asia, and South America. Zohary and Hopf (2000) documented the geographical history of Old World crops, and Sauer (1994) detailed the geographical areas of origin of most of the world's major crops. These references indicate clearly that North America north of Mexico is of very minor importance as a site of origin of the world's leading food crops (note Figure 1.4 and Table 1.5).

**TABLE 1.4**
**World's Billion Dollar Food Crops and Their Geographical Sites of Origin**

| Species or Crops | Mean Annual Value (2000–2009) in Millions of 2012 U.S. Dollars[a] | Geographical Origin | Food Class | Edible Representative of the Genus Among the Top 100 Indigenous Crops of North America (Indicated by NA) |
|---|---|---|---|---|
| Almond (*Prunus dulcis* (Mill.) D.A. Webb) | 4330 | Mediterranean Basin | Nut | NA |
| Apple (*Malus domestica* Borkh.) | 31,473 | Kazakhstan | Fruit | |
| Apricot (*Prunus armeniaca* L.) | 2124 | Asia | Fruit | NA |
| Asparagus (*Asparagus officinalis* L.) | 7529 | Eurasia, North Africa | Vegetable | |
| Avocado (*Persea americana* Mill.) | 2241 | Mexico and Central America | Vegetable | |
| Banana + plantain (cooking banana) (*Musa ×paradisiaca* L.) | 23,404 | Malesia | Fruit (banana) and vegetable (plantain) | |
| Barley (*Hordeum vulgare* L.) | 20,371 | Middle East | Cereal | |
| Beans (dry + "string beans") (*Phaseolus vulgaris* L. + others) | 18,310 | South America | Vegetable | NA |
| Broad bean (*Vicia faba* L.) | 1512 | Near East | Vegetable | |
| Cabbages and other brassicas (*Brassica oleracea* L., *B. rapa* L.) + others) | 11,957 | *B. oleracea*: Europe<br>*B. rapa*: China | Vegetable | |
| Carrot (*Daucus carota* L.) + turnip (*Brassica rapa* L. subsp. *rapa*) | 7683 | Carrot: western Asia Turnip: Europe | Vegetable | |
| Cashew (*Anacardium occidentale* L.) | 2415 | Tropical America (probably northeastern Brazil) | Nut | |
| Cassava (*Manihot esculenta* Crantz) | 20,268 | Brazil and Paraguay | Vegetable | |
| Cauliflower (*Brassica oleracea* L. var. *botrytis* L.) + broccoli (*B. oleracea* var. *italica* Plenck) | 4627 | Europe | Vegetable | |
| Cherries (*Prunus* species, especially sweet cherry, *P. avium* (L.) L. and sour cherry, *P. cerasus* L.) | 2736 | Sweet cherry: Eurasia and North Africa<br>Sour cherry: southwestern Asia and southeastern Europe | Fruit | NA |

*(Continued)*

**TABLE 1.4 (*Continued*)**
**World's Billion Dollar Food Crops and Their Geographical Sites of Origin**

| Species or Crops | Mean Annual Value (2000–2009) in Millions of 2012 U.S. Dollars[a] | Geographical Origin | Food Class | Edible Representative of the Genus Among the Top 100 Indigenous Crops of North America (Indicated by NA) |
|---|---|---|---|---|
| Chestnut (*Castanea* species, especially Chinese chestnut, *C. mollissima* Blume, European chestnut, *C. sativa* Mill., and Japanese chestnut, *C. crenata* Siebold & Zucc.) | 1203 | Eurasia | Nut | NA |
| Chickpea (Garbanzo bean; *Cicer arietinum* L.) | 3337 | Turkey | Vegetable | |
| Chile peppers (*Capsicum* species, especially *C. annuum* L.) | 16,616 | Mexico, Central America, South America | Vegetable | |
| Cocoa bean (chocolate, cacao; *Theobroma cacao* L.) | 4077 | South America, West Indies | Flavorant | |
| Coconut (*Cocos nucifera* L.) | 5229 | Tropical and subtropical seacoast through Asia, Africa, Latin America, and Pacific region; possibly originated in Indo-Malayan region or the Pacific islands | Miscellaneous | |
| Coffee (green coffee; *Coffea* species, especially *C. arabica* L.) | 6717 | *C. arabica*: Ethiopia | Flavorant | |
| Corn (maize; *Zea mays* L.) | 105,365 | Mexico, Guatemala | Cereal | |
| Cotton (seed; *Gossypium* species, especially American upland cotton, *G. hirsutum* L.) | 10,790 | American upland cotton: Central America<br>Other species: South America, Africa, Asia | Edible oil | |
| Cowpea (black-eyed pea; *Vigna unguiculata* (L.) Walp. subsp. *unguiculata*) | 1851 | Tropical Africa | Vegetable | |
| Cucumber (*Cucumis sativus* L. and others) | 12,005 | *C. sativus*: India | Vegetable | |

| | | | | |
|---|---|---|---|---|
| Date (*Phoenix dactylifera* L.) | 3677 | Western Asia and North Africa | Fruit | |
| Eggplant (aubergine; *Solanum melongena* L.) | 8086 | Southeast Asia | Vegetable | |
| Garlic (*Allium sativum* L.) | 8722 | Western Asia | Flavorant | NA |
| Ginger (*Zingiber officinale* Roscoe) | 1151 | Tropical Asia or Indonesia | Flavorant | |
| Grape (*Vitis* species, mostly *V. vinifera* L.) | 43,507 | *V. vinifera*: Mediterranean area | Fruit | NA |
| Grapefruit (*Citrus* ×*paradisi* Macfad.) and pomelo (*C. maxima* (Burm.) Merr.) | 1212 | Grapefruit: West Indies<br>Pomelo: Southeast Asia | Fruit | |
| Hazelnut (filbert; *Corylus* species, especially European filbert, *C. avellana* L., and Turkish filbert, *C. colurna* L.) | 1594 | European filbert: Europe, western Asia<br>Turkish filbert: southern Europe and Turkey | Nut | NA |
| Leek (*Allium ampeloprasum* L.) and other alliaceous vegetables | 2054 | Eurasia, Africa | Vegetable | NA |
| Lemon (*Citrus limon* (L.) Burm. f.) and lime (*C. aurantiifolia* (Christm.) Swingle) | 5224 | Lemon: northwest India?<br>Lime: Indo-Malayan region | Flavorant | |
| Lentil (*Lens culinaris* Medik.) | 1438 | Near East | Vegetable | |
| Lettuce (*Lactuca sativa* L.) and chicory (*Cichorium intybus* L.) | 11,692 | Lettuce: Asia Minor or Middle east<br>Chicory: Eurasia and North Africa | Vegetable | |
| Mango (*Mangifera indica* L.), mangosteen (*Garcinia mangostana* L.), and guava (*Psidium guajava* L.) | 17,384 | Mango: southern Asia<br>Mangosteen: Sunda Islands, Moluccas, or Malaya<br>Guava: southern Mexico, Central America | Fruit | |
| Melon (*Cucumis melo* L.) | 6246 | Old World tropics | Fruit | |
| Millet (several grass species, especially pearl millet, *Pennisetum glaucum* (L.) R. Br.; proso millet, *Panicum miliaceum* L.; foxtail millet, *Setaria italica* (L.) P. Beauv.; finger millet, *Eleusine coracana* (L.) Gaertn.; and barnyard millet, *Echinochloa frumentacea* Link) | 5892 | Africa, Asia, Europe | Cereal | |
| Oat (*Avena sativa* L.) | 3198 | Middle East | Cereal | |

(*Continued*)

**TABLE 1.4 (*Continued*)**
**World's Billion Dollar Food Crops and Their Geographical Sites of Origin**

| Species or Crops | Mean Annual Value (2000–2009) in Millions of 2012 U.S. Dollars[a] | Geographical Origin | Food Class | Edible Representative of the Genus Among the Top 100 Indigenous Crops of North America (Indicated by NA) |
|---|---|---|---|---|
| Okra (*Abelmoschus esculentus* (L.) Moench) | 3322 | Asia, North Africa | Vegetable | |
| Oil palm (*Elaeis guineensis* Jacq.) | 12,681 | West Africa | Edible oil | |
| Olive (*Olea europaea* L.) | 15,294 | Eastern Mediterranean area | Edible oil | |
| Onion (*Allium* species, mostly *A. cepa* L.) | 15,012 | Middle East or Asia | Vegetable | NA |
| Orange (*Citrus* species, especially *C. sinensis* (L.) Osbeck) | 12,262 | Asia | Fruit | |
| Papaya (*Carica papaya* L.) | 2262 | Central or South America | Fruit | |
| Pea (*Pisum sativum* L.) | 7083 | Middle East | Vegetable | |
| Peach and nectarine (*Prunus persica* (L.) Batsch) | 9408 | Tibet and western China | Fruit | NA |
| Peanut (groundnut; *Arachis hypogaea* L.) | 13,865 | South America | Nut | |
| Pear (*Pyrus* species, especially common pear, *P. communis* L. and Asian pear, *P. pyrifolia* (Burm. f.) Nakai) | 9229 | Common pear: Eurasia<br>Asian pear: eastern Asia | Fruit | |
| Persimmon (*Diospyros* species, especially *D. kaki* Thunb.) | 1644 | *D. kaki*: Asia | Fruit | NA |
| Pigeon pea (*Cajanus cajan* (L.) Huth.) | 1060 | Probably India (often claimed to be from Africa) | Vegetable | |
| Pineapple (*Ananas comosus* (L.) Merr.) | 4624 | Southern Brazil, Paraguay | Fruit | |
| Pistachio (*Pistacia vera* L.) | 2097 | Central Asia | Nut | |
| Plum (*Prunus* species, primarily common plum, *P. domestica* L.; the sloe (*P. spinosa* L.) is included by FAO, but is a rather insignificant fruit crop) | 5182 | Common plum: Caucasus and trans-Caucasus region<br>Damson plum: Europe, western Asia<br>Sloe: Africa, Eurasia | Fruit | NA |
| Potato (*Solanum tuberosum* L.) | 54,805 | South America (Andes, Chile) | Vegetable | |

| | | | | |
|---|---|---|---|---|
| Pumpkin and squash (*Cucurbita* species: *C. pepo* L., *C. maxima* Duchesne, *C. moschata* Duchesne, *C. argyrosperma* C. Huber) and gourds (*Cucurbita* species; bottle gourd, *Lagenaria siceraria* (Molina) Standl.) | 3996 | *Cucurbita* is native to southern Mexico and northern Central America (*C. pepo* in small part originated from the United States)<br>Bottle gourd probably originated in tropical Africa and was introduced in early times to South America | Vegetable | NA |
| Rapeseed (canola; *Brassica* species, mostly turnip rape, *B. rapa* L. subsp. *oleifera* (DC.) Metzg. and Argentine rape, *B. napus* L. subsp. *napus*) | 14,336 | Eurasia | Edible oil | |
| Rice (paddy rice; mostly Asian rice, *Oryza sativa* L.; also African rice, *O. glaberrima* Steud.) | 159,643 | Asian rice: Southeast Asia<br>African rice: Niger, West Africa | Cereal | |
| Rye (*Secale cereale* L.) | 2270 | Southwestern Asia | Cereal | |
| Sesame (*Sesamum indicum* L.) | 1651 | Africa or India | Flavorant | |
| Sorghum (*Sorghum bicolor* (L.) Moench) | 9839 | Africa | Cereal | |
| Soybean (Soja; *Glycine max* (L.) Merr.) | 51,827 | China | Vegetable | |
| Spinach (common spinach, *Spinacia oleracea* L. + others) | 4597 | Common spinach: Asia | Vegetable | |
| Strawberry (*Fragaria* ×*ananassa* Duchesne ex Rozier) | 6963 | A garden-generated hybrid of Virginia strawberry, *F. virginiana* Mill.) of eastern North America and Chilean strawberry (*F. chiloensis* (L.) Mill.), found from Chile to the mountains of western North America | Fruit | NA |
| Sugar beet (*Beta vulgaris* L. subsp. *vulgaris*) | 11,034 | Eurasia, North Africa | Flavorant | |
| Sugar cane (*Saccharum officinarum* L. + hybrids) | 39,726 | Southeast Asia | Flavorant | |
| Sunflower (*Helianthus annuus* L.) | 7610 | United States | Edible oil | NA |
| Sweet potato (*Ipomoea batatas* (L.) Lam.) | 10,327 | Mexico, South America | Vegetable | |

(*Continued*)

**TABLE 1.4 (*Continued*)**
**World's Billion Dollar Food Crops and Their Geographical Sites of Origin**

| Species or Crops | Mean Annual Value (2000–2009) in Millions of 2012 U.S. Dollars[a] | Geographical Origin | Food Class | Edible Representative of the Genus Among the Top 100 Indigenous Crops of North America (Indicated by NA) |
|---|---|---|---|---|
| Tangerine and mandarin oranges (*Citrus reticulata* Blanco), clementine (*C. reticulata* × *C. aurantium* L.) | 6992 | Tangerine and mandarin oranges originated in southeastern Asia Clementine is an artificial hybrid | Fruit | |
| Taro (cocoyam; *Colocasia esculenta* (L.) Shott) | 2929 | Southern Asia | Vegetable | |
| Tea (*Camellia sinensis* (L.) Kuntze) | 5231 | Assam–Burma–Yunnan triangle | Flavorant | |
| Tomato (*Lycopersicon esculentum* Mill.) | 49,261 | Mexico, Central America, South America | Vegetable | |
| Triticale (×*Triticosecale* Wittm.) | 1760 | No natural distribution; an artificial cereal crop genus, created by crossing wheat (*Triticum*) and rye (*Secale*) | Cereal | |
| Walnut (*Juglans* species, especially Persian walnut, *J. regia* L. and Japanese walnut, *J. ailantifolia* Carrière) | 2904 | Persian walnut: southeastern Europe, western Asia Japanese walnut: Japan | Nut | NA |
| Watermelon (*Citrullus lanatus* (Thunb.) Matsum. & Nakai) | 12,960 | Kalahari Desert of southern Africa | Fruit | |
| Wheat (*Triticum* species, especially common wheat, *T. aestivum* L.) | 104,258 | Common wheat was probably selected in the Near East, likely in the Fertile Crescent | Cereal | |

| | | | |
|---|---|---|---|
| Yams (*Dioscorea* species) | 19,435 | Asia: *D. alata* L., *D. esculenta* (Lour.) Burkill, *D. polystachya* Turcz.<br>West Africa: *D.* × *cayenensis* Lam., *D. rotundata* Poir<br>South Africa: *D. elephantipes* (L'Hér.) Engl.<br>Africa, Asia, Australia: *D. bulbifera* L.<br>South America: *D. trifida* L. f. | Vegetable |

[a] This table presents information for crops with mean annual cumulative values for all reporting countries of the world, of more than 1 billion dollars. The means were calculated from data in FAOSTAT (http://faostat.fao.org/default.aspx; accessed February 2012). Note that a proportion of the harvest of some species is used for nonfood purposes, such as forage, fuel, and inedible industrial or medicinal products; the data for several classes of crops are combined; and there are some ambiguities at the FAOSTAT website regarding the species identity of crops. Also keep in mind that "farm gate" receipts (money paid to producers) does not reflect on-site usage of crops, and that because many crops are starting materials for extracts and products, their generated value-added income may considerably exceed the values reported above.

Students of crop evolution have noted the close association of major ancient civilizations and the development of the world's major crops. Although people in many areas of the world selected superior cultivated food crops from local wild plants, this process was advanced to a far greater degree in the great cultures of Eurasia and in the Americas south of the United States, where the world's leading food crops originated. By comparison, the indigenous plants of North America north of Mexico (where, coincidentally, classical civilizations comparable to the great urbanized areas of history did not develop) have not produced any of the world's leading food plants, except for sunflower. As detailed in the following, in North America north of Mexico, transfer of superior domesticates from Mexico and more southern areas in pre-Columbian times suppressed progress of domesticating indigenous North American plants, and transfer of superior Eurasian domesticates in post-Columbian times led to the abandonment of consumption of more than a thousand wild food plant species native to the North American continent north of Mexico.

## NORTH AMERICAN WILD AND DOMESTICATED INDIGENOUS FOOD CROPS

### Pre-Columbian Agriculture in North America North of Mexico

Contrary to a rather common misconception, North America north of Mexico was not sparsely populated in pre-Columbian (or "precontact") times; it probably was home to 7 to 12 million inhabitants. Also contrary to popular portrayal, at the time of the discovery of the New World, all North American Indians did not live in small migratory bands subsisting on hunting, fishing, and gathering wild plants, but in fact many had established semipermanent villages and some were farmers. However, the degree to which agriculture was adopted varied considerably. The Inuit and several Indian tribes of the Far North were almost completely dependent on hunting and fishing. In fact, animals had not been domesticated for food anywhere in the United States or Canada, and diets were usually meat-heavy, based on hunting and fishing.

Hancock (2004) pointed out that "Until the last 500 years virtually all crop dispersal was within continents and not between them." In the case of North America, the areas immediately to the south of the United States were of key influence in determining the course of agriculture. The Mesoamerican center (extending from Mexico City to Honduras) was responsible for the domestication (or at least one of the domestications) of corn (maize), amaranth, avocado, beans, chile peppers, pumpkins, and squash. The cultivation of most of these was taken up by Native Americans, especially in the southwestern United States, and in some cases as far north as southern Canada. Native North Americans cultivated corn, beans (either common bean, *Phaseolus vulgaris* L., or tepary bean, discussed in this book), and squash (collectively known as "the three sisters") together because of their complementary nutritional and agricultural advantages. As companion plants, the corn provided a platform on which the bean vines could climb; the beans provided nitrogen to the soil through their nitrogen-fixing bacteria; and the squash vines growing on the soil shaded out weeds, deterred pests through their spiny surfaces, and acted as a mulch to slow water evaporation from the soil. Corn is low in the amino acids lysine and tryptophan, essential for protein synthesis, whereas beans tend to be low in methionine; in combination, corn (a cereal) and beans (a legume or pulse) compensate for each other's deficiencies, and indeed cereal–pulse combinations are very common throughout the world (this topic is discussed additionally in the following). The importation of such very successful crops to the United States and Canada was responsible in large part for the suppression of interest in utilizing and domesticating indigenous North American crops.

### Pre-Columbian Food Crop Domestication in North America north of Mexico

Several plants were grown and possibly domesticated for medicinal and other nonfood purposes in ancient North America. The following discussion is restricted to crops that are at least partly used as food. Ancient domestication of North American food crops south of the U.S.–Mexico border

**FIGURE 1.4** Location and approximate domestication times of some of the world's major crops. (Modeled after Procheş, Ş., et al., *BioScience*, 58, 151–159, 2008; Prepared by B. Brookes.)

is very well documented, because the area is a major center of crop development. By contrast, prehistoric North American crop domestication is much less appreciated (e.g., Minnis and Elisens 2000). Probably dozens of species were cultivated (or at least managed), but good evidence of ancient domestication centers has been found in only two areas: in the southwestern United States and in the eastern United States. Crops of these centers included some that are now extinct; a few that are extant but of no or almost no importance; a few minor crops; and sunflower, the only crop of major world importance, which originated from North America north of Mexico (Heiser 1993).

One very minor "center" of domestication is the southwestern United States. Devil's claw was domesticated, at least in part, in north of Mexico (the two species of devil's claw are discussed in Chapter 34). However, the cultivated selections were domesticated primarily as material for basketry and similar applications rather than for food. An ancient cereal, Sonoran panic grass (*Panicum sonorum* Beal, also known as Mexican panic grass and sauwi), seems to have been domesticated and grown at least partly in the southwestern United States (Nabhan and de Wet 1984), although particularly used in southwestern Chihuahua. Tepary bean (treated in Chapter 96) and amaranth (*Amaranthus hybridus* L.) may have undergone some domestication north of the U.S.–Mexico border, but perhaps mainly in Mexico.

A more substantial center of ancient North American domestication is the eastern United States, in the area now known as Alabama, Arkansas, Illinois, Kentucky, Missouri, Ohio, and Tennessee (Ford 1985; Smith 1989, 2006; Minnis 2003; Smith and Yarnell 2009; Price 2009), a region that has been termed "the Eastern Woodlands." In summarizing the evidence, Smith (2007) stated: "the eastern woodlands of the United States is now widely acknowledged as one of the ten or so world regions where human societies independently brought wild species of plants under domestication between 1000 and 4000 years ago." There is convincing evidence of domestication of sunflower (*Helianthus annuus*) and squash (*Cucurbita pepo*), important indigenous crops that are still grown today and are discussed in detail in this book. There is also clear evidence of domestication of chenopod (*Chenopodium berlandieri* Moq., also known as goosefoot) and annual marsh elder (*Iva annua* L., also known as marshelder and sumpweed). There is evidence of cultivation (and hence at least the suggestion that domestication may have occurred) of erect knotweed (*Polygonum erectum* L.), little barley (*Hordeum pusillum* Nutt.), maygrass (*Phalaris caroliniana* Walter, also known as Carolina canarygrass), and giant ragweed (*Ambrosia trifida* L.). Except for squash, all of these were harvested for their seeds, rich in oil or fat. After AD 800, the practice of cultivating corn (maize), imported from Mexico, came to dominate eastern North America (by contrast, cultivation of corn in the southwestern United States may date back 4000 to 5000 years ago). Of the seven crops, only sunflower and squash are still cultivated. Until recently, North America north of Mexico was usually not considered to have been a site of crop domestication. Today, while it is now recognized that eastern North America was a site of domestication, the fact that sunflower is the only crop of world importance to have been domesticated there (squash was independently domesticated in South America) rather diminishes the significance of the area.

## Most Prominent Indigenous North American Food Plants

Of the 100 crops featured in this book, only sunflower (*H. annuus*) is among the several dozen leading domesticated food crop species of the world. Other quite significant indigenous North American domesticated crop plants include numerous species of *Rubus* (primarily blackberry, dewberry, and raspberry), *Vaccinium* species (primarily blueberry and cranberry), grape (*Vitis* species), pecan, and wild rice, with substantial contributions to modern cultivars of such crops as hop (*Humulus*) and strawberry (*Fragaria*) from North American germplasm.

Several significant domesticated crops are native not just to the United States and Canada but to other regions of the world, and the U.S. and Canadian plants may have played a minor role as the ancestors of the cultivars. These crops include acerola, aronia, blue honeysuckle, cactus pear, and chive. Some successful crops of the Old World are based on species that occur there and in North America, but cultivars have been developed only or primarily from plants of the Old World. This is the case for chufa and elder.

Ginseng is the most important cultivated wild crop (i.e., the undomesticated plant is grown). This is a very unusual situation (major cultivated crops are almost always highly domesticated, with specialized cultivars). Ginseng is cultivated because the wild supply is endangered in most of the natural geographical range. It has not been domesticated because the largely Asian market is based on usage as a tonic, and tradition demands that the product be based on wild plants (albeit usually no longer from plants actually growing in the wild).

Sugar maple, the source of maple syrup, is the most important wildcrafted crop (i.e., based on collection of uncultivated, undomesticated plants). Piñon nuts are next in importance in this category. Relatively minor wildcrafted crops are wild rice (which is now mostly obtained from cultivars), blueberries (also obtained from cultivars), huckleberries, bilberry, cloudberry, persimmon, and wild leek (ramp). Mexican oregano extends into the United States, but is mainly obtained from wild plants in Mexico.

## Importance of North American Wild Plants

The information presented in this chapter indicates that plant species native to North America north of Mexico have had a limited contribution to modern plant foods, by comparison with several areas of the world that have produced the world's leading food crops. It is important that the impression not be left that indigenous North American plants are unimportant economically, either specifically for food, agriculture, or other purposes. Indeed, this book is dedicated to the most important indigenous food plants and points out that many of these have substantial potential for increased value.

The commercial importance of wild North American plants is reflected by other considerations than value for human food (for an extensive review, see Prescott-Allen and Prescott-Allen 1986). The most important harvested product from wild-growing plants is lumber for construction, pulp, industrial extracts, and other purposes. Also of immense value are wild grasslands, particularly those of the extensive North American prairies and plains, which maintain vast herds of livestock that produce considerable food for society. A particularly valuable category of wildcrafted plant (and other) materials is referred to as "nonforest products (NFPs)," and the harvest and processing of NFPs provides considerable employment and generates very large income from value-added processing (for reviews, see Emery and McLain 2001; Jones et al. 2002; Wetzel et al. 2006). NFPs include items used for food, health and personal care (particularly obtained from medicinal plants), landscaping, gardens, decorations, and other aesthetic products. "Floral" or "greenery" products include plants and parts of trees, frequently dried and used in craft items such as wreaths, potpourri, wildflower arrangements, fragrance items, ornaments, and table decorations. Essential oils are obtained not only from cultivated crops but also from wild plants such as balsam fir, birch, and spruce; these are used in cosmetics, insect repellants, insecticides, household cleaners, disinfectants, and other products.

## Categories of Food Usage with Particular Reference to North America

An analysis of common food categories of the top 100 American food plants in comparison to the world's billion dollar crops is shown in Table 1.5. Fruits make up more than a third of the American food plants and almost a quarter of the billion dollar crops. Indigenous cereals are very rare in North America. Other categories (vegetables, flavorants, nuts, edible oils) are not notably different.

## Principal Vascular Plant Families from which Important Food Plants Have Been Derived, with Particular Reference to North America

The number of plant families that should be accepted is currently unclear, but usually 300 to 400 are recognized, about 200 of which are native to the United States and Canada. Procheş et al. (2008) found that plant families are mostly comparable in proportion of "edible" species (i.e., although families differ greatly in numbers of species, they have about the same percentage of edible species). However, this should not obscure the fact that the most important food plants occur particularly in several families (see Ladizinsky [1998] for an extensive list of botanical families

**TABLE 1.5**
**Percentages of Principal Food Uses of the 100 American Food Plants Compared with the 78 Crops with World Values of over $1 Billion Annually**

| Food Class | Percentage of the 100 American Food Plants (Table 1.2) | Percentage of the 78 Billion Dollar Crops (Table 1.4) |
|---|---|---|
| Fruit | 36 | 23.7 |
| Vegetable | 28 | 36.5 |
| Flavorant (herb, spice, sweetener, stimulant beverage) | 19 | 11.5 |
| Nut | 8 | 9.0 |
| Edible oil | 6 | 6.4 |
| Cereal or pseudocereal | 2 | 11.5 |
| Miscellaneous | 1 | 1.3 |

**TABLE 1.6**
**Principal Plant Families That Have Produced the World's Dominant Crops and Their Representation in the List of Billion Dollar Food Crops and the List of Top 100 Indigenous North American Food Plants**

| Plant Family | Species Generating at Least $1 Billion Annually (Table 1.4) | Species Included in the Top Indigenous 100 North American Food Plants (Table 1.2) |
|---|---|---|
| Poaceae (Gramineae; grass family) | Barley, corn (maize), millets, oat, rice, rye, sorghum, sugar cane, triticale, wheat | Reed, wild rice |
| Fabaceae (Leguminosae; pea family) | Broad bean, common bean, chickpea, cowpea, lentil, pea, peanut, pigeon pea, soybean | Groundnut, hog peanut, mesquite, prairie turnip, tepary bean |
| Brassicaceae (Cruciferae; cabbage or mustard family) | Broccoli, cabbages, cauliflower, turnip, rapeseed | Scurvy grass |
| Rosaceae (rose family) | Apple, apricot, cherries, nectarine, peach, pear, plums, strawberry | Aronia, blackberries and dewberries, cloudberry, cherries, hawthorns and mayhaws, plums, raspberries, roses, salmonberry, saskatoon, strawberries |
| Solanaceae (potato or tomato family) | Chile peppers, eggplant, potato, tomato | Dwarf cape gooseberry |
| Apiaceae (Umbelliferae; carrot family) | Carrot | Honewort, Scotch lovage, yampah |
| Cucurbitaceae (cucumber family) | Cucumber, melon, pumpkins and squash, watermelon | Buffalo gourd, squash |
| Liliaceae (lily family) | Asparagus, garlic, leek, onion | Canada garlic, chive, nodding onion, wild leek |
| Araceae (arum family) | Taro | |

*Source:* After Harlan, J.R., *Crops and Man*, second edition, American Society of Agronomy, Crop Science Society of America, Madison, WI. p. 68, 1992.

containing major crop plants). Table 1.6 provides information on the plant families with food species of world importance and compares the representation of these families in North America. As discussed later, the representation of cereals (of the grass family) and pulses (of the legume family) is very much reduced in North America, compared to other locations, and this has very important significance. The rose family has produced numerous temperate zone fruits and is very well represented in Canada and the United States (Table 1.6). However, other families known for having

**TABLE 1.7**

**Plant Families Represented in North America North of Mexico with the Largest Proportion of Species Present That Were Used as Food by Native North American Indians**

| Plant Families with the Largest Proportion of Species Used as Food (After Moerman 1996) | Species Included in the Top Indigenous 100 North American Food Plants |
|---|---|
| Liliaceae (lily family) | Canada garlic, chive, nodding onion, wild leek |
| Rosaceae (rose family) | Aronia, blackberries and dewberries, cherries, cloudberry, hawthorns and mayhaws, plums, raspberries, roses, salmonberry, saskatoon, strawberries |
| Ericaceae (heath family) | Bilberry, blueberries, cranberry, crowberry, huckleberries, Labrador tea, lingonberry, salal |
| Apiaceae (carrot family) | Honewort, Scotch lovage, yampah |
| Chenopodiaceae (goosefoot family) | Epazote |
| Fagaceae (beech or oak family) | American chestnut, chinkapin |
| Grossulariaceae (currant family) | Golden currant, northern gooseberry |
| Cactaceae (cactus family) | Cactus pear, saguaro |
| Pinaceae (pine family) | Piñon pine |
| Polygonaceae (knotweed family) | Sea grape |

*Source:* Based on Moerman's analysis of 1625 species that were consumed as food. Moerman, D.E., *J. Ethnopharmacol*, 52: 1–22, 1996. [Note that the analysis includes introduced species and that some species may be extremely important for food, but their family may not be represented here because most of the species were not consumed as food.]

produced major food plants are not well represented in North America (Table 1.6). Moerman (1996) listed the top plant families with respect to proportion of species used as food by indigenous people (Table 1.7). Of these, only the rose family and blueberry families have substantial representation in the top 100 food plants discussed in this book. Once again, plant families known for having produced many of the world's most important crops are poorly represented in North America.

## Plant Genera with Food Crops of Major World Importance That Are Represented by Food Species in North America

Eleven of the genera of crops represented in the 78 billion dollar crops of the world (Table 1.2) are also represented by species in the top 100 North American indigenous food species (Table 1.2). However, only one of the genera, *Helianthus*, has a species (*H. annuus*, sunflower), which is both an indigenous North American crop and one of the world's billion dollar crop species. The remaining top 100 North American indigenous plants are of much less importance or have merely contributed genes to the species that are important. The top 100 North American indigenous crops include 5 relatively minor species (or species groups) of *Prunus* (cherries, plums, etc.) and 3 relatively minor species of *Allium* (garlic, onion, etc.). Similarly, North America has relatively minor species of genera that contain species of world importance: the grape genus *Vitis*, the bean genus *Phaseolus*, the squash genus *Cucurbita*, the strawberry genus *Fragaria*, the persimmon genus *Diospyros*, the chestnut genus *Castanea*, the hazelnut genus *Corylus*, and the walnut genus *Juglans*.

The apparent limited importance on a world scale of indigenous North American food crop species is somewhat misleading. Many fruit crops (including blackberry/dewberry, cherry, grape, plum, raspberry, and strawberry) have utilized North American species as important sources of germplasm for cultivar improvement. Other North American food plants that have also been used significantly to breed improved cultivars include evening primrose, hazelnut, hop, and squash. Another important contribution of North American species is that some of the tree fruit species are used as rootstocks on which cultivars are grafted.

## RELATIVE HUMAN CONSUMPTION OF ANIMALS AND PLANTS: CRITICAL CONSIDERATIONS

The much-quoted Old Testament phrase "All flesh is grass" reflects the fact that the animals that we consume ultimately trace to consumption of plants. Another biblical phrase, "milk and honey," points out two of mankind's most desired foods of animal origin, which would not exist but for the pastures available to cows and bees. The herbivores that we humans eat transform plant tissues into forms that are exceptionally nutritious, often much more so than available plant tissues, and this has had profound influences on the evolution of humans, as explained in this section. The comparative availability of animal and plant foods has been a major determinant of agriculture and civilization, as explained in the following.

### Comparative Human Adaptation to Consuming Plants and Animals

Both animals and plants are of great importance for very valuable nonfood resources. Hides of wild animals are useful for clothing, bedding, and shelter, and the bones, horns, and sinews are useful for making weapons, implements, and decorative objects. Wild plants have also been used as materials for shelters, tools, and fabrics. Some uses are more or less unique to plants (most medicines have come from plants in the past), and some uses are more or less unique to animals (for example, horses used for transportation and dogs used to assist in hunting). However, animals and plants have always been valued primarily as food, and domestication is best understood in this regard.

Dietary adaptations are critical to understanding the past, present, and future of humans, and as explained in this section, the relative consumption of plant foods and meat may be the most important consideration, both from theoretical and practical perspectives. Humans have demonstrated amazing dietary flexibility, capable of thriving on either meat or plant staples. Nevertheless, as argued below, we have inherited a genetic legacy of adaptation to and preference for eating meat, and this has hindered the domestication of indigenous North American food crops. Moreover, as also discussed in the following, excessive consumption of meat in modern times is a major contributor to the obesity and correlated health epidemic, and the growing world demand for more meat and less plant foods in the diet is a major contributor to the deterioration of the ecology of the planet and its natural biodiversity.

Humans are naturally omnivorous, adapted to consuming both plants, animal, fungi, and indeed most groups of living things currently recognized. We are also generalist feeders, suited to a very varied diet; indeed, by a considerable margin, humans consume more species than any other organism that exists or has ever existed. Flowering plants (angiosperms) constitute 50% to 90% of the volume of food of most humans, with up to 75,000 species being edible (if not necessarily palatable) and about 7000 commonly eaten somewhere (Procheş et al. 2008). Most of the remaining human diet is furnished by vertebrate animals, with minor contributions from algae, fungi, and other groups of animals and plants (Procheş et al. 2008). The average world diet is composed of about 14% meat (including fish, poultry, and products such as eggs and milk, but excluding animal fats such as lard) and virtually all of the remainder are plant-based foods (Small 2009, p. 4). However, the modern human diet is not necessarily the same as in the past, and as discussed below, past food consumption likely was quite different.

### Evolution of Human Meat Consumption

In this section, a rather widespread view is adopted that the evolutionary origin of humans, commonly accepted as having occurred in east Africa and marked by the transformation of ape-like, forest tree-dwellers to bipedal human-like species adapted to hunting in savannas, was accompanied by a substantial increase in meat-eating from a former mostly vegetarian lifestyle. The associated changes in human anatomy, physiology, and psychology led directly to human dominance of the world. In short, meat turned man's primate ancestors (of the genus *Homo*) into humans. This phenomenon is dealt with at some length here, because it explains much of the human relationship with plant-based foods in North America.

The anthropoid primates from which humans evolved extend back over 25 million years, and many of their dietary adaptations likely have been conserved in humans (Milton 2000). Species of the primate line that gave rise to hominids (great apes) are (or seem to have been) herbivores primarily, with some omnivory, and so hominids in general may have tended toward herbivory (Johns 1990; Milton 2000; Barnicott 2005). Certain nutrients, such as vitamin C, carotenes (vitamin A and several antioxidants), fiber, and energy-rich carbohydrates, are only or mostly obtainable by humans from plant foods, and all primates, including humans, have retained adaptations for consumption of plants. (Carnivores often consume the gastrointestinal tract of herbivores, and so benefit from plant tissues that are present.) Humans have notably higher concentrations of amylase (the enzyme which digests starch) by comparison with related primates, indicating that we have shifted somewhat from the fruit and foliage diet prevalent in apes to one that has more starch, present in numerous plant storage organs and in many seeds (Perry et al. 2007; Lucas 2011).

Hominids (relatively advanced primate genera descended from gorillas, apes, and orangutans) first appeared perhaps 7 million years ago in east Africa, from which the genus *Homo* arose perhaps 2.5 million years ago. There is considerable support for the theory that the evolution of the brain during the early evolution of *Homo* (about 2.0–1.7 million years ago) was necessarily correlated with a shift from eating mostly plants to adding a very substantial proportion of meat to the diet (e.g., Bogin 1997; Gibbons 1998; Ungar 2006; Leonard et al. 2007). The anatomy and physiology of modern humans indicate striking adaptation to what has been termed "endurance hunting." Although humans are slow compared to most terrestrial mammals, it has been claimed that our skeleton, musculature, and bare skin (facilitating dissipation of heat by evaporation) is ideal for persistently following animals until they overheat and succumb (in short, we seem to be the world's best long-distance runners). There is good reason to believe that the immediate ancestors of humans evolved large brains to carry out the complex activities needed to acquire and process a wider variety of foods than any other species, by cooperative hunting and gathering. Humans have much larger brains in relation to body weight by comparison with other primates, and indeed in relation to other terrestrial mammals. Moreover, by comparison with other primates, humans allocate a much larger share of their energy budget to brain metabolism (Leonard et al. 2007). The human gastrointestinal tract, especially the colon, is reduced by comparison with other primates, and human teeth are also less robust, and so humans would have great difficulty in supplying the needed energy for the brain on the kind of plant-based diet that maintains most primates and probably sustained early hominids. It has been concluded that high-quality food, likely animal food, was necessary, and that in the early evolution of humanoids there indeed was a substantial shift to meat consumption. The fact that humans evolved in tropical Africa, where animals tend not to develop fat stores, has been suggested as an indication that plants were still necessary to provide essential fats (Milton 2000), but it has been shown that wild plant foods like tubers and nuts on the African savanna (where hominids evolved) contain only trace amounts of the key fatty acids required by the brain (docosahexaenoic acid and arachidonic acid), but muscle and organ tissues of wild African ruminants provide sufficient levels. It has been noted that fatty cuts of meat are often prized in hunter–gatherer societies (Farrington and Urry 1985), but this could be simply because of the higher energy level.

The theory that animals were essential to supply brain development in humans is weakened by the fact that processing of plant foods, particularly by cooking, can also increase the supply of nutrients and lower the need for chewing and digestion of tough materials, compensating for the relatively small human gastrointestinal tract and puny jaws in comparison with apes. Wrangham (2009) championed the view that cooking (by using fire), not meat consumption, was the stimulus responsible for the evolution of big brains in early species of *Homo* (and the correlated reduction of big jaws, freeing the head to house a larger brain). Although building a fire requires energy expenditure, this is usually more than compensated by energy saved in digestion. Wrangham claims that cooking led to sexual division of labor (males hunting, females gathering), and even to pair-bonding (and marriage). The evidence for cooking, however, extends back not much more than 1.5 million years ago, which is after the time that big-brained hominids came into existence (Leonard et al. 2007). Brain tissue requires an excellent supply of glucose, most easily obtained from plant carbohydrates, suggesting that some plant intake is desirable. In any event, it seems clear that the human brain does require excellent nutrition, and that either or both of animals or high-quality plant foods are required. The fact remains, however, that meat from most wild animals is relatively easy to digest, while most wild plant tissues are not. Moreover, vitamin $B_{12}$, indispensable for life, may not be obtainable from flowering plants (it is available from bacteria and animals). Additionally, the vitamin riboflavin and the minerals calcium, zinc, phosphorus, and iron are more easily assimilated from animals. Still another reason why animal tissues are often preferable to plant tissues is the relative potential for toxicity. Sedentary species, which includes most plants and some animals, have tended strongly to develop toxins as protective agents. Vertebrate animals (especially mammals, birds, and fish) constitute almost all of the animal food consumed by humans. The flesh of vertebrates, which overwhelmingly are motile (and so can usually escape predators), is very rarely toxic. However, many animals can harbor diseases and parasites that can be transmitted to humans, and it may be that these have contributed to the many more taboos associated with meat by comparison with plant foods.

## Lack of a Prominent Indigenous Cereal–Pulse Combination in North America North of Mexico and How This Is Consistent with Dependence on Meat

Today, cereals are the most important category of food. Cereals furnish more than half of the food calories consumed in the world. Rice, wheat, and corn (maize) provide almost 90% of the world's cereal crop and are staple foods of over 4 billion people. Ten cereals are among the 78 crops with an annual world value of more than a billion dollars (Table 1.4). In the past, cereals were "the principal crops of most civilizations" (Zohary and Hopf 2000). In pre-Columbian times in North America, wild rice was the only important wildcrafted cereal. Today, it is the only domesticated cereal to have originated north of Mexico, but it remains a fairly minor plant. By contrast, corn, of Mexican origin, became the leading pre-Columbian crop of North America. Wild rice is an extremely attractive and nutritious gourmet food, but is dwarfed in efficient productivity by corn. Once corn became available to indigenous peoples, interest in developing wild rice would have been minimized. Pseudocereals are nongrass species that similarly furnish a considerable harvest of small, nutritious fruits (seeds). Earlier in this chapter, it was pointed out that several crops were harvested for their small nutritious seeds, including domesticated sunflower (usually classified as an oilseed, not a pseudocereal), chenopod, and marsh elder and the possibly domesticated erect knotweed, little barley (a true cereal), maygrass (another true cereal), and giant ragweed. Corn may have been instrumental in suppressing the development of these plants (except sunflower).

Legumes (species of the Fabaceae), particularly pulses (usually defined as legumes with seeds consumed by humans), are second only to the grasses (Poaceae) as sources of food. Nine species of pulses have a world value of more than a billion dollars (Table 1.4). Pulses have been of critical food value because they furnish most amino acids necessary for protein synthesis, and together with cereals provide all of the amino acids required. Agronomically, pulses are superb rotation crops for cereals, because the associated nitrogen-fixing bacteria in the root nodules enrich the soils. However, the only North American pulse treated in this book is tepary bean, a very minor domesticated crop. (Some people have expanded the definition of pulse to include all edible legumes; other edible legumes among the top 100 North American species include groundnut, hog peanut, mesquite, and prairie turnip, none of which provides a significant edible seed crop.)

The lack of a prominent domesticated cereal–pulse combination in North America is of huge significance. Together, cereals and pulses provide a balance of essential amino acids to synthesize protein, indispensable for survival, and especially needed when adequate meat is unavailable.

Anthropologists have noted that societies dependent on gathering make particular efforts to acquire both cereals and pulses, and indeed the same is true of advanced agricultural societies. In the Mediterranean area, the cereals wheat and barley have been combined with the pulse lentil. In Asia, the cereal rice has been combined with several Asian legumes, notably soybean and mung bean. In Africa, the cereal sorghum is consumed with several legume vegetables. In Mexico and South America, the cereal corn (maize) is combined with the pulse common bean and this combination was imported into North America north of Mexico. The cereal wild rice was important in eastern North America, and the legume vegetable prairie turnip was important in the west, but nowhere was there a prominent combination of cereal and legume that could function as a meat replacement, until corn and beans were imported from Mexico. The prominence of meat in the diet, discussed in detail below, is clearly consistent with the lack of an indigenous cereal–pulse combination.

## Why Hunting Wild Animals Was Preferable to Gathering Wild Food Crops, and How This May Have Retarded Crop Domestication in North America

As noted earlier in this chapter, various hypotheses have been advanced to explain the differential inception of agriculture in various areas and times. Agriculture was in fact practiced in the United States and Canada in pre-Columbian times. However, domestication of animals for food did not occur, and the selection of domesticated plants was not nearly as advanced as in other areas of the

**FIGURE 1.5** Comparative merits of hunting and cultivation of domesticated crops in post-Columbian North America. So long as game was plentiful, hunting (primarily by men) provided most nutrients required, but as the availability of wild animals decreased and key domesticated crops became available, cultivated plants were progressively substituted. (a) "Buffalo Hunt on the Southwestern Prairies" by the American painter and explorer John Mix Stanley (1814–1872). Photograph of the painting housed in the Smithsonian American Art Museum by AgnosticPreachersKid (CC by 3.0). (b) New York State Museum diorama showing Iroquois family planting beans, corn, and squash. As discussed in the text, these three domesticated crops, a combination known as "The Three Sisters," provided an adequate replacement for meat protein.

world. Why? The answer probably depends on the comparative availability and nutritional merit of meat and edible plants. Analyses of the diets of the few, modern hunter–gatherer societies, which frequently are high in meat, have suggested to some anthropologists that meat was important in the past. Extant hunter–gatherers may be unrepresentative of those of the past (Milton 2000), and it is probable that relative availability of plant foods and animals foods determined how much of each category was consumed. Nevertheless, there is good reason to believe that meat was critical to the survival of many, perhaps the majority of humans in the past.

Meat, especially the meat of mammals, is nearly identical in composition to human flesh, and accordingly is remarkably suited as food for humans. In contrast with almost all plant tissues, meat provides a perfect (complete) or near-perfect balance of the essential amino acids that make up protein. Meat is also a very concentrated source of nutrients (although nuts and some other forms of plant tissue are also very nutrient dense). Fat is the chief storage reserve in animals; carbohydrates are much more frequent in plants, but contain less energy than fats. Although fats are viewed very negatively today, in previous times the large supply of calories from fats would have been essential for survival when people were often subject to feast/famine cycles. In the past, North America was teeming with wild animals, especially large herbivorous mammals, birds, and fish. Moreover, people were willing to consume a much wider array of animals than marketed today, especially invertebrates such as shellfish provided by the extremely extensive river systems of the continent. (Several of the indigenous people cookbooks cited in the Appendix to this book provide recipes for preparing virtually every animal that most people encounter today only as roadkill.) The Inuit of the north, where food plants are in very short supply, existed almost exclusively on a diet of meat, compensating for the lack of calcium by consuming the bones and compensating for the deficiency of vitamin C by eating fresh liver. (Unlike humans, most animals synthesize their own vitamin C, which particularly accumulates in the liver. Nevertheless, meat rarely supplies sufficient vitamin C, and in the case of the Inuit, local herbs rich in the vitamin were consumed, such as scurvy grass discussed in this book.) In preagricultural times, except for many fruits and some nuts, the taste of plants would have been highly inferior (indeed, often somewhat repulsive) compared to modern crop cultivars, making meat all the more preferable. Although food plants were dried and stored for winter use, very few plant foods are available fresh during the winter, while many animals can still be hunted. In precivilization times at the hunter–gatherer stage of progress, people everywhere would certainly have preferentially killed animals for meat wherever and whenever they were available and the risk seemed worthwhile.

The nutritional limitations of many native food plants are indicated by an analysis of the food value of plants used by eastern Canadian indigenous peoples (Kuhnlein and Turner 1991). It was found that only the nuts and legumes were good sources of protein (several nuts are also excellent sources of fat). Good sources of indigenous legumes are very limited in the United States and Canada. Not surprisingly, acorns were very widely consumed in North America (indeed, throughout the world) in past times, although removing the toxic tannins is very time-consuming and troublesome.

Animal tissue ("meat" in the broadest sense) is generally the most desired food, but today it has become viewed negatively in some circles. In modern times, excessive consumption of meat contributes to health problems, particularly obesity. The trend for the production of beef marbled with saturated fat has led to cholesterol clogging arteries. Excessive harvesting of animals from the wild has endangered biodiversity. Growing affluence in developing countries, notably China and India, is associated with growing demand for a diet with more meat, and this is severely straining the world's food production capacity (Rask and Rask 2011). The planet's natural grasslands have largely been converted to cereal production or grazing pastures for livestock. Agricultural production of animals is associated with pollution of soil, water, and the atmosphere; degradation of ecosystems; and shrinking availability of fresh water.

A diet based predominantly on plant tissues is now known to contribute to health; conversely, dietary diseases are associated with avoidance of plant foods. For a minority, vegetarianism (or more strictly veganism, which excludes not only animals but also their products, such as milk and eggs) represents an ethical path to avoiding harm to animals and a way of diminishing mankind's harm to the world (since crops require much less resources than livestock to produce equivalent food material). However, as discussed earlier, anatomically, physiologically, and behaviorally *Homo sapiens* are adapted to an omnivorous diet, and our ancestors normally preferred meat over plants. It is important that the modern prejudice against meat not obscure its attractiveness and importance in the past.

Unlike animals, most of which can be consumed at almost any stage of their development and at any season, plant tissues that are edible are developed at certain seasons. Fruits and nuts (which technically are also fruits) are usually available only at a certain time, and underground storage organs are usually most nutritional during the inhospitable season. Vegetative tissues (stems, leaves) tend to be tougher and less nutritional as plants age (animal tissues tend to become tougher with age, but lose nutritional value much less). People learned to collect and store plant foods for times that they were unavailable. However, succulent fruits frequently do not store well, and primitive conditions made food storage difficult, especially for itinerant people. Meat on the hoof was sometimes also only available seasonally (e.g., the migrating American bison), but a variety of animal species could supply meat throughout much of the year.

Sufficient local availability of wild animals and/or plants would have limited the motivation of hunter–gatherer societies to raise livestock or cultivate crops (necessity is the mother of invention). In particular, since animals would have always been a preferable food, so long as wild game was available, people would have been less likely to progress toward agriculture. Indeed, it does seem that many preagricultural societies practiced considerable meat consumption, and that this was true for much of North America. The reasons why crops were domesticated in certain areas of the world but not others are complex and explanations are rather speculative. Nevertheless, the availability of meat has been advanced as a principal reason why plant domestication was limited in certain areas of the world, notably Australia and North America.

**FIGURE 1.6** Comparison of health effects of eating meat by hunter–gatherer societies and in modern times. Top panel: In the past, hunting demanded considerable physical effort, promoting fitness. Because it was dangerous, hunting (usually conducted by men) was intended to meet minimal (subsistence) needs, discouraging overeating. More easily obtained plants (usually harvested by women) provided health benefits not or minimally provided by meat. Bottom panel: Today, overabundance and overconsumption of processed, refined, nutritionally deficient prepared food products has led to an epidemic of obesity and associated health issues, worsened by sedentary, stressful lifestyles. (Illustrations courtesy of B. Brookes.)

## TASTE PREFERENCES IN RELATION TO THE DEVELOPMENT OF OLD AND NEW CROPS

> During a lifetime, a human being will eat thousands of pounds of food. The body will use this food to grow, to repair damaged tissue, and to maintain organs such as the brain and heart. Some of these foods will be enjoyable to eat because they are perceived to look appetizing and taste delicious. Other foods may not be enjoyable to eat, but will be consumed anyway because they are 'good for the body or the spirit'. Biochemically, the body does not distinguish between foods that are liked or disliked, for the human body does not use food, rather the body requires the biological nutrients contained in food.
>
> **Barry Bogin (1997)**

Students of food crop evolution have generally ignored the importance of taste, which is in fact a principal determinant of food plant domestication. Humans have natural preferences and aversions for certain food constituents, and these reflect adaptations for acquiring nutritional materials and avoiding toxins. The intrinsic, adaptive preferences discussed in the following section are responsible for why humans have domesticated certain species to provide staple foods, and accordingly it is important to understand how this constrains the future development of new food crops. The predominant importance of taste is emphasized here, but other sensory aspects of food—color, smell, and texture—may also have determined why some plants were or were not adopted as staples.

Human taste buds are mainly located on the tongue, but are also found in the mouth and upper throat. The five basic tastes are sweet, sour, bitter, salty, and (unknown to most people) unami (savory). Of these, sweet is by far the most attractive. For the most part, sugars that are responsible for sweetness are found naturally in fleshy fruits, so clearly their presence is a lure to animals that reward the plants by distributing their seeds. Many of our primate relatives are primarily frugivores (fruit eaters); fruits make up more than 75% of the diet of chimpanzees, bonobos (pygmy chimps), and orangutans (Lindeberg 2010). It seems obvious that humans have inherited a love of fleshy, sweet fruits. Such fruits are produced in numerous wild plants in most terrestrial areas of the world, and historically they have been harvested wherever they are found. Almost everywhere in the world, indigenous plants with edible fruits constitute the largest numbers of desired food plant species. As noted in Table 1.5, species with fleshy fruits make up the largest proportion (36%) of the categories of the top 100 North American food plants. This makes sense because fleshy fruits are designed by nature (at least most of the time) to provide nutrition to animals, but virtually all other parts of wild plants are designed for avoiding being eaten (although every plant is susceptible to some herbivore species). Nature's extensive bounty of fruits is evident by the variety available in most supermarkets.

Humans lack taste receptors for the fundamental classes of "macronutrients" (those that need to be consumed regularly in large amounts): protein, fat, and carbohydrate (although many simple carbohydrates and a few natural proteins are detected by sweet taste buds). However, texture clearly plays a role in attracting humans to macronutrients, especially fat. Meat, which is largely animal muscle rich in protein, is generally much easier to chew than fibrous plant tissues. Fats, whether from animals or plants, and carbohydrates (like starch) from plants, also tend to be pleasantly soft. Fat and oil (most fats are solid at room temperature, and most oils are liquid) are higher in energy content on a unit weight basis in comparison to protein and carbohydrate (fat has about 9 cal/g, and carbohydrate and protein have about 4 cal/g), and the pleasant creamy mouth feel of butter and chocolate is attractive to humans because it correlates with high intake of calories, necessary for survival (at least in moderation).

Plant tissues are often quite tough to withstand the rigors of life. Herbivores have specialized, robust teeth to harvest and grind plant tissues and complicated guts usually with special microorganisms that can digest cellulose and other plant constituents from which plants are constructed. Humans are not well adapted to consuming tough plant tissues, and indeed the rather widespread low esteem for vegetables probably reflects a lack of enthusiasm for chewing. Whole-grain cereals and salads have gained increased popularity in recent times, because of the realization of the value of fiber and plant foods in the diet, not because of taste considerations. Many food plants have become softer because of breeding to reduce fibrous tissues; "string beans," for example, are no longer stringy because the fibers have been eliminated in modern cultivars.

Modern Western medicines are mostly synthetic (made from raw chemicals) or semisynthetic (chemically synthesized at least in part from biochemicals, i.e., chemicals extracted from plants). In most of the developing world, medicines are mostly based on plant materials. In the modern Western world, a clear distinction is usually made between food and medicine. In much of the developing world, and indeed everywhere in the past, foods and medicines were often the same and were often plant based. (Moerman [2010] discusses at length how Native Americans equated food and medicine.) The saying "medicine has to taste bad to be good" reflects the reality that medicinal substances are frequently very unattractive in taste. The saying "a spoonful of sugar makes the medicine go down" reflects the aversion that people have to consuming bad-tasting medicines. On a world basis, at least tens of thousands of plants have been used medicinally (Small and Catling 1999). In North America, more than 2600 species of plants have been used as medicines by indigenous peoples (Moerman 1996, 2009). These medicinal plants normally taste bad and have not been consumed with a spoonful of sugar. The lesson is clear: in past times (and indeed in modern times in most of the world still reliant on herbal medicine), people became habituated to bad-tasting plant material. This habituation to bad-tasting medicinal plants would have reinforced the tolerance to bad-tasting food plants (indeed, as pointed out above, many plants were consumed for both medicine and food).

In times of famine, people have eaten plants that are repugnant in taste and sometimes quite toxic. In India, driven by hunger, grass pea (*Lathyrus sativus* L.) is still widely consumed, a plant that results in permanent paralysis of the legs (Small and Catling 2004). The plant foods that were consumed in the past in North America are for the most part extremely unpalatable by modern standards, and one is tempted to assume that they were eaten not by choice but of necessity. Certainly, there must have been times when people turned to unpalatable food plants because other more desirable foods were temporarily unavailable. However, it does appear that in much of North America in the past, hunting, gathering, and in some cases cultivation provided an adequate, probably often a very attractive diet (see comments in Harlan [1992] regarding the probable misperception that life in hunter–gatherer societies was largely a desperate, all-consuming search for food).

Judgment about taste is confounded by the phenomenon of acquired preferences from personal experience. People acquire a desire (even an addiction) for substances such as caffeine (in tea and coffee), isohumulones (which give a bitter taste to hop-flavored beer), and capsaicin (in pungent chile pepper). Black pepper, mustard, and horseradish are common condiments that seem attractive to most people, but when first consumed the experience was likely unpleasant. Babies will try to eat almost anything that is placed in their mouths, so humans are easily conditioned to accept a wide variety of foods. The foods that were consumed in infancy and childhood usually seem very attractive for the rest of one's life, whereas the unique cuisines of foreign cultures or of different ethnic groups often seem unattractive. Possibly the most dramatic example of this is the consumption of insects such as ants, termites, and beetle larvae in parts of Africa and South America, a practice that is nutritionally desirable but which most people find disgusting. Sometimes the desire for foods that are surprisingly bitter, sour, or otherwise objectionable become firmly entrenched in a society or region (Johns 1994). It may be that some bad-tasting herbs have been selected because they have (usually unappreciated) medicinal properties (Johns 1990). Epazote (see Chapter 39) is a very popular condiment served with beans in Mexico, but people unfamiliar with it are discouraged by its medicinal taste and turpentine odor (this herb is somewhat toxic, but has been widely used in the past to expel intestinal parasites). Bitter and sour tastes and unpleasant odors are usually signs that organic material is toxic or decayed, and unfamiliar foods with these characteristics typically require a prolonged period of consumption before they seem pleasant. Indeed, the need for such conditioning or habituation is a critical barrier preventing unfamiliar foods from becoming established. Curiously, the ethnic foods that seem to evoke the most disgust by people not conditioned to eating them are mostly animal foods (e.g., haggis, chitlins, insects, snails, frog's legs). Nevertheless, it is probable that the taste of the numerous plants that were consumed in past times in North America was not nearly as objectionable as judged by modern tastes.

Finally, even if foods taste bad, are monotonous, and are poisonous, humans are amazingly talented in improving the flavor and eliminating toxicity by cooking and various processing procedures. Taste modifiers in the form of culinary herbs and spices are widely used in most societies (Small 2006). Sugar is the world's most widely used taste modifier, indeed constituting an astonishing 30% of the calorie consumption in the United States. Refined sugar was not available in North America in pre-Columbian times, but maple syrup was, at least in the east. Several other plant species were used as sweeteners, as was honey, but on the whole in precontact times, North Americans did not make extensive use of sweeteners. There is no shortage of plants that can be used to improve flavor (as noted in Table 1.5, 19 of the 100 plants examined are flavoring plants), and very likely in most parts of North America dating back thousands of years, dinner was a pleasure, not an ordeal. Culinary herbs were commonly used to improve the flavor of meals (Kuhnlein and Turner 1991). However, it appears that most indigenous peoples north of Mexico consumed their food without extensive use of spices, unlike in Mexico and Central America, where meat was less available and more plant species were incorporated into dishes.

## INDIGENOUS NORTH AMERICANS AND INDIGENOUS FOOD PLANTS

A collective term for aboriginal peoples is often a sensitive issue. "Indigenous North Americans" refers to the pre-Columbian inhabitants of North America, who comprised hundreds of diverse cultures and their descendants (often including those with partial Amerindian ancestry). Recent census data indicate that the following population percentages of people self-identified as indigenous: Canada: 0.8% (5.4% when partial ancestry is included); United States: 0.9% (1.5% when partial ancestry is included). The term "Indians" is almost always assumed to trace to Columbus' label for the people he encountered in the New World, based on an allusion to the people of India. "Indians" is problematic when applied to people of North America; although entrenched in much legislation and accepted by some indigenous North Americans, the term is often perceived to be pejorative. "Native" is acceptable or pejorative, depending on the context. In Canada, the phrase of choice is "First Nations" (or "First Peoples)" or occasionally "Native Canadians," and in the United States "Native Americans" is widely used. "Amerindians" and "aboriginals" are correct but are now considered to be academic terms. In reference to aboriginal people, ambiguity arises from the practices of using the word "American" to refer to, on the one hand, the United States exclusively, or on the other hand, to various areas of North and South America and related regions. Phrases like "Indigenous Americans" and "Indigenous American Peoples" may refer to indigenous peoples of any areas of the Americas.

Native North Americans creatively, ingeniously, and practically utilized and indeed relied on the wild resources of the entire continent in pre-Columbian times. However, reliance on natural resources decreased starting in the early sixteenth century, as Europeans progressively usurped land and confined indigenous tribes to reservations. In modern times, land ownership or at least territorial rights remain contentious issues, tracing back to ambiguous and typically unfair treaties. Despite considerable reduction of populations, assimilation, and suppression of cultural practices, many Native North Americans have retained traditional interests in sustainable use of wild resources, and indeed often control harvests from lands in which particular economic species are wildcrafted in significant quantities.

This book is written from the point of view of maximizing the profit potential of North America's indigenous plants, a perspective which is not always shared by indigenous peoples, who are often much more concerned with respecting nature by limiting harvests to sustainable levels. The conflict between narrow economic interests and conservation of traditional "Indian lands" is unfortunately evident today in struggles between indigenous North Americans and some lumber companies, mining interests, fossil fuel pipeline initiatives, and land developers.

European empires throughout the world have left a legacy of suppressed indigenous peoples who in recent times are asserting intellectual property rights to plant resources that are or were under their control. Knowledge acquired over millennia and indeed the existence of most crops trace to indigenous peoples, but to date very little financial gain has accrued to native peoples from these considerations. A specific example of this follows.

In recent times, the wild North American crop that has had the greatest potential for expansion has been wild rice, the only significant indigenous cereal. Several decades ago, the harvest was entirely under the control of eastern indigenous groups, who supplied a small world market based on traditional wildcrafting from canoes of the plants from native stands. The crop was then domesticated and grown in paddy culture (like rice), first in Minnesota and now primarily in California, which supplies a very large world market. Although traditional wild-collected wild rice still has a niche market (the natural crop does have a romantic appeal), it cannot compete price-wise with the cultivated crop. This situation illustrates that when large profits are possible, wildcrafting is likely to be replaced by cultivation, and traditional "ownership" of the crop is likely to be ignored.

Modern urban life is often extremely stressful, conducted in a "concrete jungle," with the result that many yearn for the simpler rural life of past times. Collecting and preparing wild food plants has become a popular escape from today's hectic lifestyles. Curiously, the movement to return to previous culinary practices has included many indigenous peoples, with the realization that modern-refined, high-caloric but nutritionally deficient foods are particularly harmful to many native groups. The harm occurs because indigenous peoples often live in poverty, and like poor people in general are susceptible to an unhealthy diet; and because of better physiological/genetic adaptation to a traditional diet low in simple carbohydrates and higher in protein and fats, and correspondingly poorer adaptation to highly processed and/or junk foods (Turner and Turner 2008). In North America and several parts of the world, indigenous peoples recently descended from hunter–gatherer societies appear to have superior ability to store fat acquired during times of plenty so that they can survive during times of famine. However, this adaptation is counterproductive today. Unfortunately, the result has been that indigenous people suffer inordinately from obesity, heart disease, and type-2 diabetes. In addition to the possible nutritional benefits from returning to the harvest of traditional food plants, there may be associated benefits: "the production and harvesting of traditional food is also generally healthful, since it promotes activity and exercise, which is itself recognized to be beneficial to health in the broadest sense: physically, spiritually, emotionally, and mentally" (Wong, cited in Turner and Turner [2008]).

## DEVELOPMENT POTENTIAL OF WILD, INDIGENOUS NORTH AMERICAN FOOD PLANTS

As shown in Table 1.3, more than half of the food plants discussed in this book are harvested only or predominantly from the wild. The most important product harvested exclusively from wild plants is sugar maple, which has an annual value ranging between $500 million and $1 billion. Sugar maple is a rather unique food species; it is in high demand, but its growth for harvest of food is confined to North America, and the industry is virtually certain to remain in its present location because the tree is not productive in other areas. For most crops in high demand, the wild supply becomes inadequate, and the crop is cultivated and often domesticated. Several North American wild plants that are now wildcrafted are very attractive as food plants, but are still quite common in nature (particularly bilberry, cloudberry, fiddlehead fern, and huckleberries); nevertheless, such plants have high potential for cultivation and domestication.

Native North Americans learned to consume more than 1500 plant species (cf. Yanovsky 1936; Fernald and Kinsey 1943; Moerman 1996, 2010), including virtually all of those discussed in this book. Well over 90% of these failed to enter the modern commercial marketplace. As noted above, berries (and nuts, to a much lesser extent) have been the most successful category of North American food crops. Unfortunately, most berries and nuts are borne on long-lived perennials (trees and shrubs) that take many years to produce their first seeds, which means that plant breeding is extremely slow, discouraging investment in crops that are not already proven successes. By contrast with fruits, there is very poor representation of commercial cereals, vegetables, and spices among the top 100 native food plants of North America. Of the plants that have not yet achieved even slight market penetration, the chances of becoming successful now seem extremely limited.

However, there is one circumstance where a plant is not palatable or may even be toxic, but yet still has promise for development as a food crop species. This situation prevails when a plant has become widely cultivated, perhaps domesticated, for a nonfood purpose, but nevertheless has a history of minor food usage. In this case, there is normally a wealth of research, knowledge, and technology specific to the species. Consequently, it is very much easier to improve the plant by breeding or simply to use technology to extract constituents that have food value. Jojoba turned into a very widely grown crop because its normally inedible oil meets unique industrial requirements. It is

**FIGURE 1.7** Indigenous North American plants once used as sources of edible crops. All of these are significantly toxic. (a) Florida arrowroot (*Zamia integrifolia* L.f.). (From Jacquin, N.J., *Icones plantarum rariorum,* B. White et filium, London, UK, Vol. 3, Plate 635, 1786–1793.) (b) Sweet flag (*Acorus calamus* L.). (From Köhler, H.A., *Köhler's Medizinal Pflanzen,* Verlag von F.E. Köhler, Germany, Vol. 2, 1887.) (c) Wintergreen (*Gaultheria procumbens* L.). (From Köhler, H.A., *Köhler's Medizinal Pflanzen,* Verlag von F.E. Köhler, Germany, Vol. 3, 1898.) (d) American beech (*Fagus grandifolia* Ehrh.). (Courtesy of Rob Routledge, Sault College, and Bugwood.org [CC By 3.0].) (e) Bracken fern (*Pteridium aquilinum* (L.) Kuhn). (Courtesy of Rasbak [CC By 3.0].)

included in this book because of the possibility that it can be transformed into a food crop. Although not as important as jojoba, other plants grown for industrial oil but with food potential include blue waxweed, coast tarweed, and dwarf glasswort.

Sometimes a crop is grown for a nonfood purpose, but produces edible parts that can be harvested. Crops with such flexibility are usually viewed favorably. May-apple would not normally be included in a book dedicated to food plants because, except for the mature fruit, it is extremely poisonous. However, it is cultivated for medicinal products, and so the fruits could be considered as a secondary product. Ginseng is also cultivated mainly for extracted medicinal preparations. However, there is a small market for the root as a vegetable. Black walnut is one of the most valuable of lumber trees; the nuts are a bonus. Jerusalem artichoke periodically has been advanced as having potential to generate great wealth for the production of alcohol as fuel and for other industrial chemicals from its edible tubers, which have received only limited market interest to date, since they must compete with the much cheaper potato. Should demand for growing the plant for industrial chemicals increase, the food usage could also increase.

## CAN OBSOLETE COMMERCIAL NORTH AMERICAN CROPS BE RESURRECTED?

Some indigenous North American food plants achieved notable success in recent centuries, but are now obsolete. Chapters are presented on camas, prairie turnip, and yampah, perhaps the very best of the root crops that were important and widely harvested by North American indigenous people. None of these was domesticated, or indeed ever entered the modern commercial marketplace (they were very significant as trade and barter items). By virtue of the ability of these crops to produce substantial amounts of palatable food that stores well, they deserve to be highlighted. Certainly with dedicated research and development, they could be transformed into domesticated crops. However, given the enormously competitive potato, and indeed numerous other very productive root crops, the chances of these old Indian specialties becoming significant are very limited.

A chapter is devoted to American chestnut, once extensively harvested for its nuts. It has almost vanished from the landscape because of decimation by a blight fungus. The nuts are so delicious that they would easily find a commercial niche today but for the fact that the fungus continues to prevent the tree from reestablishing its former prominence. However, should technology be discovered to make the tree resistant to the blight, it could certainly become a significant nut crop once again.

A very few food plants stimulated a brief period of commercial interest and then vanished from the economic market. Perhaps the most notable was *Zamia integrifolia* L. (*Z. floridana* A. DC.; "*Z. pumila* L." of some authors, a name applicable to another species; Figure 1.7a), called wild sago, comtie, and coontie, which occurs in Florida and southern Georgia and in areas south of the United States. Its tuberous roots yield "Florida arrowroot starch," sometimes also called sago starch (although "sago starch" is usually from *Metroxylon sagu* Rottb., one of several species called sago palm). In the nineteenth century, there was a short-lived industry based on the production of Florida arrowroot starch, used to make baby food, spaghetti, biscuits, and laundry starch (Gearhart 1952). The plant contains the toxin cyasin, and the extracted starch requires treatment to render it nontoxic. From a commercial perspective, the world is abundantly supplied with starch from other crops, obtained far more cheaply than from wild sago, and so the prospects for reviving the Florida arrowroot industry are virtually nil. The species has become a popular ornamental, which is fortunate, because its habitats are being destroyed.

Wintergreen (*Gaultheria procumbens* L; Figure 1.7b) is a small, low-growing shrub of northeastern North America. The foliage was once harvested commercially as a source of methyl salicylate, used in very low concentrations (because of its toxicity) as a flavoring agent and at higher concentrations for medicinal applications. Today, the compound is synthesized chemically, making the natural source obsolete. The plant is valued by wild food collectors for its minty, edible fruit and its leaves are used to make tea.

Sweet flag (*Acorus calamus* L; Figure 1.7c) is a semiaquatic herb of temperate and subtemperate regions of Eurasia and the Americas. Essential oil from the rhizome has been used primarily for medicinal purposes. The oil has also been used as a condiment or flavoring agent for alcoholic beverages, aromatic vinegar, and confectionery, and the rhizome has been candied. Because of the presence of a carcinogen, in the late twentieth century its use was forbidden in food, and indeed generally for oral medicinal use. There is evidence, however, that the North American plants are not as toxic as the Asian plants (Small and Catling 1999), and there is at least a remote possibility that the species could regain commercial significance.

The American beech tree (*Fagus grandifolia* Ehrh; Figure 1.7d) is native to the eastern half of the United States and southeastern Canada. Its fruit (beech nuts) are gathered and eaten fresh, dried, or roasted by wild food collectors (they are somewhat toxic, and consumption should be limited). The nuts have occasionally been sold commercially, and oil expressed from the nuts was best known for flavoring beechnut gum, a mint-flavored gum popular in the mid-twentieth century. The European beech (*F. sylvatica* L.) of the Old World has been used for similar purposes. However, beech nuts have almost no commercial importance today and are very unlikely to become significant in the future.

Almost all of the world's major crops were domesticated by 5000 BP, so the prospects of finding or creating new major crops would seem to be dismal. Many writers have pointed out how difficult it is to predict the future; at the same time, many other writers have noted the old wisdom that the best guide to the future is the past. Certainly judging by at least the last century, it is very rare for North American food crops that are not now commercially successful to become so, whether or not they were popular in past times. Nevertheless, this is a time of great advances in biotechnology, and it could be that abandoned North American food crops are simply awaiting new discoveries to realize their potentials.

## Chief Obstacles to Transforming Wild Plants into Profitable Crops

The introduction of new domesticated crop species that achieve notable market success is rare. Prescott-Allen and Prescott-Allen (1986) concluded that for North America north of Mexico, blueberry and wild rice are the only important food crops that were domesticated during the twentieth century.

Most of the crops presented in this book are harvested from nature. There is an undeniable romanticism and nostalgia associated with life in past times, although very few people realize how uncertain and difficult it usually was to acquire, prepare, and indeed sometimes to digest the foods of the past. Two centuries ago, most people in North America lived a rural existence, consuming crops and animals they raised themselves or food they harvested from nature. In pre-Columbian times, indigenous people living in what is now the United States and Canada relied on nature for most of their food needs, with relatively limited agriculture practiced. Except for fishing, North Americans now rely on agriculture to supply virtually all of their foods. Of the more than 1000 species that were once consumed by Native Americans, almost none is eaten today; most of these taste bad by current standards and have undesirable constituents.

Philosopher Francis Bacon (1561–1626) wrote "Acorns were good until bread was found." In many areas of the world, prodigious quantities of acorns (the fruits of oaks and related trees) are produced. These were once widely eaten by people, providing a caloric content per unit weight than cereal grains. To this day, numerous recipes can be found on the Web detailing how to extract flour and make bread from acorns. But removing the toxic tannins is very time-consuming and laborious, and using wheat and other cereals is much more efficient.

Many of the plants discussed in this book are wild (undomesticated), and most of these are either not cultivated or cultivated for nonfood purposes. Asa Gray (1874) observed that had civilization started in America instead of Asia, "our ground-nut [*Apios americana*] would have been the first developed esculent tuber and would probably have held its place in the first rank along with potatoes

**FIGURE 1.8** Asa Gray (1810–1888), the foremost nineteenth-century North American botanist. (From Deane, W., *Bull. Torrey Bot. Club* 15: 59, 1888.)

and sweet potatoes of later acquisition." Although Gray overlooked the fact that both potato and sweet potato were domesticated in the Americas south of the United States, not in the Old World, his observation is profoundly important: once a given class of food crop has achieved wide success in satisfying a market niche, it drastically suppresses the development of similar crops based on wild species. This might seem counterintuitive, since the free market thrives on competition, and new products regularly replace older ones. However, crops are unlike most marketed products in important respects. Several factors combine to suppress or at least discourage the development of new crops: (1) Wild species that are attractive to the marketplace (in their present form) have virtually all been discovered (mostly in the distant past) and have already achieved market saturation. (2) Remaining wild species have serious drawbacks (e.g., they are toxic or have poor taste) and developing them into profitable crops requires considerable time, resources, and investment. (3) Intellectual property rights for crops have been nonexistent or negligible until recent times, and even today they are much weaker than for patentable inventions and other creations. Once a new crop has been created, there is little to prevent others from using that material or slightly altering it; in most cases, royalties are not paid to the creator (plant variety rights do provide protection, but basically for selected cultivars not for the crop in general). Accordingly, there is very limited incentive for the private sector to invest in creating new crops from wild species. (4) The extremely conservative nature of agriculture makes introduction of new crops difficult. Farmers are generally risk-averse, since the consequences of crop failure can be devastating. (5) Established crops are associated with extensive research knowledge, technologies including specialized machinery, value-added processes, and political influence. Accordingly, competing directly with an established crop can be extremely expensive.

Given the above difficulties, it is essential to consider market potential for an undeveloped wild plant before investing in its domestication. Cereals, the most important class of food plants, so dominate the food world that domesticating wild grasses as new cereals today seems very unlikely; wild rice has been successfully domesticated in recent decades, but no other wild grass north of Mexico seems to have such potential. Oilseeds are the second most important class of food plant, but except for the already developed sunflower, the several oilseeds included in this book (blue waxweed, buffalo gourd, coast tarweed, jojoba) do not seem to have great promise for food purposes. Tree fruits (apple, peach, etc.) and berries (blueberries, strawberries) represent an extremely competitive market, and new fruits need competitive advantages (for example, saskatoon and aronia are achieving success in part because they are hardy enough to be grown in quite northern areas). Leafy vegetables, root crops, nuts, and plants used for flavor are categories that are well represented already in the food market. There is room for new offerings, but basically for niche markets interested in novelties.

## Technological Innovations and the Possibility of Rapid Domestication of North America's Wild Plants

Just when "modern" plant breeding started is open to debate. Nonintentional, so-called "unconscious selection" has been important in the domestication of plants (Heiser 1988), but hardly qualifies as scientific. Certainly, scientific breeding had started by 1900 with the rediscovery of Gregor Mendel's laws of inheritance. Some authorities accept the middle or late nineteenth century as the time when plant breeding had reached a level of systematic application sufficient to qualify as "modern." Modern plant breeders have enormously advanced the domestication of the world's major crops, for which very advanced cultivars are now available. Nevertheless, it has been claimed that most of the work of domesticating the world's major crops was carried out thousands of years ago, well before the era of modern plant breeding (Simmonds 1979). It might appear that our most important domesticated crops are so extensively changed from their wild ancestors that it is foolish to domesticate wild plants today, because the work required is so extensive and time consuming. Indeed, most current plant breeders prefer to utilize cultivars or at least land races to create new cultivars; although wild relatives are invaluable sources of genes that can improve available cultivars, wild plants also have genes that are very deleterious to productivity, and using wild plants in hybridization programs requires much work to get rid of these unwanted genes while retaining the desired ones.

The idiom "low-hanging fruit" is based on the natural cleverness (or laziness) of people to minimize the work required to fulfill a need. In early historical times, the need for certain kinds of edible plant (cereal, fruit, spice, etc.) was met by local indigenous plants. With growing knowledge of superior foreign crops, interest in most indigenous food plants usually ceased or became minimal. Reviving such interest in seemingly anachronistic food sources is a formidable, but not a lost cause.

Mankind's choice of species to favor (or enslave) as domesticates was not guided by systematic, logical analysis, but rather by the intuition of small groups of people in scattered locations over the last 12,000 years; our domesticated species meet the world's needs remarkably well, but surely other species could meet the same and other needs even better, were it not for the work and time needed. To some, it might seem that there is no more need to invent new fruits and vegetables than there is to reinvent the wheel, but the comparison is not fair. Imagine that you are in the fresh fruit and vegetable area of a large supermarket; yes, the offerings are diverse and impressive, but the quality is usually variable, almost nothing is perfect, and choosing the best is a tiresome exercise. Moreover, as anyone who has visited the tropics knows, the choice of fresh foods is much more impressive, but the shelf life of most species is so short that they cannot be exported. In short, by no means have the possibilities been exhausted of introducing new and superior offerings, and the normal course of marketplace economics is for the new to replace or at least compete with the old.

Several discoveries and technologies suggest that rapid domestication of wild plants may be possible in the future, and these considerations give North America's indigenous flora, particularly the wild food species highlighted in this book, potential significance as important new crops for the future. Improved understanding of the genetic and molecular basis of domestication (e.g., Burger et al. 2008) may facilitate applying this knowledge directly to the domestication of wild plants. Genetic engineering has the potential to transfer desirable traits into plants in ways that were impossible in the past.

## How the Modern Era of Synthetic Foods Can Stimulate the Selection of New Food Domesticates

At least for babies and infants, the most perfect food is mother's milk. With the domestication of cows and some other mammals for dairy production, humans created a substitute for mother's milk. However, cow's milk has proven to be a rather imperfect food, and indeed so have other substitutes for human milk. Mother's milk is perfect because millennia of selection have removed undesirable components and included a very complex array of constituents in exactly the right balance; milk substitutes are imperfect because decades of scientific studies have not yet identified all the factors that are desirable and/or undesirable, and technology has not yet proven capable of producing a completely acceptable alternative. In a similar vein, nutrition based on consumption of natural or "whole" foods is increasingly being replaced by complex, processed products of technology, which are significantly unsatisfactory (for a comprehensive review, see Lindeberg 2010). However objectionable this is, it is a reality of modern times, and in the following analysis an attempt is made to show how this might influence the future of plant domestication, with special reference to North American indigenous food plants.

For better or worse, technology is transforming the familiar natural commodities of life. In the past, all medicines originated from nature; today, at least in the Western world, most are synthetic, prepared from raw chemicals. In the past, wood and stone were the standard materials from which buildings were made; today, they are increasingly replaced by composites. In the past, clothing and other textiles were manufactured from plant and animal fibers; today, they are usually prepared from petroleum. The foods of today are also being synthesized from a variety of ingredients. Synthetic dyes and texturizing techniques are being used to create imitation foods such as "milk shakes" that are devoid of milk. We live in an era of "prepared," "fast," and "convenient" food, or less charitably "junk," "counterfeit," or "artificial" food. The traditional bases on which staple foods have been selected—taste and nutritional qualities—are no longer essential criteria, since taste and nutritional molecules can simply be added.

Horrific food preparation and marketing practices in the past, such as adding lead, urine, or rust to foods to improve appearance or simulate an expensive spice, were terminated by food purity, quality, and safety legislation. Today, some practices of the food synthesis industry seem equally reprehensible (Table 1.8), contributing substantially to the current obesity epidemic (half of people in the Western world are overweight, one in six is obese), and the resulting health problems and

**TABLE 1.8**
**Seven Deadly Sins of the Synthetic Food Industry**

| Transgression | Harm |
|---|---|
| 1. Too many calories resulting from excessively rich foods served in oversize portions | Obesity, heart disease, diabetes |
| 2. Too much bad (saturated) fat, too little good (unsaturated, essential) fat | Cardiovascular disease, unbalanced nutrition |
| 3. Too much sugar and other refined carbohydrates | Obesity, diabetes, dental caries |
| 4. Too much salt | Cardiovascular disease, stroke |
| 5. Too many additives (preservatives, taste enhancers, colorants, etc.) | Toxicity |
| 6. Too much meat, too little plant-based food | Dietary lack of minerals, vitamins, antioxidants, fiber, and other natural plant components that are inadequately substituted by additives |
| 7. Too little taste; too monotonous | Diminished enjoyment of life, diminished appreciation of the true taste of food, diminished attention to the social and societal values of the meal ritual |

suffering. Many major chronic diseases characterize modern industrialized societies, notably heart disease, hypertension, obesity, adult-onset diabetes, some kinds of cancer, and dental caries, and most foods currently marketed are responsible. The "freedom of choice" defense offered by the industry is unconvincing, since foods laced with calories, fat, sugar, and salt are highly addictive. Moreover, high-tech food is astonishingly cheap, ubiquitous, effectively advertised and marketed, and so it is very difficult for healthy foods to compete. The enormous financial clout of the prepared food industry is accompanied by political power, making it difficult to initiate corrective legislation. This is a shame because it should be possible to manufacture synthetic foods that are genuinely healthy and to market them in a responsible fashion that does not encourage overeating. Humans require about 50 essential nutrients for growth, maintenance, and repair of the body, including protein, carbohydrate, fat, vitamins, minerals, and water (fiber is beneficial but not essential). Unfortunately in the Western world and increasingly everywhere, food is being marketed primarily for sensory gratification, not for sustenance. In the past, a potential new food crop had to compete against established crops, particularly their taste; today, potential new food crops need to compete against foods that have been artificially flavored, strongly limiting the likelihood of success. Regardless of the undesirable aspects of most modern synthetic foods, the need for raw food components raises the possibility of creating new domesticated plants as sources.

It is logical that very productive food plants that were abandoned historically because of their taste or because of the presence of toxins could be new sources of food components, since only the desired constituents would be extracted. For example, jojoba and buffalo gourd (both taste awful) are potential sources of edible extractives, since it should be possible to separate nutritional components from the other constituents, many of which have considerable nonfood industrial values. The world's major crops in many instances are already superior sources of food extracts because they have been selected for high productivity, but there are good reasons for seeking out new sources. Potatoes are a natural source of starch, but require huge inputs of herbicides, fungicides, fertilizer, and irrigation. Wild crops like cattail (Chapter 24) and reed (Chapter 79) can be enormously productive of starch while growing naturally on land unsuitable for agriculture. Jerusalem artichoke has been viewed as a source of food and industrial constituents for the last half century, but deserves increased attention. Abandoned root crops that once were so productive that they became staples of North America's indigenous people (see Chapter 22 on Camas) need to be reexamined for their potential to become sources of food constituents.

## LITERATURE CITED


Arnason, T., Hebda, R.J., and Johns, T. 1981. Use of plants for food and medicine by Native Peoples of eastern Canada. *Can. J. Bot.* 59: 2189–2325.

Barnicott, N.A. 2005. Human nutrition: Evolutionary perspectives. *Integr. Physiol. Behav. Sci.* 40: 114–117.

Black, M.J. 1978. Plant dispersal by native North Americans in the Canadian subarctic. *Anthropol. Pap. Mus. Anthropol. Univ. Mich.* 67: 255–262.

Bogin, B. 1997. The evolution of human nutrition. In *The anthropology of medicine: From culture to method.* Third edition. Edited by L. Romanucci-Ross, D.E. Moerman, and L.R. Tancredi. Bergen and Garvey, South Hadley, MA. pp. 996–142.

Burger, J.C., Chapman, M.A., and Burke, J.M. 2008. Molecular insights into the evolution of crop plants. *Am. J. Bot.* 95: 113–122.

Candolle, A. de. 1882 (imprinted "1883"). *Origine des plantes cultivées.* Germer Baillière, Paris, France. pp. 379. (English translation: 1884. *Origin of cultivated plants.* Kegan Paul, London, UK. pp. 468.)

Darlington, C.D. 1973. *Chromosome botany and the origins of cultivated plants.* Third edition. Allen & Unwin, London, UK. pp. 237.

Diamond, J. 1987. The worst mistake in the history of the human race. *Discover* (May): 64–66.

Emery, M.R., and McLain, R. (Editors). 2001. *Non-timber forest products: Medicinal herbs, fungi, edible fruits and nuts, and other natural products from the forest.* Food Products Press, Binghamton, NY. pp. 176.

Facciola, S. 1998. *Cornucopia II. A source book of edible plants.* Kampong Publications, Vista, CA. pp. 713.

Farrington, I.S., and Urry, J. 1985. Food and the early history of cultivation. *J. Ethnobiol.* 5: 143–157.

Fernald, M.L., and Kinsey, A.C. 1943. *Edible wild plants of eastern North America*. Idlewild Press, Cornwall-on-Hudson, NY. pp. 452.

Ford, R.I. (Editor). 1985. *Prehistoric food production in North America*. Museum of Anthropology, University of Michigan, Ann Arbor, MI. pp. 411.

Gearhart, E.G., Jr. 1952. South Florida's first industry. [Coontie, *Zamia pumila*.] *Tequesta* 12: 55–57.

Gibbons, A. 1998. Solving the brain's energy crisis. *Science* 280(5368): 1345–1347.

Gray, A. 1874. Were fruits made for man or did man make the fruits? *Amer. Nat.* 8: 116–120.

Hancock, J.F. 2004. *Plant evolution and the origin of crop species*. Second edition. CABI, Wallingford, Oxon, UK. pp. 313.

Harlan, J.R. 1992. *Crops and man*. Second edition. American Society of Agronomy, Crop Science Society of America, Madison, WI. pp. 284.

Hawkes, J.G. 1983. *The diversity of crop plants*. Harvard University Press, Cambridge, MA. pp. 184.

Hawkes, J.G. 1998. Back to Vavilov: Why were plants domesticated in some areas and not in others? In *The origins of agriculture and crop domestication*. Proceedings of the Harlan symposium. Edited by A.B. Damania, J. Valkoun, G. Willcox, and C.O. Qualset. International Center for Agricultural Research in the Dry Areas, Aleppo, Syria, 5–8.

Heiser, C.B. 1988. Aspects of unconscious selection and the evolution of domesticated plants. *Euphytica* 37: 77–81.

Heiser, C.B. 1993. Ethnobotany and economic botany. In *Flora of North America*, Vol. 1, Introduction. Edited by Flora of North America Committee. Oxford University Press, New York, NY. pp. 199–206.

Johns, T. 1990. *With bitter herbs they shall eat it: Chemical ecology and the origins of human diet and medicine*. University of Arizona Press, Tucson, AZ. pp. 356.

Johns, T. 1994. Ambivalence to the palatability factors in wild food plants. In *Eating on the wild side: The pharmacologic, ecologic, and social implications of using noncultigens*. Edited by N.L. Etkin. University of Arizona Press, Tucson, AZ. pp. 46–61.

Johns, T., and Kubo, I. 1988. A survey of the traditional methods employed for the detoxification of plant foods. *J. Ethnobiol.* 8: 81–129.

Jones, E.T., McLain, R.J., and Weigand, J. (Editors). 2002. *Nontimber forest products in the United States*. University Press of Kansas, Lawrence, KS. pp. 445.

Kuhnlein, H.V., and Turner, N.J. 1991. *Traditional plant foods of Canadian indigenous peoples. Nutrition, botany and use*. Gordon and Breach Science Publishers, Philadelphia, PA. pp. 633.

Ladizinsky, G. 1998. *Plant evolution under domestication*. Kluwer Academic, London, UK. pp. 254.

Leonard, W.R., Snodgrass, J.J., and Robertson, M.L. 2007. Effects of brain evolution on human nutrition and metabolism. *Annu. Rev. Nutr.* 27: 311–327.

Lindeberg, S. 2010. *Food and Western disease. Health and nutrition from an evolutionary perspective*. Wiley-Blackwell, Chichester, UK. pp. 354.

Lucas, P.W. 2011. Cooking clue to human dietary diversity. *PNAS*. 108: 19101–19102.

Milton, K. 2000. Hunter-gatherer diets—A different perspective. *Am. J. Clin. Nutr.* 71: 665–667.

Minnis, P.E. (Editor). 2003. *People and plants in ancient eastern North America*. Smithsonian Books, Washington, DC. pp. 423.

Minnis, P.E., and Elisens, W.J. (Editors). 2000. *Biodiversity and native America*. University of Oklahoma Press, Norman, OK. pp. 310.

Moerman, D.E. 1996. An analysis of the food plants and drug plants of native North *America. J. Ethnopharmacol.* 52: 1–22.

Moerman, D.E. 2009. *Native American medicinal plants: An ethnobotanical dictionary*. Timber Press, Portland, OR. pp. 799.

Moerman, D.E. 2010. *Native American food plants: An ethnobotanical dictionary*. Timber Press, Portland, OR. pp. 455.

Morton, J.F. 1963. Principal wild food plants of the United States, excluding Alaska and Hawaii. *Econ. Bot.* 17: 319–330.

Nabhan, G., and de Wet, J.M.J. 1984. *Panicum sonorum* in Sonoran desert agriculture. *Econ. Bot.* 38: 65–82.

National Academy of Sciences. 1975. *Underexploited tropical plants with promising economic value. Report of ad hoc panel of Advisory Committee on Technical Innovation*. National Academy of Sciences, Washington, DC. pp. 188.

Perry, G.H., Dominy, N.J., Claw, K.G., Lee, A.S., Heike, F., Redon, R., Werner, J., et al. 2007. Diet and the evolution of human amylase gene copy number variation. *Nat. Genet.* 39: 1256–1260.

Peterken, G.F. 1981. *Woodland conservation and management*. Chapman and Hall, London. pp. 328.

Pickersgill, B. 2007. Domestication of plants in the Americas: Insights from Mendelian and molecular genetics. *Ann. Bot.* 100: 925–940.

Prescott-Allen, C., and Prescott-Allen, R. 1986. *The first resource: Wild species in the North American economy*. Yale University Press, New Haven, CT. pp. 529.
Prescott-Allen, C., and Prescott-Allen, R. 1990. How many plants feed the world? *Conserv. Biol.* 4: 365–374.
Price, T.D. 2009. Ancient farming in eastern North America. *PNAS* 106(16): 6427–6428.
Procheş, Ş., Wilson, J.R.U., Vamosi, J.C., and Richardson, D.M. 2008. Plant diversity in the human diet: Weak phylogenetic signal indicates breadth. *BioScience* 58: 151–159.
Rask, K.J., and Rask, N. 2011. Economic development and food production–consumption balance: A growing global challenge. *Food Policy* 36: 186–196.
Ratcliffe, D.A. (Editor). 1977. *A nature conservation review*. Cambridge University Press, Cambridge, UK. 2 vols.
Rindos, D. 1984. *The origins of agriculture: An evolutionary perspective*. Academic Press, New York, NY. pp. 325.
Sauer, J.D. 1994. *Historical geography of crop plants. A select roster*. CRC Press, Boca Raton, FL. pp. 309.
Simmonds, N.W. 1979. *Principles of crop improvement*. Longman, London, UK. pp. 408.
Small, E. 2006. *Culinary herbs*. Second edition. NRC Press, Ottawa, ON. pp. 1036.
Small, E. 2009. *Top 100 food plants: The world's most important culinary crops*. NRC Press, Ottawa, ON. pp. 636.
Small, E. 2011a. *Top 100 exotic food plants*. Taylor & Francis/CRC Press, Boca Raton, FL. pp. 658.
Small, E. 2011b. *Alfalfa and Relatives: Evolution and Classification of Medicago*. NRC Research Press, Ottawa, ON, Canada, and CABI, Wallingford, Oxon, UK. pp. 727.
Small, E., and Catling, P.M. 1999. *Canadian medicinal crops*. NRC Press, Ottawa, ON. pp. 240.
Small, E., and Catling, P.M. 2004. Blossoming treasures of biodiversity 14. Grass pea (*Lathyrus sativus*). Can a last resort food become a first choice? *Biodiversity* 5(4): 29–32.
Small, E., Catling, P.M., and Brookes, B. 2012. *Official plant emblems of Canada: A biodiversity treasure*. Public Works and Government Services Canada, Ottawa, ON. pp. 428.
Smith, B.D. 1989. Origins of agriculture in eastern North America. *Science* 246(4937): 1566–1571.
Smith, B.D. 2006. Eastern North America as an independent center of plant domestication. *PNAS* 103(33): 1223–1228.
Smith, B.D. 2007. *Rivers of change: Essays on early agriculture in eastern North America*. New edition. University of Alabama Press, Tuscaloosa, AL. pp. 302.
Smith, B.D., and Yarnell, R.A. 2009. Initial formation of an indigenous crop complex in eastern North America at 3800 B.P. *PNAS* 106(16): 6561–6566.
Thorne, R.F. 1993. Phytogeography. In *Flora of North America*. Volume 1, Introduction. Edited by Flora of North America Committee. Oxford University Press, New York, NY. pp. 132–153.
Turner, N.J. 1981. A gift for the taking: The untapped potential of some food plants of north American Native Peoples. *Can. J. Bot.* 59: 2331–2357.
Turner, N.J., and Turner, K.L. 2008. "Where our women used to get the food": Cumulative effects and loss of ethnobotanical knowledge and practice; case study from coastal British Columbia. *Botany* 86: 103–115.
Vavilov, N.I. 1926. Studies on the origin of cultivated plants. *Bull. Appl. Bot.* 26: 1–248.
Vavilov, N.I. 1949–1950. *The origin, variation, immunity and breeding of cultivated crops*. Chronica Botanica, Waltham, MA. pp. 364.
Ungar, P. (Editor). 2006. *Evolution of the human diet: The known, the unknown, and the unknowable*. Oxford University Press, New York, NY. pp. 413.
Weisdorf, J.L. 2005. From foraging to farming: Explaining the neolithic revolution. *J. Econ. Surveys* 19: 561–586.
Wetzel, S., Duchesne, L.C., and Laporte, M.F. 2006. *Bioproducts from Canada's forests: New partnerships in the bioeconomy*. Springer, Dordrecht, the Netherlands. pp. 257.
Wiersema, J.H., and León, B. 1999. *World economic plants. A standard reference*. CRC Press, New York, NY. pp. 749. (2nd edition, 2012, 1336 pp).
Wrangham, R. 2009. *Catching fire: How cooking made us human*. Basic Books, New York, NY. pp. 320.
Yanovsky, E. 1936. *Food plants of the North American Indians*. USDA Misc. Publ. 237. U.S. Gov. Printing Office, Washington, DC. pp. 84.
Zohary, D., and Hopf, M. 2000. *Domestication of plants in the Old World: The origin and spread of cultivated plants in West Asia, Europe, and the Nile Valley*. Third edition. Oxford University Press, Oxford, UK. pp. 316.

# 2 Acerola

Family: Malpighiaceae (malpighia family)

## NAMES

Scientific name: *Malpighia emarginata* Sessé & Moq. ex DC. ("*M. punicifolia*," "*M. glabra*"; frequently confused with *M. glabra* L., a related plant with a similar natural geographical range)

- The English name "acerola" is based on the American Spanish word *acerola* used for the plant, which in turn is based on the Spanish *azarole*, referring to a quite different species, the azarole (*Crataegus azarolus* L.), which has berries of similar appearance. Acerola is pronounced ah-see-ROLL-ah and as-uh-ROH-luh.
- Acerola is also known as acerola cherry, Amazon cherry, Antilles cherry, Barbados cherry, chercese, chereese, French cherry, garden cherry, Mexican myrtle, native cherry, Puerto Rican cherry, Surinam cherry, West Indian cherry, and wild crapemyrtle.
- Early Spanish explorers called the acerola *cereza*, Spanish for cherry, which the fruit resembles. "Barbados" in the name "Barbados cherry" is based on the region where the plants were observed in early times.
- The acerola has sometimes been confused with the Surinam cherry, *Eugenia uniflora* L.
- The genus name *Malpighia* commemorates Marcello Malpighi (1628–1694), an Italian naturalist, physician, and professor at Bologna, Italy.
- *Emarginata* in the scientific name *M. emarginata* is Latin for "with a shallow notch" (at the tip), a description that is normally applied to leaves, but is not accurate for most leaves of this species.

## GEOGRAPHY AND ECOLOGY OF WILD PLANTS

It has been suggested that acerola originated in the Yucatan Peninsula of Mexico, but it seems to be naturally distributed from South Texas, through Mexico and Central America, to northern South America and throughout the Caribbean (Bahamas to Trinidad). Acerola has also been collected in southern Florida, but it is uncertain whether the plant is native to the state. The species has been introduced to tropical and subtropical areas of the world, particularly in Southeast Asia, India, and South America. It grows well in Puerto Rico and Hawaii.

Acerola is adapted to tropical and subtropical lowland climates. Mature plants can tolerate slight freezing (–2°C, –28°F) for a brief period, but young plants are susceptible. The plants grow well in a variety of soils, preferring well-drained soils. Some shade is tolerated.

## PLANT PORTRAIT

The acerola is a large, bushy, evergreen shrub or small tree 2–6 m (6.6–20 feet) in height, with a trunk 10 cm (4 inches) in diameter. The plant has clusters of red or rose flowers, each 2–2.5 cm (about an inch) wide. The fruits are borne singly or in clusters of two or three, and are somewhat flattened to round. They are cherry-like but more or less three-lobed (three furrows are evident on the outside of the fruit), 1.25–2.5 cm (0.5–1 inch) wide, thin-skinned, bright red (rarely yellow-orange), with orange, very juicy pulp. The fruits contain three small, rounded seeds with wings.

**FIGURE 2.1** Acerola (*Malpighia emarginata*). (a) Branches with fruit; photo by Wilfredor (CC By 3.0). (b) Branches with fruit; photo by sapienssolutions/Leonardo Dasilva (Flickr/CC-attribution). (c) Tree with fruit; photo by Agriculturasp/Secretaria de Agricultura e Abestecimento do Estado de São Paulo Agriculturasp (Flickr/CC-attribution). (d) Flowering branch; painting from Curtis, W. 1805. *Botanical Magazine*, Vol. 21, Plate 813. (e) Distribution map.

Acerola is widely cultivated on a commercial scale in the tropics for fresh fruit and juice. Some of the largest plantations are in Brazil. Sweet varieties have been established in home gardens in Florida, and there has been very minor commercial production in the state. Acerola is more often cultivated as an ornamental shrub than a source of edible fruit in Florida and other regions of the southeastern United States. Dwarf forms are available, and the plant is also often grown as bonsai.

Acerola fruit has one of the highest known vitamin C contents of all fruits. An acerola fruit can have up to 4.5% vitamin C compared to 0.05% in a peeled orange. Vitamin C is an acid (ascorbic acid), and the higher the content of vitamin C, the more acidic the fruit. One or two of the cherry-sized fruits can provide the recommended daily allowance of vitamin C. In the late 1940s and 1950s, there was an explosion of enthusiasm for acerola fruit because of its extraordinarily high content of vitamin C, and breeding programs were undertaken in several countries. However, interest plummeted when it was realized that a fruit could not become a superstar because of its vitamin C content alone. Synthesized vitamin C is much cheaper and, despite the claims that "natural" is better, synthetic vitamin C has become dominant in the marketplace. Vitamin C is an antioxidant and a free radical scavenger, that is, it deactivates the harmful chemicals called free radicals that are produced by the body. Acerola concentrates can now be found in many over-the-counter multivitamin supplements. Recent research in cosmetology suggests that vitamin C is also useful for skin, and acerola extracts are now appearing in skin care products claiming to fight cellular aging.

## CULINARY PORTRAIT

The flavor of the sweet, red juicy pulp of acerola is reminiscent of raspberries, and when cooked, the fruit tastes like a tangy apple. Acerola is often eaten fresh in areas where it is cultivated, although it is too tart for many people. The fruit is also stewed with sugar as a dessert. The seeds need to be separated from the pulp in the mouth and returned by spoon to the dish, which is a nuisance, and the cooked fruits must be strained to remove the seeds (eating large quantities of whole fruits containing the seeds has produced illness). Acerola sauce or purée is employed as a topping on cake, pudding, ice cream, and fruit, and is used in gelatin desserts, punch, and sherbet. The fruits make excellent syrup, jelly, jam, preserves, and pies, and have also been used to produce baby food, popsicles, and wine. Acerola juice is as popular in Brazil as orange juice is elsewhere, but does not store well. Cooking causes the bright-red color to change to brownish-red. Frozen fruit falls apart when thawed. Ripe fruit bruises easily, is highly perishable, and loses flavor and nutritional content very rapidly after harvest. As a result of poor keeping and shipping qualities, acerola is not commonly cultivated for commercial fresh fruit exportation.

### Culinary Vocabulary

- "Acerola powder" is a dried, powdered form of acerola fruit that is used to boost the nutritional level of food consumed by groups with special needs (infants, elderly, invalids, etc.). It is often marketed in health-food outlets.

## PROSPECTS

Acerola is one of the few native North American plants that are cultivated for fruit on a major scale. However, its adaptive range reaches its limit in southernmost continental United States, and most cultivation is in tropical and subtropical areas, where it can be produced much more efficiently. Although acerola is a significant commercial fruit on a world basis, it has not achieved much market recognition outside of the tropics. Because it has a short storage life and does not ship well, acerola tends to be consumed locally. It is possible that there is a market niche available in the southern United States for production of the fruit for local sale, especially for a product that can be reliably certified as organic.

## CURIOSITIES OF SCIENCE AND TECHNOLOGY

- Most genus names commemorate people. Spanish botanist Antonio José Cavanilles (1745–1804) named more than 100 genera, but in 1799, a suitable person to recognize with a new genus name did not seem to occur to him, and he coined the genus *Galphimia* simply as an anagram of *Malpighia*.
- Fresh acerola juice applied to peeled or sliced bananas keeps them from darkening. Acerola juice is also useful for preventing the oxidation of a variety of other fruits.
- Acerola trees without adequate pollination often set seedless fruit.

## KEY INFORMATION SOURCES

Alves, R.E., Chitarra, A.B., and Chitarra, M.I.F. 1995. Postharvest physiology of acerola (*Malpighia emarginata* D.C.) fruits, maturation changes, respiratory activity, and refrigerated storage at ambient and modified atmospheres. *Acta Hortic.* 370: 223–229.

Alves, R.E., Filgueiras, H.A.C., Mosca, J.L., and Menezes, J.B. 1999. Brazilian experience on the handling of acerola fruits for international trade: harvest and postharvest recommendations. *Acta Hortic.(ISHS)* 485: 31–36. (ActaHort CD-rom format.)

Asenjo, C.F. 1980. Acerola. In *Tropical and subtropical fruits*. Edited by S. Nagy and P.E. Shaw. AVI Publishing, Westport, CT. pp. 341–374.

Carrington, C.M.S., and King, R.A.G. 2002. Fruit development and ripening in Barbados cherry, *Malpighia marginata* DC. *Scientia Hortic.* 92: 1–7.

Cerezal-Mezquita, P., and Garcia-Vigoa, Y. 2000. Acerola—a neglected American fruit with high ascorbic acid content. *Alimentaria (Chile)*. 37(309): 113–125. (In Spanish)

Cooper, F.L. 1971. The acerola comes to California loaded with vitamin C. *Calif. Rare Fruit Growers Yearb.* 3: 2–8.

Duarte, O., and Paull, R.E. 2008. *Malpighia emarginata*, acerola. In *The encyclopedia of fruit & nuts*. Edited by J. Janick and R.E. Paull. CABI, Wallingford, Oxfordshire, U.K. pp. 461–466.

Du-Preez, R.J. 1997. The acerola: a natural source of ascorbic acid (vitamin C). *Neltropika Bull.* (*South Africa*) 298: 32–33.

Gonzaga-Neto, L., and Soares, J.M. 1994. *Acerola fruit for exports: technical production prospect.* Publicacoes Tecnicas Fruplex (Brazil) no. 10. Brasilia, DF, Brazil. pp. 43. (In Portuguese)

Harjadi, S.S. 1991. *Malpigia glabra* L. In *Plant resources of South-East Asia, 2, edible fruits and nuts*. Edited by E.W.M. Verheij and R.E. Coronel. Pudoc, Leiden, the Netherlands. pp. 198–200.

Ledin, R.B. 1958. *The Barbados or West Indian cherry*. University of Florida Agricultural Experiment Station, Gainesville, FL. pp. 28.

Maciel, M.I.S., Melo, E. de A., de Lima, V.L.A.G., da Silva, M.R.F., and da Silva, I.P. 1999. Processing and storage of acerola (*Malpighia* sp.) fruit and its products. *J. Food Sci. Technol.* 36: 142–146.

Marino-Netto, L. 1986. *Malpighia glabra Linn. The tropical cherry*. Nobel, Sao Paulo, SP, Brazil. pp. 94. (In Portuguese)

Miyashita, R. K., Nakasone, H. Y., and Lamoureux, C.H. 1964. Reproductive morphology of acerola (*Malpighia glabra* L.). *Hawaii Agric. Experiment Station (Univ. Hawaii, Honolulu) Tech. Bull.* 63: 31.

Morton, J. 1987. Barbados cherry. In *Fruits of warm climates*. Edited by J.F. Morton. Creative Resource Systems, Winterville, NC. pp. 204–207.

Moscoso, C.G. 1956. West Indian cherry—richest known source of natural vitamin C. *Econ. Bot.* 10: 280–294.

Pino, J.A., and Marbot, R. 2001. Volatile flavor constituents of acerola (*Malpighia emarginata* DC.) fruit. *J. Agric. Food Chem.* 49: 5880–5882.

Teixeira, A.H.de C., and Azevedo, V. de. 1995. Climate limit-indexes for acerola crop growth. *Pesquisa Agropecuaria Brasileira (Brazil)* 30: 1407–1410. (In Portuguese)

Yamane, TG.M., and Nakasone, H.Y. 1961. Pollination and fruit set studies of acerola, *Malpighia glabra* L. in Hawaii. *Proc. Am. Soc. Hort. Sci.* 78: 141–148.

### SPECIALTY COOKBOOKS

Foley, R. 2008. *The rum 1000: the ultimate collection of rum cocktails, recipes, facts, and resources.* Sourcebooks, Naperville, IL. pp. 320.

Keys, J. 1986. *We the women of Hawaii cookbook: favourite recipes of prominent women of Hawaii.* Revised edition. Press Pacific, Kalui, HI. pp. 297.

Ortiz, E.L., and Caistor, N. 1998. *A taste of Latin America: recipes and stories*. Interlink, New York. pp. 121.

Ortiz, Y. 1997. *A taste of Puerto Rico: traditional and new dishes from the Puerto Rican community*. Plume, New York. pp. 288.

Valldejuli, C.A.1984. *Puerto Rican cookery*. Pelican, Gretna, LA. pp. 408.

# 3 American Chestnut

Family: Fagaceae (beech family, sometimes called the oak family)

## NAMES

Scientific name: *Castanea dentata* (Marshall) Borkh. (*C. americana* (Michx.) Raf.)

- The word "chestnut" is from the fourteenth-century Middle English *chasteine*, derived from the Old French *chastaigne*, which in turn is from the Latin *castanea* (see the explanation of the genus name *Castanea*).
- Chestnuts are not related to water chestnuts (*Eleocharis dulcis* (Burm. f.) Trin. ex Hensch. and *Trapa natans* L.). More importantly, chestnuts should not be confused with "horse-chestnuts." The European horse-chestnut, *Aesculus hippocastanum* L., is commonly planted as an ornamental in North America. Its fruit, as well as those of the native American species of *Aesculus*, are poisonous and should not be consumed (although there has been medicinal use of horse-chestnut). Horses often refuse to eat horse-chestnuts, and it has been suggested that the name came about either with the Old English practice of using the plant medicinally to cure horse maladies or with the intention of indicating that the nuts were inedible (i.e., "fit only for horses").
- The genus name *Castanea* is from the Latin, *castanea*, and Greek, *kastanea*, names for the chestnut tree. The name is commonly stated to be based on the city of Kastana in the region of Thessaly, northern Greece, but has also been attributed to the city of Kastanis (Castanea) in Pontus, Asia Minor, Turkey. Although the name may have originated from one or both cities, it has also been suggested that the cities were named for the tree or its nuts. The genus name is also the same as the name of the city of Castanea in Italy, the home of Pope Urban VII (1521–1590). A chestnut tree is shown on the heraldic symbol of that city.

## GEOGRAPHY AND ECOLOGY OF WILD PLANTS

The American chestnut was once well established in the eastern United States and southern Ontario, and persists over most of its original area. The species occurs from southern Ontario and adjacent southern Michigan to Maine and the southern states, and is indigenous to 24 American states. In western Michigan, chestnuts have become established from introduced plants. Similar establishment from cultivation undoubtedly occurred in other places, but documentation is incomplete.

Before chestnut blight, American chestnut dominated much of the eastern deciduous forests, often comprising 25% of the trees, especially in the old-growth Appalachian forests. It has been said that an ambitious squirrel could have traveled from Maine to Georgia on the branches of chestnut trees. The species usually occurred in mixtures with other hardwoods, particularly oaks, and occasionally hemlocks and pines. Chestnuts tolerated light to moderate shade until they became part of the canopy and overtopped their associates. The tree was most common on moderate slopes, avoided limestone while occupying acidic soils, and was rare or absent on poorly drained sites, preferring sandy and gravelly somewhat dry conditions.

**FIGURE 3.1** American chestnut (*Castanea dentata*). (a) Leafy twig, spiny fruits, and nuts; photo of materials from 12-year-old trees in New Jersey (which subsequently died of blight). (Courtesy of T. van Vliet [CC By 3.0].) (b) Professor D. Jacobs in front of a chestnut tree near West Salem, Wisconsin. Note footwear intended to prevent disease transfer. Photo reproduced with his permission. (c) Spiny fruits. (Courtesy of Bob MacInnes [Flickr/CC-attribution].) (d) Distribution map of geographical range before the tree was mostly exterminated by blight.

## PLANT PORTRAIT

The American chestnut was once an incredibly magnificent and dominant tree of the eastern forest of North America. So impressed were early America settlers with the stature of the American chestnut that they called it "the king of trees" and "the queen of trees." Four billion American chestnut trees are believed to have once been present in North America. One tree out of four in the old-growth Appalachian forests was an American chestnut. Some trees grew to monstrous size, leading to the description of the tree as the "redwood of the east." Trees often approached 40 m (130 feet) in height and 3 m (10 feet) or more in diameter. Early American settlers called

**FIGURE 3.2** American chestnut (*Castanea dentata*). (a) A row of trees killed by chestnut blight; photographed in 1943 by USDA Forest Service, online at Bugwood.org (CC By 3.0). (b) Chestnut blight fungus on a tree trunk. (Courtesy of J. O'Brien, USDA Forest Service, online at Bugwood.org [CC By 3.0].) (c) An American chestnut log cabin at Portville, New York, constructed in the nineteenth century. (Courtesy of Pubdog.)

the American chestnut "the farmer's friend" because it furnished large amounts of nuts and wood for various purposes. The accidental introduction of chestnut blight destroyed almost all of the trees. This beautiful tree is now represented mostly by small stump sprouts. The roots are large and highly rot resistant, so stump sprouts can appear many decades after the trunks have been killed by blight. Sprouts sometimes live for several years, occasionally developing into small trees producing nuts before becoming reinfected and dying from blight. With the possible exception of some varieties of the Chinese chestnut, the nuts were described as the finest tasting of all chestnuts. The elimination of the American chestnut is believed to have been associated with the demise of the wild turkey in the eastern United States, since the nuts were one of the bird's staples.

FIGURE 3.3 Gigantic American chestnut trees before the species was almost exterminated by blight. (Courtesy of the U.S. Department of the Interior, Office of Surface Mining Reclamation and Enforcement.)

The flowers of American chestnut occur in catkins. The male flowers are in quite prominent catkins on the ends of branches, whereas the catkins near the bases of the branches have bisexual flowers, and when fertilized produce clusters of nuts. A protective spiny husk ("burr") encloses one to three nuts. The nuts are 5–8 cm (2–3 inches) wide and are mostly dropped while the burr is still on the tree.

The American chestnut produced a large harvest of nuts. Indigenous peoples of North America made good use them. The Iroquois from New York consumed them raw, pounded and boiled, or

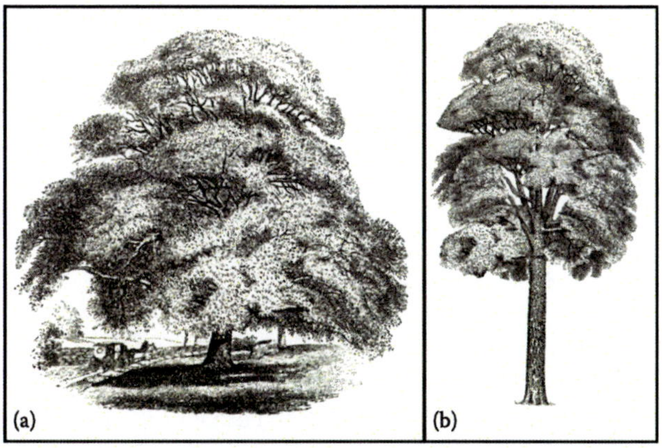


**FIGURE 3.4** American chestnut (*Castanea dentata*). (a) Open-grown tree. (From Harter, J. (Editor), *The plant kingdom compendium. A definitive volume of more than 2400 copyright-free engravings*. Bonanza Books, New York, 1988.) (b) Forest-grown tree. (Drawing courtesy of B. Brookes.)

cooked in corn bread. Chestnuts were an important cash crop in Appalachia, and boxcar loads were shipped by rail to the large cities to be sold by street vendors as roasted nuts. Those that could not be sold were often squirreled away in attics filled to the rafters. The hardships of the Depression were made much worse in Appalachia when the supply of chestnuts was ended by the blight. In Europe, chestnuts are sometimes fed to animals, particularly to pigs, as the resulting pork is highly esteemed. With amazing similarity, before the Appalachian trees were destroyed by blight, hogs were regularly turned loose to fatten on American chestnuts. Chestnut-fed pork is soft, has a stronger flavor than corn-fed pork, and has a taste that is less attractive to North Americans.

Chestnuts are probably the most important nut crop in the temperate zone, and only rank behind coconut and peanut in importance. The species native to Europe and Asia have been cultivated for many years, especially in China, Korea, Japan, and the Mediterranean basin. In the Mediterranean region, chestnuts have been grown for at least 3000 years. The ancient Greeks are thought to have been among the first to cultivate the nut, and they introduced the European chestnut from Asia Minor to Europe and Africa. In Asia, the Japanese chestnut has been raised since at least the eleventh century, and the Chinese chestnut possibly as long ago as 6000 years.

Today, world commerce in chestnuts is dominated by three species: Chinese chestnut (*C. mollissima* Blume), European chestnut (Spanish chestnut, sweet chestnut; *C. sativa* Mill.), and Japanese chestnut (*C. crenata* Siebold & Zucc.). Most of the world's chestnuts are grown and consumed in Asia. China, which specializes in cultivation of the Chinese chestnut, is currently the world's largest producer and exporter of chestnuts. Three hundred different varieties are grown in China. The world's second largest producer of chestnuts is Korea. Japan is the biggest chestnut importer. Because it is the only commercially important species of chestnut that is resistant to chestnut blight, the Chinese chestnut has become the most important chestnut species in the world. So commonly is this nut used as food in the Orient that it is said to almost take the place of the potato. The Japanese chestnut is also somewhat resistant to chestnut blight, but does not produce as high quality nuts. In Europe, the chestnut industry has long been based on the European chestnut. Italy is the largest chestnut producer, followed by France and Spain. About a century ago, the fungus chestnut blight was introduced, and ever since, most of the traditional chestnut producing areas of Europe, particularly France and Italy, have shown a progressive decline in chestnut production. In Japan, Italy, and France, the three largest consuming countries, chestnuts have often been the principal food of the poor. Just as in English a diet of bread and water connotes the simplest of food, so in French the comparable expression is "fasting on water and chestnuts" (*jeuner à l'eau et à la châtaigne*). Another French idiom, from Corsican France, is "they eat out of the drawer," referring to people who eat cheaply, and recalling the days when wild chestnuts were gathered and dried and stored in bureau drawers as survival food. Today, there is a small chestnut industry in the United States, principally in California, Oregon, and Washington, based on hybrids of European and Asian species. Most American consumption of chestnuts is based on imports, particularly from Italy. The European chestnut was planted in North America, before it proved to be highly susceptible to chestnut blight. Many regard the European chestnut as the most striking of all European trees, but unfortunately, it is no longer a good choice for cultivation, either as a nut source or as an ornamental tree.

## CULINARY PORTRAIT

The following discussion is academic, since American chestnuts are unavailable in commerce and are very rarely encountered in nature. The hope is that one day the American chestnut will once again become available.

Chestnut fruits are spiny, fibrous burs, which partially open at maturity. They contain up to five angular nuts. The "shell" or covering of the kernel is rather thin and tough, leathery rather than woody. Freshly harvested chestnuts usually contain mostly starch and very little sugar, but as they dry they "cure," some of the starch transforming to sugar. Curing or drying of the nuts is important to prepare them for eating. Because of the very low oil content of chestnuts, they do not deteriorate due to rancidity like most other nuts. As well, the low oil content makes the kernels more digestible than other nuts. Chestnuts are too bitter to be eaten raw and almost impossible to peel until they have been cooked. Cooking hydrolyzes the starch into more digestible components, just as when one cooks a potato. The shell and thin brown bitter skin are discarded. Chestnuts may be boiled, roasted, steamed, or grilled, and then served whole, mashed, or puréed. The most traditional use of chestnuts is roasted whole, with the shell and peel. A large part of the chestnut market in North America is centered on the Thanksgiving and Christmas holidays, with considerable use of roasted chestnuts. Chestnuts are associated with Christmas and New Year in Europe, and are traditionally served with game and poultry. In France and Italy, the nuts are used as a vegetable like potatoes. Chestnuts are also used in soups, fritters, porridges, stuffings, and stews; as preserves with alcohol, sugar, or syrup (as in the French delicacy marons glacés); as dressings for poultry and meats; and in confections such as puddings and cakes. Jams and purées are also prepared, the latter used to flavor puddings and cream pies. In Europe, dried, blanched nuts are often ground into flour and used to make bread, cakes, pancakes, waffles, and porridge. Indeed, in Italy, the chestnut is known as "bread of the mountains." As is the case in some North American cities, steamed chestnuts are commonly sold on the streets in Europe by vendors. Larger, selected kinds of chestnuts produced in Italy, France, and Germany are often known as "marones" (*marrons* in French) in the Old World and are traded under this term. The European chestnut may contain as much as 15% sugar, which can be extracted as a thick syrup. Starch and oil have also been extracted from the nuts.

### CULINARY VOCABULARY

- "Nesselrode pudding" or just "Nesselrode" (pronounced NEHS-uhl-rohd) is one of several dishes commemorating Count Karl Robert Nesselrode (1780–1862), chancellor of the Russian Empire, who was influential in Europe during the reign of Napoleon. Nesselrode is said to have dined and lived lavishly, and a number of rich dishes are named after him, most famously Nesselrode pudding. His chef, Monsieur Mouy, created Nesselrode pudding using chestnut purée, custard flavored with maraschino, and chopped candied fruits that had been macerated in Málaga wine. Later versions add gelatin to stabilize the dessert, rum for flavor, and whipped cream as a topping, and the preparation is often used as a filling for pie shells.

- "Mont Blanc" is a French dessert prepared with sweetened chestnut purée and whipped cream. The Italian *monte bianco* is essentially the same thing.
- The flowers of the European chestnut produce a flavorful honey, and beehives are often maintained in chestnut orchards and stands in Europe during the flowering period. *Miele di Castanea* is a popular item in specialty food shops in Italy and other European countries.

## PROSPECTS

The American chestnut was not subjected to modern breeding or selection before it was decimated by blight, so that modern cultivars are not available as they are for many other nut trees. Resistance to the blight has been the overwhelming concern for the last century, and attempts are underway to conserve germplasm for a recovery. American chestnut is grown in various arboreta, and volunteer planting programs are also maintaining some trees outside of its indigenous range. Chestnut blight attacked the European chestnut in Italy in 1938 with almost the same devastation as on American chestnut, but in the 1950s, it was observed that some trees were recovering through a natural process of weakening of the virulent form of the virus through the evolution of weaker, so-called "hypovirulent" strains. Several approaches have been taken to reestablish the American chestnut in its original range. Hybridization of the American chestnut with its blight-resistant Asian relatives, the Chinese and Asian chestnuts, has been attempted to introduce resistance genes. A second approach to overcoming the blight problem has been to exploit hypovirulent strains of the fungus. It might be possible to displace the virulent strain of the blight fungus with highly weakened strains, and indeed there has been some success using weakened strains of the fungus as a biological control agent, but much less than the case with European chestnut. A third method has been the application of ionizing radiation in hopes of generating favorable mutations. Fourth has been the attempt to genetically engineer resistance into the trees, particularly antifungal proteins such as those found in Chinese chestnut. A fifth technique is classical selective breeding for the most resistant genotypes. A sixth practice is simply to manage the plants in the light of the ecology of infection. For example, low-altitude populations in the Appalachians seem to escape the blight to a greater degree than high-altitude plants, suggesting that new plantations would best be established in low areas. Although progress has been slow, there is a reasonable prospect that in time, resistant strains of the tree will be available.

Although much smaller than the Old World commercial chestnuts, American chestnut produces what many consider to be the best-tasting nuts. The tree can also provide extremely valuable wood. Should blight-resistant trees become available, the American chestnut would be a very attractive potential new crop. There is reason to believe that the species may regain at least some of its importance.

## CURIOSITIES OF SCIENCE AND TECHNOLOGY

- Indigenous peoples of North America made good use of the American chestnut. The Iroquois from New York called the tree "O-heh-yah-tah," meaning "prickly burr." They consumed the nuts raw, or pounded and boiled to make lumps of doughy bread, or cooked in corn bread.
- In 1913, while American chestnut trees were being destroyed in large number by blight, there were widespread reports of people and animals becoming sick or dying after eating chestnuts. No cause has ever been verified. There is no evidence that the blight itself produced poisonous substances, and the fungus has been fed to rats without harmful effects. Chestnuts are not considered harmful in any respect.
- A sixteenth century custom in Britain was to roast chestnuts over a fire, making a slit or puncture in the skin of every nut except one. When that nut exploded, it indicated that the

other chestnuts were ready. The practice is the basis of Petruchio's speech in Shakespeare's *Taming of the Shrew*, Act I, Scene 2:

And do you tell me of a woman's tongue
That gives not half so great a blow to the ear
As will a Chestnut in a farmer's fire.

- The Physiocrats were a group of French economic philosophers of the 1760s who argued that government policy should be minimal, but geared to maximizing the value and output of the agricultural sector. Their writings were voluminous and, although hampered by pompousness and mysticism, influenced the future course of economic thinking in the world. Although the Physiocrats had little influence in France, one of their practical recommendations was that chestnut trees, which they viewed as supporting lazy peasants, should be cut down and wheat grown instead, and as a result, the chestnut industry was significantly repressed in the country.
- In Corsica, a tradition calls for 22 different types of dishes made from chestnut flour to be served on a wedding day.
- The French word for chestnut is *marron*, which is the basis of the dark color maroon, inspired by the claret hue of the European chestnut.
- In the Creek Indian calendar, September and October were respectively called "Little Chestnut" and "Big Chestnut" months.
- The chestnut blight that was caused by an Asian fungus produced the greatest ecological disaster in the history of the world's forests. The first cases of infected trees were reported in New York City in 1904. Half a century later, over 99.99% of the chestnut population was dead.
- American chestnut wood was iron-tough, heavy, water-resistant, as rot-resistant as redwood, and much easier to grow than oak. The wood was particularly valued for construction of furniture and similar products, and many of these objects are now collector's items. It was said that chestnut wood "carried man from cradle to grave, in crib and coffin."
- Before they were attacked by blight, American chestnut forests were very attractive when in flower: the trees developed such spectacular masses of white flowers (on the male catkins) that the floral display came to be called "summer snow." But the odor seems not to have been as attractive. Old-timers who remembered the smell of chestnuts compared it to Clorox and to "underarms without deodorant."
- Settlers learned to avoid using American chestnut as firewood, because of an annoying crackling sound made while the wood was being burned. Air spaces in the wood that became superheated and snapped were responsible for the sounds. Nevertheless, because it burned slowly and completely without producing much smoke that could attract the law, American chestnut was the favorite fuel wood at moonshine stills in the Appalachians.

## KEY INFORMATION SOURCES

(For additional literature, see Chapter 26.)

Ambrose, J.D., and Aboud, S.W. 1990. *Status report on the chestnut,* Castanea dentata *(Marsh.) Borkh., in Canada*. Committee on the Status of Endangered Wildlife in Canada, Ottawa, ON. pp. 19.

Anagnostakis, S.L. 1978. *The American chestnut: new hope for a fallen giant*. Bulletin 77. Connecticut Agricultural Experiment Station, New Haven, CT. pp. 9.

Arno, J. 1992. The demise of American chestnut: tragic loss of a great American timber. *Fine Woodworking* 92: 70–73.

Brewer, L.G. 1995. Ecology of survival and recovery from blight in American chestnut trees (*Castanea dentata* (Marsh.) Borkh.) in Michigan. *Bull. Torrey Bot. Club* 122: 40–57.

Burnham, C.R. 1988. The restoration of the American chestnut. *Am. Sci.* 76: 478–487.

Burnham, C.R., Rutter, P.A., and Treveli, D.W. 1986. Breeding blight-resistant chestnuts. *Plant Breed. Rev.* 4: 347–397.

Cochran, M. F. 1990. Chestnuts—making a comeback? *National Geographic* 177(Feb.): 128–140.

Double, M.L, and MacDonald, W.L. (Editors). 1994. *Proceedings of the international chestnut conference, West Virginia University, Morgantown, WV, July 10–14, 1992.* West Virginia University Press, Morgantown, WV. pp. 215.

Fauve-Chamoux, A. 2000. Chestnuts. In *The Cambridge world history of food.* Edited by K.F. Kiple and K.C. Ornelas. Cambridge University Press, Cambridge, U.K. pp. 359–364.

Forest, H.S., Bebee, C.N., and Cook, R.J.SO. 1990. *The American chestnut: a bibliography of references to* Castanea dentata *and other chestnut species, and on chestnut blight and its causal pathogen of the fungal genus* Endothia *or* Cryphonectria. U.S. Department of Agriculture, National Agricultural Library, Beltsville, MD. pp. 117.

Griffin, G.J. 1986. Chestnut blight and its control. *Hort. Rev.* 8: 291–336.

Griffin, G.J. 2000. The American chestnut: a story of loss and recovery. *J. Forestry* 98(2): 10–27.

Heiniger, U. 2001. *Castanea sativa*: pathology, genetic resources, ecology and silviculture. *Forest Snow Landscape Res. (Switzerland)* 76(3): 329–517.

Heiniger, U., and Rigling, D. 1994. Biological control of chestnut blight in Europe. *Ann. Rev. Phytopath.* 32: 581–599.

Huang, H., Dane, F., and Kubisiak, T.L. 1998. Allozyme and RAPD analysis of the genetic diversity and geographic variation in wild populations of the American chestnut (Fagaceae). *Am. J. Bot.* 85: 1013–1021.

Jaynes, R.A. 1974. *Genetics of chestnut.* U.S. Forest Service Research Paper WO-17. U.S. Department of Agriculture, Washington, DC. pp. 13.

Jaynes, R.A.1975. Chestnuts. In *Advances in fruit breeding.* Edited by J. Janick and J.N. Moore. Purdue University Press, West Lafayette, IN. pp. 490–503.

Jaynes, R.A. 1979. Chestnuts. In *Nut tree culture in North America.* Edited by R.A. Jaynes. The Northern Nut Growers Association Inc., Hamden, CT. pp. 111–127.

MacDonald, W.L., Cech, F.C., Luchok, J., and Smith, C. (Editors). 1978. *Proceedings of the American chestnut symposium, West Virginia University, Morgantown, WV, Jan. 4–5, 1978.* West Virginia University Books, Morgantown, WV. pp. 122.

McCarthy, M.A., and Meredith, F.I. 1988. Nutrient data on chestnuts consumed in the United States. *Econ. Bot.* 42: 29–36.

McKay, J.W. 1972. Pollination of chestnut by honey bees. North. *Nut Grow. Assoc. Ann. Rep.* 63: 83–86.

Newhouse, J.R. 1990. Chestnut blight. *Sci. Am.* 262(July): 106–111.

Nixon, K.C. 1997. *Castanea.* In *Flora of North America North of Mexico.* Vol. 3. Edited by Flora of North America Editorial Committee. Oxford University Press, Oxford, U.K. pp. 439–442.

Payne, J.A., Jaynes, R.A., and Kays, S.J. 1983. Chinese chestnut production in the United States: practice, problems and possible solutions. *Econ. Bot.* 37: 187–200.

Powell, G.H. 1898. The European and Japanese chestnuts in the eastern United States. *Delaware College Agric. Exp. Sta. Bull.* 42: 1–16.

Rutter, P.A., Miller, G., and Payne, J.A. 1990. Chestnuts (*Castanea*). In *Genetic resources of temperate fruit and nut crops.* Edited by J.N. Moore and J.R. Ballington, Jr. International Society of Horticultural Science, Wageningen, the Netherlands. pp. 761–788.

Scharf, C.S., and DePalma, N.K. 1981. Birds and mammals as vectors of the chestnut blight fungus (*Endothia parasitica*). *Can. J. Zool.* 59: 1647–1650.

Small, E., and P.M. Catling. 2002. Poorly known economic plants of Canada—33. American chestnut, *Castanea dentata* (Marsh.) Borkh. *Bull. Can. Bot. Assoc.* 35(2): 15–21.

Smith, A.H. 1976. The chestnut. *Calif Rare Fruit Growers Yearb.* 8: 15–51.

Stebbins, R.L. 1990. Requirements for a United States chestnut industry. In *Advances in new crops.* Edited by J. Janick and J.E. Simon. Timber Press, Portland, OR. pp. 324–327.

United States Agricultural Research Service. 1965. *Chestnut blight and resistant chestnuts. Farmers' Bull. 2068.* Revised edition. U.S. Department of Agriculture, Washington, D.C. pp. 21.

Volk, T. 1991. American chestnut's decline and the forest's response. *Cornell Plantations* 46: 14–18.

Wallace, R.D., and Blokland, P.J. van. 1989. *An introduction to production, marketing and economics of growing chestnuts in the southeast United States.* Institute of Food and Agricultural Sciences, University of Florida, Gainesville, FL. pp. 20.

### Specialty Cookbooks

Bhagwandin, A. 2006. *The chestnut cook book: recipes, folklore, and practical information regarding the most diverse of culinary nuts*. Third edition. Hats Off Books, Tucson, AZ. pp. 127.

Geri Camporesi, C., and Masciullo, P. 1993. *Tempo di castagne: 140 ricette [Chestnut time: 140 recipes]*. Maria Pacini Fazzi editore, Lucca, Italy. pp. 209. (In Italian)

Huntwork, J.E. 1995. *The majestic American chestnut: a history and cookbook*. Princeton Hall, Cadillac, MI. pp. 59.

Lenz, A.N., Blake, J.A., and Burkart, A.C. 1965. *How to store and use chestnuts for your eating pleasure*. College of Agriculture, Rutgers University, New Brunswick, NJ. pp. 4.

Loohuizen, R. 2006. *On chestnuts: the trees and their seeds: the horse chestnut and sweet chestnut for healing and eating*. Prospect Books, Blackawton, U.K. pp. 136.

Mulloy, A., and Lewis, E. 1999. *Celebrating the chestnut: recipes and lore from a Virginia plantation*. Marcom Publishers, Orange, VA. pp. 113.

YWCA of Seattle-King County. 1998. *Celebrity chestnut cookbook: 30 acclaimed chefs from the Puget Sound offer fabulous recipes for holiday entertaining*. YWCA of Seattle-King County and Nutty Chestnut Roasters Organization, Seattle, WA. pp. 36.

Berglund and Bolsby (1971, cited in the Appendix to this book) provide the following instructions for preparing American chestnut "if you are lucky enough to come across a tree": place about 2.5 cm (1 inch) of coarse salt on a baking sheet, put the nuts in the salt, bake in an oven for 30 minutes at 177°C (350°F), and serve with butter.

# 4 American Ginseng

Family: Araliaceae (ginseng family)

## NAMES

Scientific name: *Panax quinquefolius* L.

- The name "ginseng," first applied to Asian ginseng (*P. ginseng* C.A. Mey.), is derived from the Chinese *ren-shen* (in standard Chinese piryin; also rendered *jen-sheng*, *jin-hsien*, *shen seng*, and *shinseng*). This is usually translated as "man-shaped root," but has also been said to mean "man-essence." The name originates from (1) the fancied resemblance of ginseng root to the human form or (2) from the belief that the root represents the essence of the earth crystallized in a human form. Interestingly, an American Indian name for American ginseng, *garantoquen*, also means manlike.
- Pronounce the "gin" in ginseng as in gin, the alcoholic beverage.
- American ginseng is also known as Canadian ginseng, five-fingers (for the five leaflets), ginseng, man's health, North American ginseng, occidental ginseng, red-berry (red berry), sang (a contraction or corruption of ginseng), seng, sheng, strawberry ginseng, and ninsin (a bastardization of the Japanese name for ginseng).
- In Hong Kong, the center of ginseng commerce, ginseng from North America is called American ginseng, while commercial ginsengs from China, Korea, and Japan are simply called ginseng. However, the Chinese often use the name Chinese ginseng, while in Korea ginseng is commonly called Korean ginseng. These names are reminiscent of such place-indicating phrases as "British Columbia apples" or "Washington apples," but are often used as if they reflected quite distinct kinds.
- The term "ginseng" has been associated with at least 22 species, most of these fraudulently represented as ginseng to cheat customers. One of these is canaigre, the root of *Rumex hymenosepalus* Torr., which has been deceptively marketed as "wild red American ginseng" and "wild red desert ginseng." In China, "woman's ginseng" (*dong quai*) is derived from the root of species of *Angelica* and is esteemed for regulating menstruation and expelling afterbirth.
- The genus name *Panax* comes from the Greek *pas*, all, and *akos*, cure; or *panakes*, all healing, referring to the medicinal properties of the ginseng plant. Panacea was a goddess in Greek mythology who could heal all diseases and who found a remedy for maintaining good health.
- *Quinquefolius* in the scientific name *P. quinquefolius* is Latin for five-leaved.

## GEOGRAPHY AND ECOLOGY OF WILD PLANTS

American ginseng occurs wild in Canada in southern Ontario and southwestern Quebec. In the United States, it is found from Maine to Minnesota and South Dakota, and south to Oklahoma, Louisiana, and northern Florida. The species occurs in colonies of a few to hundreds of plants, in rich, shady, deciduous forests, in deep leaf litter. Sites are frequently on northern or northeastern, cool rocky slopes, commonly in areas with limestone outcrops, in damp but well-drained soils. American ginseng thrives in 75% shade, and even shadier locations in the southern limits of its range. Clear-cutting of virgin forests and overharvesting have drastically reduced the size of wild populations.

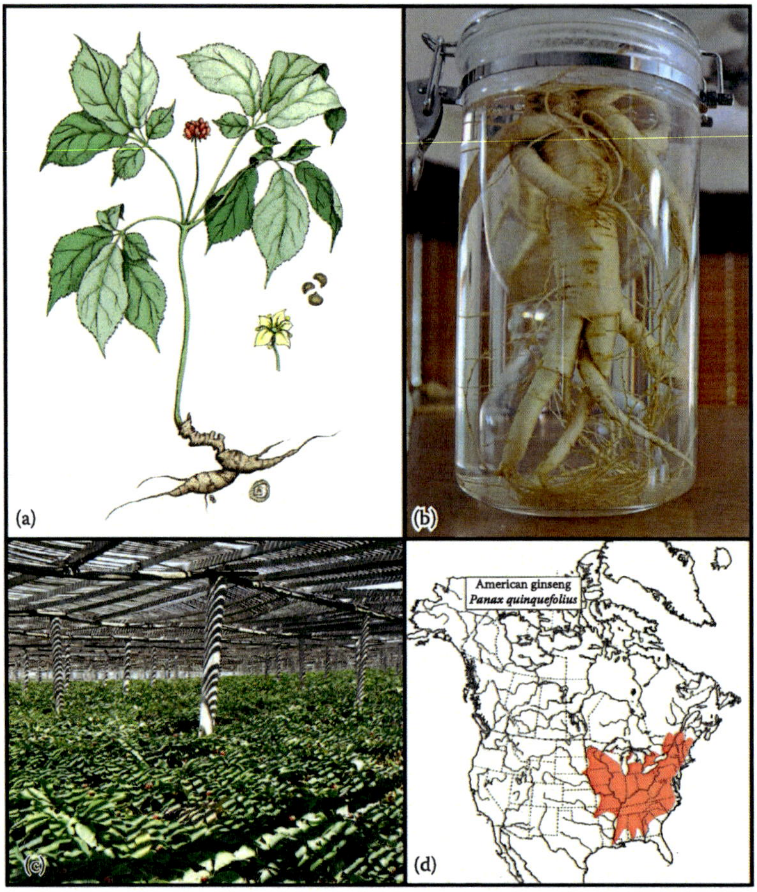


**FIGURE 4.1** American ginseng (*Panax quinquefolius*). (a) Painting by B. Brookes. (b) Large root preserved in jar. (Courtesy of Drginseng [CC By 3.0].) (c) Cultivation in a lath house; note the zebra-like effect on the support poles, the result of sunlight passing through the lath slats of the roof. (Courtesy of Drginseng [CC By 3.0].) (d) Distribution map.

## PLANT PORTRAIT

The true ginsengs are perennial herbs belonging to the genus *Panax*, of which there is one commercial species in North America (American ginseng) and a very uncertain number, perhaps four or five, in Asia. The principal Asian species is called "Asian ginseng" or simply ginseng. In most ginseng plants, an unbranched, erect stem arises from a short rhizome (underground stem) that is attached to an elongated tuberous root. One-year-old plants usually have one leaf (with three leaflets), 2-year-old plants usually have two leaves, and thereafter the plants generally have three or more leaves, each made up of three to five (occasionally seven) leaflets. Plants may not flower and produce berries for the first several years of growth. All ginseng species grow naturally in the shade of hardwood forests. The commercial species are cultivated in deep shade, mostly in special shade houses with a lath roof.

American ginseng plants are 20–70 cm (8–28 inches) tall, with a whorl of three or four long-stalked leaves, each generally with five large leaflets, of which the upper three are larger than the two lower ones. In mid-summer, 6–20 small, yellowish flowers are borne on a short stalk. The fruits mature to a deep red in the fall. The roots of ginseng have been likened to gnarled, shriveled parsnips. The principal portion of the root of American ginseng is 1–3 cm (0.4–1.2 inches) thick, and

**FIGURE 4.2** Ginseng in commerce and food (both Asian and American ginsengs). (a) A ginseng shop in Seoul, South Korea. (Courtesy of Vmenkov [CC By 3.0].) (b) A ginseng beverage. (Courtesy of Marcelo Träspel [CC By 2.0].) (c) A ginseng shop in Chinatown, Queens, New York City. (Courtesy of Jazz Guy [Flickr/CC-attribution].) (d) A traditional Korean vegetarian dish of ginseng and gingko. (Courtesy of Joe Mabel [CC By 3.0].) (e) A ginseng display in a shop in Seoul, South Korea. (Courtesy of Damien Dempsey [Flickr/CC-attribution].) (f) Chicken-ginseng soup. (Courtesy of LWY [Flickr/CC-attribution].)

5–10 cm (2–4 inches) long (sometimes much longer), spindle-shaped, and often forked. The roots of older plants become branched and acquire prominent circular wrinkles. American ginseng may live as long as 60 years. Ginseng roots are slightly aromatic, and have a sweetish, somewhat bitter taste. Wild American ginseng was apparently used by many indigenous North American Indian tribes for increasing the fertility of women; as a tonic to increase mental powers; and to treat headache, cramps, fevers, rheumatism, and cough.

The commercial significance of American ginseng was not discovered until knowledge of Asian ginseng was disseminated to the West. Asian ginseng was first described in Western literature in 1714 by Père Jartoux, a missionary in China. Jartoux conjectured that ginseng would be found in similar habitats in North America, and this information was transmitted by the Jesuits in Paris to their Canadian outposts. In 1704, Michael Sarrazin (or Sarrasin), the King's Physician to Canada, discovered American ginseng in Quebec and brought some roots to Paris. However, not until Father Lafiteau, a Jesuit priest and missionary among the Iroquois in Quebec, read Jartoux's paper, and discovered American ginseng near Montreal in 1716, did trade in the New World species begin. By 1718, the Jesuits were shipping dried roots to China. In the early 1700s, ginseng became second only to fur as a trading commodity between New France and China. During the middle of the eighteenth century, after serious depletion of local ginseng populations, very inferior material was shipped from Canada to China, resulting in a bad reputation for Canadian exports and cessation of Canadian trade with China. The problem was that roots were dug out of season, hastily harvested, and improperly dried rapidly in ovens, rather than by the time-honored method of air-drying. The Canadian trade was replaced by roots obtained in the United States. The years 1858–1896 have been described as a "green gold rush," as Americans took advantage of the Chinese interest in American ginseng. The practice of harvesting wild ginseng in North America has continued to the present, especially by rural people in southern Appalachia, who gather the roots for shipment to Asia and Europe. Overcollection has made American ginseng rare as a native wild plant.

The major centers of cultivation of American ginseng today are in southern Ontario, British Columbia, and Wisconsin (mostly Marathon County, where more than 90% of U.S. ginseng is grown). In Canada, ginseng is also raised in Quebec, Nova Scotia, New Brunswick, Prince Edward Island, and Newfoundland. In the United States, ginseng is also grown in Kentucky, North Carolina, Tennessee, and other states. The North American crop is valued at about 100 million dollars annually, about half from Canada and half from the United States. There have been attempts, generally unsuccessful, to grow American ginseng in other countries, but China is having some success in growing the American species.

Although employed for culinary purposes to a small extent, American ginseng is mostly exported for use as a medicinal tonic in Asian countries. American ginseng is sometimes incorporated in such consumer goods as hair tonic, shampoos, and facial creams. Ginseng is the world's most widely used herbal medicine, a consequence of its popularity in Asian, especially Chinese medicine. The market for ginseng extends throughout Asia, and there is minor usage as well in Western nations. Five to six million people in the United States alone use (mostly imported) ginseng regularly; ginseng sales are greater than $40 million per year in the United States; perhaps 20 million people in Western nations have used ginseng. Hong Kong is the exchange center of international commerce in ginseng, and substantial profits are made there. American ginseng tends to obtain a higher price than Asian, older material a higher price than younger, and wild roots a higher price than cultivated. For wild roots, those that are man-shaped are especially desired in Eastern culture. Ginseng brokers and graders are expert in distinguishing the value of different kinds of root. Asian ginseng is particularly grown in China, Korea, Japan, Russia, and the Ukraine.

The therapeutic value of ginseng (both American and Asian) has been a subject of continuing controversy, with Western scientists generally rejecting the claims of Asian medicine that ginseng has manifest benefits in the treatment of numerous illnesses. Ginseng has been said to enhance digestion; stimulate blood circulation; relieve fatigue, nervous disorders, and blood diseases; and, in general, have a stimulating tonic effect. Probably hundreds of millions of Asians from historical times to today have concluded that there are considerable health benefits in consuming ginseng.

## CULINARY PORTRAIT

Ginsengs are somewhat bitter, although an acquired taste often develops. The roots can be eaten like potatoes, and boiled for tea. Ginseng is used in Asia in dietary therapy as a health-giving vegetable. It is occasionally served in chicken-ginseng soup, and a slice of fresh ginseng with honey may be served to a special guest as an hors d'oeuvre. In Asia, ginseng is made into many culinary products, including extracts, tea, tablets, capsules, tinctures, syrups, drinks, wine, and candy. It is also used to flavor chewing gum, soft drinks, candy, chicken essence, jams, marmalade, and liqueurs. Some Chinese culinary recipes for ginseng include "Revitalizing ginseng tea," "Four gentlemen soup," and "Stop hunger feeling." There is also a recipe called "Stop vomiting," which might also have culinary relevance.

### Culinary Vocabulary

- "White ginseng" refers to natural ginseng root that has not been processed, except for thorough washing. In fact, the root acquires a tan color when dried.
- "Red ginseng" refers to ginseng that has been processed using steam and heat to preserve it. The roots turn red. Superior, older roots are generally required to withstand the heat, and it is often claimed that red ginseng is superior or more potent than white ginseng.

## PROSPECTS

Ginseng use is firmly established, but market prices in recent decades have often fluctuated. Most ginseng cultivation in North America is carried out by a few growers as large-scale, high-capital, and labor-intensive operations. Ginseng cultivation requires considerable knowledge, and crops are grown on long-term schedules, during which devastating problems may arise. Although potentially very profitable, ginseng cultivation is very competitive and somewhat hazardous financially. American ginseng exports from North America are an important source of revenue. It remains to be seen whether this will remain the case in the future as China invests in the cultivation of the crop in China.

Wildcrafting of ginseng in North America has a limited future. In 1988, American ginseng was officially listed as "threatened" in Canada by the Committee on the Status of Endangered Wildlife in Canada, and since then, exports of wild American ginseng have been discontinued. The Convention on International Trade in Endangered Species requires countries to show that exporting indigenous ginseng will not endanger the plant's survival. In the United States, collection and sale of wild American ginseng are subject to registration, permits, and an official season. In recent years, it has become progressively difficult to find harvestable amounts of wild ginseng in the United States. American ginseng provides an encouraging example of a plant that was seriously reduced in numbers followed by both national protection and expansion of cultivation, thus reducing pressure on natural populations. These natural populations are potentially important sources of genetic variation for the improvement of the crop.

## CURIOSITIES OF SCIENCE AND TECHNOLOGY

- In 1788, American pioneer Daniel Boone (1734–1820) collected about 13.6 metric tons (15 tons) of ginseng roots in what is now West Virginia and Kentucky to sell in Philadelphia. At a typical eighteenth-century price of 10 cents/pound, this would have fetched about $3000, an enormous sum of money in the 1780s. Unfortunately, while transporting the ginseng in a boat, the cargo was damaged by flooding and had to be redried. Adding to Colonel Boone's woes, during the delay required to dry the ginseng again, the price of ginseng in Philadelphia fell.
- In 1908, Mr. L.J. Wilson of Pennington Gap, Virginia, became totally exasperated at the prospect that the thieves who had robbed his ginseng garden the previous 2 years would return. He set up a series of shotguns with fine trip wires in his garden, with the result that a thief was shot to death and his companion wounded. A coroner's jury exonerated him from criminal blame.
- The largest wild American ginseng root collected was found in woods near Benzonia, Michigan, by Custer Higgins; it weighed 1.2 kg (2.6 pounds).
- The U.S. Food and Drug Administration once approved the addition of water to ginseng, but not ginseng to water. The latter would technically have made ginseng a food additive or drug, uses that were not approved.
- In Korea, ginseng has been fed to race horses to enhance their performance on the track.
- Manchurian wild ginseng has been considered so valuable that it has fetched as much as 250 times its weight in silver. (According to legends, ginseng has been purchased

on occasion for its weight in gold.) Almost all herbarium (plant museum) specimens of Manchurian wild ginseng, therefore, lack their roots.

- Historically, ginseng (mostly Asian) has fetched absurdly inflated prices—sometimes thousands of dollars a kilogram. A supposedly 600-year-old root found by a farmer near Beijing in 1993 reportedly sold for $282,000.
- Each season an aerial shoot is produced, a scar is left on the rhizome (underground stem), generally on different sides of it, producing a zigzag pattern. By counting these, one can determine the age of the plant. (Each season, a "wrinkle" tends to be left on the root by its contraction, but this does not seem to be as reliable as the scars left each year on the rhizome for ascertaining the age of plants.) By this method, some Asian ginseng plants have been alleged to live for about 400 years.
- Many ancient cultures advocated a "doctrine of signatures" for plants, whereby resemblances of portions of plants to human parts of the body were interpreted as a "signature" or sign of healing or beneficial qualities for the part resembled. In the case of ginseng, the roots often are forked (especially as a consequence of transplanting), and mimic the lower half of the human body. This appears to have given rise to the interpretation that the root has erotic value as an aphrodisiac, and indeed ginseng is still widely considered today to be a powerful sexual stimulant.
- Asian and American ginseng exhibit a well-known classical plant geographical distribution pattern, with one species in eastern Asia and a closely related species in eastern North America. There are numerous counterparts among other plant species.
- Ginseng has a "contractile root"—a kind of root possessed by many perennial herbaceous species, which is designed to pull the root down into the ground. This is necessary to maintain the growing tip, which regenerates the stem each year, at ground level, where it is protected. Were it not for the contractile root, each season the growing rhizome would extend the growing tip higher into the air. To counteract the vertical growth of the rhizome, the ginseng roots contracts yearly at the same rate at which the rhizome grows upward, pulling the plant downward.
- Siberian or Russian ginseng is *Eleutherococcus senticosus* (Rupr. & Maxim.) Maxim., a usually prickly shrub growing up to 6 m (20 feet) tall, found in forests. This species is native to eastern Siberia, South Korea, and parts of China. A member of the ginseng family, it is alleged to have some properties similar to those of ginseng. Its roots, and less commonly its leaves, are used medicinally in China, Russia, and elsewhere. It was described as the mystery herb taken by Soviet athletes that allowed them to win the 1980 Olympics. Millions of Russians have consumed Russian ginseng in one or another form in the belief that it has desirable medicinal effects. Siberian ginseng is now widely available in North America as an imported herbal product.
- The virtues ascribed to ginseng could be explained by a unifying concept of ginseng as an adaptogen. This has been defined as a substance that is "innocuous, causing minimal physiological disorder; non-specific in action, increasing resistance to the adverse influences of a wide range of physical, chemical and biological factors; and capable of a normalizing action irrespective of the direction of the pathological change." Although one may question whether such a definition really explains anything, studies have found that ginseng has favorable effects in the face of stress from changes in temperature, diet, restraint, exercise, and similar factors. There have also been useful pharmacological effects demonstrated for such conditions as anemia, atherosclerosis, depression, diabetes, edema, hypertension, radiation sickness, and ulcers. It has been postulated that these diverse effects could be due to alteration of the levels of pituitary, gonadal, or adrenal hormones.
- The potential use of ginseng as an aphrodisiac has attracted considerable research, but rather ambiguous and contradictory conclusions. Some experiments with animals have produced changes in hormone balance and have indeed increased sexual activity; however,

it is unclear what these effects in rats, rabbits, frogs, and hens have to do with human behavior. Some studies have shown increased sperm production in humans with low levels of production, and even improvement in impotence, but the design of these experiments was not definitive.

- Chinese herbal medicine is based on the concept of *yin* and *yang* forces of Daoist herbal theory. Yang represents masculinity, strength, and heat, and yin by contrast is feminine, mild, and cold. Ginseng is generally considered yang food. However, American ginseng is "yin of the yang" and is considered good for the respiratory or digestive system to "reduce heat." The "cooler" American ginseng also is considered more desirable in hot climates. American ginseng supposedly also has more aphrodisiac effect.
- As a rule, ginseng is rarely replanted in the same soil because it tends to die off in 2 or 3 years (this is called "the replant problem" or "replant disease"). The pathogens responsible have not been clearly identified.

## KEY INFORMATION SOURCES


Anderson, R.C., Fralish, J.S., Armstrong, J.E., and Benjamin, P.K. 1993. The ecology and biology of *Panax quinquefolium* L. (Araliaceae) in Illinois. *Am. Midl. Nat.* 129: 357–372.

Bae, H.-W. 1975. *Abstracts of Korean ginseng studies (1687–1975)*. The Research Institute, Office of Monopoly, Seoul, Republic of Korea. pp. 254.

Bae, H.-W. (Editor). 1978. *Korean ginseng*. Second edition. Korea Ginseng Research Institute, Ini-Dong, Chongro-ku, Seoul, Republic of Korea. pp. 317.

Bailey, W.G., Skretkowicz, A.L., Sawchuk, A.M., Proctor, J.T.A., Clark, L., and Lefebvre, C.M. 1994. *International ginseng conference—Vancouver 1994. Program, abstract and trade show booklet*. Simon Fraser University, Burnaby, BC. pp. 78.

Bailey, W.G., Whitehead C., Proctor, J.T.A., and Kyle, J.T. (Editors). 1995. The challenges of the 21st century. In *Proceedings of the International Ginseng Conference—Vancouver 1994*. Simon Fraser University, Burnaby, BC. pp. 645.

Baranov, A.I. 1966. Recent advances in our knowledge of the morphology, cultivation and uses of ginseng (*Panax ginseng* C.A. Meyer). *Econ. Bot.* 20: 403–406.

Chandler, R.F. 1988. Ginseng—an aphrodisiac? *Can. Pharm. J.* 121: 36–38.

Charron, D., and Gagnon, D. 1991. The demography of northern populations of *Panax quinquefolium* (American ginseng). *J. Ecol.* 79: 431–445.

Cronin, F. 1982. Elixir or not, ginseng is a lucrative cash crop. *Can. Geogr.* 102(6): 60–63.

Cruse-Sanders, J.M., and Hamrick, J.L. 2004. Spatial and genetic structure within populations of wild American ginseng (*Panax quinquefolius* L., Araliaceae). *J. Heredity* 95: 309–321.

Cruse-Sanders, J.M., Hamrick, J.L., and Ahumada, J.A. 2005. Consequences of harvesting for genetic diversity in American ginseng (*Panax quiquefolius* L.): a simulation study. *Biodivers. Conserv.* 14: 493–504.

Curran, D.F. 1983. *The complete ginseng grower's manual.* D.F. Curran Productions, Missoula, MT. pp. 146.

Curran, D.F., and Curran, P.A. 1985. *The ginseng disease and pest reference guide*. D.F. Curran Productions, Missoula, MT. pp. 118.

Duke, J.A. 1989. *Ginseng: a concise handbook.* Reference Publication, Inc., Algonac, MI. pp. 273.

English, J. 1985. The cultivation of ginseng. *Herb Spice Med. Plant Dig.* 3(2): 1–2, 5–7.

Evans, B.L. 1985. Ginseng: root of Chinese-Canadian relations. *Can. Hist. Rev.* 66: 1–26.

Goldstein, B. 1975. Ginseng: its history, dispersion, and folk tradition. *Am. J. Chinese Med.* 3: 223–234.

Hara, H. 1970. On the Asiatic species of the genus *Panax*. *J. Jpn. Bot.* 45: 197–212.

Hardacre, V. 1974. *Woodland nuggets of gold. The story of American ginseng cultivation.* Holland House Press, Northville, MI. pp. 331.

Hu, S.-Y. 1976. The genus *Panax* (ginseng) in Chinese medicine. *Econ. Bot.* 30: 11–28.

Hu, S.-Y. 1977. A contribution to our knowledge of ginseng. *Am. J. Chin. Med.* 5: 1–23.

Kelly, J. 1977. *Herb collector's manual and marketing guide. Ginseng growers and collectors handbook.* U.S. Department of Agriculture, Looneyville, WV. pp. 97.

Kitts, D.D. 2000. Chemistry and pharmacology of ginseng and ginseng products. In *Herbs, botanicals & teas.* Edited by G. Mazza and B.D. Oomah, Technomic, Lancaster, PA. pp. 23–44

Konsler, T.R. 1987. Woodland production of ginseng and goldenseal. *Stn. Bull. Purdue Univ. Agric. Exper. Sta.* 518: 175–178.

Lee, F.C. 1992. *Facts about ginseng—the elixir of life*. Hollym Press, Seoul, Korea. pp. 104.
Lewis, W.H. 1986. Ginseng: a medical enigma. In *Plants in indigenous medicine & diet: biobehavioral approaches*. Edited by N.L. Etkin. Redgrave Publishing Co., Bedford Hills, NY. pp. 290–305.
Lewis, W.H., and Zenger, V.E. 1982. Population dynamics of the American ginseng *Panax quinquefolium* (Araliaceae). *Am. J. Bot.* 69: 1483–1490.
Li, C.P., and Li, R.C. 1973. An introductory note to ginseng. *Am. J. Chin. Med.* 1: 249–261.
Li, T.S.C. 1992. *Ginseng culture methods in China, Korea and Japan*. Agriculture Canada Research Station, Summerland, BC. pp. 18.
Li, T.S.C. 1995. Asian and American ginseng—a review. *HortTechnology* 5: 27–34.
Li, T.S.C., Mazza, G., Cottrell, A.C., and Gao, L. 1996. Ginsenosides in roots and leaves of American ginseng. *J. Agric. Food Chem.* 44: 717–720.
Lindsay, K.M., and Cruise, J.E. 1975. Ginseng—native plant now rare. *Ont. Nat.* 15(2): 16–19.
Nadeau, I., and Olivier, A. 2003. The biology and forest cultivation of American ginseng (*Panax quinquefolius* L.) in Canada: a review. *Can. J. Plant Sci.* 83: 877–891. (In French)
Nantel, P., Gagnon, D., and Nault, A. 1996. Population viability analysis of American ginseng and wild leek harvested in stochastic environments. *Conserv. Biol.* 10: 608–621.
Oliver, A., Van Lierop, B., and Buonassisi, A. 1990. *American ginseng culture in the arid climates of British Columbia*. British Columbia Ministry of Agriculture and Fisheries, Victoria, BC. pp. 37.
Paseador y Barrios, C. 1901. *Ginseng; the crop that's worth its weight in sterling silver*. The Chinese-American Ginseng Co., Joplin, MO. pp. 34.
Persons, W.S. 1986. *American ginseng: green gold*. Bright Mountain Books, Asheville, NC. pp. 172. (Revised edition 1994, pp. 203.)
Persons, W.S. 1994. Growing American ginseng in its native woodland habitat. In *Herbs '94. Proceedings ninth national herb growing and marketing conference*. Edited by A. Kestner, J.E. Simon, and A.O. Tucker. International Herb Growers & Marketers Association, Mundelein, IL. pp. 32–38.
Proctor, J.T.A. 1992. The ginseng industry and the Canadian Ginseng Research Foundation. *The Grower* 42(10): 1–6.
Proctor, J.T.A. 1996. Ginseng: old crop, new directions. In *Progress in new crops*. Edited by J. Janick. ASHS Press, Arlington, VA. pp. 565–577.
Proctor, J.T.A., and Bailey, W.G. 1987. Ginseng: industry, botany, and culture. *Hort. Rev.* 9: 187–236.
Proctor, J.T.A., Lee, J.C., and Lee, S.-S. 1990. Ginseng production in Korea. *HortScience* 25: 746–750.
Proctor, J.T.A., Wang, T.-S., and Bailey, W.G. 1988. East meets West: cultivation of American ginseng in China. *HortScience* 23: 968–973.
Tanaka, O. 1994. Ginseng and its congeners—traditional Oriental food drugs. In *Food phytochemicals for cancer prevention II—teas, spices, and herbs*. Edited by C-T. Ho, T. Osawa, M-T. Huang, and R.T. Rosen. American Chemical Society, Washington, DC. pp. 335–341.
Thompson, G.A. 1987. The field cultivation of American ginseng. *Stn. Bull. Purdue Univ. Agric. Exp. Sta.* 518: 179–185.
Williams, L.O., and Duke, J.A. 1978. *Growing ginseng*. U.S. Department of Agriculture Farmers' Bull. 2201. USDA, Science and Education Administration, Washington, D.C. pp. 8.

## Specialty Cookbooks

(*Note:* numerous ginseng recipes are provided on the Web.)

Du, S. 2001. *Creative recipes of American ginseng*. Hai bin tu shu gong si, Xianggang, China. pp. 92. (In Chinese, with some English)
Durden, G. 1992. *North American cooking with ginseng*. Cookbook Publishers, Inc., Olathe, KS. pp. 66.
Taylor, D.A. 2006. *Ginseng, the divine root*. Algonquin Books, Chapel Hill, NC. pp. 272. (Mainly a review of ginseng, this book contains sample recipes.)

# 5 American Persimmon

Family: Ebenaceae (ebony family)

## NAMES

Scientific name: *Diospyros virginiana* L.

- The English word "persimmon" is of American Indian derivation, specifically of Algonquian origin, approximating the Cree *pasimian*, meaning dried fruit. American Indians dried the fruit of the American persimmon for future consumption.
- The American persimmon is also known as common persimmon, eastern persimmon, Florida persimmon, and possumwood. Obsolete names include Indian date and Virginian date plum.
- The genus name *Diospyros* is derived from the Greek *Dios*, for the god Zeus (= Jupiter), + *pyros*, wheat or grain, alluding to the edible fruit, indicating that the persimmon is the "fruit of the gods."
- *Virginiana* in the scientific name *D. virginiana* means "Virginian" and is a part of many plant names. However, the "Virginia" in question is not the modern states of Virginia or West Virginia, but the territory of Virginia, granted by the English King James I (1566–1625) in three charters, the first of which was issued in 1606. The region designated depends on the charter that was issued, but in any event included a very large region of the eastern United States.

## GEOGRAPHY AND ECOLOGY OF WILD PLANTS

The American persimmon is found throughout southeastern North America, from New England to Florida, west to Texas, Oklahoma, and Kansas. It is most common in the south Atlantic and Gulf states, developing best in the rich bottomlands of the Mississippi River and its tributaries, and in coastal river valleys. A race with 60 chromosomes and slender, smooth leaves is centered in the southern Appalachian Mountains and adjacent regions, whereas a race with 90 chromosomes and hairy leaves is in more northern and western locations. The plants have been found in clearings, old fields, prairies, thickets, open woods, and along streams and roadsides. American persimmon grows best in full sun but is very tolerant of shade. Although usually near water, the tree is sometimes in dry circumstances. The plants prefer to grow on clay or heavy loam but develop well on a very wide range of soils, including rocky soils.

## PLANT PORTRAIT

Of the approximately 400 species of *Diospyros*, several are cultivated for their edible fruit, called persimmons or date plums. The Oriental or Japanese persimmon (*D. kaki* Thunb.) has been described as the most popular fruit in the world in terms of the quantity consumed, with over one million metric tons produced annually. It is cultivated predominantly in subtropical and warm temperate regions, particularly China, Japan, Brazil, Korea, and Italy. In the United States, the tree is best adapted to central and southern California, Arizona, Texas, Louisiana, Mississippi, Georgia, Alabama, southeastern Virginia, and northern Florida. Most commercial production of Oriental

**FIGURE 5.1** American persimmon (*Diospyros virginiana*). (a and b) Tree and fruit, respectively. (Courtesy of Steven J. Baskauf [http://bioimages.vanderbilt.edu].) (c) Florida Scrub-Jays feasting on American persimmon fruit. (From Audubon, J.J., *Birds of America*, R. Havell Jr., London, Vol. 4, Plate 87, 1827–1838.) (d) Leafy twig, fruit, and seed. (From Michaux, F.A., *Histoire des arbres forestiers de L'Amérique septentrionale*, L. Haussman, Paris, France, Vol. 2, Plate 93, 1862.) (e) Distribution map.

persimmons in the United States is in California. Most persimmons available in stores in North America are oriental persimmons.

The American persimmon is usually a medium-sized tree, typically 6–20 m (20–66 feet) tall, but trees in Missouri and South Carolina taller than 40 m (131 feet) have been recorded. The scaly bark is often referred to as "alligator bark." The leaves are about 7.5 cm (2.5 inches) long, characteristically becoming smaller from the tip to the base of the branch. The trees are either male or female. Of course females are needed for fruit production, but not always males, because occasional trees will produce fruit without being pollinated (but most will not). The plum-like fruit is round or oval; variable in size; 1.5–5 cm (0.6–2 inches); green before ripening; at maturity usually orange or orange-yellow but sometimes green, blue, or black; and usually with a heavy waxy bloom on the skin. Unripe fruits are astringent and inedible, but they become very sweet and pleasantly flavored when fully ripe in the fall. The better varieties have larger, tastier fruit. The fruits should be allowed to ripen on the trees, as they do not ripen well when picked early. The heartwood of the tree is nearly black, and the wood is extraordinarily strong and heavy (the black heartwood of several species of *Diospyros* is the "ebony" of commerce). The hard, durable wood of American persimmon has been used to make furniture, flooring, paneling, and also pool cues and other sporting goods requiring shock resistance. The American persimmon is cultivated commercially to a small extent.

## CULINARY PORTRAIT

American Indians ate the fruits fresh or dried them like prunes. Early North American settlers made a tea from the fruit, which was used as a gargle for sore throats and to treat heartburn, diarrhea, and upset stomach. Persimmon beer was popular in nineteenth century America. Today, however, American persimmons are very rarely found in the marketplace and usually need to be collected from the wild. Their pulp makes delicious nut bread, puddings, and pies and is occasionally made into vinegar, molasses, beer, and brandy by home cooks. In southern Indiana, where American persimmon trees are common, restaurants and home cooks may feature such delicacies as persimmon bread and rolls, and persimmon pudding. Persimmons or their purée can be added to ice cream, custards, sorbets, puddings, cookies, cakes, pies, and the like, and they go well with rice, cheese, seafood, and poultry. The taste can be made more interesting by squeezing lemon or lime juice over the fruit. The seeds (which are inedible) should be removed (some fruits are seedless). In the past, dried, roasted, ground seeds have been used as a coffee substitute.

Because of the very distressing astringency of immature fruits, they must be allowed to ripen before consumption. A common misconception is that American persimmon fruits need to be exposed to freezing to ripen and should, therefore, be harvested after they have been exposed to heavy frost; in fact, this lowers the quality of the fruit.

"Bezoars" are rounded, layered stones found in internal organs such as the stomach, gall bladder, and kidneys of large herbivorous mammals, primarily sheep and goats. A bezoar is most often formed from an undigested mass of ingested material that remains in the stomach. Such stones can obstruct the passage of food into the intestine, and surgery is usually necessary to remove them. Bezoars of vegetable origin are most frequent in areas where persimmons are popular. So common is the persimmon as a cause of the formation of bezoars that the term "diospyrobezoar" was once proposed for persimmon bezoars. Persimmons seem extraordinarily able to cause bezoars because their tannins, when liberated in the stomach, can consolidate vegetable material into a mass, especially when the stomach is empty. Fortunately bezoars are relatively rare, although people who have had stomach surgery or suffer from slow stomach action seem prone to developing them. Because of the possibility of developing bezoars, it has been suggested that gorging on persimmons is not a good idea, especially on an empty stomach.

### Culinary Vocabulary

- Persimmon fruits are commonly called "simmons" in the American South and Midwest.
- "Persimmon pudding" is a traditional American dessert made with persimmons, notable for its high moisture content that makes it store well under refrigeration for several days. It is one of the two dessert specialties of the state of Indiana (the other being sugar cream pie).

## PROSPECTS

The American persimmon is a locally popular fruit, which has been improved to some extent. The chief impediment to its development is the Oriental persimmon, which is much superior and is established as a leading fruit. It is unlikely that the American persimmon will ever be competitive with the Oriental persimmon, but it is an interesting fruit, capable of considerable improvement by breeding, and deserving of at least some investment in research to produce better cultivars.

## CURIOSITIES OF SCIENCE AND TECHNOLOGY

- The "Victorian language of flowers" was a secret-coded language in Victorian times in England, with flowers and plants symbolic of certain messages, so when the flower or plant was mentioned in a letter those who knew the code could understand the hidden information. "Persimmon" meant "bury me amid nature's beauties."
- In American folklore, persimmon seeds were cut in half to predict weather in the winter. If the resulting seeds were spoon- or shovel-shaped, it foretold of a hard winter with deep snow.
- Shortages during the American Civil War resulted in Southerners using the large seeds of American persimmon as buttons.
- The town of Mitchell in southern Indiana is the home of what has been claimed to be the world's only persimmon-pulp canning factory, known as Dymple's Delight (named for Dymple Green).
- Astringency of many fruits is due to "phenolic" chemicals called tannins. Tannins are useful for "tanning" animal hides because these chemicals have the ability to bind protein molecules together. The dry, puckery sensation caused by tannins in the mouth results from the compounds cross-linking the proteins on the surface of the tongue and palate and in the saliva, causing the surfaces to constrict and their lubrication to fail. Persimmons have considerable tannin, but tannins are also present in peaches, bananas, dates, and other fruits when they are immature, presumable discouraging animals from eating them until the seeds are mature, at which point the fruit becomes ripe and loses its astringency.
- The wood of the American persimmon was once famous for its use in golf clubs, especially in driver-type golf club heads. Today, persimmon clubs are still available at premium prices, but synthetic materials are commonly used instead.
- Numerous colors have been defined by reference to the colors of plant parts. As a color, "persimmon" closely resembles the "medium orange-red" of a ripe persimmon fruit.
- American persimmons are relished by white-tailed deer, and in the autumn knowledgeable hunters often position themselves near trees bearing many fruits.

## KEY INFORMATION SOURCES


Ames, G.K. 1999. *Persimmon production*. Appropriate Technology Transfer for Rural Areas (ATTRA), Fayetteville, AR. (Irregularly paginated)

Baldwin, J.T., and Culp, R. 1941. Polyploidy in *Diospyros virginiana* L. *Am. J. Bot.* 28: 942–944.

Benharroch, D., Krugliak, P., Porath, A., Zurgil, E., and Niv, Y. 1993. Pathogenetic aspects of persimmon bezoars. A case-control retrospective study. *J. Clin. Gastroenterol.* 17: 149–152.

Briand, C. 2005. The common persimmon (*D. virginiana* L.): The history of an underutilized fruit tree (16th–19th centuries). *Huntia* 12: 71–89.

Choi, Y.-A., Tao, R., Yonemori, K., and Sugiura, A. 2003. Genomic distribution of three repetitive DNAs in cultivated hexaploid *Diospyros* spp. (*D. kaki* and *D. virginiana*) and their wild relatives. *Genes Genet. Syst.* 78: 301–308.

Cristoffer, C. 1993. Fruit removal and interplant distance in the persimmon *Diospyros virginiana*. *Florida Scientist* 56: 223–225.

English, J. 1982. A practical review of persimmons. *Annu. Rep. Northern Nut Grow. Assoc.* 73: 36–38.

Fletcher, W.F., and Gould, H.P. 1942. *The native persimmon*. Revised edition. U.S. Department of Agriculture, Washington, DC. pp. 22.

Gazit, S., and Levy, Y. 1963. Astringency and its removal in persimmons. *Israel J. Agr. Res.* 13(3): 125–132.

Glasgow, L.L. 1977. Common persimmon, *Diospyros virginiana*. In *Southern fruit-producing woody plants used by wildlife*. Technical Report SO-16. Edited by L.K. Halls. Forest Service, U.S. Department of Agriculture, Washington, DC. pp. 103–104.

Goodell, E. 1982. Two promising fruit plants [*Diospyros virginiana* and *Actinidia arguta*] for northern landscapes. *Arnoldia* 42: 103–133.

Griffith, E., Griffith, M.E., and McDaniel, J.C. 1982. *Persimmons for everyone*. North American Fruit Explorers, Distributed by Dorothy Nichols, Arcola, MO. pp. 145.

LaRue, J.H., Opitz, K.W., and Beutel, J.A. 1982. *Growing persimmons*. Cooperative Extension, University of California, Berkeley, CA. pp. 11.

Maisenhelder, L.C. 1971. *Common persimmon (Diospyros virginiana)*. Forest Service, U.S. Department of Agriculture, Washington, DC. pp. 6.

McDaniel, J.C. 1973a. Persimmon cultivars for northern areas. *Fruit Var. J.* 27: 94–96.

McDaniel, J.C. 1973b. American persimmon: An emerging horticultural crop. *Fruit Var. J.* 27: 116–119.

Olson, D.F. Jr., and Barnes, R.L. 1974. *Diospyros virginiana* L. – common persimmon. In *Seeds of woody plants in the United States. Agricultural Handbook No. 450*. Technical coordination by C.S. Schopmeyer. Forest Service, U.S. Department of Agriculture, Washington, DC. pp. 373–375.

Raymond, J. 2006 (September). The horticultural improvement of the American persimmon. *Nutshell*: 20–27.

Skallerup, H.R. 1953. The distribution of *Diospyros virginiana*. *Ann. Missouri Bot. Gard.* 40: 211–225.

Spongberg, S.A. 1979. Notes on persimmons, kakis, date plums, and chapotes. *Arnoldia* 39: 290–310.

Sweedman, R. 1989. *Persimmon growing*. Second edition. N.S.W. Agriculture & Fisheries, Sydney, Australia. pp. 12.

Troop, J., and Hadley, O. 1895. The American persimmon: Its habits and distribution in Indiana, methods of propagation and cultivation, and suggestions concerning its improvement. *Trans. Indiana Hort. Soc.* 1895: 116–123.

Troop, J., and Hadley, O.M. 1896. *The American persimmon*. Bulletin No. 60. Purdue University Agricultural Experiment Station, Lafayette, IN. pp. 43–54.

Warren, T.O. 1979. Comparison of northern and southern races of *Diospyros virginiana*. *North American Pomona* 12: 127–128.

Watts, R. 1899. Persimmons. *Bull. Univ. Tenn. Agr. Exp. Sta.* 11: 193–223.

Whitsom, M. 2007. The pleasing persimmon. *The Lady-Slipper (Kentucky Native Plant Society)* 22(1): 2–4.

Yonemori, K., Sugiura, A., and Yamada, M. 2000. Persimmon genetics and breeding. *Plant Breed. Rev.* 19: 191–225.

## Specialty Cookbooks

Bear Wallow Books. 1978. *Old fashioned persimmon recipes*. Bear Wallow Books, Nashville, IN. pp. 32.

California Fruit Exchange. 1964. Fresh persimmon recipes. *Blue Anchor* 40(4): 21–22.

Filut, M.A. 1985. *People pleasing persimmon recipes*. Circulation Service, Leawood, KS. pp. 38.

Francis, W.O. 1959. *Persimmon pudding*. Ohio Valley Folk Research Project, Ross County Historical Society, Chillicothe, OH. pp. 2.

Green, D. 1996. *Persimmon goodies*. Dymple's Delight Persimmon Products, Mitchell, IN. pp. 23.

Greene, M. 1983. *Persimmon recipe book*. Ozarks Mountaineer, Branson, MO. pp. 32.

Hazelton, J.W. 1999. *Persimmons (kaki) from seed to supper: For good growing, good eating and good health*. Jack's Bookshelf, Inc., St. Petersburg, FL. pp. 56.

Lum, L.A. 1992. *The French-Icarian persimmon tree cookbook*. Pollard Pr., Jacksonville, FL. pp. 189.

Euell Gibbon's classic *Stalking the Wild Asparagus* (1962) is a particularly good source of recipes for the American persimmon. Brill (2002) and Schufer (2011) each have 10 recipes featuring American persimmon fruit. Also based on American persimmon, Beatty (1987) has 4 recipes, Freitus (1980) has 3, Freitus and Haberman (2005) have 6, Johnson (1989) has 4, Kluger (1973) has 13, Robe-Terry (1997) has 8, Russell (1975) has 4, Tatum (1976) has 17, and Tull (1987) has 2. (See the Appendix to this book for details of the publications cited.)

# 6 Anise Hyssop

Family: Lamiaceae (Labiatae; mint family)

## NAMES

Scientific name: *Agastache foeniculum* (Pursh) Kuntze

- True to its name, anise hyssop is anise-scented (anise is the culinary herb *Pimpinella anisum* L.). It acquired the "hyssop" part of its name through resemblance to the true hyssop (another culinary herb, *Hyssopus officinalis* L.).
- Anise hyssop is also called blue giant hyssop, elk mint, fennel giant hyssop, fragrant giant hyssop, giant fennel hyssop, giant hyssop, lavender giant hyssop, and licorice mint. Hyphens are often found in these names, such as anise-hyssop. "Fennel," a word used in some of the above names is best reserved for *Foeniculum vulgare* Mill. (as noted below, the odor of anise hyssop is reminiscent of fennel).
- Anise hyssop is neither an anise nor a hyssop. Also known as licorice mint, it is neither licorice nor mint. And, to belabor the point, it is known as elk mint and certainly not related to the elk.
- The genus name *Agastache* comes from the Greek *agan*, very much, + *stachys*, spike of wheat, literally "many ears of wheat," an allusion to the many tiny blossoms of the flowering stalk.
- *Foeniculum* in the scientific name *A. foeniculum* is the genus name of fennel (*foeniculum* is the diminutive form of the Latin *foenum*, hay, presumably an allusion to the abundant thread- or straw-like segments of the leaves of the fennel plant) and reflects the fennel-like odor of anise hyssop.

## GEOGRAPHY AND ECOLOGY OF WILD PLANTS

Anise hyssop is essentially a prairie plant, centered in the Great Plains of the United States and Canada. In Canada, it has been collected from British Columbia and the District of MacKenzie east to southern Ontario. In the United States, anise hyssop has been collected from Washington to Illinois, although it very rarely grows wild in or west of the Rocky Mountains. The plant has escaped from cultivation in eastern Ontario, Quebec, and New Brunswick, and in parts of the eastern United States. Lint and Epling (1945) stated that the native range includes Wisconsin, Minnesota, Iowa, North and South Dakota, to Wyoming and Colorado, and in Canada from western Ontario to Alberta, and that other collections could be escapes. The habitats include dry thickets, fields, and waste ground, especially on prairies and plains. The species grows best in cool weather. It is moderately cold tolerant, occasionally dying out during cold winters in USDA Zone 6, but able to survive in USDA Zone 4. It grows relatively poorly in areas as far south as USDA Zone 9. Anise hyssop prefers full sun (but tolerates light shade) and well-drained, fertile ground (although it grows in a wide variety of soils). It stops flowering under drought conditions, and cultivated plants need to be provided with adequate moisture.

## PLANT PORTRAIT

Anise hyssop is a stiffly erect, attractive, perennial, aromatic herb, 80–150 cm (31–59 inches) tall. Its leaves are softly white underneath due to a covering of white, felty hairs. The plant usually smells strongly of anise, with overtones of mint. Blue or blue-purple flowers are produced in the summer,

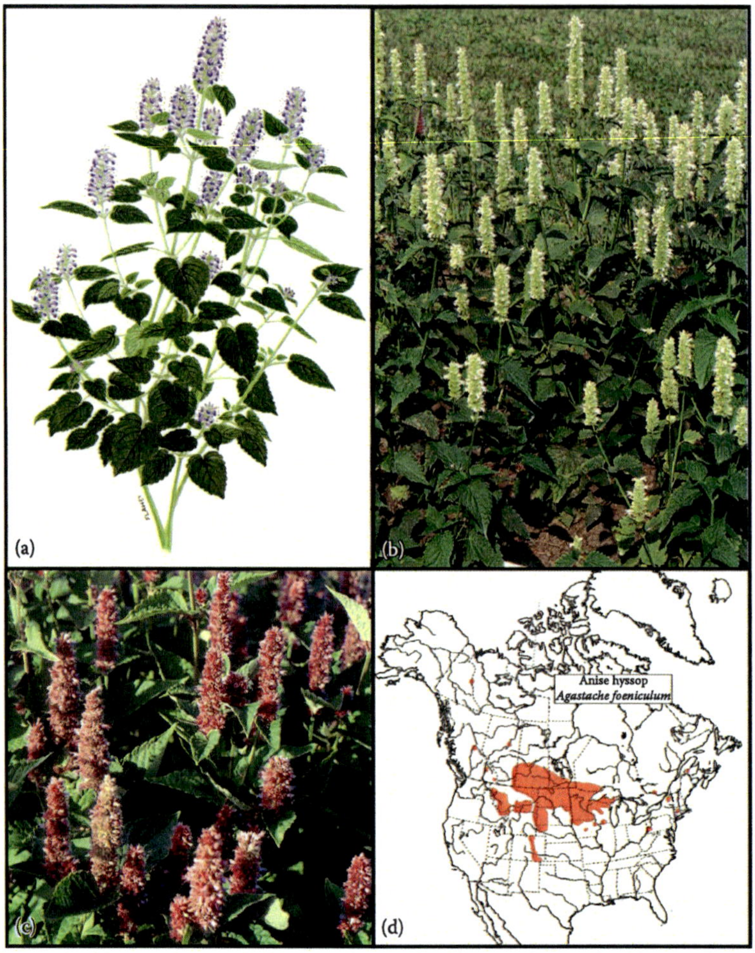


**FIGURE 6.1** Anise hyssop (*Agastache foeniculum*). (a) Painting by B. Flahey. (b) White-flowered cultivar 'Alba'. (Courtesy of Sharon and Gerry Channer.) (c) Normal, purple-flowered form. (Courtesy of Sharon and Gerry Channer.) (d) Distribution map.

with blooming lasting for many weeks. This North American herb was used by Native Americans to make beverages and as a condiment in foods, as well as medicinally. The leaves were used for tea by Woods Cree of Saskatchewan, and both for tea and as a flavoring by Montana Indians. The species is cultivated primarily as a source of methyl chavicol, which is used in the manufacture of perfumes, as well as for some food purposes, as noted below. Methyl chavicol can be chemically modified to the chemicals anethole and anisaldehyde, which are useful in color photography and in soaps, dentifrices, and other preparations.

## CULINARY PORTRAIT

Methyl chavicol from anise hyssop is used in liqueurs, some foods, and root beer. However, anise hyssop has limited commercial importance for culinary purposes. It is found occasionally in tea mixtures.

Grown in the home garden for its beautiful appearance, anise hyssop also serves as an extraordinarily useful culinary herb. The fresh or dried leaves complement lamb, pork, poultry, and fish and can be added to flavor fruit salads, cookies, cakes, breads, muffins, beverages, and jellies. Both the leaves and flowers go well with fresh or baked fruit, mushrooms, and a variety of plant foods such as squash, sweet potatoes, peas, carrots, and rice. The anise-flavored flowers are extremely tasty and can be used as a seasoning in baking and in teas, as well as in salads and as garnishes. The flowers tend to have a lighter flavor than the leaves. In some cultivated varieties, the flowers bloom for a very long period, and so are available for home use when grown in the garden. The flowers may contain tiny insects called thrips, which are impossible to remove, and so provide an extra serving of protein!

The blooms of anise hyssop are so attractive to honeybees for preparing honey that the herb has been called the "wonder honey plant," and it is cultivated in the United States specifically for bees, producing an excellent light-colored honey. Some tests have shown that bees prefer anise hyssop to white clover. While anise hyssop is used commercially to generate honey, store-bought honey can be flavored with it. A small jar can be filled with fresh anise hyssop leaves and flowers (or one-third filled with the dried herb), and liquid honey (or honey liquified by warming) can be added, and the mixture left for a month before using.

## PROSPECTS

Anise hyssop has achieved limited commercial significance, primarily as a source of the chemical methyl chavicol, which is used in part as a flavorant. This chemical and preparations of the leaves are used to a minor extent in tea preparations. The plant is also particularly useful as forage for honey bees and is widely grown in home gardens as an ornamental and flavoring herb. There are about two dozen related species, all but one native to North America, which could be used in breeding programs to produce new cultivars. There have been efforts in recent decades to develop the crop commercially, but there has been limited success. This herb appears destined to remain quite minor in importance.

## CURIOSITIES OF SCIENCE AND TECHNOLOGY

- Anise hyssop was often used for medicinal purposes by native Indians of North America. The Cheyenne used an infusion (a boiled tea) of the leaves for colds, chest pains from coughing, and a weak heart; the leaves in a steam bath to induce sweating; and powdered leaves rubbed on the body for high fevers. The Chippewa used a root infusion for colds, chest pain, and cough and a poultice of the leaves or stalks for burns. Flowers of anise hyssop were frequently included in Cree medicine bundles.
- Native North Americans chewed anise hyssop as a breath freshener.
- Anise hyssop nectar strongly attracts butterflies, and also hummingbirds, to a garden. Wild birds, especially finches, are also drawn by the seeds that form in the autumn.
- Distantly related plant species often produce similar, if not identical, chemical scents of anise, lemon, orange, pineapple, cinnamon, and a host of other flavorings so valuable to our cuisine. It is not at all surprising, therefore, that anise hyssop, a member of the mint family, has the aroma of anise, a member of the carrot family.

## KEY INFORMATION SOURCES

Ayers, G.S., and Widrlechner, M.P. 1994a. The genus *Agastache* as bee forage: A historical perspective. *Am. Bee J.* 134: 341–348.

Ayers, G.S., and Widrlechner, M.P. 1994b. The genus *Agastache* as bee forage: An analysis of reader returns. *Am. Bee J.* 134: 477–483.

Charles, D.J., Simon, J.E., and Widrlechner, M.P. 1991. Characterization of essential oil of *Agastache* species. *J. Agric. Food Chem.* 39: 1946–1949.

Fuentes-Granados, R.G., Widrlechner, M.P., and Wilson, L.A. 1998. Allozyme inheritance in anise hyssop (Lamiaceae). *Am. Soc. Hortic. Sci.* 123: 868–874.

Fuentes–Granados, R.G., Widrlechner, M.P., and Wilson, L.A. 2000. Inheritance studies of aromatic compounds in *Agastache foeniculum* (Pursh) Kuntze. *J. Essent. Oil Res.* 12: 581–594.
Galambosi, B., and Galambosi-Szebeni, Z. 1992. Studies on the cultivation methods of *Agastache foeniculum* (Pursh) Kuntze in Finland. *Acta Agron. Hung.* 41(1–2): 107–115.
Henning, J. 1994. After July—*Agastache. Am. Horticult.* 73(8): 32–35.
Henning, J. 1995. Aromatic *Agastache. Am. Nurseryman* 181(9): 59–67.
Lint, H., and Epling, C. 1945. A revision of *Agastache. Am. Mid. Nat.* 33: 207–230.
Mallavarapu, G.R., Kulkarni, R.N., Baskaran, K., and Ramesh, S. 2004. The essential oil composition of anise hyssop grown in India. *Flavour Fragrance J.* 19: 351–353.
Mazza, G., and Kiehn, F.A. 1992. Essential oil of *Agastache foeniculum*, a potential source of methyl chavicol. *J. Essent. Oil. Res.* 4: 295–299.
Nykänen, I., Holm, Y., and Hiltunen, R. 1989. Composition of the essential oil of *Agastache foeniculum. Plant Med.* 55: 314–315.
Omidbaigi, R., Kabudani, M., and Khoorang, M. 2008. Nitrogen fertilizer affecting herb yield, essential oil content and compositions of *Agastache foeniculum* Pursh. *J. Essent. Oil-Bearing Plants* 11: 261–266.
Tirillini, B.B., Menghini, A.A., and Pellegrino, R.R. 1997. Constituents of the leaf secretory hairs of *Agastache foeniculum* Kuntze. *J. Essent. Oil Res.* 9: 19–21.
Vogelmann, J.E., and Gastony, G.J. 1987. Electrophoretic enzyme analysis of North American and eastern Asian populations of *Agastache* section *Agastache* (Labiatae). *Am. J. Bot.* 74: 385–393.
Wilson, L.A., Senechal, N.P., and Widrlechner, M.P. 1992. Headspace analysis of the volatile oils of *Agastache. J. Agric. Food Chem.* 40: 1362–1366.

## Specialty Cookbooks

No cookbook was found that is dedicated only to anise hyssop. However, recipes are common in books discussing how to grow an edible herb garden and consume the herbs. For example, Creasy (1988) has two anise hyssop recipes, Barash (1993) has four, and Kublick (1990), Belsinger (1991), and Stewart (2000) each have one. (See the Appendix to this book for details of the publications cited.) Numerous recipes are also available on the Web.

# 7 Aronia

Family: Rosaceae (rose family)

## NAMES

Scientific name: *Aronia melanocarpa* (Michx.) Ell. (*Aronia arbutifolia* var. *nigra* (Willd.) Seymour, *Aronia nigra* (Willd.) Koehne, *Photinia melanocarpa* (Michx.) K.R. Robertson & J.B. Phipps)

- The name *Aronia* is a modification of *Aria*, beam tree of Europe.
- Wild plants of aronia are usually known as black chokeberry. The wild plants are less commonly called barrenberry and chokepear. Chokeberries are so named for their astringent fruit, due in part to a high content of tannins. Black chokeberry normally has black fruit. Commercial marketers of the fruit initiated using the genus name *Aronia* as the common name (i.e., aronia) in lieu of "black chokeberry" as the latter name is suggestive of an unappealing fruit.
- Sometimes chokeberries (some *Aronia* spp.) and chokecherries (*Prunus virginiana* L.) are confused. Adding to the confusion, there is a chokecherry cultivar called 'Melanocarpa'.
- The word *melanocarpa* in the scientific name is based on *melas* for "dark" and *carpos* for fruit."

## GEOGRAPHY AND ECOLOGY OF WILD PLANTS

Aronia grows wild from southern Labrador and coastal Maine west through the Great Lakes to Wisconsin and south in the Appalachians to Alabama. It occurs in a variety of habitats, including meadows, bogs, lake and seashore edges, borders of woods, roadsides, barrens, and dunes. Aronia occupies both wet and dry areas, preferring neutral to slightly acid soils, and tolerating infertile soils. The plant is usually found in sites exposed to full sunlight.

## PLANT PORTRAIT

Aronia is a deciduous, multistemmed, suckering shrub, typically only 0.5–1 m (1.6–3 feet) tall, occasionally up to 3 m (10 feet), rarely 4 m (13 feet) high. It has lustrous, dark green, nonpubescent leaves and white flowers. The species reproduces both sexually and vegetatively. New shoots arise from the rhizomes (underground stems), and the plant spreads, tending to form dense thickets and large colonies. The fruits are black or occasionally dark red, up to 1 cm (0.4 inch) in diameter, and occur in clusters. The berries often fall off the plant at maturity but are sometimes persistent in a withered condition into the early winter. Some cultivars have been selected that have larger fruit, and large size is certainly desirable in edible berry crops intended for fresh consumption. However, aronia has primarily been an "industrial" crop, harvested for extracts, and so total yield of berries/unit area cultivated, not berry size, has been considered to be most important. Aronia is more cold-hardy than most other bush fruits, and although not considered as desirable as most fresh berries, it can be grown in cold climates. Russia has more extensively cultivated aronia than other countries, but it is also commercially raised in Scandinavia and other parts of eastern Europe and very recently in the United states, particularly in Oregon.

In North America, aronia hybridizes widely with the related red chokeberry, *A. arbutifolia* (L.) Pers. (*Photinia pyrifolia* (Lam.) K.R. Robertson & J.B. Phipps), and the hybrid is often called

**FIGURE 7.1** Aronia (*Aronia melanocarpa*). (a) Fruiting branche. (Courtesy of BotBin [CC By 3.0].) (b) Flowering branch. (From House, H.D., *Wild flowers of New York*, Part 1. State Museum, Albany, NY, 1918.) (c) Fruit cluster. (Courtesy of Miraculix H.B.) (d) Distribution map.

*A.* × *prunifolia* (Marsh.) Rehd. (*P.* × *floribunda* (Lindl.) K.R. Robertson & J.B. Phipps). Red chokeberry is found on the eastern coastal plain from Newfoundland and Quebec south to Florida and eastern Texas. Because of the widespread hybridization, plants in North America are often difficult to identify.

## CULINARY PORTRAIT

Aronia is used in the food processing industry to produce a very wide range of edible products, including alcoholic beverages (wine, spirits), nonalcoholic drinks (especially mixed juices), jam, sauce, soft spreads, syrup, fillings for confectionery and bakery products, and flavorings for yogurt, cream desserts, and teas.

Aronia extracts are now widely used as red color for foods. Prominent in the extracts are anthocyanins, a subclass of flavonoids. Anthocyanins make up to 1% of the dry weight of the berries. These are water-soluble plant pigments, responsible for blue, purple, red, and black coloration. Flavonoids act as antioxidants and have anti-inflammatory effects. Because of these health-promoting properties, aronia anthocyanins have impressive economic potential as a food additive, supplementing the usefulness of the juice for flavor purposes. Recent publicity that highly colored berries (as well as vegetables) contain bioactive pigments that can fight cancer, heart disease, and even the aging process has stimulated interest in the use of aronia as a source of health-promoting dietary extracts (nutraceuticals), which can be consumed directly or used to fortify foods (functional foods). Aronia juice is considered to be healthful, and its piquant taste makes it commercially attractive, at least when combined with other juices.

Wild food enthusiasts sometimes consume aronia raw, following the traditional practice of some native inhabitants of North America, but the fruits are very astringent, much like chokecherries. Raw aronia fruits are generally too puckery, sour, and strong in taste and are usually combined with other berries such as blueberries and black currants. Wild fruits can be collected and stewed or made into jelly. They are rich in pectin and can be added to pectin-deficient fruits to produce mixtures that jell readily. Some cultivars like 'Nero' and 'Viking', selected for fruit (rather than ornamental) characteristics, have better flavor and may even be eaten out of hand.

## PROSPECTS

Aronia is inferior in taste to familiar popular bush fruits such as blueberry. However, it is increasingly becoming known for its healthful properties and is being combined with other juices for effective marketing. Several other berry-producing shrub species are currently competing for development as new crops based on the health-promoting value of their juice, notably elderberry and sea buckthorn. Aronia is naturally adapted to cold climates and so constitutes an excellent candidate for increased cultivation in colder areas of the world. The species is easy to grow, hardy, and disease resistant, and there is considerable potential for using wild relatives as breeding material to create new, superior cultivars. It is ironic that Europeans took the initiative to import and develop this North American crop, and their success provides an excellent basis for North Americans to expand on their accomplishments. Although the small-fruit marketplace is extremely competitive, aronia does seem capable of continuing to increase its market niche. This is a worthwhile crop for research and development by both government and the private sector.

## CURIOSITIES OF SCIENCE AND TECHNOLOGY

- Aronia is the second most popular ice cream flavor in Poland.
- Although birds will eat fresh, wild aronia fruits and are a significant threat to aronia orchards, the fruit is not a preferred food and tend to be ignored during the summer. In the winter, however, when food is much less available, shriveled berries that persist on the plants become a significant source of nutrition for birds, as well as for other wildlife. Relatively few woody plants retain their fruit after they mature, and it appears that the phenomenon in aronia represents an adaptation for distributing the seeds during the winter, which compensates for the lack of palatability compared to other species consumed during the summer.
- As noted in the preceding text, anthocyanins derived from aronia are being widely used to color food. However, food scientists must carefully control the acidity of foods to express a given color. Anthocyanins tend to shift hues from pinks and reds at pH 3.0 to purple-violets at pH 5.0 and blues at pH 7.0.
- Several species of *Aronia* develop spectacular red fall foliage. Red coloration in autumn leaves of most deciduous plants is primarily a result of the production of the anthocyanin cyanidin-3-glycoside, and leaf anthocyanins often increase in concentration during senescence. Several hypotheses have been advanced to explain the possible adaptive value of the development of such leaf pigments in the fall. Postulated explanations include a protective

function against high irradiance, particularly ultraviolet light; prevention against damage in sensitive tissues by photooxidation (although late-season leaves are unlikely to add much photosynthate, maintaining the biochemistry of the cells in good working order could contribute to resorption of nutrients from the leaves before they are discarded); increasing stress tolerance, particularly to cold temperatures; discouraging herbivores (at least those that overwinter or lay eggs in the vicinity); and attracting seed dispersers. (For additional information, see Lee, D.W., O'Keefe, J., Holbrook, N.M., and Feild, T.S. 2003. Pigment dynamics and autumn leaf senescence in a New England deciduous forest, eastern USA. *Ecol. Res.* 18: 677–694.)

## KEY INFORMATION SOURCES

Gätke, R., and Wilke, K. 1991. Is it possible to harvest aronia by machine? *Gartenbau* 38: 37–38. (In German)

Hall, I.V., Wood, G.W., and Jackson, L.P. 1978. The biology of Canadian weeds 30. *Pyrus melanocarpa* (Michx.) Willd. *Can. J. Plant Sci.* 58: 499–504.

Hardin, J.W. 1973. The enigmatic chokeberries (*Aronia*, Rosaceae). *Bull. Torrey Bot. Club* 100: 178–184.

Hirvi, T., and Honkanen, E. 1985. Analysis of the volatile constituents of black chokeberry (*Aronia melanocarpa* Ell.). *J. Sci. Food Agric.* 36: 808–810.

Ismail, A.A. 1974. Selective thinning of black barrenberry fruit in lowbush blueberry fields with ethephon. *Hortsci.* 9: 346–347.

Jeppsson, N.J. 1998. Evaluation of black chokeberry, *Aronia melanocarpa*, germplasm for production of natural food colourants. *Acta Hortic.* 484: 193–198.

Jeppsson, N. 2000. The effects of fertilizer rate on vegetative growth, yield and fruit quality, with special respect to pigments, in black chokeberry (*Aronia melanocarpa*) cv. 'Viking'. *Sci. Hortic.* 83: 127–137.

Jeppsson, N. 2000. The effect of cultivar and cracking on fruit quality in black chokeberry (*Aronia melanocarpa*) and hybrids between chokeberry and rowan (*Sorbus*). *Gartenbauwissenschaft* 65(2): 93–98.

Jeppsson, N. 2008. *Aronia melanocarpa*, black chokeberry. In *The encyclopedia of fruit & nuts*. Edited by J. Janick and R.E. Paull. CABI, Wallingford, Oxfordshire, UK. pp. 622–623.

Jeppsson, N., and Johansson, R. 2000. Changes in fruit quality in black chokeberry (*Aronia melanocarpa*) during maturation. *J. Hortic. Sci. Biotech.* 75: 340–345.

Kaack, K., and Kuhn, B.F. 1992. Black chokeberry (*Aronia melanocarpa*) for manufacture of food colorant. *Tidsskrift Planteavl.* 96: 183–196.

Kask, K. 1987. Large-fruited black chokeberry (*Aronia melanocarpa*). *Fruit Var. J.* 41(1): 47.

Matsumoto, M., Hara, H., Chiji, H., and Kasai, T. 2004. Gastroprotective effect of red pigments in black chokeberry fruit (*Aronia melanocarpa* Elliot) on acute gastric hemorrhagic lesions in rats. *J. Agric. Food Chem.* 52: 2226–2229.

Niedworok, J., Jankowska, B., Kowalczyk, E., Charyk, K., and Kubat, Z. 1997. Antiulcer activity of anthocyanin from *Aronia melanocarpa* Elliot. *Herba Polonica* 43: 222–227.

Oszmianski, J., and Sapis, J.C. 1988. Anthocyanins in fruits of *Aronia melanocarpa* chokeberry. *J. Food Sci.* 53: 1241–1242.

Persson Hovmalm, H.A., Jeppsson, N., Bartish, I.V., and Nybom, H. 2004. RAPD analysis of diploid and tetraploid populations of *Aronia* points to different reproductive strategies within the genus. *Hereditas* 141: 301–312.

Petrovic, D.M., and Jacimovic-Plavsic, M.M. 1992. *Aronia melanocarpa* and propagation in vitro. *Acta Hortic.* 300: 133–135.

Robertson, K.R., Phipps, J.B., Rohrer, J.R., and Smith, P.G. 1991. A synopsis of genera in Maloideae (Rosaceae). *Syst. Bot.* 16: 376–394.

Rossell, I.M., and Kesgen, J.M. 2003. The distribution and fruiting of red and black chokeberry (*Aronia arbutifolia* and *A. melanocarpa*) in a southern Appalachian fen. *J. Torrey Bot. Soc.* 130: 202–205.

Skvortsov, A.K., and Maitulina, I.K. 1982. Distinction of cultivated black chokeberry from its wild ancestors. *Biull. Gl. Bot. Sada. Moskva* 126: 35–40. (In Russian)

Slimestad, R., Torskangerpoll, K., Nateland, H.S., Johannessen, T., and Giske, N.H. 2005. Flavonoids from black chokeberries, *Aronia melanocarpa*. *J. Food Comp. Anal.* 18: 61–68.

Small, E., and Catling, P.M. 2005. Poorly known economic plants of Canada—46. Black chokeberry (*Photinia melanocarpa*). *Bull. Can. Bot. Assoc.* 38(3): 29–34.

Strigl, A.W., Kraemer, A., Fuchs, H., Scharf, M., Leitner, E., and Pfannhauser, W. 1996. Analysis and extraction of the anthocyans in black chokeberry (*Aronia melanocarpa* Michx. Ell.). In *Natural colorants for food, nutraceuticals, beverages, confectionery and cosmetics. Proceedings of the second international symposium, Acapulco, Mexico, January 1996*. Edited by A.C. Mexico and Hereld Organization. SIC Publishing Company, Hamden, CT. pp. 161–164.

Strik, B., Finn, C., and Wrolstad, R. 2003. Performance of chokeberry (*Aronia melanocarpa*) in Oregon, USA. *Acta Hortic*. 626: 439–443.

Uttal, L.J. 1984. Nomenclatorial changes, lectotypification, and comments in *Aronia* Medikus (Rosaceae). *Sida* 10: 199–202.

Yarborough, D.E., and Ismail, A.A. 1979. Effect of endothall and glyphosate herbicides on a native barrenberry *Aronia melanocarpa*, Maine and lowbush blueberry stand. *Can. J. Plant Sci.* 59: 737–740.

## Specialty Cookbooks

Cookbooks dedicated to aronia do not seem to have been prepared, but dozens of recipes for aronia jam, bread, and other preparations are available in books dedicated to wild food plants of North America, and on the Web. Marrone (2009, cited in the Appendix to this book) has six aronia recipes.

# 8 Azolla (Mosquito Ferns)

Family: Azollaceae (mosquito fern family; also placed in Salviniaceae, water fern family)

## NAMES

Scientific names: *Azolla* species

- Azolla, mosquito fern (Carolina azolla, Carolina mosquito fern, eastern mosquito fern)—*A. caroliniana* Willd.
- Azolla, mosquito fern (large mosquito fern, red water fern)—*A. filiculoides* Lam.
- Azolla, mosquito fern (Mexican mosquito fern, Pacific mosquito fern)—*A. microphylla* Kaulf. (including *A. mexicana* C. Presl; *A. mexicana* is recognized as a separate species in most literature, but molecular and other studies have shown that plants so labeled are not different from *A. microphylla*).
- *Azolla* species are also frequently called duckweed fern, fairy moss (the plants have a moss-like appearance), mosquito fern (often "mosquito-fern"), mosquito plant, and water fern. *Salvinia* species are more frequently called water ferns, so this name should be avoided.
- The name "mosquito fern" is said to have arisen because thick coats of floating plants have a reputation for preventing mosquitoes from laying eggs. As noted below, *Azolla* species are partially effective in this regard and are also useful as a mosquito larvicide.
- The genus name *Azolla* is based on the Greek *azo*, to dry, + *ollyo*, to kill, an allusion to death from drought which occurs should these aquatic plants lose their supply of water. The vernacular (common English) name azolla is based on the genus name.
- *Caroliniana* in the scientific name *C. caroliniana* is Latin for of Carolina; *filiculoides* in the name *A. filiculoides* is based on the Latin *filicum*, fern, + *oides*, resembling or like; *microphylla* in the name *A. microphylla* is Latin for small-leaved, a rather uninformative epithet since all *Azolla* species have tiny leaves.

## GEOGRAPHY AND ECOLOGY OF WILD PLANTS

*Azolla* species are tiny freshwater aquatic ferns of temperate and tropical regions. Generally, they occur in stagnant or slow-moving waters, such as ponds, lakes, marshes, swamps, and some streams. Occasionally, the plants are stranded on wet mud. Their leaves float on the surface and the roots hang down into the water. The plants have limited tolerance to freezing temperatures and saline water (some species can survive in about 1% salt water, and *A. filiculoides* can grow in 2.5% saline solution).

The species of *Azolla* are very difficult to identify, and so the following summary of the distribution of the North American species should be considered to be tentative, based on available knowledge.

*A. caroliniana* is native to southeastern Canada, the eastern half of the United States, the Caribbean, Central America, and South America. It has also become naturalized elsewhere. It is the most cold-tolerant of *Azolla* species, capable of surviving hard frosts and prolonged coverage by ice. It is also the most adapted to survival on mud.

**FIGURE 8.1** Azolla (*Azolla* species). (a) Clump of *A. caroliniana*. (b) Clump of *A. caroliniana* against a finger. (c) *Anabaena azollae*, the nitrogen-fixing cyanobacterium, forms a symbiotic relationship with *Azolla*. The large cells within the filament are "heterocysts" (arrows point to three of these), special cells that synthesize organic nitrogen compounds from the nitrogen in the air. The smaller cells are photosynthetic. (d) Floating mat of *A. caroliniana*. (Photos (a) through (d) contributed by S. Darbyshire.) (e) Floating mat of pink phase of *A. caroliniana*. (Courtesy of David Perez [CC By 3.0].)

*A. filiculoides* is native to western North America, generally in areas near the Pacific Ocean. It is also native to South America and has been become naturalized in much of the remainder of the world. This species is also cold-tolerant and can survive under thin ice.

*A. microphylla* is native to southwestern Canada, the western United States, Central America, South America, and the West Indies.

*Azolla* species are associated with a filamentous bacterium in a symbiotic relationship. This bacterium is widely identified as *Anabaena azollae* Strasb., a member of an important group of photosynthetic bacteria called cyanobacteria (these bacteria are also known as blue-green algae because of the color of the pigments responsible for photosynthesis). Controversy regarding the identification of the symbiotic associate of *Azolla* species as *Anabaena*, and whether there are other bacterial associates, is discussed below. The fern furnishes protection and possibly some nutrients

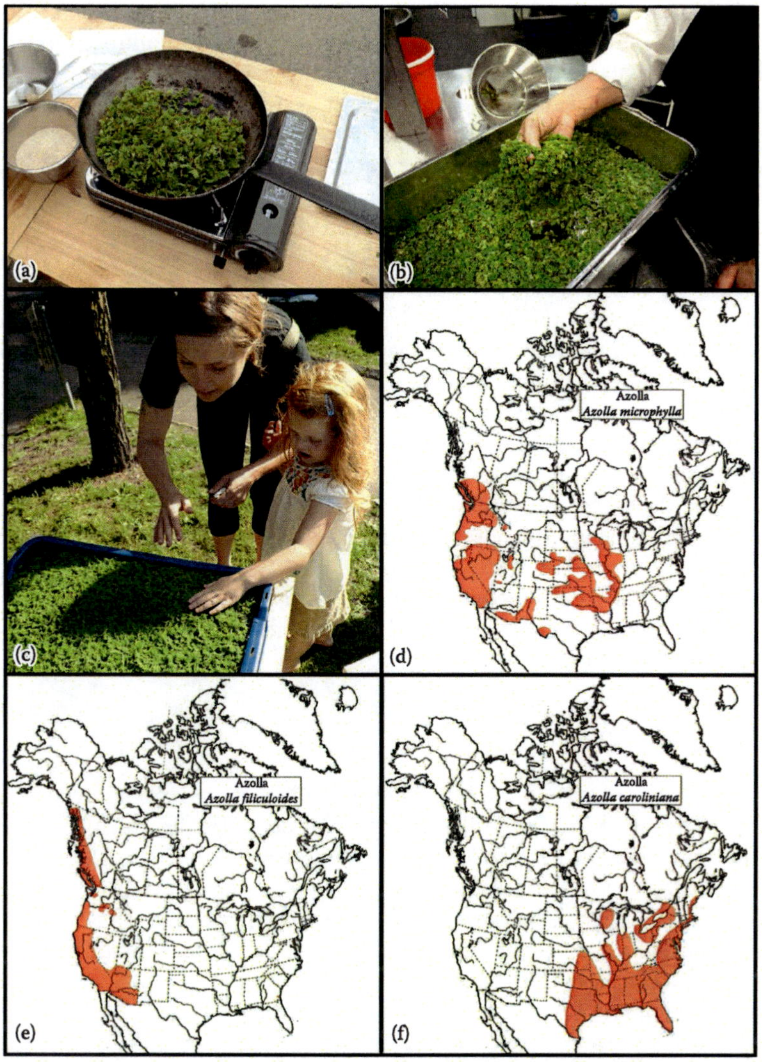


**FIGURE 8.2** Azolla (*Azolla* species). (a and b) Photos showing experimental culinary preparations of azolla. (Courtesy of E. Sjödin.) (c) A young girl being introduced to azolla; photo courtesy of E. Sjödin. (d) Distribution map of *A. microphylla*. (e) Distribution map of *A. filiculoides*. (f) Distribution map of *A. caroliniana*.

to the bacterium, whereas the latter provides nitrogen to the fern, as discussed below. The bacterium occurs in cavities in the leaves, in the growing points of the stems, and on other parts of the plant. When the plant reproduces by spores, the bacterium becomes associated with a specific area (the "indusial cap") on top of the megaspore (the female reproductive structure, described below). Symbiotic associations of plants and nitrogen-fixing bacteria are common, but the *Azolla–Anabaena* association is unique in that the bacterium cannot reproduce in nature outside of its host, the fern. No other plant–cyanobacterium symbiosis is known with a comparably precise mechanism (described above) that guarantees transfer from one plant to its offspring (in other symbioses, plants developed from seeds must acquire their bacterial partner from free-living bacteria).

The bacterium fixes atmospheric nitrogen (i.e., converts inert gaseous nitrogen into a chemically combined form that the plants can metabolize). Since nitrogen is a critical nutrient for living things and is almost always insufficiently available in soil (or water) for optimum plant growth, this provides azolla ferns with a tremendous advantage. There are as many as 10,000 species of ferns, but only *Azolla* species are known to be symbiotically associated with nitrogen-fixing organisms. (Nitrogen-fixing symbionts are best known in association with species in the pea family, but a variety of unrelated plants and even some animals have such relationships.) As a result, *Azolla* species can grow very rapidly, under ideal conditions doubling their biomass every 2–5 days. Limited phosphorus in water is usually the chief limiting factor for azolla growth, and when chemical runoff (such as from fertilizers and sewage) enriches water with phosphorus, the result may be spectacular carpeting of bodies of water with the fern (algal blooms similarly result from such "eutrophication" of water).

Most literature has identified the symbiotic nitrogen-fixing associate of *Azolla* as *Anabaena azollae*. However, since the 1990s, this has been challenged. It has been claimed that *Nostoc*, another genus of blue-green algae, performs the nitrogen-fixing function in at least some *Azolla* species. Other claims are that the cyanobacterium species present is neither *Anabaena azollae* nor *Nostoc*. Moreover, it has been demonstrated that additional cyanobacteria are present in small quantities in some *Azolla* species, and some kinds of nonphotosynthetic bacteria are also present; the roles, if any, of these other bacteria is uncertain.

In some areas of the world, azollas are very significant weeds. In parts of the United States, the introduced *A. pinnata* R. Br. (called mosquito fern, water velvet, and ferny azolla) has been declared to be a noxious weed.

Most species of *Azolla* produce reddish pigments when stressed, for example, by temperature extremes or feeding damage by herbivores, and this may result in a pinkish or reddish carpet of plants covering expanses of water.

## PLANT PORTRAIT

*Azolla*, with about six or seven species, is the only genus in the family Azollaceae. It is sometimes combined with the genus *Salvinia* in the family Salviniaceae. The delimitation of the species of *Azolla* and determination of their correct names has been rather unsettled. Recent DNA studies (cited in the bibliography) support the separation of the species recognized here, but determination of their correct names is still somewhat uncertain. The species reproduce to a considerable extent by budding off new plants and usually lack sexual reproductive structures (described below), without which they are very difficult to identify with certainty.

*Azolla* species are diminutive, delicate, free-floating, annual plants. The extremely small leaves (technically "fronds" in ferns), no larger than 2 mm or 0.08 inch across, are attached alternately to the branching stems, often overlapping in two ranks. The leaves have an upper green (i.e., photosynthetic) lobe with the bacterial partner housed in a pouch on its lower side and a smaller, usually colorless lobe that is buoyant and often submersed. Thread-like, unbranched roots are produced from the axils of branches. Under good conditions, the plants can grow over each other in layers and develop mats up to 5 cm (2 inches) thick.

*Azolla* species are true ferns, but are extremely reduced in form, and do not exhibit the highly dissected (ferny) foliage of most ferns. Ferns reproduce by spores, not seeds. Most fern species produce just one kind of spore, but *Azolla* produces two kinds, in reproductive structures termed "sporocarps," which are developed on the branches. Male sporocarps are extremely small (about 2 mm or 0.08 inch in diameter) and produce male spores, which adhere to each other in clumps; female sporocarps are much smaller, each producing just one viable, large female spore. Arrowhead-like barbs on the clumps of male spores seem to assist in adhering to the female spore. Each male spore produces eight swimming sperm, some of which succeed in fertilizing the females.

In China and Vietnam, particular strains of *A. pinnata* are used in rice culture systems. Whether such strains have been "domesticated" (changed genetically from wild ancestors) or simply represent selections of forms that exist in the wild is undetermined. The North American species of

*Azolla* have not been used to any appreciable extent in agriculture, but it is possible that strains that grow well in aquaria have been unconsciously selected.

*Azolla* species are sometimes mistaken for duckweeds (Lemnaceae), a family of about five genera and 40 species of flowering plants, which are also tiny aquatic plants of worldwide distribution. Some species of duckweeds are floating, and others are submersed in the water. Duckweeds are the smallest of all flowering plants, but they rarely produce flowers and seeds and reproduce mostly by budding off new plants. In addition to their small size and aquatic nature, duckweeds are reminiscent of *Azolla* species in several practical ways: they are useful as livestock feed and occasionally as human food, they are employed in wastewater purification, they are potentially useful as a renewable energy source, and they are sometimes detrimental as aquatic weeds. *Lemna gibba* L., which is native to North America, is cultivated in Israel as a salad and vegetable plant. *Wolffia* species (the world's smallest flowering plants) are consumed as a vegetable in Southeast Asia.

*Azolla* species are most important as a "green manure" or "biofertilizer," especially in Southeast Asia and in other rice-producing regions of the world. Shortly after rice is planted, the rice paddies are inoculated with *Azolla*, which multiples rapidly to cover the water, serving to suppress weeds. More importantly, as the ferns die and their bodies disintegrate, the nitrogen that their bacterial partners have fixed is released to the rice plants, contributing greatly to their growth. Rice and wheat are the world's most important food crops, but rice is a staple food for more people than any other crop. In China alone, over 1.2 million ha (3 million acres) of rice are grown with *Azolla*, which is thought to more than double productivity. Harvested *Azolla* is also used to some extent as a mulch and as a fertilizer for crops grown conventionally in soil. Most nitrogen fertilizers today are chemically synthesized, requiring the use of fossil fuels; the use of *Azolla* in place of synthetic fertilizers has great potential to reduce the world's reliance on fossil fuels.

*Azolla* ferns are also agriculturally important as fodder for livestock, including cattle, pigs, goats, and chickens. Harvested *Azolla* is simply added as a nutritional supplement to the regular feed of mammals and poultry.

"Aquaponics" is the combined use of aquatic plants and animals to produce food (and sometimes other useful products). In Asia, *Azolla* is commonly grown in rice fields not just to increase rice yield but also simultaneously to feed aquatic species such as fish (notably tilapia), shrimp, ducks, and geese, which simply consume the living plants growing in the water.

Another common use of *Azolla* species is as a destroyer of mosquito larvae. Mosquito larvae are aquatic organisms, but they must come to the water surface to obtain oxygen. A thick carpet of *Azolla* reduces the ability of the mosquito larvae to reach the surface to breathe. It has also been shown that a thick carpet of the plants reduces deposition of eggs by mosquitoes.

Still another common use of *Azolla* species is as a component of aquaria, where the plants function as ornamentals and also provide hiding places for small fish. Plants discarded from aquaria account for the occasional appearance of *Azolla* species outside of their native distribution areas. In the United States, shops that market aquatic plants are subject to penalties for distributing, even inadvertently, species such as *A. pinnata* that are listed as noxious weeds.

*Azolla* species are sometimes additionally used to purify waste water, produce hydrogen and biogas, and for industrial and medicinal extracts.

## CULINARY PORTRAIT

*Azolla* is thought to be fit for human consumption, although to date only experimental preparation of food products has been attempted, such as vegetable meats ("azollaburgers"), omelets, soups, cookies, and noodles. The adventurous cook can experiment with utilization of azolla ferns in the same way that alfalfa sprouts and other sprouts are used. However, the liberal use of spices and herbs is recommended, and the most palatable way of eating azolla is as a cooked preparation in combination with other ingredients.

Collecting plants directly from lakes and ponds is hazardous because of the probability of water contamination. Growing one's own supply is the only feasible way of obtaining this plant at present. Aquarium supply stores are the most convenient source of starting material.

Nitrogen is essential in proteins, and plant species that are associated with nitrogen-fixing bacteria are usually rich in protein. Indeed, azolla ferns are very rich in protein, constituting as much as a quarter of their dry weight. The amino acid profile (a measure of protein quality) is good, although there are minor deficiencies of methionine, cysteine, and lysine. On a long-term basis, it may be expected that the ferns or more likely their extracts will be combined with other foods to improve their protein quality.

For the development of commercial human food products, the name "azolla" is obviously preferable to "mosquito fern," the former name having a rather attractive exotic nature and the latter name recalling a detested pest.

High oxalic acid content has been found in *A. pinnata* (also in *Lemna gibba* and *Wolffia* of the related Salviniaceae family, mentioned above as used for food). Oxalic acid is found in high amounts in several commonly consumed foods, notably rhubarb, spinach, and sorrel. High-oxalate foods can interfere with calcium absorption (a problem that can be overcome by the consumption of calcium-rich foods), and individuals susceptible to developing medical conditions from consumption of oxalic acid should avoid eating large quantities. This includes people susceptible to calcium oxalate kidney stones, and possibly also those with arthritis or gout. Glass and stainless steel cookware and utensils are advisable for cooking high-oxalate foods, while untreated aluminum and cast iron pots and pans should not be used.

### Culinary Vocabulary

- "Water fern cake" or beo cake (*bánh bèo*) is a Vietnamese specialty, which is not made with azolla ferns, despite the name. It is, however, named for its resemblance to azolla ferns floating on water. Water fern cake is a thin, steamed rice cake, the top stuffed with minced shrimp, scallions, green bean paste, and fried shallots.

## PROSPECTS

*Azolla* ferns are extremely important for generating human food, as biofertilizers for increasing crop production, and as feed for production of meat, poultry, and fish. Although not yet used directly as food for humans, the prospects for the development in this regard are very promising. As the human population continues to outpace increases in food production, *Azolla* is likely to become increasingly significant as a direct protein source.

## CURIOSITIES OF SCIENCE AND TECHNOLOGY

- Particular pigments in photosynthetic cells primarily capture a small range of photosynthetically active light. One of the secrets of the very high productivity of *Azolla* is the fact that the light-capturing pigments in the plants (chlorophylls a and b, and carotenoids) complement the light-capturing pigments in the bacterial partner (chlorophyll a and phycobilins) to capture most of the visible light in the 500–700 nm spectrum. This allows the partners to capture a wider range of the sunlight's wavelengths than either could alone.
- According to a Vietnamese legend, 300 years ago a Vietnamese woman named Ba Heng discovered that the presence of azolla ferns in rice paddies increased production. In her honor, an annual festival in the fall is still celebrated in the country.
- The "Azolla Event" is a spectacular climate change thought to have been triggered by *Azolla* in ancient times, when several dozen now extinct species occurred. Following the elimination of the dinosaurs about 65 million years ago, the climate of Earth became very different from that of today. Fifty-five million years ago, Earth had a hot (up to 12°C or 22°F warmer than

today), tropical climate, with palm trees growing at both poles. This is believed to have resulted from increased carbon dioxide in the atmosphere, caused in part by extensive volcanism, and consequent heating by the greenhouse effect. The Arctic Ocean was almost completely enclosed by land. It is hypothesized that extensive, dense patches of *Azolla* developed in freshwater lakes in the Arctic area, and perhaps also in the Arctic Ocean, the surface of which at the time was less saline than today because of the inflow of water from rivers. Because of the heat and the very high levels of carbon dioxide, incredible growth of azolla ferns took place. It has been estimated that as the plants died, layers of dead plants up to 8 m (26 feet) accumulated on the ocean floor (cold water at the bottom of deep waters is thought to have prevented decay of the mosquito ferns). This removal (sequestration) of carbon dioxide by the ferns reduced the greenhouse effect and so cooled the planet. Carbon dioxide level in the atmosphere is thought to have dropped from 3500 parts per million (ppm) to about 650 ppm, the world turned cold, ice caps formed at both poles, and sea level dropped. This cold phase has been maintained until today, but now global warming is threatening to reverse the situation.

- In the flowering or seed plants, nonswimming sperm are delivered to the egg by a tube that develops from the pollen grain, penetrates the stigma, and grows directly to the egg. By contrast, animals, including humans, have sperm that swim to reach and fertilize eggs. Swimming sperm are also present in the more primitive classes of plants: algae, mosses, cycads, Ginkgo, and ferns (including azolla ferns).
- *Azolla* is one of the fastest-growing plants on Earth, capable of producing 9 tonnes of protein per hectare (36 tons per acre). It has been suggested that growing plants like *Azolla* for human food could require only about 200 $m^2$ (239 square yards) of land to feed an average human–less than one hundredth of the area now required for the average American. Such remarkable growth has made *Azolla* one of the principal plant species that is being considered for food production during space voyages.
- Although *Azolla* species and *Anabaena azollae* are permanently associated with each other (the bacterium has not been found outside of the plant, and the plant is almost always accompanied by the bacterium), azolla roots have retained the capacity to extract nitrogen from their environment. Interestingly, when the bacterium has been experimentally killed off (using antibiotics), the plant compensates for the loss of bacterial nitrogen normally supplied by developing extra roots.
- Since azolla ferns are aquatic plants, it is not surprising that the water content of the plants is high: 85%–95%.
- In the 1970s, it was discovered that *Anabaena azollae* uses light energy to release hydrogen from water. Under natural conditions, this hydrogen is immediately combined with the nitrogen that is fixed by the bacterium to produce ammonia, which is subsequently provided to the host fern as its nitrogen source. It has been shown that when *Azolla* and its symbiotic partner are grown under certain conditions (e.g., in an atmosphere that excludes nitrogen), the cyanobacterium produces stable hydrogen gas from water. Hydrogen gas can be used as a nonpolluting, high-energy fuel. Indeed, hydrogen can produce more energy than any other nonnuclear fuel, and if produced on a large scale, could be extremely important in addressing the problem of dwindling petroleum reserves. This phenomenon has not yet resulted in a practical application, but the possibility of this is intriguing.

## KEY INFORMATION SOURCES

Bennicelli, R., Stepniewska, Z., Banach, A., Szajnocha, K., and Ostrowski, J. 2004. The ability of *Azolla caroliniana* to remove heavy metals (Hg(II), Cr(III), Cr(VI)) from municipal waste water. *Chemosphere* 55: 141–146.

Cagauan, A.G., Branckaert, R.D., and Van Hove, C. 2000. Integrating fish and azolla into rice-duck farming in Asia. *Naga* 23(1): 4–10.

Collinson, M.E., Barke, J., van der Burgh, J., van Konijnenburg-van Cittert, J.H.A., Heilmann-Clausen, C., Howard, L.E., and Brinkhuis, H. 2010. Did a single species of Eocene *Azolla* spread from the Arctic Basin to the southern North Sea? *Rev. Palaeobot. Palynol.* 159: 152–165.

Darbyshire, S.J. 2002. Ephemeral occurrence of the mosquito fern, *Azolla caroliniana*, at Ottawa, Ontario. *Can. Field-Nat.* 116: 441–445.

Dawar, S., and Singh, P.K. 2002. Comparison of soil- and nutrient-based medium for maintenance of *Azolla* cultures. *J. Plant Nutr.* 25: 2719–2729.

Dela Rosa L.S. 1992. *Leaf protein from Azolla: Its preparation, quality, and utilization in food.* Thesis, 106 leaves. Philippine Council for Advanced Science and Technology Research and Development (PCASTRD) Library.

Ekman, M., Tollbäck, P., and Bergman, B. 2008. Proteomic analysis of the cyanobacterium of the *Azolla* symbiosis: Identity, adaptation, and NifH modification. *J. Exp. Bot.* 59: 1023–1034.

Evrard, C., and Van Hove, C. 2004. Taxonomy of the American *Azolla* species (Azollaceae): A critical review. *Syst. Geogr. Plants* 74: 301–318.

Furuno, T. 2001. *The power of duck: Integrated rice and duck farming.* Tagari Publications, Sisters Creek, Tasmania, Australia. pp. 94.

Kamalasanana Pillai, P., Premalatha, S., and Rajamony, S. 2005. Azolla: A sustainable feed for livestock. *Leisa* 21(3): 26–27.

Kannaiyan, S., and Kumar, K. 2005. *Azolla biofertilizer for sustainable rice production.* Daya Publishing House, Delhi, India. pp. 452.

Katayama, N., Yamashita, M., Kishida, Y., Liu, C.-C., Watanabe, I., and Wada, H. 2008. Azolla as a component of the space diet during habitation on Mars. *Acta Astronautica* 63: 1093–1099.

Khan, M.M. 1983. *A primer on Azolla production & utilization in agriculture.* University of the Philippines at Los Baños, Los Baños, Philippines. pp. 143.

Kushari, D.P., and Watanabe, I. 1991. Differential responses of *Azolla* to phosphorus deficiency. 1. Screening methods in quantity controlled-condition. *Soil Sci. Plant Nutr.* 37: 271–282.

Lejeune, A., Cagauan, A., and Hove, C. van. 1999. *Azolla* research and development: Recent trends and priorities. *Symbiosis* 27: 333–351.

Lumpkin, T.A. 1993. Azollaceae. In *Flora of North America.* Volume 2. Edited by Flora of North America Editorial Committee. Oxford University Press, New York. pp. 338–342.

Lumpkin, T.A., and Plucknett, D.L. 1980. *Azolla*: Botany, physiology, and use as a green manure. *Econ. Bot.* 34: 111–153.

Lumpkin, T.A., and Plucknett, D.L. 1982. *Azolla as a green manure: Use and management in crop production.* Westview Tropical Agriculture Series No. 5. Westview Press, Boulder, CO. pp. 230.

Metzgar, J.S., Schneider, H., and Pryer, K.M. 2007. Phylogeny and divergence time estimates for the fern genus *Azolla* (Salviniaceae). *Int. J. Plant Sci.* 168: 1045–1053.

Moore, A.W. 1969. *Azolla*: Biology and agronomic significance. *Bot. Rev.* 35: 17–35.

Papaefthimiou, D., Van Hove, C., Lejeune, A., Rasmussen, U., and Wilmotte, A. 2008. Diversity and host specificity of *Azolla* cyanobionts. *J. Phycol.* 44: 60–70.

Peters, G.A., and Meeks, J.C. 1989. The *Azolla-Anabaena* symbiosis: Basic biology. *Ann. Rev. Plant Physiol. Plant Molec. Biol.* 40: 193–210.

Qiu, Y.-L., and Yu, J. 2003. *Azolla*—A model organism for plant genomic studies. *Geno. Prot. Bioinfo.* 1: 15–25.

Reid, J.D., Plunkett, G.M., and Peters, G.A. 2006. Phylogenetic relationships in the heterosporous fern genus *Azolla* (Azollaceae) based on DNA sequence data from three noncoding regions. *Int. J. Plant Sci.* 167: 529–538.

Sandoval, M.E.C. 1984. Azolla (*Azolla pinnata* R. Brown) as a possible human food source. Master's Thesis. pp. 66. Faculty Graduate School, University of Los Banos, Philippines.

Sandoval, M.E.C., and Mabesa, L.B. 1985. *Azolla* (*Azolla pinnata*, R. Brown) as a possible human food source. 1. Proximate composition, ascorbic acid, provitamin A and oxalic acid. *Philippine J. Nutr.* 38: 83–87.

Saunders, R.M.K., and Fowler, K. 1992. A morphological taxonomic revision of *Azolla* Lam. section *Rhizosperma* (Mey.) Mett. (Azollaceae). *Bot. J. Linn. Soc.* 109: 329–357.

Saunders, R.M.K., and Fowler, K. 1993. The supraspecific taxonomy and evolution of the fern genus *Azolla* (Azollaceae). *Plant Syst. Evol.* 184: 175–193.
Shi, D.-J., and Hall, D.O.1988. The *Azolla-Anabaena* association: Historical perspective, symbiosis and energy metabolism. *Bot. Rev*. 54: 353–386.
Small, E., and Darbyshire, S.J. 2011. Blossoming treasures of biodiversity—35. Mosquito ferns (*Azolla* species)—tiny 'super plants'. *Biodiversity* 12: 119–128.
Smith, W.H., and Cervantes, E.P. (Editors). 1987. Azolla utilization. Proceedings of the workshop on azolla use, Fuzhou, Fujian, China, March 31–April 5, 1985. International Rice Institute, Los Baños, Laguna, Philippines. pp. 296.
Speelman, E.N., van Kempen, M.M.L., Barke, J., Brinkhuis, H., Reichart, G.J., Smolders, A.J.P., Roelofs, J.G.M., et al. 2009. The Eocene Arctic *Azolla* bloom: Environmental conditions, productivity and carbon drawdown. *Geobiology* 7: 155–170.
Van Cat, D., Watanabe, I., Zimmerman, W.J., Lumpkin, T.A., and de Waha Baillonville, T. 1989. Sexual hybridization among *Azolla* species. *Can. J. Bot.* 67: 3482–3485.
Van Hove, C. 1989. *Azolla and its multiple uses with emphasis on Africa.* Food and Agriculture Organization of the United Nations, Rome, Italy. pp. 53.
Wagner, G.J. 1997. *Azolla*: A review of its biology and utilization. *Bot. Rev*. 63: 1–26.
Whaley, J. 2007. The *Azolla* story: Climate change and Arctic hydrocarbons. *GeoExPro* 4(4): 66–72.
Zimmerman, W.J., Lumpkin, T.A., and Watanabe, I. 1989. Classification of *Azolla* spp., section *Azolla*. *Euphytica* 43: 223–232.

## Specialty Cookbooks

Sjödin, E. 2012. *The azolla cooking and cultivation project*. E. Sjödin, Stockholm, Sweden. pp. 47. (Also available as Kindle edition from Amazon Digital Services Inc.; available online at http://www .eriksjodin.net/publications.html; accessed on November 2012.)

*Note:* Azolla ferns are rarely used as human food at present, and recipes are rarely available. The best source of recipes is the book cited above by Erik Sjödin of Sweden, who has conducted experimental preparations of foods based on this plant. Sjödin's recipe for "azollaburgers" (reproduced with permission) follows:

### Ingredients

100 g of fresh azolla (or 50 g of fried azolla)
Breadcrumbs (a required binder)
Salt, pepper, fresh herbs

### Preparation

1. Fry fresh azolla in a pan for 1–2 minutes.
2. Chop the fried azolla and mix it with spices and herbs.
3. Mix in breadcrumbs until the mixture can be formed into burgers.
4. Fry the burgers in vegetable oil.

# 9 Bergamot

Family: Lamiaceae (Labiatae; mint family)

## NAMES

Scientific names: *Monarda* species

- Bergamot—*M. didyma* L.
- Dotted monarda—*M. punctata* L.
- Lemon bergamot—*M. citriodora* Cerv. ex Lag. [including Mexican bergamot—*M. citriodora* Cerv. ex Lag. var. *austromontana* (Epling) B.L. Turner (*M. austromontana* Epling)]
- Wild bergamot—*M. fistulosa* L.

### Bergamot

- The name "bergamot" was first used for *Citrus bergamia* Risso, a tropical Asian citrus tree grown almost exclusively in a narrow coastal strip in the southern part of Calabria, Italy. Oil from its peel is used to make "Earl Grey tea." The resemblance of the pungent, lemony scent of *M. didyma* to the aroma of the former led to it also being called bergamot.
- Bergamot has also been called American melissa, bee balm (bee-balm), crimson bee balm, fragrant balm, horse mint, Indian('s) plume, mountain balm, mountain mint (better reserved for *Pycnanthemum* species, treated in this book as "mountain mint"), North American swamp plant, Oswego tea, red balm, red bergamot, scarlet monarda, and sweet bergamot.
- The name "bee-balm" is just as often used for *Melissa officinalis* L. (this culinary herb is best called lemon balm) as it is for *Monarda* species.
- Bergamot is sometimes called Oswego tea after the tribe of Native Americans, who resided in what is now upstate New York, and used the plant extensively.
- A very old English pear variety is known as Bergamot (said to be a corruption of the Turkish *beg-armudi*, Prince's pear).
- *Didyma* in the scientific name *M. didyma* is Latin for "twin." Linnaeus (1707–1778), who coined the name, used the word in many of his scientific names to indicate that the stamens of the flowers were in pairs.

### Dotted Monarda

- Dotted monarda is also known as spotted bee balm and horsemint. *Punctata* in the scientific name *M. punctata* is Latin for dotted, so named because the petals have dots.

### Lemon Bergamot

- The name "lemon bergamot" reflects the lemony taste of the plant.
- Lemon bergamot has also been called lemon bee-balm (lemon bee balm) and lemon mint.
- *Citriodora* in the scientific name *M. citriodora* is Latin for citrus-odored or lemon-scented.

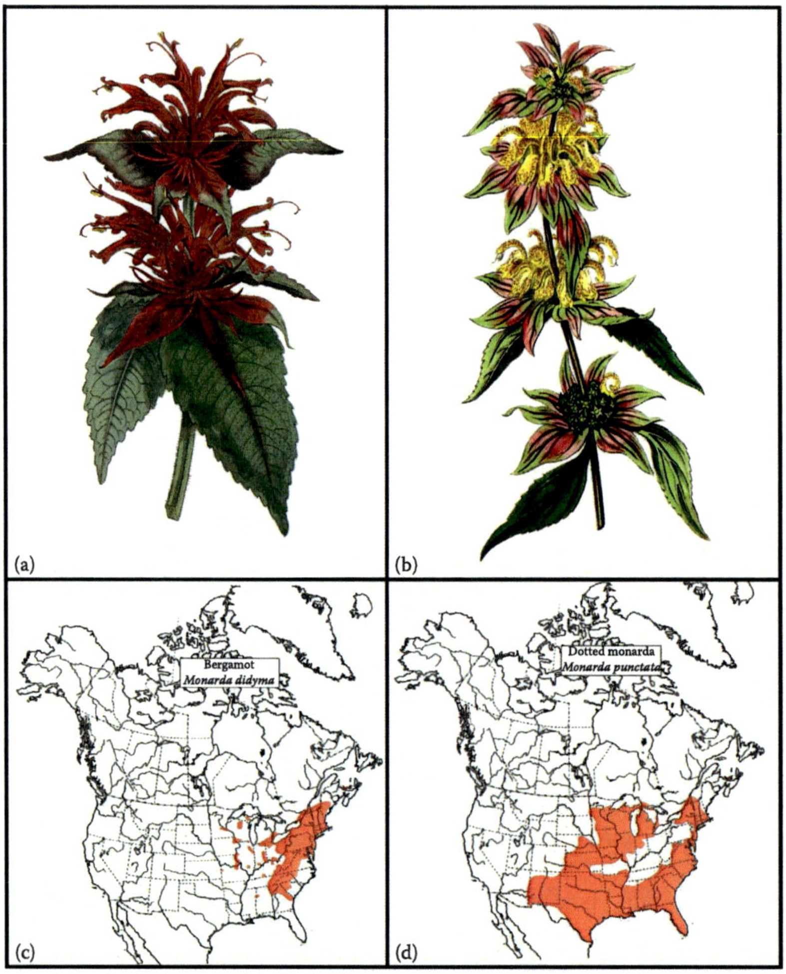


**FIGURE 9.1** Bergamot (*Monarda* species). (a) Bergamot (*M. didyma*). (From Curtis, W., *Botanical Magazine*, Vol. 15, Plate 546, 1801.) (b) Dotted monarda (*M. punctata*). (From Edwards, S.T., *Botanical Register*, Vol. 1, Plate 87, 1815.) (c) Distribution map of *M. didyma*. (d) Distribution map of *M. punctata*.

- Mexican bergamot (*M. citriodora* var. *austromontana*) is known in Mexico as napaka.
- *Austromontana* in the scientific name *M. citriodora* var. *austromontana* is based on the Latin *australis*, south or southern + *montana*, from the mountains), reflecting that the species comes from mountains, including those in Mexico and the states of New Mexico and Arizona.

## Wild Bergamot

- The "wild" in "wild bergamot" reflects the widespread culinary usage of wild forms of the species.
- Wild bergamot has also been called American horse mint, bee balm, bergamot, dry-land mint, fern mint, horse mint (horsemint), plains bee balm, and purple bee balm.
- Crushed leaves of wild bergamot were used by North American Indians to reduce the sting or itching associated with insect bites, hence the common name beebalm.

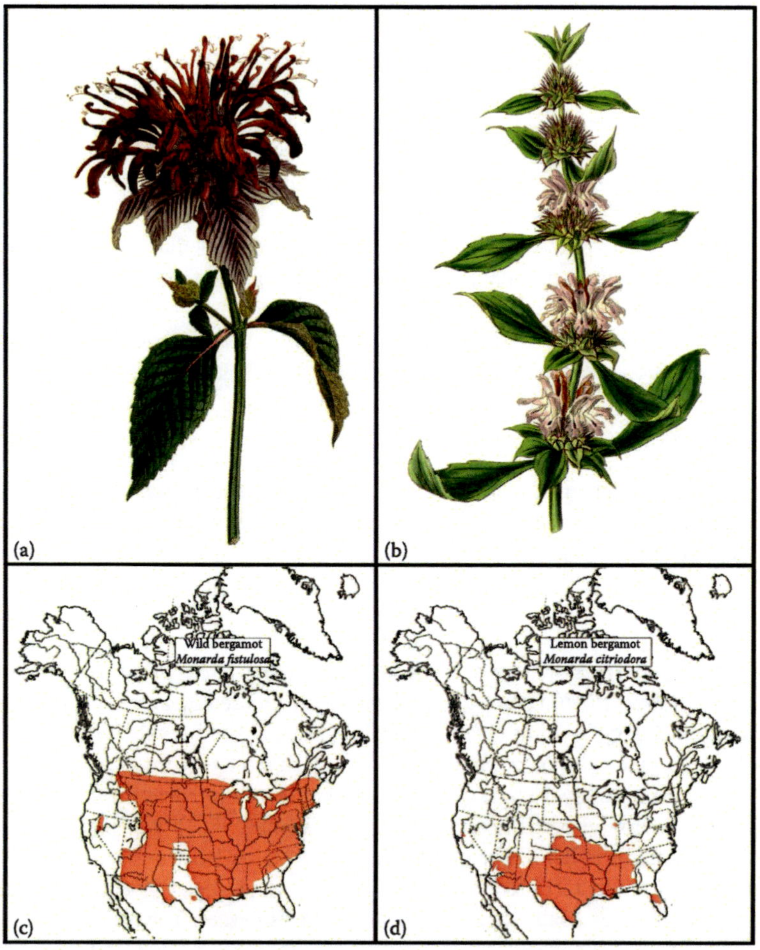


**FIGURE 9.2** Bergamot (*Monarda* species). (a) Wild bergamot (*M. fistulosa*). (From Curtis, W., *Botanical Magazine*, Vol. 5, Plate 145, 1792.) (b) Lemon bergamot (*M. citriodora*). (From Curtis, W., *Botanical Magazine*, Vol. 63, Plate 3526, 1836.) (c) Distribution map of *M. fistulosa*. (d) Distribution map of *M. citriodora*.

- The name beebalm (or bee balm) has sometimes been taken to suggest that the genus *Monarda* is good bee forage. However, the flowers of many of the species, particularly wild bergamot, have floral tubes that are too long to allow most bees to extend their proboscis to reach the nectar. The flowers are very attractive to hummingbirds.
- Wild bergamot is sometimes called "Mexican oregano," a name more usually associated with *Lippia graveolens* Kunth (see Chapter 39 on Epazote).
- *Fistulosa* in the scientific name *M. fistulosa* is Latin for hollow-stalked.

## Other

- The genus *Monarda* is named after a Spanish medical botanist, Nicholas de Monardes (1493–1588), who in 1569 authored the first medicinal flora from North America, a book whose title was translated to the delightful "Joyfull Newes out of the Newe Founde Worlde."
- "Bergamot mint" does not refer to bergamot; the name pertains to selections of several species of *Mentha* (true mint).

## GEOGRAPHY AND ECOLOGY OF WILD PLANTS

Bergamot is native to the eastern United States from Michigan and New York in the north to Georgia and Tennessee in the south. It is often found in thickets and along stream banks.

Wild bergamot is found in dry thickets, clearings, prairie hillsides, pastures, roadsides, occasionally in open woods, and usually in rocky soil. It is native to North America, occurring from northern Canada to northern Mexico. Subspecies *menthifolia* (Graham) Fern., noted in this chapter, is found in western North America.

Lemon bergamot is native to central and southern parts of the United States, from Missouri to Nebraska and southward to northern Mexico. It occurs in prairies, roadsides, and other open habitats, often in soils with high clay content.

Dotted monarda is a native of a large region of North America, from the eastern and central United States to northern Mexico. It occurs in prairies, coastal plains, and sandy areas.

## PLANT PORTRAIT

The genus *Monarda* is made up of 16 species of aromatic annual or perennial herbs native to North America. Several of these are used as culinary herbs. Many of the species and their hybrids are grown as ornamentals.

### Bergamot

Bergamot is a perennial, thick-stemmed, herbaceous plant, 60–120 cm (2–4 feet) tall, usually with crimson or scarlet flowers in wild plants, or with variously colored flowers in cultivars. The flowering head is a rather shaggy-looking assemblage of trumpet-shaped florets. Often the upper leaves and surrounding bracts are red or bronze. The plants form clumps, sometimes 1 m (about a yard) across. Details regarding early Amerindian uses of bergamot are sometimes difficult to separate from information about wild bergamot, since both are commonly called monarda and bee balm, and the latter sometimes is called Oswego tea, like the former. Both species were used medicinally by North American Indians to treat a variety of problems, and bergamot was subsequently used in Europe as a treatment for colds, sore throats, and other illnesses. The tall plants with red, pink, purple, or white flowers are particularly attractive to hummingbirds, butterflies, and bumblebees. Bergamot has also been cultivated experimentally in the Crimea for its essential oil used in flavoring and in perfumes. Hybrids of bergamot and wild bergamot with distinctive fragrances were developed in Canada for use in the perfume industry.

### Dotted Monarda

Dotted monarda was grown in the 1920s as a source of the chemical thymol. This species has been used medicinally, and leaf extracts have reportedly been used in nonalcoholic beverages. Compared to the other species discussed in this chapter, it is only occasionally used for culinary purposes.

### Lemon Bergamot

Lemon bergamot is an annual herb, 30–100 cm (12–39 inches) tall, with white, purplish, or pink and purple-dotted flowers in pagoda-like whorls. Entire young plants may be rather pink when young. Leaves were used by the Hopi Indians of the southwestern United States to flavor wild game, while the Tewa Indians used lemon bergamot as a cooked green.

Mexican bergamot is considered here to be a subspecies of lemon bergamot. This low-growing annual has a flavor like that of oregano, and indeed is locally called oregano. It has been used for many years by indigenous peoples and others of northern Mexico and the southwest of the United States to season food and as a tea considered good for the stomach.

### Wild Bergamot

Wild bergamot is a perennial herb 30–120 cm (1–4 feet) high, growing from creeping rhizomes (underground stems). Subspecies *menthifolia*, known as Oregano de la sierra, wild oregano, mint-leaved bergamot, horsemint, and bee balm, was used by Native Americans along the Rio Grande, notably the Pueblo people of the southwest, to season beans and stews, to make tea, and for medicinal purposes. The Apache, Tewa, and Hopi also used wild bergamot as a seasoning. Wild bergamot was reportedly cultivated for its greens by the Hopi, who dried the plants for winter use. It is grown locally in the southwestern United States as a culinary and medicinal plant, but has little commercial value except as a source of the chemical geraniol.

## CULINARY PORTRAIT

### Bergamot

The main culinary use of bergamot is in making tea. It can be used to make an "Earl Grey tea," but the oil of bergamot used commercially to make Earl Grey tea comes from the "bergamot orange," *Citrus bergamia*, as noted in the preceding. The leaves can either be mixed with other plants as a hot tea or combined with a slice of lemon to make a refreshing cold tea. The fresh leaves and tender sprigs have a pungent, lemon-like aroma and bitter flavor. They make a delicious addition to wine drinks, lemonades, fruit beverages, jellies, fruit ice cups, and cheese. Tender sprigs and flowers are added to tossed salads and fruit salads. Bergamot leaves can serve as a substitute for sage in stuffings and meat recipes. Young bergamot leaves go well with pork and veal.

### Lemon Bergamot

The taste of lemon bergamot is indeed lemony, although some find the aroma to be reminiscent of thyme. The scented leaves are a tasty addition to tea. The leaves can also be tossed in salads, floated in punches, or used as a garnish. Lemon bergamot makes an attractive ornamental plant, and it has been grown as a source of essential oil.

### Wild Bergamot

The twigs of wild bergamot can be used as a potherb, cooked with other foods such as meats, or dried for future use. The early leaves can substitute for mint tea, and make a good oregano-tasting seasoning, while the older leaves are the most pungent and can be used in a hot sauce. The herb has an oregano-like taste that is favored by chefs in the southwestern United States for flavoring semi-soft ricotta cheese, meat, and wild game. The flowers make an attractive edible garnish in salads.

## CURIOSITIES OF SCIENCE AND TECHNOLOGY

### Bergamot

- Tea made from bergamot became known as one of the "freedom teas" that were used by early American colonists during their conflict with Britain over a tax on imported tea.
- Early American settlers used bergamot oil to perfume hair tonics made from bear grease.

### Lemon Bergamot

- Some Native Americans seem to have used the leaves of lemon bergamot to repel insects and maggots on curing meat, in the same way that wild bergamot was used for this purpose. This is reminiscent of the widespread use of herbs and spices in Europe and Asia in past times to prevent deterioration of meat in the absence of refrigeration.

### Wild Bergamot

- The fragrance of wild bergamot was exploited by several groups of Native Americans in North America. The Cheyenne employed the chewed leaves to scent the manes and tails of their horses. The Crow mixed wild bergamot with other sweet-smelling herbs and applied the mixture as a perfume, often with a drop of castorium (a secretion produced by beavers to protect their fur from being soaked by water). The Omaha and Ponca used wild bergamot to perfume hair pomade. The Kootenay and Flathead employed the dried pulverized leaves to repel flies and maggots on curing meat.
- The Pawnee recognized four varieties of wild bergamot, including *tsusahtu* ("ill-smelling"), *parakaha* ("fragrant"), *tsostu* (not translated), and *tsakus tawirat* ("shot many times still fighting").
- *Monarda fistulosa* was the first species from the province of Ontario to be recognized by a scientific name, and also the first species introduced to European horticulture from Ontario. Gabriel Sagard, a lay brother at an early mission to the Hurons, first sent the plant to France, apparently in the early seventeenth century. He noted that the Hurons used the leaves to season boiled meat.

## PROSPECTS

The culinary significance of bergamot species is very limited, and the plants are only occasionally grown commercially as food plants, primarily for incorporation into tea mixtures. Bergamot oil production has also been limited and sporadic in recent decades. Given the very competitive nature of the market for flavorant herbs and essential oil plants, it is probable that bergamot species will remain minor economically. *Monarda* species are quite attractive to bees, and there is perhaps some potential for their use as honey plants.

## KEY INFORMATION SOURCES


Chubey, B.B. 1982. Geraniol-rich essential oil from *Monarda fistulosa* L. *Perf. Flav.* 7(3): 32–34.

Collicutt, L.M. 1989. "Marshall's Delight" monarda. *Hortscience* 24: 525.

Cresswell, J.E. 1990. How and why do nectar-foraging bumblebees initiate movements between inflorescences of wild bergamot *Monarda fistulosa* (Lamiaceae). *Oecologia* 82: 450–460.

Cruden, R.W., Hermanutz, L., and Shuttleworth, J. 1984. The pollination biology and breeding system of *Monarda fistulosa* (Labiatae). *Oecologia* 64: 104–110.

Epling, C. 1935. Notes on *Monarda*: the subgenus *Cheilyctis*. *Madroño* 3: 20–31.

Gill, L.S. 1977. A cytosystematic study of the genus *Monarda* L. (Labiatae) in Canada. *Caryologia (Italy)* 30: 381–394.

Goode, J. 1984. Bee balm [*Monarda didyma*]. *Am. Hort.* 63(6): 25.

Hood, S.C. 1916. *Commercial production of thymol from horsemint (Monarda punctata)*. U.S. Dep. Agric. Bull. 372. U.S. Department of Agriculture, Washington, DC. pp. 12.

Lawrence, B.M. 1983. Progress in essential oils—anise oil, bergamot oil. *Perf. Flav.* 8(3): 65–74.

Lawrence, B.M. 1987. Progress in essential oils. [anise oil, bergamot oil, cumin oil, cypress oil.] *Perf. Flav.* 12(2): 67–72.

Marshall, H.H. 1983. *Monarda* for geraniol production. *West. Can. Soc. Hortic. Rep. Proc.* 39: 42–44.

Mazza, G. 1991. Mint, monarda, and dill. In *Special "Cropportunities" (conference)*. University of Saskatchewan, Saskatoon, SK. pp. 85–91.

Mazza, G., Chubey, B.B., and Kiehn, F. 1987. Essential oil of *Monarda fistulosa* L. var. *menthaefolia*, a potential source of geraniol. *Flavor Fragrance J.* 2: 129–132.

Mazza, G., Kiehn, F.A., and Marshall, H.H. 1993. *Monarda*: a source of geraniol, linalool, thymol and carvacrol-rich essential oils. In *New crops—exploration, research, and commercialization*. Edited by J. Janick and J.E. Simon. John Wiley and Sons, New York. pp. 628–631.

Mazza, G., and Marshall, H.H. 1992. Geraniol, linalool, thymol and carvacrol-rich essential oils from *Monarda* hybrids. *J. Essent. Oil Res.* 4: 395–400.

McClintock, E., and Epling, C. 1942. A review of the genus *Monarda* (Labiatae). *Univ. California Publ. Bot.* 20(2): 147–194.
Morton, J.K. 1987. *Monarda punctata* L. *Atlas of the rare vascular plants of Ontario*. National Museum of Natural Sciences, Ottawa, ON. p. 1. [Unpaginated loose-leaf series]
Prather, L.A., Monfils, A.K., Posto, A.L., and Williams, R.A. 2002. Monophyly and phylogeny of *Monarda* (Lamiaceae): evidence from the internal transcribed spacer (ITS) region of nuclear ribosomal DNA. *Syst. Bot.* 27: 127–137.
Pringle, J.S. 1987. This native plant—wild bergamot, *Monarda fistulosa*. *Pappus (Hamilton Royal Botanical Gardens)* 7(2): 4–5.
Scora, R.W. 1967. Study of the essential leaf oils of the genus *Monarda* (Labiatae). *Am. J. Bot.* 54: 446–452.
Simon, D.Z., Beliveau, J., and Aubé, C. 1986. Extraction by hydrodiffusion of the essential oil of *Monarda fistulosa* grown in the province of Quebec: assay of geraniol in the hydrodiffused oil. *Int. J. Crude Drug Res.* 24: 120–122.
Straley, G.B. 1986. Wild bergamot, *Monarda fistulosa* (Lamiaceae), new to the Northwest Territories. *Can. Field-Nat.* 100: 380–381.
Turner, B.L. 1994. Taxonomic treatment of *Monarda* (Lamiaceae) for Texas and Mexico. *Phytologia* 77: 56–79.
Whitten, W.M. 1981. Pollination ecology of *Monarda didyma*, *Monarda clinopodia*, and hybrids (Lamiaceae) in the southern Appalachian mountains. *Am. J. Bot.* 68: 435–442.

## Specialty Cookbooks

No cookbook appears to be devoted to bergamot species, but several cookbooks (see the Appendix to this book for details of the following cited publications.) provide recipes, primarily for bergamot. Michael (1980) gives recipes for bergamot in salad, bergamot sauce, and bergamot tea. Barash (1993) has recipes for bergamot flowers entitled Monarda snapper, bee balm ice cream, bee balm pound cake, bee balm tea, and bee balm tartlets. Other recipes are bee balm tea (Crowhurt 1973); bergamot jelly (Garland 1979); bergamot tea (Marcin 1983; Phillips 1986 [under the title "Oswego tea"]); tomato and bergamot loaf (Hemphill and Hemphill 1984); blue cheese pasta with bergamot (flowers) (Leggatt 1987); herbal "Earl Grey" tea blend (Bremness 1988); peach shortcakes with bergamot flowers (Belsinger 1991), and bee balm tea (Stewart 2002). Turner and Szczawinski (1978), Phillips (1986), Tatum (1976), and Schufer (2011) give recipes for making tea from the leaves of wild bergamot.

# 10 Bilberry

Family: Ericaceae (heath family)

## NAMES

Scientific name: *Vaccinium myrtillus* L.

- The name "bilberry" is derived from the Danish *bollebar*, meaning dark berry. However, not all berries are dark, and a white-fruited ornamental form is known. The "red bilberry" is lingonberry (see Chapter 58).
- Bilberry is also known as blaeberry (a Scottish name meaning blue berry), dwarf bilberry, huckleberry, mountain bilberry, myrtle whortleberry, tracleberry, whinberry, whortleberry, and whortles. Most of these common names have also been applied to other species of *Vaccinium*, and so are frequently misinterpreted (see Chapters 15 and 52 on Blueberries and Huckleberries, respectively, for additional information).
- The name "whinberry," popular in England, is based on "whin," meaning "heath," that is, uncultivated, open land with soil that is poorly drained, coarse, and with considerable peat.
- See the chapter on blueberries for the interpretation that the name "whortleberry" is a corruption of "hurtleberry." It has also been contended that whortleberry is a corruption of "myrtle berry," due to resemblance of the berries of bilberry and myrtle (*Myrtus communis* L.).
- See Chapter 15 on Blueberries for information on the genus name *Vaccinium*.
- *Myrtillus* in the scientific name *V. myrtillus* is a reference to the myrtle-like leaves and fruits.

## GEOGRAPHY AND ECOLOGY OF WILD PLANTS

The bilberry is found in most of Europe, but only on mountains in the south. The species also occurs in Asia. There are some populations in southwestern Greenland that are thought to be ancient European introductions. In North America, bilberry grows wild in two areas of the Rocky Mountains, one in southeastern British Columbia and southwestern Alberta through Washington to central Oregon, the other in central Colorado, adjacent Utah, northcentral New Mexico, and southern Arizona.

Bilberry occurs in various climates in damp woodlands, heaths, and moors. It prefers filtered shade and moist, fertile, acidic soil. In North America, it is found in open, moist coniferous woods, hillsides, hummocky seepage slopes, and moraines above 1600 m (5250 feet). The plants are cold-tolerant at least to –20°C (–4°F), and much lower when protected by snow cover.

## PLANT PORTRAIT

Bilberry is a small, somewhat spreading, perennial, deciduous, shrub with slender angular branches arising from a creeping rhizome (underground stem) system. The plants are 5–30 cm (2–12 inches) tall, occasionally as high as 60 cm (2 feet), and usually shorter at higher elevations. The flowers are small, globular and waxy, with pale-green or pinkish petals. The berries are bluish-black (occasionally reddish, bluish, or blackish), globose, flat-topped, and 5–10 mm (0.2–0.4 inch) in diameter.

**FIGURE 10.1** Bilberry (*Vaccinium myrtillus*). (a) Berries. (Courtesy of Bateau ivre.) (b) Fruiting branch. (From Hallier, E., *Flora von Deutschland* [edition 5 by D.F.L. von Schlechtendal et al.], F.E. Köhler, Gera-Untermhaus, Germany. Vol. 20, Plate 2036, 1885.) (c) Fruiting branches. (From Losch, F., *Les plante médicinales*, Vigot Frères, Paris, France. Plate 56, 1906.) (d) Container of bilberries. (Courtesy of Meri-Tuuli.) (e) Distribution map.

The berries are sometimes covered when ripe with a delicate grey bloom, and are slightly acid, with a sweet flavor. The berries may contain up to 40 seeds, although generally only half this number are viable.

Bilberry is harvested mostly from native stands in North America and Europe, although it is occasionally cultivated. It has received much less attention from plant breeders than the cultivated species of *Vaccinium*, including the cranberries and blueberries. Most cultivated blueberries are stimulated to produce many young flowering and fruiting branches by heavy pruning of the shrubs. However, this practice is unsuitable for bilberry, which produces most of its fruit at the base of older branches.

The bilberry has been used in folk medicine for centuries, perhaps even for millennia. Bilberry extracts are popular and are commonly marketed in Europe, as well as by numerous North American supermarkets, drug stores, and health stores. Diverse medicinal claims are made for extracts of bilberry. It is believed to strengthen capillaries, allowing them to stretch without breaking or leaking, so that red blood cells can squeeze through tighter vessels, thereby benefiting blood flow. This is said to reduce bruising, and also varicose veins and "spider" veins, which result from leakage of blood from capillaries. Bilberry has been shown to improve vision, especially night vision. This is claimed to result from better blood flow to the capillary-rich retina of the eye. The resulting improved eye function has been said to reduce eye fatigue and lessen the incidence of cataracts and common myopia. The berries and especially the leaves of bilberry have been used in folk-medicine treatment of diabetes. Bilberry leaf has a reputation as a "blood-sugar-reducing" drug, useful in "antidiabetic" (against diabetes) teas. It is not widely used to treat diabetes today (insulin therapy has become standard treatment). Bilberry (both leaf and berry) tea has been used as a treatment for diarrhea and as a relief for nausea and indigestion. Bilberry has considerable tannin content, and the astringent action of the tannins is believed to be responsible for its effectiveness in treating digestive disorders, as well as for topical (surface) treatment of mild inflammation of the mucous membranes of the mouth and throat. It should be noted that the leaf is toxic and prolonged consumption is hazardous. Bilberry has at least some of the beneficial properties of cranberry (see Chapter 31) for maintaining urinary tract function and curing infections. Chemicals called flavonoids are found in bilberry fruits, and are natural antioxidants, which are medicinally effective because they disarm damaging free radicals. Free radicals are highly reactive chemical fragments produced during metabolism in the body, which can impair cell function.

Bilberry has at least seven North American relatives in *Vaccinium* section *Myrtillus* centered in the Pacific Northwest. Of these, the mountain bilberry (*V. membranaceum* Dougl.) has the largest fruit, ranging to 20 mm in diameter, and thus could be used in breeding to increase yield. The Cascade bilberry (*V. deliciosum* Piper) is also of potential interest in breeding and crop development because of its excellent fruit flavor, cold-hardiness, and blossom frost resistance (see Chapter 52 on Huckleberries).

## CULINARY PORTRAIT

Culinary uses for the bilberry are much the same as for blueberries (which see). The berries were extensively eaten by Native Americans, including the Kootenay, Carrier, and Shuswap tribes. North American Indians preserved berries of *Vaccinium* species in various ways, most commonly by drying in the sunshine or by a fire. In northern areas, *Vaccinium* berries were placed in seal oil, or stored in leather bags deposited in the permafrost. As with other *Vaccinium* species, especially the blueberries, bilberry fruit has long been popular in both Europe and North America for jams, jellies, preserves, pie and tart fillings, liqueurs, and wines. The fruit also can be eaten raw, and goes well with cream and sugar.

### Culinary Vocabulary

- In Norway, bilberry pancakes (*blåbaerpannekake*) and bilberry soup (*blåbaersuppe*) are traditional dishes.
- In Sweden, *blåbärsglass*, a bilberry confection much like ice cream, is prepared.

## PROSPECTS

Bilberry has a well-deserved reputation as a useful herbal medicine, as well as being a rather attractive jam and berry plant, and therefore promises a growing market. North American wild bilberry is not readily accessible (the plant grows in mountainous areas), so cultivation is desirable. Although not extensively harvested in North America, natural stands are known to produce up to 100 kg (fresh weight) of berries/ha (89 pounds/acre); a much higher production is likely possible under cultivation.

There is considerable information in the European (foreign language) literature on cultivation and management techniques. While there is a large available supply from Europe, the development of a North American source is a worthwhile enterprise.

## CURIOSITIES OF SCIENCE AND TECHNOLOGY

- During the latter part of World War I, England experienced a shortage of aniline dye, formerly imported from Germany, and the pigments from blue-black berries were substituted. So much of the bilberry crop was purchased by dye manufacturers that there was little available for making jam.
- In past times in Europe, bilberry was used prior to surgery to lessen bleeding, and even today some European doctors prescribe bilberry extracts before operations to reduce post-operative bleeding. This practice is consistent with the reputation that bilberry has for strengthening blood vessels.
- During World War II, pilots and navigators of the English Royal Air Force consumed large quantities of bilberry preserves, particularly before night missions, in the belief that this improved eyesight. As noted above, bilberry has actually been shown to be capable of improving vision.

## KEY INFORMATION SOURCES


Anisimova, G.M., and Shamrov, I.I. 2000. Ovule and seed development in *Vaccinium myrtillus* L. (Ericaceae). *Acta Biol. Cracov. (Series Botanica)* 42(1): 47–54.

Badescu, G., and Badescu, L. 1977. Basic elements in the cultivation technology of bilberries (*Vaccinium myrtillus*). *Prod. Veg. Hortic.* 26(2): 54–60. (In Romanian)

Bounous, G., Zanini, E., and Rissone, M.T.I. 1977. Bilberry clones behaviour in native stands and in culture. *Acta Hortic.* 61: 325–329.

Camire, M.E. 2000. Bilberries and blueberries as functional foods and nutraceuticals. In *Herbs, botanicals and teas*. Edited by G. Mazza and B.D. Oomah. Technomic Publishing Co., Lancaster, PA. pp. 289–319.

Camire, M.E. 2002. Phytochemicals in the *Vaccinium* family: bilberries, blueberries, and cranberries. In *Phytochemicals in nutrition and health.* Edited by M.S. Meskin, W.R. Bidlack, A.J. Davies, and S.T. Omaye. CRC Press, Boca Raton, FL. pp. 19–40.

Dierking, W., and Dierking, S. 1993. European *Vaccinium* species. *Acta. Hortic.* 346: 299–304.

Flower-Ellis, J.G. 1971. *Age structure and dynamics in stands of bilberry (*Vaccinium myrtillus *L.).* Royal College of Forestry, Stockholm, Department of Forest Ecology and Forest Soils, Research Note No. 9., Stockholm, Sweden. pp. 108.

ICON Health Publications. 2003. *Bilberry: a medical dictionary, bibliography, and annotated research guide to internet references.* ICON Health Publications, San Diego, CA. pp. 104.

Ingestad, T. 1973. Mineral nutrient requirements of *Vaccinium vitis-idaea* and *Vaccinium myrtillus. Physiol. Plant.* 29: 239–246.

Jacquemart, A.L., Mahy, G., Raspe, O., and De Sloover, J.R. 1994. An isozyme study in bilberry (*Vaccinium myrtillus*) 2. Mating system and genetic structure. *Belgian J. Bot.* 127: 105–114.

Kuusipalo, J. 1988. Factors affecting the fruiting of bilberries, an analysis of categorical data set. *Vegetatio* 76: 71–78.

Ley, B.M. 2001. *Bilberry & lutein: the vision enhancers!: protect against cataracts, macular degeneration, glaucoma, retinopathy & other health problems.* Second edition. BL Publications, Detroit Lakes, MN. pp. 39.

Morazzoni, P., and Bombardelli, E. 1996. *Vaccinium myrtillus* L. *Fitoterapia* 67: 3–29.

Raatikainen, M. 1993. *Vaccinium myrtillus* yields and environmental factors. *Aquilo. Ser. Botanica (Finland)* 31(0): 21–26.

Ritchie, J.C. 1956. Biological flora of the British Isles: *Vaccinium myrtillus* L. *J. Ecology* 44: 290–298.

Selas, V. 2000. Seed production of a masting dwarf shrub, *Vaccinium myrtillus*, in relation to previous reproduction and weather. *Can. J. Bot.* 78: 423–429.

Sjors, H. 1989. *Vaccinium myrtillus*, a plant portrait. *Svensk Botanisk Tidskrift* 83: 411–428. (In Swedish.)

Small, E., and Catling, P.M. 1999. Poorly known economic plants of Canada—21. Bilberry, *Vaccinium myrtillus* L. *Bull. Can. Bot. Assoc.* 32(2): 19–21.

Szmeja, K. 1993. Abundance and spatial organization dynamics of *Vaccinium myrtillus* and *V. vitis-idaea* populations. *Fragmenta Florist.-Geobot.* 38: 569–580.

Tolvanen, A., Laine, K., Pakonen, T., Saari, E., and Havas, P. 1994. Responses to harvesting intensity in a clonal dwarf shrub, the bilberry (*Vaccinium myrtillus* L.). *Vegetatio* 110: 163–169.

Vander Kloet, S.P. 1983. Seed and seedling morphology of *Vaccinium myrtillus. Nat. Can.* 110: 285–292.

Woodward, F.I. 1986. Ecophysiological studies on the shrub *Vaccinium myrtillus* L. taken from a wide altitudinal range. *Oceologia* 70: 580–586.

## Specialty Cookbooks

Cookbooks dedicated to bilberry were not located. Recipes for blueberries (see Chapter 15) will generally be suitable for bilberry.

# 11 Black Walnut

Family: Juglandaceae (walnut family)

## NAMES

Scientific name: *Juglans nigra* L.

- The word "walnut" has been said to come from the Old English *wealhhnutu*, literally "foreign nut" or "Gaul nut" (the Gauls or Welsh were considered foreigners) because the walnut was imported from Europe to England during the old Roman period at which time it was indeed considered foreign. In Old German, *welsche Nuß* means Welsch nut or nut of southern Europe, and this use may simply have been transferred to English. Another interpretation is that Germanic peoples considered the walnut of the Romans and Gauls to be "foreign," so the Old German name, rather than the Old English name, was the first to capture the idea of "foreign nut." Still another interpretation is that the name walnut comes from the drying of the nuts on top of English garden walls.
- The genus name *Juglans* is Latin for "nut of Jupiter," the classical king of the gods, an indication that the ancient Romans considered the nuts extremely tasty.
- The black walnut *J. nigra* is so named for the dark color of the bark. *Nigra* in the scientific name is Latin for black.
- The black walnut is also occasionally (but rarely) known as American black walnut, American walnut, Eastern walnut, Eastern black walnut, gumwood, and Virginia walnut.
- American horticulturist Luther Burbank (1849–1926) was a genuine but erratic genius (often called "the plant wizard") when it came to creating amazing new hybrid varieties of plants. In 1891, he apparently crossed the black walnut (*J. nigra*) and Hind's black walnut (*J. hindsii* (Jeps.) Rehd.) to produce the "royal walnut"—a huge tree that received sensationalistic press coverage. The name royal walnut is actually a translation of the Latin name of the English walnut, *Juglans regia* L. The name "Burbank walnut" is occasionally applied to *J. nigra*, but this may be a misinterpretation of Burbank's hybrid tree.

## GEOGRAPHY AND ECOLOGY OF WILD PLANTS

The black walnut is a native of eastern North America, from Massachusetts west through southern Ontario to southern Wisconsin and south to eastern Texas and the Florida Panhandle. It was much more abundant and dominant in the eastern North American forests in presettlement times. Black walnut occurs in mixed forest stands, generally as scattered single trees in open areas. Because it is intolerant of shade, it does not form large, pure stands. It does best on deep, well-drained but moist, rich, nearly neutral soils, and thrives on alluvial soils or bottomlands and in limestone areas.

## PLANT PORTRAIT

There are about 20 species of *Juglans*, all of which produce edible nuts. The English (or Persian) walnut, a native of southeastern Europe and western Asia, is the most widely cultivated of the walnuts and is the chief walnut of commerce. The United States (especially California), China, Iran, and Turkey are leading producers. The Japanese walnut (*J. ailantifolia* Carrière [*J. sieboldiana*

(a)

(d)

Black walnut
*Juglans nigra*

(e) (f)

**FIGURE 11.1** Black walnut (*Juglans nigra*). (a) Bowl of fruit. (Courtesy of E. Small.) (b) Photo of tree. (From Herman, D.E. et al., *North Dakota tree handbook*, United States Department of Agriculture, Bismark, ND, 1996.) (c) Base of large tree. (Courtesy of Jean-Pol Grandmont [CC By 3.0].) (d) Sectioned nuts; left, edible kernel (embryo) removed; right, with kernel. (Courtesy of Ferlut [CC By 3.0].) (e) Illustration plate showing leafy branch with drooping male catkin, fruit, nut, and kernel. (From Jacquin, N.J., *Icones plantarum rariorum*, B. White & Filum, London, U.K, Vol. 1, Plate 191, 1781–1786.) (f) Distribution map.

Maxim.]) and the black walnut are also often found in world commerce. There are many cultivars of these species (mostly of the English walnut), as well as varieties that are of hybrid origin. Walnuts are deciduous trees. The black walnut and English walnut are the tallest, and the longest-lived of the species of *Juglans*, and they are particularly attractive and valuable landscape trees.

Black walnut sometimes reaches 50 m (164 feet) in height, and may develop a trunk diameter of 2 m (6.6 feet). The trees occasionally live as long as 250 years. Black walnut produces a long, smooth trunk and a small rounded crown when growing in forests, but in the open, the trunk forks much nearer the base, producing a few major ascending, spreading branches. Male and female flowers develop on the same trees, in the spring, along with the foliage. The male flowers occur on drooping catkins from twigs of the previous year. Female flowers occur in small erect clusters on new branches.

More than 400 cultivars of black walnut have been selected but most of these are for timber quality, rather than nut yield and quality. Black walnut is prized for its durable, easily worked wood, which has an attractive straight grain pattern. The dark brown wood has been extensively used for fine furniture, including china cabinets and pianos, as a luxurious interior finish of buildings, and for coffins, boats, and gunstocks.

Native North Americans used black walnut extensively as food, and the nuts have been found at several archeological sites in the upper Great Lakes region dating back to 2000 BC. They also made walnut milk by grinding kernels and adding water. Oil from the nuts was employed in food preparations, as well as for rubbing on body and hair. The sap of the trees was also used in prepared foods.

More than 10,000 metric tons (more than 11,000 tons) of black walnuts are harvested annually in North America. The entire harvest is from native stands. Nuts are taken to buying stations, hulled and bagged for shipment to a sheller. Approximately 50% is retailed, and the remainder is sold to baking and ice cream manufacturers. Lower kernel yield in black walnut compared to English walnut due to the much thicker and tougher shell (a black walnut can be 90% shell) has stimulated the development of products from the shells and kernel fragments. The shells are ground for use in metal cleaning and polishing, oil well drilling, and as an ingredient in paints, explosives, and cosmetics.

## CULINARY PORTRAIT

The English walnut is one of the most widely consumed of nuts, and among dessert nuts only the almond (*Prunus dulcis* (Mill.) D.A. Webb) and the Brazil nut (*Bertholettia exelsa* Humb. & Bonpl.) exceed walnuts in quantities traded in the world's markets. China is the largest producer of English walnuts, followed by the United States, which harvests about 25% of the world supply of walnuts (most English), currently about 300,000 metric tons (330,000 tons). The United States is the largest exporter of walnuts. Almost all cultivated English walnuts in North America are grown in California. English walnuts are eaten directly; used in stuffings, sandwiches, sauces, pastries, and desserts; and as a flavoring for ice cream, syrup, and other food preparations. Maple-walnut ice cream and walnut fudge are favorite treats. During the seventeenth century, immature ("green") black and English walnuts, which are sour, became popular as ingredients in pickles, jams, and marmalades.

The nuts of the black walnut are enclosed in a very hard shell that is difficult to open (for most people, the easiest way is to drive over the nuts with a car). Unlike many other nuts, black walnuts retain their flavor well when cooked. Black walnuts are very strong in flavor and can overpower a recipe, so that some cooks blend one part black walnut to three parts English walnuts in recipes. The slightly bitter, distinctive taste of black walnuts is strongly desired by many gourmets. Black walnut is used particularly in candies, baked goods, and ice cream. The sap has also been obtained in the spring like maple sap, although not collected commercially.

Walnuts sold in vacuum-packed jars or cans are usually very fresh in taste. The freshness of shelled walnuts sold in bulk is uncertain, and buying a small sample to taste is probably the best assurance of quality. The kernels should be crunchy, not soft, shriveled, or rancid in taste. Unshelled walnuts should be free of cracks and holes, and should appear to be heavy and full, but once again their quality cannot be judged with certainty until they are tasted. A paint smell from cracked or shelled walnuts is an indication that the product is too old. Shelled nuts should be brittle and snap easily, and should not be soft and pliable. Once purchased, walnuts should be stored in a sealed container. Away from heat and humidity and in a cool area, they should remain fresh for several months. The nuts should not be shelled until they are about to be eaten. Shelled walnuts can be stored in a refrigerator for 6 months, and in a freezer for up to 1 year. Unless they are in a tightly sealed container, they can absorb odors from other foods.

Walnuts are rich in heart-healthy omega-3 fatty acids, which are believed to reduce the risk of heart disease by making the blood less sticky (i.e., platelets tend not to clump and form dangerous clots). Recent experimental studies have shown that the cholesterol-lowering capacity of walnuts exceeds that of olive oil, considered to be a key dietary contributor of the "Mediterranean diet" that is enormously healthier than the artery-clogging foods rich in saturated fats that are typical of North America. It does seem clear that eating a handful of walnuts a day can reduce the risk of heart disease. Of course, nuts are a high source of calories, and if not eaten in moderation will produce weight gain.

Walnuts are among the most allergenic of nuts, and can cause serious or life-threatening reactions in persons with walnut allergies. Food products sold as free of nuts, but accidentally contaminated with walnuts, are recalled fairly regularly.

Many people prefer the flavor and crispness of fresh walnuts, and harvest them directly from trees for consumption. If this is done, it should be noted that the pellicle or kernel skin of fresh walnuts is often very bitter, and should be removed. As the nuts age, this bitterness disappears. It should also be kept in mind that walnuts leave a brown stain on bare hands that some have described as "impossible to remove." To avoid staining hands, rubber gloves should be worn. Freshly collected nuts should be allowed to dry and mature in their shells for a week or more on newspaper, at which time the staining potential should be very limited.

### Culinary Vocabulary

- "Nougat," the French confection, literally means "made from walnuts," and indeed the walnut was originally used in its preparation.
- "Baklava" is a well-known Middle Eastern delicacy. This rich dessert is prepared with alternate layers of buttered filo dough and ground walnuts, topped with sweet-spiced syrup.
- Several European liqueurs are prepared using walnut husks as the base. Most famous of these is the Italian *Nocino*, a renowned cordial made from green walnuts. The nuts are picked on the Festival Day of St. John (24 June), cracked, steeped in alcohol for 2 months, and filtered.
- The culinary expression "butter the size of a walnut" is a traditional measure of butter volume, about equal to 2 tablespoons (approximately 58 mL), which is about half the size of the volume indicated by the expression "butter the size of an egg."

## PROSPECTS

Black walnut has been described as a "money tree," and its future is assured, not because of its value as a nut source, but for wood. It is the most valuable timber species in temperate North America. Although there has been dreadful overexploitation and decimation of wild black walnut trees, this is an underexploited crop from the point of view of cultivation. The demand for both black walnuts and black walnut timber exceeds the supply. Thin-shelled cultivars of black walnut, from which the kernels could easily be extracted, have not been adequately developed. At present, most of the supply of nuts comes from wild trees, but this is a vanishing resource. Integrated forestry-farming (agroforestry), coupling both walnut and timber production, and additionally combining such a system with intercropping of agricultural crops, has been described as "one of the best investment opportunities available in the profession of forestry."

## CURIOSITIES OF SCIENCE AND TECHNOLOGY

- In primitive times, medicinal practice was more often than not based on the "Doctrine of Signatures"—the idea that plants with characteristics reminiscent of human organs could cure diseases of those organs. The ancient Greeks interpreted the walnut's shell as a human skull (their word *caryon* for the walnut meant head) and the nut as a human brain (the joined halves of a kernel of a walnut certainly look like the brain). Accordingly, they believed that walnuts cured headaches. Others have thought that low intellect and madness could be cured by eating walnuts. The association of walnuts and the human brain is also reflected in the Afghanistani word for walnut, *charmarghz*, literally "four brains."

- A paste of ground English walnuts mixed with white lead was a recommended cure for baldness in China. In North America, burnt kernels of black walnut in red wine was supposed to stop hair from falling out. (Again, note the association of the walnut with the human head.)
- The nuts and leaves of the black walnut are thought to repel insects to some degree, and have been used, both by Native Americans and colonists, as an insect repellent. Walnut leaves were sometimes rubbed on the faces of cattle and horses to discourage flies. It was once customary to plant a walnut in the area of a stable, in the expectation that the smell from the very aromatic leaves would keep flies away from the animals.
- The "mound builders" were peoples who built earth mounds in a large area from the Great Lakes to the Gulf of Mexico and from the Mississippi River to the Appalachian Mountains. The greatest concentrations of mounds are found in the Mississippi and Ohio valleys. It is thought that the mound builders were the ancestors of Native Americans, and were sedentary farmers who lived in permanent villages. The mounds vary considerably in shape and size (from less than 0.4 ha or 1 acre to more than 40 ha or 100 acres). The mounds were used chiefly as burial places but also as foundations for buildings. Mounds also vary in age, some dating back to the sixth century, while others were built after Europeans began colonizing North America. Black walnut shells carved in the form of birds and pierced to serve as earrings were discovered in the mounds of Ohio and Indiana.
- The early Pennsylvania Dutch (who were from Germany and Switzerland, despite the "Dutch" in their name) judged land to be especially suitable for farms when stands of walnut were present—a good indication that the soil was rich.
- Walnut wood, to this day, is considered excellent for gunstocks. "Shouldering walnut" once meant enlisting in the military. In 1806, 12,000 walnut trees were reportedly needed annually in France for manufacturing muskets. Kaiser Wilhelm II (1859–1941) purchased large quantities of black walnut wood as a stockpile for gunstocks in anticipation of World War I.
- The shells of walnuts were a valuable war material. In World War I, the shells were converted to high-quality charcoal that was used extensively in gas mask filters. In World War II, ground nutshells proved excellent as an abrasive for cleaning aircraft pistons and cylinder heads.
- Toyo Tire & Rubber Co. Ltd. has manufactured winter tires (notably its Observe tires) with ground walnut shells in the tread compound to increase grip on ice. The shells act as sandpaper grit to improve traction.
- At one time, airplane propellers were made of walnut wood.
- Actors once used yellow dye from walnut fruits and a dark brown dye from the tree's roots to stain their skins.
- The flesh of black walnuts has a food value equal to the best beef. During the fifteenth century, the aboriginal city of Cahokia Mound on the Mississippi River was larger than London, England, at the same time. Its large population was sustained by the protein value of black walnuts.
- A black walnut tree in Tennessee produced more than 6000 nuts in one season.
- A single black walnut tree on the Lower Missouri was said to have furnished 200 fence posts.
- Black walnut is a bad neighbor for other plants. The toxic substance juglone is present in the roots, leaves, and seed husks, and this has been reported to kill several species of pines when they grow nearby. Such crops as alfalfa, potatoes, and tomatoes are easily injured by juglone, and fruit trees such as apple may not bear fruit if planted too close to walnut trees. Members of the heath (Ericaceae) family, such as rhododendrons, azaleas, blueberries, and mountain laurel, are particularly susceptible. The toxic zone from a mature tree extends on average over a 15–18 m (50–60 feet) radius from the trunk, but can be up to 24 m (80 feet)

away. English walnut also exhibits similar toxicity, and this was recorded by the Roman scholar Pliny (23–79).

- Horses hauling logs or lumber at sawmills or bedded on sawdust or wood shavings from black walnut sometimes suffered from lameness owing to an inflammation of the hoof (laminitis). Still an occasional problem today, this is believed to be caused by juglone and other chemicals present in large amounts in the heartwood.
- Walnut lumber is highly valued. Walnut veneers may be sliced less than 1 mm (0.04 inch) in thickness after the wood has been softened by hot water and steam. A veneer company paid $39,000 for a black walnut tree growing near Johnson City, Iowa. There have been cases of highly organized "rustling" of mature trees from their owners' properties.
- As much as 90% of a black walnut can be shell, and this has intricate internal cavities. It has been suggested that these are an adaptation to confuse rodents trying to locate the edible nut inside the shell.

## KEY INFORMATION SOURCES


Aletà, N., Girona, J., and Tasias, J. (Editors). 1993. *Proceedings international walnut meeting, IRTA–Generalitat de Catalunya, Tarragona, Spain, Oct. 21–25, 1991*. International Society for Horticultural Science, Wageningen, the Netherlands. pp. 315.

Beineke, W.F. 1983. The genetic improvement of black walnut for timber production. *Plant Breed. Rev.* 1: 236–266.

Brooks, M. 1951. *Effect of black walnut trees and their products on other vegetation*. Agricultural Experiment Station, College of Agriculture, Forestry and Home Economics, West Virginia University, Morgantown, WV. pp. 31.

Chenoweth, B. 1995. *Black walnut: the history, use, and unrealized potential of a unique American renewable natural resource*. Sagamore Publishing, Champaign, IL. pp. 334.

Davis, E.F. 1928. The toxic principle of *Juglans nigra* as identified with synthetic juglone, and its toxic effects on tomato and alfalfa plants. *Am. J. Bot.* 15: 620.

Dillow, M.K., and Funk, D.T. 1975. *Bibliography of walnut, supplement no. 2*. North Central Forest Experiment Station, Forest Service, U.S. Department of Agriculture, St. Paul, MN. pp. 26.

Dillow, M.K., Hawker, N.L., and Funk, D.T. 1971. *Annotated bibliography of walnut, supplement no. 1*. North Central Forest Experiment Station, Forest Service, U.S. Department of Agriculture, St. Paul, MN. pp. 23.

Forde, H.I. 1975. Walnuts. In *Advances in fruit breeding*. Edited by J. Janick and J.N. Moore. Purdue University Press, West Lafayette, IN. pp. 439–455.

Funk, D.T. 1966. *Annotated bibliography of walnut and related species*. North Central Forest Experiment Station, Forest Service, U.S. Department of Agriculture, St. Paul, MN. pp. 48.

Funk, D.T. 1970. *Genetics of black walnut*. U.S. Department of Agriculture, Washington, DC. pp. 13.

Gergely, I., and Szentiványi, P. (Editors). 1990. *First international symposium on walnut production, Budapest, Hungary, Sept. 25–29, 1989*. International Society for Horticultural Science, Wageningen, the Netherlands. pp. 369.

Germain, E. (Editor). 2001. *Proceedings of the fourth international walnut symposium, Bordeaux, France, Sept. 13–16, 1999*. International Society for Horticultural Science, Leuven, Belgium. pp. 616.

Gomes Pereira, J.A., Martins, J.M.S., and Pinto de Abreu, C. (Editors). 1997. *Third international walnut congress, Alcobaça, Portugal, June 13–16, 1995*. International Society for Horticultural Science, Leuven, Belgium. pp. 445.

Kessler, K.J. Jr., and Weber, B.C. (Coordinators). 1979. *Walnut insects and diseases: workshop proceedings, June 13–14, 1978, Carbondale, Illinois*. North Central Forest Experiment Station, Forest Service, U.S. Department of Agriculture, St. Paul, MN. pp. 100.

Malhotra, S.P. 2008. *World edible nuts economy*. Concept Publishing., New Delhi, India. pp. 538.

McGranahan, G., and Leslie, C. 1990. Walnuts (*Juglans*). In *Genetic resources of temperate fruit and nut crops*. Edited by J.N. Moore and J.R. Ballington, Jr. International Society for Horticultural Science, Wageningen, the Netherlands. pp. 907–949.

North Central Forest Experiment Station. 1966. *Black walnut culture; papers presented August 2 and 3, 1966*. North Central Forest Experiment Station, St. Paul, MN. pp. 94.

North Central Forest Experiment Station. 1973. *Black walnut as a crop*. U.S. Department Of Agriculture, Forest Service general technical report NC 4, North Central Forest Experiment Station, St. Paul, MN. pp. 114.

North Central Forest Experiment Station. 1982. *Black walnut for the future*. U.S. Department of Agriculture, Forest Service, North Central Forest Experiment Station, St. Paul, MN. pp. 151.

North Central Forest Experiment Station. 1985. *How to start walnut trees from seed*. U.S. Forest Service, North Central Forest Experiment Station, St. Paul, MN. pp. 6.

Ramos, D.E. 1985. *Walnut orchard management*. Cooperative Extension, University of California, Division of Agriculture and Natural Resources, Oakland, CA. pp. 178.

Ramos, D.E. 1997. *Walnut production manual*. University of California, Division of Agriculture and Natural Resources, Oakland, CA. pp. 320.

Rink, G. 1985. *Black walnut*. Revised edition. Forest Service, U.S. Department of Agriculture, Washington, DC. pp. 7.

Sambeek, J.W. Van. (Editor). 1997. *Knowledge for the future of black walnut: proceedings of the fifth black walnut symposium, Hammons Conference Center, Springfield, Missouri and Sho-Neff Plantation, Stockton, Missouri, July 28–31, 1996*. North Central Forest Experiment Station, U.S. Department of Agriculture Forest Service, St. Paul, MN. pp. 256.

Schlesinger, R.C., and Funk, D.T. 1977. *Manager's handbook for black walnut*. U.S. Department of Agriculture, Forest Service, North Central Forest Experiment Station, St. Paul, MN. pp. 22.

Thompson, B.S. 1976. *Black walnut for profit*. Second edition. Timber Press, Forest Grove, OR. pp. 285.

Thomsen, M.E., Davis, E.G., and Rush, B.R. 2000. Black walnut induced laminitis. *Vet. Hum. Toxicol.* 42: 8–11.

U.S. Department of Agriculture. 1978. *Growing black walnuts for home use*. Revised edition. U.S. Department of Agriculture, Science and Education Administration, Washington, DC. pp. 9.

University of California Integrated Pest Management Program. 1993. *Integrated pest management for walnuts*. Second edition. University of California, Statewide Integrated Pest Management Project, Division of Agriculture and Natural Resources, Oakland, CA. pp. 96.

Whittemore, A.T., and Stone, D.E. 1997. *Juglans*. In *Flora of North America north of Mexico*. Vol. 3. Edited by Flora of North America Editorial Committee. Oxford University Press, New York. pp. 425–428.

Williams, R.D. 1974. *Planting methods & treatments for black walnut seedlings*. North Central Forest Experiment Station, Forest Service, U.S. Department of Agriculture, St. Paul, MN. pp. 6.

## Specialty Cookbooks

Bough, C. 1993. *Cooking with black walnuts: recipes collected by Carole Bough*. Missouri Dandy Pantry, Stockton, MO. pp. 151.

California Walnut Growers Association. 1930. *100 delicious walnut recipes*. Diamond Walnut Growers and California Walnut Growers Association, Los Angeles, CA. pp. 31.

California Walnut Growers Association. 1936. *To win new cooking fame just add walnuts*. California Walnut Growers Association, Los Angeles, CA. pp. 31.

California Walnut Growers Association. 1950–1959? *Around the kitchen clock with walnuts*. California Walnut Growers Association, Los Angeles, CA. pp. 39.

Carroll, J.P. 1993. *California walnuts 7 day meal plan*. Walnut Marketing Board, Sacramento, CA. Unpaginated.

Cunningham, M. 1984. *California walnuts: talk of the town: a collection of delicious recipes, menus and tips featuring America's favorite nut*. Walnut Marketing Board, Sacramento, CA. pp. 35.

Diamond Walnut Growers. 1950–1959? *The new walnut cook book*. Diamond Walnut Growers, Stockton, CA. pp. 28.

Diamond Walnut Growers. 1978. *Diamond walnut recipe favorites*. Diamond Walnut Growers, Stockton, CA. pp. 112.

Diamond Walnut Growers. 1990. *The Diamond walnut collection*. Diamond Walnut Growers, Stockton, CA. pp. 80.

Ryan, F. 1993. *The walnut diet and fitness plan*. Swift, Bromsgrove, Worcestershire, U.K. pp. 62.

Toussaint, J.-L. 1998. *The walnut cookbook*. Ten Speed Press, Berkeley, CA. pp. 206.

Turner, M. 1978. *The Napa Valley walnut cookbook: from soup to nuts*. M. Turner, Rutherford, CA. pp. 56.

Walnut Marketing Board. 1993–2002? *Cooking with taste: expanding your taste with California walnuts*. Walnut Marketing Board, San Francisco, CA. pp. 8.

West Virginia Black Walnut Festival, Inc. 1980. *Black walnut festival baking competition: award winning recipes*. West Virginia Black Walnut Festival, Inc., Spencer, WV. pp. 130.

Brill (2002) presents 15 recipes featuring black walnut, Freitus (1980) has six, Kluger (1973) has five, Robe-Terry (1997) has five, and Tatum (1976) has eight (See the Appendix to this book for details of the publications cited.).

# 12 Blackberries and Dewberries

Family: Rosaceae (rose family)

## NAMES

Scientific name: *Rubus* species and hybrids

- "Blackberries" are of course named after their fruit color.
- The origin of the name "dewberry" is obscure. It has been suggested that the word is a corruption of "dove berry," the name in Germany for centuries.
- The Loganberry originated about 1881 in the garden of Judge J.H. Logan (1841–1928), an amateur horticulturist in California. It seems to be a hybrid between a blackberry and a red raspberry. The new berry was introduced commercially in 1882.
- The Youngberry was introduced in 1926 by B.M. Young of Morgan City, Louisiana, and is a hybrid of blackberry, red raspberry, and Loganberry.
- The Phenomenal Berry (also called Burbank's Logan) is a second-generation cross between blackberry and raspberry, and was introduced in 1905 by Luther Burbank (1849–1926), an American plant breeder and horticulturist, who was very fond of such exaggerated names for his new varieties.
- The Boysenberry was developed in 1923 by horticulturist Rudolph Boysen (died in 1950) of Anaheim, California, and is apparently a cross between a Loganberry, red raspberry, and blackberry. Walter Knott (of the now famous Knott's Berry Farm in Buena Park, California) sold and popularized the Boysenberry in the 1930s.
- Marionberries are blackberries named for Marion Country, Oregon, which are grown in the Northwest of the United States.
- See Chapter 77 on Raspberries for information on the genus name *Rubus*.

## GEOGRAPHY AND ECOLOGY OF WILD PLANTS

There appear to be hundreds of *Rubus* species throughout the world, which are called blackberries and dewberries, and generally local species have been domesticated in various countries to produce locally grown cultivars. As noted in the following, the most important North American blackberries are derived from several wild species of *Rubus* of eastern North America. Dewberries are native mostly to the southeastern United States. Blackberries are natural colonizers of disturbed and open areas, including fields, forest edges, railroad tracks, fence lines, and roadsides. The fruit and seeds are very attractive to birds, and so are easily and widely dispersed. Moreover, the tips of branches often "layer" (i.e., bend over naturally to contact the soil, and develop into new plants), spreading the plants once they get into an area.

## PLANT PORTRAIT

Varieties of blackberries and their relatives are often of complex hybrid origin, and frequently cannot be reliably identified in terms of their species composition. Several wild species of *Rubus* of Europe, Asia, and North America are ancestors of the cultivated blackberries. Many of the cultivars grown for the fresh market are largely derived from several eastern American species, such

**FIGURE 12.1** Blackberries and dewberries (*Rubus* species). (a) Commercial variety of blackberry (unidentified). (b) Blackberry, cultivar 'Eldorado'. (c) Blackberry, cultivar 'Kittatinny'. (d) Loganberry (thought to be a hybrid of the dewberry *R. ursinus*, and raspberry). ([a] Courtesy of S. Bauer, U.S. Department of Agriculture; [b–d] From Hedrick, U.P., *The small fruits of New York*. Department of Farms and Markets, New York State, 1925.)

as Allegheny blackberry (*R. allegheniensis* Porter), sawtooth blackberry (*R. argutus* Link), and Pennsylvania blackberry (*R. pensilvanicus* Poir., currently regarded as including Yankee blackberry, *R. frondosus* Bigelow), while the trailing cultivars that account for much of the processing industry are largely derived from the western American dewberry (or "trailing blackberry," *R. ursinus* Cham. & Schltdl.). Some wild blackberry species develop separate male and female plants, but as fruiting plants are desired in cultivation, the plants have been selected to have both male and female flowers. Blackberry fruits are compounds of small, juicy fruits called drupelets, each containing a tiny seed. The compound fruits are like raspberries, but the fleshy "core" (central receptacle) is retained inside the fruit when it is harvested, in contrast to raspberries, in which the core is not retained by the fruit. The fruits are nearly globose or cylindrical in shape, 1.3–2 cm (0.5–0.8 inch) in diameter and 2–3.8 cm (0.8–1.5 inches) long. The compound fruit of cultivated blackberries are large, juicy, and often seedless, whereas wild species are often too seedy to consume as fruit (although they can be used for jelly and alcoholic beverages). Blackberry cultivation began in the early nineteenth century, mostly in North America, where the majority of varieties were selected. Most of the crop is produced in Texas, Oklahoma, and Arkansas, with considerable harvesting of

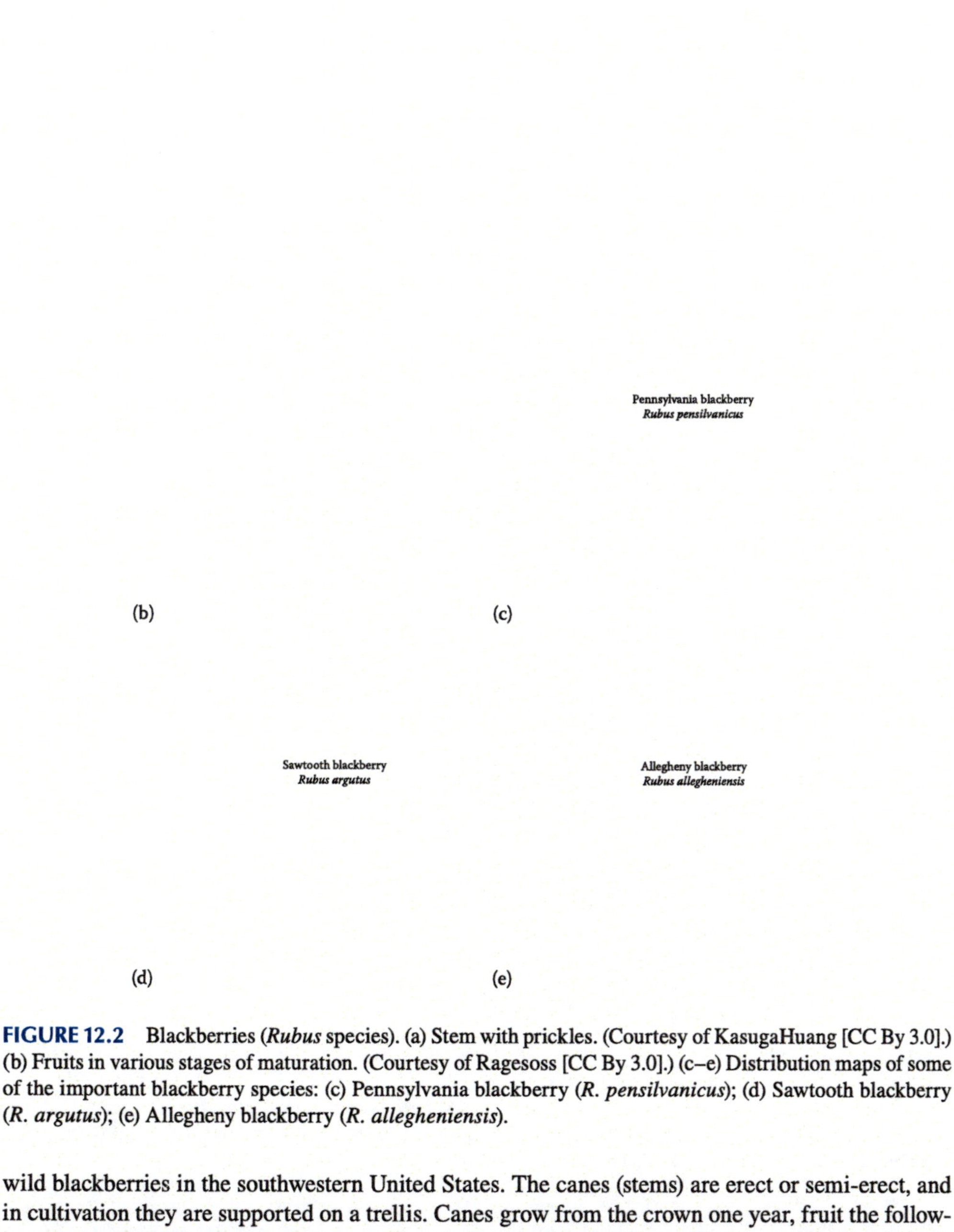


**FIGURE 12.2** Blackberries (*Rubus* species). (a) Stem with prickles. (Courtesy of KasugaHuang [CC By 3.0].) (b) Fruits in various stages of maturation. (Courtesy of Ragesoss [CC By 3.0].) (c–e) Distribution maps of some of the important blackberry species: (c) Pennsylvania blackberry (*R. pensilvanicus*); (d) Sawtooth blackberry (*R. argutus*); (e) Allegheny blackberry (*R. allegheniensis*).

wild blackberries in the southwestern United States. The canes (stems) are erect or semi-erect, and in cultivation they are supported on a trellis. Canes grow from the crown one year, fruit the following season, then die. They may grow to 5 m (16 feet) in length. Most varieties are heavily thorned, but thornless kinds are now available.

Dewberries are simply trailing blackberries. Usually dewberries ripen earlier than blackberries and the flowers are few and scattered, while blackberries usually have dense clusters of flowers. However, the two groups have been hybridized and these distinctions do not always hold. The canes of dewberries are slender and trailing, and generally are less thorny than those of blackberries. Thornlessness is of particular importance in trailing varieties, because they require trellising, and thorns tend to injure workers and damage fruits when they are picked. Some kinds, such as Loganberry, apparently have some raspberry in their ancestry. The fruits are generally cylindrical in shape, about 1.3 cm (0.5 inch) in diameter and 2.5–3.8 cm (1–1.5 inches) in length. Leading

FIGURE 12.3 Dewberries (*Rubus* species). (a) Flowers of Western American dewberry (*R. ursinus*). (Courtesy of Walter Siegmund [CC By 3.0].) (b) Child picking dewberries in Missouri. (Courtesy of U.S. Fish and Wildlife Service.) (c) Dewberry cultivar 'Lucretia' (*R. roribaccus* (L.H. Bailey) Rydb.). (From Hedrick, U.P., *The small fruits of New York*, Department of Farms and Markets, New York State, 1925.) (d) Distribution map of the Western American dewberry.

varieties are Boysenberry, Loganberry, Olallie, Lucretia, Bingleberry, Mammoth Blackberry, Cory, Lowberry, Phenomenal berry, Rossberry, Lavaca, Nectar, and Youngberry. Most of the crop of dewberries comes from western North America, and is canned or frozen.

## CULINARY PORTRAIT

Blackberries are an excellent fruit when perfectly ripe but are unfit for use when picked before they are mature. Some varieties become black before they are fully ripe, and are picked too soon. Such unripe blackberries are sometimes sold, giving the fruit an undesirable reputation. The fruit is marketed fresh, canned, frozen, in preserves, and in the form of wines, cordials, and a variety of other processed food products. Blackberries may be eaten in the same manner as raspberries—fresh alone, with cream, or cooked in tarts, pastries, and preserves. They go well with ice cream, fresh cream, and yogurt, and complement fruit salads, tarts, crêpes, and breakfast cereals. Blackberries are also used to make jams, jellies, syrup, juice, wine, and brandy. In Denmark, a blackberry soup is prepared. Blackberries are delicate fruits, and are not tolerant of heat, handling, and shipping. They tend to spoil rapidly. Blackberries may be stored in a refrigerator for several days, preferably unwashed, loosely packed, and after damaged berries have been removed. In addition to the fruit, there is a small market for blackberry leaves, used as a herbal tea. The red and purple (anthocyanin) pigments have made blackberries an occasional food colorant in past times, and such extracts, known for their antioxidant properties, have potential for use in various products marketed for health promotion.

### Culinary Vocabulary

- "Kentucky Jam Cake" (also known as Tennessee Jam Cake) is a spice cake usually prepared with blackberry jam added to the batter.
- A "Polish Sidecar" is a cocktail prepared with gin, lemon juice, blackberry liqueur or blackberry brandy, and fresh blackberries.

## PROSPECTS

Blackberries and dewberries are tasty and extremely rich in antioxidants, compounds known to be beneficial to health. Accordingly, they are increasing in popularity in the early part of the twenty-first century. Once available primarily as frozen and processed products, these berries have become a staple fresh fruit in supermarkets. Breeding of improved blackberries is occurring, with the goals of producing thornless, machine harvestable, winter-tolerant forms with fruit that is firmer and more flavorful. Increasingly, blackberries and dewberries are being grown in the Southern Hemisphere to provide fresh, off-season fruit to the Northern Hemisphere. While overseas blackberries and dewberries are marketed in North America, most in-season fruit available on the continent is grown in the United States, particularly in California and Oregon, although Mexico has become a major producer as well. Accordingly, the fresh fruit market is likely to continue to be important, and to be supplied domestically. Processed fruit, however, is likely to be supplied increasingly to North America by foreign producers, particularly China.

## CURIOSITIES OF SCIENCE AND TECHNOLOGY

- Thorny blackberry hedges reportedly were used as barriers to repel invading forces as long as 2000 years ago.
- "Batology" (from the Greek *batos*, bramble), is the science of blackberry classification (species of the genus *Rubus* are called brambles). Batologists are botanists devoted to this study, which is rather difficult because the genus *Rubus* has evolved in odd ways to produce as many as 1000 recognizably different kinds, which some experts recognize as species. The study of bats (the flying mammals, not baseball bats) is called chiropterology, and chiropterologists are zoologists who study bats.
- Many species of *Rubus* produce shoots that bend over, with the tips rooting in the soil. The loop thus formed was once believed to have restorative properties. In Europe, children with hernia and whooping cough were passed forward and backward through the arch to alleviate the conditions. This treatment was also used for boils, rickets, and other conditions. It was said to be good for curing sick cows as well, although one may doubt that a loop large enough for this purpose could be located.

- Hair dye was once made from blackberries. The famous English herbalist Nicholas Culpeper (1616–1654) recommended that blackberries be boiled in a lye solution to "maketh the hair black."
- Blackberry tea was used in the past as a cure for dysentery. During the American Civil War, temporary truces were sometimes declared when there were outbreaks of dysentery on both sides, so that both Union and Confederate soldiers could "go blackberrying" to treat the disease.
- "Blackberry winter," a term dating at least to 1900, refers to the period of cool weather in early spring when blackberries are in bloom.
- There is a white (albino) blackberry.
- Both Alabama and Kentucky declared the blackberry to be their official state fruit in 2004.
- Blackberries contain salicylates, which can cause a reaction in aspirin-sensitive people.
- The "BlackBerry" is a well known, hand-held wireless communication device used for telephone and e-mail messages and Internet access. It was developed by Research in Motion Ltd. (Waterloo, Ontario, Canada), which contracted the task of adopting a new name for it to Lexicon Branding (California). The name BlackBerry was adopted because the tiny buttons were reminiscent of the seeds of a blackberry, and indeed the device at the time was black. As shown by the BlackBerry, product naming has become a very specialized industry, with hundreds of naming companies established in the last 15 years. Memorable trademarks such as BlackBerry, Apple (computer), and Yahoo (Internet search engine) are termed "fanciful" or "arbitrary" names, with no functional connection to the product.

## KEY INFORMATION SOURCES

(For additional references, see Chapter 77 on Raspberries.)

Brennan, R.M., Gordon, S.L., and Williamson, B. (Editors). 2002. *Proceedings of the eighth international Rubus and Ribes symposium, Dundee, Scotland, July 2001*, Vol. 2. International Society for Horticultural Science, Leuven, Belgium. pp. 696.

Card, F.W. 1925. *Bush fruits: a handbook of raspberries, blackberries, dewberries, currants, gooseberries, and other small fruits*. New edition. Macmillan Co., New York. pp. 411.

Clark, J.R., Stafne, E.T., Hall, H.K., and Finn, C.E. 2007. Blackberry breeding and genetics. *Plant Breed. Rev.* 29: 19–144.

Colby, A.S., Anderson, H.W., and Flint, W.P. 1940. *Bramble fruits: raspberries, blackberries, dewberries: how to grow in Illinois*. College of Agriculture, University of Illinois, Agric. Exp. St., Urbana, IL. pp. 72.

Coyner, M.A., Skirvin, R.M., Norton, M.A., and Otterbacher, A.G. 2005. Thornlessness in blackberries: a review. *Small Fruits Rev.* 4(2): 83–106.

Crandall, P.C. 1995. *Bramble production: the management and marketing of raspberries and blackberries*. Food Products Press, New York. pp. 213.

Darrow, G.M. 1942. *Blackberry growing*. U.S. Govt. Print. Off., Washington, D.C. pp. 18.

Daubeny, H.A. 1996. Brambles. In *Fruit breeding, Vol. 2. Vine and small fruits*. Edited by J. Janick and J.N. Moore. Wiley, New York. pp. 109–190.

Deighton, N., Brennan, R., Finn, C., and Davies, H.V. 2000. Antioxidant properties of domesticated and wild *Rubus* species. *J. Sci. Food Agric.* 80: 1307–1313.

Finn, C. 2008. *Rubus* spp. Blackberry. In *The encyclopedia of fruit & nuts*. Edited by J. Janick and R.E. Paull. CABI, Wallingford, Oxfordshire, U.K. pp. 748–751.

Finn, C.E. and Clark, J.R. 2011. Emergence of blackberry as a world crop. *Chron. Hort.* 51: 13–18.

Hall, H.K. 1990. Blackberry breeding. *Plant Breed. Rev.* 8: 249–312.

Jennings, D.L. 1980. A hundred years of Loganberries. *Fruit Var. J.* 35: 34–37.

Jennings, D.L. 1988. *Raspberries and blackberries: their breeding, diseases and growth*. Academic Press, London, U.K. pp. 230.

Jennings, D.L., Daubeny, H.A., and Moore, J.N. 1990. Blackberries and raspberries (*Rubus*). In *Genetic resources of temperate fruit and nut crops*. Edited by J.N. Moore and J.R. Ballington, Jr. International Society for Horticultural Science, Wageningen, the Netherlands. pp. 331–389.

Kalkman, C. 1991. *Rubus* L. In *Plant resources of South-East Asia, 2, edible fruits and nuts*. Edited by E.W.M. Verheij and R.E. Coronel. Pudoc, Leiden, the Netherlands. pp. 277–278.
Knight, R.L. 1972. *Abstract bibliography of fruit breeding and genetics, 1956–1969, Rubus and Ribes*. Commonwealth Agricultural Bureau, Wallingford, Oxon, U.K. pp. 449.
La Vina, P.D. 1982. *Growing boysenberries and olallie blackberries*. Division of Agricultural Sciences, University of California, Berkeley, CA. pp. 15.
Logan, M.E. 1955. *The loganberry*. Mary E. Logan, Oakland, CA. pp. 20.
Moore, J.N., and Skirvin, R.M. 1990. Blackberry management. In *Small fruit crop management*. Edited by G.J. Galletta and D.G. Himelrick. Prentice Hall, Englewood Cliffs, NJ. pp. 214–244.
Ourecky, D.K. 1975. Brambles. In *Advances in fruit breeding*. Edited by J. Janick and J.N. Moore. Purdue University Press, West Lafayette, IA. pp. 98–129.
Pritts, M.P., Handley, D., Eames-Sheavly, M., and Ohira, A. 1989. *Bramble production guide*. NRAES-35. Northeast Reg. Agr. Eng. Serv., Ithaca, NY. pp. 189.
Seelig, R.A., and Bing, M.C. 1990. Blackberries. In *Encyclopedia of produce*. [Loose-leaf collection of individually paginated chapters.] Edited by R.A. Seelig and M.C. Bing. United Fresh Fruit and Vegetable Association, Alexandria, VA. pp. 12.
Strik, B.C., Clark, J.R., Finn, C.E., and Bãnados, P. 2007. Worldwide blackberry production. *Hort Technology* 17: 205–213.
Swartwout, H.G. 1929. *Blackberries, raspberries and dewberries*. Bull. Univ. Missouri Agric. Exp. St. 265, Columbia, MO. pp. 31.
Thomas, R.H., Woods, F.M., Dozier, W.A., Ebel, R.C., Nesbitt, M., Wilkins, B., and Himelrick, D. 2005. Cultivar variation in physicochemical and antioxidant activity of Alabama-grown blackberries. *Small Fruits Rev*. 4(2): 51–71.

## Specialty Cookbooks

Cookbook Publications. 1993. *Taste the treasure: the Wild Blackberry Festival, Elma, Washington*. Cookbook Pub., Olathe, KS. pp. 52.
Dufour, R.M., Bowes, D., and Göritz, U. 2000. *Blackberries: recipes from Power River*. Powell River & Region Transition House Society/Works Consulting, Powell River, BC. pp. 84.
Fleer, J. 2002. *Blackberry basics: recipes from Blackberry Farm*. Blackberry Farm Press, Walland, TN. pp. 88.
Fulton, K. 1994. *Wild & free: a cookbook and guide to Northwest blackberries*. K. Fulton, Kirkland, WA. pp. 108.
Harter, L.L. 1989. *Blackberry delights!* Ideas Unique Publishers, North Bend, OR. pp. 98.
Hendrickson, R. 1981. *The berry book. The illustrated home gardener's guide to growing and using over 50 kinds and 500 varieties of berries*. Doubleday & Company, Garden City, NY. (Chapter 7: Backyard blackberrying, pp. 104–127; Chapter 8: Boysenberry bonanzas—north and south, pp. 128–135; Chapter 9: Dewberry delights—including loganberries, nectarberries, and youngberries.)
Thompson, S. 1993. *Recipes from a Kentucky blackberry patch*. Windstone Farms, Carlisle, KY. pp. 63.

Baird (1980) has 12 "wild blackberry" recipes, Brill (2002) has 14, Freitus and Haberman (2005) have 11, Kluger (1973) has 13, Marie (2008) has 2, Marrone (2009) has 9, Mogelon (2001) has 2, Robe-Terry (1997) has 4, Schufer (2011) has 5), Snell (2006) has 1, and Watts and Watts (2007) have 12. Hibler (2004) has several dozen recipes either specifically for blackberries or which can be used with the fruit (See the Appendix to this book for details of the publications cited.).

# 13 Blue Honeysuckle

Family: Caprifoliaceae (honeysuckle family)

## NAMES

Scientific name: *Lonicera caerulea* L. (*L. caerulea* var. *edulis* Turcz. ex Herder)

- The name "honeysuckle" is based on the Old English *hunigsuge* (honeysuck), referring to the practice of sucking the "honey" (nectar) from the tube of honeysuckle flowers. The name is sometimes applied to species of different genera but is best reserved for species of the genus *Lonicera*.
- Blue honeysuckle is named for its blue flowers.
- Blue honeysuckle is also known as bearberry honeysuckle, blue-berried honeysuckle, blueberry honeysuckle, bluefly honeysuckle, edible honeysuckle, fly honeysuckle, haskap, honeyberry, mountain fly honeysuckle, northern fly honeysuckle, swamp fly honeysuckle, sweet-berried honeysuckle, and sweet-berry honeysuckle.
- The name honeyberry has become quite popular for edible berries of *L. caerulea* in recent times. Traditionally, the name has been applied to two tree species with sweetish berries: *Melicoccus bijugatus* Jacq. (also known as Spanish lime, and employed as a fresh fruit in the West Indies and South America), and *Celtis australis* L. (also known as Mediterranean hackberry, and native to the Mediterranean area).
- The name haskap is an ancient name for the plant used by the Ainu—indigenous people in Japan (the name has also been transliterated as hasakapu, haskappu, hascap, hascup, and hasukappu). The name is being promoted by growers for market purposes.
- The genus name *Lonicera* commemorates German botanist, physician, and professor Adam Lonitzer (also known as Adamus Lonicerus and Adam Lonicer; 1528–1586).
- *Caerulea* in the scientific name is Latin for dark sky blue, a reference to the flower color.

## GEOGRAPHY AND ECOLOGY OF WILD PLANTS

Blue honeysuckle is a circumpolar, widely ranging species, native to northern boreal forests of Asia, Europe, and North America. In Canada, the species occurs in the territory of Nunavut and all Canadian provinces. In the United States, it is found in the western states (Washington, Oregon, Idaho, Montana, Wyoming, California, Nevada), and in the northeast.

*Lonicera caerulea* occupies low-lying wet areas, marshes, and high mountains. Although it grows well in boggy areas and in seasonally wet ground, the plant grows better in well-drained soil. The species tolerates semi-shade and a variety of soil types. The northernmost forms are capable of overwintering in very cold regions.

## PLANT PORTRAIT

Blue honeysuckle is a very polymorphic species and is known in various areas of the world by at least a dozen different species names. Many of the kinds previously considered to be different species are now often recognized as different varieties of *L. caerulea*. The plants of eastern North America have been called *L. villosa* (Michx.) Schult. var. *villosa* (when considered to be a different

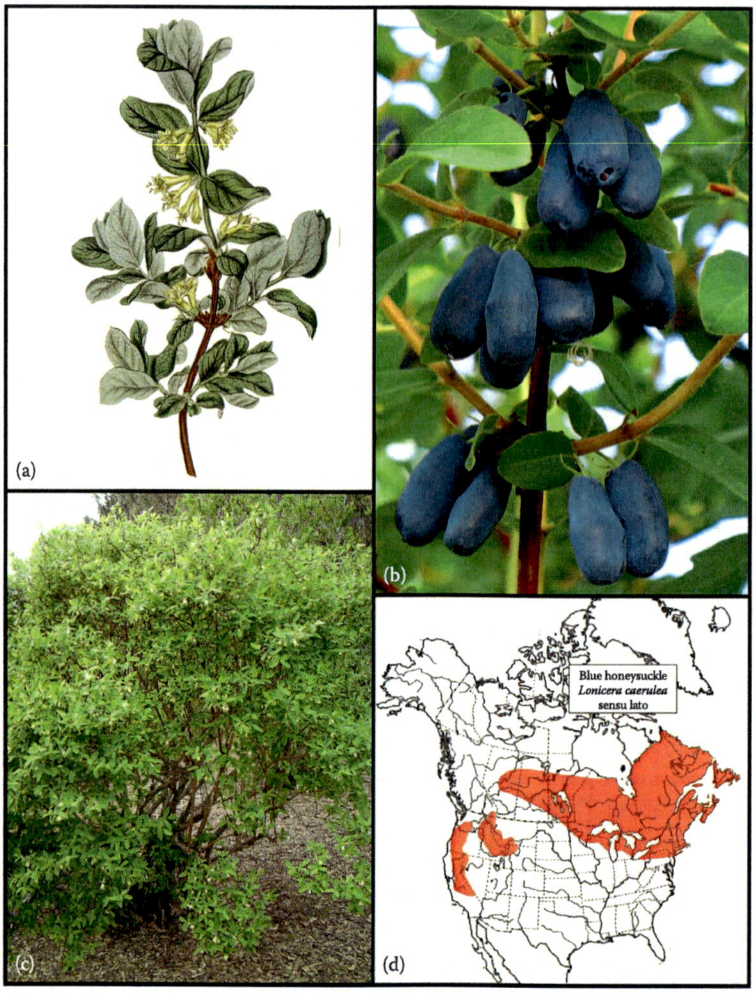


**FIGURE 13.1** Blue honeysuckle (*Lonicera caerulea*) (a) Flowering branch. (From Curtis, W., *Botanical Magazine*, Vol. 45, Plate 1965, 1818.) (b) Fruit of a cultivar. (Courtesy of K. Wiederholdt and North Dakota State University.) (c) Shrub cultivated in Roger Van den Hende Botanical Garden, Quebec. (Courtesy of Louis-M. Landry.) (d) Distribution map.

species) and *L. caerulea* L. var. *villosa* (Michx.) Torr. & A. Gray (when considered to be a variety of *L. caerulea*). The plants of western North America have been called *L. cauriana* Fernald (when considered to be a different species), and *L. caerulea* Schult. var. *cauriana* (Fernald) B. Boivin (when considered to be a variety of *L. caerulea*). The kind of blue honeysuckle that has been most important as the starting material for cultivated plants has been called *L. caerulea* var. *edulis* Turcz. ex Freyn), although other Russian varietal names have also been discussed. The Japanese variety, which has been used to generate cultivated plants, has been called *L. caerulea* var. *emphyllocalyx* (Maxim.) Nakai. A comprehensive taxonomic analysis of the species is required to determine an appropriate classification and correct names of variants within *L. caerulea*. For purposes of this chapter, *L. caerulea* is considered to be one species, including all of the variants mentioned earlier.

This species is a bush, generally 0.3–2 m (1–6.6 feet) tall. It has yellowish or yellowish-white, elongated flowers 8–16 mm (0.3–0.6 inch) long, the flowers occurring in pairs. The blue fruit is reminiscent of blueberries but is often cylindrical and elongated in Russian varieties. The berries may be 5–40 mm (0.2–1.6 inches) long, with a diameter of 5–15 mm (0.2–0.6 inch). Wild fruits are frequently the size of peas, cultivated fruits are often larger. The fruits ripen extremely early in the season. Fruits contain up to 20 small seeds, occasionally more.

Blue honeysuckle is a relatively modern crop. It has been harvested from the wild in regions of China, Japan, and Russia for hundreds of years. Breeding programs began in Russia in the 1950s, in Japan in the 1980s, and in the late 1990s at Oregon State University and the University of Saskatchewan. Many advanced cultivars have been selected in Eurasia and a few in North America. For the most part, cultivars have been created starting with wild material from Eurasia, but recently efforts are being made to also employ North American plants in breeding programs.

Many small berries are being touted as wonderful sources of vitamins and health-promoting constituents, and blue honeysuckle is no exception. The fruit is high in antioxidant chemicals, including ascorbic acid (vitamin C), anthocyanins, and phenolics. Antioxidants are thought to promote health by countering free radicals—substances resulting from metabolism that are harmful. Extracts of the fruit are also being examined for anti-inflammatory and antibacterial activities.

## CULINARY PORTRAIT

Of the 200 or so species of *Lonicera*, most have mildly poisonous berries. However, those of *L. caerulea* are edible. Wild forms and ornamental varieties of blue honeysuckle may have fruits that have a poor taste, and the culinary merit of the species should be judged only by reference to advanced fruit cultivars. The berries are sweet and can be eaten fresh. They can also be made into a wide variety of processed products, including jam, jelly, juice, wine, ice cream, yogurt, candies, sauces, and pastries. The taste varies somewhat among cultivars, and has been compared to blueberry, kiwi, and raspberry.

### Culinary Vocabulary

- "Honeysuckle wine" is a traditional wine prepared from honeysuckle flowers (not fruits). Honeysuckle flowers of various species have also sometimes been employed to flavor teas, vinegars, jams, and jellies. Whether the flowers of species known to have toxic berries also have toxic constituents is unclear. Recipes for honeysuckle wine often caution that only the petals should be used. Blue honeysuckle berries can be used to make an excellent wine, but this should not be confused with the traditional meaning of "honeysuckle wine," as the use of the berries of some species is likely to lead to poisoning.

## PROSPECTS

The blue honeysuckle is rather similar in appearance and taste to blueberry, which is well established in the marketplace, making it difficult for this new crop to become competitive. However, blue honeysuckle is extremely winter hardy and can grow in regions where blueberry cannot. Because of the early maturation of the berries, blue honeysuckle may find a market niche in the early season, before the small fruits that currently dominate the marketplace (such as strawberry, blueberry, and raspberry) become available. *Lonicera caerulea* is a quite attractive small fruit crop, and although the fruit industry is extremely competitive and difficult to penetrate, blue honeysuckle has very good potential.

## CURIOSITIES OF SCIENCE AND TECHNOLOGY

- As with many ornamental plants, superstitious beliefs are associated with honeysuckles. In Scotland, honeysuckle was usually considered to have protective properties. In the Scottish highlands, the presence of honeysuckle in a dairy on May 1 was believed to ward off evil influences from the cows and their milk and butter. It was also thought in Scotland that honeysuckle growing around the entrance to the home would prevent a witch from entering. In England, many believed that bringing a bouquet of honeysuckle into the house would result in prosperity. However, during Victorian times, teenage girls were forbidden to bring honeysuckle into the home because it was thought to induce erotic dreams. Such beliefs are usually harmless, and in the past when life often was uncertain, they provided people with reassurances that misfortune could be avoided.

- Nepetalactone, the chemical that makes catnip so attractive to cats, is also present in the wood of Tatarian honeysuckle (*L. tatarica* L.), shavings of which are often used in cat toys. This honeysuckle species has become established in many parts of North America and is weedy or invasive in some regions.

## KEY INFORMATION SOURCES

Anikina, E.V., Syrchina, A.I., Vereshchagin, A.L., Larin, M.F., and Semenov, A.A. 1989. Bitter iridoid glucoside from the fruit of *Lonicera caerulea*. *Chem. Nat. Compd.* 24: 512–513.

Chaovanalikit, A., Thompson, M.M., and Wrolstad, R.E. 2004. Characterization and quantification of anthocyanins and polyphenolics in blue honeysuckle (*Lonicera caerulea* L.). *J. Agric. Food Chem.* 52: 848–852.

Gasiewicz, E. 1999. Usefulness of honeysuckle fruits for juice production. In *Fruit and vegetable juices and drinks today and in the XXI century*. Edited by L. Michalczuk and W. Plocharski. Research Institute of Pomology and Floriculture, Skierniewice, Poland. pp. 251–260.

Heinrich, J., Svarcova, I., and Valentova, K. 2008. *Lonicera caerulea*: a prospective functional food and a source of biologically active compounds. *Chemicke Listy*. 102: 245–254. (In Czech, English summary)

Hummer, K. 2006. Blue honeysuckle: A new berry crop for North America. *J. Am. Pomol. Soc.* 60: 3–8.

Huo, J.-W. Yang, G.-H., Sui, W., and Yu, Z.-Y. 2005. Review of studies on blue honeysuckle (*Lonicera caerulea* L.) germplasm resources. *Acta Hort. Sinica* 32: 159–164. (In Chinese, English summary)

Khasbagan, Soyolt, and Geng X.-H. 2006. Analysis of nutritional components in dry fruits of wild plant *Lonicera caerulea* var. *edulis*. *J. Plant Resour. Environ.* 15: 77–78.

Leveau, A.M., Durand, M., and Paris, R.R. 1977. The toxicity of fruits of various *Lonicera* Caprifoliaceae. *Plantes Medicinales Phytotherapie* 11: 94–105.

Ochmian, I., Grajkowski, J., and Skupień, K. 2008. Field performance, fruit chemical composition and firmness under cold storage and simulated "shelf-life" conditions of three blue honeysuckle cultigens (*Lonicera caerulea*). *J. Fruit Ornam. Plant Res.* 16: 83–91.

Oszmianski, J., Kucharska, A., and Gasiewicz, E. 1999. Usefulness of honeysuckle fruits for juice production. In *Fruit and vegetable juices and drinks today and in the XXI century*. Edited by L. Michalczuk and W. Plocharski. Research Institute of Pomology and Floriculture, Rytro, Poland. pp. 251–260.

Palikova, I.H., Bednar, J., Marhol, P., Kren, P., Cvak, V., Valentova, L., Ruszicka, K., et al. 2008. Constituents and antimicrobial properties of blue honeysuckle: a novel source for phenolic antoxidants. *J. Agric. Food Chem.* 56: 11883–11889.

Plekhanova M.N. 1995. *Blue honeysuckle: a new berry crop in Russia*. Balsgaard Annual Report for 1992–1994, Department of Horticulture and Plant Breeding, Swedish University of Agricultural Sciences, Balsgaard, Sweden. pp. 180–182.

Plekhanova, M.N. 2000. Blue honeysuckle (Lonicera caerulea L.)—a new commercial berry crop for temperate climate: genetic resources and breeding. *Acta Hort.* 538: 159–164.

Rehder, A. 1909. Note on the morphology of the fruit of *Lonicera caerulea*. *Rhodora* 11: 209–211.

Skupień, K., Oszmiański, J., Ochmian I., and Grajkowski J. 2007. Characterization of selected physico-chemical features of blue honeysuckle fruit cultivar 'Zielona'. *Polish J. Nat. Sci. Suppl.* 4: 101–107.

Skupień, K., Oszmiański, J., Ochmian I., and Grajkowski J. 2009. Influence of ripening time on fruit chemical composition of two blue honeysuckle cultigens. *J. Fruit Ornam. Plant Res.* 17: 101–111.

Svarcova, I., Heinrich, J., and Valentova, K. 2007. Berry fruits as a source of biologically active compounds: the case of *Lonicera caerulea*. *Biomed. Pap. (Olomouc)* 151: 163–174.

Takada, M., Nakano, H., Hoshino, Y., and Sato, H. 2003. Evaluation of eating qualities and some horticultural characteristics for selection of elite lines in *Lonicera caerulea* L. *Research Bull. Hokkaido Univ. Farm* 33: 21–38. (In Japanese, English summary)

Tanaka, T., and Tanaka, A. 1998. Chemical composition and characteristics of hasukappu berries in various cultivars and strains. *J. Jap. Soc. Food Sci. and Tech.* 45: 129–133. (In Japanese, English summary)

Theis, N., Donoghue, M.J., and Jianhua Li, J. 2008. Phylogenetics of the Caprifolieae and *Lonicera* (Dipsacales) based on nuclear and chloroplast DNA sequences. *Syst. Bot.* 33: 776–783.
Thompson, M.M. 2006. Introducing haskap, Japanese blue honeysuckle. *J. Am. Pomol. Soc.* 60: 164–168.
Thompson, M.M., and Barney, D.L. 2007. Evaluation and breeding of haskap in North America. *J. Am. Pomol. Soc.* 61: 25–32.
Thompson, M., and Chaovanalikit, A. 2003. Preliminary observations on adaptation and nutraceutical values of blue honeysuckle (*Lonicera caerulea*) in Oregon, USA. *Acta Hort.* 626: 65–72.
Vereschchagin, A.L., Anikina, E.V., Syrchina, A.I., Lapin, M.F., Azin, L.A., and Semenov, A.A. 1989. Chemical investigations of the bitter substances of the fruit of *Lonicera caerulea. Chem-Nat-Compd.* (Consultants Bureau, New York), 25: 289–292. (Translated from Russian from *Khimiia Prirodnykh soedinenii* 25: 338–342.)

## Specialty Cookbooks

Blue honeysuckle is basically an experimental crop at present in North America, and recipes have not been compiled. Recipes for blueberries (see Chapter 15) may be used.

# 14 Blue Waxweed

Family: Lythraceae (loosestrife family)

## NAMES

Scientific name: *Cuphea viscosissima* Jacq. (*C. petiolata* (L.) Koehne)

- The term "waxweed" is applied to *Cuphea* species and reflects the fatty oil obtainable from some of them ("wax" often means hard, ungreasy fat). The "weed" in the name indicates the weedy nature of some of the species. "Blue" in the name is a mischaracterization, as the flower color is generally purplish.
- Blue waxweed is also known as clammy cuphea, fluxweed (a name given to various species that have been used to treat dysentery; one of the medical meaning of flux is material discharged from the bowels), and tarweed (tar-weed).
- The genus name *Cuphea* is based on the Greek *kyphos*, bent, curved, or humped, referring to the curved calyx, which is often slightly swollen on its upper side at the base.
- In agronomic journals, the genus name *Cuphea* is sometimes transformed to "cuphea" as a generic term indicating any of the several species or hybrids being grown for oil harvest.
- *Viscosissima* in the scientific name is based on Latin *viscos*, sticky + *issim*, the most, pointing out the extreme stickiness of the plant. This stickiness is also the basis for the names clammy cuphea and tarweed.

## GEOGRAPHY AND ECOLOGY OF WILD PLANTS

*Cuphea viscosissima* is a native of east central and southeastern North America and is the most widely distributed species of *Cuphea* north of Mexico. The plant is found along roadsides and in fields, usually in dry places. Although the species normally occurs in open locations, it occasionally occurs in low woods.

## PLANT PORTRAIT

Blue waxweed, one of the 260 or so species of *Cuphea*, is an annual, 15–60 cm (6–24 inches) tall. The plant is herbaceous, but becomes woody at the base in older plants. The stems commonly turn deep purple-red. The leaves are arranged in pairs on opposite sides of the stem, and one or two, pink to purple flowers, 8–12 mm (0.3–0.5 inch) long, arise from between the leaf stalks of a pair of leaves. The stems, leaves, and calyxes are covered with sticky, glandular hairs. The species is sometimes grown as an ornamental but is mainly of interest as a very recent source of oil from the seeds.

Up to a third of the weight of the seeds is oil, with a high percentage of "medium-chain fatty acids," especially caprylic, capric, and lauric acids. More than two-thirds of the oil is capric acid, which is used primarily in the food industry. Capric acid is also used in feed stocks and for perfumes, lubricants, and cosmetics. Compounds (triglycerides) of caprylic and capric acids are employed in specialized clinical diets to alleviate certain digestive disorders. Lauric acid is used primarily in the manufacture of soaps and detergents; it is more important for industrial purposes, but normally occurs in the species in amounts too small for commercial use. Commercial production has focused on an interspecific hybrid population derived from the Mexican *C. lanceolata* W.T. Aiton and *C. viscosissima*.

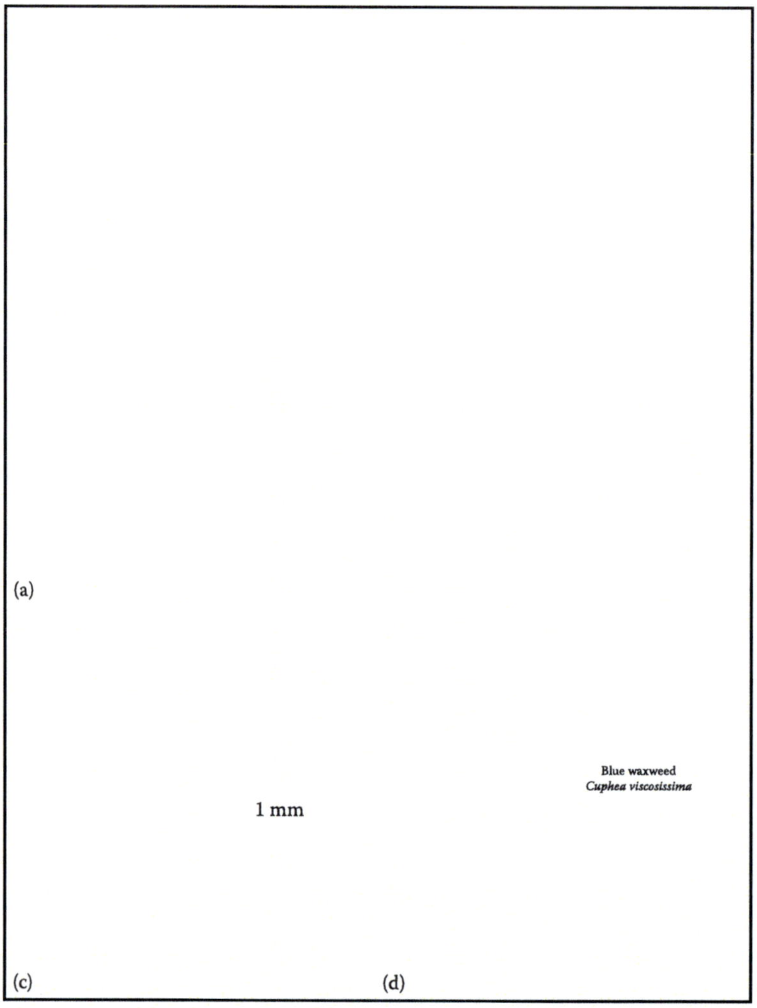


**FIGURE 14.1** Blue waxweed (*Cuphea viscosissima*). (a) Flowering plant. (From Jacquin, N.J., *Icones plantarum rariorum*, B. White & Filum, London, U.K., Vol. 2, Plate 177, 1781–1786.) (b) Flowering branch. (Courtesy of K. Weller, U.S. Department of Agriculture, online at Bugwood.org [CC By 3.0].) (c) Seeds. (Courtesy of Jose Hernandez, U.S. Department of Agriculture.) (d) Distribution map.

## CULINARY PORTRAIT

Blue waxweed is significant as an experimental source of edible oil for the marketplace. The oil is incorporated into commercial products prepared on an industrial scale and has not yet been made available to the general consumer.

### Culinary Vocabulary

- The distinction of "saturated fats" and "unsaturated fats" has become a critical nutritional issue in recent times because the latter is considered to be much healthier. Saturated fat becomes solid at low temperatures. Polyunsaturated fat remains as an oil in liquid form even at low temperatures. Unsaturated fat more easily turns rancid, especially when exposed to heat, light, and oxygen. The stability and permanent liquid nature of saturated fats, and their generally lower cost, led to their preferential use in the food industry for many years. Unfortunately, saturated fats are the biggest dietary cause of high levels of LDL (low-density lipoprotein; "bad cholesterol"). Saturated fats are found in animal products such as butter, cheese, whole milk, and fatty meats. They are also present in some vegetable oils, notably coconut palm and oilseed palm ("palm kernel") oils (the palm oil industry has waged a vigorous campaign arguing that their oils are beneficial). Most other vegetable oils contain unsaturated fat and are healthy. Unfortunately, capric acid, the chief component of blue waxweed oil, is saturated.

## PROSPECTS

Coconut (*Cocos nucifera* L.) and oil palm (*Elaeis guineensis* Jacq.) oil are the principal sources of medium-chain fatty acids. Extensive efforts have been made during the last three decades to develop selected *Cuphea* species as an alternative source of these compounds for industrial and food uses. As well, breeding may result in cultivars with additional economically desirable properties. Other species of *Cuphea* are also potential oil sources, but grow in and are adapted to the tropics of South America. *Cuphea viscosissima*, by contrast, is adapted to North American conditions. As with numerous other wild species, there are difficulties that need to be overcome to make blue waxweed a productive crop. The uneven ripening of seeds is probably the chief problem. The U.S. government has invested heavily in blue waxweed development, mainly to establish a domestic alternative to imported lauric acid from coconut and oil palm for use in detergents, soaps, cosmetics, lubricants, and biodiesel. Although the food use of the plant has not been the primary objective, the edible product industry associated with blue waxweed is also likely to be established.

## CURIOSITIES OF SCIENCE AND TECHNOLOGY

- Numerous wild species such as blue waxweed have two principal seed deficiencies from the point of view of agriculture. The seeds "shatter" (fall off the plant irregularly, so that efficient collection is not possible) and they are dormant (germinate irregularly, so that plants of uniform age cannot be generated). "Domestication" of many species that are valuable for their seeds particularly involved selection of nonshattering and nondormant forms. Indeed, the discovery of nonshattering populations of seed crops was responsible for the shift from hunting and gathering to agrarian cultures, and the establishment of civilization (see Chapter 1). Frequently, single gene mutations are all that is required to change a wild species of very limited usefulness to one that can be a staple crop. There has been appreciable selection to date, and "semi-domesticated" oilseed forms of *C. viscosissima* are available.
- The sticky hairs of *Cuphea* species almost certainly represent a defense against insects. The plants are frequently covered with dead insects that were trapped on the hairs. It has been suggested that the plants may benefit when the insects disintegrate, producing nitrogen fertilizer that the plants perhaps may use, in the manner of insectivorous plants. However, there are numerous plant species that have sticky hairs and it seems clear that most of these are not benefiting from the release of elements from insects that become trapped.

## KEY INFORMATION SOURCES

Amarasinghe, V., Graham, S.A., and Graham, A. 1991. Trichome morphology in the genus *Cuphea* (Lythraceae). *Bot. Gaz.* 152: 77–90.

Barber, J.C., Ghebretinsae, A., and Graham, S.A. 2010. An expanded phylogeny of *Cuphea* (Lythraceae) and a North American monophyly. *Plant Syst. Evol.* 289: 35–44.

Ben-Salah, H., and Roath, W.W. 1994. Somaclonal variation in *Cuphea viscosissima* Jacq. for plant improvement. *Indust. Crops Prod.* 2: 239–244.
Berti, M.T., and Johnson, B.L. 2008. Growth and development of *Cuphea. Indust. Crops Prod.* 27: 265–271.
Berti, M.T., and Johnson, B.L. 2008. Physiological changes during seed development of *Cuphea. Field Crops Res.* 106: 163–170.
Geller, D.P., Goodrum, J.W., and Knapp, S.J. 1999. Fuel properties of oil from genetically altered *Cuphea viscosissima. Indust. Crops Prod.* 9: 85–91.
Gesch, R.W., Cermak, S.C., Isbell, T.A., and Forcella, F. 2005. Seed yield and oil content of *Cuphea* as affected by harvest date. *Agron. J.* 97: 817–822.
Graham, S.A. 1988. Revision of *Cuphea* section *Heterodon* (Lythraceae). *Syst. Bot. Mono.* 20:1–168.
Graham, S.A. 1989. *Cuphea*: a new plant source of medium chain fatty acids. *Crit. Rev. Food Sci. Nutr.* 28: 139–173.
Graham, S.A., Freudenstein, J.V., and Luker, M. 2006. A phylogenetic study of *Cuphea* (Lythraceae) based on morphology and nuclear rDNA ITS sequences. *Syst. Bot.* 31: 764–778.
Gesch, R.W., Archer, D.W., and Forcella, F. 2010. Rotational effects of *Cuphea* on corn, spring wheat, and soybean. *Agron. J.* 102: 145–153.
Hirsinger, F. 1985. Agronomic potential and seed composition of *Cuphea*, an annual crop for lauric and capric seed oils. *J. Am. Oil Chem. Soc.* 62: 76–80.
Hirsinger, F., and Knowles, P.F. 1984. Morphological and agronomic description of selected *Cuphea* germplasm. *Econ. Bot.* 38: 439–451.
Knapp, S.J. 1993. Breakthroughs toward domestication of *Cuphea.* In *New crops*. Edited by J. Janick and J.E. Simon. Wiley and Sons, New York. pp. 372–379.
Knapp, S.J. 1993. Modifying the seed oils of *Cuphea*–a new commercial source of caprylic, capric, lauric, and myristic acid. In *Seed oils for the future*. Edited by S. MacKenzie and D. Taylor. American Oil Chemists Society, Monoarauh, Champaign, IL. pp. 142–154.
Knapp, S.J., Tagliani, L.A., and Roath, W.W. 1991. Fatty acid and oil diversity of *Cuphea viscosissima*: a source of medium-chain fatty acids. *J. Am. Oil Chem. Soc.* 68: 515–517.
Knapp, S.J., Crane, J.M., Tagliani, L.A., and Slabaugh, M.B. 1997. *Cuphea viscosissima* mutants with decreased capric acid. *Crop Sci.* 37: 352–357.
Roath, W.W., and Widrlechner, M.P. 1988. Inducing germination of dormant *Cuphea* seed and the effects of various induction methods on seedling survival. *Seed Sci. Technol.* 16: 699–703.
Roath, W.W., Widrlechner, M.P., and Kirkbride, J.H. 1993. Collecting *Cuphea* in Brazil, Mexico, and the United States. *FAO/IBPGR Plant Genetic Resour. Newsl.* 93: 29–33.
Roath, W.W., Widrlechner, M.P., and Kleiman, R. 1992. Morphological and agronomic variability in *Cuphea viscosissima* Jacq. *Indust. Crops Prod.* 1: 5–10.
Roath, W.W., Widrlechner, M.P., and Kleiman, R. 1994. Variability in *Cuphea viscosissima* Jacq. collected in east-central United States. *Indust. Crops Prod.* 3: 217–223.
Thompson. A.E. 1981. *Cuphea* a potential new crop. *Hort. Sci.* 19: 352–354.
Wolf. R.B., Graham. S.A., and Kleiman. R. 1983. Fatty acid composition of *Cuphea* seed oils. *J. Am. Oil Chem. Soc.* 60: 27–28.

## Specialty Cookbooks

Recipes for blue waxweed are not available.

# 15 Blueberries

Family: Ericaceae (heath family)

## NAMES

Scientific names: *Vaccinium* species

- Highbush blueberry—*V. corymbosum* L.
- Lowbush blueberry—*V. angustifolium* Aiton
- The English name blueberry is named for the blue (-black) color of the fruit, and originates from the Scottish "blaeberry," used to denote European species. The "blue" or "blae" traces to the Old High German word for blue, *blao*.
- Blueberries are known by many names, principally because early settlers noted that they resembled the European berries they knew as blaeberries, hurtleberries, trackleberries, whinberries, and whortleberries (the Scots associated them with the blaeberry, the Irish with whortleberries, the Danes with bilberries). "Whortleberry" is a dialectal variant of "hurtleberry," derived from Middle English *hurtilberi*, and is perhaps based on *hurt*, an azure-colored ball (from Old French *heurte*) + *berye*, berry (see the third following bullet entry for additional information, and see Chapter 10 for an alternative interpretation). Other interesting names for blueberry species include deerberries, farkleberries, southern gooseberries, and sparkleberries.
- The highbush blueberry is also known as swamp blueberry and whortleberry.
- The rabbiteye blueberry is a very vigorous form of the highbush blueberry, cultivated in the southeastern United States, especially Florida. It is often recognized as a separate species, *V. ashei* Reade.
- The lowbush blueberry is also known as late sweet blueberry, low sweet blueberry, and sweet-hurts. ("Hurts" is from the archaic West-Country (English) verb "to-go-a hurting," i.e., collecting berries; "hurts" and "hurtleberries" are used in parts of England to designate the bilberry, *V. myrtillus* L.)
- The genus name *Vaccinium* is the classical Latin name for the blueberry. It is presumed to derive from the Latin *vaccinus*, of cows, since cows seem to love the berries, a fact that was first recorded by English navigator and explorer Captain James Cook (1728–1779). The genus name has also been interpreted as originating from *bacca*, berry, in allusion to the numerous berries.
- *Corymbosum* in the scientific name *V. corymbosum* is based on a Latin term meaning bunch or cluster, and refers to the nature of the flower arrangement.
- *Angustifolium* in the scientific name *V. angustifolium* is Latin for narrow-leaved, in allusion to the characteristically narrow leaf.

## GEOGRAPHY AND ECOLOGY OF WILD PLANTS

The highbush blueberry is native in North America from eastern Canada to Florida: from northeastern Illinois, northern Indiana, and south central Michigan, northward along the St. Lawrence Valley to Quebec City, then east to southwestern Nova Scotia, south to Florida, and west to northeastern

**FIGURE 15.1** Highbush blueberry (*Vaccinium corymbosum*). (a) Clump of berries. (Courtesy of Opiola Jerzy [CC By 3.0].) (b) Bowl of berries. (Courtesy of Steve Hillebrand, U.S. Fish and Wildlife Service.) (c) Flowering branch. (From Curtis, W., *Botanical Magazine*, Vol. 62, Plate 3433, 1835.) (d) Distribution map.

Texas and nearby Oklahoma. It occurs in open swamps and bogs and along sandy margins of lakes, ponds, and streams, and occasionally in woods and abandoned farmland.

The lowbush blueberry is native to eastern Canada and the northeastern United States: from Labrador and Newfoundland, west to southern Manitoba and Minnesota, south to northern Illinois, Pennsylvania, and Delaware, and in the mountains to Virginia and West Virginia. It occurs in dry sandy areas, peaty barrens, rocky outcroppings, and woods, always in quite acidic soils.

## PLANT PORTRAIT

There are dozens of species of the genus *Vaccinium* in North America, and many more in other continents, and many of these are known as "blueberries" (some are known as "huckleberries," and are treated in a separate chapter). All are edible and are collected from the wild locally, but only a few have reached the status of commercial importance. The highbush and lowbush blueberries are, by far, the most important blueberries in North America, and so are featured in this chapter. Blueberries are the second most popular berry in the United States (strawberries are first.)

**FIGURE 15.2** Lowbush blueberry (*Vaccinium angustifolium*). (a) Flowering shrub. (Courtesy of Thomas Kent, FloraFinder.com. With permission.) (b) Fruiting shrub. (Courtesy of J. Dumais [CC By 3.0].) (c) A "red sea" of shrubs in autumn coloration in Nova Scotia. (Courtesy of Dennis Jarvis [CC By 2.0].) (d) A blueberry harvester. (Courtesy of Dr. Wilson [CC By 3.0].) (e) 'Chanticleer', a hybrid cultivar with some lowbush blueberry parentage. (Courtesy of U.S. Department of Agriculture.) (f) Distribution map.

The highbush blueberry is the most important of all the blueberries. It is a very variable shrub. The plants range between 1 and 7 m (3–23 feet) in height, but usually are 2–3 m (6–10 feet) tall. The berry is dull black or blue, between 4–18 mm (0.2–0.7 inch) in diameter, the larger size found in dozens of cultivated varieties. The highbush blueberry was not domesticated until the early twentieth century. It is especially cultivated in the United States, Canada, Europe, Australia, New Zealand, and Chile. Many highbush varieties are the result of hybridization with the lowbush blueberry. There is much variation in taste.

The lowbush blueberry is a depressed shrub, varying from 20–90 cm (8–36 inches) in height, and is generally about 30 cm (1 foot) tall. The bluish-black berries are 5–13 mm (0.2–0.5 inch) in diameter and are sweet. The lowbush blueberry is mostly managed in wild stands, rather than being planted as a horticultural crop like the highbush blueberry. Native Americans in Maine apparently realized that the wild blueberry does not bear fruit well when overgrown with brush and weeds, and burned the area to encourage vigorous regrowth. Although the lowbush blueberry is still not planted today, the plants are managed much like garden plants—they are fertilized and pests are controlled. Most lowbush blueberries are produced in Maine and eastern Canada (especially in Nova Scotia). Only a few cultivated varieties are available.

## CULINARY PORTRAIT

Blueberries ship well and are a favorite fresh dessert eaten alone, topped with cream or liqueurs, mixed in fruits salads and cereals, and used as a topping for waffles and crêpes. As well, syrup, jams, jellies, ice cream, tarts, pies, muffins, cheesecakes, and dozens of other blueberry confections and breakfast pastries are widely enjoyed. To prevent bleeding of the color into other ingredients during cooking, blueberries should be added at the last minute, and stirred gently, to keep the skins from breaking. More than half of the crop is processed by quick freezing. The fruit may also be canned and freeze-dried, and a pleasant wine can be manufactured. Blueberries are fragile and should be handled carefully. When purchased, they should be consumed promptly. If stored in a refrigerator, where they will keep for a few days, damaged berries should first be removed to prevent the spread of mold.

### Culinary Vocabulary

- Fungy (fungee) is a deep dish blueberry pie, popular for example in Nova Scotia.
- A "grunt" (also known as a slump) is a colonial American dessert, typically prepared with fresh fruit topped with biscuit dough and steamed in a closed container. Blueberry grunt is a specialty of Nova Scotia and adjacent regions, where wild blueberries grow.

## PROSPECTS

Blueberries are esteemed, well-established fruits in great demand for both fresh and prepared foods. Considering that the domestication of blueberries began only about a century ago, there has been remarkable progress in the improvement and popularization of this crop. In recent times, blueberries have acquired a considerable reputation for healthful properties because of the content of antioxidant constituents. Blueberries are a seasonal crop with a relatively short shelf life, but international trade has extended the season, the imported fruit often selling at premium prices. Reminiscent of the rather tasteless tomatoes that have been bred to withstand shipping well, highbush blueberries that are impressively large but with limited taste have appeared on the market—an unwelcome development that could harm the reputation of the fruit. Nevertheless, it seems likely that blueberries will continue to increase in popularity. Wild blueberries remain exceptionally tasty, and their widespread collection and sale represent the most profitable wildcrafted food crop in North America. The future of the wild crop is dependent on wise management of the natural landscape.

## CURIOSITIES OF SCIENCE AND TECHNOLOGY

- North American Indians preserved blueberries in various ways. In northern areas, the Inuit placed the berries in seal oil, or stored them in leather bags deposited in the permafrost. The berries were also dried in the sunshine or by a fire.
- Native North Americans taught European settlers how to prepare various blueberry dishes. In 1615, the French explorer Champlain observed Native Americans gathering wild blueberries to make a dish called "Sautauthig." The native people would dry the wild blueberries and beat them into a powder before adding it to parched meat. American explorers, Lewis and Clark, during their trip to the northwestern United States (see Chapter 71 on Pawpaw for details), witnessed Native Americans smoke-drying wild blueberries to use in soups and stews. Venison was cured and flavored by pounding blueberries into the meat and smoke-drying it.

- Native American people used blueberries as a source of dye for coloring clothing and basketry.
- Early American colonists made grey paint by boiling blueberries in milk.
- Lowbush blueberries are harvested in one picking operation. They can be gathered by hand raking, using a metal rake that was invented by Abijah Tabbutt of Maine in 1822. This looks like a dust pan, the bottom of which is made up of many close-set knitting needles. Most wild blueberries were handpicked using this device until recently, but mechanical harvesters are now becoming popular.
- The color pigments of blueberries are chemicals called anthocyanins, and these include antioxidants. Antioxidants are needed by the body to fight compounds called free radicals that are formed as a byproduct of breathing, digesting, and exercising. Free radicals are believed to be harmful, increasing the risk of cancer, heart disease, other diseases, and premature aging. Blueberries were found to have the highest level of antioxidants in a study of 40 fruits and vegetables by the U.S. Department of Agriculture. Blueberries are very strongly recommended by nutrition experts.
- In Alaska, red-backed voles (*Clethrionomys rutilus*) are so fond of blueberries that most of them have blue teeth when the berries are ripe.
- Blueberries are widely used as official political emblems. The "Nova Scotia wild blueberry" (presumably *V. angustifolium*) became Nova Scotia's "Provincial Berry" in 1996. In 2001, the blueberry (no particular species) became the official State Blue Berry of North Carolina (its official State Red Berry is the strawberry; red and blue are the official state colors of North Carolina). Highbush blueberry was declared to be the state fruit of New Jersey in 2004. The blueberry muffin was made Minnesota's "State Muffin" in 1988. The "wild blueberry" (interpreted as *V. angustifolium*) was declared to be Maine's "State Berry" in 1991. Blueberry pie made with wild Maine blueberries was declared to be the "state dessert" of Maine in 2011. The province of Benguet of The Philippines adopted the "native highbush blueberry" (identified by the local names ayosi and alumani) as its Official Provincial Fruit in 1987. Web documents produced by the government of Benguet identify the species by the names highbush blueberry (*V. corymbosum*) and mountain blueberry (*V. membranaceum*), species that are not native to The Philippines. The species appears to be *V. myrtoides* (Blume) Miq. (personal communication, D.A. Madulid, Philippines National Herbarium).
- Blueberries will not ripen after they are picked.
- About 85% of a blueberry is water.

## KEY INFORMATION SOURCES

(For additional references, see Chapter 31.)

Austin, M.E. 1994. *Rabbiteye blueberries: development, production, and marketing*. Agscience, Auburndale, FL. pp. 160.

Barker, W.G., Hall, I.V., Aalders, L.E., and Wood, G.W. 1964. The lowbush blueberry industry in eastern Canada. *Econ. Bot.* 18: 357–365.

Bertelsen, D.R., Harwood, J.L., and Zepp, G.A. 1995. *The U.S. blueberry industry*. U.S. Department of Agriculture, Economic Research Service. Commercial Agriculture Division, Washington, DC. pp. 39.

Camp, W.H. 1945. The North American blueberries with notes on other groups of Vacciniaceae. *Brittonia* 5: 203–275.

Caruso, F.L., and Ramsdell, D.C. (Editors). *Compendium of blueberry and cranberry diseases*. American Phytopathological Society Press, St. Paul, MN. pp. 87.

Cuddy, J.D., and Cuddy, T.M. 2000. *Growing blueberries: a guide for the serious gardener & commercial grower*. Second edition. Rush River Publications, Maiden Rock, WI. pp. 139.

Darrow, G.M. 1957. *Blueberry growing*. U.S. Government Printing Office, Washington, DC. pp. 33.

Eck, P. 1986. Blueberry. In *CRC handbook of fruit set and development*. Edited by S.P. Monselise. CRC Press, Boca Raton, FL. pp. 75–85.

Eck, P. 1988. *Blueberry science*. Rutgers University Press, New Brunswick, NY. pp. 284.

Eck, P., and Childers, N.F. (Editors). 1966. *Blueberry culture*. Rutgers University Press, New Brunswick, NY. pp. 378.
Forney, C.F., and Eaton, L.J. (Editors). 2004. *Proceedings of the ninth North American blueberry research and extension workers conference (Halifax, Nova Scotia)*. Food Products Press, Binghamton, NY. pp. 452.
Gough, R.E. 1994. *The highbush blueberry and its management*. Food Products Press, New York. pp. 272.
Gough, R.E., and Korcak, R.F. (Editors). 1995. *Blueberries: A century of research*. Food Products Press, Binghamton, NY. pp. 245.
Hall, I.V. 1979. *Lowbush blueberry production*. Canada Department of Agriculture (Public. 1477), Ottawa, ON. pp. 39.
Hancock, J.F. 1995. Blueberry, cranberry, etc. In *Evolution of crop plants*. Second edition. Edited by J. Smartt and N.W. Simmonds. Longman Scientific & Technical, Burnt Mill, Harlow, Essex, U.K. pp. 121–123.
Hutchinson, U. 1995. *Blueberries*. Amana Publications, Beltsville, MD. pp. 39.
ICON Health Publications. 2003. *Blueberries: A medical dictionary, bibliography, and annotated research guide to Internet references*. ICON Health Publications, San Diego, CA. pp. 112.
Kevan, P.G. 1977. Blueberry crops in Nova Scotia and New Brunswick: pesticides and crop reductions. *Can. J. Agric. Econ.* 25: 61–64.
Luby, J.L., Ballington, J.R., Draper, A.D., Pliszka, K., and Austin, M.E. 1990. Blueberries and cranberries (*Vaccinium*). In *Genetic resources of temperate fruit and nut crops*. Edited by J.N. Moore and J.R. Ballington, Jr. International Society for Horticultural Science, Wageningen, the Netherlands. pp. 393–456.
Lyrene, P.M., and Sherman, W.B. 1979. The rabbiteye blueberry industry in Florida—1887 to 1980—with notes on the current status of abandoned plantations. *Econ. Bot.* 33: 273–243.
Maine Agricultural Experiment Station. 1950. *Producing blueberries in Maine. Bull. 479*. Maine Agricultural Experiment Station, Orono, ME. pp. 42.
Moore, J.N. 1994. The blueberry industry of North America. *HortTechnology* 4: 96–102.
Shutak, V.G., and Gough, R.E. 1982. *Grow the best blueberries*. Garden Way, Charlotte, VT. pp. 32.
Trehane, J. 2004. *Blueberries, cranberries, and other vacciniums*. Timber Press, Portland, OR. pp. 256.
Vander Kloet, S.P. 1980. The taxonomy of the highbush blueberry, *Vaccinium corymbosum*. *Can. J. Bot.* 58: 1187–1201.
Vander Kloet, S.P. 1988. *The genus Vaccinium in North America*. Research Branch, Agriculture Canada, Ottawa, ON. pp. 201.
Wild Blueberry Commission of Maine. 2002. *Maine wild blueberries, America's native berry*. Wild Blueberry Commission of Maine, Orone, ME. pp. 62.

## Specialty Cookbooks

(*Note:* there are at least a hundred cookbooks dedicated to blueberries; the following is a selection.)

Barton, E.W. 1974. *The complete blueberry cookbook*. Phoenix Publishing, Canaan, NH. pp. 156.
Buszek, B.R. 1980. *The blueberry connection: blueberry cookery with flavour, fact, and folklore, from memories, libraries, and kitchens of old and new friends and strangers*. Second edition. S. Greene Press, Brattleboro, VT. pp. 209.
Elliot, E., and Lee, V. 2002. *Blueberries: 40 recipes for fine dining at home*. Formac, Halifax, NS. pp. 64.
Hillcrest Publications. 1986. *Newfoundland's favourite blueberry recipes*. Hillcrest Publication, St. John's, NL. pp. 64.
Matthews, K. 1997. *Great blueberry recipes*. Storey Publishing, Pownal, VT. pp. 32.
Millang, T. 2003. *The joy of blueberries cookbook*. Adventure Publications, Cambridge, MN. pp. 224.
Morrison, C.D. 1972. *The country art of blueberry cookery: 185 mouth-watering recipes*. Publisher not identified. Middletown, NY. pp. 119.
Periwinkle Publications. 1980s–1990s? *The Maine wild blueberry cookbook: a collection of fine recipes and history of the Maine wild blueberry*. Periwinkle Publications, Bar Harbor, ME. pp. 128.
Schmidt, R.M. 1993. *The best of blueberry cookery*. Pine Barrens Press, Barnegat Light, NJ. pp. 76.
Siegrist, J. 1988. *Blueberry sampler: a collection of fresh blueberry recipes*. New England Press, Shelburne, VT. pp. 48.
Walker, F. 1992. *Basically blue: a collection of blueberry recipes*. Creative Bound, Carp, ON. pp. 76.

Freitus and Habeman (2005) have 11 wild blueberry recipes, Krumm (1991) has 9, Mogelon (2001) has 4, and Stanek and Butcher (2007) have 29. (See the Appendix to this book for details of the publications cited.)

# 16 Buffaloberries

Family: Elaeagnaceae (oleaster family)

## NAMES

Scientific names: *Shepherdia* species

- Canadian buffaloberry—*S. canadensis* (L.) Nutt.
- Silver buffaloberry—*S. argentea* (Pursh) Nutt.
- The vernacular name "buffaloberry" has occasionally been explained as a reflection of the fondness of the plains bison (buffalo) for the berries. In fact, the name reflects the practices of Plains Indians and pioneers of making a sauce from the berries for bison meat and also of using the berries in pemmican made with bison meat and the fruit.
- Buffaloberry is often spelled "buffalo berry" and "buffalo-berry."
- The name buffaloberry has also been applied to *Solanum rostratum* Dunal, a weed native to North America.
- The Canadian buffaloberry has also been called bitter buffaloberry, Canada buffaloberry, and russet (red) buffaloberry.
- Another popular name for the Canada buffaloberry is soapberry or occasionally foamberry. The names soopolallie and soapolallie simply mean soap berry in Chinook jargon, the Northwest U.S. trade language. The name soapberry indicates the soapy feel of the berries and the content of soapy chemicals called saponins.
- The silver buffaloberry is so named because of the silvery appearance of the leaves and young twigs.
- Other names for the silver buffaloberry are beef suet tree, bull berry (so named for the bulls of the buffalo), Nebraska currant (the flavor has been said to be reminiscent of the currant), rabbit berry, silver berry, silverleaf, sour buffaloberry, thorny buffaloberry (named for the thorns), and wild oleaster.
- The Stoney Indians of Alberta used the phrase "slave berry" for the Canadian buffaloberry. This was because the berries were thought to be used in great quantities by the Piegan Blackfoot women in southern Alberta. The Stoney believed that the Piegan women were ill-treated by their husbands and so called them slave women.
- The genus *Shepherdia* was named for British botanist John Shepherd (1764–1836), a curator of the Liverpool Botanic Garden.
- *Canadensis* in the scientific name *S. canadensis* means "of Canada." Carl Linnaeus (1707–1778) was the first to recognize the species by using the term *canadensis*; at that time "Canada" was understood to include parts of the northeastern United States, where Linnaeus knew the species occurred.
- *Argentea* in the scientific name *S. argentea* means "silvery."

## GEOGRAPHY AND ECOLOGY OF WILD PLANTS

Canadian buffaloberry reaches its northern limits north of the Arctic Circle and is found in all provinces of Canada except Prince Edward Island. It occurs from Newfoundland to Alaska and follows the Rocky Mountains south to Arizona and New Mexico. Canadian buffaloberry is rare in Maine

**FIGURE 16.1** Canadian buffaloberry (*Shepherdia canadensis*). (a) Fruiting branch. (Courtesy of Knottyboy/Wayne Noffsinger [Flickr/CC-attribution].) (b) "Indian ice cream." (Courtesy of Matyáš Havel [CC By 3.0].) (c) Distribution map.

and Illinois, where it has "endangered" status. It has been reported at elevations up to 1600 m (5000 feet) in Alberta and 2500 m (8000 feet) in Idaho. Canadian buffaloberry grows in the understory of conifer and aspen forests and in grasslands and shrub associations. The species is generally found on sandy, gravelly, or rocky soils. It grows well on shores, riverbanks, dry slopes, rocky woods, and occasionally in calcareous marshes. Although it can develop well in moist sites, it often thrives in very dry circumstances. The species is considered very desirable for revegetation of disturbed sites, since it is a native plant, provides food and cover for wildlife, and its nitrogen-fixing ability allows it to grow in poor soils while improving them. It is used by wildlife managers for watershed management and habitat improvement.

Silver buffaloberry is native in Canada in the prairie and southern parklands of the Prairie Provinces. In Canada, it is most abundant in southwestern Saskatchewan and southeastern Alberta. In the United States, it is best developed in the Great Plains, extending east to Minnesota and Iowa, and

**FIGURE 16.2** Silver buffaloberry (*Shepherdia argentea*). (a) Fruiting branch. (Courtesy of Julia Adamson [http://en.wikipedia.org/wiki/User: SriMesh], photographer in the Saskatoon area [CC By 3.0].) (b) Cluster of fruit. (Courtesy of Louis-M. Landry, reproduced with permission.) (c) Distribution map.

south to California, Arizona, New Mexico, and Oklahoma. Silver buffaloberry particularly occupies sandy soils that are often poorly drained and frequently somewhat calcareous. Although it is tolerant of poorly drained soils and some flooding, it is intolerant of prolonged flooding and permanently high water tables. The species is found in low-lying somewhat marshy areas, wet meadows, floodplains, and near streams, rivers, lakes, springs, and ponds. It is usually in open areas but will tolerate some shading. Silver buffaloberry tends to be a member of more or less dense shrub communities. Vegetative reproduction from the rootstocks and roots is pronounced in silver buffaloberry, and this commonly produces thickets, all of the "plants" representing the same clone. As with Canadian buffaloberry, silver buffaloberry is browsed by deer and elk, and the berries are consumed by bears and birds (especially grouse), and it is also a useful plant for soil stabilization and erosion control.

## PLANT PORTRAIT

Buffaloberry plants are shrubs or small trees with separate male and female plants. Canadian buffaloberry is a spreading shrub between 0.5 and 4 m (20 inches–13 feet) in height, whereas silver buffaloberry is a shrub or small tree 1–6 m (3–20 feet) high, with stems that may exceed 10 cm (4 inches) in diameter. The fruits are yellowish-red, reddish orange, or bright red when ripe, and have an insipid, sour, or bitter taste. Silver buffaloberry has dangerous large thorns unlike Canadian buffaloberry. Native peoples of North America made extensive use of buffaloberries. The berries were often pressed into cakes that were smoked and eaten. They were also a component of pemmican, a combination of buffalo meat and berries that was a staple food. The berries are very rich in vitamin C (ascorbic acid) and contributed to keeping Indians and pioneers healthy.

## CULINARY PORTRAIT

It is almost impossible to find buffaloberries or buffaloberry jams in stores, but both Canadian buffaloberry and silver buffaloberry are grown as ornamental shrubs, particularly where hardiness is required, so people can pick their own berries. Silver buffaloberries are generally regarded as much more palatable than Canadian buffaloberries. The fruits of both species do not taste good enough to eat fresh, but make good jams and pies. Buffaloberries are in fact collected for home use for pies, jams, jellies, and wine, and there is a small North American cottage industry for the production of buffaloberry jams and jellies. The fruits contain abundant pectin and make an excellent jelly. Like most fruits, pectin content decreases somewhat in the later stages of ripening, and slightly immature, sour fruit is best for preparing jelly. For eating fresh and for pies, it is recommended that picking be delayed until after the first frost, when the flavor of the berries is sweetest. The crop may even be harvested in winter, if spared by birds. The fruits preserve very well when dried and can be stored for future use as North American Indians did traditionally. However, it is preferable to process the berries immediately into food preparations. The juice of buffaloberries has a puckering quality not unlike chokecherry (*Prunus virginiana* L.) and has been used to make a cool, sweet, summer drink.

The Canadian buffaloberry was called "ice cream bush" by some indigenous peoples of western North America, particularly the Salishan and Athapaskan. They made a confection called "Indian ice cream" by mixing Canadian buffaloberries with water and a sweetener (usually sweet berries) and beating them to make a frothy dessert. Specially carved, paddle-like wooden spoons were often used to eat this dish, and in some households each person had his or her own spoon, which was carefully hung up when not in use. Buffaloberry froth is sour and bitter and is an acquired taste. For those who want to try making Indian ice cream, it should be borne in mind that care needs to be taken in harvesting and preparing the berries so they do not come in contact with oil or grease of any kind, or they will not whip. An eggbeater is a convenient tool to prepare the dish. The "ice cream" should be swished in and out of the mouth to get rid of the air before swallowing (ingestion of the considerable air that can accompany a large amount of Indian ice cream has been observed to cause discomfort).

## WARNING

Canadian buffaloberry has a high content of chemicals called saponins in the berries, which are responsible for the foaming properties of the fruit, as noted above. Saponins have the undesirable properties of irritating the digestive system and causing nausea, vomiting, and diarrhea. It has been recommended that only limited quantities of the berries be consumed to avoid possible toxicity.

### Culinary Vocabulary

- A "hairy buffalo" has nothing to do with buffaloberries nor indeed with buffalos. It is a cocktail prepared in bars from the remains of the almost-empty bottles at the end of the day, often combined and distributed gratis.

## PROSPECTS

Buffaloberries are rather sour and small, and collection is tedious. The thorns of silver buffaloberry make picking difficult, although some have described the berries as superior in taste (like a cross between wild grape and currant). The silver buffaloberry seems to have been first introduced to the horticultural trade as a fruit-bearing plant in 1890 by G.J. and L.E.R. Lambriger of Big Horn City, Wyoming. Since then a number of horticulturists made attempts to improve and popularize it. In the early part of the twentieth century, N.E. Hansen of the South Dakota Experiment Station at Brookings had about 7500 seedlings under observation. However, few other individuals have put much effort into selecting and breeding better buffaloberries. The plants are widely recommended for marginal agricultural areas in cold climates, where only the hardiest of fruit plants will survive, especially as a substitute for currants and gooseberries (of the genus *Ribes*). Buffaloberries are also well suited to moderately steep slopes, where most bushes have difficulty in securing a foothold. However, their popularity has not extended to regions where the major domesticated fruits can be grown.

Today, Canadian buffaloberry and silver buffaloberry are grown primarily as ornamental shrubs, particularly where climatic hardiness and tolerance of poor soils is required. Cultivars of both Canadian and silver buffaloberry are available in the horticultural trade, but these are intended mainly for ornamental use. Some limited breeding has resulted in improved selections, but true commercial fruit cultivars are not yet available.

The buffaloberries, particularly the silver buffaloberry, were considered to be potential new fruits for most of the twentieth century. They have not yet been developed as new crops, although they remain the subject of continuing research in the United States and Canada. The desirable characteristics of buffaloberries include high content of vitamin C, extreme winter hardiness and drought tolerance, and high productivity. Their disadvantages include small size of the berries, sour or bitter taste, and the difficulty of growing a crop that has male and female plants. The names "buffaloberry" and "soapberry" are hardly suitable for marketing, and more attractive names, such as "ice cream berry" (explained above), would be helpful. Buffaloberries have limited prospects for development into a significant fruit crop, but they remain one of the more interesting new fruit possibilities.

## CURIOSITIES OF SCIENCE AND TECHNOLOGY

- In British Columbia, Canadian buffaloberries were an important trade item among First Peoples throughout the province, used much like currency. A tradition arose, maintained to the present, of using the berries as gifts during exchanges of presents, especially between coastal and interior tribes.
- Buffaloberries were often used by indigenous peoples of North America for medicinal purposes. For example, the Salish and Kootenai tribes used Canadian buffaloberry branches to make an eyewash; the Sioux used the boiled roots to treat diarrhea; and the Wet'suwet'en used the berries to induce childbirth. Silver buffaloberry was much less frequently used medicinally but was employed by the blackfoot to treat stomach troubles and as a laxative, and by the Navajo to treat fevers.
- Silver buffaloberry spreads extremely efficiently by its underground root system. New plants are produced from the roots and often remain connected to the founder or "mother plant." All of the plants living within a 12 m (40 feet) diameter circle may still be connected through the root system that produced them.
- Although the Canadian buffaloberry is known as soapberry because of its saponin content, the plants with the most valid claim to this name belong to the genus *Sapindus*, especially the tropical American evergreen soap tree, *S. saponaria* L., which grows as far north as Missouri. The fleshy part of the fruit has a very high content of saponin and is used instead of detergent for washing clothes (unfortunately some people develop a skin rash from contact with saponins). Although saponins have been considered to be

extremely undesirable in food plants, recent research suggests that these chemicals bind onto cholesterol and may be useful for lowering cholesterol levels, and perhaps even for fighting cancer. Although saponins are toxic in large quantities to many animals, they are commonly found in small amounts in most of the plants eaten by herbivores and in many plant foods eaten by humans.

- Although nitrogen makes up 80% of the atmosphere, this is in an inert form that is unavailable to plants. Three classes of bacteria engage in symbiotic (mutually helpful) associations with plants, whereby they "fix" the nitrogen (convert it to a form that can be used by the plants). Most plants do not have bacteria to help them acquire nitrogen, and have to survive on the very small amounts of nitrogen that have been fixed and are present in the soil. The bacteria fix nitrogen for certain privileged plants, mostly in swellings of their roots called nodules. The three bacterial groups are, respectively, called rhizobial bacteria, cyanobacteria, and actinomycetes. Rhizobial bacteria have long been considered to belong to the "true" bacteria (Eubacteria). Actinomycetes form long, thread-like branched filaments like fungi, which led to their being considered as related to the fungi in the past (they are now included in the Eubacteria). The cyanobacteria, unlike the Eubacteria, carry out photosynthesis in very much the same way as do flowering plants, and while these were called blue-green algae by botanists, they are increasingly being recognized as a separate group. It is well known that members of the pea family (Leguminosae or Fabaceae) generally are provided with fixed nitrogen by rhizobia in root nodules. But what other plants have nitrogen delivered to them by bacteria? Cyanobacteria fix nitrogen for the following: the aquatic fern *Azolla* treated in Chapter 8 of this book (the only fern that can fix nitrogen, and it does so with the assistance of the cyanobacterium *Anabaena azollae*); some bryophytes (mosses); some lichens; some cycads (palm-like tropical plants); and some flowering plants that do not belong to the pea family, including *Gunnera*, which has the largest leaves in the world (an unusual angiosperm with N-fixing cyanobacteria in pockets at the base of the petioles; often said to be the only angiosperm that forms a N-fixing symbiosis with cyanobacteria, although some tropical angiosperms have cyanobacterial films on their leaf surfaces). About 25 genera of nonlegume flowering plants have actinorhizal bacteria, including *Shepherdia* (the buffaloberries discussed here), and the genera *Hippophae, Alnus, Ceanothus, Myrica, Cercocarpus, Purshia*, and *Casuarina*.
- In the northern Rocky Mountains, the Canadian buffaloberry is extremely common, and when the berries ripen in mid-August, it becomes the most important food source for bears, which eat up to 16 hours every day, with 80% of their diet being plants. Buffaloberries have a single seed, so the number of seeds in bear dung exactly measures the number of berries eaten. Biologists have carefully counted buffaloberry seeds in bear scat to determine daily intake. An adult grizzly may eat upwards of 200,000 buffaloberries every day. Bears become so engrossed in feeding that they may not hear hikers approach, and this can be the cause of deadly encounters. Accordingly, hikers are often advised to avoid areas where buffaloberries are abundant and in season. Some campgrounds where buffaloberries are common have been closed in August because many bears are attracted.

## KEY INFORMATION SOURCES

(*Note:* at least six recent PhD theses have been conducted on *Shepherdia* in North American universities; they can be located using online bibliographic databases.)

Arohonka, T., and Rousi, A. 1980. Karyotypes and C-bands in *Shepherdia argentea* and *Elaeagnus angustifolia*. *Ann. Bot. Fenn.* 17: 258–263.

Ayer, W.A., and Browne, L.M. 1970. Alkaloids of *Shepherdia argentea* and *Shepherdia canadensis*. *Can. J. Bot.* 48: 1980–1984.

Batzli, J.M., Zimpfer, J.F., Huguet, V., Smyth, C.A., Fernandez, M., and Dawson, J.O. 2004. Distribution and abundance of infective, soilborne *Frankia* and host symbionts *Shepherdia*, *Alnus*, and *Myrica* in a sand dune ecosystem. *Can. J. Bot.* 82: 700–709.

Boboreko, E.Z., Shapiro, D.K., Anikhimovskaia, L.V., and Narizhnaia, T.I. 1978. *Shepherdia argentea* (Pursh.) Nutt., a promising vitamin-containing and ornamental plant. Izv. *Akad. Nauk. BSSR, Ser. Biol. Nauk. (Minsk)* 4: 89–91. (In Russian, English summary)

Grubb, B.H. 2007. Propagation protocol for bareroot silver buffaloberry (*Shepherdia argentea*). *Native Plants J.* 8: 233–235.

Hamer, D. 1996. Buffaloberry [*Shepherdia canadensis*] fruit production in fire-successional bear feeding sites. *J. Range Manage.* 49: 520–529.

Hayes, P.A., Steeves, T.A., and Neal, B.R. 1989. An architectural analysis of *Shepherdia canadensis* and *Shepherdia argentea*: Patterns of shoot development. *Can. J. Bot.* 67: 1870–1877.

Hayes, P.A., Steeves, T.A., and Neal, B.R. 1990. An architectural analysis of *Shepherdia canadensis* and *Shepherdia argentea* (Elaeagnaceae): The architectural models. *Can. J. Bot.* 68: 719–725.

Huguet, V., Batzli, J.M., Zimpfer, J.F., Gourbiere, F., Dawson, J.O., and Fernandez, M.P. 2004. Nodular symbionts of *Shepherdia*, *Alnus*, and *Myrica* from a sand dune ecosystem: Trends in occurrence of soilborne *Frankia* genotypes. *Can. J. Bot.* 82: 691–699.

Huguet, V., Batzli, J.M., Zimpfer, J.F., Normand, P., Dawson, J.O., and Fernandez, M.P. 2001. Diversity and specificity of *Frankia* strains in nodules of sympatric *Myrica gale*, *Alnus incana*, and *Shepherdia canadensis* determined by rrs gene polymorphism. *Appl. Environ. Microbiol.* 67: 2116–2122.

Hyland, F. 1976. *Canada buffalo-berry,* Shepherdia canadensis *(L.) Nutt., in Maine and its relevance to the Critical Areas Programs*. Maine State Planning Office, Augusta, ME. pp. 18.

Kjosen, H., and Liaaen-Jensen, S. 1969. Major carotenoids of *Shepherdia canadensis*. Isolation and synthesis of methyl apo-6'-lycopenoate. *Phytochemistry* 8: 483–491.

Knudson, J.J., Haas, R.J., Tober, D.A., Darris, D.C., and Jacobson, E.T. 1990. Improvement of chokecherry, silver buffaloberry, and hawthorn for conservation use in the Northern Plains. General Technical INT 276. In *Proceedings symposium "Cheatgrass invasion, shrub die-off, and other aspects of shrub biology and management," Las Vegas, NV, April 5–7, 1989*. Intermountain Research Station, U.S. Department of Agriculture, Forest Service, 291–299.

Krishnan, S., and Hughes, H. 1991. Asexual propagation of *Shepherdia canadensis* and *S. rotundifolia*. *J. Environ. Hortic.* 9: 218–220.

Looman, J. 1984. The biological flora of Canada. 4. *Shepherdia argentea* (Pursh) Nutt., buffaloberry. *Can. Field-Nat.* 98: 231–244.

Nelson, A. 1935. Rocky mountain herbarium studies. III. The Elaegnaceae: A mono-generic family. *Am. J. Bot.* 22: 681–683.

Richard, P. 1974. Presence of *Shepherdia canadensis* (L.) Nutt. in the Laurentides Park region, Quebec, during the late glacial period. *Nat. Can.* 101: 763–768. (In French, English summary)

Small, E., and Catling, P.M. 2001. Poorly known economic plants of Canada—32. Buffaloberry, *Shepherdia canadensis* (L.) Nutt. and *S. argentea* (Pursh) Nutt. *Bull. Can. Bot. Assoc.* 35(1): 4–9.

Smith, R.C. 1987. *Shepherdia argentea. Am. Nurseryman* 166(10): 134.

Steeves, T. 1997. Buffalo berry crop is ripe for the pickin'. *Blue Jay* 55: 169–170.

Thilenius, J.F., Evans, K.E., Garrett, E.C., and Chester, E. 1974. *Shepherdia* Nutt. buffaloberry. In *Seeds of woody plants in the United States. Agric. Handb.* 450. U.S. Dept. Agric. Forest Service, Washington, DC. pp. 771–773.

Turner, N.J. 1981. Indian use of *Shepherdia canadensis*, soapberry, in western North America. *Davidsonia* 12(1): 1–14.

United States Department of Agriculture. 1985. *'Sakakawea' silver buffaloberry*. Program Aid, U.S. Department of Agriculture, Washington, DC. pp. 5.

Visser, S., Danielson, R.M., and Parkinson, D. 1991. Field performance of *Elaeagnus commutata* and *Shepherdia canadensis* (Elaeagnaceae) inoculated with soil containing *Frankia* and vesicular-arbuscular mycorrhizal fungi. *Can. J. Bot.* 69: 1321–1328.

## Specialty Cookbooks

Holm, D. 1969. *The old-fashioned Dutch oven cookbook; complete with authentic sourdough baking, smoking fish and game, making jerky, pemmican, and other lost campfire arts*. Caxton Printers, Caldwell, ID. pp. 106.

Krumm (1991) has seven buffaloberry recipes. Stewart (2002) provides a recipe for buffaloberry lemonade. Freitus and Haberman (2005) present six recipes, and Snell (2006) has four recipes. Gray (2011) has the following instructions for preparing "Indian ice cream": "Add 1 cup (250 mL) berries to ¼ cup (60 mL) water and add 6 tablespoons (90 mL) white or brown sugar. Mix or beat ingredients until you have a stiff and pink foam." Genest (2010) has a similar recipe entitled "Soapberry ice cream." (See the Appendix to this book for details of the publications cited.)

# 17 Buffalo Gourd

Family: Cucurbitaceae (gourd family)

## NAMES

Scientific name: *Cucurbita foetidissima* Kunth (The name is often given as *Cucurbita foetidissima* H.B.K., for Humboldt, Bonpland, & Kunth.)

- The name "buffalo gourd" is a pejorative reference to the plant, suggesting it is fit to eat only by buffalos (i.e., American bison).
- In addition to the name buffalo gourd, the species has also been called calabazilla, chilicote, chilli coyote, coyote ear, coyote gourd, fetid gourd, Missouri gourd, prairie gourd, stink gourd, stinking cucumber, stinking gourd, stinky gourd, wild gourd, and wild pumpkin.
- In Spanish and Italian, buffalo gourd is sometimes called *calabacilla loca*, "crazy squash."
- The genus name *Cucurbita* is a classical Latin name for some gourd. *Cucurbitare* in old Latin had the meaning "to commit adultery"; it is not clear, however, if this sexual meaning has any relevance to the derivation of the name *Cucurbita*, although some of the species do interbreed promiscuously.
- *Foetidissima* in the scientific name *C. foetidissima* is Latin for very fetid, a reference to the pungent, bad smell of the species (although occasional plants lack the bad odor).

## GEOGRAPHY AND ECOLOGY OF WILD PLANTS

Buffalo gourd is native to the southwestern United States and northern Mexico. In the United States, the plants grow particularly well in Arizona, New Mexico, and Texas. The species grows naturally in Nebraska, Missouri, Kansas, Colorado, Utah, Nevada, Texas, New Mexico, and California, but has been collected from other states where it has been distributed by people. The plants occur in dry waste land, prairies, shrub associations, and old washes, in open and sunny locations, often in sandy soils. They very frequently grow as weeds in disturbed soils, particularly along roadsides, fences, and railways, and in abandoned fields and pastures.

## PLANT PORTRAIT

Buffalo gourd is a vigorous, perennial, low-growing, herbaceous vine. It produces very large, fleshy, storage taproots, which after three or four seasons of growth may be 2 m (about 2 yards) long and 30 cm (1 foot) wide, and can reach weights of over 100 kg (220 pounds). These tuberous roots serve as overwintering organs, from which runners (i.e., vine-like shoots that spread horizontally on the ground) are produced at the beginning of each new season. The vines grow up to 12 m (39 feet) in length. They are cold-sensitive and are killed by frost (sometimes the vines also die back to the roots during prolonged drought). Roots develop from the nodes of the runners (i.e., where the leaves are attached). From a single new plant that has arisen from a seed, hundreds of meters of vines can be produced, resulting in an extensive, homogenous colony that covers considerable ground with a dense mat of vines. The roots can extend down 5 m (16 feet) to reach water. Male and female flowers are borne singly at many of the nodes of the stems (some plants bear female flowers, whereas others

**FIGURE 17.1** Buffalo gourd (*Cucurbita foetidissima*). (a) Mature fruits. (Courtesy of R.B. Lewis, III, BonTerra Consulting.) (b) Immature fruits. (Courtesy of U.S. Department of Agriculture.) (c) Vines. (Courtesy of U.S. Department of Agriculture.) (d) Flower. (Courtesy of R.B. Lewis, III, BonTerra Consulting.) (e) Distribution map.

have male as well as female flowers). The orange-yellow flowers are similar to those of pumpkins and squashes: bell-shaped and 5–12.5 cm (2–5 inches) long. The flowers open just before daylight and remain open just for one morning. Fertilized female flowers produce fruits (gourds) that are hard and usually round, with diameters of 5–7 cm (2–3 inches). The immature fruits have alternating yellow and green stripes (and are reminiscent of miniature watermelons), but dry to a yellowish-brown. Each fruit contains 200–300 seeds. A large plant can produce more than 200 fruits in one season. The plants are foul-smelling, with an aroma that has been described as a mixture of human body odor and sulfurous sewage. The leaves are large (up to 30 cm or a foot in length), triangular-elongated, and rather abrasive because of the presence of prickly hairs. Stepping on the foliage is likely to result in stinky shoes. However, some plants are smellier than others, and occasional plants lack the repulsive odor.

The seeds contain 30%–40% edible oil and 30%–35% protein. About two-thirds of the oil may be linoleic, a polyunsaturated fatty acid that is very desirable nutritionally. Starch in the roots varies seasonally and with age and may constitute more than half of the dry root weight. The starchy roots have potential for food, fuel ethanol (the starch can be fermented into alcohol), and other industrial uses, such as incorporation into diesel fuel.

Although related to the edible pumpkins and squashes, the foul odor and extremely bitter taste of buffalo gourd fruits make them inedible. Nevertheless, Native Americans used the seeds as food for perhaps 10,000 years. The young fruits may also have been boiled repeatedly to make them palatable. Indians also used the seed oil as a cosmetic, the green fruit and root as a scouring, laundry detergent (saponin chemicals that are present have cleansing properties), and the dried fruits as rattles in ritualistic gourd dances.

It has been speculated that the long association of buffalo gourd and Native Americans may have resulted in some domestication of the plant, but it remains basically a wild plant. Attempts have been made to domesticate buffalo gourd in the southern United States as a source of seed oil, seed meal, and perhaps root starch.

## CULINARY PORTRAIT

Unlike pumpkins and squashes, the flesh of buffalo gourds is much too bitter to eat. Most parts of the plant are bitter and obnoxious in taste and smell, and Native Americans commonly treated young fruits by various cooking methods to render them palatable. Wild gourds mature from late spring through the fall and should be collected with gloves to avoid the irritating, prickly hairs on the plant. The seeds are the principal edible part of the plant because they lack the bitter chemicals (cucurbitacins) found in the flesh of the fruit and in the leaves and roots. However, fruit pulp that clings to the seeds can cause them to taste bitter. Native Americans dried and roasted the seeds, and their practices can be emulated. To prepare the seeds for consumption, wild food enthusiasts are advised to first dry the gourds until they turn yellow or tan. The gourds are cut open, the seeds removed, washed, and thoroughly sun-dried. The seeds are roasted for 15 to 30 minutes or sautéed (heat destroys digestive inhibitors that are present). The roasted seeds can be consumed with a little salt, like pumpkin seeds. They can also be ground into a flour or meal, and boiled to prepare a mush. The seeds coats are tough, and if one is not prepared to chew them thoroughly, they should be strained away.

## WARNINGS

In Texas, the buffalo gourd could be confused with some other wild gourds that may be poisonous, including species of *Melothria* (which have tiny, green, cucumber-like fruit) and species of *Momordica* (balsam apple, which have yellow or orange spiny fruit).

The folk medicinal usage of buffalo gourd is inadvisable. Buffalo gourd was sometimes used by Native Americans as an abortifacient, the chemicals present thought to cause uterine contractions intense enough to kill an early-stage fetus. Native Americans also used the plant as a laxative, an unsafe practice that can cause diarrhea and irritation of the digestive tract.

### CULINARY VOCABULARY

- "Gourds" have been defined as members of the Cucurbitaceae with durable, hard-shelled fruit grown for ornament, utensils, or general interest (a few kinds occasionally called gourds have soft flesh and nondurable rinds). Most gourds are either inedible or not used primarily as food. The two species most commonly encountered in North America as gourds, and sometimes consumed, are *Lagenaria siceraria* (Molina) Standl. (bottle gourd) and *Cucurbita pepo* L. subsp. *ovifera* (L.) D.S. Decker, often known as the white-flowered and yellow-flowered gourds, respectively.

## PROSPECTS

Buffalo gourd is a potential new crop for arid lands, used primarily for human food (extracted seed oil and root starch) and liquid fuels, and possibly also for chemical extracts and forage. The species is a potentially useful crop of hot, arid, or semiarid lands, to which it is naturally adapted. It is considered suitable for development as a new crop because it can grow well on poor lands with little rainfall. Buffalo gourd also is suited to mechanical harvest. An extensive program of research and development of buffalo gourd was conducted at the University of Arizona (Tucson), but interest in it waned by 1990. Thompson (1990) concluded "It appears that the lack of a truly unique, high-valued specialty product is a major constraint. In addition, the currently minimal economic prospects for the conventional oils, protein, and starch produced by the buffalo gourd have not generated sufficient interest by industry to push development and commercialization." Nevertheless, buffalo gourd remains a very interesting potential new crop because the world has essentially run out of new sources of arable land and of water for agriculture. The needs for exploiting marginal land for food and fuel production are becoming more urgent, and arid, tropical, and subtropical areas of the world are in great need of crops with the properties of buffalo gourd.

## CURIOSITIES OF SCIENCE AND TECHNOLOGY

- The expression "camp followers," as applied to plants, refers to species that thrive near human settlements. Such plants are often weeds and have characteristics that adapt them to the often disturbed soils, dung deposits, cleared areas, and other conditions that develop where humans are established. Numerous crops are believed to have evolved as camp followers that humans found to be useful and subsequently domesticated. Buffalo gourd is thought to have been a camp follower that was associated with Native Americans for thousands of years.
- Native Americans used juice from the root of buffalo gourd to treat toothache. Although toxic and not recommended, the antibacterial properties of the juice may have been beneficial for some dental problems.
- Some Indian mothers weaned their infants from nursing by smearing bitter juice from buffalo gourd on their nipples.
- Some Native Americans thought that the roots of buffalo gourd possessed mystical powers. They took special care when unearthing the root in the belief that an injury to the root would lead to injury of the person digging it up or to a member of one's family.
- The Navajo used buffalo gourds as ceremonial rattles.
- The large leaves of buffalo gourd tend to fold upward on both sides of the midrib, an orientation that is advantageous in substantially exposing the leaves to the sun in the morning and evening when sunlight is low or moderate, but reducing exposure to the intense midday sunlight.
- Cucurbitacins are the chemicals responsible for the rank smell of buffalo gourd. These occur mostly in species of the gourd family. Although toxic, these compounds have some medicinal properties, notably acting against tumors. They repel or poison many insects but curiously cause some beetles (including corn rootworms and cucumber beetles) to feed compulsively.

## KEY INFORMATION SOURCES


Anon. 1992. Buffalo gourd extract fights corn rootworm. *AgBiotechnology News* 9(2): 22.

Bemis, W.P., Berry, J.W., and Weber, C.W. 1978. The buffalo gourd, a potential crop for arid lands. *Arid Lands Newslett.* 8: 1–7.

Bemis, W.P., Berry, J.W., and Weber, C.W. 1979a. Domestication studies with the feral buffalo gourd. In *Arid land plant resources*. Edited by J.R. Goodin and D.K. Northington. Texas Tech University, Lubboch, TX. pp. 371–383.

Bemis, W.P., Berry, J.W., and Weber, C.W. 1979b. The buffalo gourd, a potential arid land crop. In *New agricultural crops*. Edited by G.A. Ritchie. Westview Press, Boulder, CO. pp. 65–87.

Bemis, W.P., Berry, J.W., Weber, C.W., and Whitaker, T.W. 1978. The buffalo gourd: A new potential horticultural crop. *HortScience* 13: 235–240.

Bemis, W.P., Curtis, L.C., Weber, C.W., and Berry, J.W. 1978. The feral buffalo gourd, *Cucurbita foetidissima. Econ. Bot.* 32: 87–95.

Berry, J.W., Scheerens, J.C., and Bemis, W.P. 1978. Buffalo gourd roots: Chemical composition and seasonal changes in starch content. *J. Agric. Food Chem.* 26: 354–356.

Daghir, N.J., and Sell, J.L. 1980. Buffalo gourd (*Cucurbita foetidissima*) seed and seed components for growing chickens. *Nutr. Rep. Int.* 22: 445–452.

DeVeaux, J.S., and Schultz, E.B. 1985. Development of buffalo gourd (*Cucurbita foetidissima*) as a semiarid-land starch and oil crop. *Econ. Bot.* 39: 454–472.

Dreher, M.L., Scheerens, J.C., Weber, C.W., and Berry, J.W. 1981. Nutritional evaluation of buffalo gourd *Cucurbita foetidissima* root starch. *Nutr. Rep. Int.* 23: 1–8.

Hogan, L., and Bemis, W.P. 1983. Buffalo gourd and jojoba: Potential new crops for arid lands. *Advances Agron.* 36: 317–349.

Horak, M.J., and Sweat, J.K. 1994. Germination, emergence and seedling establishment of buffalo gourd (*Cucurbita foetidissima*). *Weed Sci.* 42: 358–363.

Icerman, L., and Schulz, E.B. Jr. 1985. A preliminary economic analysis of buffalo gourd as a diesel fuel and ethanol feedstock in the high plains of New Mexico, Texas and Oklahoma. *Energy* 10: 901–909.

Khoury, N.N., Dagher, S., and Sawaya, W. 1982. Chemical and physical characteristics, fatty acid composition and toxicity of buffalo gourd oil, *Cucurbita foetidissima. J. Food Tech.* 17: 19–26.

Kohn, J.R. 1989. Sex ratio, seed production, biomass allocation, and the cost of male function in *Cucurbita foetidissima* HBK (Cucurbitaceae). *Evolution* 43: 1424–1434.

Lancaster, M., Storey, R., and Bower, N. 1983. Nutritional evaluation of buffalo gourd: An elemental analysis of seed. *Econ. Bot.* 37: 306–309.

Merrick, L.C. 1995. Squashes, pumpkins and gourds. In *Evolution of crop plants.* Second edition. Edited by J. Smartt and N.W. Simmonds. Longman Scientific & Technical, Burnt Mill, Harlow, Essex, UK. pp. 97–105.

Nelson, J.M., Scheerens, J.C., Berry, J.W., and Bemis, W.P. 1983. Effect of plant population and planting date on root and starch production of buffalo gourd *Cucurbita foetidissma* grown as an annual. *J. Am. Soc. Hort. Sci.* 108: 198–201.

Saade, L., and Hernández, S.M. 1994. Cucurbits. (*Cucurbita* spp.). In *Neglected crops, 1492 from a different perspective.* Edited by J.E. Hernández-Bermejo and J. Léon. Food and Agriculture Organization of the United Nations, Rome, Italy. pp. 63–77.

Scheerens, J.C., and Berry, J.W. 1986. Buffalo gourd: Composition and functionality of potential food ingredients. *Cereal Foods World* 31: 183–184.

Scheerens J.C., McGriff, T.L., Ralowicz, A.E., Wilkins, M.H., and Nelson, J.M. 1989. Variation in seed yield components of buffalo gourd (*Cucurbita foetidissima* HBK.) grown as an annual. *Euphytica* 40: 55–62.

Thompson, A.E. 1990. Arid-land industrial crops. In *Advances in new crops.* Edited by J. Janick and J.E. Simon. Timber Press, Portland, OR. pp. 232–241.

Toufeili, I., and Baydoun, E. 1995. Some selected physiochemical properties of buffalo gourd (*Cucurbita foetidissima* HBK) root starch. *Starch (Staerke)* 47: 413–415.

Tu, M., Eustace, W.D., and Deyoe, C.W. 1978. Nutritive value of buffalo gourd seed protein. *Cereal Chem.* 55: 766–772.

Vasconcellos, J.A., Bemis, W.P., Berry, J.W., and Weber, C.W. 1981. The buffalo gourd, *Cucurbita foetidissima* HBK, as a source of edible oil. *Am. Oil Chem. Soc.* 9: 55–68.

Vasconcellos, J.A., and Berry, J.W. 1982. Characteristics of laboratory-processed *Cucurbita foetidissima* seed oil. *J. Am. Oil Chem. Soc.* 59: 79–84.

Vasconcellos, J.A., Berry, J.W., Weber, C.W., Bemis, W.P., and Scheerens, J.C. 1980. The properties of *Cucurbita foetidissma* seed oil. *J. Am. Oil Chem. Soc.* 57: 310–313.

Young, P.G., Morgan, R.P., and Shultz, E.B. 1982. Buffalo gourd: Potential as a fuel resource on semi-arid lands. *Am. Soc. Agric. Engineers Proceedings of International Conference on Plant and Vegetable Oils as Fuels, Fargo, ND, August 2–4.* 4-82: 63–69.

## Specialty Cookbooks

Buffalo gourd is generally not considered to be of value for conventional meals. Tull (1987) and Niethammer (1999) provide advice on how to harvest and prepare the seeds. Niethammer (1999) also provides a recipe entitled "Buffalo gourd mush" based on the seeds. (See the Appendix to this book for details of the publications cited.)

# 18 Butternut

Family: Juglandaceae (walnut family)

## NAMES

Scientific name: *Juglans cinerea* L.

- European pioneers in North America named the butternut for its buttery, oily seeds (it is also called the oilnut).
- "Kisky-Thomas nut" is an obsolete name, used more for the nuts than for the tree.
- "White walnut" is an occasional name for butternut, and very uncommonly for shagbark hickory (see Chapter 48 on Hickories). The name is a reference to the light-colored bark of the tree.
- Other uncommon names for butternut include American white walnut, butternut walnut, lemon walnut (for the oblong-lemon shape), and white butternut.
- "Buartnuts" are hybrids of butternut and Japanese walnut (*J. ailantifolia* Carr.).
- The word *cinerea* in the scientific name of the butternut is Latin for "ashes" and refers to the characteristic ash-gray color of the trunk and limbs.
- For information on the genus name *Juglans*, see Chapter 11 on Black Walnut.

## GEOGRAPHY AND ECOLOGY OF WILD PLANTS

The butternut is native to the area bounded by southwestern New Brunswick west to Georgian Bay in Ontario and Minnesota, and south to Virginia, Georgia, Arkansas, and Kansas. It is a relatively infrequent species that does not develop large stands and tends to grow among other trees. Butternut grows best in valleys, along river banks, and on hillsides, but occurs occasionally in dry, stony habitats, particularly if the soil is chalky. It is intolerant of shade. Heavy production of nuts occurs on an irregular basis, once in several years. Butternut is sometimes a weed of open flower beds wherever a nut-producing tree is nearby. The species often colonizes old fields, forest clearings, and gaps in forests. Squirrels that bury nuts in open prairie at some distances from the trees are responsible for colonization of open areas. Butternut was much more abundant and dominant in the eastern North American forests in presettlement times.

## PLANT PORTRAIT

The butternut is a small-to-medium-sized deciduous tree, usually 12–18 m (40–60 feet) tall and 30–60 cm (12–24 inches) in diameter at breast height. Some reach 30 m (about 100 feet) in height. The trees develop male flowers (in pendulous, slender catkins) and female flowers (in short clusters). The nuts (which are technically fruits), produced singly or in clusters of 2–6, are oblong, 3–6 cm (1.2–2.4 inches) long, covered with sticky hairs, and thick-husked. The shell is hard and tough, but can usually be broken to release the kernel. A mature tree can produce several bushels of hulled nuts, but the majority of trees are less productive most of the time. More than 40 cultivated varieties are known, selected from natural populations for their relatively thin shells and high yield. The kernels in some of these can be readily cracked into halves. Butternut

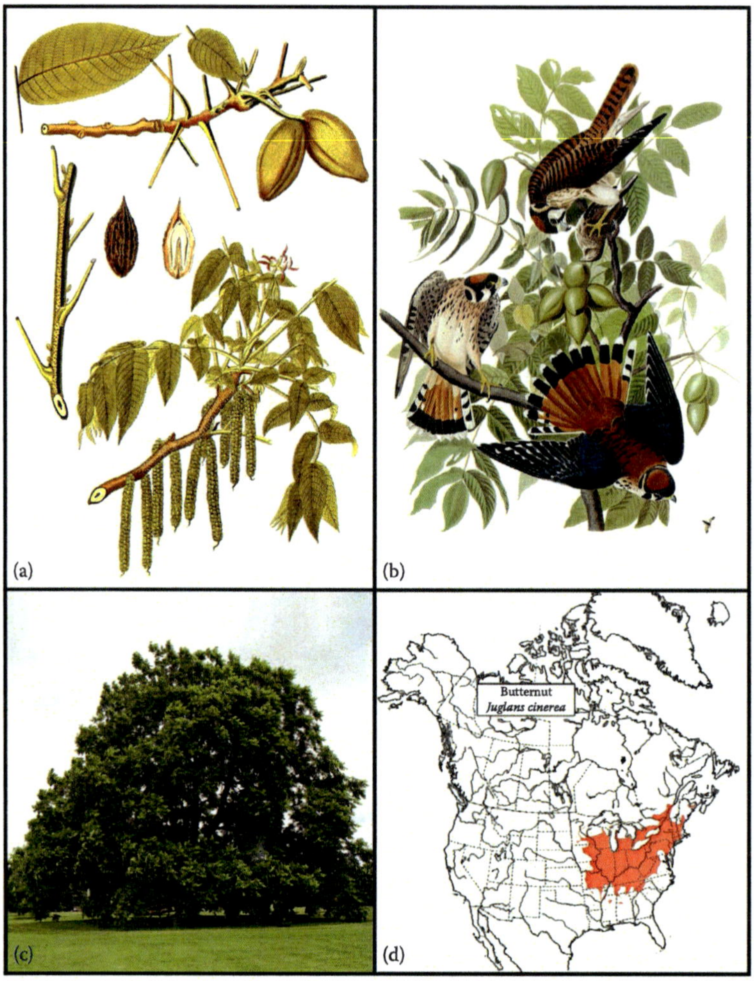


**FIGURE 18.1** Butternut (*Juglans cinerea*). (a) Plate showing fruiting branch (above) and male branch with catkins (below). (From Jacquin, N.J., *Icones plantarum rariorum*, B. White & Filum, London, UK, Vol. 1, Plate 192, 1781–1786.) (b) American kestrels in a butternut tree. (From Audubon, J.J., *Birds of America*, R. Havell Jr., London, 4 Vols, Plate 142, 1827–1838.) (c) A large, open-grown tree in Ottawa, Canada. (Courtesy of E. Small.) (d) Distribution map.

wood is relatively light, soft, and weak, and has limited use in construction. It has been used in furniture, often in church pews. The relatively light wood was advantageous for use in building rafts and boats.

The butternut was an important source of food for Native Americans in regions where it grew. The nuts have been found at archeological Indian sites in the Great Lakes region dated back to 4000 years ago. Nut milk was made by grinding the kernels and adding water. Oil from the nuts was used in food preparations, as well as for rubbing on body and hair. The Native Americans also used the bark in medicines to treat toothache, muscular pains, and wounds.

Butternut trees rarely live longer than 70 years, and unfortunately many are dying prematurely because of a disease called butternut decline or butternut canker, caused by the fungus *Sirococcus clavigignenti-juglandacearum* Nair, Kostichka & Kuntz. Spores from dying branches are spread by rainwater to the stems of healthy trees, and after infestation stem cankers develop 1 to 3 years later. The disease is killing trees throughout the range of the butternut and is so serious that commercial cultivation of butternuts is a bad risk. More than 90% of all butternut trees in Ontario and western Quebec are now infected, and the disease is also serious in the United States (especially in Wisconsin and the Carolinas). The species has achieved endangered status in Canada and similar rank in some states. Some trees have evidently survived the infection suggesting the presence of natural genetic resistance, and it is possible that these could be used to breed resistant trees. However, there is no known control for the fungal disease. Although butternut is particularly susceptible, black walnut and black walnut hybrids are also vulnerable to the disease.

## CULINARY PORTRAIT

The flavor of the butternut is agreeable, although some people object to its strong, oily taste. The nut is used in cooking and, like the black walnut, in the manufacture of confectionary. Butternuts have been used in candy, ice cream, cakes, cookies, and even in pickles and ketchup. In New England, the butternut is popularly combined with maple sugar to produce maple–butternut candy, which is sold at the roadside. However, this nut has limited commercial significance. Butternuts are not as tasty as most walnuts, but the plant is very cold hardy and therefore is useful as a nut source in cold climates. Since butternuts are rarely marketed commercially, those wishing to obtain them will usually have to do so from wild tree or trees cultivated for ornament. The fruits mature in late summer or early fall, at which time they are very sticky and are advisedly collected with gloves. They should be allowed to dry for several weeks, turning brown. The nuts can be cracked with a heavy-duty nutcracker, hammer, or a vice, and the nutmeat removed with a nutpick, and consumed raw. Alternatively, prior to extraction, the nuts may be roasted in a pan in a preheated oven at 120°C (250°F) for about 40 minutes, with occasional stirring.

### Culinary Vocabulary

- "Butternut squash," which has nothing to do with butternuts, was so-named for its butter-colored shell and its nutty-tasting flesh.

## PROSPECTS

The butternut is seriously threatened by disease and, pessimistically, its future could be as bleak as that of the American chestnut and American elm. Butternuts have never achieved much commercial importance and would not seem to be a good choice for economic development under the circumstances. Nevertheless, should control measures be found for the disease problem, the species does have sufficient potential that it remains a possibility for development.

## CURIOSITIES OF SCIENCE AND TECHNOLOGY

- The French explorer Samuel de Champlain (1567–1635) wrote that Indians constructed dugout canoes in what is now Massachusetts by charring and scraping one side of a large butternut log. In the state of New York, Native Americans were commonly observed making river canoes from one or two butternut logs.
- North American homespun cloth pants of the 1800s, stained brown with butternut husks, were called "butternut jeans." The word jean is derived from Genoa, Italy, and was first applied to a fabric made there of a blend of cotton, linen, and/or wool. By the eighteenth century, rugged workpants called jeans were made completely out of cotton, and by the early nineteenth century, blue became a popular color.
- The famous gray coats of the Confederate Army in the American Civil War were not always gray. Soldiers dressed in light-brown or butternut-colored uniforms began showing up in Confederate ranks after Union blockades shut off virtually all commerce to the southern

states. The Southerners were nicknamed "Butternuts" because the dye used on their often homemade uniforms for a time was made of butternut, creating a light brown khaki color.

- Several North American Indian tribes used butternut bark as a laxative, and so did many European settlers in North America before the 1900s.The French botanist André Michaux (1746–1802 (estimated)) called the tree *Juglans cathartica* to point out the laxative qualities of the tree.
- The small village of Butternut in northern Wisconsin (first officially recognized in 1877 by the Wisconsin Central Railway) is situated beside Butternut Lake. Both landmarks were indeed named for the local butternut trees. According to local history, pioneers there made a delectable relish of the fruit of the butternut by pickling the nuts with spices and vinegar, and usually served this with wild game, fish, and fowl.

## KEY INFORMATION SOURCES

(For additional references, see Chapter 11 on Black Walnut.)


Betts, H.S. 1945. *Butternut* (Juglans cinerea). Forest Service, U.S. Dept. of Agriculture, Washington, DC. pp. 4.

Catling, P.M., and Small, E. 2001. Poorly known economic plants of Canada—30. Butternut (*Juglans cinerea*) and black walnut (*J. nigra*). *Bull. Can. Bot. Assoc.* 34(3): 34–39.

Fegluel, V.R. 1996. *A literature review of butternut and the butternut canker.* Eastern Ontario Model Forest Information Report 20. Ontario Ministry of Natural Resources, Kemptville, ON. pp. 32.

Fjellstrom, R.G., and Parfitt, D.E. 1995. Phylogenetic analysis and evolution of the genus *Juglans* (Juglandacaeae) as determined from nuclear genome RFLPs. *Plant Syst. Evol.* 197: 19–32.

Furnier, G.R., Stolz, A.M., Mustaphi, R.M., and Ostry, M.E. 1999. Genetic evidence that butternut canker was recently introduced into North America. *Can. J. Bot.* 77: 783–785.

Gunter, L.E., Kochert, G., and Giannasi, D.E. 1994. Phylogenetic relationships of the Juglandaceae. *Plant Syst. Evol.* 192: 11–29.

Hill, L. 1977. Make it butternuts for the North. *Org. Gard. Farm.* 24(9): 92–94.

Hopkin, A., Innes, L., and Harrison, K. 2001. Distribution of butternut canker (*Sirococcus clavigignenti-juglandacearum*) in Eastern Canada. *Can. Plant Dis. Surv.* 81: 154–157.

Manning, W.E. 1978. The classification within the Juglandaceae. *Ann. Missouri Bot. Gard.* 65: 1058–1987.

McDaniel, J.C. 1979. Other walnuts including butternut, heartnut, and hybrids. In *Nut tree culture in North America.* Edited by R.A. Jaynes. Northern Nut Growers' Association, Hamden, CT, 98–110.

McIlwrick, K., Wetzel, S., Beardmore, T., and Forbes, K. 2000. Ex situ conservation of American chestnut (*Castanea dentata* (Marsh.) Borkh.) and butternut (*Juglans cinerea* L.), a review. *Forest Chron.* (Ottawa) 76: 765–774.

Millikan, D.F., and Stefan, S. 1983. Buttering up the butternut. *Ann. Rep. Northern Nut Growers Assoc.* 74: 48–53.

Millikan, D.F., Stefan, S.J., and Shaffer, W.H. Jr. 1985. Butternuts worth propagating. *Ann. Rep. Northern Nut Growers Assoc.* 76: 103–105.

Millikan, D.F., Stefan, S.J., and Rigert, K.S. 1990. Selection and preservation of butternut, *Juglans cinerea* L. *Ann. Rep. Northern Nut Growers Assoc.* 81: 22–25.

Morin, R., Beaulieu, J., Deslauriers, M., Daoust, G., and Bousquet, J. 2000. Low genetic diversity at allozyme loci in *Juglans cinerea. Can. J. Bot.* 78: 1238–1243.

Nair, V.M.G., Kostichka, C.J., and Kuntz, J.E. 1979. *Sirococcus clavigignenti-juglandacearum*: An undescribed species causing canker on butternut. *Mycologia* 71: 641–646.

Oboussier, J. 1929. *Contribution to the study of Juglans cinerea.* Frères & Rious, Lyon, France. pp. 95. (In French)

Ontario Ministry of Natural Resources. 1996. *Butternut.* O.M.N.R., Toronto, ON. pp. 6.

Ostry, M.E., and Pijut, P.M. 2000. Butternut: An underused resource in North America. *HortTechnology* 10: 302–306.

Ostry, M.E., Mielke, M.E., and Skilling, D.D. 1994. *Butternut: Strategies for managing a threatened tree.* North Central Forest Exp. St., U.S. Dep. Agric., Forest Service, St. Paul, MN, 7 pp. (Also in *Ann. Rep. Northern Nut Growers Assoc.* 85: 63–67.)

Ostry, M.E., Moore, M.J., and Worrall, S.A.N. 2003. *Butternut* (Juglans cinerea) *annotated bibliography.* United States For. Serv., North Central Res. St., St. Paul, MN. pp. 28.

Pijut, P.M., and Moore, M.J. 2002. Early season softwood cuttings effective for vegetative propagation of *Juglans cinerea*. *HortScience* 37: 697–700.
Rink, G. 1990. *Juglans cinerea* L. Butternut. In *Silvics of North America*, Vol. 2. Hardwoods. Edited by R.M. Burns and B.H. Honkala. United States For. Serv. Agric. Handb. 654, Washington, DC. pp. 386–390.
Ross-Davis, A., Huang, Z., McKenna, J., Ostry, M., and Woeste1, K. 2008. Morphological and molecular methods to identify butternut (*Juglans cinerea*) and butternut hybrids: Relevance to butternut conservation. *Tree Physiol.* 28: 1127–1133.
Schroeder, J.G. 1972. *Butternut.* U.S. For. Serv. Agric., Washington, DC. pp. 6.
Tisserat, N., and Kuntz, J.E. 1983. The etiology and epidemiology of butternut canker. *Ann. Rep. Northern Nut Growers Assoc.* 74: 30–36.

## Specialty Cookbooks

*Note:* A search for "butternut recipes" on the Web will produce recipes for butternut squash, not for butternuts (instead, search "nut recipes" + butternut* - "butternut squash").

Boorman (1969) has three recipes for butternuts, Brill (2002) has five, Freitus (1980) has eight, and Robe-Terry (1997) has three. Freitus and Haberman (2005) have a recipe for salted recipes and advice on how to dry the nuts. (See the Appendix to this book for details of the publications cited.) Also see the recipe sources in Chapter 11 on Black Walnut.

# 19 Cabbage Palmetto

Family: Arecaceae (Palmae; palm family)

## NAMES

Scientific name: *Sabal palmetto* (Walter) Lodd. ex Schult. & Schult. f.

- Cabbage palmetto has also been called blue palmetto, cabbage palm, cabbage tree, Carolina palmetto, common palmetto, palmetto, palmetto palm, sabal palm, sabal palmetto, serpent palm, swamp cabbage, and swamp cabbage palm.
- For an explanation of the word "cabbage" in the name of the plant, see the section "Culinary Vocabulary."
- The phrase "cabbage palm" is used for several palm trees with edible hearts (as explained below); it is also employed for ornamental species of the genus *Cordyline*. The expression "palm tree cabbage" is used for some kales (certain *Brassica* species), and this also sometimes leads to confusion.
- The genus name *Sabal* is of unclear origin. It has been suggested that the botanist who coined the name in 1763, Michel Adanson, used a South American vernacular name for palms of the genus, perhaps based on the Mayan word *sibul* for a black bird.
- *Palmetto* in both the scientific and common names is a diminutive of the Latin *palma*, palm tree, probably derived through the Spanish *palmito*, meaning small palm.

## GEOGRAPHY AND ECOLOGY OF WILD PLANTS

In the southeastern United States, cabbage palmetto is native to Florida, Georgia, southeastern North Carolina, and South Carolina. In the Caribbean, it is native to the Bahamas and Cuba. The species is grown throughout the southern United States.

Cabbage palmetto occurs in hammocks, tidal flats, coastal dunes, coastal marshes, sandy shores, river banks, and seacoast woodlands. This subtropical and warm temperate palm is very cold hardy compared to most other palms and can withstand short periods of freezing. It is extremely salt-tolerant and can grow in brackish water but not seawater. It also tolerates salty winds, waterlogging, drought, and a wide range of soil conditions, and is very resistant to fire and hurricanes.

Phytoplasmas are specialized bacteria, discovered in 1967. They especially infect plants of tropical and subtropical areas, including some important crops, causing mild to fatal symptoms. Phytoplasmas are normally transmitted from plant to plant, by sap-sucking insects. A deadly phytoplasma named Date Palm Lethal Decline and Texas Phoenix Palm Decline affects several palm species, including cabbage palmetto. The disease was found in palms in Florida in 2006 and confirmed in cabbage palmetto in 2008. Quarantine regulations govern the sale and movement of palm species affected by the disease.

## PLANT PORTRAIT

*Sabal palmetto* is a slender tree with a single, unbranched trunk growing up to 20 m (65 feet) in height, occasionally to 28 m (92 feet). The trunk diameter is up to 60 cm (2 feet). At maturity, the trees tend to develop a dense, rounded crown up to 5.5 m (18 feet) across, made up of long, curved leaves. As with many palms, the crown is usually wider when grown in shade and more compact when grown in

**FIGURE 19.1** Cabbage palmetto (*Sabal palmetto*). (a) Trees. (Courtesy of Mmeknight4.) (b) Crown of tree. (Courtesy of Forest & Kim Starr [CC By 3.0].) (c) Fruit. (Courtesy of Forest & Kim Starr [CC By 3.0].) (d) Distribution map.

full sun. The evergreen leaves are up to 3.7 m (12 feet) in length, with 40–60 leaflets each up to 80 cm (2.6 feet) long. The tiny flowers are creamy to yellowish-white and produced in large branching clusters up to 2.5 m (8.2 feet) in length. The mature fruits are dark blue or black spherical berries, 5–13 mm (0.2–0.5 inch) wide, and contain one hard, brown seed. As with numerous other species of palms, the lower leaves are gradually shed. In *S. palmetto*, the base of the leaves persists for some time on the lower trunk, producing a spiky appearance. The persistent bases are called "bootjacks" or "boots" for short because they resemble the Y-shaped devices used to help remove boots from feet.

The Seminole Indians made extensive use of *S. palmetto*. The leaves were used to thatch their huts and to construct baskets, and the fruits and palm hearts were eaten. The trunks were employed to make house poles, food paddles, staffs, arrows, and a variety of utensils.

The trunks and leaves of cabbage palmetto have also had some limited usage in modern times. Fibers from the bark and leaf sheaths have been made into scrubbing brushes. The young leaves are shipped worldwide every spring for use on Palm Sunday. Ornamental table tops have been prepared using polished

**FIGURE 19.2** Hearts of palm technology. Palm hearts, the soft inner parts of the upper stems, are harvested from many species of palms, and come in various sizes. Those shown here are from South American species. (a) Extracting palm hearts outdoors. (Courtesy of Preifutura Municipal de Itanhaém [Flickr/CC-attribution].) (b) Extracting palm hearts in a factory. (Courtesy of Ricardo Eliezer de Souza and Silva Maas.) (c) Commercial containers of palm hearts. (Courtesy of Valter Campanato [CC By 3.0].) (d) Palm heart salad. (Courtesy of Lablascovegmenu [Flickr/CC-attribution].)

cross sections of the trunk. Young stems have been used as canes, and hollowed out stems have been employed as pipes. Logs were used for wharf pilings and docks because they are resistant to sea worms.

Cabbage palmetto is often grown as a low-maintenance garden and street tree in semitropical areas. In Florida, there have been extensive plantings along freeways. The tree is also widely cultivated in Hawaii.

## CULINARY PORTRAIT

There are several thousand species in the palm family, distributed around the tropical world, and all have edible "hearts," although many do not have a pleasant taste and contain chemicals that may be unhealthy to consume. Palm plants develop new leaves at a growing point (meristem) located at the top of the stem. The very young leaves are mostly just leaf stalks packed together, and typically form a cylinder of tissue in the center of the uppermost part of the stem. The cylinder of tender young leaves, before they emerge from the center of the stem (and become green), together with the stem tissue just below the growing point, make up the "heart." Only about 100 palm species have hearts that are large enough to be worth collecting, and most of these are used in local cuisines as a vegetable. International trade in palm hearts is based to a considerable degree on wild *Euterpe oleracea* C. Mart. (manicole palm, assai palm) native to the estuary of the Amazon River, and large amounts are exported from Brazil. Many palm trees die when their growing point is harvested, and it is hard to justify killing a palm tree merely to eat a small portion of it. However, multistemmed palms can be sustainably collected from the wild, so long as one does not destroy too many of the stems by harvesting their growing points. Unfortunately, many palm species are being destructively harvested from the wild, including *Sabal* species in Mexico and Central America. The peach palm (*Bactris gasipaes* Kunth), a native of Amazonian areas of Colombia, Ecuador, Peru, and Brazil, is perhaps the most important source of palm hearts. This species has been domesticated and is cultivated in Central and South America specifically for the harvest of palm hearts. There is some export of palm hearts from Asia, based on the coconut palm (*Cocos nucifera* L.), which is widely grown, and from *Daemonorops schmidtiana* Becc. (which lacks a widespread English common name). Often, when palms are cut down for timber or because of age, the edible hearts are harvested.

The Seminole Indians ate the raw fruits of cabbage palmetto, reduced them to syrup, and used them to prepare bread meal. The berries have a sweet flavor reminiscent of prunes, but generally people have not consumed the fruits (the huge fruit production is extensively eaten by wildlife). The hearts of cabbage palmetto used to be regarded as a food for poor people in Florida and were widely harvested during the Depression era. The hearts have been commercially canned and sold. They are still used in southern cooking to make "swamp cabbage" and hearts of palm salad. Because so many people were harvesting sabal palmetto for the hearts, thereby killing the trees, Florida enacted a law to protect the species.

Hearts of palm can be used raw in salads and in main dishes, or deep-fried. They can be boiled and mixed with vegetables and eggs to make a casserole. Central and South Americans like their hearts of palm very soft, and the processors cook the product longer to obtain the softer texture. The majority of Americans like their hearts of palm more crunchy, as they are when newly harvested. By comparison with the major commercial hearts of palm, cabbage palmetto hearts are relatively bland.

Cabbage palmetto is a significant source of nectar used by bees to make honey that is commercially sold.

### Culinary Vocabulary

- Because of some resemblance in form and texture to cabbages, palm hearts are often called "cabbages" or "palm cabbages," especially in Europe. The hearts are also occasionally called "palmettos" and "palmitos," especially in the New World.
- "Millionaires' salad" is a phrase often applied to edible palm hearts. The name is based on the facts that palm hearts of some species are very rare delicacies and harvesting them is very expensive because it may require sacrificing an entire tree.
- "Cabbage stew" in rural areas of Florida may mean a stew made with cabbage palmetto, not cabbage.

## PROSPECTS

The development of cabbage palm as a food crop faces two major obstacles. First, harvesting the edible growing tips kills the trees, which represents excessive waste. As a consequence, Florida legislation limits harvest of wild trees from public lands. Second, harvest of hearts of palm from both wild and cultivated species in Central and South America is already a thriving industry. At present, the major opportunity for large-scale harvest of cabbage palm hearts occurs when plantings along highways and natural areas need to be cleared for legitimate reasons. However, this does not provide a sufficient basis for providing a reliable and regular supply. The advantage that cabbage palm has over the palm species currently harvested for palm hearts is its exceptional adaptation to the northern limits of palm growth. Although currently relatively unlikely to become a significant new crop, identification of mutant forms with superior taste and, perhaps, technology that could stimulate multiple-trunked forms could make cabbage palmetto a promising possibility.

## CURIOSITIES OF SCIENCE AND TECHNOLOGY

- In 1776, during the Revolutionary War against the British, American patriots in Charleston, Florida, under the leadership of General William Moultrie (1730–1805) constructed Fort Moultrie on Sullivan's Island in Charleston Harbor and defended it successfully against the British fleet. The fortifications were made with cabbage palmetto trunks. These logs were particularly suitable because they did not produce lethal splinters when struck by cannonballs and absorbed the shock of impact.
- Cabbage palm was declared to be the state tree of Florida in 1953 and the state tree of South Carolina in 1939. The tree is on the state flag of South Carolina, which has the nickname "Palmetto State."

## KEY INFORMATION SOURCES


Bombard, M.L. 1950. *Palm trees in the United States*. Agriculture Information Bulletin 22. U.S. Department of Agriculture, Washington, DC. pp. 26.

Brown, K.E. 1976a. Ecological studies of the cabbage palm. *Principes* 20: 3–10.

Brown, K.E. 1976b. Ecological studies of the cabbage palm, *Sabal palmetto*. II. Dispersion, predation, and escape of seeds. *Principes* 20: 49–56.

Brown, K.E. 1976c. Ecological studies of the cabbage palm, *Sabal palmetto*. III. Seed germination and seedling establishment. *Principes* 20: 98–105.

Brown, K.E. 1976d. Ecological studies of the cabbage palm, *Sabal palmetto*. IV. Ecology and geographical distribution. *Principes* 20: 148–157.

Brown, K.E. 1982. Observations on the natural history of the cabbage palm, *Sabal palmetto*. *Principes* 26: 44–48.

Castano, F., Crevecoeur, M., and Stauffer, F.W. 2009. Contribution to the knowledge of the floral structure in *Sabal palmetto* (Walter) Lodd. ex Schult. (Arecaceae: Coryphoideae). *Phytomorphology* 59: 85–91.

Landry, G.P., and Reese, W.D. 1996. Sabal palmetto naturalized in western Louisiana. *Principes* 40: 177–178.

Lopez, J.C., and Dirzo, R. 2007. Floristic diversity of sabal palmetto woodland: an endemic and endangered vegetation type from Mexico. *Biodiv. Conserv.* 16: 807–825.

Martín-Alemán, N., León-Hernández, A.-M., and Rodriguez-Pérez, J.A. 1999. Effect of salinity on germination of *Phoenix canariensis* and *Sabal palmetto* (Arecaceae). *Acta Hortic.* 486: 209–213.

McPherson, K., and Williams, K. 1996. Establishment growth of cabbage palm, *Sabal palmetto* (Arecaceae). *Am. J. Bot.* 83: 1566–1570.

McPherson, K., and Williams, K. 1998. Fire resistance of cabbage palms (*Sabal palmetto*) in the southeastern USA. *Forest Ecol. Manage.* 109: 197–207.

Menge, J.T., and Brown, K.E. 1992. Commercial transplanting of wild cabbage palms, *Sabal palmetto* in Florida. *Principes* 36: 94–98.

Moegenburg, S.M. 1996. *Sabal palmetto* seed size: causes of variation, choices of predators, and consequences for seedlings. *Oecologia* 106: 539–543.

Perry, L., and Williams, K. 1996. Effects of salinity and flooding on seedlings of cabbage palm (*Sabal palmetto*). *Oecologia* 105: 428–434.

Reynolds, C.J. 1998. An unappreciated native: *Sabal palmetto*. *Principes* 32: 84–85.

Riefer, R., and Zona, S. 2005. A new cultivar of *Sabal palmetto*. *Palms* 49: 46–47.

Small, J.K. 1923. The cabbage tree: *Sabal palmetto*. *New York Bot. Gard. J.* 36: 261–270.

Wade, D.D., and Langdon, O.G. 1990. *Sabal palmetto* (Walt.) Lodd. ex J.A. and J.H. Schult. Cabbage palmetto. In *Silvics of North America*. Vol. 2, Hardwoods. Agric. Handbook 654. Technical coordination by R.M. Burns and B.H. Honkala. U.S. Department of Agriculture, Forest Service, Washington, DC. pp. 762–767.
Zona, S. 1990. A monograph of *Sabal* (Arecaceae: Coryphoideae). *Aliso* 12: 583–666.
Zona, S. 1997. The genera of Palmae (Arecaceae) in the southeastern United States. *Harvard Pap. Bot.* 2: 71–107.
Zona, S.A. 2000. Sabal. In *Flora of North America North of Mexico*. Vol. 22. Edited by Flora of North America Editorial Committee. Oxford University Press, Oxford, U.K. pp. 107–110.

## Specialty Cookbooks

(*Note:* Palm hearts have been characterized as a "national ingredient" of Brazilian food, and Brazilian cookbooks are an excellent source of recipes.)

Bogosian, F.S. 2008. *Delicious Brazil: gastronomy and tourism*. Create Space, Scotts Valley, CA. pp. 150.
Botafogo, D. 1993. *The art of Brazilian cookery*. Hippocrene Books, New York. pp. 240.
Calvo, I.M. 1981. *Usos culinarios del chontaduro*. Fundación Educación Superior, Instituto Vallecaucano de Investigaciones Científicas. Cali, Columbia. pp. 73. (In Spanish. Presents 40 recipes for palm hearts, mostly for baked products such as bread, cake, and cookies.)
Mendelson, A., and Martinez, Z. 2004. *Zarela's Veracruz: Mexico's simplest cuisine*. Houghton Mifflin Harcourt, Boston, MA. pp. 400.

# 20 Cactus Pear

Family: Cactaceae (cactus family)

## NAMES

Scientific name: *Opuntia ficus-indica* (L.) Mill.

- The word "cactus" is the root of the family name Cactaceae and also applies to the genus name *Cactus*. This name was used by the ancient Greeks for some spiny plant quite unrelated to cactus plants. Today, a "cactus" may be any of the 1700 or so species in the Cactaceae.
- The cactus pear has also been called pear apple and tuna or tuna cactus (in Latin America), and Barbary fig and Indian fig (in many European countries). The names with "Indian" (for India) and "Barbary" (for the Barbary Coast, extending from the Egyptian border to the Atlantic, and including Morocco, Algiers, Tunis, and Tripoli) reflect the past tendency of Europeans to give exotic names to unfamiliar fruit (inappropriate names in this case, as the fruit comes from the New, not the Old World).
- The cactus pear has been called the "prickly pear" for many years. Unfortunately, some 20 different species of *Opuntia* are called prickly pear, so the name can be misleading.
- The phrase "prickly pear," the traditional name for the cactus pear and its fruit, evokes a negative image. To overcome this, in 1990 several people, notably Frieda Caplan, a California marketer of exotic fruits and vegetables, argued that the name "cactus pear" be used, and this recommendation was adopted by representatives of 10 countries at the "Second International Conference on Tuna and Cochineal" in Santiago, Chile, in 1992. (See the section "Curiosities of Science and Technology" for explanation of "cochineal.")
- "Tuna," a much-used Mexican term for cactus fruit, traces to the Caribbean word *tun*, fruit or seed. In Aztec, Mexico, the word *nochtli* was used for the fruits of cactus pears, but the Spanish conquistadors took up and established the word tuna.
- In Israel, the cactus pear fruit is called *sabra* (from the Hebrew *tzabar*), a term that was coined after the cactus was brought to the Middle East for use as a living fence. Sabra is also used to denote the cactus pear in Arabic, and in northern Africa and southwestern Asia. The word has also come to be applied to native-born Israelis (sometimes to Arab women as well), by analogy meaning tough on the outside, sweet on the inside.
- Cactus pear plants are sometimes called "mission pears" because they were often grown about missions in Mexico and the southwestern United States. However, "mission pears" usually refer to very old pear varieties brought over from Europe by missionaries, especially from France, and often by Jesuits.
- In Europe, especially in Sicily, there are very tasty varieties of cactus pears with whitish or yellowish fruits. In Sicily, these are called *bastardi* or *bastadoni.*
- In Australia, where introduced *Opuntia* species are pernicious weeds, they are called "pest pears."

**FIGURE 20.1** Cactus pear (*Opuntia ficus-indica*). (a), (b) Fruits. (Courtesy of U.S. Department of Agriculture.) (c) Flowers. (Courtesy of G.M. Stolz, U.S. Fish and Wildlife Service.) (d) Distribution map. Areas in red are native distributions; areas in green are introduced distributions.

- The genus name *Opuntia* is based on the Greek name of a non-cactus plant that grew around the ancient town of Opus or Opuntia in Greece, the homesite of a tribe called the Locri Opuntii.
- *Ficus-indica* in the scientific name *O. ficus-indica* means "fig of India," a rather imaginative description for a New World cactus. As noted below, the taste of some varieties is reminiscent of figs.

## GEOGRAPHY AND ECOLOGY OF WILD PLANTS

The cactus pear is native to the desert zones of northwestern Mexico and possibly also the southwestern United States. There is some evidence that the species originated as a cultivated crop in Mexico and that its native range is based on escapes from ancient cultivation. Accordingly, ascertaining its original indigenous range is probably impossible. In the United States, it has been collected as a wild (i.e., uncultivated) plant in many places, particularly in California and Arizona.

**FIGURE 20.2** Cactus pear (*Opuntia ficus-indica*). (a) A commercial fruit (note that spines have been removed). (Courtesy of E. Small and B. Brookes.) (b) Sliced fruit. (Courtesy of E. Small and B. Brookes.) (c) Cactus pear pads (nopales) for sale in Mexico City. (Courtesy of Pro_bug_catcher.) (d) A large plant. (Courtesy of U.S. Department of Agriculture.)

Apparently wild-growing plants have been reported in many other locations, as shown on the distribution map (Figure 20.1d), but these are likely to have originated as introduced plants. In the United States, cactus pear has been collected from coastal chaparral, sage scrub, arid uplands, washes, canyons, and disturbed sites, at altitudes of 0–300 m (0–1000 feet).

## PLANT PORTRAIT

The fruits of more than 40 species of cacti can be purchased in Latin America, although only a few of these are grown on a major scale, and the plant highlighted here is the most important. It has been estimated that the average Mexican citizen eats as much cactus as the average American eats cauliflower. The cactus pear or prickly pear has the appearance of a fleshy bush or small tree 3–5 m (10–16 feet) in height. The plant was brought to Europe by the first Spanish colonists from Mexico and has been cultivated along the Mediterranean coast since the late seventeenth century. Numerous species of *Opuntia* are commonly called prickly pear, and in Mexico, they are cultivated and harvested for food. *Opuntia tuna* (L.) Mill., commonly called "tuna" and also known as "elephant ear prickly pear," is particularly popular in Mexico, for although it produces rather spiny, small fruit, the plants make an ideal hedge. ("Tuna" is also a frequent term in Mexico for cactus fruit, as discussed in the section "Names.") However, *O. ficus-indica* is the chief species of interest and the main cactus grown outside Mexico. A mature plant can yield 100–200 fruits in a season. There are dozens of cultivated varieties. The cactus pear is an ideal fruit plant to raise in arid regions because it requires far less water than conventional crops. Native Americans dried cactus fruit in the sun, and these could be stored for at least a year. Cactus pears are grown for two edible commodities, as noted in the following.

The cactus pear fruit is generally pear-shaped and has a number of small spines (except in spineless varieties). The skin of commercial cactus pear fruits can be red, pink, orange, purple, green, or yellow. The fruits vary in weight from 100 to 200 g (3.5–7 ounces), and are generally 5–10 cm (2–4 inches) in length. The skin or rind is thick and fleshy (and represents 30%–40% of the weight of the fruit). The center of the fruit is filled with a soft juicy pulp (making up 60%–70% of the total fruit weight; about 85% of the pulp is water). The flesh may be green, light yellow-green, orange-yellow, deep golden, or dark red depending on variety. The sugar content of the fruits is about 15%. There are many hard-coated, small, black seeds in the pulp (composing 5%–10% of the pulp weight). These seeds are considered very objectionable to new consumers, and there are efforts underway to breed seedless varieties.

In *Opuntia* species, the flattened branches or stem sections are technically called cladodes and are popularly called pads (or sometime paddles) and joints (although cladodes are often joined together by what appear to be joints). These have the function of leaves (although some anatomists have equated the spines on the stems with the leaves of most plants). Although *Opuntia* species reproduce sexually by seeds, they also reproduce vegetatively when cladodes fall off the plant to the ground and produce roots and new daughter cladodes (indeed, commercial plantations are established entirely with cladodes or portions of cladodes). Tender young prickly pear pads, called nopales (and usually referred to as nopalitos when cut up for culinary use), have been consumed as a vegetable in central Mexico since pre-Hispanic times. Nopalitos are mostly water (over 90%). In Mexico, most nopalitos come from *O. ficus-indica*, but in the United States, mostly in southern California and Texas, considerable amounts of nopalitos are also obtained from *O. cochenillifera* (L.) Salm-Dyck [*Nopalea cochenillifera* (L.) Salm–Dyck], called the cochineal cactus and nopal cactus.

## CULINARY PORTRAIT

The mild, pleasant flavor of a ripe cactus pear fruit, depending on variety, may resemble strawberries, watermelons, honeydew melons, figs, bananas, or citrus. The texture of the flesh is somewhat more granular than that of watermelon. Cactus pears may be eaten raw, at room temperature or chilled, and alone or with lemon or lime juice (desirable since the bland fruit lacks acidity). The rind is not edible. There are numerous hard seeds present in the edible flesh, and these make eating it uncomfortable. The seeds can be chewed and eaten, swallowed whole, or spit out. (Eating the seeds does not cause harm; indeed, Native Americans once collected the seeds, dried them in the sun, and ground them into a flourlike meal used for cooking.) The seeds can be sieved out if the fruit is to be cooked. Fresh pulp may be added to salads and various desserts. The flesh can be cooked into jams and preserves or cooked down into a syrup as a base for jelly and candy—the "cactus candy" in some Mexican food stores. This syrup can be reduced even further into a dark red or black paste that is fermented into a potent alcoholic drink called "coloncha" (see the section "Culinary Vocabulary"). The fruit pulp can be dried and ground into flour for baking into small sweet cakes or stored for future use.

Cactus pears bought in supermarkets should have had the glochids (hairlike spines) removed (by scraping, singeing, or washing with a high-pressure nozzle). These may be almost invisible but can be very painful if they pierce the hand, fingers, or tongue, and once in the skin, the irritation may last for several days.

The best fruits have deep, even color. Fruits should be ripened at room temperature and are ripe when they give slightly to palm pressure. Mature fruit may be stored in the refrigerator for up to a week, but exposure to temperature below 5°C (41°F) for several days may result in injury and decay.

Some Mexican companies candy nopalitos or process them for export as pickles, sauce, or jam. The pads are said to taste something like green beans or between that of green pepper and asparagus, although nopales sold in jars have a tart, pickle-like taste. Fresh nopales are sometimes available in supermarkets in North America as "cactus leaves" (although botanically they are flattened stems). These are generally harvested when 15–20 cm (6–8 inches) long. Nopales are highly perishable. The smaller young pads in the early spring are the most succulent, tender, and delicate in flavor, crunchy when fresh, and have the fewest spines. Immature pads also have less oxalic acid, which can be toxic in large amounts. Nopales are slippery because they exude a mucilaginous substance similar to that in okra. Because of the mucilage, diced nopalitos can help thicken soups and stew stock. (To avoid contact with the sticky fluid that oozes from the nopales, they can be steamed whole, just long enough for their color to change from bright green to olive drab. Once the color changes, they should immediately be plunged into a bowl of cool water, then sliced on a cutting board. The fluid can be mixed into dishes to enhance them.) Fresh pads are full of water and should be bright green and firm. To prepare the pad, simply hold its base and scrape the skin on both sides with a blunt knife until all the spines are removed. (Use tongs to avoid getting spines or glochids [tiny, fuzzy spines] in your fingers, and take care to remove the areoles [the places where spines develop on the pad].) Then peel the pads with a sharp knife or vegetable peeler, and cut them into shoestring strips or dice them according to the needs of the recipe. They can be eaten raw in salads, boiled and fried like eggplant, pickled with spices, or cooked with shellfish, pork, chillies, tomatoes, eggs, coriander, garlic, and onions. Omelets are commonly prepared with young cactus pear pads throughout the southwestern United States. Cactus pie can be made from cactus nopalitos. The pie may taste apple-like because both the cactus and apples contain high levels of malic acid.

In Mexico, the petals from the flowers of cactus pear are added to meat dishes. Native Americans traditionally celebrated a "cactus moon" in the early spring, by collecting flower buds of wild *Opuntia* species and cooking them as a special treat.

### Culinary Vocabulary

- *Huevos rancheros con nopalitos* is ranch-style eggs with cactus pads.
- *Calonche* or *colonche* (pronounced kahl-ohn-chay) is an alcoholic beverage made from the fermented juice of certain cacti. Native Americans once used cactus pear fruits to prepare this as well as other alcoholic drinks.
- *Queso de tuna* (pronounced KEH-soh day too-nah) is a sweet paste prepared from fermented prickly pear juice, used in Mexican confections.
- Several alcoholic beverages are produced from cactus pear in the Old World. *Ficodi* is a cactus-pear-flavored liqueur produced in the village of Gagliano Castelferrato in central Sicily. *Bajtra* (the Maltese name for cactus pear) is the name of a liqueur produced from cactus pear growing wild in Malta. *Tungi Spirit* is a locally distilled liqueur produced using cactus pear on the island of Saint Helena.

## PROSPECTS

Cactus pear has developed into an important crop in semitropical climates that are arid or semiarid. The species has become the most widely cultivated cactus grown for crops in the world. The plants are cultivated for fruit production on all continents except Antarctica, with at least 100,000 ha

(247,000 acres) of orchards (in addition, considerable fruit is harvested from wild plants and home gardens). It is grown in about 30 countries, mostly in Mexico, but with notable crops also in Chile, Bolivia, South Africa, Italy, Argentina, the United States, and Israel. Given the dramatically decreasing supplies of water in the world, and cactus pear's natural adaptation to dry, hot climates, the species is likely to become more important in the future.

## CURIOSITIES OF SCIENCE AND TECHNOLOGY

- Cactus pear seeds have been found in human coprolites (fossilized feces) dated at 9000 years of age from caves in Mexico, providing proof that indigenous peoples were using cacti as food in ancient times.
- The first published illustration of cactus pear appeared in 1535 in the Spanish book *La Historia General* by Oviedo y Valdés. His erroneous report that eating the fruits turned one's urine red was widely circulated.
- Cacti are native to North and South America and the West Indies (some species of the primitive genus *Rhipsalis* are apparently native to Madagascar and a few other parts of the Old World), and characteristically grow in hot, dry, and hostile desert areas. The family has 122 genera and about 1600 species, nearly all of which have succulent, spiny stems. The literature shows that Native Americans tasted virtually every species! Cacti are favorite ornamentals, and many people specialize in collecting and cultivating these unusual plants. It has been claimed that there are more species of cacti growing in homes (especially in cold regions) than species of any other plant family (about 300 species of cacti are available as ornamentals).
- Certain opuntias and barrel cacti can live for up to 3 years on water stored in their stems, in the absence of water uptake from the soil.
- So-called "rain roots" are produced by *Opuntia* species within a few hours of rain falling on the plants. These quickly absorb available water and die off as soon as the soil dries.
- The mucilaginous sap from the pads of the cactus pear can be used in first aid similar to the *Aloe vera* plant. Simply cut off a portion of a pad, crush it, and squeeze the juice onto a cut, burn, or bruise. In Mexico, the cactus pear is commonly used on burns and swellings.
- In Argentina, the slimy juice from cactus pear stems was traditionally used to make whitewash more adhesive. In the construction of Spanish mission buildings in California during the 1700s and 1800s, the mucilage was used to bind adobe bricks.
- Before cheap synthetic aniline dyes were developed from coal tar in 1856, cactus plantations (including the cactus pear) were established for the production of cochineal (carminic acid), a dye. This is obtained by extracting the dye that the cochineal insects produce (from the females of the genus *Dactylopius*, which are much larger than the males). About 32,000 female insects weigh 1 kg (70,000 weigh 1 pound). The scale insect, which is native to Central America and Peru, feeds on the pads and fruit and develops a large quantity of the stain within its body. The dye was once used to color the robes of Aztec emperors, including those of Montezuma (1466–1520), a deep royal red. Taxes were sometimes paid in cochineal insects, and indeed the value of the dye once exceeded its weight in gold. In the sixteenth century, the export of cochineal from Mexico was second in importance and monetary value only to silver. The Spanish traders who introduced the dye to Europe kept the source secret, but Anton van Leeuwenhoek (1632–1723), the Dutch scientist who refined microscopes and first described cells, noted insect parts in the cochineal stain he was using, and deduced that the dye came from insects. The dye was used for the robes of European royalty, the red jackets of British soldiers (made famous by American patriot Paul Revere when he warned in 1775 that "the redcoats are coming"), and the crimson jackets of the Northwest Mounted Police or "Mounties" (now the Royal Canadian Mounted Police). Cochineal is used to make a modern dye called carmine. Cochineal or carmine is

still used in botanical stains and as a cloth dye, and has some use as an edible (but tasteless and odorless) dye for foods such as maraschino cherries, icings, creams, jellies, cakes, candies, wines, and liqueurs, as well as in lipsticks. The dye has reacquired importance since coal-tar (aniline) dyes used for these purposes have been linked to cancer in laboratory animals.

- Some forms of wild cactus pear form massive thickets. These have been observed to survive fires in southern California by regenerating from live stems in the center of the thickets where fire was unable to penetrate.
- Prickly pear cacti are frequently grown into hedges and fences by planting them a foot (30 cm) or so apart. Within several years, the plants grow together to form a spiny barrier that will repel any intruder larger than a rabbit.
- The American plant breeder and horticulturist Luther Burbank (1849–1926) was a very controversial genius, who introduced many new plants of great value, but also made some whose virtues were considered very dubious. Burbank helped develop a spineless form of cactus pear, and in 1911, he claimed that it was so good that it "promises to be of as great or even greater value to the human race than the discovery of steam." This led to numerous large plantings, especially around Los Angeles, and great disappointment when the food value of the plants proved to be much lower than expected.
- In Texas and Mexico, ranchers commonly use propane torches to singe the spines of wild *Opuntia* species so that cattle can forage on them. Cattle readily eat the singed plants, consuming cactus up to 10% of their body weight daily. Because the cacti typically contain about 90% water, the cattle can be sustained for many months of drought by relying entirely on the water in the plants.
- The introduction of *Opuntia* cacti to Australia in 1832 as hedging plants caused an ecological disaster. By 1925, several species were spreading through Australia like wildfire. A variety of control measures proved ineffective until 1925, when the Argentinean Moth *Cactoblastis cactorum* was introduced, the larvae of which feed on the cacti. By 1933, 90% of the cacti had been eradicated. A similar situation occurred in South Africa with *O. ficus-indica* covering a huge territory before suitable biological control agents (a weevil and a beetle, as well as the moth used in Australia) were found in the 1930s. (Ironically, cactus pear is now considered a valuable crop in South Africa, and farmers growing it there have great difficulties controlling the biological control agents.)
- Cacti are highly desirable as ornamental plants, and unfortunately many wild species have been badly overcollected from nature. There is legislation that now protects cacti. The Convention on International Trade in Endangered species of Wild Fauna and Flora (CITES) necessitates appropriate permits and documentation for moving plants across international boundaries, and this applies to all cacti except certain species used in commerce. Arizona law requires that native cacti must be tagged and a fee paid before they can be moved. California also has strict requirements that deal with the movement of cacti.
- The cactus fruit is naturally protected by tufts of glochids (small, barbed spines). However, the biological purpose of fleshy fruits such as that of the cactus pear is to attract animals to eat the flesh, so that the seeds will be distributed (the seeds of the cactus pear are distributed principally by being consumed by birds and mammals, which spread the seeds in their droppings). As the fruit ripens, the glochids drop off, so that animals will not be discouraged from eating the fruits.
- Plants acquire carbon dioxide through pores (stomates), simultaneously losing water through the pores. Most plants keep their stomates open during the day, when sunlight is available, because the usual photosynthetic machinery requires a continuous influx of carbon dioxide while the sun is shining. Plants regulate the opening of the pores, and most plants close them at night to prevent unnecessary loss of water (keeping the pores open during the day is the price paid to acquire carbon dioxide). In arid environments,

so much water can be lost when pores are open during the day that plants may die. Six to seven percent of the world's plant species, including cacti, have special photosynthetic machinery ("crassulacean acid metabolism" [CAM]) that permits the plants to keep their stomates closed during daylight. During the much cooler nights, when water loss is much lower, the pores open, taking in carbon dioxide and storing it in a fixed chemical form (as "organic acids") until light is available during the day to process the organic acids further by photosynthesis. Although there are other adaptations to drought, CAM is the most significant mechanism adapting plants to dry environments. The average plant loses five times as much water as cactus pear on a surface area basis. Moreover, because the surface area of cactus pear is far less than that of plants that have true leaves, it is able to survive under conditions that would kill most plant species. Cactus pear produces dry matter on a surface area basis at the same rate as the average plant species, despite using only 20% as much water, and this remarkable water use efficiency is why it is so useful with respect to water conservation.

- As noted above, the peculiar photosynthetic mechanism of cacti involves storage of acidic compounds during the night, which are metabolized into sugars when sunlight is available. As a result, cactus tissue harvested during the night will taste more acidic than the same tissue harvested during the day. Depending on the time of day (which affects the amount of light available), the taste of nopales may differ considerably.

## KEY INFORMATION SOURCES


Annecke, D.P., and Moran, V.C. 1978. Critical reviews of biological pest control in South Africa. 2. The prickly pear, *Opuntia ficus-indica* (L.) Miller. *J. Ent. Soc. Sth. Afr.* 41: 161–188.

Barbera, G., Inglese, P., and Pimienta-Barrios, E. (Editors). 1995. *Agro-ecology, cultivation and uses of cactus pear.* FAO Plant Production and Protection Paper 132. Food and Agriculture Organization of the United Nations, Rome. pp. 216.

Curtis, J.R. 1977. Prickly pear farming in the Santa Clara Valley, California. *Econ. Bot.* 31: 175–179.

Donkin, R. 1977. Spanish red: an ethnogeographical study of cochineal and the *Opuntia* cactus. *Transact. Am. Phil. Soc.* 67: 1–77.

Griffith, M. P. 2004. The origins of an important cactus crop, *Opuntia ficus-indica* (Cactaceae): new molecular evidence. *Am. J. Bot.* 91: 1915–1921.

Hanselka, C.W., and Paschal, J.C. 1991. Prickly pear cactus: a Texas rangeland enigma. *Rangelands* 13: 109–111.

Inglese, P. (Editor), and Brutsch, M.O. (Conference convenor). 1997. *Third international congress on cactus pear and cochenille.* International Society for Horticultural Science, Wageningen, the Netherlands. pp. 180.

Inglese, P., Barbera, G., and La Mantia, T. 1995. Research strategies for the improvement of cactus pear (*Opuntia ficus-indica*) fruit quality and production. *J. Arid Environ.* 29: 455–468.

Jacobo, C.M. 2001. Cactus pear domestication and breeding. *Plant Breed. Rev.* 20: 135–166.

Meyer, N.B., and McLaughlin, J.L. 1981. Economic uses of *Opuntia. Cactus Succulent J.* 53: 107–112.

Mondragon-Jacobo, C., and Perez-Gonzalez, S. 1996. Native cultivars of cactus pear in Mexico. In *Progress in new crops.* Edited by J. Janick. ASHS Press, Arlingon, VA. pp. 446–450.

Nefzaoui, A., and Inglese, P. 2002. *Fourth international congress on cactus pear and cochineal.* International Society for Horticultural Science, Wageningen, the Netherlands. pp. 356.

Nobel, P.S. 1988. *Environmental biology of agaves and cacti.* Cambridge University Press, Cambridge, U.K. pp. 270.

Nobel, P.S. (Editor). 2002. *Cacti. Biology and uses.* University of California Press, Los Angeles. pp. 280.

Pareek, O.P., Singh, R.S., Nath, V., and Vashishtha, B.B. 2001. *The prickly pear (*Opuntia ficus-indica *L. Mill.).* Agrobios, Jodhpur, India. pp. 76.

Pimienta-Barrios, E. 1993. Vegetable cactus (*Opuntia*). In *Pulses and vegetables.* Edited by J.T. Williams. Chapman and Hall, London, U.K. pp. 177–191.

Pinkava, D.J. Opuntia. In *Flora of North America North of Mexico.* Vol. 4. Edited by Flora of North America Editorial Committee. Oxford University Press, Oxford, U.K. pp. 123–148.

Russell, E.C., and Felker, P. 1987. Comparative cold-hardiness of *Opuntia* spp. and cvs. grown for fruit, vegetable and fodder production. *J. Hortic. Sci.* 62: 545–550.

Russell, E.C., and Felker, P. 1987. The prickly pears (*Opuntia* spp. Cactaceae): a source of human and animal food in semiarid regions. *Econ. Bot.* 41: 433–445.

Saenz, C. 2000. Processing technologies: an alternative for cactus pear (*Opuntia* spp.) fruits and cladodes. *J. Arid Environ.* 46: 209–225.

Savio, Y. 1987. Prickly pear cactus: the pads are "nopales", and the fruits are "tunas"— they are easy to grow and wonderful to eat. *Cactus Succulent J.* 59(3): 113–117.

Savo, Y. 1989. *Prickly pear cactus*. University of California, Davis, CA. Family Farm Series. 6 pp.

Small, E., and Catling, P.M. 2004. Blossoming treasures of biodiversity: 11. Cactus pear (*Opuntia ficus-indica*)—Miracle of water conservation. *Biodiversity* 5(1): 27–31.

Stintzing, F.C., Schieber, A., and Carle, R. 2001. Phytochemical and nutritional significance of cactus pear. *Food Res. Technol.* 212: 396–407

## Specialty Cookbooks

Niethammer, C.J. 2004. *The prickly pear cookbook.* Rio Nuevo, Tucson, AZ. pp. 84.

Tate, J.L. 1972. *Cactus cook book; succulent cookery international.* Second edition. Cactus & Succulent Society of America, Reseda, CA. pp. 126.

Latorre (1977) has two recipes for cactus leaves, Russell (1975) has three, and Tull (1987) has three recipes for cactus leaves and one for the fruit. (See the Appendix to this book for details of the publications cited.)

# 21 California Bay

Family: Lauraceae (laurel family)

## NAMES

Scientific name: *Umbellularia californica* (Hook. & Arn.) Nutt.

- The English word "bay" is derived from the Latin *baca*, berry, originally applied to the berries of the much better known European bay or laurel tree (*Laurus nobilis* L.), not the leaves used in commercial bay preparations.
- California bay is also known as bay (likely to lead to confusion with European bay), California bay laurel, California laurel, California olive, California pepper, California spice tree, cajeput tree, laurel (likely to lead to confusion with other species), cinnamon bush, Oregon bay, Oregon myrtle, myrtlewood, peppernut tree, Pacific myrtle, pepperwood, sassafras laurel, spicebush (but see Chapter 90 on Spicebush [*Lindera benzoin*]), and spice tree. There are also numerous other local names, many applied to other species as well, and best avoided.
- The names pepperwood and California pepper are based on the strong peppery-menthol scent given off when the leaves are crushed.
- The odor released from the foliage of California bay can cause headaches, dizziness, and nausea in susceptible individuals; as a result, the plant has acquired the name "headache tree." (Ironically, California Native Americans used the foliage to treat headaches, and the practice was taken up by some European settlers.)
- California bay or California laurel should not be confused with plants with similar sounding names, such as bayberry (*Myrica pensylvanica* Loisel.), a native of eastern North America that provides bayberry oil used in perfumes and in candle making; California bayberry (*Myrica californica* Cham.), a native of the U.S. Pacific states; mountain laurel (*Kalmia latifolia* L.), a poisonous, eastern North American bog-loving plant sometimes grown as an ornamental (several other species of *Kalmia* are also called laurel; the constituent andromedotoxin in *Kalmia* is so potent that it can poison honey made by bees visiting its flowers); West Indian bay tree (*Pimenta racemosa* (Mill.) J.W. Moore, a relative of allspice, *Pimenta dioica* (L.) Merr.), a native of the Caribbean region, with oil used in perfumery and to make bay-rum; and cherry (or English) laurel (*Prunus laurocerasus* L.), a native tree of southeastern Europe and Asia Minor. Cherry laurel has sometimes been mistaken for common European bay laurel with disastrous results; cherry laurel produces prussic acid, a virulent poison. Despite its toxicity, cherry laurel leaves were once occasionally used to flavor milk in England. Portuguese (or Portugal) laurel (*Prunus lusitanica* L.) is another toxic plant that has caused poisoning by virtue of being mistaken for bay laurel.
- The genus name *Umbellularia* is based on the Latin word *umbella*, sunshade or parasol, an allusion to the umbrella-like (umbelliform) arrangement of the flowers on stalks that converge to a point (like the fabric supports of an umbrella).

**FIGURE 21.1** California bay (*Umbellularia californica*). (a) Semiedible fruit, some split open to show large, somewhat edible nut. (Courtesy of Nimmolo.) (b) Huge street tree. (Photo reproduced with permission of Urban Forestry Division, Los Angeles.) (c) Fruiting branches. (Courtesy of A. Barra [CC By 3.0].) (d) Distribution map.

- *Californica* in the scientific name is Latin for Californian, an appropriate description since almost all of the distribution range of the species is in the state.

## GEOGRAPHY AND ECOLOGY OF WILD PLANTS

This species is native to the West coast of the United States, from southwestern Oregon south through California to northern Baja California in Mexico. It occurs widely in the Coast Ranges, less frequently in inland valleys and in the foothills of the western Sierra Nevada. Occasionally it develops pure stands, but mostly occurs with other trees and shrubs. It grows both in cool-humid and hot-dry climates. It tolerates drought and shade but does not grow large in such conditions. California bay is often in moist soils in alluvial flood plains, semishaded valleys and canyons. It is found in oak woodland, mixed evergreen forest, redwood forest, and chaparral.

## PLANT PORTRAIT

California bay, the only species in the genus *Umbellularia*, is a broad-leaved evergreen tree. It commonly grows to heights of 12–24 m (39–79 feet), occasionally up to 30 m (98 feet) with trunk diameters of 46–76 cm (18–30 inches), very rarely up to 305 cm (10 feet). One tree grew to a record height of 53.3 m (175 feet). Near the Pacific Ocean and in very dry, rocky habitats, the species is often merely a shrub. The flowers are small and pale yellow. The fruits resemble miniature avocados the size of olives; they are 1–2.5 cm (0.4–1 inch) long, slightly longer than broad, greenish when immature, developing various colors at maturity, including speckled yellow-green, pale yellow, and yellow-green tinged with dull red or purple. The fruits contain a single, large, thin-shelled seed. The leaves are 2.5–10 cm (1–4 inches) long, 1.5–3 cm (0.6–1.2 inches) wide, generally widest in the lowest third. The leaves are thinner and more elongated than those of Mediterranean bay.

The leaves, fruits, and seeds are rich in chemicals (particularly umbellone) that can cause medical issues. Leaf vapors have produced various problems, including sneezing, headache, sinus pain, and even unconsciousness. Menthol and umbellone in the foliage can interfere with respiration, heartbeat, and blood circulation. Umbellone is a fungicide and germicide. Volatile oil from the leaves has been used for various applications. Native Americans used the leaves internally and externally for medicinal purposes, and for vermin control. Oil from the leaves, seeds, and wood has been adopted for a range of pharmaceutical purposes.

Small amounts of California bay timber are harvested, the wood often called myrtlewood. (Numerous unrelated species have "myrtle" in their names. The true myrtle, *Myrtus communis* L., a native of the Mediterranean region, produces spicy leaves and berries which are used in cooking like bay leaves.) The hard, fine, beautifully grained wood is in demand for furniture and paneling, and for specialty products, including gunstocks, guitars, novelty carving, and turnery (bowls and the like).

The species is grown as an ornamental evergreen tree for landscaping. It is cultivated well beyond its natural range, for example, in southern British Columbia and western Europe, and sometimes as a decorative potted plant for patios and lobbies. It tolerates moderate pruning, and the dense, aromatic foliage can be molded into pleasing shapes.

## CULINARY PORTRAIT

Native Americans consumed considerable amounts of fruits and seeds of California bay, and used root bark to prepare a beverage. The seed is reminiscent of an avocado pit, and Native Americans roasted it to remove much of the pungency before consumption. Today, wild food collectors often roast California bay pits. The pits contain a waxy fat reminiscent of cocoa butter; and ground, roasted pits with added sugar are reminiscent of chocolate. The fleshy portion of the fruit is palatable for a brief period when mature (afterward it decays like an overripe avocado) and was eaten raw by Native Americans. The sun-dried flesh was also consumed, usually just the lower third of the fleshy part of the fruit, as the upper, thinner flesh is high in the objectionable chemicals of the plant.

The expression "bay leaves" normally refers to the dried leaves of the Mediterranean bay (*Laurus nobilis*), one of the major commercial spices of the world, very widely used in cookery to flavor a considerable range of dishes. By contrast, California bay leaves are a product that has entered the commercial marketplace only to a minor extent. Nevertheless, California bay leaves are somewhat comparable to Mediterranean bay leaves, and can be used for the same purposes. Bay leaves are very frequently used in combination with other spices, such as oregano, marjoram, rosemary, and thyme, and the same can be done with California bay. California bay produces more potent-tasting aromatic leaves than the common European bay, with a stronger camphor and cinnamon flavor. California bay needs to be used in smaller amounts (no more than half as much) in recipes calling for conventional bay. Overusing California bay can make a dish inedible. Most cooks who are familiar with both European and California bay prefer the former, as California bay tends to be relatively bitter or harsh, and produces a menthol-like note. Umbellone, the primary flavorant of California bay, is a central nervous system toxin, and unlike conventional bay, California bay has not been given GRAS status ("generally recognized as safe") by the U.S. Food and Drug Administration.

### Culinary Vocabulary

- "Bay nuts" are the pits (seeds) of California bay, popularly roasted by wild food collectors. Technically, nuts are "large" achenes (an achene is a kind of single-seeded, dry, indehiscent fruit in which the seed coat is not part of the fruit coat), and bay nuts are not true nuts in this sense. In a nontechnical sense, nuts are hard-textured, usually oily, kernel-like edible plant materials housed in a frequently hard shell. In this popular sense, nuts are usually seeds or fruits, and bay nuts are, in fact, seeds, although they do not occur in a hard shell, like walnut (which is a nut both in the technical and nontechnical senses).
- "Turkish bay leaves" is a phrase often used for the leaves of European bay (*Laurus nobilis*).
- "Mexican Bay" (*Litsea glaucescens* Kunth of the Lauraceae) grows in Mexico and Central America. Its leaves resemble Mediterranean bay in appearance and flavor. While the spice is used south of the United States, it is rarely encountered north of Mexico.

## PROSPECTS

The California bay produces leaves that are reasonably comparable to those of the commercially successful Mediterranean bay. Although the flesh and pits of the fruits are edible, their palatability is insufficient to justify commercial development. Most Mediterranean bay leaves are collected from wild undomesticated plants, which are plentiful. Although somewhat attractive, and the leaves are marketed to a small extent as a food seasoning, California bay is not available in sufficient quantity or quality to justify a new spice industry. However, selection of superior trees together with a program of cultivation could result in the California bay developing into a minor commercial success.

## CURIOSITIES OF SCIENCE AND TECHNOLOGY

- The aromatic chemicals of strongly scented plants are likely present in many (perhaps most) cases because they repel damaging insects. The foliage of such plants is often used as insect repellents where modern technology is not being applied, and California bay is an example of this. The leaves were used as a fumigant by Native Americans to repel biting insects, especially fleas. A decoction (watery tea) was used to remove head lice.
- Allelopathy is the inhibition of competing plants by the release of chemicals into the soil by a plant species. This is a kind of chemical warfare waged in the plant kingdom. Chemicals that California bay leaves release into the soil (water-soluble terpenes) strongly inhibit seedlings of many species from establishing nearby.
- Since 1995, California bay has become a major host of a disease-producing fungus (*Phytophthora ramorum*) suspected of having been imported to California. Numerous other species have been found to also act as hosts (carrying the infection but not dying from it), but generally not as extensively as California bay. A range of species has been found to develop disease when infected by spores from the fungus growing on California bay and other hosts, but California oaks are especially susceptible, and hundreds of thousands have succumbed to "Sudden Oak Death." It has been observed that oak trees growing within 5 m (16 feet) of a California bay tree are in grave danger of becoming infected and dying. This deadly fungus problem is now the subject of major scientific and landscape management research.
- In a remarkable analysis of adaptive animal behavior, it has been shown that dusky-footed woodrats (*Neotoma fuscipes*) place California bay leaves on or near their sleeping nests with the result that volatile substances in the leaves repel skin parasites that otherwise would harm the rats.

## KEY INFORMATION SOURCES


Benemei, S., Appendino, G., and Geppetti, P. 2010. Pleasant natural scent with unpleasant effects: cluster headache-like attacks triggered by *Umbellularia californica*. *Cephalalgia* 30: 744–746.

Bergen, J.H. 1971. The Oregon myrtle, *Umbellularia californica*. *Arboretum Bull.* 34(3): 13–15.

Buttery, R.G., Black, D.R., Guadagni, D.G., Ling, L.C., Connolly, G., and Teranishi, R. 1974. California bay oil, 1: constituents, odor properties. *J. Agric. Food Chem.* 22: 773–777.

DiLeo, M.V., Bostock, R.M., and Rizzo, D.M. 2009. *Phytophthora ramorum* does not cause physiologically significant systemic injury to California bay laurel, its primary reservoir host. *Ecol. Epidemiol.* 99: 1307–1311.

Frankel, S.J., Kliejunas, J.T., and Palmieri, K.M. (Technical Coordinators). 2008. *Proceedings of the sudden oak death third science symposium.* Gen. Tech. Rep. PSW-GTR-214, Pacific Southwest Research Station, Forest Service, U.S. Department of Agriculture, Albany, CA. pp. 491.

Goralka, R.J.L., and Langenheim, J.H. 1995. Analysis of foliar monoterpenoid content in the California bay tree, *Umbellularia californica*, among populations across the distribution of the species. *Biochem. Syst. Ecol.* 23: 439–448.

Goralka, R.J.L., and Langenheim, J.H. 1996. Implications of foliar monoterpenoid variation among ontogenetic stages of the California bay tree (*Umbellularia californica*) for deer herbivory. *Biochem. Syst. Ecol.* 24: 13–23.

Goralka, R.J.L., Schumaker, M.A., and Langenheim, J.H. 1996. Variation in chemical and physical properties during leaf development in California bay tree (*Umbellularia californica*): Predictions regarding palatability for deer. *Biochem. Syst. Ecol.* 24: 83–103.

Granholm, J.W. 1985. An evergreen from the West Coast offers beauty and versatility: *Umbellularia californica*. *Am. Nurseryman* 162(9): 65–67.

Hemmes, R.B., Alvarado, A., and Hart, BL. 2002. Use of California bay foliage by wood rats for possible fumigation of nest-borne ectoparasites. *Behav. Ecol.* 13: 381–385.

Kasapligil, B. 1951. Morphological and ontogenetic studies of *Umbellularia californica* Nutt. and *Laurus nobilis* L. *Univ. Calif. Publ. Bot.* 25(3): 115–239.

Kelsey, R.G., McCuistion, O., and Karchesy, J. 2007. Bark and leaf essential oil of *Umbellularia californica*, California bay laurel, from Oregon. *Nat. Prod. Comm.* 2: 779–780.

Kepner, R.E., Ellison, B.O., Breckenridge, M., Connolly, G., Madden, S.C., and Muller, C.J. 1974. Volatile terpenes in California bay foliage. Changes in composition during maturation. *J. Agric. Food Chem.* 22: 781–784.

MacGregor, J.T., Layton, L.L., Buttery, R.G. 1974. California bay oil, 2: Biological effects of constituents. *J. Agric. Food Chem.* 22: 777–780.

McClintock, E. 1993. California bay and Mediterranean laurel. *Pacific Hort.* 54(1): 10–12.

Munger, T.T. 1966. Trees in hazard: Oregon's myrtle groves. *Oregon Histor. Quart.* 67: 41–53.

Neville, H.A., and Bohm, B.A. 1994. Flavonoids of *Umbellularia californica*. *Phytochemistry* 36: 1229–1231.

Reynolds, T., Dring, J.V., and Hughes, C. 1991. Lauric acid-containing triglycerids in seeds of *Umbellularia californica* Nutt. (Lauraceae). *J. Am. Oil. Chem. Soc.* 68: 976–977.

Sargent, C.S. 1892. *Umbellularia*. *Gard. Forest* 5: 349–350, 355.

Stein, W.I. 1958. *Silvical characteristics of California-laurel.* Silvical Ser. 2. U.S. Department of Agriculture, Forest Service, Pacific Northwest Forest and Range Experiment Station, Portland, OR. pp. 16.

Stein, W.I. 1974. *Umbellularia* (Nees) Nutt., California-laurel. In *Seeds of woody plants in the United States. Agriculture Handbook 450.* Technical coordination by C.S. Schopmeyer. U.S. Department of Agriculture, Forest Service, Washington, DC. pp. 835–839.

Stein, W.I. 1990. *Umbellularia californica* (Hook. & Arn.) Nutt., California-laurel. In *Silvics of North America. Agricuture Handbook 654.* Vol. 2, Hardwoods. Technical Coordination by R.M. Burns and B.H. Honkala. U.S. Department of Agriculture, Forest Service, Washington, DC. pp. 826–834.

Trussell, M.E. 1984. The changeling [*Umbellularia californica*]. *Am. Forests* 90: 43–44.


### SPECIALTY COOKBOOKS


Evans, H. 1996. *Basil, bay & borage.* JG Press, North Dighton, MA. pp. 61.


Farnsworth (1999; see the Appendix to this book) provides advice on the culinary use of California bay, and two recipes.

# 22 Camas

Family: Asparagaceae (asparagus family; placed in the past in the Liliaceae, the lily family, which has recently been split into several families, and the composition of some of these has varied according to different authorities; of these segregate families, *Camassia* has also been placed in Agavaceae, Asparagaceae, and Hyacinthaceae)

## NAMES

Scientific name: *Camassia quamash* (Pursh) Greene

- The name camas is based on the name *camas* or *quamash*, used in the Nez Perce (Nez Percé) language; so are the genus name *Camassia* and the epithet *quamash* in the scientific name.
- Camas is also called black camas, blue camas, common camas, common camash, comosh, Indian hyacinth, quamash, and small camas. Camas is sometimes spelled camass. Blue camas and quamash are the most frequent alternative names.

## GEOGRAPHY AND ECOLOGY OF WILD PLANTS

The six species of the genus *Camassia* are native to North America, and at least three were consumed by native people. Camas, the most widely consumed species and the subject of this chapter, is indigenous to western North America. It occurs in southern British Columbia and southern Alberta in Canada and the states of California, Idaho, Montana, Nevada, Oregon, Utah, Washington, and Wyoming. It has been suggested that native groups transplanted bulbs of the species to places beyond their natural range, where they became established.

Camas grows in soils that are moist in the spring but often dry out as the season progresses. It commonly occurs near seasonal pools and streams. The species is found on the edges of grassy prairies, in meadows and depressions, on bluffs and rocky outcroppings, and in open areas of forests. In the past, camas stands were often so extensive that in flower they carpeted meadows with blue. The once-extensive stands have been greatly reduced by livestock grazing and other forms of agriculture. There are efforts underway to restore camas in prairies and other areas.

## PLANT PORTRAIT

Camas is a very variable perennial herb with long, narrow, grass-like leaves 10–60 cm (4–24 inches) in length, growing in a clump from the base. From the center of the clump of leaves, a flower stalk 20–90 cm (8–35 inches) tall develops, bearing star-shaped, pale to deep blue or bluish-violet (rarely white) flowers 12–35 mm (0.5–1.4 inches) long. The fruit is a capsule 6–19 mm (0.2–0.7 inch) long, with 5–10 seeds. The bulbs have thin, black or brown "tunics" (coats). The bulbs generally are 5–20 cm (2–8 inches) below the soil surface. They are egg-shaped, reminiscent of onion bulbs, made up of concentric layers of condensed leaf tissue. Most bulbs are small, 1–3 cm (0.4–1.2 inches) in diameter. However, descriptions and old photos indicate the past harvest of larger bulbs, 5–7.5 cm (2–3 inches) in diameter, and the bulbs can become even larger in

**FIGURE 22.1** Camas (*Camassia quamash*). (a) Native Americans collecting camas bulbs. (Painting courtesy of Barry Flahey.) (b) Long section of camas bulb in hand. (Painting courtesy of Barry Flahey.) (c) Distribution map.

cultivation (some have been described as "the size of tangerines"). Camas is mostly reproduced by seeds, but is vegetatively propagated to a small extent by offset "bulblets."

*Camassia* species were among the most important foods of Native Americans west of the Rocky Mountains. Usage of camas as food in the Pacific Northwest dates back at least 7000 years. It has been estimated that camas supplied 20% to 50% of the caloric needs of some native people of the northern Pacific Coast. Apparently, in some areas where there were vast meadows of camas, tens of thousands of collectors were sometimes busy gathering and preparing the bulbs. Camas stands were often managed by native peoples, who weeded out the toxic death camas (see the section "Warning"), burned off grass, and removed invading weeds. No other plant food was more widely traded than camas bulbs. Indicative of its importance, some tribes and families asserted hereditary property rights over camas stands. It has been estimated that 2.7 ha (6.7 acres) of a typical wild camas meadow would have supplied 1 metric ton (0.907 ton) of camas bulbs, which would have been sufficient to supply about 20% of the yearly caloric requirement for a family of five (assuming an average intake of 2500 calories per person per day).

**FIGURE 22.2** Camas (*Camassia quamash*). (a) Bulbs (Courtesy of T. Abe Lloyd.) (b) Flowering plant. (Courtesy of Walter Siegmund [CC By 3.0].) (c) Flower. (Courtesy of Walter Siegmund [CC By 2.5].)

Camas is grown as an ornamental, and several cultivars are available, including a white-flowered variety. The plant is suitable for flower borders, beds, ground cover, and rock gardens.

## CULINARY PORTRAIT

Camas was an important food for native peoples. The bulbs were harvested in the fall. They do not keep well, so they were frequently baked and roasted for consumption, or dried and pounded into flour, or simply sun-dried for later use. Dried bulbs could be reconstituted by soaking overnight. Camas contains considerable inulin instead of starch as the storage protein. Inulin is substantially indigestible, but breaks down into sweet fructose when cooked. (See Chapter 53 on Jerusalem Artichoke for information on inulin.) Before the European introduction of sugar and molasses into North America, sweeteners were in short supply, and baked camas was valued for its sweetening property. The cooked bulbs have been said to taste like chestnuts.

## WARNING

In western North America, wild food collectors may confuse edible *Camassia* plants with very toxic (sometimes lethal) death camas species, particularly *Toxicoscordion venenosum* (S. Watson) Rydb. (*Zigadenus venenosus* S. Watson).

### Culinary Vocabulary

- The bulbs (which sometimes naturally have a black covering) are sometimes called "black camas" because slow pit cooking turns the bulbs black.
- "Pit cooking" was a widely used method of food preparation used by indigenous peoples of the world, and it is still used. A hole is dug in the ground, stones are placed in the hole (more elaborately, the pit is lined with stones) along with wood fuel, the wood is burned, allowing the fire to thoroughly heat the stones, the food to be cooked is placed on the stones, the pit is covered to starve the fire of oxygen, and the hot stones cook the food, the cooking time sometimes extending for days. The Lewis and Clark expedition of 1803–1806 (see Chapter 85 on Sassafras for details) began in St. Louis and traveled over the Rocky Mountains and through Oregon country to the Pacific Ocean. In 1805, the explorers descended from the Bitterroot Mountains (of the Rocky Mountains) and were met by members of the Nez Perce tribe who gave them a meal that included camas. Clark described the method of cooking. He noted that the Indians dug a deep pit, lined it with split wood and stones, and placed more wood on top. A fire was lit in the pit and allowed to burn until the stones were hot. The fire was then extinguished and the camas roots were placed between layers of grass atop the stones to cook.

## PROSPECTS

Camas was one of the most important food and trade commodities of native people of North America in past times, and collecting and consuming it remains an important cultural heritage. However, it does not appear to have been domesticated, and never became a commercial modern food. One Web document commented, "It's not apt ever in the modern age to become a food crop." The species is slow-growing, and seems adapted to best growth in quite moist soils, and these are not agriculturally desirable characteristics. Moreover, camas bulbs do not keep well fresh, becoming desiccated and soft. Nevertheless, camas is an intriguing possibility for domestication. It is naturally good-tasting, and does not seem to have a particular competitor in the marketplace (although numerous much more competitive "root crops" are now commonly sold). The name "camas" is neutral (unlike names such as "gopher potato" and "mole's carrot"), and would not require wordsmithing. This is the kind of vegetable that could become very attractive as a new crop, and despite its unsophisticated background, would likely have snob appeal.

## CURIOSITIES OF SCIENCE AND TECHNOLOGY

- Because of its exceptional sustenance value to indigenous peoples, camas was often included in key cultural rituals marking transitions of life. The Nez Perce people provided presents of dried camas in corn husk bags during marriages, and at funerals widows gave gifts of camas to relatives and friends.
- Gophers are quite fond of camas bulbs, and sometimes their efforts to collect and store them for future use went unrewarded: some of the cached bulbs simply sprouted and produced new plants; and occasionally Native Americans sought out the stored bulbs and robbed them from the animals.

## KEY INFORMATION SOURCES

Beckwith, B.R. 2004. *"The queen root of this clime": ethnoecological investigations of blue camas (*Camassia leichtlinii *(Baker) Wats.,* C. quamash *(Pursh) Greene; Liliaceae) and its landscapes on southern Vancouver Island, British Columbia.* PhD thesis, University of Victoria, Victoria, BC. pp. 263.

Bryce, C. 2005. *Kwetlal "camas"*: Camas harvest and pit cook. Songhees Land Department, Victoria, BC. (Flyer)

DeLane, T.M., and Sharp, W.H. 1976. The blue camas, *Camassia quamash*, a plant new to Waterton Lakes National Park, Alberta. *Can. Field-Nat.* 90: 79–80.

Fishbein, M., Kephart, S.R., Wilder, M., Halpin, K.M., and Datwyler, S.L. 2010. Phylogeny of *Camassia* (Agavaceae) inferred from Plastid *rpl16* intron and *trnD–trnY–trnE–trnT* intergenic spacer DNA sequences: implications for species delimitation. *Syst. Bot.* 35: 77–85.

Gould, F. W. 1942. A systematic treatment of the genus *Camassia*. *Am. Midl. Nat.* 28: 712–742.

Gross, B.L., Rollenhagen, K., Mellor, J., and Kephart, S.R. 2001. Species boundaries in *Camassia*: analysis of a putative hybrid zone. *Am. J. Bot.* (Suppl.) 88: 157.

Konlande, J.E., and Robson, J.R. 1972. The nutritive value of cooked camas as consumed by Flathead Indians. *Ecol. Food Nutr.* 2: 193–195.

Plew, M.G. 1992. Nutritional analysis of camas (*Camassia quamash*) from southern Idaho. *Northwest Anthropol. Res. Notes* 26: 217–223.

Ranker, T.A., and Hogan, T. 2002. *Camassia*. In *Flora of North America*. Vol. 26. Edited by Flora of North America Editorial Committee. Oxford University Press, New York, NY. pp. 303–307.

Statham, D.S. 1975. A biogeographical model of camas and its role in the aboriginal economy of the Northern Shoshoni in Idaho. *Tebiwa* 8(1): 59–80.

Stevens, M.L., and Darris, D.C. 1999. *Ethnobotany, culture, management, and use of great camas. Plant Materials Technical Note No. 23.* U.S. Department of Agriculture Natural Resources Conservation Service, Portland, OR. pp. 10.

Stevens, M., Darris, D.C., and Lambert, S.M. 2001. Ethnobotany, culture, management and use of common camas. *Native Plants J.* 2: 47–52.

Stevens, M.L., and Darris, D.C. 2000. *Plant guide for common camas: Ethnobotany, culture, management, and use. Plant Materials No. 25.* U.S. Department of Agriculture Natural Resources Conservation Service, Portland, OR. pp. 10.

Thoms, A.V. 1989. *The northern roots of hunter-gatherer intensification: Camas and the Pacific Northwest.* Ph.D. thesis, Washington State University, Pullman, WA. 521 leaves.

Tomimatsu, H., Kephart, S.R., and Vellend, M. 2009. Phylogeography of *Camassia quamash* in western North America: postglacial colonization and transport by indigenous peoples. *Molec. Ecol.* 18: 3918–3928.

Turner, N.J., and Kuhnlein, H.V. 1983. Camas (*Camassia* spp.) and riceroot (*Fritillaria* spp.): Two liliaceous "root" foods of the Northwest Coast Indians. *Ecol. Food Nutr.* 13: 199–219.

Uyeda, J.C., and Kephart, S.R. 2006. Detecting species boundaries and hybridization in *Camassia quamash* and *C. leichtlinii* (Agavaceae) using allozymes. *Syst. Bot.* 31: 642–655.

Walker Rose, S. 1997. Earth play–camas day–pit bake way. *Bull. Primitive Technol.* 13: 41–42.

### SPECIALTY COOKBOOKS

Krockmal and Krochmal (1974) provide three recipes, as does Mohney (1975). (See the Appendix to this book for details of the publications cited.)

# 23 Canada Garlic

Family: Liliaceae (lily family; sometimes placed in Alliaceae, the onion family)

## NAMES

Scientific name: *Allium canadense* L.

- The word "garlic" is derived from the Anglo-Saxon *gar*, lance or spear (after the shape of the stem) and *leac*, leak or pot-herb. The name "garfish," referring to a fish with a long pointed snout, is also based on the similarity to a spear.
- Canada garlic is also known as American wild onion, Canada onion, Canadian garlic, meadow garlic, meadow leek, rose garlic, rose leek, wild Canada garlic, wild garlic, wild onion, and wild shallot.
- The name "wild garlic" is ambiguous, since it is applied to other species, such as *A. oleraceum* L. and *A. vineale* L., which are Eurasian species naturalized in North America. *Allium vineale* is usually known as "field garlic," a name confusingly sometimes also applied to *A. canadense*. *Allium vineale* and *A. canadense* often grow together, look similar (but the leaves of *A. vineale* are round while those of *A. canadense* are flat), and are frequently confused with each other.
- The name "tree onion" is sometimes used for Canada garlic, but tree onion is usually one of the alternative names of Egyptian onion, *A.* × *proliferum* (Moench) Schrad.ex Willd. (*A. fistulosum* × *A. cepa*), an oddity of gardens that, like Canada garlic, also produces edible bulbils in the flower head.
- According to Kindscher (1987; see the Appendix to this book), *kha-a'mot-ot-ke-wat*, the Cheyenne name for *A. canadense*, means skunk testes, "probably a polite translation for something more derogatory."
- The genus name *Allium* is the classical Latin name for garlic.
- *Canadense* in the scientific name is Latin for Canadian. However, Canada garlic is rare in Canada and the bulk of its geographical occurrence is in the United States. When Linnaeus coined the scientific name, "Canada" was understood to include parts of the northeastern United States, where Linnaeus knew the species occurred.

## GEOGRAPHY AND ECOLOGY OF WILD PLANTS

Canada garlic occupies the eastern half of the United States and southeastern Canada, from Ontario to New Brunswick in the north, south to Texas and Florida. The species is found in thickets, low, open woods, glades, meadows, prairies, ditches, lawns, and disturbed sites. It occurs in both moist and well-drained ground, in a variety of soils, particularly sand and sandy loam.

Canada garlic is considered to be a noxious weed when found in commercial seed, in the legislation of at least 12 states. The species is also considered to be an invasive weed in Canada and some regions of the United States. In crops, especially winter wheat in the Midwestern United States, its presence reduces the quality of the harvest, as the grain-like bulbils can become intermixed in the harvested grains, imparting an onion odor. Indeed, when cultivated crops, lawns, or roadsides where Canada garlic has become established are mowed, the odor becomes very evident. When consumed by livestock, Canada garlic can produce a garlic flavor in dairy and meat products.

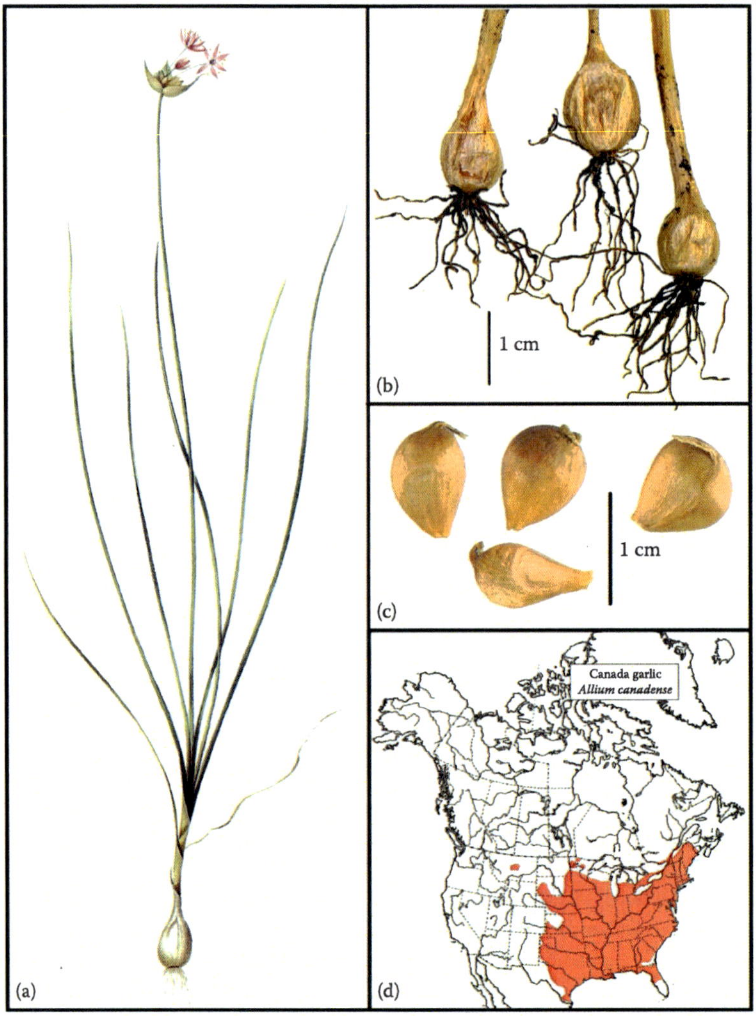


**FIGURE 23.1** Canada garlic (*Allium canadense*). (a) Painting of plant in flower. (From Redouté, P.-J., *Les liliacées*, Redouté, Paris, France. Vol. 4, Plate 240, 1808.) (b) Dried, underground bulbs, on a herbarium specimen. (Courtesy of E. Small and B. Brookes.) (c) Aerial bulblets (developed from flowers). (Courtesy of S. Hurst, U.S. Department of Agriculture.) (d) Distribution map.

## PLANT PORTRAIT

Canada garlic is a perennial herb with narrow, grass-like leaves up to 40 cm (16 inches) in length, arising from the base. A flowering stalk up to 60 cm (2 feet) in height arises in the spring or early summer, and bears a cluster of up to 60 star-shaped white, pink, or lavender flowers 4–8 mm (0.2–0.8 inch) long. One or several bulbs are developed at the base. These are ovoid, up to 3 cm (1.2 inches) long and 2.5 cm (1 inch) wide. The bulbs are covered with a thick, brown or gray network of intermeshed fibers.

Some plants develop small "bulbils" instead of flowers, or a combination of flowers and bulbils in the flowering head. The bulbils are miniature bulbs, which can fall off the plant and develop into new plants. Since they are produced without sexual fertilization, they represent a kind of vegetative reproduction. Forms of the species that sets bulbils in the flower head rather than producing fruits with seeds are thought to have been distributed beyond their natural native range by indigenous peoples and European traders, as evidenced by presence along historically important travel routes and campsites.

As with other species of *Allium*, Canada garlic was extensively used as medicine by both Native Americans and early settlers. The juice is naturally antiseptic, and was often applied directly to wounds and burns, or used in a poultice to treat boils, bee stings, insect bites, and even snakebites. Preparations were also consumed to treat coughs, internal parasites, and a variety of illnesses. The juice of the plant was sometimes rubbed on exposed parts of the body to repel biting insects.

Canada garlic was widely eaten in the past by Native Americans and European settlers. It is still collected as a wild food, and is also sometimes cultivated.

## CULINARY PORTRAIT

Garlic (*A. sativum* L.) is one of the world's most important condiments. By comparison, Canada garlic is scarcely known. The bulbs smell and taste more like onion than garlic. Canada garlic can be used both like garlic and leek, although it is not exactly the same as either. The bulbs of Canada garlic are crisp, mild, and pleasant, and become sweeter after boiling.They can be consumed raw or cooked, used as a vegetable or as a flavoring in soups and stews, and pickled. The leaves are mild in flavor and can be eaten raw or cooked, preferably before flowering occurs. They may be used in salads, as greens (like chives), and as a flavoring in cooked food. The flowers are somewhat stronger in flavor than the leaves, but can also be consumed raw, used as a flavoring, and as a garnish on salads. The bulbils produced by some plants make an excellent onion-flavored pickle. In general, the bulbs and leaves can be used in any recipe calling for onions or chives, but as they are more pungent in flavor, the amounts used should be reduced.

## WARNINGS

(1) *Allium* species are often collected from the wild and used as food. In western North America, amateurs may confuse edible *Allium* species with very toxic (sometimes lethal) death camus species, particularly *Toxicoscordion venenosum* (S. Watson) Rydb. (*Zigadenus venenosus* S. Watson), well known for poisoning livestock. There are also other poisonous plants in North America (sometimes called "poison onions") that have been mistaken for edible *Allium* species, such as false garlic or crow poison (*Nothoscordum bivalve* (L.) Britton). Experienced wild food collectors advise that unless onion-like plants have the very distinctive smell of onions or garlic, they should not be considered edible. (2) Some compounds in garlic (*A. sativum*) affect blood coagulation by forming the antithrombotic compound ajoene, which is thought to inhibit aggregation of blood platelets. This blood-thinning effect can be beneficial to people in danger of heart attack and stroke. However, individuals regularly taking aspirin or other anticoagulating drugs, or susceptible to low blood clotting conditions, are sometimes advised to avoid consuming large amounts of garlic. Canada garlic has similar compounds, and the same advice is probably worth considering. (3) Despite being used as food, some references state that Canada garlic can be mildly toxic to humans and other animals, causing vomiting, diarrhea, and nausea. (4) Consumption of large amounts of raw *Allium* species has poisoned livestock and pets, including dogs, and while not especially dangerous, caution should be exercised when animals have access to the plants. (5) Alliums have caused contact dermatitis in some susceptible individuals. (6) Extreme overconsumption of *Allium* species has been linked to development of goiter.

### CULINARY VOCABULARY

- The terms shallot, onion, and leek are used in the alternative names of Canada garlic. Many people confuse true shallots, green onions, and leeks. All but the leek are members of the onion species, *A. cepa*. Green onions have a definite bulb with the typical concentric

arrangement of scale leaves that the dry onion has, although they are mostly grown for their edible leaves. Leeks (*A. ampeloprasum* L. var. *porrum* (L.) J. Gay) lack a bulb, but do develop a swollen stem base just above the roots ("shallot" is occasionally used to designate leeks—a confusing and undesirable usage). The true shallot is distinguished by its distinctive bulbs that are made up of a clump of cloves (or small bulbs), reminiscent of garlic, but unlike garlic, the individual bulbs are not closely compacted together by a common membrane.
- "Alliaceous" (pronounces ALL-ley-AY-shus) means smelling like garlic or onions. The word can apply to anything, but is particularly used for food and breath.

## PROSPECTS

Canada garlic is only occasionally cultivated, and is not favored in agricultural circles because of its objectionable weedy tendencies. As a food plant, it is known mostly to wild food enthusiasts. The market is well supplied with other *Allium* vegetables and condiments, notably onion, garlic, and leek species, with which Canada garlic would have to compete should it be developed as a commercial crop. Nevertheless, the species has very good taste and grows very vigorously, indicating that it has the basic requirements to be considered for domestication. Except for the wild leek (*A. tricoccum*; see Chapter 97) and nodding onion (*A. cernuum*; see Chapter 65), this is the most attractive of the many wild North American species of *Allium*, and is certainly deserving of attention as a possible new vegetable crop.

## CURIOSITIES OF SCIENCE AND TECHNOLOGY

- Major Stephen Long (1784–1864) of the U.S. Army led a pioneering scientific expedition into the Great Plains. While exploring Nebraska and Colorado in the 1820s, about 100 men in his party died of scurvy. Local Indians provided a wild garlic to the survivors, who did not perish due to the high vitamin C content of the herb. A decade later, Prince Maximilian of Wied (1782–1867), a German naturalist and explorer, was conducting a 2-year expedition of the Missouri River Valley, and also fell ill with scurvy. A cook who had been with Major Long remembered the garlic cure, acquired wild garlic from Native Americans, and healed the prince.
- Many *Allium* species are infamous for their distinct, objectionable odor, often causing tears. The volatile oils that help to give species of *Allium* their unique flavors contain a class of organic molecules known as amino acid sulfoxides. Peeling, cutting, or crushing *Allium* tissue releases enzymes called allinases, which convert these molecules to sulfenic acids. The sulfenic acids, in turn, spontaneously rearrange to form a volatile gas (syn-propanethial-S-oxide), the chemical that triggers tears. The gas mixes with water in the eyes, forming a weak acid that causes the tear ducts to flood the eyes to get rid of the irritant. Rubbing the eyes is a natural reaction, but if juices from the plant are present on the hands it will probably worsen the irritation. Some solutions are as follows: cut the bulbs under water or run the tap over them as they are sliced; chill them in the refrigerator prior to cutting (this somewhat reduces the release of the gas); wear goggles.

## KEY INFORMATION SOURCES

(For additional references, see Chapter 97 on Wild Leek.)

Carson, J.F. 1987. Chemistry and biological properties of onions and garlic. *Food. Rev. Internat.* 3: 71–103.
Defelice, M.S. 2003. Wild garlic, *Allium vineale* L.— Little to crow about. *Weed Technol.* 17: 890–895.
Dore, W.G. 1964. The Canada onion. *Greenhouse-Garden-Grass* 4: 1–3.
Dore, W.G. 1970. The wild Canada onion. *The Herbarist* 36: 34–38.
Dore, W.G. 1971 Canada onion. Its method of spread into Canada. *Nat. Can.* 98: 385–399.

Gregory, M., Fritsch, R.M., Friesen, N.W., Khassanov, F.O., and McNeal, D.W. 1998. *Nomenclator Alliorum. Allium names and synonyms: a world guide.* Royal Botanic Gardens, Kew, U.K. pp. 83.
Hanelt, P., Hammer, K., and Knüpffer, H. (Editors). 1992. *The genus Allium— taxonomic problems and genetic resources*. Inst. Pflanzen. Kulturpflanzenfors., Gatersleben, Germany. pp. 359.
Hardcastle, W.S. 1976. Chemical control of wild *Allium. Agron. J.* 68: 144–145.
Jones, H.A., and Mann, L.K. 1963. *Onions and their allies. Botany, cultivation and utilization.* Leonard Hill, London, U.K. pp. 286.
Krochmal, A. 1960. Germination studies of aerial bulblets of *Allium vineale* L. and *A. canadense* L. *Am. Midl. Nat.* 63: 497–508.
Ownbey, M. 1950. The genus *Allium* in Texas. *Res. Stud. State Coll. Washington* 18: 181–222.
Ownbey, M., and Aase, H.C. 1955. Cytotaxonomic studies in *Allium*. I. the *Allium canadense* alliance. *Washington State Univ. Res. Stud.* 23(suppl.): 1–106.
Ownbey, M., and Aase, H.C. 1959. *Allium speculae*, a new species of the *Allium canadense* alliance from Alabama. *Rhodora* 61: 70–72.
Pierce, K.R., Joyce, J.R., England, R.B., and Jones, L.P. 1972. Acute hemolytic anemia caused by wild onion poisoning in horses. *J. Am. Vet. Med. Assoc.* 160: 323–327.
Van der Meer, Q.P. 1997. Old and new crops within edible *Allium. Acta Hortic*. 433: 17–31.

## Specialty Cookbooks

The recipes in many of the cookbooks suggested in Chapters 65 and 97 on Nodding Onion and Wild Leek, respectively, can be used for Canada garlic.

Cotter, D. 2007. *Wild garlic, gooseberries, and me: a chef's story and recipes from the land.* HarperCollins, New York. pp. 320.
Herb Society of America, Northern Illinois Unit. 1997. *The wild onion cookbook.* Jumbo Jack's Coobooks, Audubon, IA. pp. 379.
Tolleson, P. 2007. *Wild garlic. Culinary memoirs*. BPR Publishsers, North View, Creswell, Morpeth, U.K. pp. 128.

Freitus (1980) has three recipes for Canada garlic (identified as "wild garlic"), and Tatum (1976) has one. (See the Appendix to this book for details of the publications cited.)

# 24 Cattails

Family: Typhaceae (cattail family)

## NAMES

Scientific name: *Typha* species

- Broadleaved cattail—*T. latifolia* L.
- Glaucous cattail—*T.* × *glauca* Godr. (*T. angustifolia* × *T. latifolia*).
- Narrow-leaved cattail—*T. angustifolia* L.
- Southern cattail—*T. domingensis* Pers.
- The "cat tail" that is the basis of the name cattail is the furry elongated seed stalk, which also looks like a fuzzy, brown, bottle brush.
- Cattails are also called cat-tail, cat tail, cat's tail, and cat-'o-nine-tail. "Cattail" is almost universally used in North America. The following names are used in Europe: cattail-flag (cattail flag), soft flag, flag tule, flag, reed-mace (reed mace or reedmace; for the perceived similarity to both a reed and the club-like weapon called a mace), nail-rod (nailrod), water torch, candlewick, bulrush (bullrush; a name likely to cause confusion with *Scirpus* species), and clubrush.
- Broad-leaved cattail (*T. latifolia*; *latifolia* is Latin for broad-leaved) is also called common cattail, broad-leaved cattail (broadleaf cattail), great cattail, Cossack asparagus, and great reedmace.
- Narrow-leaved cattail (*T. angustifolia*; *angustifolia* is Latin for narrow-leaved) is also called narrowleaf cattail, small cattail, lesser cattail, small bulrush, lesser bulrush, lesser reedmace, and narrow-leaved reedmace.
- "Glaucous" in the name glaucous cattail (*T.* × *glauca*) means waxy bluish-green in appearance. This is also called hybrid cattail (although other species of cattails hybridize), blue cattail, and blue flag (a name usually applied to *Iris versicolor* L.).
- The genus name *Typha* has been said to trace to ancient Greek. It may have been a name for cattails, or perhaps for other aquatic plants. Alternatively, *Typha* has been interpreted as originating from *typhos*, which means marsh or bog, or from *typhein*, which means to smoke or emit smoke, in allusion to the use of the spike for maintaining smoky fires, or to the smoky brown color of the fruiting spikes.
- *Domingensis* in the scientific name *T. domingensis* means "of Santo Domingo" (the island in the West Indies divided between Haiti and the Dominican Republic where the species was first collected and described; Santo Domingo also refers to a city and port in the Dominican Republic).

## GEOGRAPHY AND ECOLOGY OF WILD PLANTS

*Typha latifolia* occurs in Alaska, all provinces and territories of Canada, all of the conterminous United States, Mexico, and Guatemala. It also occurs in Eurasia, East Africa, India, and (by introduction) Australia. *Typha angustifolia* is more narrowly distributed, occurring from southeastern Canada to southern Saskatchewan, extending along the Atlantic coast to South Carolina and Florida, occurring in the midwest and north-central Great Plains to Oregon and California, and also

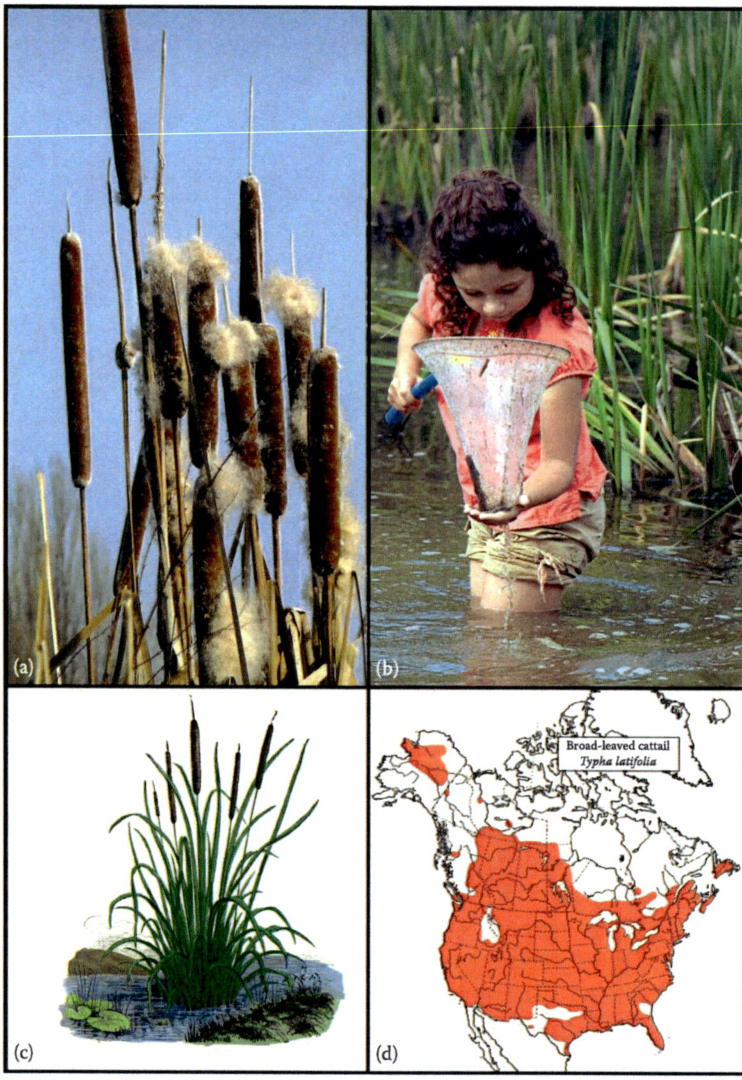


**FIGURE 24.1** Broad-leaved cattail (*Typha latifolia*). (a) Mature, disintegrating fruiting heads. (Courtesy of Pavel Šinkyřík.) (b) Child beside cattails. (Courtesy of U.S. Fish & Wildlife Service.) (c) Drawing of cattail clump (colored by B. Brookes). (From Harter, J. (Editor), *The plant kingdom compendium. A definitive volume of more than 2400 copyright-free engravings*, Bonanza Books, New York, 1988.) (d) Distribution map.

in Europe. Some authors have suggested that *T. angustifolia* is not native to North America, but is an alien, introduced with early European settlement, subsequently migrating westward; others have suggested that *T. angustifolia* of North America and of Europe are different species. *Typha* × *glauca* is generally found where *T. latifolia* and *T. angustifolia* occur together, but is also known from a few localities north of the northern range limit of *T. angustifolia*, such as Anticosti Island, central Manitoba, and southwestern Saskatchewan. It is often found in disturbed areas such as roadside ditches. Apparently during the last century, *T. angustifolia* spread strongly inland and westward across southern Canada and the northern United States, and hybrids with *T. latifolia* became more common in these regions. *Typha domingensis* is found in the southern half of the United States and in warm-temperate and tropical regions in much of the world.

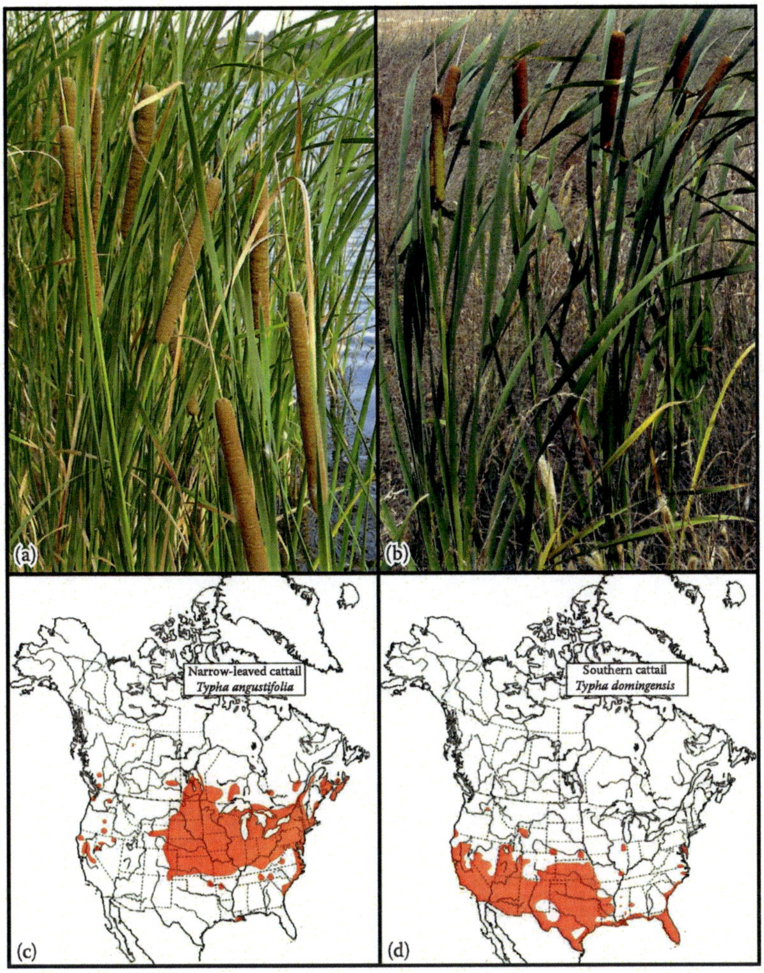


**FIGURE 24.2** Cattail (*Typha* species). (a) Narrow-leaved cattail (*T. angustifolia*). (Courtesy of Le.Loup. Gris [CC By 3.0].) (b) Southern cattail (*T. domingensis*). (Courtesy of Javier Martin.) (c) Distribution map for narrow-leaved cattail. (d) Distribution map for southern cattail.

Cattails usually grow with their base in standing water. Air channels in the plant allow oxygen to circulate to the underwater parts. The seeds sprout readily on mud, but also germinate completely underwater. Submersed cattail seedlings have flaccid leaves, the upper portions of which may float. The pointed seeds of *T. latifolia* can become embedded in the skin of fish, and accordingly transported. The plants are normally rooted in soil, although floating mats can be formed. *Typha angustifolia* can grow in deeper water than *T. latifolia* (the former may tolerate a depth of up to 1.5 m or 5 feet, the latter up to 1 m or about 3 feet), and is competitively superior in deeper water, but often outcompeted in shallower water.

*Typha latifolia* occurs from tropical to northern areas and from sea level to over 2000 m (6600 feet) in altitude. *Typha angustifolia* and *T.* ×*glauca* are found in a narrower range of climate and habitats; both are frequently found in disturbed wetland sites. *Typha latifolia* grows in almost any soil that remains wet or saturated most of the growing season, and tolerates some salinity and acidity (it has been said to be the only species of *Typha* in acidic areas). It is common in marshes, fens, wet meadows, and roadside ditches, and along lakeshores, seacoast estuaries, and pond margins. *Typha angustifolia* occurs in similar soils but can tolerate more saline and/or alkaline substrates, and frequently inhabits brackish marshes. *Typha* ×*glauca* has been said to be intermediate in its ecology, but competitively superior to its parents under unstable water conditions. *Typha domingensis* grows in both brackish and fresh water as well as on muddy soil, and is much less cold tolerant than the other North American species.

Cattails are ecologically useful, the rhizomes and roots playing a major role in stabilization of shorelines and river banks. Nevertheless, they can be extremely serious weeds of irrigated agricultural lands and managed aquatic systems. Cattails are often considered undesirable by wildlife managers because of their ability to take over marshes and other freshwater environments, excluding other plants. They tend to block or silt up irrigation and drainage canals, and obstruct commercial and recreational activities in lakes, ponds, and marshes.

## PLANT PORTRAIT

The genus *Typha* consists of about a dozen species, and is widely distributed in temperate and tropical regions of all continents except Antarctica. Cattail plants are tall (sometimes over 4 m or 13 feet), stout, erect herbs with perennial fleshy rhizomes (underground or underwater stems). The long narrow leaves are flattish, alternate in two ranks, each leaf covered for much of its length by overlapping sheaths of lower leaves. In the spring and early summer, single flowering spikes are produced. The flowers are minute, unisexual, and lack petals and sepals. The flowering spike is divided into two portions. The upper section of the flowering spike contains male flowers, while the lower section bears densely packed female flowers, which mature into an elongated, fuzzy, seed-bearing structure. Fluffy white hairs attached to the seeds carry them away on the slightest breeze (cattails have aptly been called "the dandelions of the swamps").

Broad-leaved cattail was exceptionally important to indigenous North Americans, and some authors have contended that it was their most important food plant. Early European settlers sometimes survived on it in times of need. Indeed, cattail species worldwide have been eaten during famines, and it has been said that no one should starve or even go hungry in an area where cattails are abundant. Broad-leaved cattail attracted extensive use because of its widespread occurrence and the availability of edible components throughout much of the year. The tubers were used as a starchy food staple throughout the late summer, autumn, and winter (although more palatable in the early summer, starch content is highest in the fall). Native Americans spread ripe, opened cattail heads on flat stones, ignited the floss, and swept up and ate the parched seeds that remained.

Cattail marshes were so important to North American Indians that tribal wars took place over their control. Some societies in Central and South America used cattail plants in diverse ways and developed a very high level of dependence on them. Cattails are still widely used in some countries for roofing, bedding, basketry, shoes, ropes, and paper, but are employed mostly in cottage industries for making handicrafts. Although quite combustible, cattails were used by pioneers for bedding and insulating houses. The fluff from fruiting spikes has been used for stuffing mattresses, quilts, pillows, life-preservers, dolls, and diapers. Although the plants can be very invasive, they are cultivated as ornamentals on the borders of ponds or small pools.

Cattails have been used in folk medicine to treat a wide variety of illnesses. North American native peoples widely employed the pollen to control bleeding. The pollen is still so used in Chinese medicine, and preparations are also marketed to treat coronary heart disease, to lower blood lipid level, and for other purposes. However, cattails are not used in modern Western medicine.

## CULINARY PORTRAIT

Euell Gibbons, author of the 1962 classic *Stalking the Wild Asparagus*, called cattails "the supermarket of the swamps" because they provide so many kinds of food. All parts of cattails can be used as food, and have been eaten by indigenous peoples of countries wherever the species are found. The rhizomes can be consumed as a cooked vegetable, or dried and ground into flour. Young shoots are eaten raw or cooked like asparagus. The young flowering stems resemble corn on the cob, and can be eaten raw, boiled, steamed, or made into soup. Young stem sections can be pickled. The seeds are also edible, and have an attractive nutlike flavor.

Cattails are one of the very few plant species that produce pollen in sufficient quantity that it is used as food. Although not used historically as food in Europe, cattail pollen has been eaten in India, China, New Zealand, South Africa, the United States, Canada, and in many tropical regions around the world. In the great marshes of southern Iraq and Iran, pollen of cattail (*T. domingensis*) is harvested in large quantities, mixed with sugar and steamed in a bag, and sold in marketplaces. The pollen also provides a protein-rich flour, used for bread, gruel, and a thick soup. For the most part, the pollen is mixed with water, and sometimes with wheat or corn flour.

Fresh rhizomes contain about 75% water. About 80% of the dried rhizome is carbohydrate (higher than that of potato), including 30%–40% starch and about 25% sugars. About 8% of the dried root is protein. The protein content (6%–8%) of cattail flour is comparable to flour from wheat, corn, and rice. The content of calcium, iron, and potassium—nutrients important to human health—has been evaluated as far superior to that of potato or rice. The pollen contains about 15%–20% of each of starch and protein. About 20% of the seed is made up of an edible oil, of which about 70% is linolenic acid, an important nutritional fatty acid.

If the water in which cattails grow is polluted by sewage or poisoned by heavy metals or other industrial contaminants, the resulting plants could be toxic. Cattails can absorb substantial amounts of heavy metals, and have been used to help purify metal-contaminated water. Wild food gatherers should also be aware that roadside collection of plants can be hazardous because of toxins, and roadside ditches in which cattails grow are especially susceptible to contamination. As in some other plants, crystals of calcium oxalate are common in cattails. These may irritate the mouth of some people, so eating cattail parts raw is not advisable. Cattail pollen is used by wild food enthusiasts as a protein-rich additive for bread, pastries, and porridge. The pollen is allergenic to susceptible individuals, but this is uncommon. It has been recommended that pregnant women should avoid eating cattail pollen because it is an emmenagogue, that is, a promoter of regular menstruation, which could interfere with a pregnancy. Another potential danger is that people unfamiliar with cattails that have not yet flowered could collect quite poisonous *Iris* species that also grow in shallow water and look somewhat similar when not in flower.

### Culinary Vocabulary

- For centuries, the Cossacks of the Don River marshes of Russia consumed young cattail shoots (developing from the rhizomes) as a raw vegetable. Before these shoots mature into familiar cattail heads, they bear some resemblance to asparagus spears. Inspired by the Russian practice, people in other countries have similarly harvested young flowering stems of local cattails as food, and applied the phrase "Cossack's asparagus" to the edible shoots.

## PROSPECTS

Cattails are widely regarded as an enormous potential resource for food, energy, and industrial products. A cattail stand is a natural extensive monoculture, comparable to a forest, but unlike a forest, the kind of technology needed for harvesting has not yet received much development. Because they grow in wet areas, cattails do not compete for land used for most crops or forests. This is important because agricultural land is in very short supply. Unlike conventional crops, cattails can be raised without pesticides, herbicides, or fertilizers. A very large and untapped land base is present in all areas of North America that could be used for the managed cultivation of cattails. Cattails have often been proposed as biomass crops for renewable energy, and as sources of industrial chemicals. More than 40% of the total biomass of cattails can be converted to fermentable material, so that ethanol production is quite feasible (estimated yields are 1500–6000 L/ha or

160–640 American gallons/acre). Also, pelletized fuel from cattail for home heating and grain drying is thought to be feasible. The annual productivity for *T. latifolia* has been found to range from 6 to 20 tonnes/ha (2.7–9 tons/acre). A biomass of 34 tonnes/ha (15 tons/acre) has been recorded for *T.* ×*glauca* in Wisconsin. Under managed conditions, it has been estimated that cattail's productivity can exceed 40 tonnes/ha (18 tons/acre) of dry matter. By comparison, wheat produces 1.3–7.0 tons/ha (0.6–3 tons/acre), Jerusalem artichoke 17 tons/ha (7.6 tons/acre), sugarbeet 30 tons/ha (13 tons/acre), and corn 7 t/ha (3 tons/acre). Cattail rhizomes are frequently less than 30 cm (1 foot) below the soil, so that the technology for harvesting is not demanding. There have been several North American economic analyses of the potential profitability of harvesting natural stands of cattails for biomass purposes (e.g., for alcohol production), and the predictions have been rather uncertain. However, with current high energy prices, the concept of using this renewable resource is increasingly promising.

The potential food use has particularly intrigued many specialists interested in developing new agricultural crops, because huge yields are possible. An edible oil has been extracted from the seeds, and cattail oil has been compared to linseed oil, but the small size of the seeds suggests that *Typha* would be difficult to develop as a seed crop. The rhizomes are edible, but not particularly palatable, and the food potential lies mostly in extracting edible components industrially, or perhaps feeding the rhizomes to livestock.

All of the dozen or so species of *Typha* worldwide have been harvested historically and have provided important staples for local inhabitants, and collection of wild plants remains important for various cottage industries. There has been occasional large-scale, commercial harvesting of natural stands, and even experimental cultivation as a crop, but developing cattails as a new crop is in its infancy. The basic problem is that cattails are wild plants, requiring considerable investment in breeding and technological research. Cattails offer incredible economic rewards, but do require the kind of long-term investment in research and development that is increasingly difficult to obtain.

Cattails are among the most dominant and common plants of marsh habitats, and in discussing their economic potential it should be remembered that aquatic habitats are fragile, and any programs of development undertaken in existing wetlands should have as their goal the creation of sustainable, ecologically friendly systems. One interesting proposal is to harvesting cattails in rows alternating with strips of natural aquatic vegetation; this could promote the preservation of wildlife and make replanting unnecessary. The creation of new wetlands for cattail cultivation could have great benefits for wildlife and biodiversity.

## CURIOSITIES OF SCIENCE AND TECHNOLOGY

- The buoyancy of cattails has been exploited for various purposes. Some inhabitants of Lake Titicaca (at the Peru–Bolivia boundary) live on huge rafts constructed of layers of cattail stems. The ancient Aztecs created stationary islands in marshes, made of cattails. Unsinkable boats made of cattails were commonly constructed by indigenous people in southern Asia and in the Americas. Cattail stems simply tied in bundles have served as floats to aid swimmers in India. Native North Americans ingeniously wove floating duck decoys out of cattails. Cattail floss is particularly buoyant, and during World War II, thousands of tons were used as a substitute for kapok in life jackets.
- Cattail floss is irritating to eyes. American Indians are said to have blown it into the eyes of their enemies during battle to temporarily blind them. Nevertheless, adolescents commonly engage in duals with cattail stems, showering each other with clouds of soft fluffy down.
- Indigenous inhabitants of Australia used cattail stems as spear handles.
- Native Americans relied on layers of woven cattail leaves to weatherproof their wigwams and lodges. The leaves swell when wet, so that they are excellent for closing up windy holes and cracks.

- In many old paintings, Jesus is depicted at his trials holding a cattail as a mock sceptre (e.g., Van Dyck's "Jesus Crowned with Thorns").
- The Old Testament's "reeds" of Exodus 2:3 and "rushes" of Isaiah 19:6 were probably cattails.
- Cattails and fire seem to be companions. The pollen of cattails is highly inflammable, and has been used to manufacture fireworks. The female flowers make excellent tinder, and can be lit by the spark of a flint. Cattail heads are often soaked in kerosene, lighted, and carried as torches.
- Of all flowering plants, cattails have been said to have the greatest number of flowers per unit volume of inflorescence. A single spike of broad-leaved cattail can produce over 200,000 seeds, while one of southern cattail can produce 700,000.
- A single broad-leaved cattail spike can produce 400 million pollen grains.
- True to its name, elephant grass (*T. elephantina* Roxb.) can be huge (up to 4 m or 13 feet in height), and is enthusiastically grazed by elephants in India.
- The long-billed marsh wren (*Cistothorus palustris*) is almost exclusively a dweller of cattail marshes. It almost always uses cattail leaves to build sack-shaped nests (characterized by a small circular opening in one side), which are suspended from cattail plants.
- Cattail heads are often gathered for decorative purposes. Mature cattail heads are designed to explode as they dry out to disperse the seeds, and it can be disconcerting if they dry out in the heat of a house and explode without warning in a fresh cut flower display, scattering fluff everywhere. Pick immature specimens early in the season, while male flowers are in bloom, to avoid this problem. A thick coat of lacquer on air-dried collections will keep the heads intact for use in dried arrangements.
- Since cattail pollen is probably the only pollen used directly by humans as food, it would seem to be an ideal food for bees, which can eat most pollens. However, experimental feeding of broad-leaved cattail pollen to honeybees showed that while the bees could digest the pollen, they only survived for 21 days on average, compared to 57 days for bees fed on a mixture of typical bee-preferred pollens. It has been suggested that the high starch content (which bees digest poorly) is the reason cattail pollen is bad for bees.

## KEY INFORMATION SOURCES


Austin, D.F. 2007. Sacred connections with cat-tail (*Typha*, Typhaceae)—Dragons, water-serpents and reed-maces. *Ethnobot. Res. Applic.* 5: 273–303.

Baltezore, J.F., Leitch, J.A., and Linz, G.M. 1994. *The economics of cattail management: assessing the trade-offs*. Department of Agricultural Economics, North Dakota Agricultural Experiment Station, North Dakota State University, Fargo, ND. pp. 67.

Fiala, K. 1978. Underground organs of *Typha angustifolia* and *Typha latifolia*, their growth, propagation, and production. *Acta Scientiarum Naturalium Brno* 12: 1–43.

Gabrielson, J.E. 1982. *Cattail rhizome derived alcohol: final report*. U.S. Department of Energy, Washington, DC. pp. 34.

Grace, J.B., and Harrison, J.S. 1986. The biology of Canadian weeds. 73. *Typha latifolia* L., *Typha angustifolia* L., and *Typha* ×*glauca* Godr. *Can. J. Plant Sci.* 66: 361–379.

Hull, D. 1984. *The feasibility of ethanol production from the cattail* Typha latifolia. California Department of Food and Agriculture, Sacramento, CA. pp. 34.

Lakshman, G. 1984. *A study to evaluate the potential of cattail as an energy crop*. Saskatchewan Research Council Tech. Rep. 162. Saskatchewan Research Council, Saskatoon, SK. pp. 89.

Linde, A.F. 1976. *Cattail: the significance of its growth, phenology and carbohydrate storage to its control and management*. Tech. Bull. 94. Wisconsin Department of Natural Resources, Madison, WI. pp. 27.

Linz, G.M. 1992. *Proceedings, cattail management symposium: Fargo, North Dakota, February 12, 1992*. North Dakota State University, Fargo, ND. pp. 47.

Liptay, A. 1989. *Typha*: review of historical use and growth and nutrition. *Acta Hort.* 242: 231–238.

Kuehn, M., and White, B.N. 1999. Morphological analysis of genetically identified cattails *Typha latifolia, Typha angustifolia*, and *Typha ×glauca*. *Can. J. Bot.* 77: 906–912.
Kuehn, M.M., Minor, J.E., and White, B.N. 1999. An examination of hybridization between the cattail species *Typha latifola* and *Typha angustifolia* using random amplified polymorphic DNA and chloroplast DNA markers. *Molec. Ecol.* 8: 1981–1990.
Mitich, L.M. 2000. Common cattail, *Typha latifolia* L. *Weed Technol.* 14: 446–450.
Morton, J.F. 1975. Cattails (*Typha* spp.)—weed problem or potential crop? *Econ. Bot.* 29: 7–29.
Penko, J.M. 1982. *A bibliography of the biology, ecology and utilization of* Typha. Bio-Energy Coordinating Office, University of Minnesota, St. Paul, MN. pp. 55.
Prendergast, H.D.V., Kennedy, M.J., Webby, R.F., and Markham, K.R. 2000. Pollen cakes of *Typha* spp.—"lost" and living food. *Econ. Bot.* 54: 254–255.
Riedl, H. 2003. *Typha* L. In *Plant resources of South-East Asia, 17, fibre plants*. Edited by M. Brink and R.P. Escobin. Backhuys Publishers, Leiden, the Netherlands. pp. 232–235.
Small, E., and Catling, P.M. 2001. Poorly known economic plants of Canada—28. Cattail, *Typha* spp. *Bull. Can. Bot. Assoc.* 34(1): 8–14.
Smith, S.G. 1967. Experimental and natural hybrids in North American *Typha* (Typhaceae). *Am. Midl. Nat.* 78: 257–287.
Smith, S.G. 1987. *Typha*: its taxonomy and the ecological significance of hybrids. *Arch. Hydrobiol. Beih. Ergebn. LLimnol.* 27: 129–138.
Smith, S.G. 2000. Typhaceae. In *Flora of North America north of Mexico*. Vol. 22. Edited by Flora of North America Editorial Committee. Oxford University Press, New York. pp. 278–285.
Thieret, J.W., and Luken, J.O. 1996. The Typhaceae in the southeastern United States. *Harvard Papers Bot.* 8: 27–56.
Yeo, R.R. 1964. Life history of common cattail. *Weeds* 12: 284–288.

## Specialty Cookbooks

Numerous recipes for preparing cattails are available on the Web, and in books on edible wild plants. Beaty (1987) has 6 recipes, Berglund and Bolsby (1971) have 10, Freitus (1980) has 6, Hunt (1977) has 2, Knutsen (1975) has 4, Marie (2008) has 1, Niethammer (1999) has 5, Russell (1975) has 4, Schufer (2011) has 1, Snell (2006) has 2, Szczawinski and Turner (1980) have 4, Tatum (1976) has 5, and Williamson (1995) has 16. (See the Appendix to this book for details of the publications cited.)

# 25 Cherries
## North American Species

Family: Rosaceae (rose family)

### NAMES

Scientific names: *Prunus* species (see Table 25.1).

**TABLE 25.1**
**Most Important North American Cherry (*Prunus*) Species for Breeding for Fruit**

| Common Name(s)[a] | Scientific Name | Distribution | Notes |
|---|---|---|---|
| Black cherry, cabinet cherry, rum cherry, mountain black cherry, timber cherry, wild black cherry, and wine cherry | *P. serotina* Ehrh. | Eastern Canada, central and eastern United States, Mexico, Guatemala | Large forest tree (a shrub in the northernmost regions). Valued for timber. Fruits rather unpalatable (but see comments in text) |
| Choke cherry (often spelled chokecherry) | *P. virginiana* L. | Very widespread in Canada and the United States | Shrub or small tree. Fresh fruit usually unpalatable, but wild fruit widely used, and some cultivars have palatable fruit |
| Pin cherry, bird cherry, fire cherry, hay cherry, northern pin cherry, pigeon cherry, wild red cherry | *P. pensylvanica* L. f. | Widespread in Canada and the United States | Shrub or small- to medium-sized tree. Fruits usually sour, and not consumed fresh. Several cultivars grown for fruit |
| Sand cherry, Appalachian cherry, eastern sand cherry | *P. pumila* L. | Eastern Canada, northeastern United States | Small shrub. Many ornamental cultivars are available. Fruit palatable |
| Western sand cherry, Bessey cherry, dwarf cherry, and Rocky Mountain cherry | *P. pumila* L. var. *besseyi* (L.H. Bailey) Gleason | Central Canada, north central United States | Small shrub. Several cultivars grown for fruit |

- "Cherry" is from the Latin *cerasus*, cherry tree, which in turn is based on the Greek word for the cherry tree, *kerasos*, derived ultimately from *karsu* from the Akkadian language of Mesopotamia.
- "Cherry" is part of the name of various plant species that are unrelated to cherry species. For example, "cherry currant" refers to a variety of the red currant (*Ribes rubrum* L.), "cherry tomato" refers to small tomatoes (forms of *Lycopersicon esculentum* Mill.), "cherry orange" is the kumquat (*Fortunella* species), and "cherry pepper" refers to various peppers (*Capsicum* species; a "hot cherry" is a round, pungent chile pepper mostly used for pickling).

- The name "black cherry" (for *P. serotina*) points out the almost-black color of the mature fruits.
- Sometimes chokeberries (See Chapter 7 on Aronia) and chokecherries (*P. virginiana*) are confused.
- The name "pin cherry" (for *P. virginiana*) is based on the arrangement of the clusters of fruit on long stems that resemble the pins in a pincushion.
- The name "fire cherry" (for *P. virginiana*) is based on the species growing quickly in burned-over areas.
- The name "hay cherry" (for *P. virginiana*) reflects ripening of the cherries during the hay season.
- The names "bird cherry" and "pigeon cherry" (for *P. virginiana*) indicate the attractiveness of the fruits of this species (indeed of most species of *Prunus*) to birds.
- "Cherry plums" are hybrids of cherry and plum species. These are discussed in this chapter. The Asian species *P. cerasifera* Ehrh. is also known as cherry plum, and this is likely to lead to confusion.
- The expression "bush cherries" refers to shrub forms (generally fruit cultivars) of cherry species, in contrast to the trees that are the sources of most commercial cherries.
- The plums, cherries, apricots, peaches, nectarines, and almonds are known as "drupe fruits." A drupe is a fleshy or pulpy fruit with the inner portion of the pericarp (fruit wall) being hard and stony, and normally enclosing one (sometimes more than one) seed. The inner hard portion is commonly the "pit" or "stone."
- See Chapter 74 on Plums for the derivation of the genus name *Prunus*.
- Epithets in the scientific names of *Prunus* mentioned earlier have the following derivations: *serotina* (Latin for late, referring to the relatively late-maturing fruit), *virginiana* (Latinized form for "of Virginia," an area much larger when the species was recognized than the present Virginia states), *pensylvanica* (Latinized form for "of Pennsylvania"; note the single n, the conventional spelling for the old colony at the time the species was named), *pumila* (Latin for diminutive), and *besseyi* (for C.E. Bessey, 1845–1915, influential professor of botany at the University of Nebraska).

## GEOGRAPHY AND ECOLOGY OF WILD PLANTS

Black cherry is native to eastern North America, ranging from southern Quebec and Ontario south to Texas and central Florida. There are also populations in Arizona and New Mexico, and in the mountains of Mexico and Guatemala. The species often grows in mixed deciduous forests, but also in open locations such as old fields. It is found on a variety of soils, preferring rich ground.

The pin cherry occurs in all provinces of Canada. In the United States it is very common in New England and the Lake States, and occurs sporadically in the Appalachian Mountains and the Rocky Mountains. It is usually found in recently cleared areas, but seldom in forests (except in clearings) as it is intolerant of shade. Pin cherry grows on a variety of dry and moist soils, but avoids clay.

The chokecherry is widely distributed across Canada (except for the northernmost regions) and the United States (except for the southernmost regions), and it reaches into northern Mexico. As reflected by its very wide distribution, it occurs in numerous habitats and on many types of soil. It often grows in shrub associations. Although it can develop in the shade of forests, it attains its best growth on the edge of forests.

The taxonomy of sand cherries is somewhat controversial, and authorities differ with respect to what to include in given species, and hence in the geographical limits of the species. As defined here, the sand cherry is widely distributed in eastern North America, from New Brunswick along the Atlantic seaboard to North Carolina, westward to Manitoba in Canada and the midwest of the United States. Sand cherry, true to its name, is most common in sandy areas. The species also grows along gravel bars of waterways, cliff faces, rocky slopes, and sometimes on calcareous, saline, or serpentine soils. It sometimes colonizes road cuts, gravel pits, and railroad beds. Forms of sand cherry occur on sand dunes. The species prefers open habitats. The western sand cherry extends from Minnesota to Kansas, west to Utah and Montana, with isolated populations as far east as Michigan. The eastern and western sand cherries are ecologically similar.

## PLANT PORTRAIT

The genus *Prunus* has more than 250 species (more than 400 according to some authorities) distributed across the northern hemisphere and into the subtropics and tropics. In addition to cherries, other economically significant species include peaches, plums, apricots, almonds, and a wide variety of ornamentals and timber species. Most cherry species are native to eastern Asia, whereas only a little more than two dozen species have been recognized in North America. The "true cherry" species have been traditionally classified into one of five subgenera of *Prunus*, subgenus *Cerasus*. Of the species discussed in the following presentation, subgenus *Cerasus* includes the Old World sweet cherry, sour cherry, and ground cherry, and the North American pin cherry. The North American sand cherries, which will also be examined, were previously included in subgenus *Cerasus*, but have recently been reassigned to a sixth subgenus, *Lithocerasus*. The North American black cherry and choke cherry, also examined in the following, are members of *Prunus* subgenus *Padus*, so-called "deciduous racemose cherries" in contrast to the "true cherries."

The most widely cultivated cherry species are sweet cherry (*Prunus avium* (L.) L.) and sour cherry (*P. cerasus* L.). These are the dominant cherry species of commerce, and indeed are among the world's most important fruits. Sweet cherry is a native of Europe, western Asia, and North Africa. Sour cherry is thought to occur wild in southwestern Asia and southeastern Europe, although some authorities think the species exists only in cultivation. Ground cherry (*P. fruticosa* Pall.) is widespread in central Europe, Siberia, and Northern Asia. It is an important fruit species in Russia and there are many hybrids with the other two species. The preceding three Old World species have been shown to be closely related.

Of the North American cherries, the western sand cherry has been particularly employed to select cultivars, and for hybridization with other species of *Prunus*, especially plum species (see Chapter 74 on Plums). These cultivars and hybrids have been of significant value for growing plums in the stressful climates of the north central and Great Plains states, and in the Prairie Provinces of Canada, where drought and cold resistance are important. The most notable hybrid of a plum species and an American cherry species is between the western sand cherry and the Japanese plum (P. *salicina* Lindl.). The hybrid plants tend to resemble the sand cherry in growth habit and do not attain much height. They are superior to the western sand cherry, and are recommended in preference to it. The western sand cherry has also been crossed with numerous other plums, and with the peach and apricot. It has also proven to be a good rootstock for all of these. Several cultivars of relatively minor importance have also been generated from pin cherry and choke cherry.

## CULINARY PORTRAIT

Aside from being eaten fresh, cherries are prepared in a variety of ways. They may be cooked, candied, dried, preserved, in syrup, or made into alcoholic beverages, including liqueurs. Cherries are widely used in the making of pies, jams, preserves, and pastries. Candied cherries are an essential ingredient of fruitcake. Cherries can be used to accompany poultry and wild game.

Cherries in baked goods sometimes produce blue discoloration around the fruit due to a chemical reaction between the fruit and baking powder or baking soda. This can be prevented by substituting buttermilk or sour cream for milk in the recipe or adding an acidic liquid such as lemon juice.

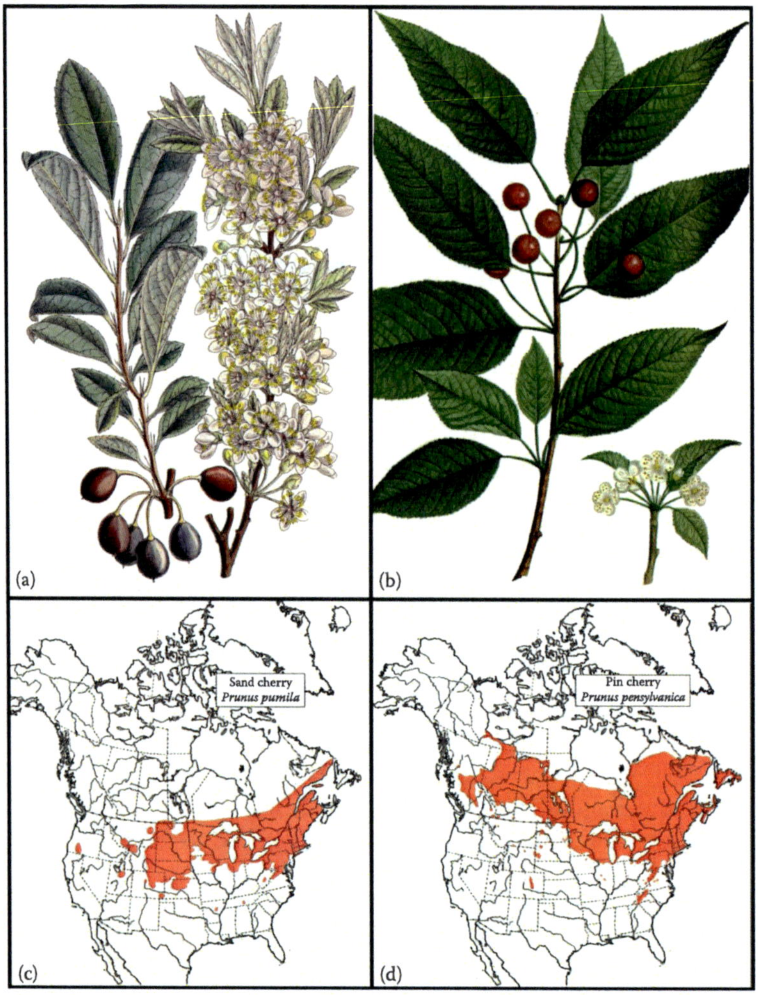


**FIGURE 25.1** American cherries (*Prunus species*). (a) Fruiting and flowering branches of sand cherry (*P. pumila*). (From Curtis, W., *Botanical Magazine*, Vol. 133, Plate 8156, 1907.) (b) Fruiting and flowering branches of pin cherry (*P. pensylvanica*). (From Michaux, F.A., *Histoire des arbres forestiers de L'Amérique septentrionale*, L. Haussman, Paris, France. Vol. 3, 1863.) (c) Distribution map of sand cherry. (d) Distribution map of pin cherry.

The black cherry, although principally of value as a source of wood for cabinet making and interior work, has several minor but interesting culinary uses. The fruits have a rich, wine-like taste that is useful for flavoring alcoholic beverages, including brandy, rum (hence the name "rum cherry"), liqueurs, and cider. An extract from the bark is used medicinally and in the food industry to flavor soft drinks, syrups, candy, and baked goods. The wood is excellent for cooking planks, which impart a pleasant woody taste to foods cooked on them. The wood is also used as a smoking wood for meat and fish.

The name "chokecherry" reflects the considerable astringency of fruits of wild plants of the chokecherry. Nevertheless, wild-collected fruits are commonly eaten raw, dried, or processed into juice, jams, jellies, syrup, pies, and (especially) wine. Several of the fruit cultivars have surprisingly palatable ("chokeless") fruit.

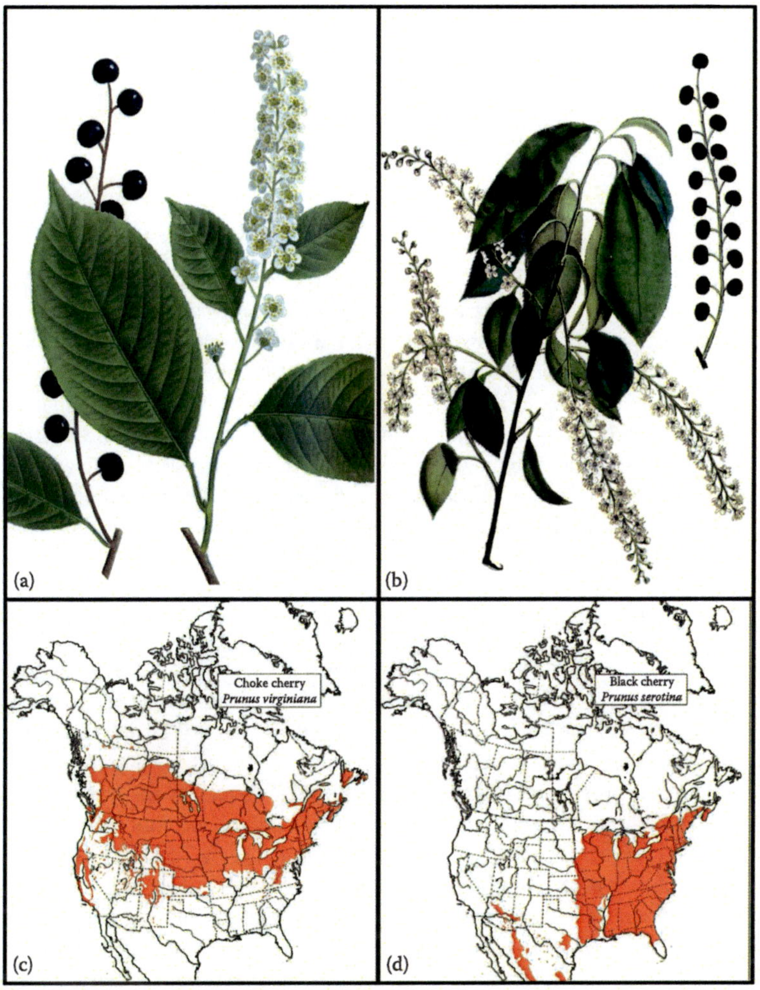


**FIGURE 25.2** American cherries (*Prunus species*). (a) Fruiting and flowering branches of choke cherry (*P. virginiana*). (From Michaux, F.A., *Histoire des arbres forestiers de L'Amérique septentrionale*, L. Haussman, Paris, France. Vol. 3, 1863.) (b) Fruiting and flowering branches of black cherry (*P. serotina*). (From Carson, J., *Illustrations of medical botany*, R.P. Smith, Philadelphia, PA. Vol. 1, Plate 35, 1847.) (c) Distribution map of choke cherry. (d) Distribution map of black cherry.

## WARNING

All members of the genus *Prunus* contain potentially deadly analogues of deadly cyanides in the form of prussic acid (hydrocyanic acid) or hydrogen cyanide. Generally among plants, cyanide is found in the form of glycosides, that is, in combination with sugars. Amygdalin in bitter almonds is an example of such a glycoside; it yields prussic acid upon hydrolysis (which may occur in the digestive system). Prussic acid and the cyanides are highly toxic, inhibiting cellular oxidative processes by combining with the red blood cells, preventing the cells from carrying oxygen to the tissues. Hydrogen cyanide is present in all parts of plants of *Prunus* except the fleshy portion of the fruit. Hydrogen cyanide is often concentrated in the seeds. Livestock are frequently poisoned by consuming the foliage of wild species of *Prunus*. Some cherry species, particularly choke cherry and black cherry, are noted for poisoning livestock that consume the foliage. Ingesting large quantities of chokecherry and black cherry fruits without removing the seeds has caused illness and death in children. Children have even been poisoned by chewing the twigs.

### Culinary Vocabulary

- "Cherry bounce" in Britain refers to cherry brandy, and in America to a traditional homemade cherry-flavored liqueur made from rum or whisky, or sometimes cider, combined with fruit, sugar, spices, and water, and fermented for several weeks. Black cherry fruit has often been used to prepare cherry bounce.

## PROSPECTS

The North American cherries highlighted in this chapter are inferior as fruits to the Eurasian cherry species, and have very limited commercial importance as direct sources of cherries. The Eurasian sweet and sour cherries are widely consumed, and it is likely that the economic significance of the American species will remain chiefly as (1) sources of breeding material for improvement of cultivars of the Eurasian species, (2) parental material for hybrids with other species of *Prunus*, and (3) as rootstocks for dwarfing, or for adapting the Eurasian species to stressful environments. The American species are quite valuable as starting material for breeding cultivars adapted to cold and dry environments, and given the attractiveness of cherries, there is substantial justification for investing in research and management to develop superior cultivars.

## CURIOSITIES OF SCIENCE AND TECHNOLOGY

- Colors are often difficult to describe and distinguish. "Cherry" is part of the names of various, rather subjectively defined colors, including cherry (a moderate red), cherry blossom (also a moderate red, but paler), cherry pink (pink-like cherry blossoms), cherry rose (a deep pink), and cherry wine (a deep red).
- Small nectar-producing glands ("extra-floral nectaries") occur on the leaf margins of some cherry species. It has been found that the predaceous red-headed ant (*Formica obscuripes* Forel) feeds on the nectaries of the foliage of black cherry. The ant seems to repay the cherry for the nectar by consuming some insects that feed on the plant.
- Overripe cherries left on a tree may ferment under hot conditions, and have been reported to produce intoxication in birds that consume them in large quantities.
- Cherries should be picked with the stems attached, as this prolongs their shelf life.
- In 1997, the cherry was declared to be the official state fruit of Utah.
- In 2007, the chokecherry became the official fruit of the state of North Dakota. A petition started by grade six schoolchildren was the key impetus that led to the legislation.

## KEY INFORMATION SOURCES

### General

Badenes, M.L., and Parfitt, D.E. 1995. Phylogenetic relationships of cultivated *Prunus* species from an analysis of chloroplast DNA variation. *Theor. Appl. Genet.* 90: 1035–1041.

Faust, M., and Suranyi, D. 1997. Origin and dissemination of cherry. *Hortic. Rev.* 19: 263–317.

Fogle, H.W. 1975. Cherries. In *Advances in fruit breeding*. Edited by J. Janick and J.N. Moore. Purdue University Press, West Lafayette, IN. pp. 348–366.

Hedrick, U.P., Howe, G.H., Taylor, O.M., Tubergen, C.B., and Wellington, R. 1915. *The cherries of New York.* J.B. Lyon Co., Albany, NY. pp. 371.

Iezzoni, A., Schmidt, H., and Albertini, A. 1990. Cherries (*Prunus*). In *Genetic resources of temperate fruit and nut crops*. Edited by J.N. Moore and J.R. Ballington, Jr. International Society for Horticultural Science, Wageningen, the Netherlands. pp. 111–173.

Looney, N.E., and Webster, A.D. (Editors).1996. *Cherries: crop physiology, production and uses.* CABI, Wallingford, U.K. pp. 513.

Marshall, R.E. 1954. *Cherries and cherry products*. Interscience Publishers, New York. pp. 283.

Mulligan, G.A., and Munro, D.B. 1981. The biology of Canadian weeds. 51. *Prunus virginiana* L. and *P. serotina* Ehrh. *Can. J. Plant Sci.* 61: 977–992.

Shaw, J., and Small, R.L. 2004. Addressing the "hardest puzzle in American pomology": phylogeny of *Prunus* sect. *Prunocerasus* (Rosaceae) based on seven noncoding chloroplast DNA sequences. *Am. J. Bot.* 91: 985–996.

Tilman, D. 1978. Cherries, ants, and tent caterpillars: timing of nectar production in relation to susceptibility of caterpillars to ant predation. *Ecology* 59: 686–692.

Wen, J., Berggren, S.T., Lee, C.-H., Ickert-bond, S., Yi, T.-S., Yoo, K.-O., Xie, L., Shaw, J., and Potter, D. 2008. Phylogenetic inferences in *Prunus* (Rosaceae) using chloroplast *ndh*F and nuclear ribosomal ITS sequences. *J. Syst. Evol.* 46: 322–332.

## Black cherry (*P. serotina*)

Carter, K.K., Cech, F.C., and Dehayes, D.H. 1983. Geographic variation in *Prunus serotina*. *Can. J. For. Res.* 13: 1025–1029.

Hough, A.F. 1960. *Silvical characteristics of black cherry (Prunus serotina)*. Archive Station Paper NE 139. U.S. Department of Agriculture, Northeastern Forest Experiment Station, Upper Darby, PA. pp. 26.

Labrecque, M., and Barabé, D. 1984. Fruit development in *Prunus serotina* (Rosaceae). *Can. J. Bot.* 62: 195–206. (In French.)

Maynard, C.A., Kavanagh, K., Fuernkranz, H., and Drew, A.P. 1991. Black cherry (*Prunus serotina* Ehrh.) *Biotech. Agric. Forest.* 16: 3–22.

McVaugh, R. 1951. A revision of the North American black cherries (*Prunus serotina* Ehrh., and relatives). *Brittonia* 7: 279–315.

Morden-Moore, A.L., and Willson, M.F. 1982. On the ecological significance of fruit color in *Prunus serotina* and *Rubus occidentalis*: field experiments. *Can. J. Bot.* 60: 1554–1560.

Pairon, M., Petitpierre, B., Campbell, M., Guisan, A., Broennimann, O., Baret, P.V., Jacquemart, A.-L., and Besnard, G. 2010. Multiple introductions boosted genetic diversity in the invasive range of black cherry (*Prunus serotina*; Rosaceae). *Ann. Bot.* 105: 881–890.

Petitpierre, B., Pairon, M., Broennimann, O., Jacquemart, A.L., Guisan, A., and Besnard, G. 2009. Plastid DNA variation in *Prunus serotina* var. *serotina* (Rosaceae), a North American tree invading Europe. *Europ. J. For. Res.* 128: 431–436.

Vanhellemont, M., Wauters, L., Baeten, L., Bijlsma, R.-J., De Frenne, P., Hermy, M., and Verheyen, K. 2010. *Prunus serotina* unleashed: invader dominance after 70 years of forest development. *Biol. Invas.* 12: 1113–1124.

## Chokecherry (*P. virginiana*)

Hall, I.V., and Gourley, C.O. 1981. Biology of *Prunus pensylvanica* L.f. *Proc. N.S. Inst. Sci.* 31: 101–108.

Labrecque, M., Barabé, D., and Vieth, J. 1985. Fruit development in *Prunus virginiana* (Rosaceae). *Can. J. Bot.* 63: 242–251. (In French.)

Majak, W., McDiarmid, R.E., and Hall, J.W. 1981. The cyanide potential of Saskatoon serviceberry (*Amelanchier alnifolia*) and chokecherry (*Prunus virginiana*). *Can. J. Anim. Sci.* 61: 681–686.

Pijoan, M. 1942. Cyanide poisoning from choke cherry seed. *Am. J. Med. Sci.* 204: 550–553.

Samolova, L.P., and Safronova N.E. 1991. Large-fruited forms of *Prunus virginiana*. *Sadovodstvo i vinogradarstvo* 6: 36–37. (In Russian.)

Vilkitis, J.R. 1974. Common chokecherry. In *Shrubs and vines for northeastern wildlife*. Edited by J.D. Gill and W.M. Healy. USDA Forest Service General Technical Report NE 9. U.S. Northeastern Forest Experiment Station, Upper Darby, PA. pp. 23–25.

Zatylny, A.M., Ziehl, W.D., and St-Pierre, R.G. 2005. Physicochemcial properties of fruit of chokecherry (*Prunus virginiana* L.), highbush cranberry (*Viburnum trilobum* Marsh.), and black currant (*Ribes nigrum* L.) cultivars grown in Saskatchewan. *Can. J. Plant Sci.* 85: 425–429.

## Pincherry (*P. pensylvanica*)

Marks, P.L. 1974. The role of pin cherry, *Prunus pensylvanica*, in the maintenance of stability in northern hardwood ecosystems. *Ecol. Monogr.* 44: 73–88.

### Sand cherry (*P. pumila*)

Catling, P.M., and Larson, B.M.H. 1997. The decline and current status of the dune race of dwarf cherry, *Prunus pumila* var. *pumila*, on the Canadian shores of the lower Great Lakes. *Can. Field-Nat.* 111: 187–193.
Catling, P.M., McKay-Kuja, S.M., and Mitrow, G. 1999. Rank and typification in North American dwarf cherries, and a key to the taxa. *Taxon* 48: 483–488.

### Western sand cherry (*P. pumila* var. *besseyi*)

Hansen, N. E. 1904. *The western sand cherry.* Bulletin 87. South Dakota Agricultural College Experiment Station, Brookings, SD. pp. 64.
Erskine, R. 1971. Varieties of western sand cherry (*Prunus besseyi*). *Arch. Fruit Var. Hort. Dig.* 25(4): 80–81.
Evreinoff, V.A. 1946. The dwarf sand cherry (*Prunus besseyi*). *Rev. Hort.* (Paris) 118: 116–117. (In French.)

### Specialty Cookbooks

The following cookbooks are based on commercially available cherries (i.e., sweet and sour cherry species); the cherry species described in this chapter can be substituted, although several of them have a harsher taste.

Bader, M. 1995. *Cherry creations: The ultimate cherry cookbook.* Northstar Pub. Co., Las Vegas, NV. pp. 224.
Cherry Cookbook Committee. 1976. *150 cherry recipes.* Cherry Cookbook Committee, National Cherry Festival, Traverse City, MI. pp. 106.
Cherry Marketing Institute. 1994. *A celebration of cherries: Delightful recipes for dining adventure.* Cherry Marketing Institute, Okemos, MI. pp. 37.
Conyers Cherry Blossom Festival Foundation. 1991. *A taste of the cherry blossom festival.* Illustrated Art & Design, Conyers, GA. pp. 173.
Eberly, C. 1984. *101 cherry recipes.* Eberly Press, Lansing, MI. pp. 47.
Eschmann, C. 1984. *Cherries galore: A cookbook.* Macon Telegraph and News, Macon, GA. pp. 206.
Gohlke, A. 1982. *Cherry delights.* Reiman Publications, Milwaukee, WI. pp. 64.
Hill, J., and Littler, J. 1993. *Cherries galore: A Door County cherry recipe collection.* Hill Orchards of Scandia Road, Siste Bay, WI. pp. 252.
Morgenroth, B. 1998. *The best 50 cherry recipes.* Bristol Pub Enterprises, San Leandro, CA. pp. 78.
Rubin, C., and Rubin, J. 1974. *Cherry cookbook.* Emporium Publications, Newton, MA. pp. 91.
Whitaker, T. 1976. *Light-sweet cherries: Over 100 recipes and serving ideas.* Whitaker, Silveton, OR. pp. 88.

Berglund and Bolsby (1971) have 3 chokecherry recipes, Lund (2005) has 2, Mogelon (2001) has two, Stanek and Butcher (2007) have 2, and Williamson (1995) has 12. Freitus and Haberman (2005) have 11 recipes for wild cherries, Krumm (1991) has 18, Marrone (2009) has 14, Robe-Terry (1997) has 4, and Snell (2006) and Marie (2008) each have 2. (See the Appendix to this book for details of the publications cited.)

# 26 Chinkapin

Family: Fagaceae (beech family, sometimes called the oak family)

## NAMES

Scientific name: *Castanea pumila* (L.) Mill.

- "Chinkapin" (also spelled chincapin, chinquapin, and chinquepin) is a word of Native American (Algonquian) derivation. It designates American chestnuts other than the American chestnut tree, *C. dentata* (Marshall) Borkh. (some specialists recognize up to eight such species, whereas others recognize only one species, *C. pumila*). The word chinkapin also refers to the nuts produced by the trees.
- *Castanea pumila* is also known as the Allegheny chinkapin, American chinkapin, bush chestnut, common chinkapin, dwarf chestnut, eastern chinkapin, golden chinkapin (an ambiguous name, also applied to *Chrysolepis*, as noted below), and tree chinkapin.
- The name chinkapin is also used for some other plant genera, including species of the Asian genus *Castanopsis* and the golden chinkapins of the Pacific United States (these are sometimes included within *Castanopsis* but are more often considered a separate but very closely related genus, *Chrysolepis*).
- Chinkapins are sometimes called dwarf or bush chestnuts.
- See Chapter 3 on American Chestnut for the derivation of the genus name *Castanea*.
- *Pumila* in the scientific name *C. pumila* is Latin for small or dwarf, indicative of the relatively small stature by comparison with other trees.

## GEOGRAPHY AND ECOLOGY OF WILD PLANTS

The American chinkapin is native to the southeastern United States, especially the Appalachian Mountain range. It is found from southern New Jersey and Pennsylvania south to central Florida, west to eastern Texas, and north to southern Missouri and Kentucky. It occurs in dry, sandy, loamy, or rocky uplands and ridges and in open woods, up to an elevation of 1000 m (about 3300 feet) in the Appalachian Mountains. The species is often found in disturbed sites, such as along railways, powerline clearings, and fences, and in old fields. It grows best on well-drained soils in full sun or partial shade. Two varieties are often recognized: the Ozark chinkapin (*C. pumila* var. *ozarkensis* (Ashe) G.E. Tucker) and the Allegheny chinkapin (*C. pumila* var. *pumila*).

The Ozark chinkapin is named for its presence in the Ozark Highlands of Oklahoma, Arkansas, and Missouri. It has been severely affected by the chestnut blight fungus that devastated the closely related American chestnut. The Ozark chinkapin was once present in Alabama but has been eliminated there by the chestnut blight. It is also being severely affected in the Ozark Plateau. This variety occurs on dry uplands in deciduous or mixed woodlands, and on cliff margins, talus slopes, and rocky ridges.

The more widely distributed Allegheny chinkapin (which is rare in the Ozark Plateau) is distributed from Pennsylvania to Florida, west to east Texas, eastern Oklahoma, southwestern Missouri, and south-central Kentucky. In contrast to the Ozark chinkapin, it is found in disturbed habitats such as along railways, powerline clearings, and fences, and in old fields and pine plantations. As well as occupying disturbed sites, it occurs in coniferous, mixed, and deciduous woodlands. This

**FIGURE 26.1** Chinkapin (*Castanea pumila*). (a) Branch with male catkins; nut and fruit at bottom. (From Michaux, F.A., *Histoire des arbres forestiers de L'Amérique septentrionale*, L. Haussman, Paris, France, Vol. 2, 1862.) (b) Tree. (Courtesy of Salvor.) (c) Nuts. (Courtesy of Steve Hurst, U.S. Department of Agriculture.) (d) Prickly fruits. (Courtesy of Alan Cressler, reproduced with permission.) (e) Distribution map.

variety is found from sea level on the coastal plain to high elevations in the Appalachians. It is fairly resistant to chestnut blight. Affected trees often continue to send out suckers that will produce fruit.

## PLANT PORTRAIT

Chinkapins are shrubs and small trees of North America, placed in one species (*C. pumila*), by many authorities, a treatment that is followed in this chapter. The chinkapin is a prolific producer of sweet, nutty-flavored, small chestnuts. *Castanea pumila* is a spreading shrub or small tree, growing 2–20 m (6–65 feet) in height at maturity, and developing trunks 15–56 cm (6–18 inches) in diameter, rarely up to 90 cm (3 feet). Both varieties occur as trees, but the Allegheny chinkapin tends to more often form shrubs than does the Ozark chinkapin. Small, pale yellow to white, male flowers are borne on erect catkins 10–15 cm (4–6 inches) long attached to the base of a leaf. Small female flowers are located at the base of some catkins. Some people consider the odor of the plants at flowering time to be unpleasant. The fruit is a golden-colored bur 2.5–3.5 cm (1–1.5 inches) long with many sharp spines, maturing in autumn (technically, the burs are clusters of fruits). The bur contains one ovoid, shiny, sweet, chestnut-brown nut 7–20 mm (0.3–0.8 inch) long. The burs split into two halves, like a clamshell, freeing the nut. A mature tree can produce about 4.5 kg (10 pounds) of edible material annually. The species forms thickets or colonies by spread of rhizomes (underground stems).

## CULINARY PORTRAIT

The nuts of American chinkapin are sweet and can be eaten raw. Some authorities claim that when they are baked they are even sweeter, with a floury texture. The nuts can be baked, roasted in the shell, ground into flour, and incorporated into candies. Chinkapin nuts are miniature versions of chestnuts and can be prepared and served in the same ways; see Chapter 3 on American Chestnut for suggestions.

### Culinary Vocabulary

- A "chestnut pan" is a shallow frying pan with a perforated bottom, used to roast chestnuts. The holes allow a heat source (usually a flame) to furnish diluted contact with the nuts. This utensil is ideal for roasting chinkapins, provided that the holes are small enough to prevent the nuts from falling through.

## PROSPECTS

Payne et al. (1993) provided the following economic analysis of chinkapin:

### Advantages

Precocious, produces nuts in 2 to 3 years
Prolific; large number of female flowers per catkin, large number of female catkins per shoot
Distinct flavor and aroma (sweet and edible)
Attractive foliage, flowers, and burs
Wildlife food and cover crop
Suitable for dry areas and for use in site reclamation

### Problems

Excessive bird and mammal feeding
Difficult to harvest
Fall germination
Small nut size
Adherence of nuts within the bur and germination inside the bur
Susceptible to blight
Multistemmed–suckering
Dwarfing rootstock possibility
Lack of pesticide registrations

Although it was once predicted that the American chinkapin would become a cultivated plant of commercial value, this has not happened. At present, the species is viewed chiefly as a source of nuts for wildlife, with some potential for development as a landscape ornamental. There has been very limited success in establishing a commercial industry based on chestnuts from chinkapins. The small size of the nuts and difficulty in harvesting because of uneven ripening have been claimed to be among the species' chief disadvantages. However, the prodigious production of some trees compensates for the small nuts, and it should be possible to select for uniform maturation. Wild-collected seeds have been sold in local markets in the United States, especially during the nineteenth and twentieth centuries, but they are very uncommonly available today. "The economic potential of this crop still remains to be adequately demonstrated" (Payne et al. 1993). Nevertheless, "it may well be our most ignored and undervalued native North American nut tree" (Payne et al. 1993), and the American chinkapin needs to be included among the wild plants of North America that should be considered for economic development.

The chinkapin is closely related to the American chestnut, a species that was once an important source of chestnuts but was almost exterminated from North America by a fungus. There remains hope that one day the American chestnut industry can be revived, and the chinkapin may play a role in this, as some forms of it are moderately resistant to the fungus and this resistance could be transferred to the American chestnut by hybridization. The chinkapin might also be used in programs of hybridization with the Eurasian chestnut species that currently dominate the chestnut market.

## CURIOSITIES OF SCIENCE AND TECHNOLOGY

- Captain John Smith (1580–1631), an English explorer who became Admiral of New England, is known for contributing to the establishment of the first permanent English settlement in North America, at Jamestown Virginia. Smith was captured by the Powhatan Confederacy, but reputedly was saved by the Virginian Indian Pocahontas who prevailed on her father, the chief, to release him. He may have recorded the first European record of the chinquapin in 1612 in writing "The Indians have a small fruit growing on little trees, husked like a chestnut, but the fruit most like a very small acorne. This they call *Checkinquamins*, which they esteem a great daintie."
- The Cherokee Indians treated headaches by blowing heated chinkapin leaves on the patients.
- The embryos within chinkapin nuts sometimes germinate while the fruits are still attached to the tree. That is, the radicle (embryonic root) may start to emerge before the nuts have fallen off the tree. This is related to the fact that the nuts germinate in the fall, shortly after they become mature, and this phenomenon is one of the drawbacks of chinkapin as a crop.

## KEY INFORMATION SOURCES

Dane, F. 2009. Comparative phylogeography of *Castanea* species. *Acta Hort.* 844: 211–222.

Dane, F., Hawkins, L.K., and Huang, H. 1999. Genetic variation and population structure of *Castanea pumila* var. *ozarkensis*. *J. Am. Soc. Hort. Sci.* 124: 666–670.

Dane, F., Lang, P., Huang, H., and Fu, Y. 2003. Intercontinental genetic divergence of *Castanea* species in eastern Asia and eastern North America. *Heredity* 91: 314–321.

Fu, Y., and Dane, F. 2003. Allozyme variation in endangered *Castanea pumila* var. *pumila*. *Ann. Bot.* 92: 223–230.

Hooker, W.V. 1967. Chinquapins in Oklahoma. *Ann. Rpt. No. Nut Growers Assn.* 58: 118–120.

Johnson, G.P. 1987. Chinquapins: Taxonomy, distribution, ecology, and importance. *Ann. Rpt. No. Nut Growers Assn.* 78: 58–62.

Johnson, G.P. 1988. Review of *Castanea* sect. *Balanocastanon* (Fagaceae). *J. Arnold Arb.* 69: 25–49.

Lang, P., Dane, F., and Kubisiak, T.L. 2006. Phylogeny of *Castanea* (Fagaceae) based on chloroplast *trn*T-L-F sequence data. *Tree Genet. Genomes* 2: 132–139.

Lang, P., Dane, F., Kubisiak, T.L., and Huang, H. 2007. Molecular evidence for an Asian origin and a unique westward migration of species in the genus *Castanea* via Europe to North America. *Molec. Phylogenet. Evol.* 43: 49–59.

Paillet, F.L. 1993. Growth form and life histories of American chestnut and Allegheny and Ozark chinkapin at various North American sites. *Bull. Torrey Bot. Club* 120: 257–268.

Payne, J.A., Dutcher, J.D., and Wood, B.W. 1982. Chinkapins: A promising nut crop in the south? *Ann. Rpt. No. Nut Growers Assn.* 73: 23–26.

Payne, J.A., Johnson, G.P., and Miller, G. 1993. Chinkapin: Potential new crop for the south. In *New crops*. Edited by J. Janick and J.E. Simon. Wiley, New York, NY. pp. 500–505.

Payne, J.A., Miller, G., Johnson, G.P., and Senter, S.D. 1994. *Castanea pumila* (L.) Mill.: An underused native nut tree. *HortScience* 29: 62, 130–131.

Senter, S.D., Payne, J.A., Miller, G., and Anagnostakis, S. 1994. Comparison of total lipids, fatty acids, sugars, and nonvolatile organic acids in nuts from four *Castanea* species. *J. Sci. Food Agr.* 65: 223–227.

Tucker, G.E. 1975. *Castanea pumila* var. *ozarkensis* (Ashe) Tucker, comb. nov. *Arkansas Acad. Sci. Proc.* 29: 67–69.

## Specialty Cookbooks

Chinkapin nuts are very rarely sold commercially and at present are basically a wild food that is found only in the native area of the species. Accordingly, there are no cookbooks dedicated to the culinary preparation of the nuts. Chinkapins are essentially miniature chestnuts, and the specialty cookbooks listed for chestnut (which see) can be consulted for recipe suggestions. Freitus (1980) presents eight recipes for chinkapins. Krochmal and Krochmal (1974) provide advice on preparing "chestnut coffee" using *Castanea* nuts. (See the Appendix to this book for details of the publications cited.)

# 27 Chive

Family: Liliaceae (lily family; sometimes placed in Alliaceae, the onion family)

## NAMES

Scientific name: *Allium schoenoprasum* L. (*A. sibiricum* L.)

- The singular "chive" is best used as the name of the plant. When referred to as a condiment, the plural "chives" is usually used.
- "Chive" is derived from the Old French *cive*, which in turn comes for the Latin *cepa*, onion.
- The genus name *Allium* is the classical Latin name for garlic.
- *Schoenoprasum* in the scientific name for common chive is formed from the Greek *skhoinos*, meaning rush, and *prason*, meaning leek, a reference to the rush-like leaves of the plant. The obsolete name "rush-garlic" also reflects a superficial resemblance to the rush genus, *Juncus*.

## GEOGRAPHY AND ECOLOGY OF WILD PLANTS

Wild chive plants are distributed in northern regions of the North Temperate Zone, extending as far south as southern Europe, Iran, India, and China. In North America, the species grows naturally as far south as northern New York to northern Colorado and Oregon. Chive grows in several habitats, but prefers moist soil conditions such as wet meadows and edges of rivers and lakes.

## PLANT PORTRAIT

Chive is a perennial herbaceous plant with slim, hollow leaves arising from thick tufts to a height of 70 cm (28 inches) or more. Slender, white-sheathed bulbs develop in dense masses at the base of the plant. Rose-purple or mauve flowers are produced in a small, round cluster at the end of a leafless stalk. The domesticated selections are little changed from the wild forms. Chive has been collected from the wild since antiquity but was probably not cultivated until the Middle Ages. Wild chive was a popular food among indigenous peoples of North America. The plant has been grown in Europe at least since the sixteenth century, and cultivation, thought to have originated in Italy, spread to Germany, and then to other European countries. During the nineteenth century, chives became very popular in Europe, particularly in French *haute cuisine*. Chive has never been of great economic importance. Although among the most universally grown of herbs, relatively little is produced commercially (worldwide estimates ranging from as little as 1000 ha/year to as much as 24,000 ha/year, or 2470–59,300 acres/year).

Chive plants are often used for edging flower beds because of their abundant foliage and beautiful lavender flowers. Dried chive flower heads are widely used in flower arrangements.

The sulfur compounds that give species of *Allium* their characteristic onion taste are considered beneficial to the circulatory, respiratory, and digestive systems. Chive contains much lower concentrations of these compounds than, for example, garlic (*Allium sativum* L.), and as a result it is used medicinally to a much lesser extent. It does, however, contain substantial amounts of vitamins A and C.

**FIGURE 27.1** Chive (*Allium schoenoprasum*). ([a and b] Courtesy of Sharon and Gerry Channer.) (c) Commercial package of chives. (Courtesy of E. Small.) (d) Painting. (From Hallier, E., *Flora von Deutschland* [fifth edition by D.F.L. von Schlechtendal et al.], F.E. Köhler, Gera-Untermhaus, Germany. Vol. 3, Plate 266, 1880.) (e) Distribution map.

## CULINARY PORTRAIT

Chives have a delicate onion flavor. The tender, mild leaves are eaten raw or cooked in many dishes, including salads, soups, vegetables, sauces, meats, eggs, fish, butter, and cream cheese. Chopped chives make an exceptional garnish for salads and potato dishes, going well with sour cream. Many cooks maintain potted chive plants on sunny windowsills, cutting off as much as is needed for a dish in preparation (dried chives are tasteless). Some cooks recommend cutting chives with scissors and never chopping them. Chives is a traditional accompaniment for vichyssoise. The flavor of chives is destroyed with prolonged heat and so it is best added to cooking foods during the last 5 to 10 minutes. The processing industry freezes, freeze-dries, and dehydrates chives for use in packaged foods such as soup mixes, salad dressings, cocktail dips, sour cream, and cottage cheese products. The bulbs are often pickled.

Fresh chives can be stored for up to 5 days by wrapping in damp paper towels, placing in a plastic bag, and kept in a refrigerator. Chives can be frozen after rinsing, drying, and cutting but should be consumed within 3 months.

Most *Allium* species can cause distress to many animals when consumed in excess. Although chive is not an especially potent member of the onion group, it has been reported to produce poisoning symptoms in horses, dogs, and other animals, and deer are said to avoid it. However, it is generally not considered to be a significant threat to animals.

### Culinary Vocabulary

- "Garlic chives" should not be confused with chives. Garlic chive, also known as Chinese chive, flowering chive, kuchai, and oriental garlic, is a different species (*A. tuberosum* Rottler ex Spreng.). This is a perennial herb native to Asia, widely grown in eastern Asia from China and Japan south into Thailand and the Philippines, with some commercial cultivation in California. The leaves of garlic chive are eaten in Western dishes much as those of chive, but with more discretion in view of their stronger taste.

## PROSPECTS

The market demand for chives as an ingredient in commercial food products is fairly stable. On the other hand, there is growing sales of fresh chives, especially out of season. There is therefore good potential for greenhouse and hydroponic producers to supply grocery stores, restaurants, and caterers with a high-quality fresh product. Chive grows very well in hydroponic culture and produces considerable material in a relatively small space. Chive cultivation is much more established in Europe, especially northern Europe, Italy, and Greece, than in North America. Since chive is more widely consumed in Europe than elsewhere in the world, the greatest efforts at breeding cultivars and preserving germplasm have been made there. However, serious breeding of chive did not begin in Europe until the early 1960s, and as a result, cultivars are relatively undeveloped. There are only a few hundred germplasm collections of chive and most of these are in Europe. Fundamental to the requirement for breeding chive cultivars is the basic need for germplasm acquisition and a better appreciation of the classification of *A. schoenoprasum*. Only a few dozen commercial cultivars are available, and in North America very frequently the chives that are grown are not named cultivars. There is a need for the breeding of improved varieties.

## CURIOSITIES OF SCIENCE AND TECHNOLOGY

- Ancient wisdom is sometimes captured in poetry, as evidenced by the following:

  He who bears chives on his breath
  Is safe from being kissed to death.

  —**Marcus Valerius Martialis** (from "Martial's Epigrams," about 80 AD; Martial was a witty Roman who invented the "epigram," a short satirical poem typically with a clever last line)

- Charlemagne (742–814) was a Frankish king (the Franks were a Germanic people) who came to be Emperor of the West. To better the economy of his realm, he composed an edict, "The Capitulare," which directed all those he governed to grow about 90 different kinds of plant wherever possible. Chive was among the edible plants that were ordered to be grown. Charlemagne's list is one of the best documented early examples of an agricultural policy.
- Romanian Gypsies (Roma) used chives as part of their fortune-telling rituals.
- In past centuries in Europe, it was believed that hanging dried bunches of chives around the house, especially suspended from ceilings or hung on bed posts, drove away evil influences and disease.

- Chives became so popular in Holland during the nineteenth century that farmers fed plants to their cattle to produce a flavored milk. Today, milk that has been tainted as a result of cattle feeding on weedy forms of chive is considered to have a bad taste.
- The juice of the chive plant has sometimes been used as an insect repellent.
- In past times, Native Americans of the Great Basin region prepared a golden-brown dye from the skin of wild chive bulbs.
- Of the approximately 600 species of *Allium*, chive is the only one that is native to both the Old and New Worlds.
- Chives are more popular than their everyday consumption would suggest, and in home cooking they are often reserved for special meals. The Texas Department of Criminal Justice until the early twenty-first century provided detailed online information on the last meal requests of executed murderers. Of 285 final meals between 1982 and 2002, the most elaborate was "two 16 oz. ribeyes, 1 lb. turkey breast (sliced thin), 12 strips of bacon, 2 large hamburgers with mayo, onion, and lettuce, 2 large baked potatoes with butter, sour cream, cheese, and chives, 4 slices of cheese or one-half pound of grated cheddar cheese, chef salad with blue cheese dressing, 2 ears of corn on the cob, 1 pint of mint chocolate chip ice cream, and 4 vanilla Cokes or Mr. Pibb."

## KEY INFORMATION SOURCES

Bougourd, S.M., and Plowman, A.B. 1996. The inheritance of B chromosomes in *Allium schoenoprasum* L. *Chromosome Res.* 4: 151–158.

Brobst, J.E. 1988. Chives for everyone. *The Herbarist* 54: 48–52.

Friesen, N., and Blattner, F.R. 2000. RAPD analysis reveals geographic differentiations within *Allium schoenoprasum* L. (Alliaceae). *Plant Biol. (Stuttgart)* 2: 297–305.

Hashimoto, S., Miyazawa, M., and Kameoka, H. 1983. Volatile flavor components of chive (*Allium schoenoprasum* L.). *J. Food Sci.* 48: 1858, 1859, 1873.

Kmiecik, W., and Lisiewska, Z. 1999. Effect of pretreatment and conditions and period of storage on some quality indices of frozen chive (*Allium schoenoprasum* L.). *Food Chem.* 67: 61–66.

McNeal, D.W., Jr., and Jacobsen, T.D. 2002. *Allium*. In *Flora of North America*, Volume 26. Edited by Flora of North America Editorial Committee. Oxford University Press, Oxford, UK. pp. 224–276.

Mercer, J. 1994. Production of mints and chives as cut flowers. In *Herbs '94. Proceedings of ninth national herb growing and marketing conference.* Edited by A. Kestner, J.E. Simon, and A.O. Tucker. International Herb Growers & Marketers Association, Mundelein, IL. pp. 17–23.

Peterson, J. 2000. Part IIC.2. Staple foods: Domesticated plants and animals: Important vegetable supplements: The *Allium* species (onions, garlic, leeks, chives, and shallots). In *The Cambridge world history of food*, Volume 1. Edited by K.F. Kiple and K.C. Ornelas. Cambridge University Press, Cambridge, UK. pp. 249–271.

Poulsen, N. 1983. *Chives (*Allium schoenoprasum *L.): A literature review.* Statens planteavlsforsøg, Copenhagen, Denmark. pp. 39. (In Danish, English summary.)

Poulsen, N. 1990. Chives, *Allium schoenoprasum* L. In *Onions and allied crops*. Volume 3, biochemistry, food science, and minor crops. Edited by J.L. Brewster and H.D. Rabinowitch. CRC Press, Inc., Boca Raton, FL. pp. 231–250.

Schoneweis, S.D. 2000. *Growing onions, shallots, and chives.* University of Nebraska Cooperative Extension, Lincoln, NE. pp. 8.

Small, E., and Catling, P.M. 2004. Poorly known economic plants of Canada—43. Chive (*Allium schoenoprasum* L.). *Bull. Can. Bot. Assoc.* 37(4): 64–67.

Stevens, J.P., and Bougourd, S.M. 1988. Genetic analysis of flower colour variation in *Allium schoenoprasum* L. (wild chives). *Heredity* 60: 253–256.

Stevens, J.P., and Bougourd, S.M. 1991. The frequency and meiotic behavior of structural chromosome variants in natural populations of *Allium schoenoprasum* L. (wild chives) in Europe. *Heredity* 66: 391–401.

Tardif, B., and Morisset, P. 1990. Clinal morphological variation of *Allium schoenoprasum* in eastern North America. *Taxon* 39: 417–429.

Tardif, B., and Morisset, P. 1991. Chromosomal C-band variation in *Allium schoenoprasum* (Liliaceae) in eastern North America. *Plant Syst. Evol.* 174: 125–137.

Tatlioglu, T. 1993. Chive, *Allium schoenoprasum* L. In *Genetic improvement of vegetable crops*. Edited by G. Kalloo and B.O. Bergh. Pergamon Press, New York, NY. pp. 3–13.

van der Meer, Q.P. 1993. *Allium tuberosum* Rottler ex Sprengel. In *Plant resources of South-East Asia*. Volume 8, vegetables. Edited by J.E. Siemonsma and K. Piluek. Pudoc Scientific Publishers, Wageningen, the Netherlands. pp. 80–82.

## Specialty Cookbooks

Batcheller, B. 1986. *Lilies of the kitchen: Recipes celebrating onions, garlic, leeks, shallots, scallions, and chives*. St. Martin's Press, New York, NY. pp. 262.

Cooper, G., and Taylor, G.I. 1981. *In celebration of chives*. Herb Society and Juniper Press, London, UK. pp. 48.

Glover, B. 2001.*Onion: The essential cook's guide to onions, garlic, leeks, scallions, shallots and chives*. Lorenz, New York, NY. pp. 256.

Glover, B. 2002. *Essential onions: Over 80 delicious recipes using onions, leeks, garlic, shallots, scallions and chives*. Southwater, London, UK. pp. 128.

Herb Society of America. 1997. *The wild onion cook book: All you want to know about the edible alliums: Chives, garlic, leeks, onions, scallions, shallots*. Jumbo Jack's Cookbooks, Audubon, IA. pp. 377.

Oldmeadow, E.J.F., Simon, A.L., and Phillimore, V. 1943. *The lilies of the kitchen: Onions, leeks, shallots, garlic and chives*. The Wine and Food Society, London, UK. pp. 22.

Shaw, D. 1992. *Sweet basil, garlic, tomatoes, and chives: The vegetable dishes of Tuscany and Provence*. Harmony Books, New York, NY. pp. 192.

Tatum (1976) has four wild chive recipes. Gray (2011) has a recipe entitled "Wild-onion spread" that is made with wild-collected chive. (See the Appendix to this book for details of the publications cited.)

# 28 Chufa

Family: Cyperaceae (sedge family)

## NAMES

Scientific name: *Cyperus esculentus* L.

- "Chufa" is a Spanish dialect name for the plant. It is said to be based on the Spanish *chufar*, to make fun of, tracing to the Latin *sufilre*, to hiss or whistle at. Alternatively, it has been claimed that the word is based on an old Spanish term meaning "trifle" or "tidbit." In addition to being used as the name of the whole plant, "chufa" is also the name for the edible underground tubers and for a nondrying oil obtained from the underground stems (rhizomes) of the plant.
- Pronunciation: CHOO-fuh.
- Chufa is also known as duck potato (a name best reserved for *Sagittaria latifolia*; see Chapter 35 on Duck Potato), earth almond, earth nut, edible rush, Florida almond, galingale, ground almond, northern nutgrass, nutgrass, rush nut, tigernut, yellow nut grass, yellow nut sedge, water grass, and Zulu nut.
- "Earth nut" is also applied to some other plants (see Chapter 45 on Groundnut).
- The resemblance of the tubers to nuts is the basis for the "nut" and "almond" in many of the names.
- The "tiger" in the name "tigernut" is based on the striping often evident on the tubers of domesticated forms of the plant.
- The names "rush nut" and "edible rush" are based on the similarity of chufa to plants of the rush family (Juncaceae), particularly to species of the genera *Juncus* and *Scirpus*. However, species of *Cyperus* are normally called "sedge."
- The Zulus of South Africa have long used chufa, hence the name "Zulu nut." As well as using the tubers for food, the Zulus chew them to relieve indigestion and to sweeten breath.
- The name galingale is easily confused with the name of several spice plants called galangale (*Alpinia galanga* (L.) Sw., *A. officinarum* Hance, and *Kaempferia galanga* L.).
- The name "water grass" is based on the similarity of chufa to grasses. Indeed, species of the sedge family (of which there are 4000–5000 species) are often mistaken for grasses (the 10,000 or so species of the grass family, the Poaceae or Gramineae). True grass species have flowering stems that are round in cross section, usually hollow, and there are nodes (swellings) on the stems, whereas sedge species have flowering stems that are usually solid and triangular in cross section (leading to the saying "sedges have edges"), and their stems lack nodes.
- The genus name *Cyperus* is based on the ancient Greek name for the plant, *cypeiros*.
- *Esculentus* in the scientific name *C. esculentus* is Latin for edible.

## GEOGRAPHY AND ECOLOGY OF WILD PLANTS

Chufa is native to the warmer parts of the Northern Hemisphere (including Europe, Asia, and North America) and parts of Africa and South America. The species is also established as a weed worldwide. It can be found both at the equator and in Alaska. In North America, chufa seems to be native to most states, as well as in southern Canada. Weedy forms often do not produce seeds and rely mostly on their tubers for reproduction. By comparison, cultivated forms of chufa rarely flower,

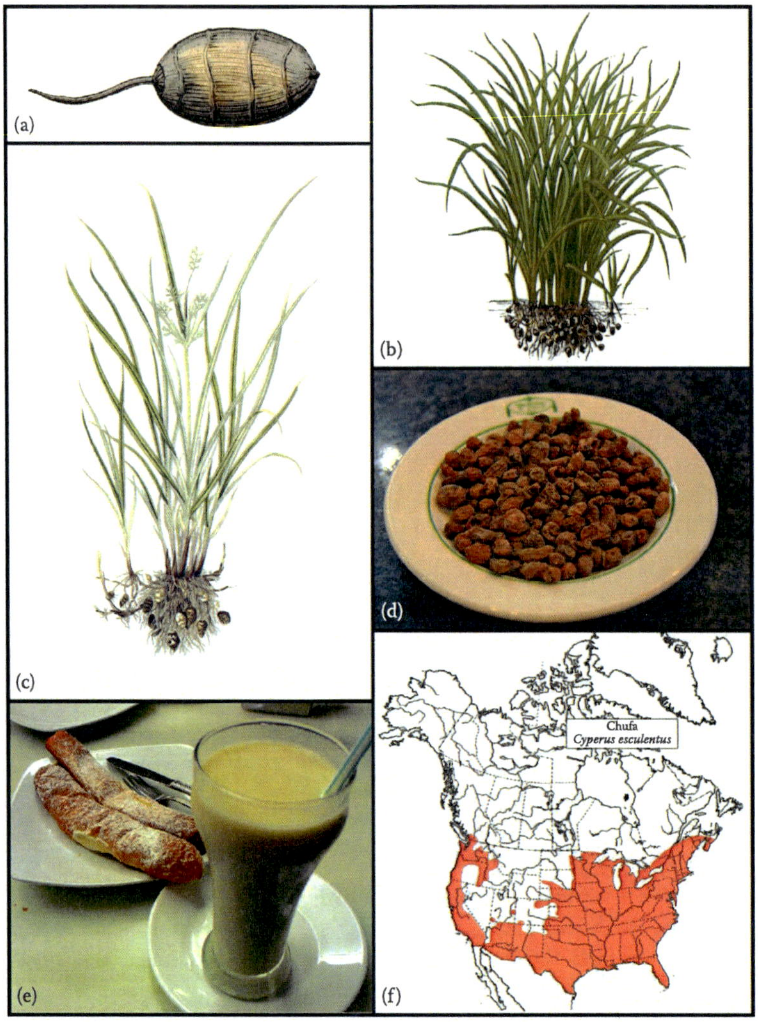


**FIGURE 28.1** Chufa (*Cyperus esculentus*). (a) Tuber (notice stripes, the basis of the name "tigernut"). (Drawings for [a] and [b] [colored by B. Brookes] from Vilmorin-Andrieux. M.M., *The vegetable garden*, John Murray, London, U.K, 1885.) (b) Clump of plants. (c) Painting of plant in flower, with underground tubers. (From Host, N.T., *Icones et descriptions graminum Austriacorum*, Vindobonae, Austria, Vol. 3, Plate 75, 1805.) (d) Plate of chufa. (Courtesy of Tamorlan [CC By 3.0].) (e) Horchata and fartons (see text). (Courtesy of Heather Cowper; required attribution: www.heatheronhertravels.com [CC By 2.0].) (f) Distribution map.

have more leaves bunched together, and more tubers crowded together at the underground base of the plant. The forms that are cultivated are said to be much less inclined to become weedy than the wild forms of the species. Chufa occurs in various habitats, especially damp soils. Weedy forms occur in moist depressions and ditches; in wet ground on the margins of streams and ponds; in pastures, old fields, and lawns; and along roadsides and railways.

## PLANT PORTRAIT

Chufa is a perennial or annual herb. Its very narrow, grass-like leaves are 5–90 cm (2 inches–3 feet) long. A few cultivated varieties are available in Spain and appear to trace to ancient Egypt, where the plant was cultivated. The edible, starchy tubers are produced at the ends of rhizomes (underground stems). The tubers are usually 10–20 mm (0.4–0.8 inch) in diameter, unevenly globe shaped or cylindrical, hard, brownish or black, with a bumpy skin and nutlike flesh. The plants are grown not from seeds but from the tubers. The tubers become sweeter after storage for a few weeks, at which time they are somewhat wrinkled. The tubers were used as food in ancient Egypt and have been cultivated since early times in Europe and western Africa. Today, the tubers are still gathered from the wild in some regions, particularly western Africa, and are cultivated for human food in some regions, for example Ghana, Nigeria, and Togo. The plants were frequently cultivated in the southern United States to provide food for pigs in the past and, although less common, such cultivation continues today. Although this is a relatively inexpensive way of feeding hogs, the meat tends to be fatty and soft in comparison to corn-fed pork. Chufa is occasionally found in home gardens. Because it is widely considered to be a noxious weed and accordingly is sprayed with herbicides, collecting the tubers from wild plants should only be done if sure that the area has not been treated. Several taxonomic varieties of *Cyperus esculentus* have been recognized, one of these in Eurasia, and the domesticated forms appear to have been derived from it rather than from New World plants.

## CULINARY PORTRAIT

Although not "nuts," the tubers have a slightly sweet, nutlike flavor, resembling the taste of almonds and chestnuts. The tubers are eaten raw or roasted. The outer rind is tough and dry, and not readily chewed, so this vegetable is commonly boiled. Chufa is an interesting addition to salads. Flour can be obtained by slowly drying the tubers until they can be broken apart by hand, ground or hammered lightly, and processed in a blender. This flour can be mixed half and half with regular flour in any recipe. The tubers are sometimes candied or made into conserves. A very unusual aspect of Spanish cooking is the extensive use of nuts. Almonds and pine nuts are used in sauces for fish, meat, and eggs; in many sweets; and are cooked with spinach and chard. Although "nuts" are usually fruits, chufa is treated as a nut for culinary purposes, and it is used more extensively in cuisine in Spain than in any other country. In the eighteenth century, chufa tubers were popular in Europe, particularly Germany and Hungary, as a substitute for coffee (although the taste is quite different), and they are still sometimes used for this purpose. Bags of dried chufas are sometimes available in Latin markets and health-food stores. Since chufa is grown in the southern United States as livestock feed, it can be purchased from some farmers' co-ops as "seed" for planting (i.e., tubers). The tubers can be stored tightly wrapped in a cool, dark place for up to a year.

In Spain and more recently in Mexico, the tubers are made into *horchata* (pronounced as hor-CHAH-tah), a refreshing milky drink with the flavor of vanilla and almonds, served cold or at room temperature. This can be prepared by soaking 227 g (0.5 pound) of tubers for 48 hours, mashing them, adding 1 L (1 American quart) of water, 115 g (0.25 pound) of sugar, and sieving the liquid. Nuts and grains are sometimes substituted for chufa (especially in Mexico), and the drink is commonly spiced with cinnamon. In France, orgeat (almond syrup) is often substituted, and the beverage made may be called orgeat instead of horchata.

### Culinary Vocabulary

- An old story accounts for how the name "horchata" came to be used for the beverage made of chufa (described above). A girl in a little town of Spain offered some of this traditional peasant drink to the visiting king of Catalonia (northeastern Spain including the city of Barcelona) and Aragon (also northeastern Spain, once an ancient kingdom). After enjoying the drink, the king asked, *Que es aixo?* (What is this?). The girl answered, *Es leche de chufa* (It's chufa milk). The king replied, *Aixo no es llet, aixo es OR, XATA*! (This is not milk, this is GOLD, CUTIE). The word *xata* in Catalan—which the king spoke—is an affectionate nickname for a child. Gradually as the drink became popular in Spain, it came to be known in Spanish as *orchata*. Later, the "h" was added to the beginning.

- In Spain, chufa beverage (horchata) is traditionally served with fartons. Should your waiter in a Spanish restaurant ask you if you would like "horchata with fartons," you should know that fartons are a kind of sweet breadstick.

## PROSPECTS

Chufa has been used historically as food for millennia, and some cultivars are available. It is an odd crop in that it is native to an extraordinarily extensive area of the planet, and there are widespread weedy forms that have been classified as pernicious weeds. The plant seems to be more useful as food for animals than for humans, but nevertheless chufa is often consumed as a gourmet specialty food, mostly in Spain. Given the fact that chufa is a very old crop, and there is hostility to it because of its harmfulness as a weed, it seems unlikely that its market potential will expand. Nevertheless, cultivars are available that are not weedy, and it is possible that the plant could become more popular, either by breeding of superior forms or simply by marketing. The tubers have considerable oil content, and it has been suggested that chufa could be developed as a new source of biodiesel.

## CURIOSITIES OF SCIENCE AND TECHNOLOGY

- Chufa has been rated as one of the world's worst 20 weeds. The closely related purple nutsedge, *C. rotundus* L., which has been introduced to North America, has been called "the world's worst weed."
- Despite its weediness, chufa is considered in the United States to be a wonderful crop for attracting and feeding wild game, particularly wild turkeys. Of course, many of those who plant chufa to feed the turkeys intend to feed the turkeys to themselves!
- Russia conducted experiments with chufa in the hope of using it to grow food to feed people in space ships and on colonies on foreign planets (for additional information see *Bioscience* 47(9), October 1997).
- Joel Wapnick, a McGill University music professor from Montreal, won $7000 and Canada's Scrabble™ championship in 1998 by playing such unusual and obscure words as reovirus, acylates, beglads, dracaena, and chufa.

## KEY INFORMATION SOURCES


Abad, P., Pascual, B., Maroto, J.V., Lopez—Galarza, S., Vicente, M.J., and Alagarda, J. 1998. RAPD analysis of cultivated and wild yellow nutsedge (*Cyperus esculentus* L.). *Weed Sci.* 46: 318–321.

Arafat, S.M., Gaafar, A.M., Basuny, A.M., and Nassef, S.L. 2009. Chufa tubers (*Cyperus esculentus* L.): As a new source of food. *World Appl. Sci. J.* 7: 151–156.

Barko, J.W., and Smart, R.M. 1978. The growth and biomass distribution of two emergent freshwater plants, *Cyperus esculentus* and *Scirpus validus*, on different sediments. *Aquatic Bot.* 5: 109–117.

Barko, J.W., and Smart, R.M. 1979. The nutritional ecology of *Cyperus esculentus*, an emergent aquatic plant, grown on different sediments. *Aquatic Bot.* 6: 13–28.

Barminas, J.T., Maina, H.M., Tahir, S., Kubmarawa, D., and Tsware, K. 2001. A preliminary investigation into the biofuel characteristics of tigernut (*Cyperus esculentus*) oil. *Bioresource Technol. (UK)* 79: 87–89.

Cantalejo, M.J. 1997. Analysis of volatile components derived from raw and roasted earth-almond (*Cyperus esculentus* L.). *J. Agric. Food Chem.* 45: 1853–1860.

Cortes, C., Esteve, M.J., Frigola, A., and Torregrosa, F. 2004. Physical and chemical properties of different commercially available types of 'horchata de chufa'. *Italian J. Food Sci.* 16: 113–121.

de Vries, F.T. 1991. Chufa (*Cyperus esculentus*, Cyperaceae): A weedy cultivar or a cultivated weed? *Econ. Bot.* 45: 27–37.

Dodet, M., Petit, R.J., and Gasquez, J. 2008. Local spread of the invasive *Cyperus esculentus* (Cyperaceae) inferred using molecular genetic markers. *Weed Res.* 48: 19–27.

Doll, J.D. 1986. *Cyperus esculentus* L.: Ecology, biology, physiology, morphology and importance. In *Ecology and control of perennial weeds in Latin America. Papers presented by panel of experts on ecology and control of perennial weeds, Santiago, Chile, November 28–December 2, 1983*. FAO Plant Production and Protection Paper no. 74. Food and Agricultural Organization of the United Nations, Rome, Italy. pp. 54–70.

Ejoh, R.A., Djomdi, and Ndjouenkeu, R. 2006. Characteristics of tigernut (*Cyperus esculentus*) tubers and their performance in the production of a milky drink. *J. Food Process. Preserv.* 30: 145–163.

Gifford, E.M., and Bayer, D.E. 1995. Developmental anatomy of *Cyperus esculentus* (yellow nutsedge). *Int. J. Plant Sci.* 156: 622–629.

Groenendael, J.M. van., and Habekotte, B. 1988. *Cyperus esculentus* L.: Biology, population dynamics, and possibilities to control this neophyte. *Zeitschrift Pflanzenkrankheiten Pflanzenschutz (Germany)* Suppl. 11: 61–69.

Jansen, P.C.M., and Aguilar, N.O. 1996. *Cyperus* L. In *Plant resources of South-East Asia, Vol. 9, plants yielding non-seed carbohydrates*. Edited by M. Flach and F. Rumawas. Backhuys Publishers, Leiden, the Netherlands. pp. 78–82.

Karababa, E., Coskuner, Y., and Nazlican, A.N. 2001. Chemical composition and technological evaluation of chufa tuber (*Cyperus esculentus* L.). *Gida (Turkey)* 26: 243–246.

Kelly, J.R. 1990. Biomass production of chufa (*Cyperus esculentus*) in a seasonally flooded wetland. *Wetlands* 10: 61–68.

Kelley, J.R., and Fredrickson, L.H. 1991. *Chufa biology and management*. U.S. Department of the Interior, Fish and Wildlife Service, Washington, DC. pp. 6.

Linssen, J.P.H., Cozijnsen, J.L., and Pilnik, W. 1989. Chufa (*Cyperus esculentus*): A new source of dietary fibre. *J. Science Food Agric.* (U.K.) 49: 291–296.

Matthiesen, R.L., and Stoller, E.W. 1978. Tuber composition in yellow nutsedge (*Cyperus esculentus* (L.)) variants. *Weed Res.* 18: 373–377.

Mitchell, W.A., and Martin, C.O. 1986. *Chufa* (Cyperus esculentus)*: Section 7.4.1, U.S. Army Corps of Engineers wildlife resources management manual*. U.S. Army Engineer Waterways Experiment Station, Springfield, VA. pp. 14.

Mokady, S., and Dolev, A. 1970. Nutritional evaluation of tubers of *Cyperus esculentus* L. *J. Sci. Food Agric.* 21: 211–214.

Mulligan, G.A., and Junkins, B.E. 1976. The biology of Canadian weeds. 17. *Cyperus esculentus* L. *Can. J. Plant Sci.* 56: 339–350.

Negbi, M. 1992. A sweetmeat plant, a perfume plant and their weedy relatives: A chapter in the history of *Cyperus esculentus* L. and *C. rotundus* L. *Econ. Bot.* 46: 64–71.

Oderinde, R.A., and Tairu, O.A. 1988. Evaluation of the properties of yellow nutsedge (*Cyperus esculentus*) tuber oil. *Food Chem.* 28: 233–237.

Pascual, B., Maroto, J.V., Lopez-Galarza, S., Sanbautista, A., and Alagarda, J. 2000. Chufa (*Cyperus esculentus* L. var. *sativus* Boeck.): An unconventional crop. Studies related to applications and cultivation. *Econ. Bot.* 54: 439–448.

Pascual-Seva, N., Pascual, B., San Bautista, A., Lopez-Galarza, S., and Maroto, J.V. 2009. Growth and nutrient absorption in chufa (*Cyperus esculentus* L. var. *sativus* Boeck.) in soilless culture. *J. Hort. Sci. Biotech.* 84: 393–398.

Sanchez-Zapata, E., Fuentes-Zaragoza, E., Fernandez-Lopez, J., Sendra, E., Sayas, E., Navarro, C., and Angel Perez-Alvarez, J. 2009. Preparation of dietary fiber powder from tiger nut (*Cyperus esculentus*) milk ("horchata") byproducts and its physicochemical properties. *Agric. Food Chem.* 57: 7719–7725.

Schippers, P., Ter Borg, S.J., and Bos, J.J. 1995. A revision of the infraspecific taxonomy of *Cyperus esculentus* (yellow nutsedge) with an experimentally evaluated character set. *Syst. Bot.* 20: 461–481.

Stribling, H.L. 1990. *Growing chufa for wild Turkeys*. Circular ANR, Alabama Cooperative Extension Service, Auburn University, Auburn, AL. pp. 2.

Taylor, J.P., and Smith, L.M. 2003. Chufa management in the middle Rio Grande Valley, New Mexico. *Wildlife Soc. Bull.* 31: 156–162.

Temple, V.J., Ojobe, T.O., and Kapu, M.M. 1990. Chemical analysis of tiger nut (*Cyperus esculentus*). *J. Sci. Food Agric.* 50: 261–263.

Tucker, G.C. 2002. *Cyperus esculentus*. In *Flora of North America north of Mexico*, Vol. 23. Edited by Flora of North America Editorial Committee. Oxford University Press, New York, NY. pp. 168–169.

Turesson, H., Marttila, S., Gustavsson, K.-E., Hofvander, P., Olsson, M.E., Bülow, L., Stymne, S., and Carlsson, A.S. 2010. Characterization of oil and starch accumulation in tubers of *Cyperus esculentus* var. *sativus* (Cyperaceae): A novel model system to study oil reserves in nonseed tissues. *Am. J. Bot.* 97: 1884–1893.

Umerie, S.C., and Enebeli, J.N. 1996. Malt caramel from tubers of *Cyperus esculentus*. *Bioresource Technol. (UK)* 57: 215–216.

Umerie, S.C., Obi, N.A.N., and Okafor, E.O. 1997. Isolation and characterization of starch from *Cyperus esculentus* tubers. *Bioresource Technol. (UK)* 62: 63–65.

Umerie, S.C., Okafor, E.O., and Uka, A.S. 1997. Evaluation of the tubers and oil of *Cyperus esculentus*. *Bioresource Technol. (UK)* 61: 171–173.

Umerie, S.C., and Uka, A.S. 1998. Brew wort from *Cyperus esculentus* tubers. *Bioresource Technol. (UK)* 66: 83–85.

Wills, G.D., Hoagland, R.E., and Paul, R.N. 1980. Anatomy of yellow nutsedge (*Cyperus esculentus*). *Weed Sci.* 28: 432–437.

## Specialty Cookbooks

Enriquez, J.O. 2010. *Horchata heaven: Mexican soft drinks from scratch.* Solar-vision, Saint George, UT. pp. 40.

Freitus and Haberman (2005) have two recipes for preparation of the tubers, Knutsen (1975) has four, and Schufer (2011) has one. Angier (1969) provides general advice on the culinary preparation of chufa tubers. (See the Appendix to this book for details of the publications cited.)

# 29 Cloudberry

Family: Rosaceae (rose family)

## NAMES

Scientific name: *Rubus chamaemorus* L.

- "Cloudberry" is from the Old English *clūd*, meaning rocky hill, + berry. Another interpretation, hundreds of years old, is that the name is based on the plant growing on the cloudy tops of mountains (mountain berry is still another name). The name dates back at least to the sixteenth century.
- The cloudberry is also referred to as baked-apple (berry), bakeapple (berry), and bake-apple (berry). Such names presumably allude to the taste, which is reminiscent of baked apples.
- Cloudberry has also been called malka, mountain berry, and yellow-berry (yellowberry, yellow berry).
- In parts of England, the archaic name "noops" was once used for the cloudberry. The word originated from the Anglo-Saxon *cnoep*, a button, for the supposed resemblance of the berries to buttons.
- Knot-berry is another old English name for the cloudberry, based on the knotty joints of the stems.
- Imaginative if far-fetched explanations of the derivations of the names cloudberry and bake-apple can be found on the Web. One interpretation claims that "cloudberry" is so named because the fruit looks like cumulus clouds. "Bake-apple" is said to have arisen when Newfoundlanders overheard French speakers ask "Baie qu'appelle?" (What do you call this berry?).
- The species is sometimes called salmonberry, a name that is usually used for *R. spectabilis* Pursh of the Pacific coast (see Chapter 83 on Salmonberry). Alaska natives called both of these species salmonberries because of the resemblance of the raspberry-like fruits to salmon eggs.
- For information on the genus name *Rubus*, see Chapter 77 on Raspberries.
- *Chamaemorus*, in the scientific name *R. chamaemorus* is from the Greek *chamae*, on the ground, a reference to the low habit, and the Latin *morus*, the mulberry tree, an allusion to the mulberry-like fruit.

## GEOGRAPHY AND ECOLOGY OF WILD PLANTS

The cloudberry is an Arctic and north-temperate species. In North America, it grows wild in cool and cold regions from Newfoundland to British Columbia and Alaska, south to the coast of eastern Maine, the mountains of western Maine and northern New Hampshire, with an isolated colony on Long Island, New York. Further west, its southern limit is near the 49th parallel in southern British Columbia and northern Minnesota. Cloudberry also grows well in areas of Finland, Norway, Sweden, England, and Russia.

Cloudberry inhabits moist, peaty, and turfy soils, including sphagnum bogs and hummocks, muskegs, mossy tundra, and black spruce bogs. The species occupies a broad spectrum of sites from

**FIGURE 29.1** Cloudberry (*Rubus chamaemorus*). (a) Flowering plant. (From Hallier, E., *Flora von Deutschland* [fifth edition by D.F.L. von Schlechtendal et al.], F.E. Köhler, Gera-Untermhaus, Germany, Vol. 25, Plate 2573, 1886.) (b and c) Plants in fruit. (Courtesy of Epukas.) (d) Fruits in hand. (Courtesy of Christoph Műller [CC By 2.0].) (e) Distribution map.

dry to wet, but is most common on wetter sites, particularly raised bogs, meadows, and freshwater marshes. It is adapted to very acidic peatlands. Cloudberry is shade tolerant and is an important understory component of both open and closed forest habitats, primarily in black spruce–sphagnum bogs. However, the plants grow best in full sun. The species is considered to be a pioneer that rapidly colonises areas that have been cleared.

Cultivation and management techniques for cloudberry have been extensively studied and developed in northern Europe, especially in Scandinavian countries. Burning of surface vegetation is sometimes practiced to increase plant density and fruit production. Control of competing plants may substantially increase productivity of cloudberry. Although it is often found in nutrient-poor substrates, productivity can be increased at least somewhat by appropriate fertilization, especially with phosphorus (but simply broadcasting fertilizer on a bog may benefit other bog species at the expense of the cloudberry). Frost damage to the flowers is sometimes controlled by flooding bogs, as is done with commercial cranberry bogs. Also, plowing furrows into a bog seems to increase the growth of cloudberries while lowering damage from seasonal freezing of flowers. Short fences have been used to increase snow cover. Of course, the above measures could upset the natural vegetation of bogs and should not be undertaken without consideration of environmental effects. In nature, male plants predominate, but 10% males have been proposed as sufficient, and this can be achieved by planting male and female clones in a 1:10 ratio.

## PLANT PORTRAIT

The cloudberry is a low-growing shrub, 10–20 cm (4–8 inches) high, with rather leathery leaves. The globular fruit is 1–2 cm (0.4–0.8 inch) in diameter and looks rather like its relative, the raspberry. The fruits are bright red when immature, quite hard and firm at this stage, becoming orange or amber, and later soft and pale translucent yellow. When the fruit ripens, it becomes juicy and yellow. Overripe fruits are a pale cream color, and the translucence has been lost.

Semicommercial to large-scale commercial harvesting occurs in Scandinavia, England, and Russia. In North America, large areas of cloudberry that could be commercially harvested are found in the bogs of northern Quebec, and Newfoundland and Labrador, but the amount of harvesting is much less than in Scandinavia. Most cloudberries on the market today come from wild plants from native sphagnum peat bogs in Scandinavia and Newfoundland. The demand for the berries far outstrips the supply. Cloudberry has a very attractive taste and interesting color, and because it is considered a gourmet item, it is quite competitive in the fruit market.

In many northern communities, particularly of the Inuit, fruit imported from the south is rare and if available may be in poor condition and extremely expensive. Wild-growing cloudberries are, accordingly, highly valued, and indeed are the favorite wild fruit of the Inuit. Native peoples of northern areas often conserved the berries in a frozen state in the snow. The leaves have occasionally been used as a tea substitute. Because of the high content of vitamin C, cloudberries were an important remedy for scurvy among Arctic hunters.

The cloudberry is afforded considerable respect in some northern areas of the world. It is the badge of the Scottish clan McFarlane. In Poland, cloudberry is considered to be a species that is threatened with extinction in the country, and is strictly protected by law. Scandinavians are so respectful of the cloudberry that the right to pick it is enshrined in some laws. In Norway, only landowners can harvest cloudberries on their property, and in northern Finland only local inhabitants can pick them in quantity. A celebration of Labrador heritage and a tribute to cloudberry called the "Bakeapple Folk Festival" is held in mid August in Labrador Straits.

## CULINARY PORTRAIT

Cloudberries are considered best when freshly picked and may be served on pancakes or waffles, or with sugar, cream, whipped cream, or ice cream. The taste and smell have been said to be unique and difficult to describe. The berries have a musky aroma and a strong sour but delicious flavor. Both the Finnish liqueur Lakka and the Quebec liqueur Chicoutai are distilled from cloudberry. Cloudberries are used in the baking of pies, puddings, and stews, and to decorate cakes and desserts. Since the berries are perishable, they are often processed into juices, preserves, liqueurs, and frozen products. Cloudberries are often found in yogurts, jams, and juice concentrate.

Cloudberries are infrequently in North American supermarkets. The plant is often grown in cool rock gardens as an ornamental, and the berries can be picked for home use. Investigators have made the following observations regarding harvesting, preserving, and preparing the berries. The best way to harvest is to pick a combination of the hard red fruit and the soft golden fruit. The hard red fruits seem to have a higher pectin content, and when cooked results in a delicious, apricot-like flavor. However, using only the hard red berries produces a preserve that is too thick and dry. The soft golden fruit is fragile, deteriorates easily, and when cooked without some hard red fruits produces a slightly runny product with an inferior flavor. When used in baking, a smoother texture results when the berries are puréed in a blender before they are cooked. The fresh berries will keep for several weeks in a refrigerator.

### Culinary Vocabulary

- "Eskimo ice cream" was once a favorite treat of the Inuit people. This is a beaten mixture of berries, seal oil, and chewed caribou tallow, or just berries and fluffy fat, frequently made with cloudberries.

## PROSPECTS

Large amounts of cloudberries are harvested in Europe, but the industry is relatively minor in North America. Cloudberry has a very attractive taste and interesting color, and because it is considered a gourmet item it is quite competitive in the fruit market. Since it is perishable, cloudberry is often processed rather than marketed fresh. Not only is the European and world market demand for this very special fruit largely unmet, even the demand by local markets, particularly in Newfoundland and Labrador, far outstrips the supply. In Norway, 200 to 300 metric tons of cloudberry are imported in some years to satisfy domestic needs. The fact that there is already a large market demand is very important when considering development of minor or wild crops. The fruit market is highly competitive today, and new fruits need to displace some of the sales of well-established fruits. Under these circumstances, there is universal reluctance to introduce new fruits. This does not apply to cloudberry since it already has a guaranteed market. Moreover, cloudberry is a suitable crop for northern areas that are rich in peatlands, but lack extensive agricultural lands of good quality. Bogs of eastern Canada and New England are particularly suitable for cloudberry, and the industry deserves to be developed.

## CURIOSITIES OF SCIENCE AND TECHNOLOGY

- North American Indians used cloudberry for several medicinal purposes. The Woodland Cree used a decoction of the root to aid women experiencing a difficult childbirth, and a decoction of the root and stem to help women become pregnant. The Micmac used the roots for cough and tuberculosis and to treat fever.
- Cloudberry fruits have been used by northern peoples as a source of a purple to dull-blue dye.
- During a productive year, a cloudberry bog may suddenly change color when the berries ripen, a consequence of fruit densities up to 1200 per square meter (13,000 square feet).
- As with the harvest of some other wild fruit, picking cloudberries is not the easiest exercise. The bog environment requires slogging through wetlands, and there is the inevitable battle against black flies and mosquitoes. When the berries are picked in the red stage, they need to be "dehulled," that is, the adherent sepals have to be removed and this results in sore fingers and ruined fingernails. The picking rate in productive Newfoundland bogs has been estimated to be 4 L/h or 4.5 kg/h (about 1 American gallon/hour or 10 pounds/hour).
- Sexual competition in cloudberry has attracted scientific research. The species is composed of male and female plants, and the males outcompete the females, hogging much more of the area where they grow. In most cloudberry bogs, less than 25% of the area is occupied by the females. Of course, this reduces the number of offspring (seeds), which seems like a very dumb thing for a species to do. It has been hypothesized that the dominance of the male sex is related to the fact that both males and females spread vigorously by vegetative reproduction, lessening the need of the species for sexual reproduction.

- Nations are very sensitive to their culinary reputation, and international diplomacy needs to be sensitive to this fact. In 2005, French president Jacques Chirac publicly stated that Finland serves the worst food in the world, and Britain the second worst. The result was that the Finnish Olympic delegation switched its vote from Paris to London, and Britain won the 2012 games by a narrow margin. Subsequently, at a European Union summit hosted by Finland in 2006, the Finns resurrected their honor by having their best chef prepare a sumptuous meal of Finlandia blinis with fish roe, reindeer, vegetables, and cloudberry mousse prepared with fresh berries from the Arctic Circle.

## KEY INFORMATION SOURCES

Agren, J. 1987. Intersexual differences in phenology and damage by herbivores and pathogens in dioecious *Rubus chamaemorus* L. *Oecologia* 72: 161–169.

Agren, J. 1988a. Sexual differences in biomass and nutrient allocation in the dioecious *Rubus chamaemorus*. *Ecology* 69: 962–973.

Agren, J. 1988b. Between-year variation in flowering and fruit set in frost-prone and frost-sheltered populations of dioecious *Rubus chamaemorus*. *Oecologia* 76: 175–183.

Agren, J. 1989. Seed size and number in *Rubus chamaemorus*: Between-habitat variation, and effects of defoliation and supplemental pollination. *J. Ecol.* 77: 1080–1092.

Agren, J., Elmqvist, T., and Tunlid, A. 1986. Pollination by deceit, floral sex ratios and seed set in dioecious *Rubus chamaemorus* L. *Oecologia* 70: 332–338.

Baardseth, P., and Russwurm, H., Jr. 1978. Content of some organic acids in cloudberry (*Rubus chamaemorus* L.). *Food Chem.* 3: 43–46.

Dumas, P., and Maillette, L. 1987. Sex ratio and reproductive success of *Rubus chamaemorus*, a herbaceous dioecious species of subarctic distribution. *Can. J. Bot.* 65: 2628–2639. (In French, English summary)

Gustafsson, M., and Kortesharju, J. 1996. Occurrence of hermaphroditism in a dioecious plant, *Rubus chamaemorus*, in northern Finland. *Aquilo Ser. Bot.* 36: 53–60.

Johansson, A.K., Kuusisto, P.H., Laakso, P.H., Derome, K.K., Sepponen, P.J., Katajisto, J.K., and Kallio, H.P. 1997. Geographical variations in seed oils from *Rubus chamaemorus* and *Empetrum nigrum*. *Phytochemistry* 44: 1421–1427.

Karst, A.L., and Turner, N.J. 2011. Local ecological knowledge and importance of bakeapple (*Rubus chamaemorus* L.) in a southeast Labrador Métis community. Ethnobiol. Lett. 2: 6–18.

Kaurin, A., and Junttila, O. 1982. Vegetative growth and frost hardiness of cloudberry (*Rubus chamaemorus*) as affected by temperature and photoperiod. *Physiol. Plant.* 55: 76–81.

Korpelainen, H. 1994. Sex ratios and resource allocation among sexually reproducing plants of *Rubus chamaemorus*. *Ann. Bot.* 74: 627–632.

Korpelainen, H., Antonius-Klemola, K., and Werlemark, G. 1999. Clonal structure of *Rubus chamaemorus* populations: Comparison of different molecular methods. *Plant Ecol.* 143: 123–128.

Kortesharju, J. 1993. Ecological factors affecting the ripening time of cloudberry (*Rubus chamaemorus*) fruit under cultivation conditions. *Ann. Bot. Fennici* (*Finland*) 30: 263–274.

Kortesharju, J. 1995. Effects of frost on the female flowers, unripe fruits and vegetative growth of the cloudberry (*Rubus chamaemorus*) in Finnish Lapland. *Aquilo Ser. Bot.* 35: 31–38.

Mäkinen, Y., and Oikarinen, H. 1974. Cultivation of cloudberry in Fennoscandia. *Rep. Kevo Subarctic Res. Stat.* (*Finland*) 11: 90–102.

Pelletier, L., Brown, A., Otrysko, B., and McNeil, J.N. 2001. Entomophily of the cloudberry (*Rubus chamaemorus*). *Entomol. Exp. Appl.* (*Dordrecht*) 101: 219–224.

Rantala, E.M. 1976. Sexual reproduction in the cloudberry. *Ann. Agric. Fenn.* 15: 295–303.

Rapp, K., Næss, S.K., and Swartz, H.J. 1993. Commercialization of the cloudberry (*Rubus chamaemorus* L.) in Norway. In *New crops*. Edited by J. Janick and J.E. Simon. Wiley, New York, NY. pp. 524–526.

Resvoll, T.R. 1929. *Rubus chamaemorus*. L. A morphological-biological study. *Nyt Mag. Naturvid.* (*Oslo*) 67: 55–129.

Saebo, S. 1968. The autecology of *Rubus chamaemorus* L. I. Phosphorus economy of *Rubus chamaemorus* in an ombrotrophic mire. *Meld. Norg. LandbrHøgsk* 47(1): 1–67.

Saebo, S. 1970. The autecology of *Rubus chamaemorus* L. II. Nitrogen economy of *Rubus chamaemorus* in an ombrotrophic mire. *Meld. Norg. LandbrHøgsk* 49(9): 1–37.

Saebo, S. 1973. The autecology of *Rubus chamaemorus* L. III. Some aspects of calcium and magnesium nutrition of *Rubus chamaemorus* in an ombrotrophic mire. *Meld. Norg. LandbrHøgsk* 52(5): 1–29.

Saeboe, S. 1977. The autecology of *Rubus chamaemorus* L. IV. Potassium relations of in an ombrotrophic mire. *Meld. Norg. LandbrHøgsk* 56(26): 1–20.
Small, E., and Catling, P.M. 2000. Poorly known economic plants of Canada-27. Cloudberry, *Rubus chamaemorus*. *Bull. Can. Bot. Assoc.* 33(4): 43–47.
Taylor, K. 1971. Biological flora of the British Isles. *Rubus chamaemorus* L. *J. Ecol.* 59: 293–306.
Warr, H.J., Savory, D.R., and Bal, A.K. 1979. Germination studies of bakeapple (cloudberry) seeds. *Can. J. Plant Sci.* 59: 69–74.

## Specialty Cookbooks

(*Note*: Scandinavian cookbooks (examples included below) are a rich source of recipes for cloudberry.)

Henderson, H. 2005. *The Swedish table*. University of Minnesota Press, Minneapolis, MN. pp. 187.
Kallio, T., and Saira, K. 2009. *Simply Scandinavian: Travelling in time with Finnish cuisine and nature*. Raikas Publishing, Helsinki, Finland. pp. 128.
Kiros, T. 2007. *Falling cloudberries and apples for jam*. Murdoch Books, London, UK. pp. 417.
Michelsen, S. 2008. *Scandinavian gourmet cooking*. AuthorHouse, Bloomington, IN. pp. 134.
Spencer, M.N., and Canter, M. 1989. *Scandinavian holiday recipes: Norway, Sweden, Demark, Finland & Iceland*. Penfield Press, Penfield, IA. pp. 159.
Viestad, A. 2003. *Kitchen of light: The new Scandinavian cooking*. Artisan, New York, NY. pp. 303.

Anonymous (1977) presents a recipe entitled "Bakeapple jam." Gray (2011) has a recipe called "Cloudberry-buckwheat pancakes," and Mogelon (2001) has a recipe for "Cloudberry muffins." (See the Appendix to this book for details of the publications cited.)

# 30 Coast Tarweed

Family: Asteraceae (Compositae; sunflower family)

## NAMES

Scientific name: *Madia sativa* Mol.

- "Tarweed" in the name is based on the sticky, glandular secretions of the plant. "Coast" reflects the West coast distribution.
- Coast tarweed is also known as Chile tarweed, Chilean tarweed, coastal tarweed, common tarweed (a name applied to other species), cultivated tarweed, madia oil plant, melosa (a Spanish name meaning "like honey," applied to other species), and tarweed (another name applied to other species).
- The genus name *Madia* is based on *madi*, the native Chilean name for *M. sativa*.
- *Sativa* in the scientific name is Latin for cultivated.

## GEOGRAPHY AND ECOLOGY OF WILD PLANTS

*Madia sativa* is native to the western seaboard of North America, Mexico, Argentina, and Chile. In North America, it occurs on the Pacific Coast from British Columbia to Baja California. The species has been mistakenly reported to be native in more eastern regions of North America, based on misidentifications. It has been introduced in Hawaii and elsewhere. Coast tarweed occurs in grasslands; open areas of shrublands and woods; stream banks; and disturbed sites such as roadsides, meadows, and fencerows. It withstands some frost (down to –5°C or 23°F) and considerable drought, and grows well in poor sandy soils.

## PLANT PORTRAIT

Coast tarweed is an annual herb, usually 20–100 cm (8–39 inches) high, rarely as tall as 240 cm (8 feet). The plant is covered by sticky, glandular, foul-smelling hairs. It has daisy-like, yellow flowers, which produce grey or black "seeds" (technically achenes, which are fruits) 4–8 mm (0.16–0.31 inch) long.

Since pre-Columbian times, *M. sativa* was extensively employed for its oily seeds by indigenous peoples of the Pacific Coast of South America, less so in North America. The seeds were crushed to a flour that was used in porridge and pinole (see the section "Culinary Vocabulary"), and the extracted oil was used for cooking. Indigenous peoples of South America sometimes prepared cold-pressed oil (produced by applying pressure without heat) by wrapping crushed seeds in a cloth and twisting the cloth to express the oil. Cold-pressed oil is resistant to becoming rancid. However, South American Indians generally boiled the seeds in water and simply skimmed off the floating oil. In addition to making extensive food use of the oil, it was also used medicinally, both internally as a purgative and externally for gout, sciatica, rheumatism, and pain in general. The oil has occasionally been used as a lamp oil for illumination and for industrial purposes, such as for preparing soap and lubrication of machinery.

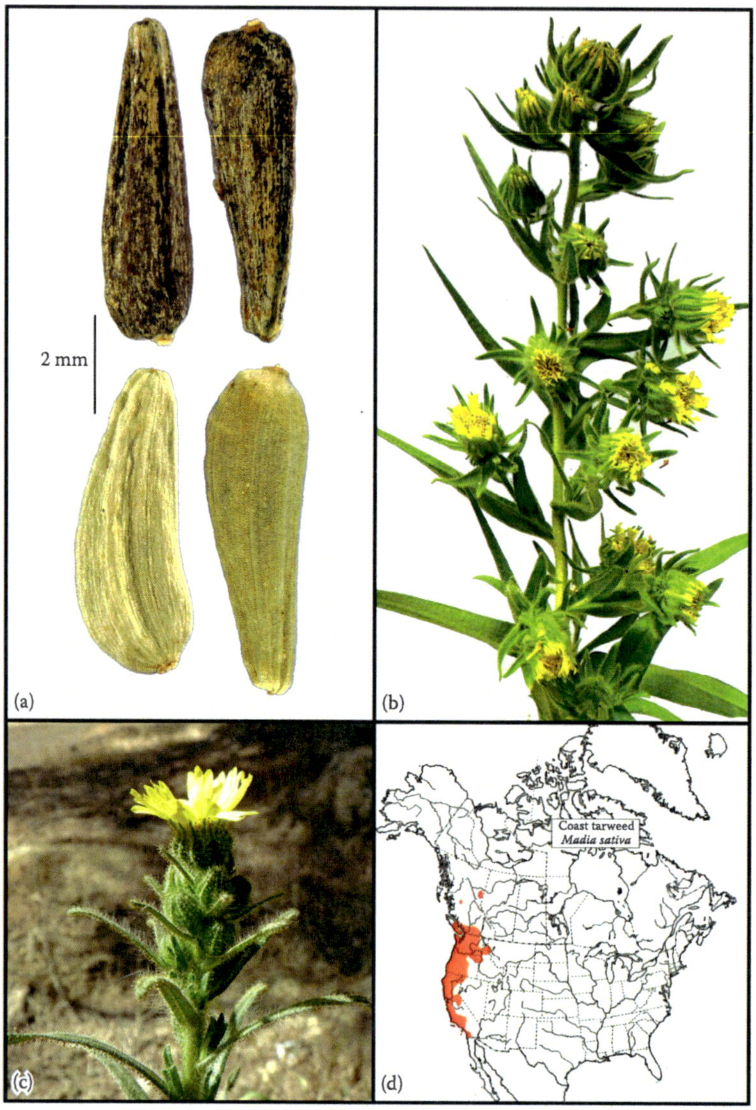


**FIGURE 30.1** Coast tarweed (*Madia sativa*). (a) Normal brown and selected white "seeds" (achenes). (Courtesy of Steve Hurst, U.S. Department of Agriculture.) (b) Flowering plant. (Courtesy of E. Small and B. Brookes.) (c) Top of flowering shoot (notice sticky hairs). (Courtesy of Pato Novoa [Flickr/CC-attribution].) (d) Distribution map.

## CULINARY PORTRAIT

In modern times, coast tarweed is grown mainly for seed oil (called "madia oil") in South America, Eurasia, and Africa. When cold-pressed, the oil is transparent, golden yellow, pleasant-smelling, with a mild, sweet, nutty, agreeable taste. The oil has been compared to olive oil. Meal remaining after oil is extracted from the seeds is high in protein and has been used to feed livestock, although there is some concern about the presence of toxic substances. The seeds are still used as an edible meal to a small extent.

### Culinary Vocabulary

- "Madia nut" refers to the seeds of *M. sativa*, used for culinary purposes.
- "Pinole" or panole is a Mexican and southwestern United States pudding prepared from ground dried corn, spices, and a sweetener.

## PROSPECTS

*Madia. sativa* is a very minor edible oil crop at present. It needs considerable improvement by breeding. Like numerous wild plants, the seeds drop irregularly from the plant, making collection very difficult. The stickiness of the leaves also complicates seed collection. The oil has some desirable fatty acids, but there are low concentrations of antinutritional chemicals present. Edible oil crops are extremely important economically and nutritionally. A small number of crops currently dominate the marketplace, but there is room for new candidates. The similarity in appearance and taste to olive oil is both an advantage and a handicap since the latter is widely considered to be the premium cooking, salad, and healthful oil, and the standard by which all edible oils are judged. Coast tarweed is a tough plant, reflected by its weediness. This annual crop requires relatively limited agricultural inputs (particularly irrigation and fertilization). It represents a worthwhile investment in breeding and management. The name "tarweed" is inappropriate for marketing purposes and would obviously have to be changed.

## CURIOSITIES OF SCIENCE AND TECHNOLOGY

- The foliage of coast tarweed is water-repellent and difficult to moisten even with the use of a wetting agent surfactant.
- There are numerous species that, like *M. sativa*, have one distribution area in North America that is well separated from a second distribution area in South America. It has been hypothesized that this pattern was produced by migratory birds transporting seeds from one area to the other. (Key reference: Raven, P.H. 1963. Amphitropical relationship in the floras of North and South America. *Quart. Rev. Biol.* 38: 151–177.)

## KEY INFORMATION SOURCES


Anonymous. 1915. *Madia sativa* seed from South Africa. *Bull. Imperial Inst.* 13: 344–346.

Baldwin, B.G., and Strother, J.L. 2006. *Madia*. In *Flora of North America north of Mexico*. Vol. 21. Edited by Flora of North America Editorial Committee. Oxford University Press, New York. pp. 303–308.

Baldwin, B.G., and Wessa, B.L. 2000. Origin and relationships of the tarweed–silversword lineage (Compositae–Madiinae). *Am. J. Bot.* 87: 1890–1908.

Baldwin, B.G., Kyhos, D.W., Dvorako, J., and Carr, G.D. 1991. Chloroplast DNA evidence for a North American origin of the Hawaiian silversword alliance (Asteraceae). *Proc. Natl. Acad. Sci. USA* 88: 1840–1843.

Barrier, M., Robichaux, R.H., and Purugganan, M.D. 2001. Accelerated regulatory gene evolution in an adapative radiation. *Proc. Natl. Acad. Sci. USA* 98: 10208–10213.

Bohlmann, F., Jakupovic, J., King, R.M., and Robinson, H. 1982. New labdane derivatives from *Madia sativa*. *Phytochemistry* 21: 1103–1107.

Carlquist, S. 1958. Structure and ontogeny of glandular trichomes of Madiinae (Compositae). *Am. J. Bot.* 45: 675–682.

Carlquist, S. 1959. Studies on Madiinae: anatomy, cytology and evolutionary relationships. *Aliso* 4: 171–236.

Celedon-Neghme, C., Gonzales, W.L., and Gianoli, E. 2007. Cost and benefits of attractive floral traits in the annual species of *Madia sativa* (Asteraceae). *Evol. Ecol.* 21: 247–257.

Clausen, J., Keck, D.D., and Hiesey, W.M. 1945. *Experimental studies on the nature of species. II. Plant evolution through amphiploidy and autoploidy, with examples from the Madiinae*. Carnegie Institution of Washington Publication, Washington, DC. pp. 174.
Crins, W.J., and Bohm, B.A. 1990. Flavonoid diversity in relation to systematics and evolution of the tarweeds. *Ann. Missouri Bot. Gard.* 77: 73–83.
Demo, P., Kozisek, Z., Rosa, J., Schofield, A.J., Stonham, T.J., Mehta, P.A., Angelini, L.G., et al. 1997. Variation in agronomic characteristics and seed oil composition of new oilseed crops in central Italy. *Indust. Crops Prod.* 6: 313–323.
Gonzalez, W.L., Negritto, M.A., Suárez, L.H., and Gianoli, E. 2008. Induction of glandular and non-glandular trichomes by damage in leaves of *Madia sativa* under contrasting water regimes. *Acta Oecol.* 33: 128–132.
Gonzales, W.L., Suárez, L.H., Molina-Montenegro, M.A., and Gianoli, E. 2008. Water availability limits tolerance of apical damage in the Chilean tarweed *Madia sativa*. *Acta Oecol.* 34: 104–110.
Haridy, M.S.A., Ahmed, A.A., and Doe, M. 2006. Microbiological transformation of two labdane diterpenes, the main constituents of *Madia* species, by two fungi. *Phytochemistry* 67: 1455–1459.
Hesse, V.F. 1968. Notes on *Madia sativa* and related species. *Madroño* 19: 210.
Kyhos, D.W., Carr, G.D., and Baldwin, B.G. 1990. Biodiversity and cytogenetics of the tarweeds (Asteraceae: Heliantheae–Madiinae). *Ann. Missouri Bot. Gard.* 77: 84–95.
Luther, E. 2012. *Madia sativa*. Acu Publishing, Saarbrücken, Germany. 84 pp. [Acu Publishing is an imprint of VDM Publishing, devoted to the on-demand reproduction of Wikipedia content.]
Ministry of Agriculture and Agrarian Reform. 1977. Oil plant cultivation: the madia oil plant. In *Activity Report 1974–1975*. Service de Phytotechnie, Ministère de l'Agriculture et de la Reforme Agraire. Rabat, Morocco. pp. 97–108. (In French)
Schmeda-Hirschmann, G. 1995. *Madia sativa*, a potential oil crop of central Chile. *Econ. Bot.* 49: 257–259.
Wollenweber, E., Dorr, M., Dorsam, M., Hassan, A.E., Ahmed, A.A., Hegazy, M.F., and Zeller, K.P. 2003. Flavonoids and terpenoids from the resinous exudates of *Madia* species (Asteraceae, Helenieae). *Z. Naturforsch. C.* 58: 153–160.
Zardini, E. 1992. *Madia sativa* Mol. (Asteraceae-Heliantheae-Madiinae): an ethnobotanical and geographical disjunct. *Econ. Bot.* 46: 34–44.

## SPECIALTY COOKBOOKS

Recipes for the culinary use of *Madia sativa* do not seem to be available, reflecting its rare culinary use at present.

# 31 Cranberry

Family: Ericaceae (heath family)

## NAMES

Scientific name: *Vaccinium macrocarpon* Aiton (*Oxycoccus macrocarpos* (Aiton) Pers.).

- "Cranberry" is usually interpreted as a corruption of "crane-berry," an early American name for the plant in use since the late seventeenth century. Dutch settlers are said to have originated the English word from their word *kranbeere*, which in turn comes from the Low German *kraanbere*, "crane berry." The flower bud was said to resemble a crane (the slender curving flower stalk, calyx, and corolla of the flower bud, before it opens, respectively simulate the neck, head, and beak of a crane). The name cranberry has also been claimed to have arisen from the French word for cranberry, *canneberge*, meaning "shore reed," the name coined for the plant by French colonists in Canada.
- The cranberry is also known as American cranberry and large cranberry. The name "large cranberry," based on the relatively large size of the fruit, serves to distinguish the species from the very similar, related, small cranberry, *V. oxycoccos* L. The small cranberry (which has smaller fruit) is a very widespread species, occurring from Greenland and Labrador to Alaska; south to New Jersey, Pennsylvania, Ohio, Indiana, and Minnesota; in the west to California; and also in Eurasia. It is not grown commercially, but is often collected by wild food enthusiasts.
- In Britain, cranberry was known by a number of now obsolete names, including fen berry, fen-wort, marsh-wort, and moss berry.
- A method that has been recommended to judge the quality of cranberries is to bounce them: the more times they bounce, the better the berry. As a result of this practice, cranberries are sometimes called bounceberries. According to legend, "Peg-Leg" John I. Webb, a New Jersey grower, initiated the first cranberry separator by simply allowing cranberries to tumble down his stairs—the ripest, firmest berries arrived at the bottom while the inferior fruit remained on the steps.
- See Chapter 15 for information on the genus name *Vaccinium*.
- *Macrocarpon* in the scientific name *Vaccinium. Macrocarpon* is based on Greek *makros* + *karpos*, large-fruited.
- The highbush cranberry or cranberry bush (see Chapter 32), *Viburnum trilobum* Marsh., is not related to *Vaccinium* although the fruit does resemble that of the true cranberries in appearance and taste; this species is a close relative of the elderberry (see Chapter 38). The Florida cranberry (better known as roselle), *Hibiscus sabdariffa* L., is also quite unrelated to *Vaccinium*, but the fleshy portion that surrounds the mature fruit is cooked and served as "cranberry." The lingonberry (see Chapter 58), *Vaccinium vitis-idaea* L., is also known as mountain cranberry and rock cranberry.

## GEOGRAPHY AND ECOLOGY OF WILD PLANTS

*Vaccinium macrocarpon* is distributed from Newfoundland to central Minnesota, south to Nova Scotia, New England, Long Island (New York), West Virginia, northern Ohio, central Indiana, northern Illinois, and rarely Arkansas, and it also occurs in the Appalachian Mountains of

**FIGURE 31.1** Cranberry (*Vaccinium macrocarpon*). (a) Flowering and fruiting branches. (From Curtis, W., *Botanical Magazine*, Vol. 52, Plate 2586, 1825.) (b) Harvesting cranberries from a flooded bog. (Courtesy of Keith Weller, U.S. Department of Agriculture.) (c) Painting (1880, by Eastman Johnson (1824–1906), "The Cranberry Harvest on the Island of Nantucket.") (d) Cranberries. (Courtesy of Cjboffoli [CC By 2.0].) (e) Distribution map.

Tennessee and North Carolina. The cranberry is grown within this native range, but also in British Columbia, Washington, and Oregon, where it has escaped and become established in the wild. Plants are cultivated in parts of northern and central Europe, and the species has also escaped and become naturalized in Britain, Germany, the Netherlands, and Switzerland. *Vaccinium macrocarpon* occurs in open acid bogs, swamps, wet shores, and damp heaths. It is adapted to highly acidic soils (thriving at pH 3.2–4.5). However, the species also occurs in circumneutral substrates.

## PLANT PORTRAIT

Red-fruited species of *Vaccinium* are almost universally called cranberries in North America. *Vaccinium macrocarpon* is the only species of commercial value and is referred to simply as cranberry in this chapter. The cranberry is an evergreen, creeping, mat-forming plant with slender, intricately forking, woody, horizontal stems 30–150 cm (1–5 feet) long. Although often referred to as a vine, it is a trailing shrub, rather than a true vine. Small, nodding, pink flowers are borne on upright shoots 5–15 cm (2–6 inches) high. The berries are 1–2.5 cm (0.4–1 inch) thick, globose, ellipsoid, or pear-shaped, with hard, shiny skin. Numerous horticultural varieties have been selected. Berries of cranberry cultivars may be bell-shaped, bugle-shaped, or cherry-shaped. Color of the berries is typically bright red, but may vary from a light yellow through dark red to almost black, and some plants produce white fruit. The berries are high in acid and pectin, and extremely tart (astringent) because of their low sugar content.

Cranberry culture was developed in the New World, and although some fruit is still collected from the wild, almost all consumed cranberries are now cultivated. This crop grows naturally in open bogs, in swamps, and on wet shores, and it is cultivated in artificially created bogs that mimic the natural habitat. Cranberries are confined to cool, moist regions, and culture of them is largely restricted to acid soils along the edges of streams, and ponds, and in bogs of temperate North America. Cranberry culture is a highly specialized form of small fruit production. Although cranberries are adapted to moist habitats, too high a water level encourages rushes, sedges, and other competing plants, which then crowd the crop and interfere with pest control and picking. Sphagnum bogs are unsuitable until drained and the substrate allowed to decay to muck. Muck soils are particularly desirable because of their moisture-holding capacity. Cranberries have also been successfully planted on somewhat dry soils near the sea, where relatively low summer temperatures retard evaporation. Artificial bogs are occasionally built on sand or clay with little or no muck, but these usually require considerable fertilization, normally supplied in nature by decaying organic matter.

Captain Henry Hall of Massachusetts was the first to cultivate the crop at the beginning of the nineteenth century. The center of cranberry cultivation and production is still Massachusetts, but considerable quantities are also raised in the peatlands of British Columbia, New Jersey, Wisconsin, Washington, and Oregon. Much cranberry culture also occurs in eastern Canada, particularly in Ontario, Quebec, and the Maritime Provinces. There are over 1000 cranberry growers in the United States, using almost 14,000 ha (34,600 acres) and producing over $1 billion in retail sales annually. Ocean Spray Inc., a growers' cooperative comprising almost a thousand cranberry and grapefruit growers in the United States and Canada, markets about 90% of the cranberries harvested in North America. Cranberry is also cultivated in parts of northern and central Europe.

European colonists appreciated the value of cranberry in warding off scurvy, a legitimate medicinal use, since the berries are high in vitamin C. Cranberry also acquired a folk reputation as a treatment for urinary tract problems, and this application has also been validated. Cranberry is a strong diuretic (promoting urination), and the juice is often prescribed as dietary treatment for urinary tract infections, kidney disorders, and other conditions where the passing of fluids is desirable. Many women suffer from cystitis, an inflammation and/or infection of the bladder, at some time in their lives. As many as one in five women are estimated to have urinary tract discomfort at least once a year. Cranberry can be used both as a preventive and as an adjunct treatment for such urinary tract problems. Cranberry juice therapy may require drinking a liter (about 1 American quart) of straight cranberry juice daily, which few are willing to do. Fortunately, cranberry capsules are easily swallowed. Consumers should be aware that some of the "cranberry juice" on the market may have considerable sugar and water added, and the pure form or a concentrate is likely to be more beneficial.

## CULINARY PORTRAIT

Cranberries have long been in demand as a food plant. The fruit was prized by Native Americans, who used it in many ways, including in pemmican, a dried mixture of animal fat and fruit, which was the precursor of the dehydrated foods used by present-day hikers during camping trips. Benzoic acid, a natural constituent of the berries, likely aided in preservation. In 1955, 20% of U.S. households used cranberry juice products. By 1985, this had increased to 70%, and the popularity of cranberry has continued to expand. In North America, a small majority of people prefer canned, jellied cranberry sauce over both canned whole berry sauce and cooked preparations (relish, chutney, etc.), despite the snob appeal of the last category. Because of their high acidity, cranberries are rarely consumed fresh. They can be cooked in a small quantity of water, the pot best left uncovered so that steam will not develop in the berries, causing them to swell and explode like popcorn. When cranberries are cooked with some sugar, the resulting product is sweet yet attractively slightly bitter. Cooked cranberries provide a pleasant sauce to accompany meat and poultry. They are also incorporated in salads and numerous products, including juice, cocktails, pies, tarts, muffins, bread, jams, jellies, and preserves. The cranberry is a beautiful fruit that is not only tasty but makes an attractive garnish, and this has led to its use especially during the Christmas and Thanksgiving festive seasons. Most cranberries are frozen for storage or marketing. Fresh cranberries should be firm to the point of hardness. Cranberries should be stored in a refrigerator as they spoil quickly at room temperature.

### Culinary Vocabulary

- The "Craisin™," a dried, sweetened cranberry, is marketed for breakfast cereals and fruit mixes by Ocean Spray Inc.
- A "Cape Cod" or "Cape Codder" is a cocktail prepared with cranberry juice, vodka, sugar syrup, and lime juice.
- A "cranberry bean" is not related to the cranberry. It is a variety of common bean (*Phaseolus vulgaris* L.) with streaks or patches of red on the skin. Known by many other names (e.g., Borlottti bean, crab eye, Fagiolo Romano, Roman, Rosecoco Saluggia, shell bean, Salugia bean), it has a mild, nutty flavor and is popular in Italian cuisine.

## PROSPECTS

There has been concern that environmentally sensitive wetlands might be eliminated or harmed by development of new cranberry bogs, but technology is available to control damage to wetland sites, and even to develop bogs on dryland sites. Demand for the fruit has almost always exceeded the supply, keeping prices high. However, the potential for increasing productivity is good. There are a number of locally adapted cultivars, and newer hybrid cultivars are expected to become increasingly available as a result of accelerated breeding programs. Yield of cranberries has increased by five times between 1909 and the present because of gradual and continuous improvements in management, including frost injury prevention, improved harvest technology, and modern control of pests. The substantial increases in productivity have been stimulated by a combination of scientific research and producer innovations. Experts have suggested that considerable improvements can still be realized with a sustained breeding program for insect resistance, as well as other traits. Current markets are reported to require several thousand hectares of new plantings in the near future, and demand is expected to continue to increase. Although the cost of initial establishment is relatively high and costs are not recovered quickly, cranberry appears to be a good investment. This high-value, high-demand crop appears destined to increase in value.

## CURIOSITIES OF SCIENCE AND TECHNOLOGY

- The Pilgrims on arriving in the New World observed cranberries growing profusely in the area around Cape Cod, Massachusetts, and noted that the Native Americans used the fruit as a source of a brilliant red dye for their clothes. (Whether the Pilgrims had cranberries at the first Thanksgiving, held in October 1621, has been debated.)
- The "Victorian language of flowers" was a secret coded language in Victorian times, with flowers and plants symbolic of certain messages, so when the flower or plant was

mentioned in a letter, those who knew the code could understand the hidden information. "Cranberry" meant "cure for heartache."

- Damaged cranberries often sink in water, so that simply washing berries in water can reveal the superior berries.
- At least 700 consumer products contain cranberries in one form or another.
- Although most cultivated cranberry is used to produce juice, about 50 million kg (227 million pounds) of cranberry sauce is consumed in the United States each year.
- A sauce was made with white cranberries as a prelude to a new marketing strategy, but it was of an amber color and had an insipid taste. Because the dark pigments of the fruit contribute substantially to the taste and medicinal quality of cranberries, the value of white cranberries is debatable.
- In 1959, many cranberry growers in the United States applied the weed killer aminotriazole prematurely, before the cranberries were harvested, rather than following the usual practice of waiting until the crop was removed. As a result, the berries were contaminated. The incident led to sensationalistic publicity after it was disclosed that the chemical was capable of producing cancer in mice, and the resulting suspension of cranberry sales led to millions of dollars of losses for the cranberry industry.
- In 1970, cranberry juice was named the "State Beverage" of Massachusetts. In 1992, a fifth-grade class adopted the cause of making the cranberry the official berry of the state. Their 2 years of lobbying, petitions, and hearings were finally rewarded in 1994.
- Cranberry was declared to be the state fruit of Wisconsin in 2003.

## KEY INFORMATION SOURCES

(Also see references in Chapter 15.)

Bomser, J., Madhavi, D.L., Singletary, K., and Smith, M.A. 1996. In vitro anticancer activity of fruit extracts from *Vaccinium* species. *Planta Med.* 62: 212–216.

Bruederle, L.P., Hugan, M.S., Dignan, J.M., and Vorsa, N. 1996. Genetic variation in natural populations of the large cranberry, *Vaccinium macrocarpon* Ait. (Ericaceae). *Bull. Torrey Bot. Club* 123: 41–47.

Cross, E.C. 1973. Cranberries—the last one hundred years. *Arnoldia* 33: 284–291.

Dana, M.N. 1990. Cranberry management. In *Small fruit crop management*. Edited by G.J. Galletta and D.G. Himelrick. Prentice Hall, Englewood Cliffs, NJ. pp. 334–362.

Eck, P. 1986. Cranberry. In *CRC handbook of fruit set and development*. Edited by S.P. Monselise. CRC Press, Boca Raton, FL. pp. 109–117.

Eck, P. 1990. *The American cranberry*. Rutgers University Press, New Brunswick, NJ. pp. 420.

Elle, E. 1996. Reproductive trade-offs in genetically distinct clones of *Vaccinium macrocarpon*, the American cranberry. *Oecologia* 107: 61–70.

Fleet, J.C. 1994. New support for a folk remedy: cranberry juice reduces bacteriuria and pyuria in elderly women. *Nutr. Rev.* 52(5): 168–178.

Fiander—Good Associates Ltd. 1993. *The feasibility of establishing a cranberry industry in New Brunswick*. Fiander–Good Associates Ltd., Fredericton. (Canada/New Brunswick Cooperation Agreement on Planning, 1993.) pp. 126 +23.

Galetta, G.J. 1975. Blueberries and cranberries. In *Advances in fruit breeding*. Edited by J. Janick and J.N. Moore. Purdue University Press, West Lafayette, IA. pp. 154–196.

Hall, I.V., Blatt, C.R., Lockhart, C.L., and Wood, G.W. 1982. *Growing cranberries*. Agriculture Canada Publication 1282/E. Ottawa, ON. pp. 28.

Hall, I.V., and Nickerson, N.L. 1986. The biological flora of Canada. 7. *Oxycoccus macrocarpus* (Ait.) Pers., large cranberry. *Can. Field-Nat.* 100: 89–104.

Jackson, B., and Hicks, L.E. 1997. Effect of cranberry juice on urinary pH in older adults. *Home Healthc. Nurse* 15(3): 198–202.

Jorgensen, E.E., and Nauman, L.E. 1994. Disturbance in wetlands associated with commercial cranberry (*Vaccinium macrocarpon*) production. *Am. Midl. Nat.* 132: 152–158.

Jukes, T.H. 1970. The global "cranberry incident." *Clin. Toxicol.* 3: 147–149.

Kevan, P.G., Gadawski, R.M., Kevan, S.D., and Gadawski, S.E. 1984. Pollination of cranberries, *Vaccinium macrocarpon*, on cultivated marshes in Ontario. *Proc. Entomol. Soc. Ontario* 114: 45–53.
Kuzminski, L.N. 1996. Cranberry juice and urinary tract infections: is there a beneficial relationship? *Nutr. Rev.* 54(11 Pt. 2): S87–S90.
Liebster, G., and Delor, H.W. 1974. *Bibliography of the international literature on the cranberry,* Vaccinium macrocarpon *Ait.* Vol. 3 of Technical University of Berlin Library Bibliographic Series. Technical University of Berlin, Berlin, Germany. pp. 75.
Luby, J.J., Ballington, J.R., Draper, A.D., Kazimierz, P, and Austin, M.E. 1991. Blueberries and cranberries. In *Genetic resources of temperate fruit and nut crops*. Edited by J.N. Moore and J.R. Ballington. International Society for Horticultural Science, Wageningen, the Netherlands. pp. 393–456.
MacKenzie, K.E. 1994. Pollination requirements of the American cranberry. *J. Small Fruit Vitic.* 2: 33–44.
MacKenzie, K.E., and Averill, A.L. 1995. Bee pollinators (Hymenoptera: Apoidea) of cranberry (*Vaccinium macrocarpon* Aiton) in southeast Massachusetts. *Annals Ent. Soc. Amer.* 88: 334–341.
Nazarko, L. 1995. Infection control. The therapeutic uses of cranberry juice. *Nurs. Stand.* 9(34): 33–35.
Oertel, B. 1996. A conservation plan for every cranberry grower. Land and Water. *Magazine Nat. Resource Manage. Restor.* 40(4): 41–43.
Porsild, A.E. 1938. The cranberry in Canada. *Can. Field-Nat.* 52: 116–117.
Rodale, J.I. 1960. Cranberries. *Org. Garden.* 7(3): 21–25.
Roper, T.R. and Vorsa, N. 1997. Cranberry: botany and horticulture. *Hort. Rev.* 21: 215–249.
Sarracino, J.M., and Vorsa, N. 1991. Self and cross fertility in cranberry. *Euphytica* 58: 129–136.
Serres, R.A., McCown, B.H., Zeldin, E.L., Stang, E.J., and McCabe, D.E. 1993. Applications of biotechnology to cranberry: a model for fruit crop improvement. *Acta Hortic.* 345: 149–156.
Sibert, I. 1996. Cranberry pollination. *Am. Bee J.* 136: 363–364.
Siciliano, A.A. 1996. Cranberry. *HerbalGram* 38: 51–54.
Stang, E.J. 1993. The North American cranberry industry. *Acta Hortic.* 346: 284–298.
Thomas, J.D. 1990. *Cranberry harvest: a history of cranberry growing in Massachusetts*. Spinner Publications, New Bedford, MA. pp. 224.
Vander Kloet, S.P. 1983. The taxonomy of *Vaccinium* and *Oxycoccus*. *Rhodora* 85: 1–44.
White, J.J. 1916. *Cranberry culture*. Orange Judd Co., New York. pp. 131.

## Specialty Cookbooks

(*Note*: There are dozens of cookbooks dedicated to cranberries. The following are examples.)

Anderson, P., Megow, E.D., and Buchda, T. 1997. *Wisconsin cranberry growers' favorite recipes*. Cranberry Expo Ltd., Warrens, WI. pp. 160.
Baker, J. 1986. *Something cranberry*. Brandison Press, Oak Park, IL. pp. 77.
Better Homes and Gardens. 1971. *Better homes and gardens five seasons cranberry book*. Des Moines, Iowa, Meredith Pub. Des Moines, IA. pp. 112.
Borlase, J. 1994. *Cranberry recipes*. Borlase Publications, Victoria, BC. pp. 90.
Buszek, B.R. 1978. *The cranberry connection: cranberry cookery with flavour, fact, and folklore, from memories, libraries, and kitchens of old and new friends and strangers*. Second edition. Greene Press, Brattleboro, VT. pp. 208.
Clark, L., and Baker, J.W. 2004. *Cranberry companion*. Brick Tower Press, New York. pp. 112.
Colizzi, J.D. 1991. *Cranberries: cranberry recipes*. Revised edition. J.D. Colizzi Publishing Company, Somerdale, NJ. pp. 176.
Johnston, J. 1996. *The cranberry cookbook: recipes from 40 years of cooking with cranberries*. J. Johnston, Bala, ON. pp. 180.
Lodi, Y. 2000. *Yolanda's cranberry country recipes*. Rock Village Publishing, Middleborough, MA. pp. 98.
Ocean Spray Cranberries, Inc. 1980–1988? *101 all-time favorite cranberry recipes*. Ocean Spray Cranberries, Inc., Hanson, MA. pp. 36.

# 32 (American) Cranberry Bush (Highbush Cranberry)

Family: Adoxaceae (moschatel family; *Viburnum* is widely placed in the Caprifoliaceae, the honeysuckle family, but recent work has shown that the Adoxaceae is a better choice)

## NAMES

Scientific name: *Viburnum opulus* L. var. *americanum* Aiton (*V. trilobum* Marshall)

- "Cranberry" in the name "cranberry bush" is based on the similarity in size and color of the berries to the true cranberry (see Chapter 31). Sometimes the cranberry bush is simply called cranberry. Although cranberry bush fruit looks and tastes like conventional cranberry, they are quite unrelated.
- "Highbush cranberry" is the name by which the American cranberry bush is most widely known. Unfortunately, this name is also often used for the lowbush cranberry (*V. edule* (Michx.) Raf.), a use of the word "highbush" that is inappropriate since lowbush cranberry is not a "high bush."
- The American cranberry bush is also called crampbark, grouseberry, mooseberry, pembina, pimbina, squawbush, summerberry, and tree cranberry (note that some of these names are also applied to *V. edule* mentioned above).
- The name crampbark reflects the old use of the bark to treat stomach and menstrual cramps. The bark of species of *Viburnum* has been used for this purpose, as well as to relieve asthma by various cultures in the Old and New Worlds for centuries. The bark contains a bitter compound called viburnine that is antispasmodic, that is, it has muscle relaxant properties, which explains why it is medicinally effective.
- The name pembina (or pimbina) is said to be derived from *anepeminan*, the Chippewa word for the fruit, supposedly a contraction of two Indian words meaning "summer berry," an allusion to the bright red fruit showing up in the depth of winter, suggesting the return of summer.
- The closely related European cranberry bush (*V. opulus* var. *opulus*) has also been called black haw, club bunches, crampbark, cranberry bush, dog rowan tree, gaitre berries, guelder rose, high cranberry, king's crown, May ball, May rose, queen's cushion, red elder, rose elder, silver bells, snowball bush, snowball tree, tisty-tosty, water elder, whitsun boss(es), whitsun rose, and whitten tree.
- The name "guelder rose" is based on *guelder* from Gueldersland, a Dutch province (the modern Gelderland of the Netherlands) where the plant was first cultivated, from where it was introduced into England under the name "Gueldres rose," an English adaptation of the original Dutch *geldersche roos*, meaning "the tree with rose-like bloom." (An alternative, less plausible explanation is that the name is a corruption of elder rose, an old European name for the plant, which was once considered to be a kind of elder.)
- The name "whitten tree" (as well as the names whitsun boss and whitsun rose) are based on "white," that is, the relatively whitish branches.

**FIGURE 32.1** American cranberry bush (*Viburnum opulus* var. *americanum*). (a, b) Branches with fruit. (From Herman, D.E. et al., *North Dakota tree handbook*, U.S. Department of Agriculture, Bismark, ND, 1996.) (c) Distribution map.

- The genus name *Viburnum* is the classical Latin name of *V. lantana* L. of Eurasia. It has been suggested that the name is derived from the classical Latin *viere*, from the Greek *vieo*, which means to tie or bind, based on the idea that the flexible branches could be tied.
- *Opulus* in the scientific name *V. opulus* is the classical Latin for a kind of maple, indicating the maple-shaped leaves of the plant.

## GEOGRAPHY AND ECOLOGY OF WILD PLANTS

The American cranberry bush grows wild in all provinces of Canada and in all northern states bordering Canada, but is most common in eastern North America. In the United States, American cranberry bush has been ranked as endangered in Indiana and Ohio, and rare in Pennsylvania. European cranberry bush, a native of Europe and western Asia, is an occasional established escape in Canada, and has become more frequently established in the northeastern United States than elsewhere in North America. The American and European varieties of *V. opulus* are known to hybridize, and with the establishment of the European variety in parts of North America, such hybridization may have resulted in the generation of intermediate plants that make distinguishing the two kinds difficult. European cranberry bush is considered to be somewhat invasive, and it is often recommended that American cranberry bush be planted instead when used as an ornamental. Another option is to plant only the cultivar *V. opulus* 'Sterilis', which does not produce seeds, and so will not spread.

American cranberry bush is adapted to the relatively fertile soils of southern Canada and the northern United States where it is native. It grows well in cool, moist conditions, and does not tolerate extended drought. It also grows well in partial shade, and indeed is often found naturally in shady habitats, but thrives in full sunlight. Typical habitats include stream banks, wet thickets, and old river bottoms.

## PLANT PORTRAIT

The fruit of species of *Viburnum* has probably been consumed by people for thousands of years, both in the Old and New Worlds, where *V. opulus* occurs. Only the American form of this species produces palatable fruit by commercial standards. However, the fruit is not currently of commercial significance and faces difficulty becoming important because it would have to compete with the similar fruits of cranberry and red currant.

The American cranberry bush is a coarse, deciduous shrub or small tree, generally 2.4–4 m (8–13 feet) in height, but depending on circumstances may be no taller than 0.6 m (2 feet), and dwarf cultivated varieties are available. The stems may be as thick as 4 cm (1.6 inches) in diameter. The leaves tend to be three-lobed, and rather reminiscent of maple leaves. The fruit is produced in clusters. The fruits are orange to glowing red at maturity, 6–15 mm (0.2–0.6 inch) long, subglobose to ellipsoid, juicy and acidic at maturity, with a round, flat, hard stone that occupies a fairly large proportion of the fruit. The berries hang indefinitely on the plants, sometimes until the next bearing period. During the twentieth century, the Canadian and American departments of agriculture introduced cultivated varieties with superior fruit characteristics.

The European cranberry bush is similar in appearance to the North American species. It is commonly cultivated as an ornamental in North America, known as the snowball, snowball bush, or snowball tree. This has very large, rounded heads of small flowers, which are generally sterile. These forms are often confused with ornamental *Hydrangea*, which also produce large white balls of sterile flowers and are also called snowball. The Old-World cranberry bush is inferior to the American cranberry bush for culinary purposes, although its comparatively bitter fruit has sometimes been marketed. Cultivars derived from the European species are basically grown as ornamentals.

The only other *Viburnum* species with palatable fruit is *V. edule*, known as squashberry (as the berries are often "squashed" in culinary preparation), lowbush cranberry, mooseberry, pimbina, and pembina. This is a straggling shrub as tall as 2.5 m (8 feet), but usually much smaller. It grows wild from Labrador to Alaska, south in the east to Pennsylvania, Michigan, and Minnesota, and in the west to Colorado and Oregon. It also occurs in northeastern Asia. Wild fruit is often consumed fresh or cooked, and often used in pies, jellies, and jams. Squashberries are smaller and not as acidic as the fruit of American cranberry bush.

## CULINARY PORTRAIT

The American cranberry bush produces fruit that is prized as a source of jelly by many familiar with it, although many others do not appreciate its unique flavor. Varieties selected for superior fruit are large, attractive shrubs that produce berries for the home gardener. The fruit is often gathered from wild stands in late August or early September for use in sauces, jellies, and juices. The fresh fruit is hard, very sour, and high in pectin. If picked after a heavy frost, the berries are softer and more palatable. The ripe fruits produce a somewhat musty odor, which is accentuated during cooking, and people find it objectionable. However, the "dirty feet" smell eventually disappears during cooking. The addition of lemon or orange peelings during cooking will help to eliminate the natural odor. The disagreeable odor is also lessened when firm berries are used. The fruit is best when picked slightly underripe and firm, and prepared as sauce or jelly. The flavor is similar to that of the true cranberry. The jelly is rich in color and as high in pectin as that made from cranberries and currants. Cranberry bush sauces and jellies are delicious condiments for meat and game. The prominent seeds make jam-making difficult, although the seeds can be strained away. Pies, and occasionally distilled spirits and wines, are also manufactured from the fruit. In general, the fruit can be used in recipes for conventional cranberry.

### Culinary Vocabulary

- The term "cranberry" mostly refers to the common cranberry, *Vaccinium macrocarpon* Aiton. However, some other edible berries are also sometimes called cranberries, including the species of *Viburnum* mentioned in this chapter. Other "cranberry" species include the Florida cranberry (better known as roselle), *Hibiscus sabdariffa* L., and the lingonberry (see Chapter 58), *Vaccinium vitis-idaea* L., which is also known as mountain cranberry and rock cranberry.

## PROSPECTS

The fruit of American cranberry bush has minor commercial significance at present, and the species has not yet acquired the status of a crop. It faces difficulty becoming important because it has to compete with the similar fruits of cranberry and red currant. However, it does seem to have potential for small-scale, local production, and for preparation of unique processed food products. Such products are currently sold at the cottage industry level and often marketed by Web advertisements. As with several other wild berry crops that are indigenous to North America, current interest in the health-promoting content of flavonoids and other natural beneficial compounds may serve to stimulate commercial development.

## CURIOSITIES OF SCIENCE AND TECHNOLOGY

- American cranberry bush was an important food for native people of the northeastern coast region of North America. Berry patches were often considered to be the property of families, with ownership passing from generation to generation. Boxes of fruit were given as a most prestigious gift, and the fruit was served at many feasts.
- The Kwakiutl Indians considered a box of American cranberry fruit to be equivalent to four blankets.
- Bears love the fruit of cranberry bush. The berries are a major food of grizzly bears, which is well to keep in mind in areas where these animals are found.
- The berries of European cranberry bush, after they turn black, have been used to make ink.
- Hundreds of plants were assigned special meanings in the "Victorian Language of Flowers," popular in Victorian times, and many delighted in sending coded messages by this means, especially for romantic purposes. Guelder rose (i.e., European cranberry bush) meant "winter" or "age."
- Cranberry bush produces flat-topped masses of white flowers up to 15 cm (6 inches) across, composed of two different types of flowers. The showy flowers around the outside of each group are sometimes 2.5 cm (1 inch) in diameter, but they are sterile, whereas the less conspicuous flowers toward the centre of the cluster are fertile. The function of the showy sterile

flowers is to lure pollinating insects so that they will pollinate the less attractive but fertile flowers. Only the inner, fertile flowers provide nectar, so that they will be visited by pollinators.

- The viburnum leaf beetle (*Pyrrhalta viburni*) is an exotic European invader that is seriously damaging several native species of *Viburnum* in eastern North America, including the cranberry bush. The adult is about the size of the head of a large matchstick, while the caterpillar-like larvae are slightly larger. The plant's leaves are often reduced to skeletons, the insects consuming everything but the veins. The insect will eat only *Viburnum* species. It has become a major pest in recent decades, and there is considerable apprehension about the extent of damage it will cause, both to wild and cultivated plants.

## KEY INFORMATION SOURCES

Berg, A.J. van den. 1988. *Growing ornamental* Prunus *and snowball tree*. Proefstation voor de Bloemisterij in Nederland, Aalsmeer, the Netherlands. pp. 75. (In Dutch)

Boiko-Alaux, V., Carnat, A., Lamaison, J.L., Bastide, J. and Bastide, P. 1988. Catechic tannins as angioprotective principles in *Viburnum opulus* L. *Plantes Med. Phytother. (France)* 22: 242–246. (In French, English summary)

Czekalski, M. 1984. European cranberry bush *Viburnum opulus* L. a shrub protected in part by conservation laws. *Rocznik Sekcji Dendrologicznej Polskiego Towarzystwa Botanicznego* 36: 73–82. (In Polish)

Darrow, G.M. 1923. *Viburnum americanum* as a garden fruit. *Proc. Am. Soc. Hortic. Sci.* 21: 44–54.

Darrow, G.M. 1975. Minor temperate fruits. In *Advances in fruit breeding*. Edited by J. Janick and J.N. Moore. Purdue University Press, W. Lafayette, IN. pp. 269–284.

Englund, R. 1993. Fruit removal in *Viburnum opulus*: copious seed predation and sporadic massive seed dispersal in a temperate shrub. *Oikos* 67: 503–510.

Fedec, P., and Knowles, R.H. 1973. Afterripening and germination of seeds of American highbush cranberry (*Viburnum trilobum*). *Can. J. Bot.* 51: 1761–1764.

Horhammer, L., Wagner, H., and Reinhardt, H. 1966. On the chemistry pharmacology and pharmacognosy of the contents of *Viburnum prunifolium*, American snowball, and *Viburnum opulus*, common snowball. *Zoological Mag. (Tokyo)* 79: 510–525.

Ivanov,V.D., and Ladygina, E.Ya. 1985. Anatomical structures of European cranberry bush *Viburnum opulus* fruit. *Farmatsiya (Moscow)* 34(6): 13–17. (In Russian)

Jones, E., and Wheelwright, N.T. 1987. Seasonal changes in the fruits of *Viburnum opulus*, a fleshy-fruited temperate-zone shrub. *Can. J. Bot.* 65: 2291–2296.

Kollmann, J., and Grubb, P.J. 2002. *Viburnum lantana* L. and *Viburnum opulus* L. (*V. lobatum* Lam., *Opulus vulgaris* Borkh.). *J. Ecol.* 90: 1044–1070.

Krannitz, P.G., and Maun, M.A. 1991 a. An experimental study of floral display size and reproductive success in *Viburnum opulus*: importance of grouping. *Can. J. Bot.* 69: 394–399.

Krannitz, P.G., and Maun, M.A. 1991 b. Insect visitors to the guelder rose, *Viburnum opulus* var. *opulus* (Caprifoliaceae), in London, Ontario. *Can Field-Nat.* 105: 13–17.

McAtee, W.L. 1956. *A review of the nearctic* Viburnum. Privately published: Chapel Hill, NC. pp. 125.

Ohashi, H. 1994. The correct name for the Asiatic variety of *Viburnum opulus* L. Caprifoliaceae). *J. Jpn. Bot.* 69: 32–33.

Ovodova, R.G., Golovchenko, V.V., Popov, S.V., Shashkov, A.S., and Ovodov, Y.S. 2000. The isolation, preliminary structural studies, and physiological activity of water-soluble polysaccharides from the squeezed berries of snowball tree *Viburnum opulus*. *Bioorganicheskaya Khimiya* 26: 61–67. (In Russian)

Schurz, J. 1972. *Viburnum opulus* and *Euonymus europaeus* [biochemistry]. *Kosmos* 86: 257–258. (In German)

Shapiro, D.K., Kisilevs'kii, I.R., Moroz, P.A., Potopal'skii, A.I., Kuznetsova, Z.P., Vasilevskaya, T.I., Dovnar, T.B., et al. 1992. Biochemical composition of fruits of *Viburnum opulus* L. forms growing in the Ukrainian Polesye and forest steppe. *Rastitel'nye Resursy* 28: 54–63. (In Russian)

Shapiro, D.K., Kisilevs'kii, I.R., Moroz. P.A., Vasilevs'ka, T.I., Gorbatsevich, V.I., Prilishch, N.P., and Vereskovs'kii, V.V. 1989. Biologically active substances of the fruits of the perspective forms of *Viburnum opulus*. *Farmatsevtychnyi Zhurnal (Kiev)* 1: 55–58. (In Ukrainian)

Small, E., and Catling, P.M. 2005. Poorly known economic plants of Canada—47. American cranberry bush (*Viburnum opulus* var. *americanum*). *Bull. Can. Bot. Assoc.* 38(4): 48–54.

Stang, E.J. 1990. Elderberry, highbush cranberry, and Juneberry management. In *Small fruit crop management*. Edited by G.J. Galletta and D.G. Himelrick. Prentice Hall, Englewood Cliffs, NJ. pp. 363–382.

Staszkiewicz, J., and Bialobrzeska, M. 1997. The variability of leaves and fruits of *Viburnum opulus* (Caprifoliaceae). *Fragmenta Floristica Geobot. Series Polonica*, Suppl. 2: 263–276. (In Polish)
Vakhrameeva, M.G. 1978. *Viburnum opulus* nomenclature, taxonomy, USSR. *Biol. Flora Mosk. Obl. Moskva, Moskovskii Universitet.* 1978(pt. 4): 191–197. (In Russian)
Velioglu, Y.S., Ekici, L., and Poyrazoglu, E.S. 2006. Phenolic composition of European cranberrybush (*Viburnum opulus* L.) berries and astringency removal of its commercial juice. *Internat. J. Food Sci. Technol.* 41: 1011–1015.
Winston, D. 2000. *Cramp bark: Viburnum opulus: analytical, quality control, and therapeutic monograph.* American Herbal Pharmacopoeia, Santa Cruz, CA. pp. 17.
Witmer, M.C. 2001. Nutritional interactions and fruit removal: cedar waxwing consumption of *Viburnum opulus* fruits in spring. *Ecology* 82: 3120–3130.
Zuzuk, B.M., Rohovs'ka, L.Ya., and Shtokalo, M.R. 1995. *Viburnum opulus* fruits are promising medicinal raw material. *Farmatsevtychnyi Zhurnal (Kiev)* 3: 72–75. (In Ukrainian)

## Specialty Cookbooks

Wilson, K.A. 1977. Lowdown on highbush "cranberry" and other native viburnums. *Flower—Gard. (West. edition)* 21(9): 34, 43. [Presents a recipe for jam.]

Freitus and Haberman (2005) have 10 cranberry bush recipes, and Marrone (2009) has 18. Gray (2011) has a recipe entitled "Highbush cranberry applesauce," which is intended for *Viburnum edule*, but should be suitable also for *V. opulus*. Stanek and Butcher (2007) have nine "highbush cranberry" (species not identified) recipes. (See the Appendix to this book for details of the publications cited.)

# 33 Crowberry

Family: Ericaceae (heath family; until recently *Empetrum* was commonly placed in the Empetraceae, but recent evidence has demonstrated it should be in the Ericaceae)

## NAMES

Scientific name: *Empetrum nigrum* L. (*E. hermaphroditum* (Hagerup) D. Löve)

- The name "crowberry" reflects the observation that birds frequently eat the berries.
- Crowberry is often called black crowberry, and is also known as blackberry (a name that should be reserved for other species; see Chapter 12 on Blackberries and Dewberries), crakeberry (crake-berry), curlew berry (curlewberry), mossberry, and mountain crowberry. In Scandinavia, the name *kraaihei* is often used.
- The genus name *Empetrum* is based on the Greek *en*, in + *petros*, rock, an allusion to the fact that the species often grows on rocky shores.
- *Nigrum* in the scientific name is Latin for black, a reference to the fruit color.

## GEOGRAPHY AND ECOLOGY OF WILD PLANTS

*Empetrum nigrum* is a circumboreal species. In North America, it occurs in Alaska, all of the provinces and territories of Canada, the three West coast states, and the northeastern states. On the Pacific coast it is found as far south as Del Norte County in California. In the United States it has also been recorded in Maine, Michigan, Minnesota, New Hampshire, New York, and Vermont. Outside of North America it is distributed throughout the Northern Hemisphere, including Greenland, Iceland, and Eurasia.

This species occurs in a very wide range of habitats, including both arctic and alpine environments, and temperate and subarctic climates. It is often near large bodies of water, and is also associated with coastal forests, moist coniferous forests, moorlands, tundra, and muskeg. Black crowberry thrives in well-drained sandy soils, wet peat soils, bogs, and rocky shores, usually showing a very strong avoidance of basic soils. Frequently, the species is quite dominant, forming large, continuous mats.

## PLANT PORTRAIT

The classification of *Empetrum nigrum* is controversial. Kinds with male and female flowers on different plants are frequently separated as a different species or subspecies from plants with bisexual flowers or both kinds of flowers on the same plant; however, this division does not appear to be justified. Kinds of plant with pink, red, or purple berries are often included in the species, but often not; this chapter includes only plants with black fruits in the species.

Black crowberry is a creeping, mat-forming, much-branched, dwarf shrub 10–50 cm (4–20 inches) in height, with very small evergreen leaves reminiscent of the needles of spruce and fir trees. Male and female flowers may occur on different plants, or plants may have bisexual flowers or both kinds of flowers. The solitary, inconspicuous, pinkish or purplish flowers are quite small, 1–2 mm (0.04–0.08 inch) long. The berries are 4–10 mm (0.2–0.4 inch) in diameter, black or dark purple when ripe, with 6–9 hard seeds.

**FIGURE 33.1** Crowberry (*Empetrum nigrum*). (a) Branch with numerous fruits. (Courtesy of Hajjotthu [CC By 3.0].) (b) Fruiting twig, fruits, and seed. (From Thomé, O.W., *Prof. Dr. Thomé's Flora von Deutschland, Österreich und der Schweiz*, H.V. Verlag, Berlin-Lichterfelde, Germany. Vol. 3, Plate 395, 1905.) (c) Fruit. (Courtesy of T. Lee Tibbitts, U.S. Department of the Interior [Agency: Geological Survey].) (d) Distribution map.

In addition to food uses of the berries, Northern indigenous people used the fruits and sometimes the branches for several medicinal purposes. Crowberries are occasionally grown as ornamentals, and there are a few cultivars available. Although the environment suitable for crowberry is rather demanding, the plant can be a good groundcover and landscape plant, especially in rock gardens.

## CULINARY PORTRAIT

Crowberry fruits are juicy, slightly acidic, varying from bland to sweet. Indigenous peoples throughout the natural range of the species harvested the berries for food in the autumn when they matured. Indeed, the berries were a staple among subarctic people. An advantage was that some berries remain on the plant overwinter, and could be gathered even under snow, and in the spring. The berries were (and continue to be) consumed raw or cooked, and used in a variety of confections. Indigenous people and wild food enthusiasts still harvest large amounts of the berries, which are mild in flavor and juicy. Crowberries can be eaten fresh or used in pies, jelly, tea, and wine. Because the berries are mild in taste, they are generally combined with other, more flavorful berries in recipes.

### Culinary Vocabulary

- "Eskimo ice cream" (*ahkootuk*, *akutaq*, *akutq*) is a combination of fish or meat fat (commonly moose, reindeer, caribou, or seal oil) and berries (often crowberries). Some modern recipes substitute vegetable shortening for the meat and add sugar. For a recipe, see Lay and Santora (2003, listed in the appendix).

## PROSPECTS

Crowberry is collected in considerable amounts from wild stands, and mostly consumed locally, but it is also sold commercially in northern climates, and used in some commercial blended juices. The wild supply is considerable, so that deliberately cultivating the plant seems unnecessary. Moreover, outside of the natural geographical range, it can be difficult to supply the soil and climate conditions that are appropriate, and other berry crops are much more competitive. Crowberry has the potential of being managed in the same manner as lowbush blueberry (see Chapter 15 on Blueberries); that is, natural stands are provided with fertilization, weeding, and other procedures to increase productivity, but planting is not necessary. Crowberry fruits have high levels of anthocyanin pigments in the skin, which have potential as nutritive extracts and natural food colorants. Indeed, there is some use of the berries as a purple dye for food products and for antioxidant benefits in nutritional supplements, and these applications have potential for expansion. In many northern areas, where agriculture is very difficult, crowberry is a natural food source, and as such deserves serious consideration for future development as a crop.

## CURIOSITIES OF SCIENCE AND TECHNOLOGY

- The berries of *Empetrum nigrum* are quite watery, and native people and knowledgeable hikers have often consumed them in the field as much to slake thirst as to enjoy the fruits.
- In subarctic areas, native people commonly used the natural refrigeration provided by the climate to preserve foods overwinter. Black crowberries harvested in the fall were frequently mixed with other berries or greens, and meat, fish, seal oil, blubber, or fat, for storage in the frozen ground until required.
- Plants of crowberry as old as 140 years have been recorded.
- From a commercial perspective, names are usually critical for market success. For example, the names "monkey peach" and "vegetable mouse" were abandoned in preference to the much more attractive "kiwi," contributing to the success of this fruit. "Crowberry" is not the kind of name that most people find attractive, and names like "winter berry" or "Arctic blackberry" would find greater acceptability.

## KEY INFORMATION SOURCES

Altan, Y., and Ozdemir, C. 2004. Morphological and anatomical studies on economically important *Empetrum nigrum* L. subsp. *hermaphroditum* (Hagerup) Bocher (Empetraceae). *Econ. Bot.* 58: 679–683.

Barclay-Estrup, P., and Nuttall, D.V. 1974. Some aspects of the distribution and ecology of crowberry *Empetrum nigrum* on the north shore of Lake Superior. *Can. Field-Nat.* 88: 171–181.

Bell, J.N.B., and Tallis, J.H. 1973. Biological flora of the British Isles: *Empetrum nigrum* L. *J. Ecol.* 61: 289–305.

Bell, J.N.B., and Tallis, J.H. 1974. The response of *Empetrum nigrum* L. to different mire water regimes with special reference to Wybuntury moss, Cheshire and Featherbed moss, Debyshire. *J. Ecol.* 62: 75–95.

Elvebakk, A. and Spjelkavik, S. 1995. The ecology and distribution of *Empetrum nigrum* ssp. *hermaphroditum* on Svalbard and Jan Mayen. *Nordic J. Bot.* 15: 541–552.

Finn, C. 2008. *Empetrum nigrum*, crowberry. In *The encyclopedia of fruit & nuts*. Edited by J. Janick and R.E. Paull. CABI, Wallingford, Oxfordshire, U.K. pp. 348.

Johansson, A.K., Kuusisto, P.H., Laakso, P.H., Derome, K.K., Sepponen, P.J., Katajisto, J.K., and Kallio, H.P. 1997. Geographical variations in seed oils from *Rubus chamaemorus* and *Empetrum nigrum*. *Phytochemistry* 44: 1421–1427.

Kallio, H., and Markela, E.0. 1982. Nonvolatile acids in the juice of crowberry, *Empetrum nigrum*. *J. Food Sci.* 47: 1466–1468.

Kim, C., Han, H., Koh, H., and Byun, G. 2005. Review on the taxonomic and biogeographic characteristics of an arctic-alpine species, *Empetrum nigrum* var. *japonicum*, in Mt. Halla, Korea. *KFRI J. For. Sci. (Seoul)* 68:128–136. (In Korean, English abstract.)

Kron, K.A. 1996. Phylogenetic relationships of Empetraceae, Epacridaceae, Ericaceae, Monotropaceae, and Pyrolaceae: Evidence from nuclear ribosomal 18s sequence Data. *Ann. Bot.* 77: 293–303.

Kron, K.A., Judd, W.S., Stevens, P.F., Crayn, D.M., Anderberg, A.A., Gadek, P.A., Quinn, C.J., and Luteyn, J.L. 2002. Phylogenetic classification of Ericaceae: molecular and morphological evidence. *Bot. Rev.* 68: 335–423.

Laaksonen, O., Sandell, M., Jarvinen, R., and Kallio, H. 2011. Orosensory contributing compounds in crowberry (*Empetrum nigrum*) press-byproducts. *Food Chem.* 124: 1514–1524.

Malinowska, H. 1991. Container cultivation of crowberry (*Empetrum nigrum* L.). *Roczniki Akademii Rolniczej w Poznaniu, Ogrodnictwo* [Annals of Agricultural University of Poznan, Gardening] 19: 51–58. (In Polish, English summary.)

Matsuura, H., Saxena, G., Farmer, S.W., Hancock, R.E.W., and Towers, G.H.N. 1995. Antibacterial and antifungal compounds from *Empetrum nigrum*. *Planta Medica* 61: 580.

Monni, S., Salemaa, M., and Millar, N. 2000. The tolerance of *Empetrum nigrum* to copper and nickel. *Environ. Pollut.* 109: 221–229.

Monni, S., Uhlig, C., Hansen, E., and Magel, E. 2001. Ecophysiological responses of *Empetrum nigrum* to heavy metal pollution. *Environ. Pollut.* 112: 121–129.

Monni, S., Uhlig, C., Junttila, O., Hansen, E., and Hynynen, J. 2001. Chemical composition and ecophysiological responses of *Empetrum nigrum* to aboveground element application. *Environ. Pollut.* 112: 417–426.

Murray, D.F., Mirré, V., and Elven, R. 2009. *Empetrum*. In *Flora of North America North of Mexico*. Vol. 8. Edited by Flora of North America Editorial Committee. Oxford University Press, Oxford, UK. pp. 486–489.

Mutikainen, P., and Ojala, A. 1993. Simulated herbivory and air pollution: growth and reproduction of an evergreen dwarf shrub, *Empetrum nigrum*. *Acta Oecol.* 14: 771–780.

Ogawa, K., Sakakibara, H., Iwata, R., Ishii, T., Sato, T., Goda, T., Shimoi, K., and Kumazawa, S. 2008. Anthocyanin composition and antioxidant activity of the crowberry (*Empetrum nigrum*) and other berries. *J. Agric. Food Chem.* 56: 4457–4462.

Puntari, I., Kallio, H., and Linko, R.R. 1985. Volatiles of crowberry, *Empetrum nigrum*. *Lebensmittel - Wissenschaft + Technologie [Food Science + Technology]* 18: 249–251.

Shevtsova, A., Haukioja, E., and Ojala, A. 1998. Growth response of subarctic dwarf shrubs, *Empetrum nigrum* and *Vaccinium vitis-idaea*, to manipulated environmental conditions and species removal. *Oikos* 78: 440–458.

Teppner, H. 1987. *Empetrum nigrum* L. s. str., new for the Alps. *Bot. Jahrb.* 108: 355–362. (In German)

Zverev, V.E., Zvereva, E.L., and Kozlov, M.V. 2008. Slow growth of *Empetrum nigrum* in industrial barrens: Combined effect of pollution and age of extant plants. *Environ. Pollut.* 156: 454–460.

## Specialty Cookbooks

Lay, J.S., and Santora, B. 2003. *What real Alaskans eat. Not your ordinary cookbook*. Epicenter Press, Kenmore, WA. pp. 129.

Freitus (1980) has two crowberry recipes, Freitus and Haberman (2005) have two, Gray (2011) has one, Moerman (1998) provides one, and Stanek and Butcher (2007) have five. (See the Appendix to this book for details of the publications cited.)

# 34 Devil's Claw

Family: Martyniaceae (unicorn plant family; sometimes this family is placed in the Pedaliaceae, the pedalium or sesame family (best known for sesame, *Sesamum indicum* L., the seeds of which are commonly sprinkled on hamburger buns); however, recent evidence suggests the two families should not be combined)

## NAMES

Scientific names: *Proboscidea* species

- Louisiana devil's claw—*Proboscidea louisianica* (Mill.) Thell. (*P. jussieui* Medik, *P. fragrans* (Lindl.) Decne., *Martynia fragrans* Lindl., *M. louisianica* Mill., *M. proboscidea* Gloxin).
- New Mexico devil's claw—*Proboscidea parviflora* (Wooton) Wooton & Standley (*M. parviflora* Wooton).
- The name "devil's claw" (and similar names like elephant's tusks) is based on the pair of prominent hooked appendages attached to the fruit. Both species discussed here are usually called "devil's claw," but the occasional names Louisiana devil's claw and New Mexico devil's claw are used here to distinguish them.
- "Devil" is often capitalized, although this convention is not followed here. "Devil's claw" is sometimes hyphenated ("devil's-claw").
- In addition to being called devil's claw, *P. louisianica* has been termed aphid trap, common unicorn plant, elephant tusks, goat's head, Louisiana devil's claw, Louisiana unicorn plant, martynia, pale devil's claw, proboscis-flower, purple-flower devil's claw, purple-flowered devil's claw, ram's horn (ram's-horn), and unicorn plant (unicorn-plant).
- In addition to being called devil's claw, *P. parviflora* has been termed doubleclaw, devil's horn, New Mexico devil's claw, red devil's claw, and unicorn plant.
- Several species which also have hooked fruits are also called devil's claw; many of these are in the Martynaceae and Pedaliaceae families. "African devil's claw" refers to *Harpagophytum procumbens* (Burch.) DC. ex Meisn. and *H. zeyheri* Decne.; these are medicinal plants of the Pedaliaceae. Frequently, when the phrase "devil's claw" is used, it refers to African devil's claw. Occasionally websites misidentify *Proboscidea* as the source of anti-inflammatory preparations used for conditions such as arthritis, when the preparations in fact are from *Harpagophytum*.
- The genus name *Proboscidea* is based on the Greek *proboskis*, elephant's nose, an allusion to the proboscis-like projection of the fruit.
- "*Louisianica*" in the scientific name *P. louisianica* is Latin relating to Louisiana, a region much larger than the present state of Louisiana when the epithet was first proposed for the species in 1768. The name "*P. louisiana*" is sometimes encountered but is erroneous (confusion began when the term "louisiana" was first coined for the plant, but the author changed this to "louisianica" on a page listing his errors).
- "*Parviflora*" in the scientific name *P. parviflora* is Latin for small-flowered, an allusion to the comparatively small flowers of this species.

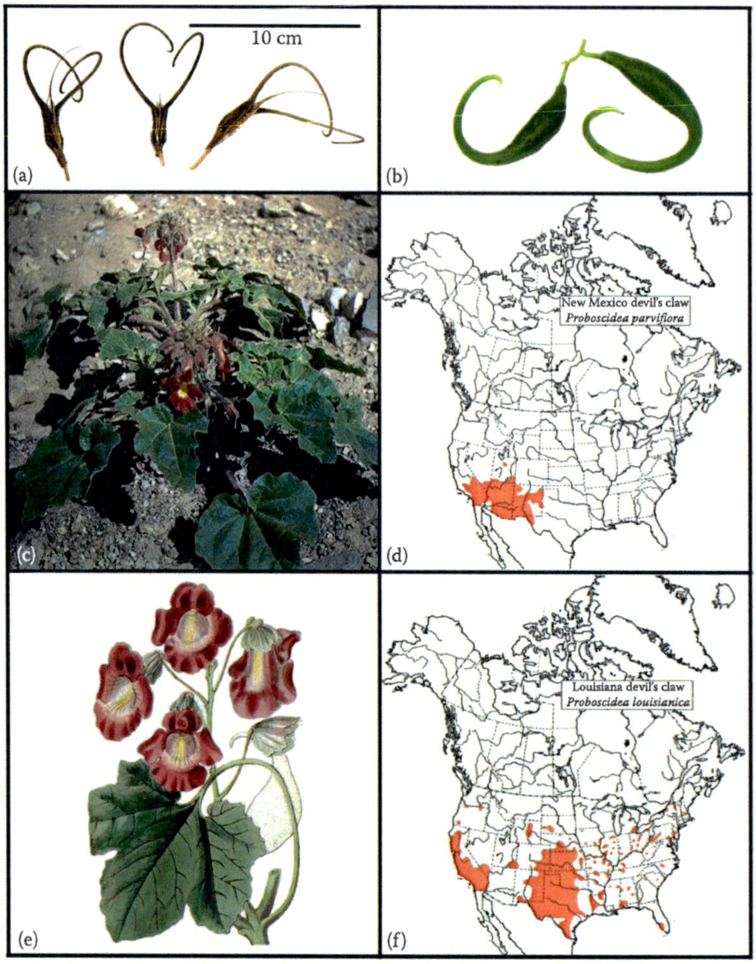


**FIGURE 34.1** Devil's claw (*Proboscidea* species). (a) Mature fruits of New Mexico devil's claw (*P. parviflora*). (Courtesy of E. Small and B. Brookes.) (b) Immature fruits of New Mexico devil's claw (note that elongated tips have not yet split into two hooks). (Courtesy of E. Small and B. Brookes.) (c) Flowering plant of New Mexico devil's claw (*P. parviflora*). (Courtesy of Clarence A. Rechenthin, U.S. Department of Agriculture.) (d) Distribution map of New Mexico devil's claw. (e) Flowering plant of Louisiana devil's claw (*P. louisianica*). (From Edwards, S.T., and Lindley, J., *Edward's botanical register*, James Ridgway, London, UK, Vol. 26, Plate 85, 1840.) (f) Distribution map of Louisiana devil's claw.

## GEOGRAPHY AND ECOLOGY OF WILD PLANTS

*Proboscidea louisianica* is native to Colorado, Kansas, Nebraska, New Mexico, Oklahoma, Texas, and Mexico. It has been introduced in other areas. The species can be quite weedy and has been collected from almost all states of the United States, from Manitoba and Ontario in Canada, and from other countries. It occurs in open areas, in a variety of substrates, but often in the fertile soils of cultivated fields, typically in moist ground in contrast to most other *Proboscidea* species, which are found in arid regions. The species occurs in meadows, playas, waste places, and on stream banks.

*Proboscidea parviflora* is native to Arizona, California, Nevada, New Mexico, Texas, Utah, and northern Mexico. As an inhabitant of deserts, it grows in open areas, in sandy or gravelly soils (or sometimes in clay), and in dry disturbed habitats, including plains, mesas, and slopes.

The hooked pods of devil's claw species seem obviously adapted to attaching to the hooves and legs of large grazing animals. When hooked to the hooves of animals, the pods are crushed open as the animals travel, dispersing the seeds. (It has been suggested that the fruits are designed to hobble and kill mammals, so that their decaying corpses will act as fertilizer for the plants that grow from the seeds. This seems unlikely.) With the possible exception of bighorn sheep, large grazing animals that are native to the distribution range of devil's claw species are absent. It has been speculated that a few thousand years ago bison, antelope, or some other North American animals were present in the range of North American devil's claw species. It has also been speculated that hundreds of thousands of years ago large grazers such as the ground sloth and mammoths, which are now extinct, served to distribute the fruits. Today, the pods are often distributed by becoming attached to introduced livestock, the shoes of people, or tires.

The copious glandular secretions of the plants probably act as protection against insects, but may also serve for allelopathy (a form of plant chemical warfare, whereby chemicals secreted into the soil inhibit the germination and/or growth of competing plants).

## PLANT PORTRAIT

The plants of both species of devil's claw highlighted in this chapter are annual taprooted herbs with large leaves and sprawling stems somewhat reminiscent of pumpkin vines. The stems and foliage are covered with sticky hairs, which secrete a gummy, gelatinous liquid that produces an odor which has been compared to moldy sneakers, rotten gym socks, cat urine, and melted rubber. The leaves of devil's claw plants are heliotropic, tracking the sun. When immature, the curious pods have a curved elongation, which has given rise to the name "unicorn plant." As the fruits mature, the soft outer "rind" degenerates, revealing a hard, woody pod. The elongated, horned portion of the pod splits into two sharp "claws," each reminiscent of a very large fish hook. The fruits are often used to construct *objets d'art*, often decorated as insects or birds. They are also used in flower arrangements. Seeds of devil's claw (advertised as "martynia") were offered by Philadelphia seedsman Bernard M'Mahon in his 1805 catalog. Devil's claw plants are still sometimes grown in gardens, but have a very unpleasant odor, the slime produced on the foliage is easily transferred to clothing, and the fruits can be dangerous for animals and children.

*Proboscidea louisianica* is a branching herb up to 1 m (about 3 feet) tall, with stems ranging up to 80 cm (32 inches). The showy flowers, somewhat reminiscent of orchids, are 3.5–5 cm (1.5–2 inches) wide and are white, creamy, pink, yellowish, or lavender, and commonly have dark spots. The flowers are violet to reddish-purple in subsp. *fragrans* (Lindl.) Bretting of Texas and central Mexico, known as Chihuahuan devil's claw. The pods of *P. louisianica* are about 10 cm (4 inches) long.

*Proboscidea parviflora* is a sprawling herb similar to *P. louisianica*, up to 50 cm (20 inches) high and up to 1.5 m (59 inches) across. The species has white, pink, or reddish-purple flowers about 2.5 cm (1 inch) wide. The fruits, exclusive of the claws, are 5–10 cm (2–4 inches) long, and the two claws are up to four times as long as the remaining fruit body. Devil's claw pods have long been used by Native Americans of the southwestern United States for basketry. The pods are cut into narrow strips, which are quite flexible, and can be employed in weaving (most of the fibrous material used is harvested from the claws). The Tohono O'odham (formerly the Papago) of southern Arizona selected a white-seeded cultivar of *P. parviflora* (devil's claw seeds are normally black) with fruits having claws over 45 cm (15 inches) long, which was advantageous for basket making since longer strips of material could be made from the fruits. The white seeds lack germination inhibitors, and the seeds germinate quickly, unlike the seeds of wild devil's claw plants. The Tohono O'odham also selected a form with three or four claws, but it is said to have fallen out of favor when the belief developed that this would result in women having twins. Several other cultivars have been selected. A tea prepared by steeping the dried seedpods was used by Native Americans as a headache medicine. The plants were once extensively cultivated for use as a food source by Native Americans of the southwestern

United States and Mexico. *Proboscidea louisianica* was cultivated in Michigan and Massachusetts in the early 1900s and the pods canned commercially. Today, the plants are grown for food only to a small extent, both by Native Americans and by gardeners elsewhere.

## CULINARY PORTRAIT

Although slimy and bitter, the immature, green fruits of the two devil's claw species discussed in this chapter are soft, tender, and somewhat fleshy, and at this stage they are consumed as a cooked vegetable, as a pickle like cucumbers, or as a thickener like okra, especially by Native Americans in the southwestern United States and Mexico. The taste has been compared to okra, with occasional undertones of morels. Devil's claw has been noted in cookbooks since the eighteenth century, mainly in pickling recipes. For culinary uses, the pods need to be picked before they are 2.5–5 cm (1–2 inches) long, after which they become quite bitter. A simple method of preparation is to wash the pods, brush off their sticky hairs, boil them for 15–20 minutes, and serve them with butter. The seeds are rich in protein and oil and are also consumed. In Mexico, the pods are shelled and the raw seeds are collected, preferably while they are still green because the immature seeds can be eaten directly without removing their seed coat. The seeds have an excellent flavor (sometimes likened to sunflower seeds) and can be toasted or ground into a meal. In some areas of Mexico, the seeds are sold in markets.

### Culinary Vocabulary

- In older cookbooks, devil's claw often appears under the name "martynia."

## PROSPECTS

Thousands of plant species have been utilized historically by indigenous peoples, and the overwhelming majority of such uses are obsolete. Several considerations suggest that devil's claw, unlike most other obsolete "Indian plants," may have commercial potential. First is the fact that, unlike most plants that were useful in the past, one devil's claw species (*P. parviflora*) was domesticated, and the cultigen is still available for further development. This means that considerable breeding has already been done, so future breeding begins with a good base. There are also several other species of *Proboscidea* that can be used in breeding programs to select improved cultivars. Second, and most importantly, *P. parviflora* is adapted to deserts and *P. louisianica* can be grown in droughty areas, so both species are suitable for dryland cultivation, a critical consideration in the development of new crops, because water for irrigation is becoming increasingly scarce and expensive. Third, devil's claw has potential to be competitive in two food market classes, niche vegetables, and oilseeds; by comparison, several other market classes—fruits, spices, cereals—are much more difficult to penetrate. The edible oilseed possibility is much more significant than the potential development of the plant as a novelty vegetable. There have been indications that devil's claw could become a dryland oilseed, and as the availability of irrigation water becomes more acute in the future, the crop will become increasingly attractive. For commercial purposes, the name "unicorn plant" would probably be much more acceptable than "devil's claw."

## CURIOSITIES OF SCIENCE AND TECHNOLOGY

- Devil's claw fruit hooks were once used as coat hooks and pot hooks.
- Hopi Indians believed that the long spines of devil's claw attracted lightning and (more importantly) rain, and so did not weed it out of their crops.
- Devil's claw species have been termed "protocarnivorous," that is, preceding the state of being truly carnivorous since their malodorous, sticky leaves attract, trap, and kill

numerous small insects such as aphids, midges, and gnats, but the plants do not seem to actually digest the insects as do insectivorous plants. True carnivorous plants have digestive enzymes to break down the bodies of trapped animals. (Reference: Płachno, B.J., Adamec, L., and Huet, H. 2009. Mineral nutrient uptake from prey and glandular phosphatase activity as a dual test of carnivory in semi-desert plants with glandular leaves suspected of carnivory. *Ann. Bot.* 104: 649–654.)

## KEY INFORMATION SOURCES


Armstrong, W.P. 1979. Unicorn plants in California. *Fremontia* 7: 16–22.

Armstrong, W.P. 1980. A gourmet's guide to unicorns. *Desert Magazine* 43: 36–39.

Armstrong, W.P. 1981. Unicorn plants. *California Garden* 72: 13–19.

Armstrong, W.P. 1985. Devil's claw: Trademark of the southwest. *Terra* 23: 16–19.

Armstrong, W.P. 1992. Devil's claws. *Pacific Hortic.* 53: 19–23.

Asai, T., Hara, N., and Fujimoto, Y. 2010. Fatty acid derivatives and dammarane triterpenes from the glandular trichome exudates of *Ibicella lutea* and *Proboscidea louisiana. Phytochemistry* 71: 877–894.

Berry, J., Bretting, P.K., Nabhan, G.P., and Weber, C. 1981. Domesticated *Proboscidea parviflora*: A potential oilseed crop for arid lands. *J. Arid Environ.* 4: 147–160.

Bretting, P.K. 1982. Morphological differentiation of *Proboscidea parviflora* ssp. *parviflora* (Martyniaceae) under domestication. *Am. J. Bot.* 69: 1531–1537.

Bretting, P.K. 1983. The taxonomic relationship between *Proboscidea louisianica* and *Proboscidea fragrans* (Martyniaceae). *Southw. Nat.* 28: 445–449.

Bretting, P.K. 1984. Folk names and uses for martyniaceous plants. *Econ. Bot.* 38: 452–463.

Bretting, P.K. 1985. Geographical intergradation in *Proboscidea parviflora* ssp. *sinaloensis* (Martyniaceae). *Southwest. Nat.* 30: 343–348.

Bretting, P.K. 1986. Changes in fruit shape in *Proboscidea parviflora* ssp. *parviflora* (Martyniaceae) with domestication. *Econ. Bot.* 40: 170–176.

Bretting, P.K., and Nabhan, G.P. 1986. Ethnobotany of devil's claw (*Proboscidea parviflora* ssp. *parviflora*: Martyniaceae) in the greater southwest. *J. Cal. Great Basin Anthropol.* 8: 226–237.

Ditomaso, J.M. 1983. Adaptations to animal dispersal in two weed species or true confessions of a happy hooker. *Weeds Today* 14(3): 3–4.

Gardner, C.A. 1932. A new noxious weed. The devil's claw (*Martynia* (*Proboscidea*) *louisiana* Mill.). *J. Dept. Agric. Western Austral.* 9: 129–131.

Ghosh, A., and Beal, J.L. 1979. Seed lipid constituents of three species of *Proboscidea. J. Nat. Prod.* 42: 287–291.

Hevly, R.H. 1969. Nomenclatural history and typification of *Martynia* and *Proboscidea* Martyniaceae). *Taxon* 18: 527–534.

Nabhan, G.P., and Rea, A. 1987. Plant domestication and folk biological change: The Upper Piman/devil's claw example. *Am. Anthropol.* 89: 57–73.

Nabhan, G.P., Whiting, A., Dobyns, H., Hevly, R., and Euler, R. 1981. Devil's claw domestication: Evidence from southwestern Indian fields. *J. Ethnobiol.* 1: 135–164.

Ortega-Nieblas, M., and Robles-Burgueno, M.R. 1993. Physicochemical characterization of crude and refined oil from the seeds of *Proboscidea parviflora. Grasas y Aceites (Spain)* 44: 30–34. (In Spanish, English summary)

Ortega-Nieblas, M., Robles-Burgueno, M.R., and Vazquez-Moreno, L. 2003. Nutritional and sensorial evaluation of *Proboscidea parviflora* (devil's claw) oil. *Grasas y Aceites (Spain)* 54: 48–52. (In Spanish)

Phillippi, A., and Tyrl, R.J. 1979. The reproductive biology of *Proboscidea louisianica* (Martyniaceae). *Rhodora* 81: 345–361.

Riffle, M.S., Waller, G.R., Murray, D.S., and Sgaramello, R.P. 1990. Devil's-claw (*Proboscidea louisianica*), essential oil and its components: Potential allelochemical agents on cotton and wheat. *J. Chem. Ecol.* 16: 1927–1940.

Riffle, M.S., Waller, G.R., Murray, D.S., and Sgaramello, R.P. 1991. Composition of essential oil from *Proboscidea louisianica* (Martyniaceae). *Proc. Okla. Acad. Sci.* 71: 35–42.

Riffle, M.S., Thilsted, W.E., Murray, D.S., Ahring, R.M., and Waller, G.R. 1988. Germination and seed production of unicorn-plant (*Proboscidea louisianica*). *Weed Sci.* 36: 787–791.

Thieret, J.W. 1976. Floral biology of *Proboscidea louisianica* (Martyniaceae). *Rhodora* 78: 169–179.

## Specialty Cookbooks

Recipes for devil's claw are very difficult to locate. Niethammer (1974, 1999; see the Appendix to this book) provides advice on cooking the pods. An early pickling recipe for devil's claw from *Vaughan's Vegetable Cookbook* (4th edition, 1919) follows:

> Gather the pods when young and tender enough to thrust a needle through them easily, later they become hard and useless for pickles. Leave half an inch of stem on each, and lay them in salt water a couple of days, then cook in weak vinegar until tender, but not so long as to break them. Drain well from this, place them in jars and prepare vinegar for them in the proportion of an ounce each of cloves, allspice and black pepper to a gallon of vinegar; scald all these together with half a teaspoonful of prepared mustard. Pour hot over the martynias, cover closely and keep in a cool place. They will soon be ready for use.

# 35 Duck Potato

Family: Alismataceae (water-plantain family)

## NAMES

Scientific name: *Sagittaria latifolia* Willd.

- The name "duck potato" for *S. latifolia*, as well as some other species of the genus, is based on the fact that ducks commonly eat the tubers.
- Occasionally, the name "duck potato" is used for chufa (see Chapter 28).
- Duck potato is also known as American arrowhead, arrowhead, arrowleaf, broadleaf arrowhead, common arrowhead, Indian potato (a phrase applied to several unrelated species), Indian swamp potato, muskrat potato, swamp potato, swan potato, wapato, and wapatoo.
- "Wapato" is the Chinook jargon trade language word for potato (the Chinook term is derived from the Cree *wapatowa*, meaning "white mushroom"). This pidgin language was used along the western North American coast from the California/Oregon border to the Alaskan Panhandle, at least since contact was first made with Europeans. The word was used both for duck potato and potato. Chinook jargon includes vocabulary from indigenous native languages of the area as well as French and English, and should not be confused with native American Chinookan languages.
- In the lower Sacramento Valley of California, duck potato has been called Tule potato and Tule root. The word "tule" traces to the Aztec word "tullin" or "tollin," meaning a group of aquatic plants (especially *Scirpus* or bulrush species). The word entered into Spanish and came to be applied to marshy land.
- The genus name *Sagittaria* is based on the Latin *sagitta*, an arrow, + *aria*, like, a reference to leaf shape of several of the species. Some, but not necessarily all, of the leaves of duck potato are indeed arrow shaped (the leaf blade appearing like an arrowhead and the leaf stalk like the shaft of an arrow).
- *Latifolia* in the scientific name is Latin for broad-leaved.

## GEOGRAPHY AND ECOLOGY OF WILD PLANTS

Duck potato is native to most of the eastern and several western states of the United States, all of the provinces of Canada, and extends southward to South America. It is also present in Hawaii. It is the most common species of *Sagittaria* found in eastern North America and has been introduced to Europe and Australia; it is an invasive weed in parts of these three continents.

The species commonly grows in swamps, bogs, ditches, the edges of lakes, ponds, and stream margins. It is found in muddy soils as well as in shallow, still, or slow-moving water. It thrives in water 15–30 cm (6–12 inches) deep.

## PLANT PORTRAIT

The genus *Sagittaria* includes about two dozen species of marsh or aquatic herbs with white, buttercup-like flowers. Similar to corn, the upper flowers are often male, and the lower ones are frequently female. There are over a dozen species in North America. The tubers of the Eurasian

**FIGURE 35.1** Duck potato (*Sagittaria latifolia*). (a) Flowering branch, mature arrow-shaped leaf, and cluster of fruits. (From New York Botanical Garden, *Addisonia*, Vol. 2, Plate 54, 1917.) (b) Intact tubers (above) and sectioned tubers (below). (Courtesy of E. Small.) (c) Flowering plant. (Courtesy of M. Ryon, Oak Ridge National Laboratory.) (d) Distribution map.

*S. sagittifolia* L. (*S. trifolia* L.) and the North American *S. latifolia* have been widely consumed. (Tubers are swollen underground plant stems, and are storage organs for plants, most familiarly represented by potato.) *Sagittaria cuneata* E. Sheldon (arum arrowhead or arumleaf arrowhead) is another widespread species of North America that is collected and used as "duck potato." In addition to their food use, arrowheads are often grown in aquaria, ponds, and bog gardens.

Duck potato is a perennial herb with deciduous (i.e., annual) arrow-shaped aerial leaves 10–30 cm (4–12 inches) long and 8–20 cm (3–5 inches) wide. White flowers 3–4 cm (1.2–1.6 inches) across are produced on the flower stalks, which grow as high as 1.5 m (5 feet). Some plants have male flowers only, some female flowers only, and some both kinds of flowers. The plant produces tubers in shallow waters, often up to a meter (about a yard) away from the main plant stalk, on the ends of the plant's rhizomes (horizontally creeping underground stems). The tubers have a purplish skin.

A single plant can produce up to 40 tubers. The larger tubers can be up to 5 cm (2 inches) across, and they contain an unpleasant-tasting milky juice.

Duck potato is very widely planted to provide habitat, food, and cover for wildlife, particularly waterfowl. It is also grown to promote aquatic environment restoration.

Native Americans used the plant medicinally: the rhizome was ground and used as a tea for indigestion, the pulverized rhizomes were also used in poultices for wounds and sores, a tea made from the leaves was used to relieve rheumatism and to wash babies with fever, and ground leaves were used in a poultice to stop milk production.

## WARNINGS

Wild food enthusiasts collecting duck potato from aquatic habitats need to be extremely careful not to accidentally harvest poisonous water plants, or to collect from polluted waters. Arrowhead tubers can cause dermatitis in people with sensitive skin.

## CULINARY PORTRAIT

Native North Americans prepared a potato-like food by roasting or broiling the starchy tubers of arrowhead species, particularly of *S. latifolia*. Duck potato was in fact a staple food for many Indian tribes. On boiling or roasting, the tubers become pleasant in taste. Some Native North Americans sliced boiled tubers and strung then on twine to dry and store for winter use (dried tubers store about as well as potatoes). According to Williamson (1995, cited in the Appendix to this book), "In the mid to late 1800s the Chinese, who worked the mining areas of the Central Rockies, not only harvested wild arrowhead tubers but cultivated them, as well. They believed that the tubers bore a resemblance in texture and flavor to the water chestnuts of their homelands." The tubers can be eaten raw (although they may taste unpleasant or bitter) or cooked (boiled, fried, or roasted). The taste has been likened to potatoes, chestnuts, and corn. The dried tubers can be ground into a flour.

### CULINARY VOCABULARY

- Duck potatoes are "tubers," but this term is widely used inconsistently. In a restricted sense, the term tuber refers to underground (sometimes above ground), fleshy storage organs that are formed from stem tissue, and so can easily develop new shoots (e.g., potato, which can produce new shoots [and roots connected to the shoots] from "eyes" [buds] on the surface; many yams [*Dioscorea* species] similarly produce stem tubers). However, similar structures formed from roots are also sometimes called tubers (e.g., sweet potato, *Ipomoea batatas* (L.) Lam.), but these typically do not produce new shoots, serving exclusively for food storage. Similarly, "tuberous roots" (as present in dahlia, i.e., species of *Dahlia*) are simply elongated storage roots that are relatively enlarged with storage tissue but do not form swollen compact structures like tubers. The tubers of duck potato are sometimes incorrectly referred to as "corms" (they have also been characterized as "corm-like tubers"). Corms like tubers are made up of swollen stem tissue, have eyes from which shoots develop, and contain energy stores (usually starch) to grow the plant anew in the spring, but unlike tubers, they have a distinctive basal area from which new roots will arise (the roots and shoots of corms thus arise from different places, whereas in tubers they both arise from the eyes).

## PROSPECTS

Duck potatoes are sometimes available in stores, but rarely. They are in the main a wild-collected food. The plants need a shallow-water or muddy soil environment to grow and therefore, harvest is not easy. This is a wild crop, and so domestication could considerably improve its marketability.

Because duck potato is widely planted to establish or maintain shallow-water aquatic habitats, there already is considerable management information available. The rather similar *S. sagittifolia* is widely used as food in Asia. The great difficulty with numerous "root crops" is that they need to compete with the king of this category, the potato, and the more similar the taste is to potatoes, the more difficult it is to gain a market foothold. Nevertheless, duck potatoes have a genuinely attractive taste, the plants are productive, and there is at least a possibility of it acquiring a market niche in the future.

## CURIOSITIES OF SCIENCE AND TECHNOLOGY

- Captain Meriwether Lewis (see Chapter 71 on Pawpaw for a description of the historic expedition of Lewis and Clark of 1804–1806), while in Oregon, recorded how Indian women collected duck potato using light, shallow canoes: "She takes one of these canoes into a pond where the water is as high as the breast, and by means of her toes, separates from the root this bulb, which on being freed from the mud, rises immediately to the surface of the water and is thrown into the canoe. In this manner, these patient females remain in the water for several hours, even in the depth of winter." The Lewis and Clark team had to subsist on wild root vegetables during the winter of 1805–1806 and liked the duck potato best of those they ate. Meriwether Lewis wrote, "They are nearly equal in flavor to the Irish potato and afford a very good substitute for bread."
- Because the tubers are difficult to collect, some Native Americans searched out and appropriated the large stores of duck potatoes collected by beavers and muskrats.
- A southeastern U.S. species of duck potato, *S. lancifolia* L., is native to the coastal plain from Florida to Texas. It was traditionally used by the Seminole Indians for treating shock following alligator bites.
- The diet of ducks differs greatly and depends in part on whether they "dabble" or "dive," with dabblers tending to feed on the surface or in shallow water, whereas divers have access to food in deeper water. Dabbling ducks (e.g., American black duck, mallard, green-winged teal, wood duck, Northern shoveler, Northern pintail, gadwall, European widgeon, and American widgeon) feed at the water surface or by submerging their head and pitching their tail up. They also feed on land in open areas, and several dabblers have bills adapted for straining tiny items out of muddy water. Their diet is diverse, including seeds of aquatic plants, crops, and small aquatic animals. Diving ducks (e.g., ring-necked ducks, common goldeneyes, buffleheads, hooded mergansers, common mergansers, greater scaups, lesser scaups, canvasbacks, and ruddy ducks) feed by diving from the surface and swimming underwater. Mergansers are ducks that specialize in eating fish, for which they have serrated bills to grasp them. The main diet of diving ducks consists of aquatic vegetation, including duck potato, that tends to grow in deep water, as well as aquatic insects, fish, and molluscs.
- *Sagittaria sagittifolia*, a species widely used as food as mentioned earlier, is one of the aquatic plants that are listed on the U.S. *Federal Noxious Weed List* because it has proven to be a troublesome introduced weed in southern areas of the United States. The U.S. Congress passed the Plant Protection Act, which provides for penalties for knowing violations, such as moving a federal noxious weed without a permit. Civil penalties range from $1,000 to $250,000 per violation. Many independent dealers, pet shops, and even some mail order firms that sell aquatic plants for use in aquaria and ornamental ponds are unaware that the plant is on the list.

## KEY INFORMATION SOURCES

Bloedel, C.A., and Hirsch, A.M. 1979. Developmental studies of the leaves of *Sagittaria latifolia* and their relationship to the leaf-base theory of monocotyledonous leaf morphology. *Can. J. Bot.* 57: 420–434.

Bogin, C. 1955. Revision of the genus *Sagittaria* (Alismataceae). *Memoirs N.Y. Bot. Gard.* 9: 179–233.

Calhoun, A., and King, G.M. 1998. Characterization of root-associated methanotrophs from three freshwater macrophytes: *Pontederia cordata*, *Sparganium eruycarpum*, and *Sagittaria latifolia*. *Appl. Environ. Microbiol.* 64: 1099–1105.

Darby, M.C. 2005. The intensification of wapato (*Sagittaria latifolia*) by the Chinookan People of the Lower Columbia River. In *Keeping it living: traditions of plant use and cultivation on the northwest coast of North America*. Edited by D. Deur and N.J. Turner. University of Washington Press, Seattle, WA. pp. 194–217.

Delesalle,V.A., and Blum, S. 1994. Variation in germination and survival among families of *Sagittaria latifolia* in response to salinity and temperature. *Int. J. Plant Sci.* 155: 187–195.

Dorken, M.E., and Barrett, S.C.H. 2003. Gender plasticity in *Sagittaria sagittifolia* (Alismataceae), a monoecious aquatic species. *Plant. Syst. Evol.* 237: 99–106.

Dorken, M.E., and Barrett, S.C.H. 2003. Life-history differentiation and the maintenance of monoecy and dioecy in *Sagittaria latifolia* (Alismataceae). *Evolution* 57: 1973–1988.

Dorken, M.E., and Barrett, S.C.H. 2004. Chloroplast haplotype variation among monoecious and dioecious populations of *Sagittaria latifolia* (Alismataceae) in eastern North America. *Molec. Ecol.* 13: 2699–2707.

Dorken, M.E., and Barrett, S.C.H. 2004. Phenotypic plasticity of vegetative and reproductive traits in monoecious and dioecious populations of *Sagittaria latifolia* (Alismataceae): a clonal aquatic plant. *J. Ecol.* 92: 32–44.

Dorken, M.E., Friedman, J., and Barrett, S.C.H. 2002. The evolution and maintenance of monoecy and dioecy in *Sagittaria latifolia* (Alismataceae). *Evolution* 56: 31–41.

Glaettli, M., and Barrett, S.C.H. 2008. Pollinator responses to variation in floral display and flower size in dioecious *Sagittaria latifolia* (Alismataceae). *New Phytol.* 179: 1193–1201.

Gordon, E. 1996. Dispersal type, germination and seedling growth of *Sagittaria latifolia* (Alismataceae). *Frag. Flor. Geobot.* 41: 657–668. (In Spanish)

Haynes, R.R., and Hellquist, C.B. 2000. *Sagittaria*. In *Flora of North America north of Mexico*. Vol. 22. Edited by Flora of North America Editorial Committee. Oxford University Press, New York, NY. pp. 11–23.

Haynes, R.R., and Holm-Nielsen, L. 1994. The Alismataceae. *Flora Neotropica Monogr.* 64: 95–98.

Hunter-Cario, L. 2007. Propagation protocol for broadleaf arrowhead *Sagittaria latifolia* Willd. (Alismataceae). *Native Plants J.* 8: 80–83.

Kane, M.E., Gillis, M.R., Philman, N., and Campbell, S. 2000. Seasonal differences in ex vitro growth and corm formation between two micropropagated *Sagittaria latifolia* ecotypes. *Acta Hortic.* 520: 229–237.

Lerman, M.M. 1989. Arrowhead, or duck potato. *Minn. Horticulturist* 108(7): 207.

Marburger, J.E. 1993. Biology and management of *Sagittaria latifolia* Willd. (broad-leaf arrow-head) for wetland restoration and creation. *Restoration Ecol.* 1: 248–255.

Rogers, G.K. 1983. The genera of Alismataceae in the Southeastern United States. *J. Arnold Arb.* 64: 383–420.

Sarkissian, T.S., Barrett, S.C.H., and Harder, L.D. 2001. Gender variation in *Sagittaria latifolia* (Alismataceae): is size all that matters? *Ecology* 82: 360–373.

Wooten, J.W. 1971. The monoecious and dioecious conditions in *Sagittaria latifolia* L. (Alismataceae). *Evolution* 25: 549–553.

Zepeda, C., and Lot, A. 1999. Acuitlacpalli or *Sagittaria macrophylla* (Alismataceae): a Mexican endemic hydrophyte and a threatened food resource. *Econ. Bot.* 53: 221–223.

## Specialty Cookbooks

Williamson (1995) has 18 recipes for preparing duck potato, Boorman (1969) has 1, Freitus (1980) has 5, Knutsen (1975) has 5, Johnson (1989) has 1, Hall and Hall (1980) have 5, and Hunt (1977) has 13 (See the Appendix to this book for details of the publications cited.).

# 36 Dwarf Cape Gooseberry

Family: Solanaceae (potato family)

## NAMES

Scientific name: *Physalis grisea* (Waterf.) M. Martínez (*P. pruinosa* of some authors)

- "Dwarf" in the name dwarf cape gooseberry is a comparison of stature (and fruit size) with the larger cape gooseberry, *P. peruviana* L. (*P. edulis* Sims). Neither of the cape gooseberries is a true gooseberry (i.e., a species of *Ribes*) nor a native of the cape. As noted in Chapter 66 on Northern Gooseberry, "gooseberry" has also been used to label other fruits quite unrelated to the true gooseberry. The name cape gooseberry was coined by Australians who imported the cape gooseberry from South Africa. The cape gooseberry was grown by early settlers at the cape of Good Hope (more generally in the Cape Peninsula) at least by the early nineteenth century.
- Unfortunately, common names are rather confused for *Physalis* species used for their edible fruit. Such species are often called ground cherry, husk tomato, Andean cherry, cap berry, and cap gooseberry, and it is often not possible to discern the identity of the species from these names alone.
- *Physalis grisea* is also called hairy ground cherry, a name also used for *P. pubescens* L., mentioned below (in Plant Portrait). "Strawberry tomato" seems to be becoming popular, but as noted below, this is also applied to *P. pubescens*. The name "Aunt Molly's ground cherry" is sometimes applied to the species, but is also often represented as a cultivar of the species. "Cossack pineapple" appears to be a recent name introduced in the trade.
- "Golden berry" is sometimes used for dwarf cape gooseberry, but is much better established for cape gooseberry, and seems to be based on confusion with the latter.
- The genus name *Physalis* is based on the Greek *physa*, a bladder, so-named for the husk that surrounds the fruit.
- *Grisea* in the scientific name *P. grisea* is Latin for grey, for the greyish-green leaves.

## GEOGRAPHY AND ECOLOGY OF WILD PLANTS

Dwarf Cape gooseberry is native to the eastern United States, with scattered introductions on the West coast and southeast of the United States. It is most likely to be encountered growing as a weed in waste places, old fields, and along roadsides, usually in open habitats, often in sandy (well-drained) soils.

## PLANT PORTRAIT

There are almost 100 species of the genus *Physalis*, most of which are native to the New World. Some are grown for their small edible fruits, while others are cultivated as ornamentals, particularly for their large, brightly colored, inflated papery husks (formed from the sepals of the flowers) that surround the fruits. Most are native to warm climates and are sensitive to cold temperatures. Some species of *Physalis* are toxic, and even unripe fruits of species grown for food may be poisonous at least to some people. Although fruits of wild species are sometimes gathered for food, this should only be done when the species are identified as edible with certainty.

**FIGURE 36.1** Dwarf cape gooseberry (*Physalis grisea*). (a)–(c) Views of fruit painted by B. Flahey: (a) Husk bracts folded back revealing fruit; (b) Sectioned fruit; (c) Left, normal fruit hidden within husk; right: two bracts removed to reveal fruit. (d) and (e) Photos by E. Small, of young fruiting plant. (d) Branch with immature, green fruit; (e) Plant. (f) Distribution map.

The cape gooseberry (*P. peruviana*) produces piquant, grape-sized fruits that are usually regarded as superior to those of the dwarf cape gooseberry. It is grown as an annual in temperate regions and a perennial in the tropics. Good crops are produced in several central and South American countries, Australia, New Zealand, China, India, Malaya, and other countries. Several cultivated varieties have been selected; one of these, 'Golden Berry', is said to produce juice that looks and tastes like orange juice.

Tomatillo (*P. philadelphica* Lam.; *P. ixocarpa* of some authors) produces fruits that are 2.5–7.5 cm (1–3 inches) in diameter, and resemble small tomatoes, except for the usual enveloping husk. These fruits are also regarded as superior to those of the dwarf cape gooseberry, and have become a staple in Mexican and some South American cuisines.

The dwarf cape gooseberry is the third most important species of *Physalis* used for food. It is a low-growing, sprawling, bushy, annual plant typically reaching 45–75 cm (18–30 inches) in height when in flower. It produces cherry-sized green to yellow-gold fruit, which drop to the ground when ripe. The husks turn brown when the fruits are mature. In gardens, well-cultivated plants have been observed to produce as much as a half bushel of fruits. It may be noted that seeds offered in the trade as dwarf cape gooseberry may be misidentified, and may actually be one of the other cultivated *Physalis* species.

Aside from *P. grisea*, *P. pubescens* (variously known as downy ground cherry, ground cherry, hairy ground cherry, husk tomato, pops, and strawberry tomato) is the only other species of *Physalis* that is native north of Mexico and has attracted some interest as a food plant. It is a native of Mexico, the United States, and South America, but has become widely naturalized in the world, and is cultivated for its edible fruit. It is grown mostly as an occasional garden vegetable, particularly in the United States but also in other countries, and is sometimes cultivated as a minor marketplace offering.

## CULINARY PORTRAIT

Dwarf cape gooseberries have a pleasing and distinctive sweet or bittersweet, acidic flavor, but the fruit is not to everyone's taste. At its best, the taste is often described as a combination of tomato and pineapple, or strawberry. The texture is similar to tomato. The fruits are most commonly used to make jam, but are also eaten raw or used in pies and other cooked desserts, stews, sauces, and preserves. They make interesting subjects when served raw in cocktails and produce good cooked sauce for cakes and puddings. One interesting serving suggestion is to coat them with chocolate. The fruit husks are inedible and should be removed before eating or cooking. A sticky substance under the husks is sometimes encountered, and can be washed off. Small seeds are present, which some consumers find slightly objectionable. Fruits are rarely found in the marketplace, and should be picked when completely ripe (otherwise the flavor is likely to be off, and some have contended that there is potential toxicity). When grown in northern areas, many unripe fruits are likely to be available at the end of the season, and these are best discarded. Ripe fruits drop off, and gardeners commonly place a tarpaulin under the plants to keep the fallen fruit from deteriorating. The fruits can be kept in a dry atmosphere for several weeks or even longer.

### Culinary Vocabulary

- "Salsa," Spanish for sauce, has come to mean a Mexican cold sauce prepared from certain vegetables, such as cilantro, chiles, onions, and especially tomatoes. In recent times, salsa has exceeded the dollar sales of ketchup in the United States (but ketchup, which is cheaper, exceeds sales by volume). A popular variation of salsa is "salsa verde," made with tomatillos (*Physalis philadelphica*) instead of tomatoes. Other *Physalis* species could be substituted, including the dwarf cape gooseberry, at least in homemade preparations.

## PROSPECTS

Little selection of superior fruits of dwarf cape gooseberry has taken place to date, so that the species has limited market appeal. The fruit does not ship well, and if picked under-ripe to improve surviving long-distance transportation, it is very likely to suffer considerably in taste. The very similar cape gooseberry is tastier and is much more widely cultivated and marketed, providing stiff competition for dwarf cape gooseberry. The dwarf cape gooseberry has limited prospects of becoming a significant crop, but is nevertheless an interesting subject for research and development and it could become economically attractive in the future.

## CURIOSITIES OF SCIENCE AND TECHNOLOGY

- An ancient Indian camp site, Shawnee-Minisink located on a terrace of the Delaware River near Stroudsburg, Pennsylvania has been found by the radiocarbon dating method to have been occupied some time between 9700 and 9500 years ago. *Physalis* fruits were recovered from several hearths, indicating that the fruits have been used by people in North America for most of the time that the continent has been populated.
- The subject of whether tomatoes are a fruit (which they are botanically) or a vegetable (which they are in supermarkets) has been much debated, and a similar argument can be made for *Physalis* fruits. For trade purposes, the European Union has ruled that all *Physalis* berries are classified as "fruit," although some species are consumed as "vegetables."
- Girls in Japan once fashioned *Physalis* fruits into a kind of whistle.

## KEY INFORMATION SOURCES

Dremann, C.C. 1985. *Ground cherries, husk tomatoes, and tomatillos*. Redwood City Seed Co., Redwood City, CA. pp. 22.

Fisher, B. 1977. "Instant" fruit—it's hard to ask more [including *Physalis grisea*]. *Org. Garden. Farm*. 24(6): 74–75.

Henry, R.D. 1990. Some embryological observations of *Physalis pruinosa* L. Solanaceae. *Phytomorphology* 40: 309–318.

Mahmoud, M.H. 1995. Seed germination of *Physalis pruinosa* (L.) as affected by salinity. *Egypt. J. Hortic*. 22(2): 127–135.

Martinez, M. 1993. The correct application of *Physalis pruinosa* L. (Solanaceae). *Taxon* 42: 103–104.

Menzel, M.Y. 1951. The cytotaxonomy and genetics of *Physalis*. *Proc. Amer. Phil. Soc*. 95: 132–183.

Menzel, M.Y. 1957. Cytotaxonomic studies of Florida coastal species of *Physalis*. Yrbk. *Amer. Phil. Soc*. 1957: 262–266.

Quiros, C.F. 1984. Overview of the genetics and breeding of husk tomato. *HortScience* 19: 872–874.

Raghava, R.P., and Raghava, N. 1994. Enhancement of fruit yield of husk tomato by seed-applied plant growth regulators and gamma rays. *Proc. Nat. Acad. Sci. India Sect. B*. 64: 305–309.

Waterfall, U.T. 1958. A taxonomic study of the genus *Physalis* in North America north of Mexico. *Rhodora* 60: 107–114.

Waterfall, U.T. 1967. *Physalis* in Mexico, Central America, and the West Indies. *Rhodora* 69: 82–120.

### SPECIALTY COOKBOOKS

Dwarf cape gooseberries are rarely specified in recipes, but can be substituted for cape gooseberries, which are widely used in the cuisines of the world, but particularly so in Mexican cooking. A selection of cookbooks with recipes for cape gooseberries follows:

Curtis, S., Hoyer, D., and Smith, R.A. 2000. *Tacos*. Gibbs Smith, Salt Lake City, UT. pp. 44.

Macpherson, N. 1954. Cape gooseberry recipes. *New Zeal. J. Agr*. 88(4): 401–404.

Padmanabhan, C. 1999. *Dakshin: vegetarian cuisine from South India*. Periplus, Singapore. pp. 176.

Reader's Digest Association. 2007. *The ultimate soup cookbook*. Reader's Digest, Pleasantville, NY. pp. 544.

Schmidt, A., and Nam, I. 1996. *The book of hors d'oeuvres and canapes*. Wiley, Somerset, NJ. pp. 320.

Wolf-Cohen, E. 2000. *The book of dips and salsas*. HP Trade, New York. pp. 96.

Russell (1975) has 3 recipes for fruit from wild-collected *Physalis* species, Mogelon (2001) has 1, and Niethammer (1999) has 2. Hall and Hall (1980) have 10 recipes for "ground cherry," which can be adapted for dwarf cape gooseberry. (See the Appendix to this book for details of the publications cited.)

# 37 Dwarf Glasswort

Family: Amaranthaceae (amaranth family; until recently, generally placed in the Chenopodiaceae, the goosefoot family)

## NAMES

Scientific name: *Salicornia bigelovii* Torr.

- The ashes of *Salicornia* plants were once widely used in glassmaking, hence the name "glasswort." The ashes are high in soda ash (mainly sodium carbonate), once a principal starting material for glass. This usage ceased in the first half of the nineteenth century, with the introduction of the (now outmoded) LeBlanc process for industrial production of glass, which used soda ash produced from chemicals.
- The occasional name "annual glasswort" points out the annual nature of *S. bigelovii* (most species of *Salicornia* are perennial).
- "Sea bean" is another name sometimes encountered for *S. bigelovii*. This is an allusion to the fruits ("beans").
- The names "pickleweed" and "Bigelow's pickleweed" are based on the appearance of the jointed stems, resembling strings of tiny pickles.
- "Drift-seed" is another name for the plant, pointing out that the seeds are dispersed by drifting away on water.
- Rock samphire (*Crithmum maritimum* L. of the Apiaceae or carrot family) is a small, perennial shrub native to European seasides, which is sometimes confused with *Salicornia* species in Europe because it was once similarly used as a vegetable. However, it has an unappealing aroma of varnish, and is now rarely encountered as a food plant.
- The genus name *Salicornia* is based on the Latin *sal*, salt + *cornu*, a horn, allusions to the salty habitats occupied and the horn-like appearance of the stems.
- *Bigelovii* in the scientific name commemorates the discoverer of *S. bigelovii*, American botanist James Bigelow (1787–1879).

## GEOGRAPHY AND ECOLOGY OF WILD PLANTS

Dwarf glasswort occurs as scattered populations in temperate and subtropical regions on the East and West coasts of North America, including the Gulf of Mexico and the Atlantic seaboard. The East coast distribution includes several Caribbean islands (Anguilla, Bahamas, Cuba, Guadeloupe, Hispaniola, Netherlands Antilles, and Puerto Rico), the Yucatan Peninsula of Mexico, coastal United States from Texas to Maine, and Nova Scotia in Canada. On the West coast of North America, the plant has been collected from Baja California Norte of Mexico and from California. *Salicornia bigelovii* thrives in moist, saline habitats. The plants grow in the intertidal zones of estuarine salt marshes and in interior saline basins. That is, it is a terrestrial plant, the base often submersed in salty water for part of the day. The species is best adapted to subtropical coastal deserts. It colonizes mud flats and tidal wetlands, establishing by prolific seed production. Dwarf glasswort uses the C4

**FIGURE 37.1** Dwarf glasswort (*Salicornia bigelovii*). (a) Plant growing on a beach in Texas. (Courtesy of Alan Cressler, reproduced with permission.) (b) Fuzzy seeds, resembling puffed wheat cereal. (c) Plant in fruit. ([b] and [c] Courtesy of J. Macdonald, Rancho Santa Ana Botanic Garden.) (d) Distribution map.

mechanism of photosynthesis, which is common in tropical and subtropical plants, adapting them to efficient use of very high levels of sunlight.

## PLANT PORTRAIT

Dwarf saltwort, is an annual, erect, branching, fleshy herb, 30–90 cm (1–3 feet) tall. The succulent stems are "articulated" or "jointed"; the joints are the nodes where the leaves arise. The stems have short internodes (the portions between the nodes). The scale-like leaves are fused into a band around the stem, but are so rudimentary that the plant appears to be leafless. Branches and the very small leaves occur in opposite pairs on the stem. The plants are usually green in early growth, but frequently become quite red in the autumn. The flowers are tiny and as inconspicuous as the leaves; they occur in clusters of about three, and appear to be sunken into the stem joints. The plants are wind-pollinated. The fruit spikes occur in the upper part of the plant. The fruit is small, succulent, and contains a single fuzzy, ovoid seed 1–1.5 mm (0.04–0.06 inch) long.

## CULINARY PORTRAIT

There is hope that extracted oil from dwarf glasswort can be developed into a new major oilseed crop for human consumption, much like the oil harvest from soybean, sunflower, canola (rapeseed), and corn. The oil is highly polyunsaturated (about 90%), and similar to safflower oil in fatty acid composition. More than half of the oil in the seeds can be linoleic acid, an essential dietary fat precursor considered extremely valuable in the human diet. The oil has a pleasant, nutlike taste and a texture similar to olive oil, and it can be used as a cooking oil. Although this edible oil application could transform dwarf glasswort into a very important food plant, its current food use is as a very minor vegetable.

A relative of *S. bigelovii*, *S. europaea* L., usually known simply as salicornia, is a popular, traditional edible plant in Europe. (It is also called marsh samphire, pickleweed, saltwort, sea asparagus, sea bean, sea pickle, and sometimes Seaphire, a commercial name associated with Seaphire International, which produces *Salicornia* crops.) In recent times, it has become a cultivated vegetable, available as a specialty item in some supermarkets, and the development of this crop is proceeding rapidly in both Europe and North America. *Salicornia europaea* has become established as an introduced plant in parts of North America. It is sometimes served as part of specialty dishes in the finest restaurants. Although occasionally added raw to salads, salicornia is usually cooked (often steamed or microwaved), coated in butter, and frequently served with fish or seafood. The flavor and texture resemble young spinach or asparagus. Small twigs that are bright green and firm, not flabby, dark, or slimy, are best. After a week, salicornia starts to lose flavor, so it should be consumed promptly. Fresh twigs can be boiled for 4 minutes at most (cooking for longer deteriorates the taste). The plant has a peculiar, salty taste, and indeed can replace the salt in dishes to which it is added. Because of the high salt content, there is no need to add salt as a condiment for this vegetable. Soaking for 12 hours in cold water in the refrigerator will take away some of the saltiness. Salicornia goes well with sea food in general, many fish dishes, particularly smoked salmon, and vegetables. It also can be added to stews, or cooked like asparagus or green beans and served with butter. The plant is also pickled in vinegar and served as an hors d'oeuvre.

Dwarf glasswort can be prepared and served like salicornia, described above. Indeed, in Mexico and China, selected strains of *S. bigelovii* are being grown and sold for consumption in the same way as salicornia. The stems have a short shelf life, generally about 6 days, but this can be extended by storage at 2°C (36°F). As with conventional salad and stewed greens, the tastiest portions are young and tender. The plant becomes woody as it ages, with a thin woody core present in the stems. The fleshy exterior can be stripped off, but it is preferable to harvest the tips of the plant when the shoots are young, and to cut the stems short enough to exclude the tougher bottom part (just as one consumes only the upper part of the shoots of asparagus). In China, desalted, dehydrated dwarf saltwort has been produced.

There are two nutritional problems associated with glassworts. First is the presence of salt, an essential compound for metabolism, but its excess consumption in the human diet is now known to be associated with major health problems. Fortunately, the extracted seed oil is free of salt. However, the edible stems have considerable salt. The second nutritional problem is the presence of saponins, which are bitter, antinutritional chemicals, common in plants and serving to protect them against many herbivores. Consumed saponins reduce the value of several nutrients, including fats and protein, and can irritate tissues. In many domesticated crop plants, the concentration of saponins has been greatly reduced by selection, and possibly this could be done for dwarf glasswort. The bitter saponins give the fresh stem tissues and whole seeds a somewhat bitter flavor, reducing their palatability. The safety of consuming the high content of salt and saponins of fresh dwarf glasswort needs to be examined, and until this has been done, it is probably advisable to limit consumption.

### Culinary Vocabulary

- *Salicornia* species are sometimes called "seaweeds," although this term is normally reserved for large (macroscopic) marine algae. Edible seaweeds are sometimes called "sea vegetables," a phrase that nicely describes edible *Salicornia* stems. According to an anonymously authored website, "Not quite a vegetable but not quite a seaweed, salicornia must have gone through a tough identity crisis as a teenager."

## PROSPECTS

Dwarf glasswort can be grown in areas that are quite unsuitable for conventional crops—extreme coastal desert environments, inland regions that have soils that are naturally high in salt, and agriculturally exploited lands that have been very seriously degraded by salinization (deposition of salts in soil resulting from inappropriate irrigation). Moreover, the plant can be irrigated with brackish water or seawater, saving precious freshwater resources. The world has virtually run out of land and freshwater that can be used for growing foodstuffs, and the prospect of expanding food and feed production is extremely attractive. In recent decades, several research and pilot commercial initiatives have been attempting to develop dwarf glasswort into a viable biofuel and livestock feed crop, particularly in the Middle East, Africa, Mexico, and the United States. These efforts have had limited success to date. However, the plant is still basically wild, with very limited selection of improved varieties. There is minor commercial success of dwarf glasswort as a novelty, salad plant, much like the European *S. europaea*. There also remains very good potential for developing dwarf glasswort into a new oilseed crop for human use. There are pressing needs to improve seed quality by breeding, improve management practices, and create optimal practices for commercial production. Despite these significant obstacles, dwarf glasswort has extraordinary long-term economic advantages for both food and nonfood applications, and is especially deserving of development.

## CURIOSITIES OF SCIENCE AND TECHNOLOGY

- On average, seawater contains 35 parts per thousand of sodium chloride. *Salicornia bigelovii* is a land plant, but it can thrive in soil water that has three times the saltiness of the ocean, and can grow with its roots in water that is up to six times saltier, an ability matched by almost no other land plant in the world. Its extraordinary tolerance of salty water has led to efforts to develop it as a new crop that can be irrigated with seawater.
- Up to 30% of the dry weight of *S. bigelovii* can be salt that the plant has accumulated.
- Experimental particle board has been manufactured using the straw from dwarf glasswort plantations. The high salt content proved useful as a fire retardant and as a termite repellent. (Reference: Dutton, G. 2000. Salty food stymies termites. *Popular Science*, 257 [2; August]: 51.

## KEY INFORMATION SOURCES

Alsaeedi, A.H., and Elprince, A.M. 2000. Critical phosphorus levels for *Salicornia* growth. *Agron. J.* 92: 336–345.

Anwar, F., Bhanger, M.I., Nasir, M.K.A., and Ismail, S. 2002. Analytical characterization of *Salicornia bigelovii* seed oil cultivated in Pakistan. *J. Agric. Food Chem.* 50: 4210–4214.

Ayala, F., and O'Leary, J.W. 1995. Growth and physiology of *Salicornia bigelovii* Torr. at suboptimal salinity. Int. *J. Plant Sci.* 156: 197–205.

Bashan, Y., Moreno, M., and Troyo, E. 2000. Growth promotion of the seawater-irrigated oilseed halophyte *Salicornia bigelovii* inoculated with mangrove rhizosphere bacteria and halotolerant *Azospirillum* spp. *Biol. Fertil. Soils*. 32: 265–272.

Glenn, E.P., and Brown, J.J. 1999. Salt tolerance and crop potential of halophytes. *Crit. Rev. Plant Sci.* 18: 227–255.

Glenn, E.P., Brown, J.J., and O'Leary, J.W. 1998. Irrigating crops with seawater. *Sci. Amer.* 279(2): 76–81.

Glenn, E., Miyamoto, S., Moore, D., Brown, J.J., Thompson, T.L., and Brown, P. 1997. Water requirements for cultivating *Salicornia bigelovii* Torr. with seawater on sand in a coastal desert environment. *J. Arid Environ.* 36: 711–730.

Glenn, E.P., O'Leary, J.W., Watson, M.C., Thompson, T.L., and Kuehl, R.O. 1991. *Salicornia bigelovii* Torr.: an oilseed halophyte for seawater irrigation. *Science* 251: 1065–1067.

Grattan, S.R., Benes, S.E., Peters, D.W., and Diaz, F. 2008. Feasibility of irrigating pickleweed (*Salicornia bigelovii* Torr.) with hyper-saline drainage water. *J. Environ. Qual.* 37(5, Suppl. S): S149–S156.

Kadereit, G., Ball, P., Beer, S., Mucina, L., Sokoloff, D., Teege, P., Yaprak, A.E., and Freitag, H. 2007. A taxonomic nightmare comes true: phylogeny and biogeography of glassworts (*Salicornia* L., Chenopodiaceae). *Taxon* 56: 1143–1170.

Kudo, N., and Fujiyama, H. 2010. Responses of halophyte *Salicornia bigelovii* to different forms of nitrogen source. *Pedosphere* 20: 311–317.

Lu, Z., Glenn, E.P., and John, M.M. 2000. *Salicornia bigelovii*: an overview. Int. *News Fats Oils Related Mat.* 11: 418–423.

Lu, D., Zhang, M., Wang, S., Cai, J., Zhou, X., and Zhu, C. 2010. Nutritional characterization and changes in quality of *Salicornia bigelovii* Torr. during storage. *Food Sci. Technol.* 43: 519–524.

Rivers, W.G., and Weber, D.J. 1971. The influence of salinity and temperature on seed germination in *Salicornia bigelovii. Physiol. Plant.* 24: 73–75.

Rozemal, J., and Flowers, T. 2008. Crops for a salinized world. *Science* 322: 1478–1480.

Rueda-Puente, E., Castellanos, T., Troyo-Diéguez, E., Diaz De León-Alvarez, J.L., and Murillo-Amador, B. 2003. Effects of a nitrogen-fixing indigenous bacterium (*Klebsiella pneumoniae*) on the growth and development of the halophyte *Salicornia bigelovii* as a new crop for saline environments. *J. Agron. Crop Sci.* 189: 323–332.

Rueda-Puente, E.O., Garcia-Hernández, J.L., Preciado-Rangel, P., Murillo-Amador, B., Tarazón-Herrera, M.A., Flores-Hernández, A., Holguin-Peña, J., et al. 2007. Germination of *Salicornia bigelovii* ecotypes under stressing conditions of temperature and salinity and ameliorative effects of plant growth-promoting bacteria. *J. Agron. Crop Sci.* 193: 167–176.

Small, E. 2010. Blossoming treasures of biodiversity. 32. Dwarf saltwort—reducing the environmental salt threat. *Biodiversity* 11(3 & 4): 65–72.

Somers, G.F. (Editor). 1975. *Seed-bearing halophytes as food plants: proceedings of a conference, University of Delaware, June 10–11, 1974*. College of Marine Studies, University of Delaware, Newark, DE. pp. 156.

Stumpf, D.K., Prisco, J.T., Weeks, J.R., Lindley, V.A., and O'Leary, J.W. 1986. Salinity and *Salicornia bigelovii* seedling establishment water relations. *J. Exper. Bot.* 37: 160–169.

Troyo-Diéguez, E., Ortega-Rubio, A., Maya, Y., and León, J.L. 1994. The effect of environmental conditions on the growth and development of the oilseed halophyte *Salicornia bigelovii* Torr. in arid Baja California Sur, Mexico. *J. Arid Environ.* 28: 207–213.

Webb, K.L. 1966. NaCl effects on growth and transpiration in *Salicornia bigelovii*, a salt marsh halophyte. *Arch. Plant Soil.* 24: 261–268.

Weete, J.D., Rivers, W.G., and Weber, D.J. 1970. Hydrocarbon and fatty acid distribution in the halophyte, *Saliconria bigelovii. Phytochemistry* 9: 2041–2045.

Zerai, D.B., Glenn, E.P., Chatervedi, R., Lu, Z.J., Mamood, A.N., Nelson, S.G., and Ray, D.T. 2010. Potential for the improvement of *Salicornia bigelovii* through selective breeding. *Ecol. Engineering*. 36: 730–739.

## Specialty Cookbooks

Cookbooks dedicated to glasswort are not available, but recipes for the very similar salicornia may be adopted. Occasional recipes are presented in European cookbooks, and in cookbooks dealing with wild food plants of North America. Szczawinski and Turner (1980) present recipes entitled Seaside potato salad, Glasswort omelette, Glasswort-sardine salad, and Glasswort pickles. Freitus (1980) presented a recipe entitled beach asparagus salad for salicornia ("beach asparagus was the name he used for salicornia). (See the Appendix to this book for details of the publications cited.) Dwarf glasswort is easily prepared as a vegetable like salicornia, and reliable recipes for the latter can simply be found on the Web, occasionally under the name salicornia, but much more often under the names marsh samphire and glasswort.

# 38 Elder

Family: Adoxaceae (moschatel family; *Viburnum* is widely placed in the Caprifoliaceae, the honeysuckle family, but recent work has shown that the Adoxaceae is a better choice)

## NAMES

Scientific name: *Sambucus nigra* L.

- European elder—*S. nigra* L. subsp. *nigra* (these plants have been recognized as the sole component of the species *S. nigra* L. in most literature, but increasingly this group is treated as one of several subspecies of *S. nigra*). This subspecies is not native to North America, where it is cultivated as an ornamental and as a minor fruit source.
- American elder—*S. nigra* subsp. *canadensis* (L.) R. Bolli (recognized as the species *S. canadensis* L. in most literature, but increasingly treated as a subspecies of *S. nigra*).
- Blue elder—*S. nigra* subsp. *cerulea* (Raf.) R. Bolli (recognized as the species *S. cerulea* Raf. in most literature, but increasingly treated as a subspecies of *S. nigra*; the "Mexican elder," "*S. mexicana*" of some authors, is a synonym; so is *S. glauca* Nutt.).
- The name "elder" comes from the Old English *aeld* and several variants of this. *Aeld* meant "fire," the name associated with the elder apparently because the pith of the young elder branches was removed and the hollow tubes thus formed were used to blow up fires.
- The European elder is called the bour tree ("bore tree") in Scotland, because of the ease with which the pith may be bored out of the stems, leaving a hollow tube.
- The American elder has also been called blackberry elder, Canadian elder, common elder, common elderberry, elderberry, elder-blow, pie elder, and sweet elder.
- The blue elder has also been called Arizona elderberry, blue elderberry, blueberry elder, hairy blue elderberry, New Mexican elderberry, and velvet-leaf elder.
- "Poison elder" is *Toxicodendron vernix* (L.) Kuntze (*Rhus vernix* L., *R. venenata* DC.). This is not an elder, but rather a relative of poison ivy, which is also called swamp sumac, poison ash, poison dogwood, and poison sumac. It occurs in swamps from Canada to the southern United States, and causes severe rashes.
- The genus name *Sambucus* is derived from the classical Latin *sambucus*, the name of the plant, and more distantly from the Greek *sambuca* or *sambuce*. It has been contended that these ancient names were based on the sackbut, an archaic stringed musical instrument, often incorrectly stated to be a wind instrument, much used among the classical Romans, and allegedly constructed of wood of the European elder. This interpretation has been challenged on the basis that the bark of the elder is easily removed in tubes and has been used to manufacture wind instruments such as flutes and whistles since ancient times, and it would seem more plausible that such instruments, rather than a stringed instrument, would have commemorated the elder.
- *Nigra* in the scientific name *S. nigra* means black, a reference to the color of the berries; *canadensis* in the scientific name *S. nigra* subsp. *canadensis* means Canadian; *cerulea* in the scientific name *S. nigra* subsp. *cerulea* is the classical Latin word for blue (the spelling *caerulea* is often encountered, and is arguably an acceptable alternative).

**FIGURE 38.1** American elder (*Sambucus nigra* subsp. *canadensis*). (a) Shrub in flower. (Courtesy of H. Zell [CC By 3.0].) (b) Cluster of fruits. (Courtesy of H. Zell [CC By 3.0].) (c) Fruiting branches. (Courtesy of Franklin Bonner, online at Bugwood.org [CC By 3.0].) (d) Flowering branch with insert of a cluster of berries. (From Millspaugh, C.F., *American medicinal plants*, Boericke & Tafel, New York. Vol. 1, Plate 75, 1887.) (e) Distribution map.

## GEOGRAPHY AND ECOLOGY OF WILD PLANTS

The American elder and blue elder are deciduous shrubs growing naturally in North America. The American elder is native to eastern and central North America, east of the Rocky Mountains, ranging from Nova Scotia south to Florida and west to Manitoba and Texas. Blue elder is native to western North America, from southern British Columbia south to California, Arizona, and New Mexico.

Both American elder and blue elder occur in semiopen plant communities. Full sun is preferred, but American elder tolerates up to two-thirds shade. Elder habitats include swamps, woodlands, roadsides, shrubby thickets, fencerows, and the edges of rivers and streams. Other habitats occupied are open montane forests, old fields, pastures, bogs, disturbed sites, and (occasionally) upland woods and forests. The plants prefer moist, well-drained, fertile soils but can tolerate both very wet and droughty soils. Both elders grow in sandy soils, heavy clays, peats, and muck, and both tolerate a wide range of climatic conditions. In the northern part of its range, American elder withstands

**FIGURE 38.2** Blue elder (*Sambucus nigra* subsp. *cerulea*). (a) Branches with fruit. (Courtesy of Keir Morse, reproduced with permission.) (b) Shrub in flower. (Courtesy of Dave Powell, United States Department of Agriculture Forest Service, online at Bugwood.com [CC By 3.0].) (c) Distribution map.

winter temperatures down to −40°C (−40°F); in the southern part of its range, summer temperatures are as high as 38°C (100°F) and frost is rare. Blue elder in its northernmost location of British Columbia is almost as cold hardy, and in its southernmost location in Mexico (as "*S. mexicana*") it withstands hotter conditions.

## PLANT PORTRAIT

The genus *Sambucus* includes about two dozen species, found particularly in the temperate regions of the world. The European elder grows widely throughout Europe, as well as in North Africa and western Asia. It is a large shrub or tree to 10 m (33 feet) in height, differing most obviously from American elder and blue elder by often growing into a tree, and tending not to develop numerous stems at the base.

The American elder and blue elder are deciduous shrubs. They are typically 1–4 m (3–13 feet) in height in the northern (Canadian) part of their ranges. In the southern portion of the range of

American elder, in the United States, the plants can grow as tall as 9 m (30 feet), and in Florida they have been reported to be evergreen.

Individual berries of the wild elders are small, about 4–6 mm (0.16–0.25 inch) in diameter, globose, with prominent seeds. The fruits of cultivars are larger (often averaging 8 mm or about 0.3 inch in diameter) than wild fruits, and often mature earlier.

In Europe, especially in Austria, the European elderberry has become an established cultivated crop. Although reported to be cultivated during early settlement in North America (American elder in 1761, blue elder in 1850), Native North American elders were not extensively grown until the twentieth century. Berries were simply collected from the wild. Commercial crops are now produced in several states, notably Ohio, Pennsylvania, Oregon, Kansas, and New York.

Although the berries of several other elder species are sometimes harvested, *S. nigra* is the most important fruit species. The European elder has been used as a food source in Europe since antiquity. 'Haschberg' is the main European elder cultivar. Native Americans harvested the berries of American elder and blue elder long before the arrival of European settlers. American elder is the main source of elderberries in North America. Blue elder generally has larger and less acid fruit than American elder. The juice and preserves of elderberries have often been mainstays of rural American pantries, and early settlers used them as a dried and medicinal crop. However, elder was seldom cultivated. High-quality cultivars of American elder were developed in Canada and the United States in the twentieth century.

There are also ornamental cultivars, which produce masses of spectacular fragrant white flowers and abundant fruits that attract birds to the garden. Most ornamental elders bear fruits that are not as palatable as those of fruit cultivars.

In addition to their food use, elderberries have been used as a source of deep red and purple colors for dye-making (the berries usually vary in color from red to blue-black). The hollowed stems are used in various crafts, especially for constructing whistles and toys. The elder has long been used medicinally as a herbal remedy for the treatment of numerous common ailments in man and animals.

## CULINARY PORTRAIT

Elderberries have a sweet, slightly acid flavor, but are not suitable for use as an uncooked dessert fruit. These are very minor fruits, generally harvested from the wild, with limited commercial cultivation. Although the characteristic flavor is not universally appreciated, the fruit of the elderberry is especially popular for pies. Elderberries are also used for jams, jellies, juice, sauces, chutney, and homemade wine. The purple-black fruits are highly regarded for coloring wines and jams. Commercially, elderberries are usually preserved frozen. Fresh elderberries do not withstand shipment well because of their softness, and are most suitable for homegrown and local market purposes.

Several species of elder have the reputation of being poisonous, but the toxic potential is largely limited to the leaves, bark, roots, and seeds. Eating the fruit of the American and the European elders is safe, although uncooked berries are believed to cause nausea in some individuals. Occasional reports of people being poisoned by consuming elder may be due to mistaking poisonous species for elder, or mixing in leaves with the fruit. In the past, children have been occasionally poisoned by using the hollow stems to make pea shooters and whistles that they held in their mouths (such crafts are now so old-fashioned that there is little probability of this problem). As with the berries, elder flowers are used in preparing wines. Elder flowers are also occasionally fried in batter to make fritters. Some people who have consumed elder flowers have reported that diarrhea resulted (curiously, to this day, flower extracts are sometimes used in herbal medicine as a cure for constipation). Extracts from the flowers are used in some commercial flavoring preparations (note references cited later in this chapter).

Almost a century ago, the idea was proposed by that the objectionable taste of elderberries could be removed simply by drying them like raisins: "Many people are still ignorant of the fact that the berries of the common elder (*Sambucus canadensis*) make excellent pies. Others who have tasted the so-called elderberry pie were inclined to call it a nauseating mixture. The trouble is not so much in the pie itself as in the way it is put together. Pies made of fresh elderberries are scarcely likely to appeal to many palates. The fruit still retains some of the rank eldery flavor possessed by the entire plant and made evident when the stem is broken; but if one will collect the berries when fully ripe and dry them in flat trays in the sun or in a warm oven he will have a cheap and appetizing material from which to manufacture pies all winter—and pies that are not inferior to huckleberry pies in flavor. The eldery flavor seems to be dissipated by drying." (W.N. Clute in *The American Botanist*, 1905; quoted in Fernald et al. 1958 [see the Appendix to this book]).

Elderberry has recently become an important source of nutritional extracts in Europe, particularly in Austria. The Haschberg cultivar of European elderberry has been reported to have an anthocyanin content of 1500 mg per 100 g (for comparison: bilberry, 300–700; cranberry, 60–170; grape, 120; black currant, 250–600; wild European elder, 450–600). Anthocyanins have been shown to have a variety of health benefits, partly based on antioxidant activity. Antioxidants are thought to counteract the harmful effects of "free radicals" that are generated in metabolism. Free radicals are associated with damage to cells and the onset of some diseases, including cancer and cardiovascular disease, as well as aging.

## WARNING

Elderberries sold in stores will rarely cause health difficulties. However, the seeds of wild elderberries (especially red-fruited species, which should never be consumed) contain hydrocyanic acid, which can produce mild cyanide poisoning, nausea, and diarrhea, if consumed in large amounts. As noted above, the American species highlighted in this chapter are considered to be basically edible, albeit sometimes capable of causing digestive difficulties in some individuals. For wild, blue-fruited elderberries, straining out the seeds or thorough cooking of the berries is believed to substantially remove possible toxicity. It is essential that wild plants be reliably identified before consumption.

### Culinary Vocabulary

- "Baga" refers to elderberry juice once used to color port wines.

## PROSPECTS

Elderberries are very productive, easily grown plants. Although the fresh berries are not attractive in taste and do not ship well, they are useful for processing into a variety of fruit products. Domestic and export demand will likely grow for all native berry crops in response to expanding clinical evidence of health benefits of a berry-rich diet (consumers are increasingly seeking such products, that prevent rather than treat disease). With selection of better fruit varieties of the native American elders, as well as development of processed products from the berries, there is good potential for elder to become a significant North American crop, as successful as it is in Europe.

## CURIOSITIES OF SCIENCE AND TECHNOLOGY

- The use of elderberries as a dye may trace to prehistoric times. They were used by some ancient Romans to color hair black, a practice that was also known in recent centuries in Europe.
- North American Indians and pioneers used hollowed elder stems as a tap for drawing sap from sugar maple trees.
- The unpleasantly scented leaves of the elder have a reputation for repelling flies. The bridles of horses were often decorated with elder flowers in the belief that this would keep flies away. Dried, powdered leaves have often been used as an insect repellent for people

and animals. A decoction of the flowers and leaves, or an ointment containing them was once applied to large wounds to prevent visits from flies.

- A remarkable range of illnesses were treated with elder in the past. In medieval Europe, elder wood was used to cure toothache, interrupt epileptic fits, and remove poison from metal vessels. In Ireland, a necklace of nine elder twigs or berries was thought to cure epilepsy. Drinking a preparation of boiled elder root was considered to be a cure for lizard bites. Berries plucked on St. John's Day were a remedy for baldness. Elder leaves picked on the last day of April were thought to be especially good at healing wounds. Warts were believed to be cured by rubbing them with elder. In Germany, it was thought that if a patient silently stuck an elder twig in the ground, his fever would go away. This is reminiscent of a Slavic cure for fever: three elder branches were bent into an arch and the patient crawled through it. In Massachusetts, a bag of elderberry pulp carried around the neck was used to cure rheumatism. In England, an elder twig with three or four knots was carried in the pocket as a charm against rheumatism.
- In the early twentieth century, elder flowers were used to make elder flower water, a common article among every lady's toiletries to keep complexions clear and remove unwanted freckles. Elder flowers are a mild astringent (so is water!), so wrinkles would, at least temporarily, be reduced.
- Wine from elderberries has often been used as an adulterant of fine wines and misrepresented as port. In 1747, the practice had reached such alarming proportion in Portugal that the cultivation of elders was forbidden.
- The "Victorian language of flowers" was a semisecret coded language in Victorian times, with flowers and plants symbolic of certain messages, so when the flower or plant was mentioned in a letter, those who knew the code could understand the hidden information. "Elder" meant "zealousness."
- Because it is nontoxic, elderberry dye is used to mark off cuts of beef, pork, and other meat on carcasses. Dye extracted from American elder has proven to be particularly resistant to light and heat.

## KEY INFORMATION SOURCES

Atkinson, M.D., and Atkinson, E. 2002. *Sambucus nigra* L. *J. Ecol.* 90: 895–923.

Belli, R. 1994. *Revision of the genus Sambucus*. J. Cramer, Berlin, Germany. pp. 227.

Braga, F.G., Carvalho, L.M., Carvalho, M.J., Guedes-Pinot, H., Torres-Pereir, J.M., Neto, M.F., and Monteiro, A. 2002. Variation of the anthocyanin content in *Sambucus nigra* L. populations growing in Portugal. *J. Herbs Spices Med. Plants*. 9(4): 289–295.

Bridle, P., and Garcia-Viguera, C. 1996. A simple technique for the detection of red wine adulteration with elderberry pigments. *Food Chem.* 55: 111–113.

Clancy, J.A., and Maguire, J.D. 1979. Seed dormancy in blue elderberry. *J. Seed Technol.* 4: 24–33.

Craig, D.L. 1878. *Elderberry culture in eastern Canada*. Publication 1280. Canada Department of Agriculture, Ottawa, ON. pp. 6.

Finn, C.E., Thomas, A.L., Byers, P.L., and Serce, S. 2008. Evaluation of American (*Sambucus canadensis*) and European (*S. nigra*) elderberry genotypes grown in diverse environments and implications for cultivar development. *HortScience* 43: 1385–1391.

Fisher, N. 1981. Blue elderberry. *Herb. Q.* (*Wilmington*) 2(8): 18–19.

Förster-Waldl, E., Marchetti, M., Schöll, I., Focke, M., Radauer, C., Kinaciyan, T., Nentwich, I., et al. 2003. Type I allergy to elderberry (*Sambucus nigra*) is elicited by a 33.2 kDa allergen with significant homology to ribosomal inactivating proteins. *Clin. Exp. Allergy*. 33: 1703–1710.

Frost, S.W. 1977. Insects associated with the extrafloral nectaries of elderberry. *Florida Entomol.* 60: 186.

Frost, S.W. 1979. A preliminary study of North American insects associated with elderberry flowers. *Florida Entomol.* 62: 341–355.

Gottfried, R. 1993. Designer drinks. Elderberry flower extract. *Food Market. Technol.* 7(3): 9–10.

Gray, A.M., Abdel-Wahab, Y.H.A., and Flatt, P.R. 2000. The traditional plant treatment, *Sambucus nigra* (elder), exhibits insulin-like and insulin-releasing actions in vitro. *J. Nutr.* 130: 15–20.

Jensen, K., Christensen, L.P., Hansen, M., Jorgensen, U., and Kaack, K. 2001. Olfactory and quantitative analysis of volatiles in elderberry (*Sambucus nigra* L.) juice processed from seven cultivars. *J. Sci. Food Agric.* 81: 237–244.
Jorgensen U., Hansen., M., Christensen, L.P., Jensen., K., and Kaack, K. 2000. Olfactory and quantitative analysis of aroma compounds in elder flower (*Sambucus nigra* L.) drink processed from five cultivars. *J. Agric. Food Chem.* 48: 2376–2383.
Kilham, C. 2000. Health benefits boost elderberry. *HerbalGram* 50: 55–57.
Koch-Munz, M., and Holyoak, M. 2008. An evaluation of the effects of soil characteristics on mitigation and restoration involving blue elderberry, *Sambucus mexicana. Environ. Manage.* 42: 49–65.
Lee, J.M., and Finn, C.E. 2007. Anthyocyanins and other polyphenolics in American elderberry (*Sambucus canadensis*) and European elderberry (*S. nigra*) cultivars. *J. Sci. Food Agric.* 87: 2665–2675.
Mershon, W.A. 1965. A tasty native, the elderberry. *Org. Gard. Farm.* 12(11): 50–51.
Novelli, S. 2003. Developments in berry production and use. *Bi-weekly Bulletin* 16(21): 1–6.
Pierquet, P. 1978. Why not grow elderberries? *Minnesota Hortic.* 106(4): 98–99.
Plewes, J. 1969. Meet the elders. *Org. Gard. Farm.* 16(8): 66–67.
Pogorzelski, E. 1982. Formation of cyanide as a product of decomposition of cyanogenic glucosides in the treatment of elderberry fruit *Sambucus nigra. J. Sci. Food Agric.* 33: 496–498.
Ritter, C.M., and Mckee, G.W. 1964. *The elderberry: history, classification and culture.* Bulletin 709. Pennsylvania State University, College of Agriculture, Agriculture Experiment Station, University Park, PA. pp. 22.
Schaefer, H. M. 2009. Reliable cues and signals of fruit quality are contingent on the habitat in black elder (*Sambucus nigra*). *Ecology* 90: 1564–1573.
Skirvin, R.M., and Otterbacher, A. 1977. Elderberry cultivar performance in Illinois. *Fruit Var. J.* 31: 7–10.
Small, E., Catling, P.M., and Richer, C. 2004. Poorly known economic plants of Canada—41. American elder (*Sambucus nigra* subsp. *canadensis*) and blue elderberry (*S. nigra* subsp. *cerulea*). *Bull. Can. Bot. Assoc.* 37(2): 20–28.
Taylor, J.R., and Fernandez-Flores, E. 1969. Chemical composition of fresh elderberries. *J. Assoc. Offic. Analyt. Chem.* 52: 643–646.
Vallès, J., Bonet, M.À., and Agelet, A. 2004. Ethnobotany of *Sambucus nigra* L. in Catalonia (Iberian Peninsula): the integral exploitation of a natural resource in mountain regions. *Econ. Bot.* 58: 456–469.
Walker, M., Nunez, J., Walkingstick, M., and Banack, S.A. 2004. Ethnobotanical investigation of the Acjachemen clapperstick from blue elderberry *Sambucus mexicana* (Caprifoliaceae). *Econ. Bot.* 58: 21–24. [As pointed out in the text, "*S. mexicana*" = *S. canadensis* subsp. *cerulea*.]
Way, R.D. 1964. Elderberry varieties and cultural practices. *Proc. 110th Annual Meeting New York State Hort. Soc.* 110: 233–236.
Way, R.D. 1966. *Elderberry growing in northeastern United States.* 57th Annual Report of the Northern Nut Growers Association. pp. 124–128.
Way, R.D. 1981. *Elderberry culture in New York State.* N.Y. Food Life Sci. Bull. 91, New York State Agricultural Experiment Station, Geneva, NY. pp. 4.
Weinsheimer, D.W. 1961. The versatile elderberry. *Org. Gard. Farm.* 8(1): 56–58.

## Specialty Cookbooks

Belt, T.E. 1981. *Winemaking with elderberries ("the British grape").* The "Amateur Winemaker" Publications, Ltd., Andover, Hants, U.K. pp. 106.
Hendrickson, R. 1981. *The berry book. The illustrated home gardener's guide to growing and using over 50 kinds and 500 varieties of berries.* Doubleday & Company, Garden City, NY. [Chapter 6: Magic elderberries, pp. 96–103.]

Wild food books of North America usually have considerable information on the culinary preparation of elderberries, which are one of the most popular of wild berries. Baird (1980) has 2 elderberry recipes, Berglund and Bolsby (1971) have 3, Boorman (1969) has 10, Freitus (1980) has 4, Freitus and Haberman (2005) have 6, Johnson (1989) has 13, Kluger (1973) has 16, Knutsen (1975) has 4, Krumm (1991) has 6, Marie (2008) has 2, Marrone (2009) has 15, Mogelon (2001) has 5, Niethammer (1999) has 6, Robe-Terry (1997) has 4, Russell (1975) has 5, Schufer (2011) has 5, Tatum (1976) has 9, Williamson (1995) has 12, and Van Atta (1991) has 8. (See the Appendix to this book for details of the publications cited.)

# 39 Epazote

Family: Chenopodiaceae (goosefoot family)

## NAMES

Scientific name: *Dysphania ambrosioides* (L.) Mosyakin & Clemants (*Chenopodium ambrosioides* L.; the name *D. ambrosioides* is recent, and almost all literature produced to date uses the name *C. ambrosioides*)

- The name "epazote" is derived from the Nahuatl words *epatl*, meaning skunk, and *tzotl*, meaning sweat, a reference to the rank smell. Nahuatl is the language of the Amerindian peoples of southern Mexico and Central America, spoken by the Aztecs before the Spanish conquest, and still a minority language in Mexico. Suggested pronunciations: EH-pah-ZOT, eh-pah-soh-teh, eh-puh ZOE-tay, e-pah-ZOH-the.
- The Spanish names of some Mexican species unrelated to epazote nevertheless have epazote in their name. *Epazote de zorrillo* refers to *Teloxys graveolens* (Willd.) W. A. Weber, a medicinal plant of Mexico. *Epazote zorrillo* (*Chenopodium graveolens* Willd., skunk saltwort) is another medical species.
- Epazote is also known as American goosefoot, American wormseed, Baltimore wormseed, bitter weed, Californian spearmint (in New Zealand), demi-god's food, epasote, herb sancti Mariae, Indian wormweed, Jesuit's tea, Mexican goosefoot, Mexican tea, mouse food, pazote, Spanish tea, skunkweed, West Indian goosefoot, worm bush, worm grass, wormseed, wormseed goosefoot, wormseed oil plant, and wormweed.
- "Hedge mustard" is used as a name for epazote, but is well established for the common weed *Sisymbrium officinale* (L.) Scop. "Stinkweed" is another name that is much more commonly used for other species, for example *Thlaspi arvense* L., and *Pluchea camphorata* (L.) DC.
- Rarely, the name "ambrosia" has been used for epazote, but ambrosia is almost always used for *Chenopodium botrys* L., a Eurasian plant that is weedy in North America, and is occasionally used as a culinary herb. Ambrosia is also known as Jerusalem oak. Rarely, epazote has been called Jerusalem oak, Jerusalem bush, Jerusalem tea, and Jerusalem parsley.
- For pharmaceutical purposes, epazote is often known as wormseed. Extracted oil has been used as an anthelmintic (a medicine for controlling internal parasites) for many years. In the early 1900s, epazote oil was one of the major anthelmintics used to treat ascarids (roundworms) and hookworms in humans, cats, dogs, horses, and pigs. Wormseed oil from epazote was replaced with other, more effective and less toxic anthelmintics in the 1940s. In a few areas in Latin America, epazote is still used to treat worm infections in livestock.
- The name Baltimore wormseed arose from the fact that the Baltimore Maryland area was the center of production of wormseed oil in North America for more than a century.
- The name Mexican tea is sometimes applied to *Ephedra nevadensis* S. Watson, a native of the southwestern United States and Mexico, more commonly called Mormon tea.
- The genus name *Dysphania* is based on the Greek *dysphanis*, visible with difficulty, an allusion to the tiny flowers.
- *Ambrosioides* in the scientific name C. *ambrosiodes* is Latin for ambrosia-like, referring to the strong odor. According to Greek mythology, Ambrosia (Greek for "not mortal") was a nourishment reserved for the Olympic gods.

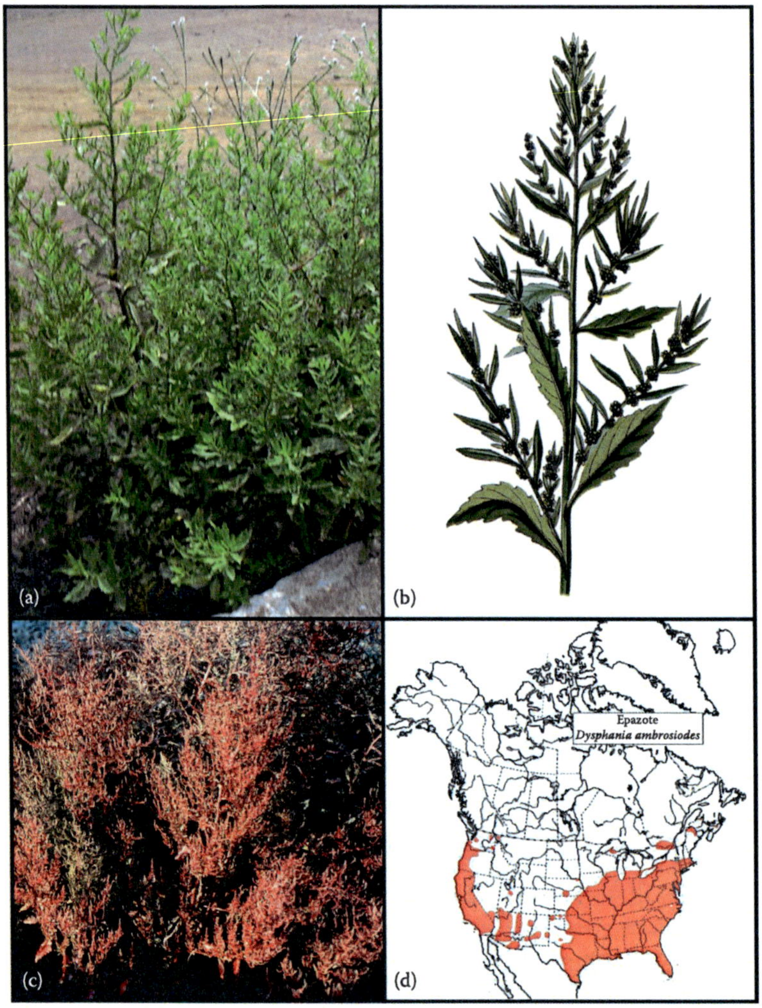


**FIGURE 39.1** Epazote (*Dysphania ambrosiodes*). (a) Group of plants. (Courtesy of Forest & Kim Starr [CC By 3.0].) (b) Plant in fruit. (From Hallier, E., *Flora von Deutschland* [edition 5 by D.F.L. von Schlechtendal et al.], F.E. Köhler, Gera-Untermhaus, Germany. Vol. 9, Plate 870, 1882.) (c) Plants in fall coloration. (Courtesy of Sharon and Gerry Channer.) (d) Distribution map.

## GEOGRAPHY AND ECOLOGY OF WILD PLANTS

Until recently, epazote has been considered to be a native of tropical and subtropical America, including Mexico, Central America, and South America. Its widespread occurrence in the United States has been interpreted as naturalization resulting from introductions. Clemants and Mosyakin (2003) concluded that the southern populations in the United States represent the northern indigenous range of the species. The United States Department of Agriculture's Germplasm Resources Information Network (http://www.ars-grin.gov/cgi-bin/npgs/html/index.pl?language=en; consulted Jan. 2011) also indicates that epazote is indigenous to the United States. Epazote has become naturalized in North America as far north as Canada, and has also become established in tropical and warm temperate regions throughout the world.

In North America, epazote grows along roadsides and in waste and cultivated grounds. In the southwestern United States, it thrives along dry river bottoms, stream beds, and lake beds.

## PLANT PORTRAIT

Epazote is an annual or a perennial depending on its situation. The plants grow from 0.5 to 1.5 m (1.6 to 5 feet) tall. Large clusters of small, yellowish-green flowers are produced in midsummer to early fall and, later, small brownish-black seeds. Plants that are cultivated to make tea and soup are sometimes put in *D. ambrosiodes* var. *ambrosioides*, while those cultivated for use as a vermifuge and source of wormseed oil are sometimes placed in var. *anthelminticum* (L.) A. Gray. The species is strongly aromatic. Wormseed oil, the chief product of interest of the plant, is produced mostly in the leaves and the seeds, the latter normally furnishing more than half of the plant's yield. The leaves are conspicuously dotted beneath with oil glands. Epazote was used for centuries as a vermifuge in the Americas, and was also used for the same purpose in post-Columbian times in Europe and Asia.

## CULINARY PORTRAIT

The availability of epazote as a wild plant has made it a "poor man's herb" for centuries in Mexico, enhancing otherwise common peasant fare. It has been described as "the most Mexican of the culinary herbs," and is also used by Mexican–Americans near the border. Its rather objectionable odor has been compared to turpentine and petroleum, with hints of citrus, savory, or mint. The taste is medicinal and pungently camphoraceous, and requires habituation. Epazote leaves are used to flavor corn, black beans, mushrooms, fish, soups, stews, chili sauces, shellfish, and freshwater snails. The foliage is also used to brew a tea, and the tender leaves are sometimes used as a potherb. Epazote added to cooking beans is said to reduce the flatulence caused by the latter. In Mexico, where the herb is well known, epazote is commonly added to bean dishes to avoid socially embarrassing consequences. Refried beans made with epazote are a Mexican specialty, but when served in Tex-Mex-style restaurants outside of Mexico and the southern United States, this dish almost never contains epazote. Outside of Mexico, epazote has some popularity as a tea in the West Indies, the United States, southern France, and Germany. Because epazote is so strongly flavored, it has been recommended that it be added only during the last 15 minutes of cooking so that the food will not become bitter. It has also been advised that this herb be used sparingly.

Epazote contains pharmacologically active principles and cases of poisoning have been recorded. Although widely used as a culinary herb, especially in Mexico, its consumption could be hazardous. If eaten at all, it should only be consumed in very limited quantities, and rarely. The culinary use of epazote outside of the Mexican and Latino communities, where it represents an acquired taste, is limited by its toxicity, peculiar taste, and unappealing aroma.

### CULINARY VOCABULARY

- *Sopa de flor de calabaza* (pronounced soh-pay day floor day cah-lah-bah-zah) is a soup prepared in Central and South America from squash blossoms, flavored with epazote, onions, and pepper.
- "Mexican tea tassies," "Mexican tea cakes," and "Mexican tea cookies" are all bakery treats made without "Mexican tea" (i.e., epazote).
- "Refried beans" is a Mexican–American dish of mashed, cooked beans, often served as a side dish or used as a filling for various tortillas. It is frequently prepared with epazote. The English name is based on a mistranslation of the Spanish name "frijoles refritos." The Spanish means well fried, not refried.

## PROSPECTS

Epazote oil is the principal commercial product from the plant, and is obtained from cultivated forms selected for oil quality and quantity. Although the oil has lost favor in Western countries, it continues to be produced and used for medicinal and insect-repellent purposes in many Third World countries. The culinary uses are largely restricted to the Latino community, much of which has become habituated to the taste of the herb. Although epazote is likely to maintain both its oil and culinary market niches, there seems very little prospect for growth, since more desirable alternatives are available.

## CURIOSITIES OF SCIENCE AND TECHNOLOGY

- In New Mexico, suppositories of dried pulverized leaves of epazote, ground spearmint, and salt have been used as an alleged remedy for appendicitis.
- One of the many health problems treated with an epazote folk remedy is athlete's foot. It has been found that epazote indeed inhibits fungi that cause this disease, confirming the wisdom of this herbal remedy.
- The popular use of epazote in New York City by Latinos has led to it becoming somewhat weedy there, and it is found growing in cracks of sidewalks and in Central Park.
- Epazote is sometimes used to repel pests. It has been used to deter insects from eating stored grains. Indochinese farmers mix the fruiting stalks with fertilizer to discourage insect larvae in their crops. In Brazil, the plant is used as a household insect repellent and insecticide.
- Beans are notorious for producing flatulence, and epazote is only one of several solutions to the problem. Water-soluble compounds (oligosaccharides) that contribute to flatulence can be reduced by a lengthy soaking in water followed by boiling, but this also removes valuable water-soluble nutrients. Fermented beans also reduces the likelihood of flatulence. The commercial produce "Beano" ("Take Beano before and there'll be no gas") contains the enzyme α-galactosidase, extracted from the fungus *Aspergillus*. This enzyme neutralizes some of the sugars that otherwise cause flatulence. Three to eight drops of Beano are taken when first consuming gas-inducing food, but they must be placed in the mouth with food that is no more than lukewarm, or it will not work, as the enzyme becomes inactive at 49°C (130°F).

## KEY INFORMATION SOURCES


Bahrman, N. 1984. Biometrical analysis of the quantitative variability of some morphological characters in *Chenopodium* [*C. ambrosioides* and other species]. *Rev. Gen. Bot. (France)* 91: 153–161. (In French, English summary.)

Boutkhil, S., El-Idrissi, M.M., Amechrouq, A., Chbicheb, A., Chakirt, S., and El-Badaoui, K. 2009. Chemical composition and antimicrobial activity of crude aqueous ethanol extracts and essential oils of *Dysphania ambrosiodes* (L.) Mosyakin & Clemants. *Acta. Bot. Gall.* 156: 201–209.

Clemants, S.E., and Mosyakin, S.L. 2003. *Dysphania*. In *Flora of North America north of Mexico*. Vol. 4. Edited by Flora of North America Editorial Committee. Oxford University Press, New York. pp. 267–275.

Coile, N.C., and Artaud, C.R. 1997. Chenopodium ambrosioides *L., (Chenopodiaceae): Mexican tea, wanted weed?* Department of Agriculture and Consumer Services, Division of Plant Industry, Gainesville, FL. pp. 6.

Gadano, A., Gurni, A., Lopez, P., Ferraro, G., and Carballo, M. 2002. In vitro genotoxic evaluation of the medicinal plant *Chenopodium ambrosioides* L. *J. Ethnopharmacol.* 81: 11–16.

Gupta, D., Charles, R., Mehta, V.K., Garg, S.N., and Kumar, S. 2002. Chemical examination of the essential oil of *Chenopodium ambrosioides* L. from the southern hills of India. *J. Essent. Oil Res.* 14: 93–94.

Jardim, C.M., Jham, G.N., Dhingra, O.D., and Freire, M.M. 2008. Composition and antifungal activity of the essential oil of the Brazilian *Chenopodium ambrosiodes* L. *J. Chem. Ecol.* 34: 1213–1218.

Jimenez-Osornio, F.M.V.Z.J., Kumamoto, J., and Wasser, C. 1996. Allelopathic activity of *Chenopodium ambrosioides* L. *Biochem. Syst. Ecol.* 24: 195–205.

Kishore, N., Chansouria, J.P.N., and Dubey, N.K. 1996. Antidermatophytic action of the essential oil of *Chenopodium ambrosioides* and an ointment prepared from it. *Phytother. Res.* 10: 453–455.

Kliks, M.M. 1985. Studies on the traditional herbal anthelmintic *Chenopodium ambrosioides*: Ethnopharmacological evaluation and clinical field trials. *Social Sci. Med.* 21: 879–886.

Kumar, R., Mishra, A.K., Dubey, N.K., and Tripathi, Y.B. 2007. Evaluation of *Chenopodium ambrosioides* oil as a potential source of antifungal, antiaflatoxigenic and antioxidant activity. Internat. *J. Food Microbiol.* 115: 159–164.

Logan, M.H., Gwinn, K.D., Richey, T., Maney, B., and Faulkner, C.T. 2004. An empirical assessment of epazote (*Chenopodium ambrosioides* L.) as a flavoring agent in cooked beans. *J. Ethnobiol.* 24: 1–12.

MacDonald, D., VanCrey, K., Harrison, P., Rangachari, P.K., Rosenfeld, J., Warren, C., and Sorger, G. 2004. Ascaridole-less infusions of *Chenopodium ambrosioides* contain a nematocide(s) that is (are) not toxic to mammalian smooth muscle. *J. Ethnopharmacol.* 92: 215–221.

Monzote, L. 2007. Essential oil from *Chenopodium ambrosioides*: A return to traditional medicine? *Internat. J. Essent. Oil Therap.* 1: 179–183.

Mosyakin, S.L., and Clemants, S.E. 2002. New nomenclatural combinations in *Dysphania* R. Br. (Chenopodiaceae): Taxa occurring in North America. *Ukrayins'k. Bot. Zhurn.* [new series] 59: 380–385.

Olajide, O.A., Awe, S.O., and Makinde, J.M. 1997. Pharmacological screening of the methanolic extract of *Chenopodium ambrosioides. Fitoterapia* 68: 529–532.

Onocha, P.A., Ekundayo, O., Eramo, T., and Laakso, I. 1999. Essential oil constituents of *Chenopodium ambrosioides* L. leaves from Nigeria. *J. Essent. Oil Res.* 11: 220–222.

Pare, P.W., Zajicek, J., Ferracini, V.L., and Melo, I.S. 1993. Antifungal terpenoids from *Chenopodium ambrosioides. Biochem. Syst. Ecol.* 21: 649–653.

Pino, J.A., Marbot, R., and Real, I.M. 2003. Essential oil of *Chenopodium ambrosioides* L. from Cuba. *J. Essent. Oil Res.* 15: 213–214.

Sagrero-Nieves, L., and Bartley, J.P. 1995. Volatile constituents from the leaves of *Chenopodium ambrosioides* L. *J. Essent. Oil Res.* 7: 221–223.

Su, H.C. 1991. Toxicity and repellency of *Chenopodium* oil to four species of stored-product insects. *J. Entomol. Sci.* 26: 178–182.

Tapondjou, L.A., Adler, C., Bouda, H., and Fontem, D.A. 2002. Efficacy of powder and essential oil from *Chenopodium ambrosioides* leaves as post-harvest grain protectants against six stored product beetles. *J. Stored Prod. Res.* 38: 395–402.

Umemoto, K. 1978. Essential oil of *Chenopodium ambrosioides* L. containing (-)-pinocarveol as a major component. *J. Agric. Chem. Soc.* Japan 52: 149–150. (In Japanese, English summary)

Weiland, G.S., Broughton, L.B., and Metzger, J.E. 1935. Wormseed oil production. *Univ. Maryland. Agric. Exp. Stn. Bull*: 315–335.

## Specialty Cookbooks

Hutson, L. 1987. *The herb garden cookbook.* Texas Monthly Press, Austin, TX. pp. 278.

Farnsworth (1999; see the Appendix to this book) provides a recipe for beans with epazote.

# 40 Evening Primrose

Family: Onagraceae (evening primrose family)

## NAMES

Scientific name: *Oenothera biennis* L.

- The evening primrose is not a "primrose," a name best applied to the genus *Primula*. It is unclear how the common name primrose came to be applied to *Oenothera*. The "evening" in the name relates to the fact that the flowers of many of the approximately 125 species of *Oenothera* open in the evening and release a scent that attracts moths for pollination.
- The name evening primrose is optionally hyphenated, as evening-primrose. The species is also known as yellow evening primrose. Archaic names include king's cureall, nightwillow herb, scabbish, scurvish, and tree primrose.
- In Texas, evening primroses are sometimes called "buttercups" (presumably because of the superficial resemblance of the cupped flowers).
- The genus name *Oenothera* is based on the Latin *oenothera*, "a plant with juice that may cause sleep," tracing perhaps to the Greek name *oinos* wine + *therea* "booty" (an expression meaning "wine catcher;" according to folklore, the plant can prevent hangovers). Alternatively (and less plausibly), it has been speculated that the genus name traces to the Greek words *onos*, "an ass" + *thera*, "hunting" or *ther*, "wild beast."
- *Biennis* in the scientific name is Latin for biennial.

## GEOGRAPHY AND ECOLOGY OF WILD PLANTS

Evening primrose, a native of North America, is found in all provinces of Canada, but is more frequent in the east than the west. It occurs in the eastern half of the United States and in scattered locations in the west. The species is widely naturalized in temperate and tropical regions, and the issue of whether it is indigenous outside of North America is unsettled.

Evening primrose is a frequent weed of roadsides, waste places, and abandoned land, often found in light sandy and gravelly soils. It commonly occurs in association with early successional, biennial, and perennial weeds.

## PLANT PORTRAIT

*Oenothera biennis* is a biennial (as the name suggests) or short-lived perennial herb (and very rarely an annual), producing thick, conical, fleshy roots and a basal rosette of leaves in the first year. In the second year, the stem grows to 1–2 m (3–6.6 feet) in height and develops yellow flowers 3–5 cm (1.2–2 inches) wide. There are ornamental forms of *O. biennis* with attractive growth habit and flowers. The fruit is a capsule containing many reddish-brown seeds that mature in the fall. The seeds are very small, but a single plant can easily produce 150,000 seeds.

The taxonomic classification of *O. biennis* and related species of both North America and Eurasia is very complex. Most texts recognize var. *canescens*, with dense grayish pubescence, as the predominant plant of western North America, while the eastern plants are referable to var. *biennis*. Races of evening primrose with distinctive chromosomal composition are sometimes segregated

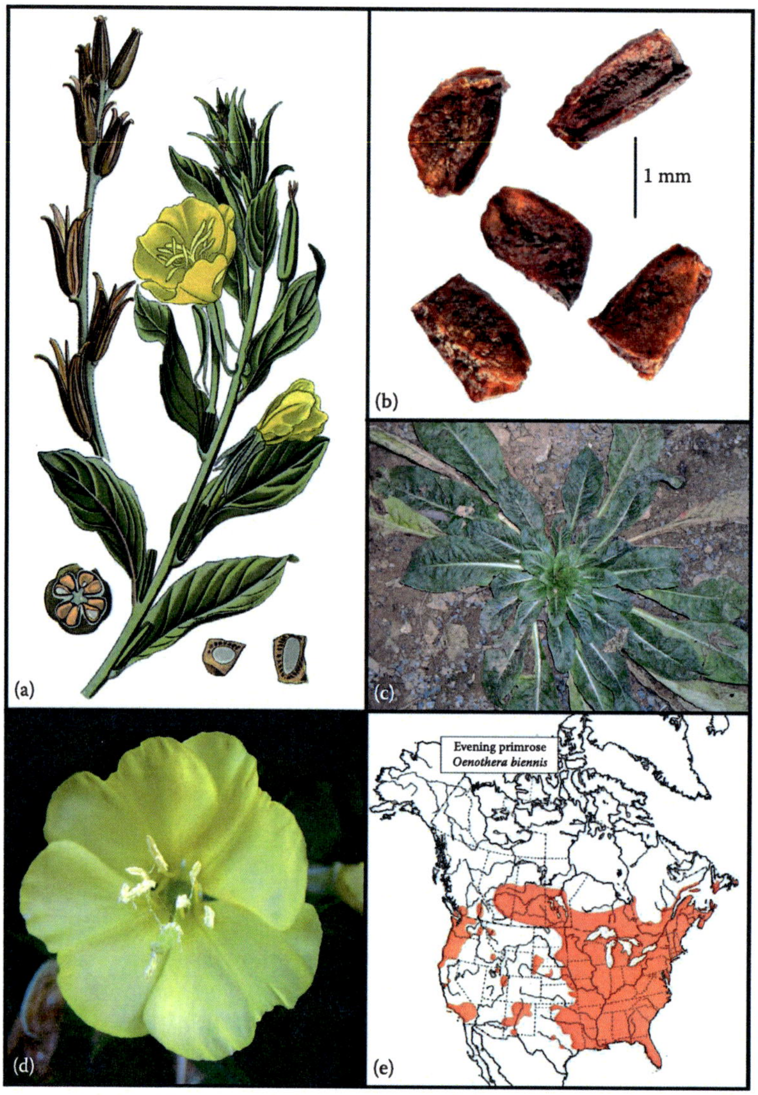


**FIGURE 40.1** Evening primrose (*Oenothera biennis*). (a) Flowering and fruiting branches, and sectioned fruit and seeds. (From Thomé, O.W., *Prof. Dr. Thomé's Flora von Deutschland, Österreich und der Schweiz*, H.V. Verlag, Berlin-Lichterfelde, Germany, Vol. 3, Plate 421, 1905.) (b) Seeds. (Courtesy of Steve Hurst, U.S. Department of Agriculture.) (c) Plant at overwintering rosette stage. (Courtesy of noebse.) (d): Flower. (Courtesy of TeunSpaans [CC By 3.0].) (e) Distribution map.

as distinct species, although these are usually difficult to distinguish morphologically. Genetically, *O. biennis* is categorized as a "complete translocation complex-heterozygote," with two sets (the "complexes") of seven chromosomes maintained by a system of balanced lethal genes. This type of inheritance is known in a few other genera, but was first described in *O. biennis*, and is the classical example of the phenomenon discussed in evolution and genetics courses. At meiosis, translocations link the chromosomes into a ring of 14, but zigzag (alternate) separation of the chromosomes generates the original parental sets. Lethal factors kill the pollen carrying one of the sets (so that there is 50% pollen fertility), and ovule lethal factors limit survival to the set of chromosomes complementary to that in the pollen. Self-pollination generates offspring with the two chromosome complements found originally in the maternal plant. The permanent hybrid vigor resulting from the combination of two quite different genomes is thought to explain the success of evening primrose as a colonizing species.

While it is clear that *O. biennis* is the chief *Oenothera* species that has been grown as a medicinal oilseed, related species have also been cultivated, often unknowingly. Other species from which cultivars have been derived include *O. glazioviana* Micheli ("*O. lamarckiana*" of many authors) and *O. parviflora* Micheli.

Evening primrose extracts were used medicinally by both Indians and early settlers. In Europe, during the early 1600s it was called "king's cure-all." An infusion of the whole plant was thought to counter asthmatic cough, gastrointestinal disorders, and whooping cough, and to reduce pain. Poultices were used to treat bruises and wounds.

Evening primrose has attracted great interest for its seed oil, used medicinally as a nutritional supplement. The health value of the seed oil resides in an unusual polyunsaturated fatty acid, gamma-linolenic acid (γ-linolenic acid) or simply GLA. The seeds contain 17%–25% oil, of which only 7%–10% is GLA, although climate and maturity affect oil content and qualitative composition, as well as overall yield. GLA is one of the so-called essential fatty acids needed by humans for maintenance of cell functions. It is a precursor in the biosynthesis of prostaglandins, especially prostaglandin $E_1$, a hormone-like substance that has been clinically shown to regulate metabolic functions in mammals; it affects cholesterol levels, dilates blood vessels, reduces inflammation, and is used to treat erectile dysfunction. GLA is thought to be important for development of brain tissue and other tissue growth, and nature seems to provide for human infants with high levels of GLA in human milk. GLA is a normal conversion product of linoleic acid, a major constituent of most vegetable oils, so that it would appear that humans should not experience a shortage. Nevertheless, some people, perhaps 10%–20% of the population, evidently do not have adequate levels, even when receiving large amount of linoleic acid. The deficiency seems due to lack of an enzyme that metabolizes GLA from linolenic acid, so that there is a deficiency of GLA in the blood. Useful for treating atopic eczema, GLA has therapeutic promise for premenstrual syndrome, diabetes, multiple sclerosis, alcoholism, inflammation, heart disease, and stroke. Rubbing GLA into the skin is thought to be an alternative route of assimilation, so cosmetic preparations often incorporate GLA.

## CULINARY PORTRAIT

Evening primrose leaves, shoots, roots, and seed pods were consumed by American Indians as food. There are also forms with fleshy edible roots, used as a vegetable, which were more commonly grown in the nineteenth century than today. The roots have a somewhat peppery flavor, and are said to resemble salsify and parsnip. The foliage is occasionally eaten as a salad, especially in China. The leaves tend to become bitter once the plant starts to flower, but identification usually requires waiting until the biennial plant flowers in the second year, so wild food enthusiasts usually do not harvest the plant.

The chief food value of evening primrose is as a source of the health-promoting food supplement GLA. Pharmaceutical and food companies are developing GLA-containing supplements and specialty foods for infants, the elderly, and people with health problems. Side effects of consumption of GLA-fortified foods and supplements have been documented (including itching, sore throat, and gassiness), so that use should be guided by doctors and pharmacists.

### Culinary Vocabulary

- The word "primrose" is very infrequently encountered in cooking, but when it is the plant in question is often the common ornamental primrose (*Primula vulgaris* Huds.), not evening primrose. Common primrose flowers are sometimes candied or used to make wine, and the edible leaves are occasionally used for tea or in salads.

## PROSPECTS

The chief economic value of evening primrose lies in its use as a crop source of GLA. Although GLA has been obtained by fermentation of some yeasts and other fungi, and from currants (*Ribes* species), the chief commercial sources are evening primrose and borage (*Borago officinalis* L.). Companies have engaged in a boastful debate about the comparative efficacy of GLA in their preparations made from evening primrose on the one hand, and from borage on the other. Whether borage or evening primrose is more competitive for GLA production depends on climatic and edaphic factors at a particular location. In North America, both species are grown. Borage is much more suitable for the Canadian prairies, where available cultivars of evening primrose do not overwinter reliably. However, it is not essential to grow evening primrose as a biennial; it is often started in greenhouses in mid-winter and transplanted to the field where it is grown as an annual.

As a cultivated plant, evening primrose is tolerant of a variety of soil types and a range of pH, but soils that are prone to crusting after rains and waterlogged soils should be avoided. If planted at too high a density (14 plants/ft$^2$; 150 plants/m$^2$) the plants may not bolt (flower), which is of course necessary for seed production.

Evening primrose crops are raised in temperate areas of northern and eastern Europe, North America, and Australasia. The United States production has been centered in North and South Carolina, Texas, and Oregon. Canadian production has been centered in Nova Scotia and Ontario. Annual world production of seed has increased at least twenty-fold in the last several decades. China alone has produced over 20,000 metric tons (22,000 U.S. tons) in some years, and has become the world's major producer. Combined U.S. and Canadian annual production is less than 200 tonnes (220 U.S. tons).

Wild evening primrose plants shed their seeds when a pod matures, and since the pods do not mature simultaneously, harvest of seeds is difficult. Nevertheless, seed is gathered from wild plants in northeast China. Most modern evening primrose cultivars have nonshedding pods, which has simplified harvest and reduced seed loss.

The future of evening primrose as a pharmacological crop is uncertain because of competition from other countries and the unreliability of the present market. Hemp (*Cannabis sativa* L.) is attracting considerable interest as a new crop in Canada, to be grown for its high-GLA seed oil. Still another source of competition is canola (*Brassica* species) and other crops genetically engineered to produce GLA.

## CURIOSITIES OF SCIENCE AND TECHNOLOGY

- Because evening primrose flowers open at dusk, it was once considered to be a witch's herb, used in casting spells.
- Hugo de Vries (1848–1935) was a world famous student of evolution who, at the beginning of the twentieth century, theorized that new species arise by spontaneous changes in individuals, called mutations (Charles Darwin had earlier learned that such changes (called "sports" at the time) occur, but did not appreciate their importance for the mechanism of evolution). Unfortunately for de Vries, he chose evening primrose to demonstrate his theory. Later, scientists learned that the odd genetic system of evening primrose was responsible for the generation of altered individuals that de Vries was labeling as mutations and that this system did not occur in many other plants. As a result, his theory was discredited, although de Vries did contribute substantially to evolutionary theory.
- Some evening primrose seeds have been shown to live up to 80 years in the soil.
- According to legend, the Hoskins Library of the University of Tennessee is haunted by a female ghost known as the "Evening Primrose." She wanders the library, knocking books off of shelves and playing with the elevators, and is said to be accompanied by the smell of cornbread.

## KEY INFORMATION SOURCES


Baker, J. 1998. Evening primrose. In *Richters second commercial herb growing conference—transcripts*. Edited by R. Berzins, H. Snell, and C. Richter. Richters, Goodwood, ON. pp. 67–85.

Baskin, C.C., and Baskin, J.M. 1993. Germination requirements of *Oenothera biennis* seeds during burial under natural seasonal temperature cycles. *Can. J. Bot.* 72: 779–782.

Belisle, D. 1991. Potential of evening primrose in Canada. *Altern. Crop. Noteb.* 4: 4–5. [Reprinted from BioOptions, Newsletter of the Center for Alternative Plant & Animal Products 1(5), 1990.]

Brandle, J.E., Court, W.A., and Roy, R.C. 1993. Heritability of seed yield, oil concentration and oil quality among wild biotypes of Ontario evening primrose. *Can. J. Plant Sci.* 73: 1067–1070.

Briggs, C.J. 1986. Evening primrose: La belle de nuit, the king's cureall. *Can. Pharm. J.* 199: 248–252.

Budeiri, D., Po, A.L.W., and Dornan, J.C. 1996. Is evening primrose oil of value in the treatment of premenstrual syndrome? *Control Clin. Trials* 17(1): 60–68.

Court, W.A., Hendel, J.G., and Pocs, R. 1993. Determination of the fatty acids and oil content of evening primrose (*Oenothera biennis* L.). *Food Res. Int.* 26: 181–186.

Cisowski, W., Zielinska-Stasiek, M., Luczkiewicz, M., and Stolyhwo, A. 1993. Fatty acids and triacylglycerols of developing evening primrose (*Oenothera biennis*) seeds. *Fitoterapia* 64: 155–162.

Deng, Y.-C., Hua, H.-M., Li, J., and Lapinskas, P. 2001. Studies on the cultivation and uses of evening primrose (*Oenothera* spp.) in China. *Econ. Bot.* 55: 83–92.

Dietrich, W.M., and Wagner, W.L. 1988. Systematics of *Oenothera* section *Oenothera* subsection *Raimannia* and subsection *Nutantigemma* (Onagracceae). *Syst. Monogr.* 24: 1–91.

Gates, R.G. 1957. A conspectus of the genus *Oenothera* in eastern North America. *Rhodora* 59: 9–17.

Gates, R.G. 1958. *Taxonomy and genetics of* Oenothera. Uitgeverij Dr. W. Junk, Den Haag, Cambridge, MA. pp. 115.

Ghasemnezhad, A., and Honermeier, B. 2008. Yield, oil constituents, and protein content of evening primrose (*Oenothera biennis* L.) seeds depending on harvest time, harvest method and nitrogen application. *Ind. Crops. Prod.* 28: 17–23.

Gross, K.L., and Kromer, M.L. 1986. Seed weight effects on growth and reproduction in *Oenothera biennis* L. *Bull. Torrey Bot. Club* 113: 252–258.

Hall, I.V., Steiner, E., Threadgill, P., and Jones, R.W. 1988. The biology of Canadian weeds. 84. *Oenothera biennis* L. *Can. J. Plant Sci.* 68: 163–173.

Hanczakowski, P., Szymczyk, B., and Wolski, T. 1993. The nutritive value of the residues remaining after oil extraction from seeds of evening primrose (*Oenothera biennis* L.). *J. Sci. Food Agric.* 63: 375–376.

Hulan, H.W., Hall, I.V., Nash, D.M., and Proudfoot, F.G. 1987. Composition of native evening primrose seeds collected from western Nova Scotia. *Crop. Res.* (*Edinburgh*) 27: 1–9.

Kerscher, M.J., and Korting, H.C. 1992. Treatment of atopic eczema with evening primrose oil: rationale and clinical results. *Clin. Investig.* 70: 167–171.

Kromer, M., and Gross, K.L. 1987. Seed mass, genotype, and density effects on the growth and yield of *Oenothera biennis* L. *Oecologia* (*Berlin*) 73: 207–212.

Lapinskas, P. 1989. Commercial exploitation of alternative crops, with special reference to evening primrose. In *New crops for food and industry*. Edited by G.E. Wickens, N. Haq, and P. Day. Chapman and Hall, London. pp. 216–221.

Levin, D.A., Howland, G.P., and Steiner, E. 1972. Protein polymorphism and genic heterozygosity in a population of the permanent translocation heterzygote, *Oenothera biennis*. *Proc. Nat. Acad. Sci. USA* 69: 1475–1477.

Levy, A., Palevitch, D., Ranen, C. 1993. Increasing gamma linolenic acid in evening primrose grown under hot temperatures by breeding early cultivars. *Acta Hort.* 330: 219–225.

Mukherjee, K.D., and Kiewitt, I. 1987. Formation of gamma linolenic acid in the higher plant evening primrose (*Oenothera biennis* L.). *J. Agric. Food. Chem.* 35: 1009–1012.

Munz, P.M. 1965. *North American Flora, Series II, Part 5—Onagraceae*. The New York Botanical Garden, New York. pp. 231.

Paccalin, J., Mendy, F., Bernard, M., Delhaye, N., and Spielmann, D. 1983. Rediscovery of an oleaginous plant: *Oenothera biennis*. The importance of gamma-linolenic acid in nutrition. *Bull. Acad. Natl. Med.* 167: 923–931. [In French]

Raven, P.E. 1979. A survey of reproductive biology in Onagraceae. *New Zealand J. Bot.* 17: 575–593.

Raven, P.E., Dietrich, W., and Stubbe, W. 1979. An outline of the systematics of *Oenothera* subsect. *Euoenothera* (Onagraceae). *Syst. Bot.* 4: 242–252.

Roy, R.C., White, P.H., More, A.F., Hendel, J.G., Pocs, R., and Court, W.A. 1993. Effect of transplanting date on the fatty acid conposition, oil content and yield of evening primrose (*Oenothera biennis* L.) seed. *Can. J. Plant Sci.* 74: 129–131.

Russell, G. 1988. Physiological restraints on the economic viability of the evening primrose crop in eastern Scotland. *Crop. Res. (Edinburgh)* 28(1): 25–33.

Simpson, M.J.A. 1994. A description and code of development of evening primrose (*Oenothera* spp.). *Ann. Appl. Biol.* 125: 391–397.

Simpson, M.J.A., and Fieldsend, A.F. 1993. Evening primrose: harvest methods and timing. *Acta Hort.* 331: 121–128.

Small, E., and Catling, P.M. 1996. Poorly known economic plants of Canada—8. Evening primrose, *Oenothera biennis* L. *Bull. Can. Bot. Assoc.* 29: 8–10.

Srivastava, A., Shukla, Y.N., and Kumar, S. 1998. Chemistry and pharmacology of the evening-primrose *Oenothera biennis* and related species: a review. *J. Med. Aromat. Plant Sci.* 20: 432–440.

Yaniv, Z., and Perl, M. 1987. The effect of temperature on the fatty acid composition of evening primrose (*Oenothera*) seeds during their development, storage and germination. *Acta Hort.* 215: 31–38.

Yaniv, Z., Ranen, C., Levy, A., and Palevitch, D. 1989. Effect of temperature on the fatty acid composition and yield of evening primrose (*Oenothera lamarckiana*) seeds. *J. Exp. Bot.* 40: 609–614.

## Specialty Cookbooks

Berglund and Bolsby (1971) have six recipes for evening primrose root, Knutsen has five, and Schufer (2011) has one. Hall and Hall (1980) have advice on how to prepare the pods for consumption, and a recipe for this. (See the Appendix to this book for details of the publications cited.)

# 41 Fiddlehead Fern (Ostrich Fern)

Family: Aspidiaceae. (Ostrich fern family. In much of the older literature, *Matteuccia* is placed in the Polypodiaceae, whereas the modern consensus is that it belongs to the Aspidiaceae.)

## NAMES

Scientific name: *Matteuccia struthiopteris* (L.) Tod. (*Osmunda struthiopteris* L.)

- The young, tightly curled leaves of ferns are called "crosiers" (for their similarity to a bishop's staff), or more familiarly "fiddleheads" for their resemblance to the curled, tuning head or scroll of a violin. The word "fiddlehead" can be used in three ways: (1) to designate the ostrich fern; (2) to refer to the young curled leaf of the ostrich fern used as a vegetable; and (3) as the name of the young curled leaf of any fern.
- The name ostrich fern is based on the large leaves resembling ostrich plumes.
- The fiddlehead fern is also known as fiddlehead, garden fern, and shuttlecock fern.
- The genus *Matteuccia* was named for Carlo Matteucci (1800–1868), an Italian physicist and politician.
- *Struthiopteris* in the scientific name *M. struthiopteris* is derived from the Greek *struthion*, ostrich + *pteris*, feather.

## GEOGRAPHY AND ECOLOGY OF WILD PLANTS

The fiddlehead fern is a wild fern, native to many north temperate and sub-boreal parts of the northern hemisphere. In Canada, it occurs in parts of all provinces and territories. In the United States, it is found in Alaska, and extends south to Virginia, Ohio, Indiana, Illinois, Missouri, and South Dakota. It also occurs in Japan, China, Siberia west to Scandinavia, Belgium, France, and parts of the Alps.

Fiddlehead fern grows in low open ground, alluvial thickets, and rich woodlands. Typically it occurs along floodplains and the margins of rivers and streams where rich alluvial soil is deposited. The species characteristically occupies moist but well-drained rich soils with the water table within 1.5 m (5 feet) of the soil surface. A pH range of 5.1–5.4 gives best growth. Fiddleheads planted as ornamentals have proven to tolerate a variety of soil conditions, but thrive in moist soil. Although they generally grow naturally in half or full shade of trees, in field trials they have been successfully grown in full sunlight.

## PLANT PORTRAIT

The fiddlehead fern is an elegant plant that grows to a height of about a meter (about a yard), occasionally 2 m (6.6 feet), in clumps sometimes exceeding a meter in width. The leaves are of two types, some large and "ferny" (highly divided), others much smaller and narrower and bearing small spore cases. The spore-bearing leaves are initially green but turn brown as they mature, and are much shorter and located in the center of the regular leaves. The regular leaves are produced in the early spring, while the leaves that bear spores are produced in the summer. Ferns do not have flowers and seeds, but like mosses and some other primitive plants, scatter spores (single-celled reproductive bodies). A single spore-bearing leaf can produce a million spores.

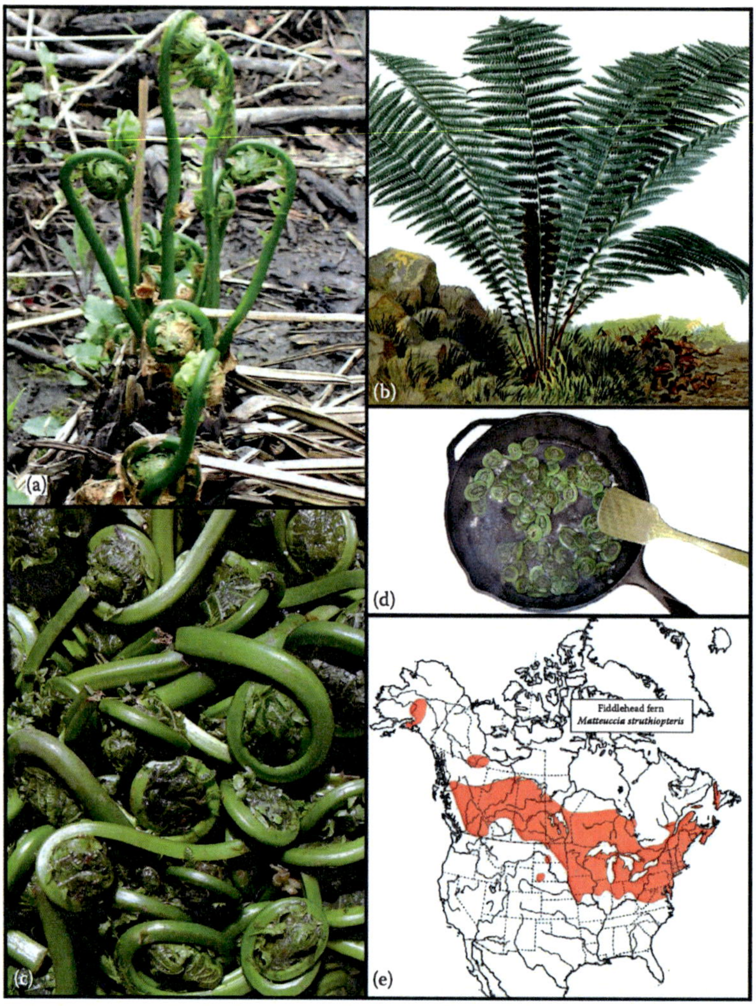


**FIGURE 41.1** Fiddlehead fern (*Matteuccia struthiopteris*). (a) Young fiddleheads emerging in the spring. (Courtesy of donhug/Donna [Flickr/CC-attribution].) (b) Mature plant in summer. (From Britten, J., *European ferns*, Cassell, Petter, Galpin, & Co., London, U.K., 1882.) (c) Freshly harvested fiddleheads. (Courtesy of New Brunswick Tourism [CC By 2.0].) (d) Fiddleheads in a frying pan. (Courtesy of John Blyberg [Flickr/CC-attribution].) (e) Distribution map.

The edible part of the plant is the short, spirally curled-up young leaf, called a fiddlehead, which first emerges and unwinds in the spring (usually April to June). The young leaves are in their coiled form for only about 2 weeks before they unfurl. Harvested fiddleheads are about 5 cm (2 inches) long and about 2.5 cm (1 inch) in diameter. Once the leaves grow beyond 7.5 cm (3 inches) they become too bitter to eat. Mature fern leaves are usually called fronds. The fiddlehead fern produces horizontal underground stems (rhizomes) with new shoots developing annually along their length. In early spring, the fiddlehead plant may have as many as 40 young leaves (i.e., fiddleheads). Repeated harvesting of all fiddleheads will kill the plants. It is advisable to remove only a few fiddleheads from a given plant, allowing some leaves on each crown to develop.

Cultivated varieties have not yet been selected. This fern is gathered from wild stands for consumption, although some experimental cultivation is in progress. Native Americans were harvesting the fiddlehead fern when European colonists arrived. In 1783 in New Brunswick, United Empire

Loyalists were poorly prepared for their first winter and had to resort to eating fiddlehead fern. The early method of consumption was to dig up and roast the entire crown before eating. The Malecite Indians of the Saint John River Valley in New Brunswick have traditionally harvested fiddleheads as a spring tonic and have sold them at local markets. In Maine, fiddleheads have been picked by the Penobscot Indians for many years, from the islands in the Penobscot River where the ferns grow. Other ferns are eaten, but the fiddlehead fern is the most economically important edible fern in the world. Its major area of use is the Maritime Provinces of Canada and neighboring Maine. It is harvested from these areas, with a small amount also from the Matapédia Valley in Gaspé, Quebec.

In commercial processing, the scales protecting the young fiddleheads are removed by a rotary washer, the fiddleheads are boiled for a few minutes, and then packaged. Processed fiddleheads are packed frozen and in cans. Recently, the proportion of the crop pickled has substantially increased.

Fiddlehead fern is also used as an ornamental plant in gardens. It is particularly attractive at the back of flower beds and serves as an excellent foundation planting beside houses. Fiddlehead fern spreads aggressively and may prove difficult to confine. It is one of the most easily cultivated ferns in temperate regions, provided moist soil is available.

## CULINARY PORTRAIT

Fiddleheads have sometimes been eaten raw, but are generally served cooked. The papery brown chaff on the outside needs to be removed by rubbing before eating (this will have been done in canned and frozen fiddleheads). The flavor has been compared to a combination of asparagus, green bean, and okra; a combination of avocado and asparagus; and artichoke. The texture has been described as appealingly chewy. The preferred method is to steam the fiddleheads until tender (about 10 minutes) and flavor them with freshly squeezed lemon, sour cream, or cheese sauce, and add salt and butter to taste. Fiddleheads go well with cheeses, tomato sauce, and East Asian cuisine, and are excellent with Hollandaise sauce. They can be used in similar ways to any firm green vegetable such as asparagus or broccoli. A wide variety of recipes can be found on the Web.

Fiddleheads are generally considered a gourmet item. They are a good source of vitamins A and C, niacin (vitamin $B_3$), and riboflavin (vitamin $B_2$), and are nutritionally comparable to asparagus and other common green vegetables. They are high in potassium and low in sodium, making them suitable for low-sodium diets. Fresh fiddleheads are available seasonally, in May and June in North America. They can be frozen after slight blanching. Frozen fiddleheads are generally available in supermarkets.

Statements appearing in the press that fiddlehead fern is carcinogenic resulted from a confusion of fiddleheads collected from fiddlehead fern with those of bracken fern (*Pteridium aquilinum* (L.) Kuhn). Studies have shown that bracken, which has been widely consumed in Japan and elsewhere, is carcinogenic, whereas fiddlehead fern is not. Some other ferns are also collected from the wild and eaten in North America, despite the fact that several are considered poisonous. Those who collect wild ferns for their fiddleheads need to be certain of the identity of the material harvested before consuming it. The fiddlehead fern has been considered to be nontoxic. However, in May 1994, a few cases of gastrointestinal illness were associated with eating raw or lightly cooked fiddlehead ferns in New York and western Canada. In the American outbreaks, the implicated ferns were eaten either raw or lightly cooked (sautéed, parboiled, or microwaved). In a similar outbreak in British Columbia in 1994, eating lightly cooked fiddleheads was also associated with gastrointestinal illness. The Canadian government conducted studies and found no toxicity problems. Although a toxin has not been identified, because of concerns that the ferns might contain a heat-labile poison, a warning was issued, advising that fiddleheads be boiled for 15 minutes or steamed for 10–12 minutes before eating. In 2002, the Canadian government recommended that fresh fiddleheads be carefully washed in several changes of cold water before cooking, and noted that fiddleheads should also be boiled or steamed prior to frying, baking, or sautéeing. All water that has come in contact with the fiddleheads during preparation should be discarded (the water used for boiling may turn dark because of the high iron content). The Center for Disease Control in the United States also investigated the outbreaks of food-borne illness associated with fiddleheads, and recommended thorough cooking before eating (boiling for at least 10 minutes). Generally, cookbooks recommend a shorter time in boiling water, often 5–7 minutes, and caution against overcooking (15 minutes in boiling water makes the vegetable mushy). The slight increase in nutrient loss and the soggier texture that occurs with longer cooking times is probably necessary, given the past occurrence of illness, and continuing rare reports of people becoming ill shortly after eating fiddleheads. However, the illnesses reported could have been due to harvesting from a polluted area or contamination during restaurant preparation. To avoid the possibility of consuming contaminated fiddleheads, it is important that they be obtained from a reliable source. It is also important to be aware of the symptoms of the illness associated with fiddleheads, which usually begin 0.5–12 hours after consumption, and may last for 24 hours. There may be diarrhea, nausea, vomiting, abdominal cramps, and headaches, and dehydration may occur, particularly among the elderly and infants. To avoid ending this section on such a negative note, keep in mind that the problem described here is very rare.

### Culinary Vocabulary

- "California cuisine" refers to a trendy "nouvelle cuisine," which was initiated in West coast restaurants of the United States in the early 1980s. Food gurus touted this type of food as healthier (at least when served with limited or no butter, cream, or cheese). Typically, portions were small (but prices were very high), composition was simple (although artistically presented), and the food often seemed exotic. Among the typical offerings were edible flowers, "baby vegetables," sundried tomatoes, and fiddleheads.

## PROSPECTS

There is potential for increased use of fiddleheads as more consumers become familiar with the vegetable. Already, demand exceeds supply of the frozen product, and the market has expanded in North America. Major limitations are the short season of availability, inaccessibility of the wild plants, and shortage of labor. It is clear that the wild source is insufficient to expand the industry (indeed, overharvesting has reduced the supply in some regions), and cultivation is, therefore, necessary. Some experimental work has been carried out on fiddlehead fern to test it as a farm crop grown in fields. In New Brunswick, 3-year-old plantings yielded 310–927 kg/ha or 277–827 pounds/acre, and the next year the yield doubled. Another study showed a yield of up to 2130 kg/ha or 1900 pounds/acre (by comparison, productive wild stands typically yield about 1300 kg/ha or 1160 pounds/acre). Cultivars are not yet available, and transplants from nature into a field setting have been very slow to establish, and, of course, seeds are not available (indeed, spore-bearing terrestrial plants are in general not grown as field crops). There is considerable opportunity for increasing quality and other desired characteristics, both by selection and development of management techniques. Another necessary step toward improving the fiddlehead industry is the development of mass propagation, perhaps by tissue culture. Recent studies have demonstrated a rapid and efficient regeneration technique for fiddlehead fern using suspension culture.

## CURIOSITIES OF SCIENCE AND TECHNOLOGY

- Fiddleheads were served at a feast hosted by the Indian Chief Chkondum, given to the French explorer Samuel de Champlain (1567?–1635, the founder of Quebec) when he reached the mouth of the Saint John River, New Brunswick.
- "Pteridomania" refers to a Victorian era craze for ferns, which were not only collected for cultivation but also used in a wide variety of decorative arts, including paintings, and both imprints and sculptures using pottery, metal, glass, textiles, and wood.

- Because ferns grow in solitary places, Christian theologians interpreted them as symbols of humility, a critical virtue to achieve a place in Heaven. Ferns often appeared on the gravestones of women in Victorian times, indicating that humility was considered much more important for women than for men.
- Ferns reproduce by microscopic spores, not by seeds. In England, "fern seeds" (which do not exist) were supposed to be visible only on St. John's Eve. Because they were not seen, those who carried them were thought to be invisible, as evidenced by the following quotation from Shakespeare's *Henry IV*: "We have the receipt of fern seed, we walk invisible."
- In Russia, fiddlehead fern has been served as a remedy for intestinal parasites.
- In Norway, the fiddlehead fern has been used as fodder for goats, and for preparing a fern beer.
- Fiddlehead fern is the only native Canadian plant that has achieved commercial success as a vegetable.
- The slim spore-bearing leaves of fiddlehead ferns are full-grown by the fall, and release their spores in the spring, before the larger, main leaves expand. This facilitates the distribution of the spores by wind, because once the main leaves grow and surround the spore-bearing leaves, they act as a barrier to the wind.
- Fiddleheads have symbolic importance in New Brunswick. The provincial coat of arms includes the fiddlehead. The literary publication of the University of New Brunswick is called "The Fiddlehead." Malecite Indians once marked their belongings and property with a symbol in the shape of the fiddlehead, and each spring they stage a "Fiddlehead Festival" on the banks of the St. John River in New Brunswick.

## KEY INFORMATION SOURCES


Blake, S.F. 1942. The ostrich fern as an edible plant. *Am. Fern. J.* 32: 61–68.

Bergeron, M.E., and Lapointe, L. 2001. Impact of one year crozier removal on long-term frond production in *Matteuccia struthiopteris. Can. J. Plant Sci.* 81: 155–163.

Bushway, A.A., Serreze, D.V., McGann, D.F., True, R.H., Work, T.M., and Bushway, R.J. 1989. *The effect of processing method and storage time on the nutrient composition of Maine fiddlehead greens.* University of Maine, Maine Agricultural Experiment Station, Orono, ME. pp. 11.

Bushway, A.A., Wilson, A.M., McGann, D.F., and Bushway, R.J. 1982. The nutrient composition of fresh fiddlehead greens. *J. Food Sci.* 47: 666–667.

Cruise, J. 1972. Spring harvest—the fiddlehead, a gourmet delight. *Ontario Nat.* 11: 18–23.

Dykeman, B.W. 1985. Effects of crozier removal on growth of the ostrich fern (*Matteuccia struthiopteris*). *Can J. Plant Sci.* 65: 1019–1024.

Hawk, R.L. 1962. The fiddlehead, New Brunswick's own vegetable. *Org. Gard.* 9(10): 47–48.

Hunter-Burley, P., Tremblay-Deveau, E., and Estabrooks, E.N. 1989. Shelf life determination of fresh fiddleheads. In *Technology development highlights 1984–1989.* Edited by G.C. Misener, and G.T. Owen. Agriculture Canada Research Station, Fredericton, NB. pp. 195–197.

Johnson, D.M. 1993. *Matteuccia.* In *Flora of North America.* Vol. 2. Edited by Flora of North American Editorial Committee. Oxford University Press, Oxford, U.K. pp. 249–250.

Lidster, P.D., Prange, R.K., Von Aderkas, P., Jackson, E.D., McDonald, J.E., and McLachlan, J. 1989. Effects of storage temperature, humidity and duration on ostrich fern (*Matteuccia struthiopteris* (L.) Todaro) quality. *J. Food Sci.* 54: 132–134.

May, L.W. 1978. The economic uses and associated folklore of ferns and fern allies. *Bot. Rev.* 44: 491–528.

Odland, A. 1992. A synecological investigation of *Matteuccia struthiopteris* dominated stands in western Norway. *Vegetatio* 102: 69–95.

Odland, A. 1995. Frond development and phenology of *Thelypteris limbosperma, Athyrium distentifolium*, and *Matteuccia struthiopteris* in western Norway. *Nordic J. Bot.* 15: 225–236.

Odland, A., Birks, H.J.B., and Line, J.M. 1995. Ecological optima and tolerances of *Thelypteris limbosperma, Athyrium distentifolium*, and *Matteuccia struthiopteris* along environmental gradients in western Norway. *Vegetatio* 120: 115–129.

Prange, R.K. 1980. Responses of the ostrich fern, *Matteuccia struthiopteris* (L.) Todaro, Pteridophyta, to lime, soil moisture and irradiance. *Proc. Nova Scotia. Inst. Sci.* 30: 171–181.

Prange, R.K., and Ormrod, D.P. 1982. Effects of ammonium and nitrate nutrition on the ostrich fern (*Matteuccia struthiopteris*), Pteridophyta. *Can. J. Plant Sci.* 62: 195–201.

Prange, R.K., and von Aderkas, P. 1985. The biological flora of Canada. 6. *Matteuccia struthiopteris* (L.) Todaro, ostrich fern. *Can. Field-Nat.* 99: 517–532.

Small, E., and Catling, P.M. 1999. Poorly known economic plants of Canada—23. Ostrich Fern, *Matteuccia struthiopteris* (L.) Todaro. *Bull. Can. Bot. Assoc.* 32(4): 54–57.

Syrchina, A.I., Pechurina, N.N., Vereshchagin, A.L., Gorshkov, A.G., Tsapalova, I.E., and Semenov, A.A. 1993. A chemical investigation of *Matteuccia struthiopteris*. *Chem. Nat. Compd.* 29: 535–536.

Thakur, R.C., Hosoi, Y., and Ishii, K. 1998. Rapid in vitro propagation of *Matteuccia struthiopteris* (L.) Todaro—an edible fern. *Plant Cell Rep.* 18: 203–208.

von Aderkas, P. 1983. Studies of the gametophytes of *Matteuccia struthiopteris* (ostrich fern) in nature and in culture. *Can. J. Bot.* 61: 3267–3270.

von Aderkas, P. 1984. Economic history of ostrich fern, *Matteuccia struthiopteris*, the edible fiddlehead. *Econ. Bot.* 38: 14–23.

von Aderkas, P., and Bird, C.J. 1983. The habitat of the ostrich fern (*Matteuccia struthiopteris*) in Nova Scotia and Prince Edward Island. *Proc. Nova Scotia. Inst. Sci.* 33: 131–135.

von Aderkas, P., and Green, P.E.J. 1986. Leaf development of the ostrich fern *Matteuccia struthiopteris*. *Bot. J. Linn. Soc.* 93: 307–321.

## Specialty Cookbooks

Comfort, J. 1985. *Spuds! Dulse! Fiddleheads!: a cook's guide to maritime foods*. Nimbus Pub, Halifax, NS. pp. 84.

Derevitzky, C. 1977. *Fiddleheads and mustard blossoms: a guide to edible plants of the forest and meadow*. Catherine Derevitzky, Morrill, ME. pp. 41.

Love, K. 1984. *Fiddleheads instead*. New Brunswick Bicentennial Commission, Saint John, NB. unpaginated. [In English and French]

Berglund and Bolsby (1971) have 8 fiddlehead recipes, Boorman (1969) has 2, Brill (2002) has 10, Freitus (1980) has 3, Gardon (1998) has 4, Hahn (2010) has 3, Kluger (1973) has 8, Mogelon (2001) has 3, Stewart (2000) has 2, and Williamson (1995) has 11. (See the Appendix to this book for details of the publications cited.)

# 42 Golden Chia

Family: Lamiaceae (Labiatae; mint family)

## NAMES

Scientific name: *Salvia columbariae* Benth.

- Chia is a Spanish name originally used for several of the species now known by the name in English. The word is derived from the Nahuatl *chian*, meaning oily (the seeds have oil, as noted later). The seeds, not just the plants of several species of *Salvia* and *Hyptis* are known as "chia," and have a long history of medicinal and culinary use in the Americas. "Chia" unqualified usually refers to the most important chia species, *S. hispanica* L., native from Mexico to Peru.
- The "golden" in golden chia is a reference to the seed color, which is usually gray or tan, by contrast with the much more commonly cultivated black-seeded chia (*S. hispanica*). (White-seeded forms are known in both species.)
- Golden chia has also been called California chia, California sage, chia, chia sage, desert chia, and desert sage.
- The genus name *Salvia* is commonly interpreted as coming from the Latin verb *salvere*, meaning to heal or to save, and is the root of the words salve and salvation. The culinary herb garden sage (*S. officinalis* L.) had a reputation for healing qualities.
- Ornamental species of *Salvia* are commonly called salvias (singular salvia). Culinary species and wild species are commonly called sage.
- When he coined the scientific name *Salvia columbariae* for golden chia, English botanist George Bentham (1800–1884) explained his unusual choice of name with the Latin phrase "habitus fere *Scahiosa columbariae*" (appearance like *Scahiosa columbaria*). *Scahiosa columbaria* L. is a Mediterranean species with a thin cluster of basal divided leaves and a head-like inflorescence, all reminiscent of golden chia.

## GEOGRAPHY AND ECOLOGY OF WILD PLANTS

Golden chia is native to much of the southwestern United States (Arizona, California, Nevada, and Utah) and parts of northern Mexico (Baja California and Sonora). The species grows from coastal scrub vegetation up to pine woodlands at an elevation of 1200 m (4000 feet), normally in open, arid, natural areas, but also in disturbed places. It is adapted to sandy and gravelly, low-fertility, dry soils. In some areas, golden chia forms extensive stands. The plant has decreased because of overgrazing, urban development, and suppression of fire. Golden chia is adapted to regular fires, and increases in abundance following fires. As noted below, Native American sometimes deliberately set fires to promote the growth of the plant.

## PLANT PORTRAIT

*Salvia columbariae* is an annual herb, sometimes as diminutive as 5 cm (2 inches) in inhospitably dry areas, or as tall as 60 cm (2 feet). Large, thick leaves are at the base of the plant, and are prominently divided, lobed, and wrinkled. When crushed, the velvety leaves produce an odor variously described as minty, sage-like, or skunky. The tiny flowers are in globular clusters (glomerules) that

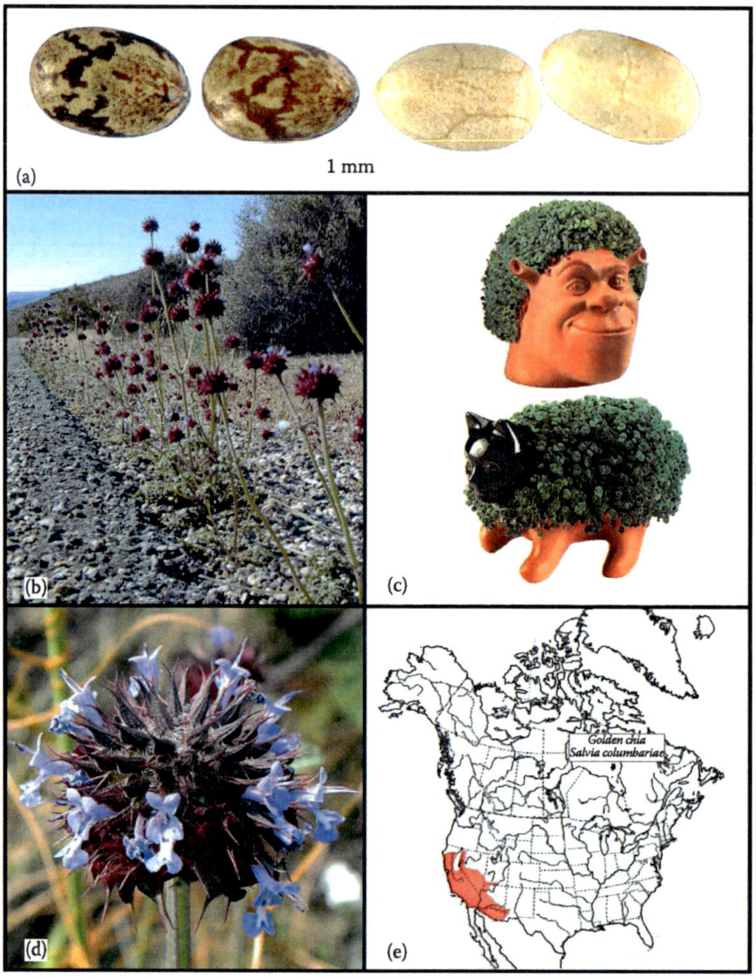


**FIGURE 42.1** Golden chia (*Salvia columbariae*). (a) Normal seeds at left, white seeds at right. (Courtesy of Steve Hurst, United States Department of Agriculture.) (b) Roadside plants. (Courtesy of Dawn Endico [CC By 2.0].) (c) Chia pets. (Photo based on a purchased commercial chia pet set.) (d) Flower head. (Courtesy of Curtis Clark [CC By 2.5].) (e) Distribution map.

are separated by about 2.5 cm (1 inch) along the main flower stalk. The pale blue of the flowers contrasts with the purplish, wine-like color of bracts (tiny leaves) at their base, so that the glomerules seem to be two-colored. The seeds are gray or light brown, flat, 1.5–2.5 mm (0.06–0.1 inch) long. Some forms of the plant have white seeds, a color that is often preferred for seeds by people, and therefore suggestive that these forms have been selected.

Chia seeds were once an important food staple for Native Americans on the southern West coast of North America. Some indigenous people continue to consume chia seeds collected from the wild, and the seeds are also marketed to a limited extent in stores. In the past, natural chia stands were managed by Indians in what is now the southwestern United States. The species was also cultivated in Mexico. Large stands of chia are believed to persist at the sites of ancient Native American villages. The harvest of chia seeds by native people is remarkably like the collection of cereals that has been practiced in premodern societies throughout the world. Seeds from the mature fruiting stalks are beaten off into a container, or the stalks are simply collected into a pile over a blanket and threshed with sticks or by trampling. Hulling (removal of the hard outer coat) is also accomplished to a considerable extent by this process. The frass (waste material) is separated and discarded by winnowing (e.g., throwing the material in the air, and fanning away the lighter, unwanted material). The seeds were subsequently parched and pulverized in a mortar, and the resulting meal was used to prepare biscuits, cakes, porridge, soup, and pinole (see "Culinary Vocabulary").

Several miscellaneous uses of chia deserve mention. Apart from the nutritional qualities of the seed, seed oil from chia species may have value as a preservative lacquer. Golden chia is often grown as an ornamental. The plant was used for several medicinal purposes by Native Americans, and today some over-the-counter, (allegedly) health-promoting preparations incorporate powdered chia seed.

## CULINARY PORTRAIT

Chia seeds, regardless of species, have several current culinary uses. The seeds are ground into meal and then processed into bread, biscuits, muffins, and cakes. The roasted seeds have a pleasantly nut-like taste. The seeds expand considerably in water, forming a thick mucilage. The gelatinous mass is used to flavor fruit juices and make a cooling drink. In Mexico, the seeds are commonly soaked for half an hour or so until a gel forms, then added to orange, lime, or other fruit juice. Gelled seeds are also made into gruel and pudding. Sprouted seeds are consumed in salads, sandwiches, soups, and stews. (However, chia seeds are too sticky for conventional sprouting jars, and Mexicans have long germinated the seeds on earthenware figurines, the predecessors of "chia pets" discussed below.) The chia seeds sold in stores and intended to be used as a nutty topping for cereals, yogurt, and salads, and to be incorporated into pasta and other prepared foods, are generally from *S. hispanica*.

Chia seeds are rich in oil and protein, and considered to be very nutritious. Golden chia is being promoted for its fatty acid content, particularly α-linoleic acid, an omega-3 fatty acid, thought to increase high-density lipoproteins ("good" cholesterol), which tend to remove low-density lipoproteins ("bad" cholesterol) which stick to the interior of arteries, increasing blood pressure and the probability of heart attacks and strokes. In North America, insufficient consumption of omega-3 fatty acids is a major health issue, although these fatty acids are available in soy, canola oils, and other foods. Some studies have suggested that adding chia to chicken feed may increase the omega-3 content and decrease the cholesterol content of the meat and eggs.

### CULINARY VOCABULARY

- Pinole or panole is a Mexican and Southwestern pudding prepared from ground dried corn, spices, and a sweetener. The word is also applicable to a mush made from the seeds of chia by Native Americans.
- Chia seeds used for culinary purposes have been said to be a "grain." A strict definition of grain restricts the term to the edible seeds of cereal species of the grass family, such as corn, rice, and wheat. More generally, however, grains can be any small edible seeds or fruits that are collected in quantity.

## PROSPECTS

In ancient times, golden chia was a cultivated crop in Mexico, and for centuries it was used as a valued food by Native Americans in the southwestern United States. Today, it has been resurrected to a small extent as a very minor, specialty food and source of nutritional extracts. For these market niches, there is considerable opportunity for expanded production, although other chia species are obvious competitors. *Salvia hispanica* has already been domesticated, and its seeds are much more readily harvested. Since golden chia is an annual, its cultivation is not necessarily restricted to its native area. Breeding and management programs could considerably improve the productivity of golden chia, and dedicated cultivars could be selected for oil, nutritional content, and food biomass. Although basically a wild plant, the species is certainly an interesting and promising investment possibility.

## CURIOSITIES OF SCIENCE AND TECHNOLOGY

- Native peoples commonly used gelatinous chia seeds to remove foreign matter from eyes, such as sand from sandstorms. One or several seeds, or mashed seeds, were placed in the eye, and rolled around or left in overnight. The gelatinous material absorbed foreign material and could then be removed.
- The Chumash once inhabited the Californian coastal region from Malibu to San Luis Obispo and inland for about 160 km or 100 miles (they still live in California and other locations). Chumash legends tell of a chia-like plant called *ilepesh* that was used to "wake the dead or the nearly dead." It is suspected that patients treated with roots of the plant had suffered from strokes or heart attacks and appeared to be nearly dead. The roots of *Salvia miltiorrhiza* Bunge from China are used in the treatment of stroke, and the roots have been shown to contain tanshinones and other compounds, which apparently are able to prevent clotting and restore blood flow in cases of stroke. It has been found that *S. columbariae* also contains tanshinones (Adams et al. 2005), perhaps explaining its reputation for "waking the dead."
- Chumash runners were known to travel 30 km (18.6 miles) or more in a day, delivering messages between villages. Eating chia seeds was thought to contribute to their endurance. Chia seeds acquired a reputation among several Native American groups for providing sustaining energy.
- The mucilaginous seeds of *S. hispanica* are widely sprouted on unglazed ceramic or other porous materials in the shape of animals, and sold as "chia pets." The sprouted seeds have the appearance of green fur. Joseph Pedott, a San Francisco-based marketing guru and his firm Joseph Enterprises, Inc., popularized chia pets in the 1980s, beginning with a terracotta ram in 1982. (The firm specializes in gadgets, including the famous Smart Clapper®, which turns lamps on and off when hands are clapped.) Chia seeds are soaked in water and then placed into the grooves of the porous clay figurines. Germination occurs in about 5 days in a sunny window. The seedlings last for 2 weeks or so before declining for lack of nutrition. The seedlings are edible.
- Why are chia seeds mucilaginous? It has been shown (Fuller and Hay 1983) that when the seeds of golden chia that are naturally scattered from the plants become wet, they tend to become covered with adhering sand grains, and that such armored seeds are significantly less consumed by insects. (Alternatively, it has been suggested that the mucilage coating chia seeds serves to keep the seeds from drying out in their often arid, desert environment. However, there is no experimental proof for this.)
- The Mexican state of Chiapas was named from the Nahuatl "chia water" or "chia river."
- The Mexican species *S. divinorum* Epling & Játiva (the Latin name translates as "salvia of the Gods") is the only one of the almost 1000 species of *Salvia*, indeed the only plant of the 4000–5000 species of the mint family, that is known for sure to be hallucinogenic. It was cultivated by Central American peoples for use in religious ceremonies before the conquest of America by Europeans. Its recent use by motion picture stars has resulted in sensationalistic reporting.

## KEY INFORMATION SOURCES

Adams, J.D., Jr., Wall, M., and Garcia, C. 2005. *Salvia columbariae* contains tanshinones. *Evidence-based Compl. Alt. Med.* 2: 107–110.

Ayerza, R., Jr., and Coates, W. 2005. *Chia: rediscovering a forgotten crop of the Aztecs*. University of Arizona Press, Tucson, AZ. pp. 215.

Ayerza, R., Jr., and Coates, W. 2007. Seed yield, oil content and fatty acid composition of three botanical sources of ω-3 fatty acid planted in the Yungas ecosystem of tropical Argentina. *Trop. Sci.* 47: 183–187.

Brayton, R.D., and Capon, B. 1980. Productivity depletion and natural selection of *Salvia columbariae* seeds. *Aliso* 9: 581–587.
Capon, B., and Brecht, P.E. 1970. Variations in seed germination and morphology among populations of *Salvia columbariae* Benth. in southern California. *Aliso* 7: 207–216.
Capon, B., Maxwell, G.L., and Smith, P.H. 1978. Germination responses to temperature pre-treatment of seeds from ten populations of *Salvia columbariae* in the San Gabriel Mountains and Mojave Desert, California. *Aliso* 9: 365–373.
Clebsch, B. 1997. *A book of salvias: sages for every garden*. Timber Press, Portland, OR. pp. 221.
Compton, J. 1994. Mexican salvias in cultivation. *The Plantsman* 15: 193–215.
Epling, C. 1938. The California salvias: A review of *Salvia*, section *Audibertia*. *Ann. Missouri Bot. Gard.* 25: 95–152.
Fuller, P.J., and Hay, M.E. 1983. Is glue production by seeds of *Salvia columbariae* a deterrent to desert granivores? *Ecology* 64: 960–963.
Gentry, H.S., Mittleman, M., and McCrohan, P.R. 1990. Introduction of chia and gum tragacanth in the U.S. In *Advances in new crops*. Edited by J. Janick and J.E. Simon. Timber Press, Portland, OR. pp. 252–256.
Hashemi, A., and Estilai, A. 1994. Seed germination response of golden chia (*Salvia columbariae* Benth.) to low temperature and gibberelin. *Indust. Crops Products* 2: 107–109.
Hashemi, A., Estilai, A., and Carapetian, J. 1993. Characterization of isozymes in *Salvia columbariae*. *Euphytica* 67: 101–104.
Hedge, I.C. 1970. Observations on the mucilage of *Salvia* fruits. *Notes Roy. Bot. Gard. Edinburgh* 30: 79–95.
Heinrich, M. 1992. Economic botany of American Labiatae. In *Advances in labiate science*. Edited by R.M. Harley and T. Reynolds. Royal Botanic Gardens, Kew, U.K. pp. 475–488.
Immel, D.L. 2003. *Plant guide: chia,* Salvia columbariae *Benth.* United States Department of Agriculture, Natural Resources Conservation Service, National Plant Data Center, Baton Rouge, LA. pp. 4.
Keeley, J.E., and Fotheringham, C.J. 1998. Smoke induced seed germination in California chaparral. *Ecology* 79: 2320–2336.
Kintzios, S.E. (Editor). 2000. *Sage: The genus Salvia*. CRC, Boca Raton, FL. pp. 440.
Luis, J.G., Grillo, T.A., Quinones, W., and Kishi, M.P. 1994. Columbaridione, a diterpenequinone from *Salvia columbariae*. *Phytochemistry* 36: 251–252.
Luis, J.G., Quinones, W., Grillo, T.A., and Kishi, M.P. 1994. Diterpenes from the aerial part of *Salvia columbariae*. *Phytochemistry* 35: 1373–1374.
Neisess, K.R., Scora, R.W., and Kumamoto, J. 1987. Volatile leaf oils of California salvias. *J. Nat. Prod.* 50: 515–517.
Reisfield, A.S. 1993. The botany of *Salvia divinorum* (Labiatae). *Sida* 15: 349–366.
Small, E. 2011. Blossoming treasures of biodiversity. 34. Chia—not just a pet. *Biodiversity* 12: 49–56.
Soulanille, E.L., and Bierzychudek, P. 1995. Can the desert annual *Salvia columbariae* adapt to global warming? *Bull. Ecol. Soc. Amer.* 76(2): 80.
Sutton, J. *The gardeners guide to growing salvias*. Timber Press. Portland, OR. pp. 160.
Timbrook, J. 1986. Chia and the Chumash: A reconsideration of sage seeds in southern California. *J. Calif. Great Basin Anthropol.* 8: 50–64.
Timbrook, J. 1990. Ethnobotany of Chumash Indians, California, based on collections by John P. Harrington. *Econ Bot.* 44: 236–253.
Timbrook, J., Johnson, J.R., and Earle, D.D. 1982. Vegetation burning by the Chumash. *J. Calif. Great Basin Anthropol.* 4: 163–186.
Visco, F.J., and Capon, B. 1970. Pollination mechanisms in three species of *Salvia* native to southern California. *Aliso* 7: 231–242.

## Specialty Cookbooks

Hoover, G. 2007. *Cooking with chia*. Lulu.com [publisher for self-published books]. pp. 110.
Scheer, J.E. 2000. *The magic of chia. Revival of an ancient wonder food.* Frog Books, Berkeley, CA. pp. 174.
Allen, D. 2010. *Chia seed.* Woodland Publishing, Salt Lake City, UT. pp. 40.

Freitus (1980; see the Appendix to this book) presents three recipes specifically for golden chia.

# 43 Golden Currant

Family: Grossulariaceae (currant family; traditionally placed in the Saxifragaceae, the saxifrage family, but recently the genus *Ribes* has been segregated as the only genus in the Grossulariaceae)

## NAMES

Scientific name: *Ribes aureum* Pursh (*R. odoratum* H.L. Wendl.)

- The word "currant" traces to the Middle English *raison of Courante*, literally "raisin of Corinth," a phrase for a small seedless raisin (*Courante* in Middle English denotes the Greek port of Corinth, the center of export). The word currant originally referred only to this dried grape (of the wine grape, *Vitis vinifera* L.). "Corinth grapes," also known as "Zante currants," are tiny, black, seedless grapes used to produce raisins, which are very popular in making pastries. The name was also applied to certain species of *Ribes* because of the similarity between their fruit clusters and those of the grapes called currants. The word currant has been used for fruit from *Ribes* only since about 1550, whereas it was used for dried raisins long before this date.
- The "golden" in the name golden currant refers to the flower color; the fruits are also sometimes gold, but more often black (and sometimes orange).
- The golden currant is also known as buffalo currant, clove currant, fragrant currant, Missouri currant, and sweet-scented currant.
- The genus name *Ribes* is said to derive from the Arabic *ribas*, rhubarb, presumably applied to the currant because of a supposed similarity in taste that was noted by Arabs while they expanded into Spain centuries ago. Alternatively, the name has been interpreted as originating from the Old Danish colloquial *ribs*, red currant, and as the Latinized form of *riebs*, an old German word for currant.
- *Aureum* in the scientific name is Latin for golden, for the flower color.

## GEOGRAPHY AND ECOLOGY OF WILD PLANTS

The golden currant occurs in the southernmost parts of Canada, from British Columbia to Prince Edward Island, and in most states of the United States (except in the southeast). In much of its range east of the Mississippi River, it is thought to have been introduced from a more western native area. The species is found in a variety of open and partially shaded areas, sometimes in woods and forests, and in both moist and dry soils. Golden currant is quite drought-tolerant and cold-hardy (tolerating temperatures down to −45°C or −45°F). It is susceptible to white pine rust.

## PLANT PORTRAIT

About 100 of the species in the genus *Ribes* are called "currants," and the remaining 50 or so are "gooseberries." The most important commercial sources of currant fruit are listed in Table 43.1. Currant species are shrubs of northern areas. They are similar to gooseberry shrubs but usually lack the spines and bristles of the latter (although spineless gooseberry varieties have been selected).

The golden currant is a deciduous, spineless shrub, growing 1–3 m (3–10 feet) in height. In the spring, yellowish flowers (turning orange or reddish in some varieties) are produced, with a distinctive odor of vanilla and cloves. The fruits are 5–10 mm (0.2–0.4 inch) in diameter, and red, orange,

**FIGURE 43.1** Golden currant (*Ribes aureum*). (a) Shrub with immature, green fruit. (Courtesy of Dave Powell, U.S. Department of Agriculture Forestry Service, online at Bugwood.org [CC By 3.0].) (b) Shrub with purple-black fruits. (Courtesy of Annelis [CC By 3.0].) (c and d) Flowering and fruiting branches (shrub with orange fruit). (Courtesy of Dan Stephens, reproduced with permission.) (e) Flowering branch. (From Edwards, S.T., *Botanical Register*, Vol. 2, Plate 125, 1816.) (f) Distribution map.

brown, black, or rarely amber or yellow in color. The species is cultivated mostly as an ornamental, not as a source of fruits, and there are several ornamental cultivars. The fruits were eaten by Native Americans and European settlers, but are rather seedy. The plants are occasionally grown as a source of berries. The plants often produce only male or only female flowers, and of course female plants are necessary for fruit production. Wild golden currant plants usually cross-fertilize with pollen from another plant to produce fruit.

About 1892, white pine blister rust was imported to North America. This fungus, which occupies *Ribes* as an alternate host, led to severe infection of five-needle (white) pine trees (some other pine species are also affected), leading to legislation preventing the cultivation of currants and gooseberries over wide areas of North America, and indeed of programs of eradication. The result was to limit the commercial significance of both currants and gooseberries in North America, most

**TABLE 43.1**
**World's Most Important Commercial Species of Currants (*Ribes* Species)**

| Common Names | Scientific Name | Distribution | Where Grown |
|---|---|---|---|
| Black currant (European black currant, cassis; not to be confused with an American black currant species that is occasionally grown) | *R. nigrum* L. | Native to Eurasia | Eurasia and North America |
| Northern red currant (downy currant, Nordic currant, red currant) | *R. spicatum* E. Robson | Native to Eurasia | Commonly grown in Europe, occasionally in North America |
| Red currant (common currant, garden currant) | *R. rubrum* L. | Native to Europe and naturalized in North America | The predominant red currant grown in North America and also grown in Eurasia |
| White currant (selected albino mutations) | Mostly hybrids of *R. rubrum* and *R. spicatum* | | Grown in temperate climates |

particularly in the United States. Today, red currants are grown in small quantities in the United States and Canada, while there is little cultivation of the black currant, which is especially sensitive to white pine blister rust. Red currants are grown mostly in Germany and Poland, with lesser amounts in other European countries. There is substantial production of black currants in Great Britain, Poland, Germany, and other European countries.

## CULINARY PORTRAIT

The Old World currants, not golden currant, are found in the marketplace today. Most fresh currants are used in jams, jellies, and preserves, and also for their juice. Red currants can be eaten raw, either directly or in salads, but because of their tartness they are usually cooked. Red currant jelly is considered a delicacy, and indeed the standard par excellence in jellies. Red and white currants are used to make the prized French Bar-le-Duc jelly (see below for additional information). Red currants are also the key ingredient of Cumberland sauce (also see below for more information). In Europe, red currant jelly is used in cooking of pastries and for making glazes and sauces for meats and poultry. Red currant jelly is a traditional accompaniment of roast venison and mutton. Red currants can also be added to cakes, pies, puddings, and mixed fruit desserts. Many find the taste of black currants objectionable, limiting the popularity of this fruit in North America. Black currants are rarely used as a dessert fruit. In the Old World, the berries are used for canning, juice, jams, jellies, and pie fillings; for black currant crême and liqueurs such as Cassis (a name not only used for liqueurs flavored with black currant juice but also for certain French wines from the Côtes de Provence region); and for converting white wine to "rosé." "Crème de cassis" is applied to a reddish-purple liqueur made from black currants, but sometimes also to a sweet syrup made from currants, which may or may not contain alcohol. Fresh currants should be consumed as soon as possible, but may be stored, unwashed, in a plastic bag in a refrigerator for a few days. Currant berries are best cooked slowly for 3–5 minutes, using a small quantity of water or juice, and adding sugar as desired after cooking.

The golden currant produces fruit that is not competitive with the commercial currant species. Home-grown and wild-collected fruit is used for pies, jams, jellies, juice, and syrup. The heirloom cultivar 'Crandall' (which has been known for more than a century) is considered very good for culinary use, and the berries are sometimes consumed fresh, with a taste that has been compared to grape.

### Culinary Vocabulary

- "Dried currants" are not dried currant berries, but are small seedless raisins, the dried fruit of several grape varieties grown in the Middle East (see derivation of the word "currant," discussed above).
- "Brownies" in North America are small, rich, chewy, squares of chocolate cake, usually with nuts. In Australia and New Zealand, the term may refer to a kind of bread made with currants.
- "Cumberland sauce" is a piquant English sweet-and-sour sauce featuring red currant jelly, and usually also with port, lemon, orange juice and zest, and mustard. It is usually served with duck, venison, or other game, typically cold. It has been speculated that the name traces to Ernest, Duke of Cumberland, brother of George IV, who in 1837 became the last independent king of Hanover, Germany; however, most authorities discount this explanation of the name, which remains of obscure origin.
- "Eccles cake" is a small round cake (a sort of cookie) prepared with buttery, flaky pastry and filled with currants. The name comes from the town of Eccles in Greater Manchester, England, and is said to have been invented by the culinary writer Elizabeth Raffald (1733–1781).
- "Election cake" is a classic English fruitcake or plum cake, originally made with molasses, spice, raisins, and currants and, later, brandy. It became a popular tradition to serve these huge cakes, each weighing about 5.4 kg (12 pounds), while waiting for election results in New England.
- Bar-le-Duc jelly (Lorraine jelly, confiture de groseille) is a famous currant preserve (mislabeled a "jelly," since as noted in the following, whole berries are present), originally from the town of Bar-le-Duc in northwestern France. It was once made from white currants, with the tiny seeds (on average, seven per berry) removed by hand, often using a sharpened goose feather, and the fruit immediately immersed in burning-hot, clear, white syrup, a method that preserved the flavor and bright color of the whole berries. The process is thought to have been created by the fourteenth century provisions chief of Duke Robert of the old province of Lorraine. The choice preserve is now produced from red and white currants as well as from other berry fruits, and the seeds are not generally removed manually.
- A "garibaldi" is a British biscuit with a layer of currants inside, which used to be popular in the first half of the twentieth century. It commemorated Giuseppe Garibaldi (1807–1882), an Italian military and national leader. Its colloquial nickname was "squashed-fly biscuit," named for the appearance of the currants.
- "Plum duff" is an old-fashioned English boiled suet pudding flavored with currants or raisins.

## PROSPECTS

Golden currant is not a commercial source of fruit at present and has limited possibilities of becoming competitive with the Old World commercial currant species. However, its considerable drought and cold tolerance, the quite attractive taste, and appreciable productivity of the cultivar 'Crandall' suggest that the species is an interesting prospect for development as a new fruit crop for North America, and perhaps for dry, cold regions of the Old World. Research to develop golden currant into a crop is in progress in the colder areas of Europe (note selected references below in foreign languages). The fact that the golden currant is popular as an ornamental plant is an additional advantage because it expands the market possibilities for the species. Considerable breeding is required, and there is limited probability of investment in golden currant as a new crop since the fruit market is already well supplied with berry crops. A major obstacle to growing currants is susceptibility to white pine blister rust, and associated legislation against growing currants in some regions so that resistant varieties would have to be bred.

## CURIOSITIES OF SCIENCE AND TECHNOLOGY

- The "Victorian language of flowers" was a secret coded language in Victorian times, with flowers and plants symbolic of certain messages, so when the flower or plant was mentioned in a letter those who knew the code could understand the hidden information. "Currant" meant "thy frown will kill me," whereas "branch of currants" stood for "you please all."
- As noted above, currant cultivation in the United States has been discouraged because the plants serve as the alternate host for a fungus that infects pines. At one time, white pine rust from currants destroyed 650 million board feet of white pine yearly. (One board foot = 2.4 L = 144 cubic inches of material, i.e., 1 foot × 1 foot × 1 inch.)
- During World War II, it was part of the war effort in England to plant black currants, which are extremely high in vitamin C, in an effort to replace oranges as the major source of the vitamin.
- Currants (and also their relatives, the gooseberries) bloom early in the spring, and severe frosts can injure blossoms and young, developing berries. Commercial planters often spray water over the plants when temperatures reach −1°C (30°F). The conversion of water to ice on the plants releases heat, which protects the blossoms and berries.

## KEY INFORMATION SOURCES


Audette, N., and Lareau, M.J. 1997. *Currants and gooseberries culture guide*. Conseil des productions végétales du Québec and Horticultural R & D Centre, Quebec, Canada. pp. 43.

Barney, D.L., and Hummer, K.E. 2005. *Currants, gooseberries, and jostaberries: A guide for growers, marketers, and researchers in North America*. Food Products Press, Binghamton, NY. pp. 299.

Berger, A. 1924. *A taxonomic review of currants and gooseberries*. New York State Agricultural Experiment Station, Geneva, NY. pp. 118.

Bird, R., and Whiteman, K. 2002. *Growing berries and currants: A directory of varieties and how to cultivate them successfully*. Lorenz, London, UK. pp. 64.

Brennan, R.M. 1990. Currants and gooseberries (*Ribes*). In *Genetic resources of temperate fruit and nut crops*. Edited by J.N. Moore and J.R. Ballington Jr. International Society of Horticulture Science, Wageningen, the Netherlands. pp. 459–388.

Crawford, M. 1997. *Currants and gooseberries: Production and culture*. Agroforestry Research Trust, Totnes, UK. pp. 48.

Crowther, D.S. 1956. *Currants and gooseberries*. W.H. & L. Collingridge, Ltd., London, UK. pp. 92.

Darrow, G.M. 1919. *Currants and gooseberries*. U.S. Dept. of Agriculture, Washington, DC. pp. 40.

Darrow, G.M., and Detwiler, S.B. 1946. *Currants and gooseberries: Their culture and relation to white-pine blister rust*. Revised edition. U.S. Department of Agriculture, Washington, DC. pp. 36.

Goldschmidt-Reischele, E. 1972a. Self-sterility in *Ribes aureum*. Part 1. Pollination, fruit setting and number of seeds. *Zeitschrift fuer Pflanzenzuechtung* 68: 225–252. (In German, English summary)

Goldschmidt-Reischele, E. 1972b. Self-sterility in *Ribes aureum*. Part 2. Pollen tube growth and seed development. *Zeitschrift fuer Pflanzenzuechtung* 68: 269–280. (In German, English summary)

Harmat, L., Porpaczy, A., Himelrick, D.G., and Galletta, G.J. 1990. Currant and gooseberry management. In *Small fruit crop management*. Edited by G.J. Galletta and D.G. Himelrick. Prentice Hall, Englewood Cliffs, NJ. pp. 245–272.

Janczewski, E. de, and Glinka, R. von. 1907. Monograph of the currants, *Ribes*. *Mem. Soc. Phys. Hist. Nat Genève* 35: 199–517. (In German; W. Kündig & fils, Geneva, Switzerland)

Keep, E. 1975. Currants and gooseberries. In *Advances in fruit breeding*. Edited by J. Janick and J.N. Moore. Purdue University Press, West Lafayette, IN. pp. 197–268.

Keep, E. 1995. Currants. In *Evolution of crop plants*. Second edition. Edited by J. Smartt and N.W. Simmonds. Longman Scientific & Technical, Burnt Mill, Harlow, Essex, UK. pp. 235–239.

Miculka, B. 1980. Chemical composition and canning property of golden currants [*Ribes aureum*]. *Zahradnictvi-UVTIZ* (Czechoslovakia) 7(1): 21–25. (In Czech)

Morin, N. 2009. *Ribes*. In *Flora of North America*, Vol. 8. Edited by Flora of North America Editorial Committee, Oxford University Press, New York, NY. pp. 9–42.

Williamson, B., Brennan, R.M., and Gordon, S.L. (Editors). 2002. *Proceedings of the Eighth International* Rubus *and* Ribes *Symposium, Dundee, Scotland, July 2001*. 2 Volumes. International Society for Horticultural Science, Leuven, Belgium.

## Specialty Cookbooks

Hendrickson, R. 1981. *The berry book. The illustrated home gardener's guide to growing and using over 50 kinds and 500 varieties of berries*. Doubleday & Company, Garden City, NY. (Chapter 12: The electrifying currant, 184–193.)

Pellowe, S. 1989. *Saffron & currants: A Cornish heritage cookbook*. Renard Productions, Aurora, IL. pp. 51.

Senn, C.H. 1906. *Currants: A few tasty recipes*. London, UK. pp. 23.

Senn, C.H. 1915. *Home recipes for cooking currants*. London, UK. pp. 32.

Senn, C.H. 1919. *Currant cookery*. London, UK. pp. 24.

Wayne, S. 1930–1939. *The currant recipe book*. Advertising Department, Central Currant Office, London, UK. pp. 32.

Freitus and Haberman (2005) have 9 wild currant recipes, Krumm (1991) has 8, Marie (2008) has 1, Marrone (2009) has 16, Niethammer (1999) has 2, Stanek and Butcher (2007) have 6, and Williamson (1995) has 11. (See the Appendix to this book for details of the publications cited.)

# 44 Grapes

Family: Vitaceae (grape family)

## NAMES

Scientific names: *Vitis* species

- Of the species native to North America, the following two are the most important:
  - Fox grape—*Vitis labrusca* L.
  - Muscadine grape—*V. rotundifolia* Michx.
- Some other native American species are of minor commercial interest, for example:
  - Summer grape (pigeon grape)—*V. aestivalis* Michx.
- The English word "grape" is from the Old French words *grape*, *grappe*, and *crape*, "bunch of grapes." The word was derived in the thirteenth century from the French term *grape*, a small billhook used to cut off grapes from the vine, this usage tracing to the Old High German *krapfo*, "a hook."
- The scientific name *Vitis* is the classical Latin word for vine.
- The fox grape is said to be named for the "foxy," aromatic, or musky odor. By 1864, Webster's Dictionary defined "foxy" as having the coarse flavor of the fox grape. Whether the grapes actually smell like a fox has been questioned, and in any event, very few people today have been close enough to a fox to smell one. It has also been said that "fox" in the name was simply a way of indicating that the species is wild. The fox grape is also known as the northern fox grape, plum grape, northern muscadine, skunk grape, and swamp grape.
- *Labrusca* in the scientific name of the fox grape is an early Latin name for European wild grapes.
- "Muscadine" is probably a variant of "muscatel," from Old French *muscadel*, resembling musk (tracing to the Latin *muscus*, musk), and applied to the Muscat grape used for wine and raisins. The muscadine grape is also known as the bullace, scuppernong, and southern fox grape.
- "Catawba" was the name of an Indian tribe of the Carolinas who were part of the Sioux nation. The word was originally *katapa*, meaning "to be separate," reflecting the separation of the Catawba from the other Sioux, who lived mostly in the west on the Great Plains. The name was passed on to the Catawba River, part of the original homeland of the tribe, and eventually to the Catawba grape.
- *Rotundifolia* in the scientific name of the muscadine grape is Latin for "rounded-leaved."
- *Aestivalis* in the name *V. aestivalis* is a Latin adjective meaning "pertaining to the summer."
- Scuppernong is the best known variety of the muscadine grape, *V. rotundifolia*. The name "scuppernong" comes from the Scuppernong River and is derived from the Algonquin Indian name *Askuponong*, interpreted as "place of the sweet bay tree" or "place of the magnolia." The first English settlement in the New World was in 1586 at Roanoke Island, off the northeast coast of North Carolina, and the colonists are credited with discovering the original vine growing there around 1584–1585. During the American Civil War, troops garrisoning Roanoke Island noted the vine was still there and growing strong. Today, the vine is more than 400 years old, has a trunk 0.6 m (2 feet) thick, covers 0.2 ha (half an acre), and is a tourist attraction. Over the years, many cuttings were taken to grow the grape commercially.

**FIGURE 44.1** Fox grape (*Vitis labrusca*). (a) Painting. (From Jacquin, N.J, *Plantarum rariorum horti caea-rei Schoenbrunnensis*, B. Apud and J. White, *Lugduni Batavorum*, Vol. 4, Plate 426, 1804.) (b) Bunch of grapes. (Courtesy of Takk [CC By 3.0].) (c) Bunch of grapes of the cultivar 'Wheeler'. (Courtesy of U.S. Department of Agriculture.) (d) Distribution map.

## GEOGRAPHY AND ECOLOGY OF WILD PLANTS

The fox grape is native to the eastern United States. It occurs in wet or dry thickets on the borders of woods and in a variety of similar habitats. The muscadine grape grows as a wild plant in the southeastern United States and in Mexico. It is found in woods, thickets, sandhills, and shores. The summer grape is native to the eastern U.S., reaching into southernmost Ontario in Canada. Summer grape occurs in open forests, woodlands, woodland borders, and thickets, and climbs whatever trees and shrubs are at hand. In Europe, several of the North American species that have been used as rootstocks have escaped from cultivation and have become invasive alien weeds, often displacing the native *V. vinifera* L.

**FIGURE 44.2** Muscadine grape (*Vitis rotundifolia*). (a) Close-up of bunch of fruit. (Courtesy of David Nance, U.S. Department of Agriculture; online at Bugwood.org [CC By 3.0].) (b) Enormous grapevine. (Courtesy of Chrishibbard7 [CC By 3.0].) (c) Painting of immature (green) bunch of grapes. (From Lindley, J., *Pomological Magazine*, J. Ridgeway, London, U.K., Vol. 1, 1827–1828.) (d) Fruits hanging on vine. (Courtesy of Carl Hunter, U.S. Department of Agriculture.) (e) Distribution map.

## PLANT PORTRAIT

Grapes are the subject of specialized areas of science and technology. Viticulture is the cultivation and production of grapes; enology is the science of wine and wine making; ampelography (pronounced am-peh-LAW-gra-fee; from the Greek *ampelos*, vine, + *graphe*, writing) is the study and classification of grapevines and grapes (which today includes DNA fingerprinting).

There are about 60 species of the genus *Vitis*, all of which produce edible, although not necessarily palatable fruits. About 18 species are native to North America north of Mexico. The woody vines of grape species climb by means of tendrils. Many grape cultivars are of hybrid origin and not simply referable to any one species. Also, many wild species and hybrids are used as rootstocks, on which the fruit-bearing portions of the principal species are grafted. Modern cultivars of *Vitis*

**FIGURE 44.3** Summer grape (*Vitis aestivalis*). (a) Norton grapes. (Courtesy of Don Kasak [CC By 2.0].) (b) Norton wine bottle. (Courtesy of Whitebox [CC By 3.0].) (c) Bunch of grapes. (Courtesy of U.S. Department of Agriculture.) (d) Distribution map.

are clones, essentially genetically uniform, and are propagated vegetatively, like apple and potato varieties. Seedless clones are of course easily reproduced without seeds. Grapes are an unusual fruit in that they must be allowed to ripen fully before harvesting, as further ripening does not occur after the fruit is picked. Accordingly, many attempts to grow grapes in regions to which they are not climatically suited end in disaster. Nevertheless, there are 2000 or more wineries in North America, and almost every state of the United States and province of Canada produces some type of wine. Mexico makes some commercial wines, but its grapes are used mostly for producing brandy.

The Old World wine grape, *V. vinifera*, is by far the most important grape species, accounting for 95% of grape production. This is not well adapted to cultivation in North America except for part of California, but hybridization with American species has increased its range of cultivation. Many of the hybrids are called "French hybrids" because French breeders did much of the work, beginning in the early 1900s. The French found that of the American species, *V. berlandieri* Planch.,

*V. riparia* Michx., and *V. rupestris* Scheele were of greatest value to their hybridization programs. Other North American species that have played important roles in generating commercial cultivars through hybridization with *V. vinifera* include *V. aestivalis*, *V. candicans* Engelm., *V. cinerea* Engelm., *V. cordifolia* Michx., *V. labrusca*, *V. lincecumii* Buckl., and *V. rotundifola*.

For many years France has been the largest producer of wine (although recently Italy has become first). In 1880, the vineyards of France were almost completely destroyed by an aphid called phylloxera or root louse phylloxera (*Phylloxera vitifoliae* (Fitch)), accidentally introduced from eastern North America about 1860. The grape phylloxera produces galls on leaves and roots. The root injury may be very destructive to European grapes, but American varieties are more or less resistant. As New World grapes are essentially immune to root injury, American species have been used as rootstocks since 1880 in Europe, effectively stopping the depredation of phylloxera, although not until 1930 had normal production resumed in Europe. However, wild *V. vinifera* of the Old World has been devastated by this insect. Today, all wines from France are made from fruit of vines grown on American roots. Chile is rather unique in still growing the wine grape on wine grape rootstocks. Most of the world's vineyards grow European *V. vinifera* vines that have been grafted onto North American species rootstock. Although phylloxera was the original and perhaps still the chief reason for growing the wine grape on American rootstocks, the practice has been found to also confer resistance to other harmful organisms, increase productivity and quality of grapes in some circumstances, and often increase tolerance of climate and soil conditions. The American species most used as rootstocks are *V. berlandieri*, *V. labrusca*, *V. riparia*, *V. rotundifolia*, and *V. rupestris*. Rootstock cultivars, often of hybrid origin, have been selected for their performance with particular wine grape cultivars grown in particular regions.

## Fox Grape (*Vitis labrusca*)

Varieties developed from the fox grape are hardier and more disease resistant than the Old World grape. The berries may be green, red, purple-black, or reddish-brown, and are 1.3–2.5 cm (0.5–1 inch) in diameter. They are sweet or astringent and "foxy" in flavor. Generally, fox grape cultivars are hybrids with other species, especially the common European grape, or at least have some other species in their parentage. Concord is the best known variety, and Catawba, Delaware, and Niagara are other well-known varieties. Concord was originated by Ephraim W. Bull (1805–1985) in Concord, Massachusetts. It grew from a seed planted in 1843, which bore fruit in 1849. Prior to the appearance of the Concord, grape growing in eastern North America had limited success. Except for cultivars of the common European grape, the Concord is the most important grape cultivar in the New World. Grape juice and jelly are typically made from Concord grapes. Catawba was first discovered in woods near the Catawba River in North Carolina in 1901. It is a hybrid of the fox grape and the wine grape, and produces a dark, rich, red wine with a taste reminiscent of muscat. The cultivation of Catawba helped to establish grape growing in the Atlantic states.

## Muscadine Grape (*Vitis rotundifolia*)

The berries of muscadine grape are 2–4 cm (0.8–1.6 inches) in diameter and have a thick, tough skin. The grapes tend to be dull purple, but a range of other colors have been recorded. As with all American species, the wild fruit was harvested by Native Americans. The species was domesticated by settlers in the New World, and several cultivars are now grown in the Cotton Belt of the United States. Commercial production of muscadine grapes is essentially limited to the U.S. southeast. Scuppernong is the oldest and best known variety, and indeed its name has become a general name for all muscadine grapes. Several dozen cultivars are now available. Like the fox grape, muscadine grapes may have strongly musk-flavored grapes. However, some modern cultivars have a unique fruity flavor with very little muskiness. The fruit of the muscadine grape is eaten fresh and is commonly made into juice, wines, pies, jellies, and other processed products.

### Summer Grape (*Vitis aestivalis*)

Two cultivars of the summer grape, 'Norton' and 'Cynthiana', that have been shown to be essentially identical, are said to represent the oldest American grape still being grown (priority is usually given to the name 'Norton'). This grape variety has been grown commercially for more than a century in the eastern United States. The cultivar is thought to be the result of hybridization between the American *V. aestivalis* and the European wine grape, *V. vinifera*. The cultivated grape lacks the distinctive (and objectionable) flavors of American grapes and so is suitable for preparing dry wines. Indeed, the cultivar has been characterized as the best native American grape variety for fine wine, and it is a special product of a number of American wineries. Norton wine has a deep dark red color and a spicy raspberry aroma.

## CULINARY PORTRAIT

### Table Grapes

Grapes used fresh should have attractive appearance (large, bright, and sometimes unusually shaped berries), and grapes used for storage or shipping should have firm, tough, fairly thick skin. Seedless table grapes have become very popular in recent decades. Grapes can be eaten out of hand; added to salads; used as an accompaniment to poultry, meat, and seafood; made into sauces, jam, or jelly; or used as a component of a very wide range of dishes, including pastries, stuffings, curries, and stews. Grapes tend to shrivel and ferment at room temperature and are best stored in a perforated plastic bag in a refrigerator. If only part of a bunch of grapes is being consumed, it is preferable to cut off small clusters from the main stem with scissors, instead of pulling off individual grapes, as this tends to cause drying out of the remaining bunch when it is stored.

### Juice Grapes

Juice grapes are used to produce unfermented juice. Grape juice is pasteurized in North America, which causes most vinifera cultivars to lose their fresh flavor. The strongly flavored American variety Concord retains its flavor very well and so is widely used for juice.

### Raisins

Grapes used for drying are table grapes, rather than wine grapes. Raisin grapes are produced primarily from the varieties Thompson Seedless, Malaga, Sultanina, Black Corinth, and Muscat, all of which are cultivars of *V. vinifera*. Desirable characteristics of raisin grapes include seedlessness; pleasing, rich flavor; soft texture; high sugar content; tender skin; and little tendency to become sticky. The large-fruited Muscat has seeds, but these are removed after drying. Black Corinth grapes, also known as Zante currants, are tiny, black, seedless grapes the raisins of which are very popular in making pastries. "Golden raisins" are produced by treating Thompson grapes with sulfites before they are dried, resulting in conservation of their light color, which varies from amber to golden yellow. Raisins are eaten directly; added to nut and dried fruit mixtures; mixed with cereals, salads, sauces, poultry stuffing, meat loaf, puddings, and other dishes; and cooked in pastries, breads, pudding, and other confections.

### Wine

Although wines can be made from various fruits and even other materials (mead, for example, is a wine made from honey), wine is mostly fermented grape juice. Today, most grapes are consumed as wine and spirits, and in the light of this it is perhaps not surprising that world production of grapes exceeds that of any other fruit. Table (dessert) grapes, raisin grapes, wine grapes, and juice grapes are often different, specialized cultivars. Cultivars suitable for table grapes are usually not suitable for making wine, and vice versa. All mature grapes may be fermented into wine, but only certain cultivars produce wine of high quality, characterized by distinctive flavor, bouquet, and general excellence. Table or dry wines are produced from grapes of high acidity and moderate sugar content, whereas dessert or sweet wines are usually made of grapes high in sugar and relatively low in acid. High yield and high quality of the resulting wine are the standards by which wine grapes are judged. Varieties based on American grape species do not make wine of high quality comparable to those produced from *V. vinifera*, but hybrids with the latter can produce excellent wines.

## Warning

Some wild grapes look similar to Canadian moonseed (*Menispermum canadense* L.), which is also a vine and grows in eastern North America, where the three species highlighted in this chapter also grow. Moonseed berries, which are very poisonous, have sometimes been mistaken and consumed as wild grapes. Moonseed fruits have a single, crescent-shaped seed; grape seeds are round.

## Culinary Vocabulary

*Note:* In the English language, there are more terms descriptive of wine than for any other food. The following represents a selection:

- The word "raisin" is from the Old French *resin*, from the Latin *racemus*, "bunch of grapes."
- The word "vinegar" comes from the French *vin aigre* (or more accurately the Old French *vyn egre*) meaning "sour wine." However, not all vinegar is made from wine, and indeed any food that can be fermented to produce alcohol can also be used to manufacture vinegar. The best vinegars are made from wine or cider. Wine vinegar is made from wine. Balsamic vinegar is made from concentrated grape juice, according to traditional methods; it is usually aged for 4 to 5 years (occasionally as long as 40 years), producing a dark brown, somewhat syrupy fluid which is only slightly acidic and has a unique taste. Malt vinegar is made from ale or beer.
- The word "toast" has been applied since Roman times to a proposal of health prior to imbibing. This tradition originated because of the Roman practice of adding spiced toast to wine to improve its flavor and absorb sediment.
- A "dry" (as opposed to "sweet") wine is one lacking in sugar. Terms for describing the dryness of champagne (French terms in italics) are *doux*: extremely sweet; *demi-sec*, dry: moderately sweet; *sec*: noticeable sweetness; extra dry: slight sweetness; *brut*: no perceptible sweetness; *brut nature*: bone dry. *Brut*, (meaning "rough" in French, reflecting the fact that long ago the French preferred sweeter champagne) indicates less than 2% sugar, and is the most widely enjoyed style, even among the French. In California wine parlance, "dry" sometimes means "unfortified."
- Under world trade agreements, a special "*Appellation d'Origine Controlée*" (AOC) is issued for certain products. An excellent AOC white wine is based on the "limited district" of Cadillac in France, and is in demand in North America by wine connoisseurs. The French name "Cadillac" was exported to the New World, most famously by French pioneer Antoine de la Mothe Cadillac (1658–1730), the founder of Detroit, and the man after whom the Cadillac car was named.
- "Cooking wine" is wine to which salt has been added—a concept that apparently traces to the Prohibition era, so that people would use the wine in cooking and not drink it. No one with any respect for food would purchase this, and regular drinking wines are used in cooking.
- "Raisin wine" is a sweet wine prepared from dried grapes.

- "Champagne" is sparkling wine made in the traditional fashion in the Champagne region of France. In North America, the word is often used as a generic term to describe all sparkling wines (a practice of course that outrages the French). To avoid confusion, most producers in the United States do not call their sparkling wine Champagne, even when it is made in the traditional method. Other generic language terms for sparkling wine are *vins mousseux* (the French term for sparkling wines produced outside France's Champagne region, *spumante* (Italian), *sekt* (German), and *cava* (Spanish). In North America, sparkling wine, especially Champagne, is often called "bubbly" and "gigglewater."
- "Jewish champagne" is an affectionate term that was sometimes used for seltzer (plain carbonated water) by New York Jews, who frequently drank it to adhere to kosher rules (for example, not drinking milk with meats).
- Wine and cheese are a natural pair of fermented foods, but wine grapes and cheese also combine well in "grape cheese," known in France as *tomme au marc* and *fondue aux raisins*. This is a cheese made with an edible crust of marc (the grape pulp that remains after pressing grapes for wine).
- "Post Grape-Nuts" has neither grapes nor nuts, but is made from wheat and barley. The cereal was created in 1897 by American inventor C.W. Post (1854–1914) who put "grape" in the name because it contained maltose (which Post called "grape sugar") and "nuts" because of its flavor.

## PROSPECTS

American grape species have been called "a rowdy and uncouth bunch," but this hardly does justice to their importance. Fox grape and muscadine grape are significant commercial crops for juice and jam in North America and are important contributors to the genetic improvement and vigor of the wine grape. The grape cultivar 'Norton'/'Cynthiana', is of much lesser significance, but does illustrate that there is a considerable range of wild American grape species with unexplored commercial value. The importance of all of these American grapes is dwarfed by the European wine grape, but the latter is the world's most important fruit (albeit, mainly for the production of wine). The market demand for grape products is steady, indeed increasing, and the future is likely to see a modest but growing expansion of the cultivation of American grapes and their hybrids. Cultivation of the wine grape in Europe is under notable threat because of global warming (wine grape quality is extremely susceptible to relatively small changes in the weather), and the hardy American grape species are likely to play an increasing role in breeding new wine grape varieties that meet future climatic stresses.

## CURIOSITIES OF SCIENCE AND TECHNOLOGY

- The Norse explorer, Leif Ericson, based in Greenland, visited the northeastern coast of what is now Canada about 1000 AD. He was so impressed by the rich growth of native grapes that he observed (likely *V. labrusca*) that he dubbed the New World "Wineland the Good." The exact location of "Vinland" as Ericson termed it has been disputed. L'Anse aux Meadows in Newfoundland is the only known Norse settlement in North America and was occupied for a short time around 1000 AD. However, the nearest wild grapes are found in New Brunswick, several hundred miles to the south. It has been suggested that L'Anse aux Meadows was Ericson's base camp, from which he explored nearby areas of North America, which is the presumed "Wineland."
- A Virginia law enacted in 1619 required every household in the colony to plant and tend 10 wine grape cuttings. Prizes were offered for the best wines.
- The first commercial winery in the United States was established in 1823 in Missouri.

- Dentist Thomas Bramwell Welch (1825–1903) of Vineland, New Jersey, created an unfermented grape sugar beverage in 1869 that he called "Dr. Welch's Unfermented Wine." In effect, he had applied the process that Louis Pasteur invented 4 years earlier to sterilize wine, beer, and milk. Today, Welch Food Inc. is one of America's largest companies, specializing in juices and other products.
- Raisin pies were traditionally prepared by Pennsylvania Dutch farm women. These and other types of pastries were stored in cool cellars until required for such occasions as wakes and weddings, and as a result raisin pies became a traditional "mourning food" served at funerals. Raisin pie became associated so often with funerals in colonial New England and Pennsylvania that an inquiry of whether a raisin pie was needed was a way of asking if someone was seriously ill.
- The Titanic carried 1500 bottles of wine, 2000 wine glasses, and 1500 grape scissors on its fateful voyage of 1912.
- In 1919, the Eighteenth Amendment to the U.S. Constitution was passed, which initiated a period of alcohol prohibition starting in 1920 and lasting until repealed by the Twenty-first Amendment in 1933 (some states had banned all alcoholic beverages as early as 1916; Maine forbade the sale of liquor in 1841). The period of Prohibition caused great hardship for U.S. vineyards and the large wine industry. Sacramental wine was allowed for Catholic and Jewish services, and much of New York's grapes were used for the purpose. Moreover, there was another loophole: section 29 of the Volstead Act of 1919, which initiated Prohibition allowed citizens to make up to 200 gallons a year of "homemade cider" or "nonintoxicating fruit juice," which was widely understood to be a euphemism for regular alcoholic wine. (Two hundred gallons of home-brewed alcoholic beverage is still the limit in many states.) This peculiarity of the legislation allowed grape growers to sell their fruit to home winemakers and resulted in a considerable increase of wine grape production in the United States during Prohibition. In 1932, wine with 3.2% alcohol was allowed and was called "McAdoo wine" after Senator William Gibbs McAdoo of California, who sponsored the bill. During Prohibition, "Grape Brick," made up of dried and pressed wine grape concentrate, was sold, along with an attached packet of yeast, and the warning, "Do not add yeast or fermentation will result." Because of these factors, wine consumption increased considerably during Prohibition, although most of it was not of high quality.
- In 2001, North Carolina declared the scuppernong grape to be its official state fruit.
- The 'Norton'/'Cynthiana' grape (partly *V. aestivalis*) was adopted as the official state grape of Missouri in 2003.
- A small winery in Arkansas sent a bottle of its prized white wine to the University of Arkansas requesting a chemical analysis. The report came back: "Your horse has diabetes."
- Researchers at Glasgow University discovered that after consuming two glasses of wine, members of the opposite sex appeared about 25% more attractive.

## KEY INFORMATION SOURCES

(*Note:* Thousands of books and tens of thousands of articles have been written about grapes and, more particularly, wine. The following is a selection.)


Adams, L.D., and Novak, B. 1990. *Wines of America.* Fourth edition. McGraw-Hill, New York. NY. pp. 528.

Alleweldt, G., Spiegel-Roy, P., and Reisch, B. 1990. Grapes (*Vitis*). In *Genetic resources of temperate fruit and nut crops*. Edited by J.N. Moore and J.R. Ballington Jr. International Society of Horticulture Science, Wageningen, The Netherlands. pp. 291–327.

Amerine, M.A., and Phaff, H. 1986. *A bibliography of publications by the faculty, staff, and students of the University of California, 1876–1980, on grapes, wines, and related subjects*. University of California Press, Berkeley, CA. pp. 244.

Arrigo, N., and Arnold, C. 2007. Naturalised *Vitis* rootstocks in Europe and consequences to native wild grapevine. *Plos One* 2: 521–529.

Brenner, L. 1995. *Fear of wine: An introductory guide to the grape*. Bantam Books, New York, NY. pp. 310.
Cass, B., and Robinson, J. 2000. *The Oxford companion to the wines of North America*. Oxford University Press, Oxford, UK. pp. 301.
Clarke, O., and Rand, M. 2001. *Oz Clarke's encyclopedia of grapes*. Harcourt, New York, NY. pp. 320.
Conner, P.J. 2009. A century of muscadine grape (*Vitis rotundifolia* Michx.) breeding at the University of Georgia. *Acta Hort*. 827: 481–484.
Cox, J. 1999. *From vines to wines: The complete guide to growing grapes and making your own wine*. Storey Books, Pownal, VT. pp. 235.
Einset, J., and Pratt, C.1975. Grapes. In *Advances in fruit breeding*. Edited by J. Janick and J.N. Moore. Purdue University Press, West Lafayette, IN. pp. 130–153.
Foulkes, C. 1994. *Larousse encyclopedia of wine*. Larousse, New York, NY. pp. 608.
Gohdes, C.L.F. 1982. *Scuppernong: North Carolina's grape and its wines*. Duke University Press, Durham, NC. pp. 115.
Gray, D.J., and Meredith, C.P. 1992. Grape. In *Biotechnology of perennial fruit crops*. Edited by F.A. Hammerschlag and R.E. Litz. CABI Publishing, Cambridge, MA. pp. 229–262.
Guttadauro, G.J. 1976. *A list of references for the history of grapes, wines, and raisins in America*. Agricultural History Center, University of California, Davis, CA. pp. 70.
Harris, J.M., Kriedemann, P.E., and Possigham, J.V. 1968. Anatomical aspects of grape berry development. *Vitis* 7: 106–119.
Iverson, J. 1998. *Home winemaking, step-by-step: A guide to fermenting wine grapes*. Second edition. Stonemark Publishing Company, Medford, OR. pp. 172.
Jackson, D., and Schuster, D. 1987. *The production of grapes & wine in cool climates*. Butterworths of New Zealand, Wellington, New Zealand. pp. 192.
Jancis, J. 1994. *The Oxford companion to wine*. Oxford University Press, Oxford, UK. pp. 1088.
Jindal, P.C. 1985. Grape. In *Fruits of India, tropical and subtropical*. Edited by T.K. Bose. Naya Prokash, Calcutta, India. pp. 219–276.
Jindal, P.C. 1999. Grape. In *Tropical horticulture*. Vol. 1. Edited by T.K. Bose, S.K. Mitra, A.A. Farooqui, and M.K. Sadhu. Naya Prokash, Calcutta, India. pp. 259–284.
Johnson, H. 1985. *The world atlas of wine: A complete guide to the wines and spirits of the world*. Third edition. Simon and Schuster, New York, NY. pp. 320.
Kasimatis, A.N. 1975. *How to produce quality raisins*. University of California, Division of Agricultural Sciences, Berkeley, CA. pp. 9.
Kaufman, W.I. 1984. *Encyclopedia of American wine, including Mexico and Canada*. J.P. Tarcher, Los Angeles, CA. pp. 564.
Kerridge, G., and Antcliff, A.J. 1999. *Wine grape varieties*. Revised edition. CSIRO Publishing, Collingwood, VIC, Australia. pp. 205.
Kolpan, S., Smith, B.H., and Weiss, M.A. 1996. *Exploring wine: The Culinary Institute of America's complete guide to wines of the world*. Van Nostrand Reinhold, New York, NY. pp. 699.
Lavee, S., and Nir, G. 1986. Grape. In *CRC handbook of fruit set and development*. Edited by S.P. Monselise. CRC Press, Boca Raton, FL. pp.167–191.
McGrew, J.R. 1993. *Growing wine grapes*. G.W. Kent, Inc., Ann Arbor, MI. pp. 91.
Moore, M.O. 1991. Classification and systematics of eastern North America *Vitis* L. (Vitaceae) north of Mexico. *Sida* 14: 339–367.
Newman, J.L. 2000. Wine. In *The Cambridge world history of food*. Edited by K.F. Kiple and K.C. Ornelas. Cambridge University Press, Cambridge, UK. pp. 730–737.
Olmo, H.P. 1995. Grapes. In *Evolution of crop plants*. Second edition. Edited by J. Smartt and N.W. Simmonds. Longman Scientific & Technical, Burnt Mill, Harlow, Essex, UK. pp. 485–490.
Parker, L., Bordallo, P., and Colova. 2006. Phylogenetic analysis of North American native Cynthia/Norton grape variety using DNA microsatellite markers. *Proc. Fla. State Hort. Soc*. 119: 61–65.
Pearson, R.C., and Goheen, A.C. 1988. *Compendium of grape diseases*. APS Press, St. Paul, MN. pp. 93.
Pratt, C. 1971. Reproductive anatomy in cultivated grapes. A review. *Am. J. Enol. Vitic*. 22: 92–109.
Priew, J. 1999. *Wine: From grape to glass*. Abbeville Press, New York, NY. pp. 256.
Ramey, B.C. 1977. *The great wine grapes and the wines they make*. Great Wine Grapes, Inc., Burlingame, CA. pp. 250.
Reisch, B., Goodman, R.N., Martens, M., and Weeden, N.F. 1993. The relationship between Norton and Cynthiana red wine cultivars derived from *Vitis aestivalis*. *Am. J. Enol. Viticult*. 50: 236–242.
Robinson, J. 1986. *Vines, grapes, and wines*. Knopf, New York, NY. pp. 280.

Robinson, J. 1996. *Jancis Robinson's guide to wine grapes*. Oxford University Press, Oxford, UK. pp. 232.
Rombough, L. 2002. *The grape grower: A guide to organic viticulture*. Chelsea Green Publishing, White River Junction, VT. pp. 289.
Simon, A.L. 1957. *The noble grapes and the great wines of France*. McGraw-Hill, New York, NY. pp. 180.
Simon, A.L., and Allen, H.W. 1967. *Wines of the world*. McGraw-Hill, New York, NY. pp. 719.
Stevenson, T. 1997. *The new Sotheby's wine encyclopedia*. DK Publishing, New York, NY. pp. 600.
Tröndle, D., Schröder, S., Kassemeyer, H.-H., Kiefer, C., Koch, M.A., and Nick, P. 2010. Molecular phylogeny of the genus *Vitis* (Vitaceae) based on plastid markers. *Am. J. Bot.* 97: 1168–1178.
Unwin, P.T.H. 1991. *Wine and the vine: An historical geography of viticulture and the wine trade*. Routledge, New York, NY. pp. 409.
Vine, R.P., Harkness, E.M., and Linton, S.J. 2002. *Winemaking: From grape growing to marketplace*. Second edition. Kluwer Academic/Plenum Publishers, New York, NY. pp. 477.
Wagner, P.M. 1976. *Grapes into wine: A guide to winemaking in America*. Knopf, New York, NY. pp. 302.
Weaver, R.J. 1976. *Grape growing*. John Wiley & Sons, New York, NY. pp. 371.
Webb, A.D. 1984. The science of making wine. *Am. Scientist* 72: 360–367.
Winkler, A.J., Cook, J.A., Kliewer, W.M., and Lider, L.A. 1974. *General viticulture*. University of California Press, Berkeley, CA. pp. 710.

## Specialty Cookbooks

### Grapes

Barnes, W. 1977. *The wine country's great grape cook book*. Toll Road Press, Napa Valley, CA. pp. 44.
California Grape Grower. 1925. *The use of grapes in the home. Tested recipes for delectable dishes. Dietetic value of the grape*. The California Grape Grower, San Francisco, CA. pp. 7.

### Raisins

Australian Dried Fruits Association. 1960–1969. *Family favourites with currants, Sultanas, and seeded raisins: Hints, suggestions, ideas, and dozens of delicious recipes*. Australian Dried Fruits Association, Melbourne, Australia. pp. 60.
California Raisin Advisory Board. 1950–1960. *Tested raisin formulas*. California Raisin Advisory Board, Fresno, CA. pp. 54.
California Raisin Advisory Board. 1967. *Bonanza of California raisin originals*. California Raisin Advisory Board, Fresno, CA. pp. 26.
California Raisin Advisory Board. 1977. *Raisins: Everything under the sun*. California Raisin Advisory Board, Fresno, CA. pp. 34.
Deacon, P. 1979. *The wrinkled grape*. Patricia Deacon, Los Angeles, CA. pp. 44. (variously paginated).
Sun-Maid Raisin Growers of California. 1950a. *Downright delicious Sun-Maid raisin recipes*. Sun-Maid Raisin Growers of California, Fresno, CA. pp. 32.
Sun-Maid Raisin Growers of California. 1950b. *More downright delicious Sun-Maid raisin recipes*. Sun-Maid Raisin Growers of California, Fresno, CA. pp. 24.

### Wine

Anderson, S.F., and Anderson, D. 1989. *Winemaking: Recipes, equipment, and techniques for making wine at home*. Harcourt Brace Jovanovich, San Diego, CA. pp. 284.
Ballard, B., and Kittler, P. 1981. *Wine in everyday cooking: "Cooking with wine for family and friends."* Wine Appreciation Guild, San Francisco, CA. pp. 122.
Bowers, W.F., and Bowers, L. 1975. *Gourmet cooking with homemade wines*. Stackpole Books, Harrisburg, PA. pp. 288.
California Wine Advisory Board. 1978. *Gourmet wine cooking, the easy way*. Revised edition. Wine Appreciation Guild, San Francisco, CA. pp. 128.
Chase, E. 1960. *The pleasures of cooking with wine*. Prentice-Hall, Englewood Cliffs, NJ. pp. 243.
Church, R.E. 1976. *Entertaining with wine*. Rand McNally, Chicago, IL. pp. 174.
Crocker, B. 1989. *Betty Crocker's cooking with American wine*. Prentice Hall, New York, NY. pp. 112.
Gaulke, J.A. 1972. *Cooking with wine*. Lane Books, Menlo Park, CA. pp. 79.
Goldstein, J.E., Williams, C., and Goldstein, E. 1999. *Food & wine pairing*. Time-Life Books, Alexandria, VA. pp. 111.

Hoffman, V., and Hoffman, R. 2003. *The great little food with wine cookbook: 76 cooking with wine recipes, pairing food with wine, how and where to buy wine, ordering wine in a restaurant.* Second edition. Rayve Productions, Windsor, ON. pp. 128.

Johnson-Bell, L. 1999. *Pairing wine and food.* Burford Books, Short Hills, NJ. pp. 194.

Logan, A.M. 1972. *Wine and wine cooking: Entertaining and cooking with American wines.* Westover Publishing Company, Richmond, VA. pp. 328.

MacDonald, B. 1976. *Wine in cooking and dining.* The Culinary Institute, Chicago, IL. pp. 96.

Sarvis, S., and Thompson, R. 1973. *American wines and wine cooking.* Creative Home Library, Des Moines, IA. pp. 182.

Smith, J. 1986. *The Frugal Gourmet cooks with wine.* Morrow, New York, NY. pp. 447.

Taylor, G.H. 1963. *Treasury of wine and wine cookery.* Harper & Row, New York, NY. pp. 278.

Time-Life Books. 1983. *Wine.* Time-Life Books, Alexandria, VA. pp.176.

Werlin, L. 2003. *All American cheese and wine book: Pairings, profiles, & recipes.* Stewart, Tabori & Chang, New York, NY. pp. 335.

Willan, A., and Clay, L. 2001. *Cooking with wine.* Harry N. Abrams in cooperation with COPIA, New York, NY. pp. 288.

# 45 Groundnut (Apios)

Family: Fabaceae (Leguminosae; pea family)

## NAMES

Scientific name: *Apios americana* Medik. (*A. tuberosa* Moench.)

- The name "groundnut" (optionally "ground nut") is descriptive of the edible part of the plant, that is, the underground tubers, which are not nuts botanically.
- Groundnut is also called American potato bean, apios, bog potato, hopniss (a Lenape Indian name, originally used by Delaware Native Americans), Indian potato, pea vine, potato bean, sea vine, Virginia potato, wild potato, and wild sweet potato. Rarely, it has been called American groundnut, a rather appropriate name.
- The genus name "apios" is often used as a common name when referring to the plant as a crop, but "groundnut" is more familiar in North America.
- When groundnuts were brought to France from Canada in the sixteenth century, they became known as "Canada."
- The peanut (*Arachis hypogaea* L.) was once called groundnut in the southeastern United States and is often known as groundnut outside of North America.
- The hog peanut or ground bean (see Chapter 49 on Hog Peanut) is often confused with groundnut. Like groundnut, this is also a native wild vine of the pea family. Its edible portion is actually a fruit, much like a peanut, produced underground or at the surface of the ground. The two species often grow together, but can be distinguished in that hog peanut has three leaflets on its leaves instead of the five to seven leaflets on the leaves of groundnut. The names Dakota pea, ground bean, and wild bean, which are best reserved for hog peanut, have been applied to groundnut, usually as a result of confusion.
- Some other unrelated plants with underground edible organs are also called groundnut. The Bambarra groundnut, *Vigna subterranea* (L.) Verdc. (*Voandzeia subterranea* (L.) Thouars. ex DC.), is widely cultivated in tropical regions for its edible seeds. The Kersting's groundnut or ground bean, *Macrotyloma geocarpum* (Harms) Maréchal & Baudet (*Kerstingiella geocarpa* Harms), is used as a vegetable in West Africa.
- The "earthnut" (or earth nut) is chufa (*Cyperus esculentus* L. see Chapter 28 on Chufa). The term is also applied to a tuberous-rooted native of Eurasia, *Carum bulbocastanum* (L.) W.D.J. Koch (*Bunium bulbocastanum* L.), which produces edible nutlike bulbs sometimes used as a wild food.
- The name "wild bean" is more often applied to wild forms of the common bean (*Phaseolus vulgaris* L.) than to groundnut. However, the name wild bean is frequently applied to groundnut when it occurs as a weed (i.e., where it is not wanted in another crop, especially in cranberries).
- In North America, the name "Indian potato," occasionally used for groundnut, may also refer to *Orogenia linearifolia* S. Watson, a western North American member of the parsley family. Outside of North America, "Indian potato" generally refers to varieties of the common potato (*Solanum tuberosum* L.) grown in India.

**FIGURE 45.1** Groundnut (*Apios americana*). (a) Dried specimen (Agriculture and Agri-Food Canada herbarium) with strings of tubers. (Courtesy of E. Small and B. Brookes.) (b) Flowers. (Courtesy of Alan Cressler, reproduced with permission.) (c) Tubers. (Courtesy of Thomas Kent, FloraFinder.com, reproduced with permission.) (d) Painting of flowering branch. (From Curtis, W., *Botanical Magazine*, Vol. 30, Plate 1198, 1809.) (e) Distribution map.

- The genus name *Apios* is derived from the Greek word *apios*, pear, referring to the pear-shaped underground tubers.
- *Americana* in the scientific name *A. americana* means American, referring to the American distribution of the species.

## GEOGRAPHY AND ECOLOGY OF WILD PLANTS

The groundnut is native to North America, from southern Canada to the Gulf of Mexico, and from the Great Plains to the east coast. The distribution ranges from New Brunswick west to Minnesota, and south to Florida and Texas. Groundnut is generally a plant of rich soil, but also is found in sand. Its habitats include woodland edges, borders of marshes and shorelines, the sides of creeks, and in low thickets, wet meadows, hammocks, floodplains, swamps, ditches, moist woodlands, and prairie ravines. Choctaw Indians called the groundnut *lukckuk ahe*, "mud potato," a charming reference to the fact that the tubers are often found in moist, muddy ground. Although the species is usually found in moist soils, it can also occur in much drier sites. Groundnut is capable of growing well in both full sun and shade. The species is common in much of its range, often forming large colonies. It is sometimes a serious weed of cultivated cranberry bogs.

## PLANT PORTRAIT

The groundnut has been described as the most famous edible wild plant in eastern North America. The plant is a twining (counterclockwise) or climbing, perennial, herbaceous vine, 1–3 m (3–10 feet), occasionally to 6 m (20 feet) long. The very slender vines sprawl over the ground and climb over shrubs and up trees. The vines are killed back to the ground by frost each year, the plants regrowing in the spring from the underground tubers. Two to 20 white-fleshed, edible tubers (swollen storage organs derived from stem tissue) are developed beneath the ground surface, in chains in which the tubers may be almost adjacent or up to 30 cm (1 foot) apart. The appearance of the tubers on the rhizomes (underground stems) is reminiscent of an oversize necklace with beads of irregular size and position. The tubers range in size from 2–10 cm (0.8–4 inches) in diameter, rarely larger. The tubers are rather small the first year, becoming larger in the second and third years of growth. Groundnut has pea-like flowers that are deep red-brown, purple-brown, or chocolate-brown, and arranged in heads. Its pods are 5–12 cm (2–4.7 inches) long and have 6–11 edible seeds.

Indigenous peoples and early explorers used the potato-like tubers and the pea-like seeds as food. The tubers were eaten raw, boiled, or ground and sometimes combined with other foods. The Iroquois collected the tubers and planted them in suitable places near their villages and are reported to have grown them in parts of New York state. Although it was apparently not a major crop, cultivation was reported by at least four early writers, and several others noted that groundnut was important when corn crops failed. Records of use by many tribes at the time of first European contact suggest that groundnut was used by native people throughout most of eastern North America. The common potato, although indigenous to the New World (Mexico southward), was only consumed by northeastern American Indians after the arrival of Europeans.

The availability of groundnut tubers contributed to the survival of the Pilgrims through their first winter, but the first European to eat the tubers was probably the scientist Thomas Harriot who accompanied Sir Walter Raleigh to Virginia in 1585 and remarked "they are very good meate." Harriot was the first to record the use of groundnut by native people and the first to bring the tubers to Europe when he brought them to England. Groundnuts were served at a feast, hosted by the Indian Chief Chkondum for the French explorer Samuel de Champlain, when he reached the mouth of the Saint John River, New Brunswick.

By the late 1800s, groundnut was well established in gardens in many parts of Europe. During the Irish potato famine of 1845, there was an effort to increase the cultivation of groundnut in Europe, but it was abandoned when growth of the familiar potato was resumed successfully with disease-resistant strains. Groundnut has been cultivated locally for more than a century in southern Europe for its edible tubers and has established itself as a wild plant in some parts of Europe. It was introduced to Japan about 100 years ago and was initially used as a medicinal crop for postnatal medication to reduce fatigue and to improve the health of children. It is still cultivated in Japan where it is called "apios."

Groundnut requires 2 to 3 years to develop full-sized tubers, and these are considerably smaller than produced by potato in a single season. But the comparison is not quite fair as groundnut is essentially wild, while potato has been selected for thousands of years. A research program in Louisiana resulted in a strain of groundnut that formed 3.7 kg (7 pounds) of tubers per plant in a single growing season, indicating that in time groundnut could be a useful new vegetable. However, that program was terminated.

Collecting wild groundnuts requires botanical expertise to be sure harmful or rare plants are not accidentally harvested. Groundnut is very seldom available commercially, so the only reliable way to experience this vegetable is to grow your own. There are a few nursery catalogue sources available that can supply tubers to be used to establish plants in the home garden.

## CULINARY PORTRAIT

After scrubbing to remove the soil, groundnut tubers can be either boiled or fried. Although the tubers are sometimes eaten raw, they contain some antinutritional factors that are destroyed with heat, so they should be cooked before being eaten. Peeling before cooking is not required (some have recommended removing the skin after the tubers have cooked for a while, which results in the "jackets" becoming loose). The tubers can be added to salads, soups, stews, and vegetable dishes. The cooked tubers have a dry, mealy, or floury texture and an appealing, mild, slightly sweet taste, reminiscent of a combination of potato, sweet potato, and peanut. In addition to starch, groundnut tubers contain a crude protein content of 12%–18%, two to three times that of potato, and the protein is of extremely high quality. The tubers make excellent chips and have desirable baking qualities when blended with corn meal or wheat flour. The tubers can be dried and ground into a powder for use as a sweetener and thickener in soups. The seeds can be cooked like beans and served as a side dish. It may be noted that a few people have shown an allergic reaction from eating groundnut.

### Culinary Vocabulary

- A "potato nail" is a large-headed, sharp-tipped nail that is inserted into a raw potato so that heat is conducted rapidly into the center of the tuber and it bakes quickly and uniformly. This can be used for the same purpose to bake groundnut.

## PROSPECTS

Groundnut is a nutritious vegetable with an appealing taste. However, it is still essentially a wild plant, with only limited domestication to date. In the 1990s, Frieda's, a distributor of specialty produce, offered groundnuts as part of its "Lost Crops of the Americas" collection of historically indigenous staples. "Apios" (certainly a preferable commercial name) was sold, washed and unpeeled in 8-ounce (227 g) bags. Unfortunately all of the Lost Crops products were discontinued for lack of sales, indicating that considerable improvement is required. The vining nature of the plant makes it difficult to grow and manage on a field scale. Growth of the tubers is slow, requiring 3 years or so to attain a reasonable size. Groundnut competes directly with potato, which is the fourth most important food crop in the world (after the cereals wheat, rice, and corn), and there is virtually no possibility that groundnut could become a staple like potato. However, groundnut has excellent potential to become a valuable niche crop. Some efforts have been made in recent years to develop groundnut as a cash crop (in the United States, Europe, and Japan), but it still requires considerable investment in breeding and production, as well as market and product development. Isoflavones have been found in groundnut that appear to have some potential for treatment of prostate and breast cancers, but this also requires more research. The British organization *Plants for a Future*, concerned with educating the public on "edible, medicinal, and useful plants for a healthier world," ranked groundnut as the fourth most important plant in its database of 7000, suggesting that the species is deserving of attention as a new crop.

## CURIOSITIES OF SCIENCE AND TECHNOLOGY

- Sir Walter Raleigh (about 1552–1629) was an English explorer, best known for allegedly introducing the potato and tobacco into England. It has been speculated that the "potato" that he apparently brought back from the New World was actually the groundnut.

- The Pilgrims of Plymouth, Massachusetts, survived in 1623 on groundnuts when their corn supply was exhausted. Groundnuts were so important to the colonists of the Connecticut River valley that in 1654, the town of Southampton passed a law that prohibited Indians from harvesting them from "English Lands." The first offence was punishable by a period in the stocks; the second by whipping. This was hardly an appropriate way of thanking the Native Americans who had helped them survive.
- Indicative of the historical importance of groundnut to Native Americans, Sag Harbor, about 160 km (100 miles) east of Manhattan on the eastern tip of Long Island, New York state, obtained its name from a place just to the south called Sagaponack, which means "the place where the big groundnuts grow" in Shinnecock, an endangered Algonquian Indian language.
- Shubenacadie, Nova Scotia, acquired its name from a Mi'kmaq language phrase for "abounding in groundnuts."
- Native Americans made such extensive use of groundnuts that one can commonly find the plants growing over old Indian village sites, the plants established from discarded tubers.
- One of the world's greatest botanists, Professor Asa Gray (1810–1888; see Figure 1.8 in Chapter 1) of Harvard University, noted that if civilization had started in America instead of in the Old World, groundnut would have been the first edible tuber to be domesticated and cultivated.
- Euell Gibbons, the famous American exponent of eating wild plants, wrote (1966) "If you want to go further afield for an exclusive perfume ..., try making some scent from the clusters of purplish-brown blossoms of the groundnut.... These blossoms have a rich, heady fragrance, and not every woman can live up to such a scent."
- Like other species in the pea family, groundnut enriches the soil by taking inert nitrogen out of the atmosphere and transforming it into nitrogen compounds that plants can use for growth. Plants in the pea family do this through the aid of bacteria housed in nodules on the roots. It has been estimated that where groundnut grows, it can add 100 kg per hectare (89 pounds per acre) of nitrogen to the soil annually.
- The flowers of groundnut are explosive. When pollinators (generally flies) contact the blooms, an explosive movement of the floral parts occurs, dusting the insects with pollen to be distributed to other plants.
- Most nectar glands occur in flowers and serve to reward pollinators. Extrafloral nectaries are sugar-secreting glands that occur outside of flowers and reward insects for services to the plant other than pollination. Extrafloral nectaries occur at the base of flower clusters in groundnut. They are visited by ants, which might assist the plant in distributing its seeds.
- A substance called genistein in the tubers of groundnut has been shown to reduce incidence of various forms of cancer, including breast cancer. Genistein also has female sex hormone properties. Some nutritionists have recommended an increase in consumption of soy foods in North America to obtain the protective effects of genistein, but it is difficult to get people to change their eating habits. It has been suggested that an increase in genistein consumption could be achieved by marketing the groundnut as an especially healthy type of potato.

## KEY INFORMATION SOURCES

Alan, H.W. 2009. Extrafloral nectaries in kudzu, *Pueraria montana* (Lour.) Merr., and groundnut, *Apios americа* Medicus (Fabaceae). *Castanea* 74: 360–371.

Beardsley, G. 1939. The groundnut as used by the Indians of eastern North America. *Pap. Mich. Acad. Sci. Arts. Lett.* 25: 507–525.

Blackmon, W.J., and Reynolds, B.D. 1986. The crop potential of *Apios americana*: Preliminary evaluations. *Hortscience* 21: 1334–1336.

Bruneau, A., and Anderson, G.J. 1988. Reproductive biology of diploid and triploid *Apios americana* (Leguminosae). *Am. J. Bot.* 12: 1876–1883.

Bruneau, A., and Anderson, G.J. 1994. To bee or not to bee? The pollination biology of *Apios americana* (Leguminosae). *Plant Syst. Evol.* 192: 147–149.
Carlisi, J., and Wollard, D. 2005. History, culture, and nutrition of *Apios americana*. *J. Nutr. Funct. Med. Foods* 4: 85–92.
Catling, P.M., and Small, E. 2000. Poorly known economic plants of Canada—24. Groundnut, *Apios americana* L. *Bull. Can. Bot. Assoc.* 33(2): 12–15.
Duke, J.A. 1987. Help rediscover an American vegetable: *Apios*. *Org. Gard.* 34: 98–101.
Gibbons, E. 1966. *Stalking the healthful herbs*. D. McKay Co., New York. NY. pp. 303.
Hoshikawa, K., and Juliarni. 1995. The growth of apios (*Apios americana* Medikus), a new crop, under field conditions. *Jpn. J. Crop Sci.* 64: 323–327.
Joly, S., and Bruneau, A. 2004. Evolution of triploidy in *Apios americana* (Leguminosae) revealed by genealogical analysis of the histone H3-D gene. *Evolution* 58: 284–295.
Juliarni, and Hoshikawa, K. 1996. Characteristics of dry matter production and yield of apios (*Apios americana* Medikus) clones. *Jpn. J. Crop Sci.* 65: 98–102.
Juliarni, Goto, Y., Nakamura, T., Takahashi, K., and Hoshikawa, K. 1997. Tuberization in apios (*Apios americana* Medikus). 1. Developmental morphology of tuber. *Jpn. J. Crop Sci.* 66: 466–471.
Krishnan, H.B. 1998. Identification of genistein, an anticarcinogenic compound, in the edible tubers of the American groundnut (*Apios americana* Medikus). *Crop Sci.* 38: 1052–1056.
Picha, D.H., Blackmon, W.J., Wilson, P.W., Hanson, L.P., and Reynolds, B.D. 1990. Compositional changes in *Apios americana* tubers during storage. [Abstract.] In *Advances in new crops*. Edited by J. Janick and J.E. Simon. American Phytopathological Society, St. Paul, MN. pp. 443.
Putnam, D.H., Heichel, G.H., and Field, L.A. 1991. Response of *Apios americana* to nitrogen and inoculation. *HortScience* 26: 853–855.
Reynolds, B.D., Blackmon, W.J., Wickremesinhe, E., Wells, M.H., and Constantin, R.J. 1990. Domestication of *Apios americana*. In *Advances in new crops*. Edited by J. Janick and J.E. Simon. American Phytopathological Society, St. Paul, MN. pp. 436–442.
Seabrook, J. 1973. *A biosystematic study of the genus Apios Fabricius (Leguminosae) with special reference to Apios americana Medikus*. Thesis. University of New Brunswick, Fredericton, NB. pp. 244.
Seabrook, J., and Dionne, L.A. 1976. Studies on the genus *Apios*. I. Chromosome number and distribution of *Apios americana* and *A. priceana*. *Can. J. Bot.* 54: 2567–2572.
Vietmeyer, N.D. 1986. The wild groundnut: The next potato? *Apios Tribune, Bull. Dept. Plant Pathol. Crop Physiol. Louisiana Agric. Exp. Stn.* 1: 2–5.
Westerkamp, C., and Paul, H. 1993. *Apios americana*, a fly-pollinated papilionaceous flower? *Plant Syst. Evol.* 187: 135–144.
Wilson, P.W., Gorny, J.R., Blackmon, W.J., and Reynolds, B.D. 1986. Fatty acids in the American groundnut (*Apios americana*). *J. Food Sci.* 51: 1387–1388.
Wilson, P.W., Pichardo, F., Blackmon, W.J., and Reynolds, B.D. 1987. Amino acids in the American groundnut (*Apios americana*). *J. Food Sci.* 52: 224–225.
Woods, M. 2005. A revision of the North American species of *Apios* (Fabaceae). *Castanea* 70: 85–100.
Zinsou, C., Desfontaines, L., and Sobesky, O. 1998. Chemical and biochemical composition of tubers of *Apios americana* Medikus grown ex situ. In *Proceedings of International Symposium on Tuberous Legumes, Celaya, Mexico, August 5–8, 1996*. Edited by M. Soerensen, J.E. Estrella, O.J. Hamann, and S.A.R. Ruiz. Kongelige Veterinaer og Landbohoejskole, Copenhagen, Denmark. pp. 557–565.

## Specialty Cookbooks

Books on edible, eastern North American wild plants will almost certainly provide recipes for groundnut. Searching "groundnut recipes" on the Web will result in some recipes for *Apios americana*, but also peanut recipes. Generally, the tubers can simply be cooked and used in recipes like potatoes. However, the taste is distinctive, and whether groundnuts can be substituted for potatoes in given recipes can only be determined by trial and error. Usually, a mixture of potatoes and groundnuts is delicious.

Thayer, S. Summer/Fall 2002. Hopniss: North America's best wild tuber? *The Forager* 2(3): unpaginated (ca. 3 pp.) (Has extensive information and suggestions on how to prepare groundnut.)

Freitus (1980) presents seven groundnut recipes. Schufer (2011) has a recipe entitled "wild rice and groundnuts." (See the Appendix to this book for details of the publications cited.)

# 46 Hawthorns (Including Mayhaws)

Family: Rosaceae (rose family)

## NAMES

Scientific names: *Crataegus* species

- The name "hawthorn" traces to the Old English *hagathorn*, based on *haga*, haw (an old word for hedge, as well as for hawthorn fruit) and *thorn*, thorn. *Hagathorn* is similar to the German *hagedorn*, also meaning hedge-thorn. This reflects the use of the spiny hawthorn to construct formidable hedges.
- The 15 or so evergreen species of the Asian genus *Rhaphiolepis* (which is also in the rose family) are also known as hawthorns. Some of the species are grown as ornamentals.
- The fruits of *C. aestivalis* and *C. opaca* tend to ripen in May, hence their name "mayhaw."
- The non-capitalization of "mayhaw" (rather than "Mayhaw") follows a frequent practice employed for many biological names (e.g., mayfly, mayweed; cf. Chapters 59 and 60). Nevertheless, mayhaw is often encountered as Mayhaw.
- The genus name *Crataegus* is based on the Greek *keratos*, strength, referring to the solidity of the wood. The hardness is due to the slow growth of the trunk. Montana Native Americans once made digging tools, fish hooks, and pins from the tough wood. Although not available in quantity, the wood is prized for turnery (making objects on a lathe).

## GEOGRAPHY AND ECOLOGY OF WILD PLANTS

Hawthorns are adapted to grow as understory trees and shrubs in temperate zone forests. With the clearing of virgin forests in North America, the plants have become frequent in open areas. Hawthorns are common along the shores of waterways, in abandoned fields, and along fences. The mayhaw species, particularly those in the Gulf states of the southeastern United States (as noted later), tend to grow in moist soils in river and creek bottoms under hardwood trees, and were once often collected from boats to prepare large quantities of homemade jelly. With increasing urbanization of the south and the associated destruction of mayhaw habitats by land clearing for forestry and agriculture, this tradition has been largely abandoned, although many rural families still prepare a batch of mayhaw jelly every year. Although the mayhaws grow well in swampy soil, they grow best in moist but well-drained soil.

## PLANT PORTRAIT

Hawthorns are deciduous shrubs or small trees, with small, crab-apple-like or rose-like fruits called haws and hawberries. The fruit flesh of most species is mealy and dry, like that of rosehips, but in some species, the fruits have the texture and sweetness of cherries. Hawthorn plants usually have thorns, clusters of white (rarely rose) flowers in the spring, and colorful orange, red, or yellow (rarely blue or black) fruits in the fall. The species are distributed in north temperate climates and are especially common in eastern North America. Over 800 *Crataegus* species have been described from

FIGURE 46.1 Hawthorn (*Crataegus* species). (a) Fruiting branch. (b) Flowers. (c) Containers of fruit for sale. (d) Tree in fruit. (e) Tree in flower. (All photos of these "mayhaws" courtesy of B.M. Talbert, Louisiana Mayhaw Association.)

North America alone, but most experts accept far fewer—perhaps, 140–200 species worldwide. Hybridization and odd pseudo-sexual methods of reproduction are known to generate a wide range of diverse plants, making identification difficult. Hawthorns are easily recognized as a group, but the species are extremely difficult to distinguish.

Hawthorns are cultivated for ornament, and the fruits of some species are used as food. Four North American species with notably edible fruits are listed in Table 46.1, and their distribution ranges are shown in Figure 46.2. Indigenous peoples in North America consumed the fruits, often drying them for winter use. There is a long folk tradition of using hawthorn medicinally, both in the Old and New Worlds, and at present hawthorn extracts (particularly glycosides) are employed to some extent for treating cardiac problems (including angina, congestive heart failure, and high blood pressure). Self-treatment with hawthorn extracts without professional medical advice could be hazardous. Whether consumption of large quantities of hawthorn fruits could be a problem for heart patients is unclear.

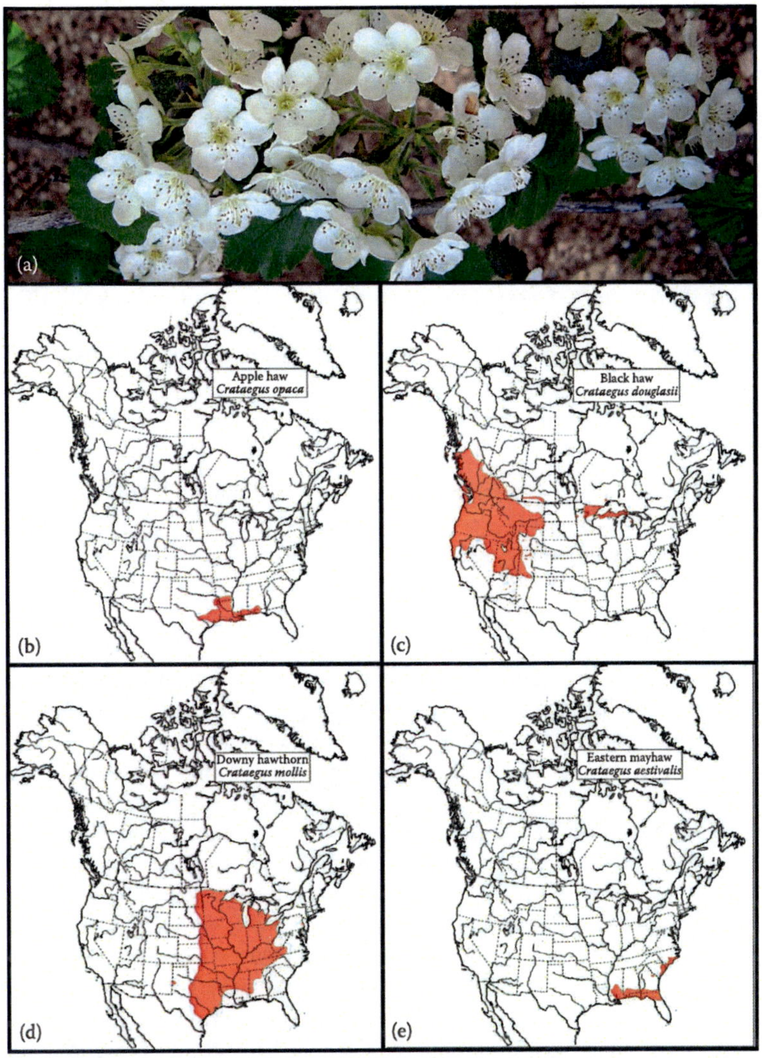


**FIGURE 46.2** Hawthorn (*Crataegus* species). (a) Flowers of downy hawthorn (*C. mollis*). (Courtesy of Nadiatalent.) (b–e) Distribution maps of the most edible species.

The azarole (*C. azarolus* L.) of western Asia and southeastern Europe produces the best fruit of all hawthorns and is one of the few hawthorns deliberately cultivated for its fruit. It is a tree up to 9 m (30 feet) in height, with few or no thorns, and orange-red to yellow fruit about 2 cm (0.8 inch) long. This species is cultivated for fruit in France, Italy, Algeria, and Spain, and is also grown in many temperate areas as an ornamental. There are varieties with yellow, red, and white fruits. Although the fruit will develop in cooler regions, it seems not to be as flavorful as when grown in warmer climates.

The mayhaws are early ripening, edible-fruited, southern U.S. species, including *C. aestivalis* and *C. opaca*. These are large shrubs or round-topped small trees (8–10 m or 26–33 feet tall). The mayhaw species have yellow to bright red fruits that are usually 8–20 mm (0.3–0.8 inches) in diameter, fragrant, acid, and juicy, resembling crabapples in taste. Fruit from the mayhaws is collected from wild plants, but in recent years, attempts have been underway to select cultivars and develop the mayhaws as orchard crops. Dozens of cultivars are now available, including thornless varieties that facilitate fruit collection.

**TABLE 46.1**
**North American *Crataegus* Species Noted for Edible Fruit**

| Scientific Name | Common Name | Native Distribution |
|---|---|---|
| *C. aestivalis* (Walter) Torr. & A. Gray (including *C. rufula* Sarg.) | Eastern mayhaw, May hawthorn, mayhaw | Southeastern United States |
| *C. douglasii* Lindl. | Black haw, black hawthorn, Douglas thornapple, river hawthorn, western hawthorn | Western North America (rare in eastern North America) |
| *C. mollis* (Torr. & A. Gray) Scheele | Downy hawthorn, red hawthorn | Eastern North America |
| *C. opaca* Hook. & Arn. | Apple haw, river-flat hawthorn, western mayhaw | South-central and southeastern United States |

Fruit of the remaining North American species listed in Table 46.1 are mostly collected from wild plants, or harvested from plants grown for ornament, for local use. The fruit of some of these species is quite extensively processed into preserves and jams.

## CULINARY PORTRAIT

The fruits of all species of *Crataegus* are probably edible, but most are not palatable. The azarole of the Old World, the most important commercial hawthorn grown for food, has fruit that is apple-like in flavor, sometimes eaten fresh, but especially used for making jam, jellies, and sweet confections, and for flavoring liqueurs. North American *Crataegus* species do not produce fruit of comparable quality.

The mayhaws have fruit that is fragrant, acid, and juicy, resembling cranberries in appearance and crabapples in taste (immature fruit is astringent and unsuitable for culinary use). However, the fruit is too tart for raw consumption and is used to make marmalades, butters, preserves, jellies, condiments, syrups, wines, and desserts. Purée can be prepared by boiling and sieving the fruits; this can be used to flavor and thicken savory dishes and stewed fruits. Mixed with vinegar, sugar, and spices, the purée can be employed to prepare ketchup. Mayhaw fruit is considered to be one of the unique culinary delicacies of southern American cuisine, and several communities hold mayhaw harvest festivals (in May, of course). Colquitt, Georgia, boasts that it is the "Mayhaw Capital of the World." Mayhaw jam is sold to some extent in local stores in the southern United States and has been produced for sale by prisoners of the Louisiana State Penitentiary.

### CULINARY VOCABULARY

- *Tanghulu* or *bīng tánghúlu* is a winter snack in northern China, particularly intended for children, and traditionally made by placing candied hawthorn fruits (from Chinese hawthorn, *C. pinnatifida* Bunge) on bamboo skewers.
- "Haw flakes" are a Chinese sweet made from the Chinese hawthorn. They are very thin discs, typically 25–40 mm (1–1.6 inches) in diameter, sometimes sold in Chinese markets in packages, stacked like Pringles potato chips.
- *Sansachun* is a liquor make in South Korea from Chinese hawthorn fruits.
- *Zalzalak* in Iran refers to the fruits of the azarole (*C. azarolus*) eaten fresh as a snack or as a jam. Similarly, in Mexico, *tejocotes* refers to the fruits of Mexican hawthorns eaten fresh or as a jam.

## PROSPECTS

Cultivated varieties of the mayhaws are currently very minor commercial fruits, basically significant as a cottage industry in the southeastern United States. Old World hawthorns (particularly the azarole and the Chinese hawthorn) produce superior fruits, and are more important commercially, albeit with quite minor market significance. The native hawthorns of North America have limited likelihood of establishing in the very competitive fruit marketplace, but it remains quite possible that new cultivars, particularly of the mayhaws, will become popular. Mayhaws can grow on land that is too wet for most crops, and so they can be useful for extending agriculture into lands that are not currently used.

## CURIOSITIES OF SCIENCE AND TECHNOLOGY

- In ancient Athens, hawthorn was considered a sign of rebirth and fertility, flowering as it does in the spring. Athenian brides carried large boughs of hawthorn while their wedding guests crowned themselves with hawthorn flowers. The flambeaux, a torch that illuminated the wedding chamber, was made from hawthorn wood.
- Various authors have commented that the smell of death is like the odor of hawthorn. It has been shown that trimethylamine, one of the first products formed when animal tissues start to decay, is present in hawthorn flowers. In older times, corpses were commonly kept in the house for up to a week before burial, so that the odor of death would have been familiar, and it is therefore not surprising that hawthorn has been long associated with death and bad luck.
- According to folklore in Serbia and Croatia, stakes used to slay vampires must be made of hawthorn wood.
- Superstitious beliefs sometimes are so entrenched that they must be respected, even in modern times. One such belief, prevalent in the British Isles, is that the hawthorn is the tree of the fairies and cutting one down is an invitation to disaster. There are numerous recorded accounts of people suffering following chopping down of a hawthorn. English workmen are said to have refused to remove a tree in the path of a new railroad, and when the track was carried over it on a trestle, the supporting earth embankments continually collapsed. As late as 1968, a road had to be realigned to pass a hawthorn that was on a new road being built from Ballintra to Rossnowlaugh. In 1982, workers in the De Lorean car plant in Northern Ireland protested that the site was cursed because a hawthorn had been destroyed during construction of the plant. To calm everyone, management ceremoniously planted another "fairy tree" on the property. On the grounds of Belfast University, a center of level-headed modern thought, people refused to trim a hawthorn for fear of offending the fairies, and it is said that when new buildings were planned, they had to be placed so that they would not disturb the tree.
- Interpreting the origin of seemingly nonsensical sayings is part of the study domain of linguists. An example is the popular rhyme "Here we go gathering nuts in May." Nuts are not available in May in England, where the expression originated. It has been determined that the rhyme was sung by young men gathering "knots" of May blossoms for the May Day celebrations. In England, hawthorn flowers are associated with May Day, and the hawthorn has long been used as a symbol of spring in English poetry.
- In September 1752, the common calendar was changed. The Gregorian calendar replaced the Old Julian calendar, with 11 days dropped (September 2 was followed immediately by September 14). Christmas, now commonly celebrated on December 25, (New Style) used to be celebrated on January 6 (Old Style, maintained by many). This affected a number of previous observances related to plant flowering dates. For example, May flowering of

hawthorns became less common in England than before the calendar change, so May Day customs based on hawthorn sometimes became more difficult to observe.

- Louisiana adopted mayhaw jelly and Louisiana sugarcane jelly as its official state jellies in 2003.
- In 2004, the oldest evidence of wine-making was published. Chemical analyses of ancient organic residues from pottery jars from the early Neolithic village of Jiahu in Henan province in China indicate that a mixed fermented beverage of rice, honey, and fruit (hawthorn fruit and/or grape) was being produced as early as 9000 years ago. [Reference: McGovern, P.E., Zhang, J., Tang, J., Zhang, Z., Hall, G.R., Moreau, R.A., Nuñez, A., Butrym, E.D., Richards, M.P., Wang, C.S., Cheng, G., Zhao, Z., and Wang, C. 2004. Fermented beverages of pre- and proto-historic China. *Proc. Natl. Acad. Sci. USA* 101(51): 17953–17958.]

## KEY INFORMATION SOURCES


Akin, J.S. 1985. Mayhaw coming out of the swamp. *Pomona* 18: 70–73.

Chapman, G.W., Jr., Horvat, R.J., and Payne, J.A. 1991. The non-volatile acid and sugar composition of mayhaw fruits (*Crataegus aestivalis, C. opaca, C. rufula*). *J. Food Qual.* 14: 435–439.

Craft, B.R., Melcher, G., and Langston, E. 1996. *Mayhaws: a guide to orchard production and propagation*. G. Melcher, Tioga, LA. pp. 72.

Graham, C.J. *Crataegus*. In *The encyclopedia of fruit & nuts*. Edited by J. Janick and R.E. Paull. CABI, Wallingford, Oxfordshire, U.K. pp. 628–634.

Horvat, R.J., Chapman, G.W., Jr., and Payne, J.A. 1991. Identification of volatile compounds from ripe mayhaw fruit (*Crataegus opaca, C. aestivalis*, and *C. rufula*). *J. Food Qual.* 14: 307–312.

Johnson, W.T., Sinclair, W.A., and Saunders, J.L. 1974. *Diseases and insects of hawthorns*. New York State College of Agriculture and Life Sciences, Cornell University, Ithaca, NY. pp. 11.

Krewer, J.W., and Crocker, T.E. 1997. Performance of mayhaws in Georgia and south Florida. *Proc. Fla. State Hort. Soc.* 110: 169–171.

Kruschke, E.P. 1965. *Contributions to the taxonomy of* Crataegus. Milwaukee Public Museum publications in botany, no. 3, Milwaukee, WI. pp. 273.

Lance, R. 1995. *The hawthorns of the southeastern United States*. R. Lance, Fletcher, NC. pp. 136.

Payne, J.A., and Krewer, G.W. 1990. Mayhaw: a new fruit crop for the south. In *Advances in new crops*. Edited by J. Janick and J.E. Simon. Timber Press, Portland, OR. pp. 317–321.

Payne, J.A., Krewer, G.W., and Fitenmiller, R.R. 1990. Mayhaws: trees of pomological and ornamental interest. *HortScience* 25: 245, 375.

Phipps, J.B. 1983. Biogeographic, taxonomic, and cladistic relationships between east Asiatic and North American *Crataegus*. *Ann. Missouri Bot. Gard.* 70: 667–700.

Phipps, J.B. 1988. *Crataegus* (Maloideae, Rosaceae) of the Southeastern United States. I. Introduction and series *Aestivales*. *J. Arnold Arbor.* 69: 401–431.

Phipps, J.B. 1997. *Monograph of northern Mexican* Crataegus *(Rosaceae, subfam. Maloideae)*. Botanical Research Institute of Texas, Fort Worth, TX. pp. 94.

Phipps, J.B. 1998. Introduction to the red-fruited hawthorns (*Crataegus*, Rosaceae) of western North America. *Can. J. Bot.* 76: 1863–1899.

Phipps, J.B. 2005. A review of hybridization in North American hawthorns – Another look at "the *Crataegus* problem". *Ann. Missouri Bot. Gard.* 92: 113–126.

Phipps, J.B., and Dvorsky, K.A. 2007. Review of *Crataegus* series *Apricae*, ser. nov., and *C. flava* (Rosaceae). *J. Bot. Res. Inst. Texas* 1: 171–202.

Phipps, J.B., O'Kennon, R.J., and Lance, R.W. 2003. *Hawthorns and medlars*. Royal Horticultural Society, Cambridge, U.K. pp. 180.

Pittler, M.H., Schmidt, K., and Ernst, E. 2003. Hawthorn extract for treating chronic heart failure: meta-analysis of randomized trials. *Am. J. Med.* 114: 665–674.

Puls, E. 1991. *Commercial mayhaw culture*. Louisiana Cooperative Extension Service Publication 2429. Louisiana State University, Baton Rouge, LA. pp. 7.

Rigelsky, J.M., and Sweet, B.V. 2002. Hawthorn: pharmacology and therapeutic uses. *Am. J. Health-Syst. Pharm.* 59: 417–422.

Trappey, A., Bawadi, H.A., Bansode, R.R., and Losso, J.N. 2005. Anthocyanin profile of mayhaw (*Crataegus opaca*). *Food Chem.* 91: 665–671.

Trappey, A.F., Johnson, C.E., and Wilson, P.W. 2006. Consumer acceptance of mayhaw (*Craaegus opaca* Hook. and Arn.) juice blended with muscadine grape (*Vitis rotundifolia* Michx.) juice. *Int. J. Fruit Sci.* 6: 53–65.
Trappey, A.F., Johnson, C.E., and Wilson, P.W. 2007. Comparison of juice extraction methods from fresh and frozen mayhaw (*Crataegus opaca* Hook.) fruit. *Int. J. Fruit Sci.* 7: 41–64.
Trappey, A.F., Johnson, C.E., and Wilson, P.W. 2008. Characterization of juice extraction methods utilizing fresh mayhaw (*Crategus opaca* Hook.) fruit. *Int. J. Fruit Sci.* 8: 318–331.
Zhang, Z., Chang, Q., Zhu, M., Huang, Y., Ho, W.K.K., and Chen, Z.Y. 2001. Characterization of antioxidants present in hawthorn fruits. *J. Nutr. Biochem.* 12: 144–152.

## Specialty Cookbooks

Craft, B.R., Melcher, G., and Langston, E. 1996. *Mayhaws, a guide to orchard production and propagation.* Morris Publishing, Kearney, NE. pp. 72. [Presents several recipes.]
Whitaker, K., and Simmons, S. 2010. *Louisiana hometown cookbook.* Great American Publishers, Kosciusko, MS. pp. 272.

Freitus and Haberman (2005) have seven recipes for hawthorn species, Krumm (1991) presents three, Marie (2008) has two, Marrone (2009) has ten, Mogelon (2001) has one, Snell (2006) has one, and Tatum (1976) has two. (See the Appendix to this book for details of the publications cited.)

# 47 Hazelnuts

Family: Corylaceae (hazel family; this is often placed within the birch family, the Betulaceae)

## NAMES

Scientific names: *Corylus* species

- American hazelnut (American filbert, American hazel)—*C. americana* Marshall (this is usually mistakenly cited as *C. americana* Walter)
- Beaked hazelnut (wild filbert, beaked hazel, beaked filbert)—*C. cornuta* Marshall
- European nuts of the genus *Corylus* are often called filberts, whereas the nuts of native species of *Corylus* of North America tend to be called hazelnuts. Some authors have employed the name filbert for any superior nuts from cultivated *Corylus*, but this usage has not been consistent. Another practice has been to call cultivated varieties filberts and wild types hazels. Hazelnut is currently preferred by the North American marketing industry. For centuries, it was common practice in England to term nuts with long husks as filberts and those with short husks as hazels, and sometimes this practice is still followed in Great Britain, but because of hybridization between the two kinds, such terminology is obsolete.
- The name "filbert" is said to have originated from the German *Vollbart*, meaning full beard, referring to the long husk of the nut. It has also been associated with St. Philbert (Phillibert, about 684), a Frankish abbot of Jumièges in Normandy, whose feast date is August 22, when the nuts of European hazelnut have matured. Still another suggestion is that "filbert" traces to the medieval King Philibert of France, who introduced new cultivated hazels.
- The word "hazel" is derived from *haesel*, the Anglo-Saxon word for hood or bonnet, descriptive of the husk enclosing the fruit. Alternatively, the root of *haesel* has been traced to the older Anglo-Saxon *haes*, or German *heissen*, "to give orders," the hazel rod representing a sign of authority for shepherd chieftains.
- "Witch hazel" is an unrelated plant, *Hamamelis virginiana* L., a shrub or small tree of North America with medicinal properties.
- "Trazels" are hybrids of Turkish filbert (*C. colurna* L.) and the European hazelnut (*C. avellana* L.).
- The name "filazel" is applied to hybrids of beaked hazelnut (*C. cornuta*) and European hazelnut.
- Hybrids of American hazelnut and European hazelnut (*C. avellana* L.) have been referred to as "hazelberts."
- The genus name *Corylus* originates from the Greek word *korys*, referring to a helmet or hood, descriptive of the husk that encloses the fruit.
- *Americana* in the scientific name *C. americana* is the Latinized form of the adjective "American."
- *Cornuta* in the scientific name *C. cornuta* is Latin for horned or beaked, a reference to the long, hornlike husk around the fruit.

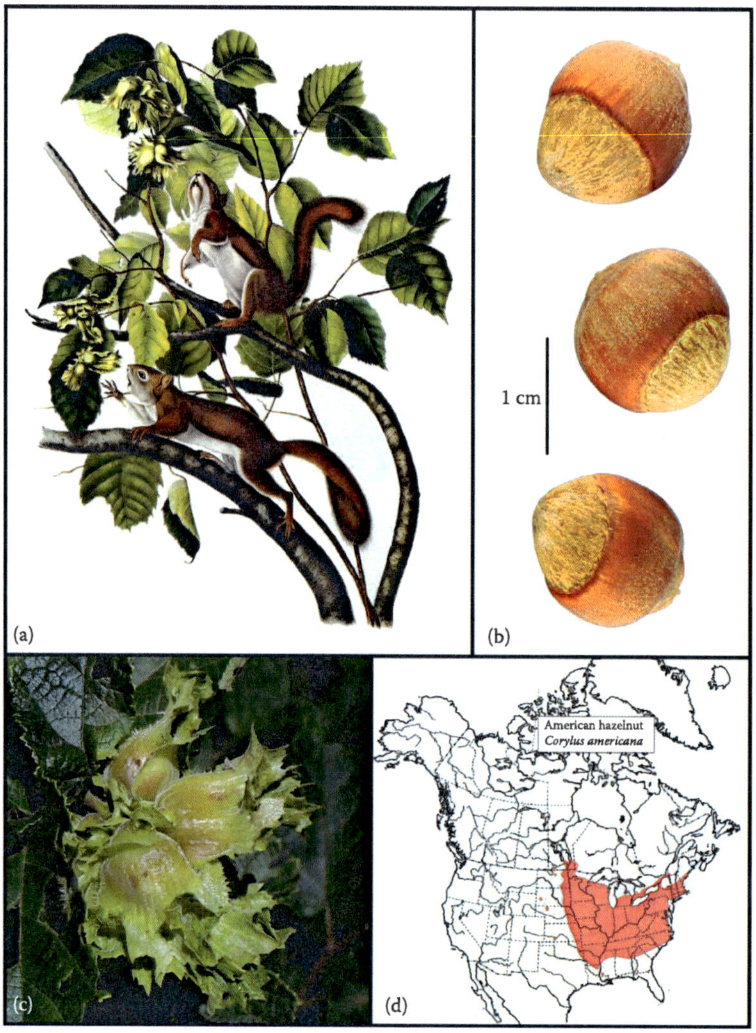


**FIGURE 47.1** American hazelnut (*Corylus americana*). (a) Red squirrels on American hazelnut. (From Audubon, J.J., *Viviparous quadrupeds of North America*, J.T. Bowen, Philadelphia, PA. Vol. 1, Plate 14, "1849" [variously dated].) (b) Nuts. (Courtesy of Steve Hurst, U.S. Department of Agriculture.) (c) Fruits with husks. (Courtesy of the U.S. Fish and Wildlife Service.) (d) Distribution map.

## GEOGRAPHY AND ECOLOGY OF WILD PLANTS

American hazelnut occurs throughout a large area of eastern and central United States from coastal New England south to Georgia and west to the Dakotas and Oklahoma. In Canada, it is confined to the southern parts of Ontario, Quebec, and Manitoba. The beaked hazelnut grows wild across Canada, in the northeastern United States, and on the West coast from California northward (the western range is often recognized as a separate subspecies, *C. cornuta* subsp. *californica* (A. DC.) E. Murray).

The two North American hazelnuts occur in dry to slightly moist woodlands, woodland edges, prairies, slopes of hills, and river banks. They appear to do best on loam and sandy loam soil but beaked hazelnut grows well in coarse, relatively dry sand and on shallow soils over either limestone or granite rocks. Both can grow well in the open or in partial shade (30% sun) but decline in well shaded situations.

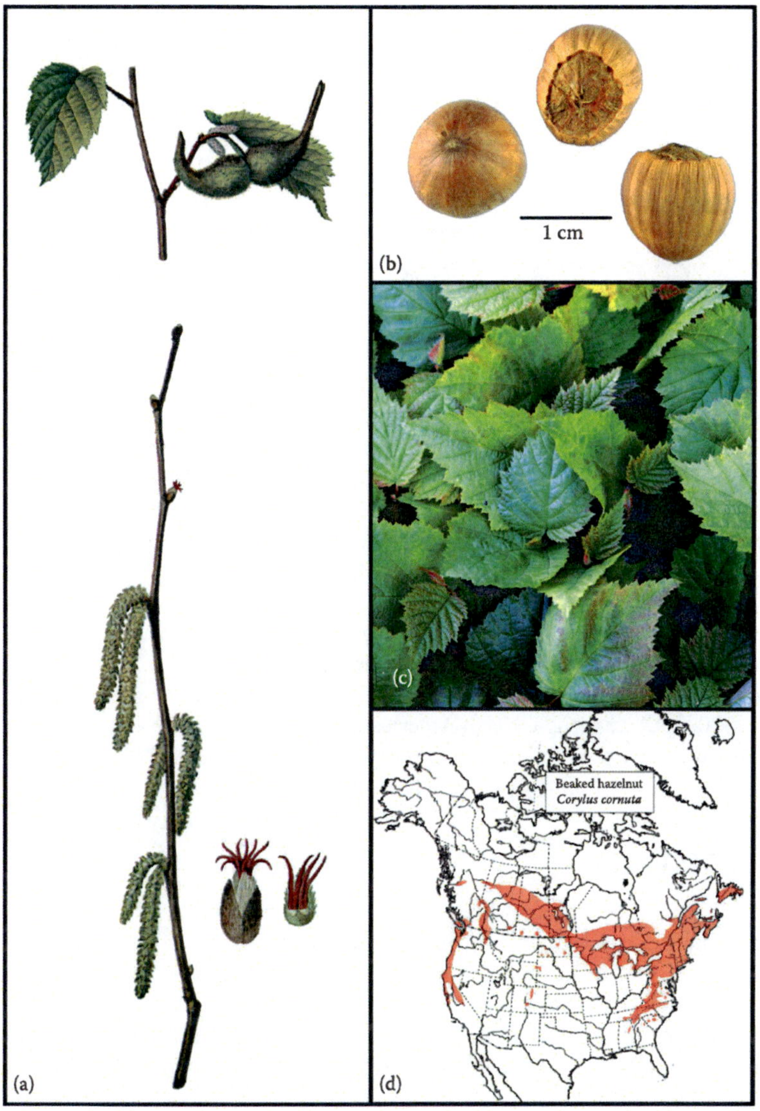


**FIGURE 47.2** Beaked hazelnut (*Corylus cornuta*). (a) Painting of female branch with a pair of fruits (above), branch bearing male catkins (below), and female flowers (at right of male branch). (From New York Botanical Garden, *Addisonia*, New York Botanical Garden, Brooklyn, NY. Vol. 5, Plate 173, 1920.) (b) Nuts. (Courtesy of Steve Hurst, U.S. Department of Agriculture.) (c) Foliage. (Courtesy of Agrosylva [CC By 3.0].) (d) Distribution map.

## PLANT PORTRAIT

The two North American species of hazelnut are deciduous shrubs growing 1–3 m (3–10 feet) tall. The fruit is a hard-shelled nut. The nuts may be single or in clusters of 2–6, each surrounded by a husk of leafy bracts. The nuts of both North American species are too small to compete with the European filberts, but there are a few varieties of American hazelnut that are much improved compared to the wild species (the most important cultivars of American hazelnut are 'Rush', a large variety from southern Pennsylvania and 'Winkler', a small, but winter-hardy and productive selection).

Both native North American species have been hybridized with the European species to produce hardy varieties of the latter. The uninitiated wild nut collector should be warned that the husk of the beaked hazel is covered with tiny, skin-piercing hairs, which can be quite irritating to the hands.

Hazelnut or filbert is second only to almond in world nut production. The main source of the nuts of commerce is the European filbert, *C. avellana*, a thicket-forming shrub or small tree up to 10 m (33 feet) in height, native to most of Europe and western Asia. Until recently, this was divided into two species, *C. avellana*, said to be restricted to Europe, and so-called giant filbert (recognized as *C. maxima* Mill.), said to occur in southeastern Europe and western Asia. Recent evidence indicates that *C. maxima* should be included as part of *C. avellana*. Some of the crop from Turkey (where most of the world's filberts are raised) is from hybrids of the European and Turkish filberts. The Turkish filbert is a deciduous shrub or small tree rarely up to 25 m (82 feet) tall, native to southern Europe eastward to Turkey. The Spanish cultivar 'Barcelona', a European filbert, is the predominant hazelnut grown in North America. It is cultivated in the Pacific Northwest, including the coastal valleys of the Cascade Range of Oregon and Washington, and the Fraser Valley in southern British Columbia. The American crop provides about 3% of the world supply, and the United States imports about 50% of the hazelnuts it consumes annually. About 70% of the world's filberts come from small Turkish farms on the southern coast of the Black Sea, about 20% from coastal regions of Italy, and about 7% from Spain's Mediterranean coastal areas. Filberts are also produced to a minor extent in other areas of Europe, northeastern Asia, Australia, New Zealand, South Africa, and eastern North America.

Hazelnuts were eaten by early Europeans during the Mesolithic period of the Stone Age 10,000 years ago, as indicated by presence of the nuts in many archeological sites. Old Chinese manuscripts indicate that filberts were harvested nearly 5000 years ago, and they were cultivated by the Greeks and Romans 2000–2500 years ago. North American wild hazelnuts were widely collected by native people throughout their areas of occurrence. Early North American explorers and settlers also ate the wild nuts.

## CULINARY PORTRAIT

Hazelnuts or filberts are consumed whole, ground, or chopped. The texture and flavor are highly appealing. The nuts are eaten extensively during the Christmas season but are also used as a substitute for flour, as a spread like butter, in toppings and coatings, and in cereals, garnishes, soups, cookies, ice cream, and chocolates. The nuts are popular in stuffed vegetables, other vegetarian and cheese dishes, and are also made into milks for desserts, and sauces and stuffings for meat, game, and fish. A superior, well-flavored salad oil (not recommended for frying) is extracted from the nuts, sometimes with over 90% unsaturated fatty acids. Like the nuts, it must be stored carefully to prevent rancidity. In western Europe, filberts are particularly important in the confectionery business. Both praline mixtures made from ground hazels and sugar, and a paste made from ground blanched filberts and chocolate are widely used in areas where rival almonds and walnuts came late, and a whole hazelnut is as likely as the latter to be found in the famous chocolate pralines of Belgium or Switzerland. The filbert is also one of the nuts commonly used in European sugar paste or nougat confections, particularly in Spain. Filberts are often substituted for almonds in the Mediterranean region.

Growers and manufacturers emphasize that hazelnuts (the Eurasian, commercial species) have many beneficial qualities. They have relatively low carbohydrate and high fiber compared to other nuts. They are a concentrated source of energy and are cholesterol-free. Over half of the weight is unsaturated fat and the fatty acid content is similar to that of olive oil (oleic and linoleic are the most abundant fatty acids). The nuts are relatively low in sugars and sodium. They contain approximately 5% water, 16% protein, 64% fat, 11%–12% carbohydrate, and 3100 calories/lb or 595 calories/100 g edible portion. The relatively high levels of vitamins E and $B_6$ may be useful against coronary heart disease and certain cancers.

European hazelnuts are marketed shelled or unshelled, and may be available whole or ground, and often roasted and salted. The American species can be similarly treated. Roasting can be carried out by placing the nuts on a cookie sheet and baking in an oven at 93°C–141°C (200°F–285°F) until golden brown, stirring occasionally. After roasting but while the nuts are still hot, the thin brown skin can be peeled by rubbing with a thick cloth. The nuts are less perishable than such fatty nuts as Brazil nuts, macadamias, and pecans, but nevertheless become bitter and dry out relatively quickly if not stored properly. Fresh nuts can be stored in a cool dry place for a month. Shelled nuts can be refrigerated for up to 4 months, or frozen for up to a year. The nuts should not be dried in the sun, as the oils will vaporize and much of the flavor will be lost.

### Culinary Vocabulary

- Nutella™ is a hazelnut preparation made with skim milk and cocoa, familiar in Europe and Canada, but little known in the United States, where the similar peanut butter almost exclusively dominates the nut spread market.
- The Italian expression *brutti ma buoni* (pronounced BROO-tee mah B'WHO-nee), meaning "good but ugly," refers to a meringue cookie prepared with hazelnuts or almonds.

## PROSPECTS

The American and beaked hazelnuts are very minor sources of nuts, with extremely small commercial significance. There is limited likelihood that breeding based on just these species will produce important new cultivated varieties. This is because the European filbert dominates the hazelnut industry of the world, and is likely to continue to do so. However, several hybrids of the American species with the European filbert have proven to have market value, indicating that both American hazelnuts will continue to contribute to the production of new cultivars.

## CURIOSITIES OF SCIENCE AND TECHNOLOGY

- After melting of the continental glaciers 7–10 thousand years ago, hazelnut was so abundant in Europe that scientists specializing in studying the past call the period the "hazel maximum."
- The famous "Ice Man" (nicknamed Ötzi, after the Ötztaler Alps) found in 1991 in the Tyrolean Alps of Italy and dated to the late Stone Age over 5000 years ago, used hazelwood to reinforce his skin quiver and backpack.
- In Greek mythology, Jupiter had two sons, Apollo (the god of harmony) and Mercury (the god of eloquence), who one day exchanged gifts. Apollo received a tortoise-shell lyre, whose music was intended to enhance the artistic spirit of mankind. Mercury got a winged wand made of hazel, supposed to assist people in expressing their thoughts through words. The symbolic winged hazel rod, entwined with two twisting serpents, became the symbol of communication, and is still used as such. It was said that whomever Mercury touched with his hazel rod gained the gift of eloquence.
- European hazelnut has had numerous medicinal uses dating from the Greek and Roman empires, for example treatment of baldness, common cold, and stomach and sexual problems. Eating 12 nuts before going to bed was employed as a cure for bed-wetting in Spanish traditional medicine. American hazelnut was used to treat ulcers and tumors by the Hurons in southern Ontario and has also been used by native people as an analgesic. Decoctions containing hazelnut were employed by the Chippewa to treat convulsions and hemorrhage and to heal wounds. There is little use of hazelnuts in modern medicine.
- Rhabdomancy, the art of finding underground riches and subterranean springs with a forked branch, was largely practiced with hazelnut twigs. Use of the Y-shaped branch,

called dowsing or divining, supposedly revealed the location of treasure, minerals, ore, and water. Divining rods were used by the ancient Greeks and were very popular during the Middle Ages. The father of biological nomenclature, Linnaeus (1707–1778), admitted to being half a convert to their use. The natural affinity of the hazel for damp places may be at the root of the idea that hazel indicates a water source.

- In the seventeenth and eighteenth centuries, hazels became popular among the wealthy for ornamental purposes. They were planted to provide privacy and color in gardens, as ornamental hedges and, in France, were trained into borders for so-called "nut walks."
- The fine oil of filberts has a reputation for improving the tone of violins when used in their varnish.
- Filberts or hazels have been associated fairly commonly with place names in Europe. For example, filberts are called *avelines* in France because they were cultivated for centuries around the Italian city of Avellino. The Gaelic word for hazel is Coll. It appears frequently in place names in the west of Scotland, such as the Isle of Coll and Bar Calltuin. It also appears in the name of Clan Colquhoun whose clan badge is the hazel. The longest place name in Britain (located in Wales) is Llanfairpwllgwyngyllgogerychwyrndrobwllllantyssiliogogogoch, which means (translated from the Welsh) "St. Mary's Church in the hollow of the white hazel near to the rapid whirlpool of Llantysilio of the Red Cave."
- A survey showed that hazelnuts are as popular as almonds in chocolate, but less popular than pecans in ice cream and walnuts in brownies.
- According to *Ripley's Believe It or Not!*, Constantin Boym and Laurene Leon of New York City invented edible pencils with almond and hazelnut flavors.
- Hazelnut (the European *Corylus avellana*) was declared to be the state nut of Oregon in 1989.

## KEY INFORMATION SOURCES


Axer, J. 1980. Filberts and the world of nuts. *Proc. Nut Grow. Soc. Oreg. Wash. B. C.* 1980: 83–87, 90–96.

Baron, L. 1985. *Growing hazelnuts in Oregon*. Revised edition. Extension Service, Oregon State University, Corvallis, OR. pp. 19.

Bergougnoux, F., Germain, E., and Sarraquigne, G.E. 1978. *Le noisetier, production et culture*. Invuflec, Paris, France. pp. 164. [*The hazelnut tree: production and cultivation*. New edition: Germain, É., Sarraquigne, J.-P., Breisch, H., Hutin, C., and Léglise, de Taffin, H. 2004. Le noisetier. Invuflec, Paris, France. pp. 296.]

Brink, V. 1964. Success with filberts. *Org. Garden. Farm.* 11(9): 28–31.

Catling, P.M., and Small, E. 2000. Poorly known economic plants of Canada—25. American hazelnut, *Corylus americana*, and beaked hazelnut, *C. cornuta*. *Bull. Can. Bot. Assoc.* 33(2): 16–20.

Commonwealth Agricultural Bureaux. 1980. *Growing* Corylus *spp. (hazels, cobnuts, filberts), 1972–1979*. Annotated Bibliographies. Plant series no. 8/80. Commonwealth Agricultural Bureaux, Farnham Royal, U.K. pp. 21.

Debor, H.W. 1978. *Bibliography of the international literature on filberts*. Technische University, Berlin, Germany. pp. 129. (In English and German)

Farris, C.W. 1972. The tree hazels. *North. Nut Grow. Assoc. Ann. Rep.* 63: 24–26.

Farris, C.W. 1978. The trazels. *North. Nut Grow. Assoc. Ann. Rep.* 69: 32–34.

Farris, C.W. 1988. Potential for genetic improvement in hazelnuts. *North. Nut Grow. Assoc. Ann. Rep.* 79: 46–47.

Furlow, J.J. 1997. *Corylus*. In *Flora of North America north of Mexico*. Vol. 3. Edited by Flora of North America Editorial Committee. Oxford University Press, New York. pp. 535–538.

Gellatly, J.U. 1966. Tree hazels and their improved hybrids. *North. Nut Grow. Assoc. Ann. Rep.* 57: 98–101.

Great Britain. Ministry of Agriculture, Fisheries and Food. 1966. *Cob-nuts and filberts*. Advisory leaflet 400. Ministry of Agriculture, Fisheries and Food, London, U.K. pp. 6.

Jona, R. 1986. Hazelnut. In *CRC handbook of fruit set and development*. Edited by S.P. Monselise. CRC Press, Boca Raton, FL. pp. 193–216.

Lagerstedt, H.B.1975. Filberts. In *Advances in fruit breeding*. Edited by J. Janick and J.N. Moore. Purdue University Press, West Lafayette, IN. pp. 456–489.

Lagerstedt, H.B. 1979. Filberts. In *Nut tree culture in North America.* Edited by R.A. Jaynes. Northern Nut Growers Association, Hamden, CT. pp. 128–147.

Lagerstedt, H.B. 1983. The American nut industry—filberts. [*Corylus avellana*, cultivars, culture, production] *North. Nut Grow. Assoc. Ann. Rep.* 74: 179–185.

Mehlenbacher, S.A. 1990. Hazelnuts (*Corylus*). In *Genetic resources of temperate fruit and nut crops.* Edited by J.N. Moore and J.R. Ballington, Jr. International Society of Horticultural Science, Wageningen, the Netherlands. pp. 791–836.

Miller, R.C., and Devlin, K.A. 1948. *Processing filbert nuts.* Oregon State College, Agricultural Experiment Station, Corvallis, OR. pp. 16.

Organisation for Economic Co-operation and Development.1981. *Unshelled sweet almonds, unshelled hazel nuts.* Organization for Economic Co-operation and Development, Paris, France, and Washington, DC. pp. 77.

Slate, G.L., and Lewis, G. 1930. *Filberts.* Bulletin New York State Agricultural Experiment Station. Bulletin no. 588. New York State Agricultural Experiment Station, Geneva, NY. pp. 32.

Tesche, W.C. 1956. *The walnut and filbert industries of the Mediterranean Basin.* United States Department of Agriculture, Foreign Agricultural Service, Washington, DC. pp. 48.

Thompson, M.M. 1977. Inheritance of nut traits in filbert (*Corylus avellana* L.). *Euphytica* 26: 465–474.

## Specialty Cookbooks

Oregon Filbert Commission, and Nut Growers Society of Oregon and Washington. 1973. *A treasury of prize winning filbert recipes.* Third edition. Oregon Filbert Commission, Tigard, OR. pp. 128.

Berglund and Bolsby (1971) present four recipes for beaked hazelnut. Boorman (1969) has two wild hazelnut recipes, Freitus (1980) has eight, Kluger (1973) has five, and Robe-Terry (1997) has two. (See the Appendix to this book for details of the publications cited.)

# 48 Hickories

Family: Juglandaceae (walnut family)

## NAMES

Scientific names: Hickories are species of *Carya*. The two species featured in this chapter are the most important nut trees.

- Shagbark hickory—*C. ovata* (Mill.) K. Koch
- Shellbark hickory—*C. laciniosa* (F. Michx.) W.P.C. Barton
- "Hickory" is an English contraction of *pikahickory*, from the Native American *powcohiccora*, a food prepared from pounded nuts, as designated in some Algonquian language of Virginia.
- The "shagbark" hickory is so-named for the peculiar, shaggy, untidy appearance of the peeling bark of mature trees.
- The degree of shagginess of shagbark hickory trees varies considerably. In the early nineteenth century, American carpenters, who used hickory wood to fashion tool handles and wheel spokes, preferred trees with limited shagginess, as it was easier to remove the bark. Nevertheless, they disrespectfully referred to the smoother-barked trees as "bastard hickories," presumably based on the idea that the parentage of such unusual trees was suspect.
- The "shellbark" is also so-named for the appearance of its bark, which seems to be peeling off in plates or "shells."
- The common names of hickory species are often highly undependable for identification. Five different species are called "pignut." The two species highlighted here are known by numerous names, some of which are the same. The shagbark hickory (*C. ovata*) is also known as bird's eye hickory, Carolina hickory, curly hickory, littlenut shagbark hickory, little pignut, little shagbark, false shagbark, mockernut hickory, red hickory, redheart hickory, scalybark hickory, shagbark, shagbark walnut, shellbark, shellbark hickory, shellbark tree, skid hickory, small pignut, small pignut hickory, sweet pignut, southern hickory, southern shagbark hickory, southern shellbark, sweet walnut, true hickory, upland hickory, white hickory, white-heart hickory, and white walnut. The shellbark hickory (*C. lacinosa*) is also known by many names: big shagbark, big shagbark hickory, big shellbark, bigleaf, bigleaf shagbark hickory, big shellbark hickory, bottom shellbark, king nut, kingnut hickory, ridge hickory, shellbark hickory, thickbark hickory, thick shellbark, thick shellbark hickory, true hickory, and western shellbark.
- The names white hickory (or white-heart hickory) and white walnut, respectively, indicate the pale color of the wood and nutshells of shellbark hickory.
- "Mockernut" is a name that was originally applied to hickory species that sometimes produced empty nuts (i.e., the nuts seemed to "mock" people looking for tasty treats). Alternatively, it has been claimed the name is derived from New York Dutch *moker-noot*, heavy-hammer nut, that is, requiring much force to break the thick shell.
- The name "king nut" (or kingnut or kingnut hickory), sometimes used for shellbark hickory, reflects the size of the nuts, sometimes reaching 7.5 cm (3 inches) in length. The shellbark hickory is often called the big shellbark hickory, apparently for the big size of the nuts (while names such as littlenut and little shagbark are applied to the shagbark because of its smaller nuts).

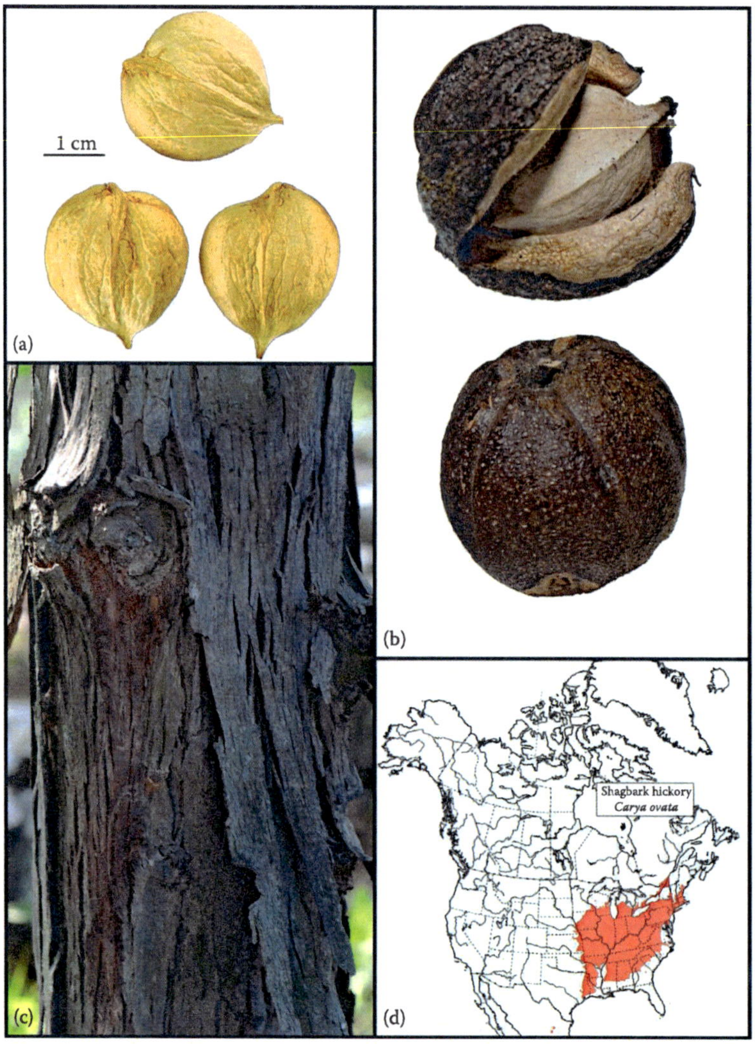


**FIGURE 48.1** Shagbark hickory (*Carya ovata*). (a) Nuts. (Courtesy of Steve Hurst, U.S. Department of Agriculture.) (b) Fruits (nuts inside husks). (Courtesy of Thomas Kent, FloraFinder.com, reproduced with permission.) (c) Shaggy, loose, older bark of trunk. (Courtesy of Thomas Kent, FloraFinder.com, reproduced with permission.) (d) Distribution map.

- For information on the genus name *Carya*, see Chapter 72.
- *Ovata* in the scientific name *C. ovata* is Latin for ovate, having an egg-like profile, interpreted by some as referring to the shape of the leaflets and by others as referring to the shape of the fruits.
- *Laciniosa* in the scientific name *C. laciniosa* is based on the Latin *lacinia*, cut in shreds, and refers to the loose plates of bark on the trunk.

## GEOGRAPHY AND ECOLOGY OF WILD PLANTS

The shagbark hickory grows wild from southern Quebec and Ontario to southwestern Maine, Michigan, and southeastern Minnesota, and south to Florida and Texas; the species also occurs in the mountains of northeastern Mexico. It is one of the most abundant hickories in the eastern and

**FIGURE 48.2** Shellbark hickory (*Carya laciniosa*). (a) Fruits (nuts visible within split husks). (Courtesy of Steven J. Baskauf [http://bioimages.vanderbilt.edu].) (b) Tree. (Courtesy of Steven J. Baskauf [http://bioimages.vanderbilt.edu].) (c) (left to right): Slice of fruit husk, compound leaf, and nut. (From Michaux, F.A., *Histoire des arbres forestiers de l'Amérique septentrionale*, L. Haussman, Paris, France. Vol. 1, 1810.) (d) Distribution map.

central states. The shagbark hickory occupies a wide range of sites over its distribution, occurring in both moist and dry woods, often along streams. In the northern part of its range, it often grows on upland slopes, generally in mixed hardwoods. At its extreme northern limit, it is confined to periodically flooded river shores. In Canada, the trees occur in valleys, on moist hillsides, and on the borders of swamps, generally in deep, moist, fertile, well-drained soils. For the most part, shagbark hickory occurs as scattered individuals or in small groups, rarely in pure stands. The shagbark is moderately shade tolerant, and saplings can persist for many years under a forest canopy, but it will be outcompeted by extremely shade-tolerant trees such as sugar maple.

The shellbark hickory grows wild as far north as southern Ontario and western New York, west to southeastern Iowa, and south to northeastern Oklahoma and Tennessee, with isolated populations

**FIGURE 48.3** Shagbark hickory (*Carya ovata*) tree. (From Harter, J. (Editor), *The plant kingdom compendium. A definitive volume of more than 2400 copyright-free engravings*, Bonanza Books, New York, 1988.)

in southern Arkansas, Mississippi, Alabama, and northern Georgia. It is found in low woods, rich bottomlands, and along streams. In much of its natural range, the trees are subject to seasonal inundation.

## PLANT PORTRAIT

There are at least 16 species of the genus *Carya*. One species occurs in eastern China, another in Indochina, and the remainder in eastern North America. The only significant commercial species of *Carya* grown for nuts is the pecan (see Chapter 72). Most species bear edible nuts, although not necessarily pleasant-tasting, but the two highlighted here have particularly tasty nuts and have been developed to a minor extent as commercial nut trees. Hickory trees grow very slowly (20 years at least for most species before they bear nuts). There is only a very limited degree of cultivation in the United States for providing a commercial supply of nuts. Hickories are notorious for attracting squirrels, and protecting the nuts from them is a major problem.

Indigenous people in North America consumed hickory nuts wherever they could be collected. Algonquin Indians in Virginia were observed to pound hickory kernels in a mortar until finely powdered, add water, and strain the mixture to produce a nourishing "hickory milk." Early settlers also valued the nuts, which could be kept for more than 2 years in a cool cellar without significant deterioration. By the early nineteenth century, a few hundred bushels of hickories were being traded among the colonies and exported to the West Indies and England. Unlike the pecan, however, only limited selection of improved varieties has occurred to date.

Next to the pecan, the shagbark hickory is the best nut tree in the genus *Carya*. The trees are 18–46 m (60–151 feet) in height, and sometimes live for 300 or more years. The bark is loose and scaly, separating into plates up to 30 cm (1 foot) long. The bark hangs down from the trunk like loose shingles and can be peeled with the bare hand. The husked nut is 3.5–6 cm (1.4–2.4 inches) long, and is somewhat longer than wide. The husk is 3–12 mm (0.1–0.5 inch) thick and splits into four separate pieces to the base of the nut. The unhusked nut is usually sharp-pointed, whitish-brown, and thin-shelled. The nut of the shagbark hickory has been esteemed for its sweetness and superior flavor since colonial times. It has been recommended as one of the best nuts for northern regions of North America, where the pecan cannot be grown. There are at least 15 selected varieties, some with large kernels and good cracking qualities.

The shellbark hickory is judged by most to be somewhat inferior to the shagbark as a source of edible nuts, although some consider it to be superior. The trees tend to be shorter than shagbark hickory (usually no taller than 30 m or 98 feet, but occasionally growing to 43 m or 141 feet) and often develop heavier branches. The bark of the tree exfoliates in plates, as in shagbark hickory, but the trunk is not as shaggy. The fruit is also similar to that of the shagbark hickory but is larger, 4.4–6.4 cm (1.7–2.5 inches) in both width and length (the fruit is round), with a very thick husk (6–13 mm; 0.2–0.5 inch). Improved nut varieties of shellbark hickory have been selected in the last few years, and there are now over 40 varieties, most originating from Iowa and Pennsylvania.

Pecans and hickories, when cross-pollinated, produce vigorous hybrids called "hicans." These tend to be poor nut producers but excellent ornamental shade trees. Hicans have some potential for use as dwarfing rootstocks, that is, for grafting pecan trees onto the rootstocks with the result that the trees grown are relatively short and the nuts are easy to reach.

Indigenous peoples used the bark of the two species highlighted here for a variety of medicinal applications, young wood and saplings for basketry, and the branches for constructing tools, bows, arrows, war clubs, and snowshoes. The husk was once ground up to produce a fish poison. A yellow dye produced from the bark was patented in the eighteenth century, but did not attain much commercial success. The size and shape of several of the hickories adapt them well as shade trees for small places. Shagbark hickory, together with pignut hickory (*C. glabra* (Miller) Sweet), furnish the bulk of commercial hickory wood. The wood has long been used by cabinet makers and is a favorite for kitchen cabinets. The Wright Brothers whittled hickory propellers for their "flying contraption." The shock and stress resistance of hickory also makes it ideal for striking tools, such as the handles of axes, hammers, and picks, for ladder rungs, and for flooring. Most hickory lumber is produced in the United States, with a limited harvest in Canada. Hickory is an excellent fuel, and during pioneer days, it was very widely used as firewood.

## CULINARY PORTRAIT

The thick-shelled nuts of hickory generally have relatively poor cracking quality, although, as noted earlier, some varieties are superior in this respect. Striking shagbark and shellbark hickory nuts at their widest center with a single sharp hammer blow is said to free the most, and often largest, pieces of kernel. But how to open hickory nuts is a subject that has been much debated, and some recommend hitting the nuts on the narrowest, not the widest, side. (No matter how the nuts are hit, the kernels tend to shatter. By contrast, a percussive blow on the ends of pecan nuts buckles their shells and produces whole kernels.) The nutmeat may be removed from the nutshells by using a nutpick, but in species with contorted shell compartments, this is labor-intensive. The buttery nuts are highly flavored, and for incorporation in ice cream, cookies, and candies, only about half as much is required as for pecans. Hickory nuts are occasionally available as a gourmet, specialty item, but to a considerable degree are simply collected from the wild or from ornamental trees for personal use. They are best harvested by allowing them to drop from the trees in the autumn. Rubber gloves may be worn to avoid staining hands. The nuts contain about 10% moisture when harvested in this way and will mold if not dried to about 5%. Properly dried nuts should keep for at least a year under cool conditions.

Hickory wood chips (especially from the bitternut, *C. cordiformis* (Wangenh.) K. Koch) are commonly used in the United States for smoking hams and bacon. Hickory once played another culinary role with respect to pork: pigs were allowed to forage on hickory nuts in the woods during the winter in the milder parts of colonial New England. There is another obsolete culinary usage of hickory: although maple trees are far superior, hickory trees were tapped by some North American settlers for their thick, sweet sap that could be boiled into sugar, although quantities were small.

### Culinary Vocabulary

- "Hickory salt" or "hickory-smoked salt" is salt that has been smoked over hickory coals, or salt or monosodium glutamate mixed with hickory ashes or the equivalent in artificial flavoring.

## PROSPECTS

In contrast to pecan (*C. illinoinensis*), which has been extensively selected for desirable traits, there has been very little selection of improved varieties of other species of *Carya*. However, most species bear edible, although not necessarily pleasant-tasting, nuts. The two species highlighted in this chapter have nuts of excellent flavor and have been developed to a minor extent as commercial nut trees. The hickories have several drawbacks as nut trees, including slow growth and a long period before bearing nuts (20 years at least for most species, 40 years until a commercial crop can be harvested), thick shells, poor cleavage or cracking qualities, small size of nuts, and the fact that the nuts are often round, whereas commercial cracking machines are adapted to oblong shapes. A few commercial plantations exist in the United States, none in Canada. Hickories are notorious for attracting squirrels, and protecting the nuts from them is a major problem. The shellbark requires a long ripening season, and in the northern part of its range, it is a less reliable nut bearer than the shagbark.

Most cultivars of hickories have originated as cuttings from wild trees. There are at least 15 cultivars of the shagbark hickory, some with large kernels and good cracking qualities. Most of these were acquired after 1950. Improved lines of shellbark hickory have been selected in recent decades, and there are now over 40 cultivars, most originating from Iowa and Pennsylvania. Several valuable germplasm collections of hickories exist in the United States. The most important of these is the National Clonal Germplasm Repository of Pecans and Hickories at Brownwood, Texas. Many named cultivars and most of the world's species of *Carya* are represented. In Canada, the University of Guelph Arboretum has a collection of big shellbark hickories from the Carolinian zone of Ontario. These collections are important for future breeding of improved cultivars.

There is a long-standing hope that hybrids between the shagbark hickory and the pecan will be the basis of thin-shelled, improved hickory cultivars suitable for Canada and the northern United States. As pointed out in the text, when cross-pollinated, pecans and hickories produce vigorous hybrids called "hicans" that tend to be poor nut producers but excellent ornamental shade trees. However, occasional hicans have been found that produce nuts combining the best characteristics of their parents. A hican called Burton has produced an abundance of good nuts in the midwest but is apparently poorly adapted to the northeastern U.S. climate.

Given that the nut market (excluding the peanut, which is not a nut botanically) is dominated by imported nuts of superior taste, the probability that hickory nut will become significantly more important is very limited. However, hickories produce very valuable lumber, and plantations may become more common by virtue of the fact that they can be harvested both for lumber and nuts.

## CURIOSITIES OF SCIENCE AND TECHNOLOGY

- Hickory species were once abundant in Europe, but were exterminated by the last ice age on that continent, about a million years ago.
- The eighteenth-century Finnish naturalist Pehr Kalm alleged that the clear strong flame of burning hickory injured the eyes of those looking at it.
- Wheelwrights commonly used shagbark hickory to form the spokes of wooden wheels for carriages and carts, especially for westward-trekking pioneers. Hickory spokes were even used for early automobiles.
- The tough, almost unbreakable wood of hickory contributed to uses considered unacceptable today. As a handle for whips, it allowed cruel owners to employ considerable force on

a victim, whether horse or slave. As a classroom ruler or stick, it caused pupils who were not learning their "readin', writin', and 'rithmetic" to cringe as the schoolmaster aimed for their heads or knuckles.

- Hickory wood is rather uniquely suited for sporting goods in its combination of strength and elasticity and is made into hockey sticks, skis, bows, baseball bats, golf club shafts, tennis rackets, heavy sea fishing rods, and the like. Almost the entire world's supply of lacrosse sticks is made in Canada from hickory, but because very little hickory lumber is produced in Canada, most of the wood is imported from the United States.
- Early American settlers commonly referred to durable things as being "as tough as hickory." American president Andrew Jackson (1767–1845) was known as "Old Hickory" because of his ruggedness as a frontier militiaman in 1813, fighting Amerindians.
- Some parrots, Staffordshire bull terriers, and the occasional human being are capable of cracking open hickory nuts with their teeth, but the nuts are nevertheless very strong. The southern U.S. expression "he can crack hickory nuts between his toes" is a way of saying that someone is a really tough character.
- Many wild trees that produce nuts and fruits tend to have consecutive years of low production, separated by occasional seasons of extremely high production, called "mast years." Black bears are particularly fond of hickory nuts, which are very nutritious, and it has been found that during hickory mast years in New England, the bears fatten up especially well and produce more cubs. By contrast, in poor years, bears tend to be more of a problem foraging for food in urban areas.
- Young red squirrels have been observed to learn to open hickory nuts more quickly by first observing experienced squirrels do so.

## KEY INFORMATION SOURCES


Althen, F.W. von. 1990. *Sowing and planting shagbark and bitternut hickories on former farmland in Southern Ontario*. Great Lakes Forestry Centre, Sault Ste. Marie, ON. pp. 11.

Boisen, A.T., and Newlin, J.A. 1910. *The commercial hickories*. U.S. Department of Agriculture, Forest Service, Washington, DC. pp. 64.

Bonner, F.T. 1976. Storage and stratification recommendations for pecan and shagbark hickory. *Tree Planters' Notes U.S. For. Serv*. 27(4): 3–5.

Campbell, R.D. 1975. The great northern hickories. *Annu. Rep. Northern Nut Growers Assoc*. 66: 146–148.

Campbell, R.D. 1980. The hickories in Ontario. *Annu. Rep. Northern Nut Growers Assoc*. 71: 35–39.

Campbell, R.D. 1987. Productivity of hickories in the North. *Annu. Rep. Northern Nut Growers Assoc*. 78: 35–36.

Campbell, R.D. 1990. Hickory nut performance in Ontario. *Annu. Rep. Northern Nut Growers Assoc*. 81: 127–128.

Crawford, M. 1999. The hickories. *WANATCA [West Australian Nut and Tree Crops Association] Yearb*. 23: 36–47.

Graney, D.L. 1990. *Carya ovata* (Mill.) K. Koch, Shagbark hickory. In *Silvics of North America*. Vol. 2. Hardwoods. Agric. Handb. 654. Edited by R.M. Burns and B.H. Honkala. U.S. Department of Agriculture, Forest Service, Washington, DC. pp. 219–225.

Grauke, L.J., and Pratt, J.W. 1986. A field guide to the hickories of North America. *Annu. Rep. Northern Nut Growers Assoc*. 77: 123–153.

Halls, L.K. 1977. Hickories/*Carya* spp. [as wildlife food]. *For. Serv. Tech. Rep. (USA)* 16: 132–133.

Hudak, J. 1968. Shagbark hickory. *Horticulture* 46(10): 10, 12.

Hyppio, P.A. 1970. *Cary ovata* 'Holden', an ornamental cultivar of shagbark hickory (Juglandaceae). *Baileya* 19: 91–96.

Loeb, R.E. 1998. Evidence of prehistoric corn (*Zea mays*) and hickory (*Carya* spp.) planting in New York City: vegetation history of Hunter Island, Bronx County, New York. *J. Torrey Bot. Soc*. 125: 74–86.

MacDaniels, L.H. 1979. Hickories. In *Nut tree culture in North America*. Edited by R.A. Haynes. Northern Nut Growers Association, Inc., Hamden, CN. pp. 4–50.

Manning, W.E. 1948. A hybrid between shagbark and bitternut hickory in southeastern Vermont. *Rhodora* 50: 60–62.

Marquard, R.D. 1991. Pollen and pollination of the hickories. *Annu. Rep. Northern Nut Growers Assoc.* 82: 21–23.
Nelson, T.C. 1961. *Silvical characteristics of shagbark hickory*. USDA Forest Service, Station Paper 135. Southeastern Forest Experiment Station, Asheville, NC. pp. 11.
Phillips, D.R 1973. *Hickory*. U.S. Department of Agriculture, Forest Service, Washington, DC. pp. 7.
Pratt, J.W. 1986. A field guide to the hickories of North America. *Annu. Rep. Northern Nut Growers Assoc.* 77: 123–153.
Robison, S.A., and McCarthy, B.C. 1999. Growth responses of *Carya ovata* (Juglandaceae) seedlings to experimental sun patches. *Am. Midl. Nat.* 141: 69–84.
Small, E., and Catling, P.M. 2004. Shagbark hickory (*Carya ovata* (Mill.) K. Koch) and shellbark hickory *C. lacinoisa* (F. Michx.) Loud.—the best native nuts? *Can. Bot. Assoc. Bull.* 37(1): 4–11.
Stone, D.E. 1997. *Carya*. In *Flora of North America north of Mexico*. Vol. 3. Edited by Flora of North America Editorial Committee. Oxford University Press, New York. pp. 417–425.
Theophilus, A., and Newlin, J.A. 1910. *The commercial hickories*. United States Forest Service Bulletin 80. U.S. Department of Agriculture, Forest Service, Washington, DC. pp. 64.
Thompson, T.E., and Grauke, L.J. 1990. Pecans and other hickories (*Carya*). In *Genetic resources of temperate fruit and nut crops*. Edited by J. Moore and J.R. Ballington, Jr. Int. Soc. Hortic. Sci. Wageningen, the Netherlands. pp. 837–904.
Totten, W.S. 1976. What can we do with [shellbark] hickories? *Annu. Rep. Northern Nut Growers Assoc.* 67: 68–71.

## Specialty Cookbooks

Flowers, A.S. 1903. *Nut menu: giving minute description of how to prepare the well-known edible nuts of our markets for the table, almonds, chestnuts, hickory nuts, pecans, peanuts, shellbarks, walnuts, etc., a treatise on the preparation of nuts for the plate*. Publisher not given, Mount Joy, PA. pp. 42.
Riotte, L. 1975. *Nuts for the food gardener. Growing quick, nutritious crops anywhere*. 2nd printing. Garden Way Publishing, Charlotte, VT. pp. 179.

Borman (1969) has one recipe featuring hickory nuts, Brill (2002) has three, Freitus (1980 has six, Freitus and Haberman (2005) have one, Johnson (1989) has three, Kluger (1973) has four, Robe-Terry (1997) has five, Russell (1975) has one, Schufer (2011) has two, and Tatum (1976) has three. (See the Appendix to this book for details of the publications cited.)

# 49 Hog Peanut

Family: Fabaceae (Leguminosae; pea family)

## NAMES

Scientific name: *Amphicarpaea bracteata* (L.) Fernald

- "Peanut" in the name hog peanut is for the peanut-like appearance of the seeds and the peanut-like characteristic of producing seeds underground. The "hog" in the name is based on the observation that hogs allowed to roam in woods often rooted up the nuts.
- Hog peanut is also known as American hog peanut, ground peanut, hog bean, hog pea, hog vine, and (wild) ground bean. The names American licorice, pea vine, wild bean, wild licorice, and wild peanut have also been used, but are inappropriate as they are best employed for other species.
- The name Dakota pea is based on frequent collecting of the hog peanut by Dakota Indians of the Missouri Valley.
- The name southern hog peanut has been employed in Texas.
- The Mexican plants of the species are called talet bean, from the Nahua language *talli*, tall + *et*, bean.
- The genus name *Amphicarpaea* is based on the Greek *amphi*, both, + *karpos*, fruit, an allusion to the plant producing two different kinds of fruit, one type aboveground and the other subterranean. The genus name has been spelled *Amphicarpa* in most older literature. Adding *ae* to *Amphicarpa* to produce the awkward spelling *Amphicarpaea* is the result of a decision to adopt the latter name that was made mandatory by the 1956 *International Code of Botanical Nomenclature*. There are two genera in the grass family with similar names: *Amphicarpon* and *Amphicarpum*. There is also an animal genus named *Amphicarpa*, belonging to the ascidians, a class of marine, sedentary, filter-feeders.
- Hog peanut should not be confused with "pignut," a name for five different species of *Carya* (see Chapter 48 on Hickories).
- Hog peanut is often confused with groundnut (*Apios americana*) (see Chapter 45). Both species are in the legume family, and both produce edible underground organs. The names Dakota pea, ground bean, and wild bean, which are best reserved for hog peanut, have been applied to groundnut, usually as a result of confusion.
- The Bambarra groundnut, *Vigna subterranea* (L.) Verd., is also sometimes called hog peanut.
- *Bracteata* in the scientific name *A. bracteata* is Latin for "small leaf," referring to the small leaves at the base of the flowers.

## GEOGRAPHY AND ECOLOGY OF WILD PLANTS

Hog peanut is native from Quebec to Montana in the north to Texas and Florida in the south. Also, there are plants growing in the Mexican states of Veracruz and Puebla. Some plants of hog peanut of Southeast Asia are so similar that some authorities have included them in *A. bracteata*, but others place them in *A. edgeworthii* Benth.

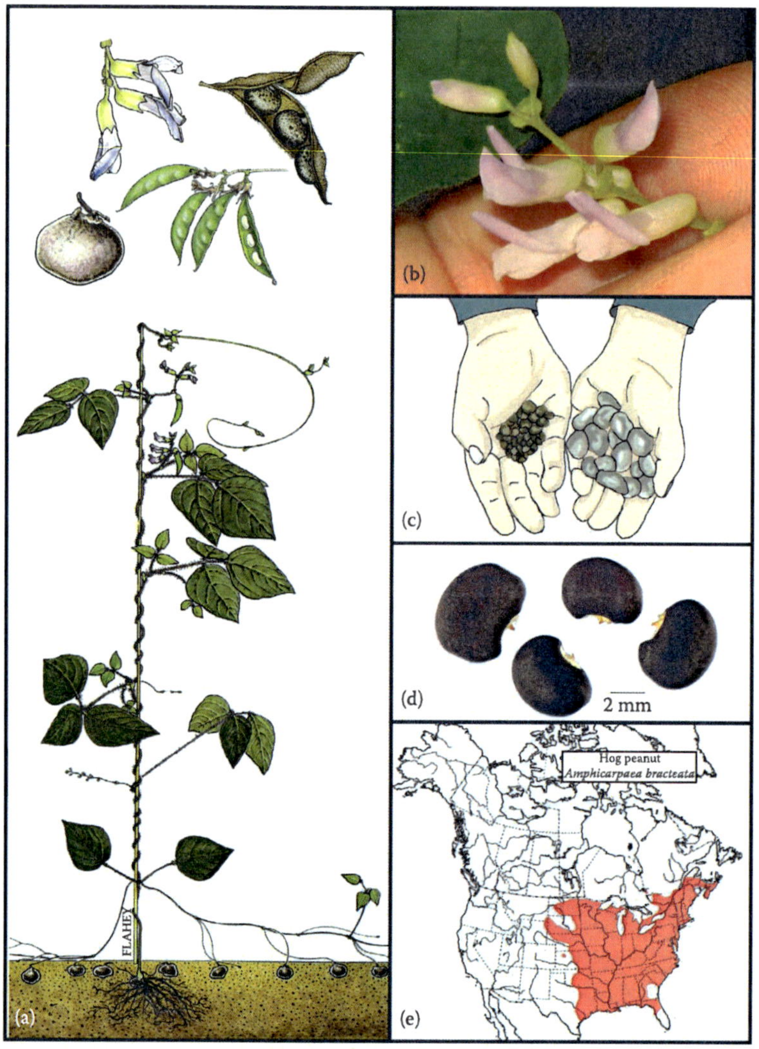


**FIGURE 49.1** Hog peanut (*Amphicarpaea bracteata*). (a) Plant (in flower, with underground one-seeded fruits) climbing on a pole; at top, views of underground fruit, young aerial fruits (pods), flowers, and mature aerial fruit. (Painting courtesy of B. Flahey.) (b) flowers. (Courtesy of Jomegat [CC By 3.0].) (c) Comparison of size of aerial fruits (left) and underground fruits (right). (Painting courtesy of B. Brookes.) (d) Aerial seeds. (Courtesy of Steve Hurst, U.S. Department of Agriculture.) (e) Distribution map.

Hog peanut tends to be found along shorelines and in alluvial soils (i.e., deposited by moving water) in disturbed sites, and usually not far from rivers and streams. The soils occupied are generally well drained and sandy. The plants occur in cool, moist, shaded deciduous woodlands; open woods; sunny meadows; thickets; the edges of forests; and prairie bottomlands. Although not considered a weed, in some areas, hog peanut grows aggressively.

## PLANT PORTRAIT

Hog peanut is an annual, twining, slender, herbaceous vine, with delicate, threadlike stems growing 30–150 cm (1–5 feet) in length, trailing on the ground or climbing over shrubs. The plant is generally branched at the base. The leaves have three leaflets (like clover leaves), which are 1.3–10 cm

(0.5–4 inches) long. In the axils of the upper leaves, there are clusters of pealike, whitish or lilac flowers, each about 1.3 cm (0.5 inch) long, which mature into small, flattened, bean-like pods 1.5–4 cm (0.6–1.6 inches) long containing 1–5 small, hard seeds. The plant also produces tiny, closed flowers with extremely small petals; these usually develop into one-seeded pods. Some of these pods are produced aboveground, others are underground at the ends of threadlike runners or at the tips of small branches of the fibrous roots. The peanut (*Arachis hypogaea* L.) similarly produces its pods underground. The underground pods of hog peanut are typically 6–12 mm (0.2–0.5 inch) in diameter and contain one edible seed resembling a peanut, although the size and shape of these differ considerably on the same plant. The pod wall is stretched over the seeds and gives the "nut" a brownish to purplish-brown appearance. When the pod wall is removed from the seed, the latter may be seen to have a white, pale blue, or purple skin. The skin loosens on cooking, leaving a cream-colored seed. These edible seeds are often abundant, frequently appearing just under the surface of the ground, and they can easily be gathered. The seeds produced below ground are much larger and more succulent than those produced aboveground, and although the latter are edible, they are not as important as a source of human food.

The underground fruits are very nutritious and were once a valuable food for Native Americans, especially in the Missouri Valley. American Indians employed the pods in various ways, cooking them separately or with cornmeal, and to make bean bread. Early explorers also used hog peanuts as food. Although a wild plant, hog peanut is said to have once been cultivated in the southern United States for its underground nuts, and in the nineteenth century, the species was grown in Scotland. In Mexico today, some Indian groups (Nahuas, Totonics, and Mestizos) allow hog peanut to grow in corn fields and while cultivating the soil, they collect the underground "beans."

## CULINARY PORTRAIT

Hog peanut is a wild plant, which has not been marketed commercially as a food plant, and its "peanuts" are obtainable only from wild-growing plants. Although it is possible for home gardeners to grow their own hog peanuts, the yield is low and picking the underground beans is tedious. The hog peanut's underground seeds resemble shelled garden beans, but are rather dry. When raw, the "bean" is sweet, pleasant-tasting, with a somewhat nutty flavor, not a "raw bean" taste. Raw hog peanuts taste much better than raw peanuts. The leathery "skin" or shell of the one-seeded pod cracks off during boiling. On cooking, the beans tend to disintegrate into a pureé, and so are suitable for soups, stews, and casserole dishes. Dried beans shrink and become somewhat harder than peanuts, although not as hard as dry beans or peas. They can be ground into flour.

The small, hard, seeds produced in the multiseeded aerial pods are also nutritious and have sometimes been consumed by Native Americans, usually boiled. They have been compared to soybeans in flavor. Typically, in Mexico, the seeds are roasted and consumed as a snack or with the daily principal meal.

As with a number of other edible legumes, antinutritional factors are present in the seeds. These include trypsin inhibitors (the subterranean seeds have levels similar to soybean), and a fairly low content of hemagglutinin (lectin). These chemicals are thought to be destroyed during cooking and are therefore not of concern, unless large quantities of raw seeds are consumed. Like other members of the bean family, hog peanuts can produce intestinal gas. It is often stated that the seeds from the fruits produced aboveground should not be eaten, but this is not true (a very few people are thought to be allergic to hog peanuts, whether aerial or subterranean). The aerial seeds are dry, whereas the fleshy underground seeds may have 50% water, although they are similar in moisture content to the aerial seeds when dried.

### CULINARY VOCABULARY

- "Bean" appears in several of the alternative names of the hog peanut. The English word "bean" has been traced with certainty only within the confines of the English language, but probably came from German. The word was spelled the same and had the same meaning a thousand years ago in English. Many, but by no means all "beans" are seeds of the legume (Fabaceae) family. A variety of seedlike products of similar size and shape are also termed beans, such as coffee beans and jelly beans.

## PROSPECTS

Hog peanut is a wild plant, which has not been marketed as a food plant. Several early explorers and botanists expressed the view that the plant should be developed as a vegetable. Although it is possible to grow hog peanuts, the yield is low and picking the small, underground beans is tedious. Keeping weeds down is very difficult because the plant is a vine and naturally grows around any nearby stems. The pods cannot be harvested by pulling up the whole plant, as one does with peanut, because the stalks are so slender that they break. Moreover, the fruits are dark in color and difficult to see in the soil. These problems can be overcome by breeding and appropriate technology. The late Canadian botanist W.G. Dore experimented with growing the plants in a sandy soil, and succeeded in harvesting 1 kg (2.2 pounds) of seed per 10 m (33 feet) of row from seeds planted 10 cm (4 inches) apart in sterilized sandy loam. The seeds can be separated out relatively easily with a 6 mm (1/4 inch) screen.

It has been suggested that hog peanut is potentially useful as a soil cover and for erosion control. There are indications in the literature that the species was cultivated in the southern United States as forage, but there is no evidence that it was ever significant as a crop. Hog peanut certainly has potential as a forage legume, which could enrich the soil through the nitrogen-fixing bacteria in the heavily nodulated roots. Hog peanut also has horticultural potential as a ground cover, since the foliage forms a dense cover and the plants remain quite low when they do not have objects to grow on.

Although the prospects of developing hog peanut as human food seem quite limited, the idea has some merit. Hog peanut grows prolifically under some circumstances, one of the basic characteristics of crop plants. As a legume, it is adapted through its symbiotic nitrogen-fixing bacterial partner to grow well in nitrogen-deficient soils, which make it a good choice for sandy soils. Coincidentally, sandy soils offer the best prospect of harvesting the edible fruits, although technology for this remains to be devised. Because hog peanut is a vine, it makes sense to intercrop it with another plant on which it can climb and which could benefit from the added soil nitrogen. However, a breeding program probably would focus on the creation of a compact type of plant that concentrates its underground seeds in a small area, to facilitate harvesting. Although the yield is low and the seeds would have to be marketed at a premium price, it is possible that specialty health food stores would be interested in stocking this most unusual bean. The seeds are easy to prepare, pleasant in flavor, and make an excellent snack. However, the unattractive name "hog peanut" is obviously unsuitable for marketing purposes. Talet bean, the name in Mexico, seems exotic and is therefore preferable. Although having one twentieth the yield of the similar peanut, talet bean may at least increase in regional importance, as other species with edible subterranean seeds and tubers already have. It has been recommended for development as a protein supplement in Mexico.

## CURIOSITIES OF SCIENCE AND TECHNOLOGY

- Native Americans used the hog peanut for a variety of medicinal purposes. The Cherokee Indians spat a boiled tea of hog peanut root onto snake bites, and also used the tea to treat diarrhea. The Iroquois used it to treat digestive problems, as well as tuberculosis. The Chippewa employed a preparation of the roots in combination with other species as a laxative. The Houma drank a boiled preparation of the leaves and stems, together with hackberry and powdered oyster shells, to treat venereal disease.
- Voles and field mice gather large quantities of the underground beans, and their nests often contain several liters (or quarts). Midwestern American Indian tribes, such as the Ojibway, Omaha, Pawnee, and Sioux, and eastern tribes, such as the Delaware, collected hog peanut

seeds from the stores of mice. The Dakota typically replaced such "stolen" beans with an equal amount of corn or other food for the rodents.

- The aboveground pods of hog peanut have an explosive mechanism that can scatter the seeds more than 2 m (6.6 feet) from the parent plant.
- The pea family contains a number of species that produce their seeds underground, where they are safe from animals that consume aboveground parts of plants. In addition to the peanut, the most important of these species are the Bambarra groundnut (noted in the section "Names") and the "groundbean," *Macrotyloma geocarpum* (Harms) Maréchal & Baudet, both cultivated mainly in Africa.
- The strategy of the hog peanut plant of producing seeds both sexually (in the aboveground pods) and asexually (especially in the underground single-seeded pods) is a clever way of ensuring its survival. Since the plant is an annual that does not overwinter by roots or tubers, the underground seeds, which are essentially genetically identical to the mother plant, ensure survival in a habitat that has proven to be suitable for the species. The sexually produced seeds are scattered widely, allowing the plant to colonize new areas.

## KEY INFORMATION SOURCES

Bibro, C.M., and Foote, B.A. 1986. Larval description of *Rivellia pallida* (Diptera: Platystomatidae), a consumer of the nitrogen-fixing root nodules of hog-peanut, *Amphicarpa bracteata* (Leguminosae). *Proc. Entomol. Soc. Wash.* 88: 578–584.

Callahan, H.S. 1997. Intraspecific differentiation in the *Amphicarpaea bracteata* (Fabaceae) species complex: varieties and ecotypes. *Rhodora* 99: 64–82.

Cheplick, G.P. 1987. The ecology of amphicarpic plants. *Trends Ecol. Evol. [TREE]* 2: 97–101.

Dore, W.G. 1970. A wild ground-bean, *Amphicarpa*, for the garden. *Greenhouse Garden Grass* 9(2): 7–11.

Gallagher, R.N., and Buhr, K.L. 1984. Plant nutrient and forage quality analysis of a wild legume [*Amphicarpa bracteata*] collected from the highland rim area of middle Tennessee. *Crop Sci.* 24: 1200–1201.

Maliarik, M.J., Roberts, D.D., and Goldstein, I.J. 1987. Properties of the lectin from the hog peanut (*Amphicarpaea bracteata*). *Arch. Biochem. Biophys.* 255: 194–200.

Marr, D.L., Devine, T.E., and Parker, M.A. 1997. Nodulation restrictive genotypes of *Glycine* and *Amphicarpaea*: a comparative analysis. *Plant Soil* 189: 181–188.

Meeson, B.W. 1974. Pollen morphology of *Amphicarpaea* (Leguminosae: Phaseoleae). *SIDA* 5: 338–340.

Parker, M.A. 1985. Local population differentiation for compatibility in an annual legume *Amphicarpaea bracteata* and its host-specific fungal pathogen *Synchytrium decipiens*. *Evolution* 39: 713–723.

Parker, M.A. 1986. Individual variation in pathogen attack and differential reproductive success in the annual legume, *Amphicarpaea bracteata*. *Oecologia* 69: 253–259.

Parker, M.A. 1988. Polymorphism for disease resistance in the annual legume *Amphicarpaea bracteata*. *Heredity* 60: 27–32.

Parker, M.A. 1994. Evolution in natural and experimental populations of *Amphicarpaea bracteata*. *J. Evol. Biol.* 7: 567–579.

Parker, M.A. 1996. Cryptic species within *Amphicarpaea bracteata* (Leguminosae): evidence from isozymes, morphology, and pathogen specificity. *Can. J. Bot.* 74: 1640–1650.

Parker, M.A., Doyle, J.L., and Doyle, J.J. 2004. Comparative phylogeography of *Amphicarpaea* legumes and their root-nodule symbionts in Japan and North America. *J. Biogeog.* 31: 425–434.

Parker, M.A., and Peters, N.K. 2001. Rhizobitoxine production and symbiotic compatibility of *Bradyrhizobium* from Asian and North American lineages of *Amphicarpaea*. *Can. J. Microbiol.* 47: 889–894.

Parker, M.A., and Wilkinson, H.H. 1997. A locus controlling nodulation specificity in *Amphicarpaea bracteata* (Leguminosae). *J. Heredity* 88: 449–453.

Peña, F.B, Villalobos, G., Martinez, M.A., Sotelo, A., Gil, L., and Delgado-Salinas, A. 1999. Use and nutritive value of talet beans, *Amphicarpaea bracteata* (Fabaceae: Phaseoleae) as human food in Puebla, Mexico. *Econ. Bot.* 53: 427–434.

Schnee, B.K., and Waller, D.M. 1986. Reproductive behavior of *Amphicarpaea bracteata* (Leguminosae), an amphicarpic annual. *Am. J. Bot.* 73: 376–386.

Small, E., and P. M. Catling. 2003. Poorly known economic plants of Canada—36. Hog peanut (talet bean), *Amphicarpaea bracteata* (L.) Fernald. *Bull. Can. Bot. Assoc.* 36(1): 8–13.

Trapp, E.J. 1988. Dispersal of heteromorphic seeds in *Amphicarpaea bracteata* (Fabaceae). *Am. J. Bot.* 75: 1535–1539.
Trapp, E.J., and Hendrix, S.D. 1988. Consequences of a mixed reproductive system in the hog peanut, *Amphicarpaea bracteata* (Fabaceae). *Oecologia* 75: 285–290.
Turner, B., and Fearing, O.S. 1964. A taxonomic study of the genus *Amphicarpaea* (Leguminosae). *Southwest. Nat.* 9: 207–218.

## Specialty Cookbooks

No cookbooks deal mainly with hog peanut, and it is difficult to find recipes for it, either on the Web or in cookbooks. Freitus (1980, listed in the Appendix to this book) presents six recipes for hog peanut.

# 50 Honewort

Family: Apiaceae (Umbelliferae; carrot family)

## NAMES

Scientific name: *Cryptotaenia canadensis* (L.) DC. (*Deringa canadensis* (L.) Kuntze)

- The word honewort, literally "hone plant," is applied to several unrelated species of the carrot family, particularly the European *Sison amomum* L., which was used to cure a swelling (especially a hard swelling in the cheek) called a hone.
- Honewort has also been called Canada honewort, Canadian honewort, stone parsley, and white chervil.
- The genus name *Cryptotaenia* comes from the Greek *cryptos*, hidden, and *taenia*, a fillet, referring to concealed oil tubes in the fruits.
- *Canadensis* in the scientific name means "of Canada." Carl Linnaeus (1707–1778) was the first to recognize the species by using the term *canadensis*; at that time "Canada" was understood to include parts of the northeastern United States, where Linnaeus knew the species occurred.

## GEOGRAPHY AND ECOLOGY OF WILD PLANTS

Honewort is found in rich woods, moist forests, thickets, and floodplains, and on the banks of streams from western New Brunswick to Manitoba, south to Georgia, Alabama, Arkansas, and Texas. It is very rare in Florida and Mississippi. The plants prefer to grow in dappled shade. They flower for about a month, from May to August depending on location, and produce seeds in late summer and autumn.

## PLANT PORTRAIT

Honewort is a succulent wild edible herb that looks like parsley and smells like celery. The plants are 30–100 cm (12–39 inches) tall after developing flowering and fruiting stalks. The leaves typically have three, sometimes five leaflets, which are sharply double-toothed on the margins, occasionally lobed. The plant bears inconspicuous white flowers in small umbels. The fruits are often curved and 4–6 mm (0.15–0.24 inch) long. *Cryptotaenia canadensis* is a perennial, flowering in the second year or later. Flowering plants die after maturing seed, but there is extensive vegetative reproduction by offshoots from the base of the stem.

Honewort is closely related to mitsuba (*C. japonica* Hassk.), a perennial herb of southeastern Asia, also known as Japanese parsley, Japanese honewort, and Japanese wild chervil. The two species are almost identical, and some authorities consider mitsuba to be a variety of honewort, *C. canadensis* (L.) DC. var. *japonica* (Hassk.) Makino (*C. canadensis* subsp. *japonica* (Hassk.) Hand.-Mazz.). Mitsuba is cultivated in Japan, Korea, China, Taiwan, and Indonesia, where it is a well-known vegetable and condiment. It is also often found in Chinese and Japanese markets in North America. Honewort has occasionally been harvested in North America and sold locally as mitsuba. Growers in the United States have occasionally exported crops of the species as a culinary herb to Japan.

**FIGURE 50.1** Honewort (*Cryptotaenia canadensis*). (a) Plant and details of fruit and flowers. (Painting courtesy of B. Flahey.) (b) Flowering plant. (Courtesy of BotBin [CC By 3.0].) (c) Distribution map.

## CULINARY PORTRAIT

In the late spring and early summer, while it is still young and tender, wild plants of honewort are collected for use as food. Like mitsuba, its cultivated Japanese counterpart, all parts of the plant are edible, and the plant is used for soup, as a potherb, in salads, as a root vegetable, and as a seasoning. Young leaves, stems, and flowers can be boiled for use as a potherb, or chopped like parsley for addition to salads and green soups. Roots can be scrubbed, boiled in salted water for 20 minutes, and served like parsnips with butter or sauce. Seeds can be used to flavor cookies, cakes, and breads. Stems can be candied in sugar.

Mitsuba is widely employed in Japanese cuisine, and serves as a guide to the uses and culinary preparation of honewort. The distinct, pungent flavor of mitsuba is reminiscent of celery, parsley, angelica, and sorrel. The leaves and leafstalks are eaten raw in salads and sandwiches; boiled or fried and consumed alone as a vegetable; used as a potherb; and added to soups, egg dishes, tempura, and a variety of fried foods. They can also simply be employed as a garnish. The leaves become bitter and lose their fragrance when simmered too long or subjected to too much heat, so mitsuba should be no more than lightly parboiled or very gently stir-fried in a scant amount of oil. Professional Japanese cooks occasionally discard the leaf blades because their cooking time is shorter than for the leaf stalks. The roots can be blanched for 5 minutes, and then sautéed, fried, or boiled. The seeds are used as a seasoning.

## WARNING

Species of the carrot family in North America are often very similar, and difficult to distinguish. Because some of these species are quite poisonous, wild food collectors should be certain of the identity of honewort before consuming it.

### Culinary Vocabulary

- Two Japanese dishes are often served with mitsuba, the domesticated relative of honewort. These are *shabu-shabu*, a traditional country dish of Japan composed of thinly sliced meat (usually beef) and vegetables, often served with dipping sauces, and *cha-wan-musha* or *chawan-mushi*, which is Japanese egg custard.

## PROSPECTS

Honewort is essentially a wild form of mitsuba, a well-known vegetable and condiment of Asia, although almost unknown in North America. The cultivation of mitsuba is very advanced in Japan, where considerable amounts are grown in hydroponic culture, cultivars have been selected, and research on cultivation and processing has been conducted. Although perennials, the plants are often grown commercially as annuals since the young vegetative plant is more tender and desirable. The usefulness of honewort lies largely as a source of breeding material for the selection of improved forms of mitsuba. These plants are easily cultivated, have an attractive and exotic flavor, and should be considered as potential minor diversification crops for markets in North America. The plants thrive in cool, shady conditions and are ideal subjects for winter glasshouse cultivation.

## CURIOSITIES OF SCIENCE AND TECHNOLOGY

- In addition to the North American *C. canadensis* and the Asian *C. japonica*, there are many other examples of plant species of eastern North America that have very closely related species in eastern Asia. Indeed, some 34 herbaceous genera of plants exhibit this pattern. The very similar deciduous forests of eastern Asia and eastern North America seem to be the major survivors of an ancestral forest that formed a continuous band around the northern hemisphere 15 to 20 million years ago. This was fragmented by climatic cooling, uplift of mountains, and continental glaciations.
- In Japan, leafstalks of mitsuba are often blanched (whitened by protection from sunlight) to increase their tenderness. This is done by hilling earth over the rootstocks in the spring, so that the young leaves have to grow through the earth, keeping the leafstalks from sunlight.

## KEY INFORMATION SOURCES

Abe, K., Sakimoto, M., and Miyazaki, A. 1994. Effects of cultivation methods and cultivar variations on characteristic aroma of Japanese honewort (*Cryptotaenia japonica* Hassk.). *J. Jpn. Soc. Hortic. Sci.* 62: 903–908. (In Japanese)

Baskin, J.M., and Baskin, C.C. 1988. The ecological life cycle of *Cryptotaenia canadensis* (L.) DC. (Umbelliferae), a woodland herb with monocarpic ramets. *Am. Midl. Nat.* 119: 165–173.

Eckenbach, U., Lampman, R.L., Seigler, D.S., Ebinger, J., and Novak, R.J. 1999. Mosquitocidal activity of acetylenic compounds from *Cryptotaenia canadensis*. *J. Chem. Ecol.* 25: 1885–1893.

Hara, H. 1962. Racial differences in widespread species, with special reference to those common to Japan and North America. *Am. J. Bot.* 49: 647–652.

Hardway, T.M., Spalik, K., Watson, M.F., Katz-Downie, D.S., and Downie, S.R. 2004. Circumscription of Apiaceae tribe Oenantheae. *South African J. Bot.* 70: 393–406.

Hawkins, T.S., Baskin, J.M., and Baskin, C.C. 2005. Life cycles and biomass allocation in seed- and ramet-derived plants of *Cryptotaenia canadensis* (Apiaceae), a monocarpic species of eastern North America. *Can. J. Bot.* 83: 518–528.

Hawkins, T.S., Baskin, J.M., and Baskin, C.C. 2007. Seed morphology, germination phenology, and capacity to form a seed bank in six herbaceous layer apiaceae species of the eastern deciduous forest. *Castanea* 72: 8–14.

Kanzaki, T. 1989. Contact dermatitis due to *Cryptotaenia japonica* Makino. *Contact Dermatitis* 20: 60.

Small, E., and Catling, P.M. 2001. Poorly known economic plants of Canada—31. Honewort, *Cryptotaenia canadensis* (L.) *DC. Bull. Can. Bot. Assoc.* 34(4): 44–46.

Spalik, K., and Downie, S.R. 2007. Intercontinental disjunctions in *Cryptotaenia* (Apiaceae, Oenantheae): an appraisal using molecular data. *J. Biogeog.* 34: 2039–2054.

Williams, C.F., and Guries, R.P. 1994. Genetic consequences of seed dispersal in 3 sympatric forest herbs. 1. Hierarchical population-genetic structure. *Evolution* 48: 791–805.

Yamaguchi, M. 1990. Asian vegetables. In *Advances in new crops*. Edited by J. Janick and J.E. Simon. Timber Press, Portland, OR. pp. 387–390.

## Specialty Cookbooks

The following cookbooks have recipes for mitsuba, for which honewort may be substituted:

Kazuko, E. 2002. *Masterclass in Japanese cooking*. Pavilion, London, U.K. pp. 92.

Kijima, N. 2006. *Healthy & tasty sushi rolls and onigiri*. Shufunotomo Company, Tokyo, Japan. pp. 62.

Staff, J. 2009. *Kawaii bento boxes: cute and convenient Japanese meals on the go*. Japan Publications Trading, Tokyo, Japan. pp. 81.

Tohata, A. 2001. *Japanese vegetable cooking*. Graph-sha, Tokyo, Japan. pp. 64.

Weinstein, K. 2007. *Japanese kosher cooking: shushi, sushi and more*. KTAV Publishing House, Jersey, NJ. pp. 193.

Yagihashi, T., and Salat, H. 2009. *Takashi's noodles*. Ten Speed Press, Berkeley, CA. pp. 168.

# 51 Hop

Family: Cannabaceae (hemp family)

## NAMES

Scientific name: *Humulus lupulus* L.

- The word "hop" traces to Old German (*hopfo*) and Dutch words for the plant. "Hop" (i.e., in the singular) refers to a hop plant; "Hops" (i.e., in the plural) is best used to refer to the hops of commerce, that is, the female cones used to flavor beer.
- Hop is also known as common hop, English hop, and European hop.
- "Spanish hop" (or Spanish hops) is a name sometimes used for dittany of Crete, *Origanum dictamnus* L., a rare ornamental and culinary herb.
- The genus name *Humulus* is derived from the Latin *humus*, soil, an allusion to their soil-hugging habit if the vines are not supported.
- *Lupulus* in the scientific name is derived from the Latin *lupus* and its diminutive *lupulus*, a wolf; in ancient Rome, hop was called *lupulus salictarius*—a "wolf among willows," suggesting that when it grew among willows it was as destructive as a wolf among sheep.

## GEOGRAPHY AND ECOLOGY OF WILD PLANTS

Hop is indigenous to the north-temperate hemisphere. There are different wild varieties of hop plants in North America, Europe, and Japan. The geography of varieties in North America is complex. Variety *neomexicanus* Nelson & Cockerell is the predominant wild hop in the western Cordillera of North America, from Mexico to British Columbia; var. *pubescens* E. Small occurs in the midwestern United States; and var. *lupuloides* E. Small is found in eastern North America. In much of Canada and the United States, the European var. *lupulus* is found as an escaped plant from past use in brewing, or as a persisting ornamental around abandoned homesteads. Hop plants frequently occur in moist thickets, slopes, river banks, alluvial woods, or along fences and hedges, often in sandy soils. The plants grow best in rich, alluvial or deep sandy or gravelly loams. Well-drained soils are especially beneficial in areas subject to frost heaving of the roots. Hop is adapted to a wide range of temperate climates. Although quite tolerant of low temperatures, good snow cover can reduce winterkill in very cold regions. Hop is somewhat shade tolerant, but prefers full sun.

## PLANT PORTRAIT

Hop vines are herbaceous, and have been known to extend more than 10 m (33 feet). The vines climb over vegetation by twining, and are assisted in holding onto surfaces by two-hooked hairs that resemble miniature grappling hooks. The annual, aboveground stem is killed by frost each year, regrowth occurring each season from perennial underground stems. The plants are male or female, pollination occurring by wind.

The European variety *lupulus* is the ancestor of most brewing cultivars used today. However, both in Japan and in North America, the local wild hop plants are hybridized with the imported European hop to produce unique cultivars. About two dozen countries raise substantial commercial crops of hops, including the United States and Canada. World centers of hop production also include

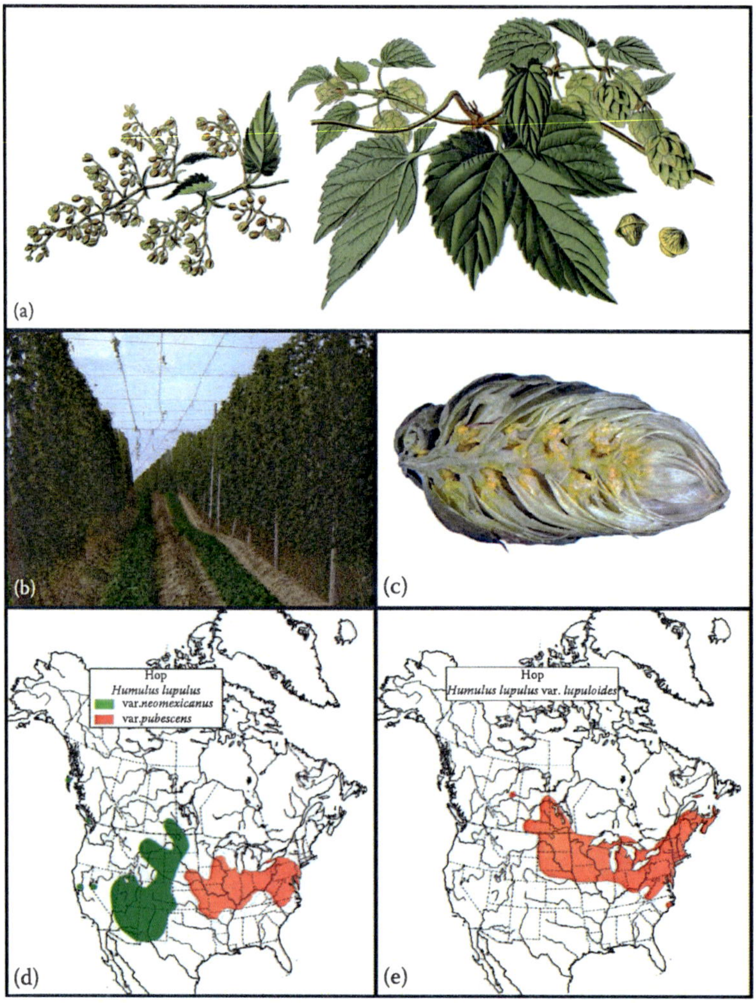


**FIGURE 51.1** Hop (*Humulus lupulus*). (a) Male branch (left) and female (fruiting) branch (right) with enlarged view of glands (bottom right). (From Köhler, H.A., *Köhler's Medizinal Pflanzen*, Verlag von F.E. Köhler, Germany. Vol. 2, 1887.) (b) Hop plantation. (Courtesy of Holledauer.) (c) Long section of cone. (Courtesy of David Gent, U.S. Department of Agriculture, online at Bugwood.org [CC By 3.0].) (d) and (e) Distribution maps of North American indigenous taxonomic varieties.

Germany, Russia, China, England, the Czech Republic, Slovak Republic, and the countries of the former Yugoslavia. The beers of different regions are often quite different in taste substantially because of the different hops used in brewing. In the United States, hop production occurs on about 15,000 ha (36,000 acres) in three Pacific Northwest states, Washington, Oregon, and Idaho, with an annual value exceeding 100 million dollars. The leading American hop varieties are Nugget, Willamette, and Galena. In North America, the vines are grown commercially on trellises 5.5 m (16 feet) high. Because the female plants produce the commercially valuable hops (the cones or fruit-clusters), and also are of greater ornamental value than the males, the latter are generally discarded as soon as they can be recognized. However, males are usually planted deliberately in England because they increase yield and it is uneconomical to grow most English cultivars to produce seedless hops. More than 100 cultivars have been named. In North America, most hops are pressed into 91 kg (200 pound) packages for commercial distribution.

**FIGURE 51.2** Hop (*Humulus lupulus*). (a) A beer tent celebrating Octoberfest in Stuttgart, Germany. (Courtesy of Annie and Andrew [Flickr/CC-attribution].) (b) Cones hanging on branches. (Courtesy of Roberto Verzo [Flickr/CC-attribution].) (c) Compressed hop pellets, the principal form of hops used in brewing. (Courtesy of Daquella manera/Daniel Lobo [Flickr/CC-attribution].)

Hop has been used through much of recorded history for culinary, medicinal, and household purposes, although it is best known as a brewing ingredient. Hops are a chief constituent of beers, which in the past have often been produced in kitchens, and not just in specialized breweries. It has been argued that hopped beer was made in prehistoric times, but the early history of the cultivated hop is obscure. Fermented liquids such as beer were important beverages throughout the ancient world. Herbs and spices were often added to improve flavor and keeping qualities, and it seems that ancient civilizations used hops for these purposes. The first unequivocal evidence for cultivation of hop comes from ninth-century Bavaria, and a tradition of usage of hops in beer has been maintained there for perhaps a millennium. A German law ("German beer purity law," *Reinheitsgebot*) states that only water, malted grains, hops, and yeast can be used to brew beer. During the Middle Ages, many monasteries were famous for their hopped beers. The brewing trade developed to a highly refined art, with beer being consumed for breakfast, dinner, and supper. The cultivation of hop was not introduced into England until the close of the fifteenth century. The hop was brought to North America and grown in the early seventeenth century. By the middle of the nineteenth century, New England and New York produced the bulk of the hops of the New World. However, by the early twentieth century, the Pacific Coast became the leading hop-producing area in North America. In the 1920s, hop growing in New York was practically wiped out by downy mildew fungus and by Prohibition (beer consumption dropped to 30% of the pre-Prohibition era). Similarly in Canada, commercial hop growing was phased out in the east by the end of the Second World War, and became established in British Columbia.

British and German traditions, respectively, account for the names of the leading breweries in Canada and the United States. The British-derived Canadian brewery labels are Molson, Carling O'Keefe, and Labatt (an Englishman, despite the French-sounding name); the German-derived American labels are Schlitz, Pabst, and Anheuser-Bush. The British tradition is to make "ale," by fermenting the yeast at the top of the brew, a method that produced the darker beer still preferred in Britain today. The German tradition was to produce "lager," a pale, light drink made with a slower, bottom-fermenting yeast. In both the United States and Canada today, lager is the principal type of beer produced. The early preference for British ale left a legacy in Canada for a somewhat darker, tangier beer than that sold in the United States.

In addition to the culinary uses mentioned below, hop extracts have been used as an aromatic bitter principle (for flavoring), and in pharmaceutical preparations. Extracts are also used (particularly in Europe) in skin creams, lotions, and shampoos for their alleged skin-softening and health-promoting properties.

## CULINARY PORTRAIT

Hops are chiefly used commercially to flavor beer. At room temperature, oxidation leads to rapid losses of the brewing value of hops, so that cold (often frozen) storage needs to be given as soon as possible after harvest. Hops are normally refrigerated until used in brewing, usually between 6 and 20 months after harvest. Hop extracts and oil have been used to flavor tobacco, yeast, beverages other than beers, frozen dairy desserts, candy, gelatins, puddings, baked goods, various confections, chewing gums, and condiments.

There are a few modern kitchen uses of hops. In home use, tea is sometimes "brewed," as an infusion of the leaves and cones. Hops can also be added to soups, stews, and other foods to produce bitterness, if desired.

Young shoots (emerging from the underground rhizome) that are 5–10 cm (2–4 inches) long are often consumed as a potherb, like asparagus. These spears can be boiled for 2–3 minutes and then boiled again in a change of water until tender. When steamed for 5 minutes and served with melted butter or cheese sauce, the shoots taste much like asparagus. In hop-producing areas of Europe, blanched hop spears are sometimes served in fine restaurants.

### CULINARY VOCABULARY

- The very young shoots of hop that are consumed in Europe as a vegetable such as asparagus spears are sometimes called "hop tops." They are also often served in Europe under their French name, *jets de houblon*.
- "Bittering hops" are hops used to produce bitterness in beer. The varieties used have high alpha acid content.
- "Noble hops" have a relatively low alpha acid content, and are used to impart aroma and flavor, but not necessarily bitterness. Varieties of this type of hop are popular in Germany and Belgium.

- "Coarse" is a term used to indicate that a beer has been overly hopped and is too bitter.
- "Hop cheese" is a German cheese that is cured by packing the fresh curd, molded into small units, between layers of hops in a cask.
- "Ale" is narrowly defined as unhopped beer (or conversely, beer is hopped ale). However, in modern practice numerous beers are called ales, and indeed because very little unhopped ale is now produced, the words ale and beer have become largely synonymous. The word "beer" comes from the Middle English *ber(e)*, and ultimately from the Latin *bibere*, to drink. "Ale" is based on the Old English, *ealu*, akin to Old Norse *ol* and the Lithuanian *alus*, words for ale.
- The most common type of beer is known as "lager" or "Pilsner" and is lighter in color, usually milder in taste, and sometimes with less alcohol than other classes of beer. Until about 1400, types of yeast were used in brewing that collected at the top of the vat (the brewing process is called "top fermentation"). Then, in Bavaria, "bottom fermentation" developed, in which yeast collects at the bottom of the vat. Bottom fermentation permits the manufacture of a lighter, milder beer. Because this type of beer spread first to Pilsen, Czechoslovakia, it became known as "Pilsner." "Lager," from the German *lager bier*, meaning "storehouse beer," was another name that became associated with this type of product. (Lager for a period became synonymous with the word beer.) Almost all North and South American beer, most European beer, the beer of Australia, and the beer of almost all major Asian countries are made in the Pilsner style. Bottom-dwelling yeasts proved suitable for the extreme temperatures of North America; because they were at the bottom of the vat they were buffered against the very hot summers and freezing winters, allowing fermentation to proceed, while top-dwelling yeasts often failed to perform well.
- By contrast with lager or Pilsner beer, darker beers that are drier and richer in taste are called porters and stouts (stout is simply a stronger, darker, sweeter porter). These are made with traditional top fermentation. The term "porter" traces to the laborers in London, England called porters who carried produce to London's Convent Garden, and popularized this type of beer. (Porter was offered in America at "porterhouses," and the proprietor of one establishment, Martin Harrison, became famous for his cut of beef, that was called after 1814 a "porterhouse steak.") Stout was first brewed by Guinness of Ireland (which, incidentally, produces the *Guinness Book of Records*). It was originally called Stout-Porter, as its recipe was a variation of that for porter beer.
- A "black and tan" is a popular English beverage made with half stout ("black") and half bitter beer ("tan"). The phrase originated apparently in the mid nineteenth century in England. It also denoted a kind of terrier with black and yellowish-brown markings. In the 1920s, the black and tans were auxiliary members of the Royal Irish Constabulary (so-named for their black and khaki uniforms) who suppressed national uprisings.
- "Red eye" (red-eye) is a very old drink made with beer and tomato juice (and sometimes flavored with Worcestershire sauce and tabasco). "Calgary Red Eye" is a mythical western Canadian beer, allegedly made with an egg and tomato juice.
- "Mead" is an alcoholic beverage based on fermented honey. "Sack mead," once somewhat popular, was made with the addition of hops.
- "Near beer" is a phrase that became popular during the Prohibition era in the United States (January 16, 1920 to December 5, 1933). The Volstead Act of 1919 that initiated Prohibition in 1920 defined a forbidden beverage as one containing 0.5% alcohol. Accordingly, the market for "near beer," which contained 0.4% alcohol, greatly expanded. When alcohol was illegally added to near beer, the product was called "needle beer" or "needled beer." In 1933, the Cullen–Harrison act that ended Prohibition raised the definition of "nonintoxication" to 3.2%, and such brews were called "3.2 beer."

## PROSPECTS

Hop is unusual as an agricultural commodity in that it is completely dependent on the brewery industry. Because consumption of beer is extraordinarily stable, so is the demand for hops. Moreover, there is really no substitute for hops. Small deviations from market demand can cause low prices in the case of oversupply, or high prices in the case of relatively small shortages, so investment in hop production can be a gamble. In recent years, hop powders, resin extracts, and compressed hop pellets have increased storage stability and decreased transportation costs considerably, so international competition is strong. Hop farming is very specialized, requiring high capital investment, and it is very labor intensive, and is increasingly an "agribusiness" venture. In North America, hop farms are usually "family corporations." These considerations suggest that there is limited prospect of increasing production in the short term. Rather, the problem for those already in the business is to retain market share.

To maintain and even increase market share, the creation of new cultivars is particularly important, since these can contribute to small changes in the taste of beer that can lead to large changes in beer preferences (albeit, beer drinkers are very loyal to their accustomed tastes). Moreover, new cultivars are essential to increasing productivity and resistance to diseases. Hop is a good example of a crop that has been substantially improved recently through incorporation of genes from wild North American plants. A wild hop from Manitoba has contributed to the improvement of many standard brewing varieties. The contribution of wild hop germplasm to the approximately $100 million of hops consumed each year in the United States has been valued at about $10 million annually. Accordingly, it is important for the future of the hop industry in North America to invest in breeding better cultivars, particularly paying attention to native American hop plants.

## CURIOSITIES OF SCIENCE AND TECHNOLOGY

- The Arabian physician Mesue, in the middle of the ninth century, claimed that the use of hops would induce sleep. There is a long Eurasian tradition of using hops as a sedative, including putting the cones in pillows, and planting hop beside bedrooms. King George III (1738–1820) of England suffered periodic bouts of mental illness and finally became permanently deranged. During his demented periods, his doctors prescribed that he sleep on a pillow stuffed with hops. American Indians independently adopted the sedative use of hops. In the days when hops were harvested using manual labor, a few people proved to be so sensitive to the hop plant that they become uncontrollably drowsy (often appearing to be drunk) after inhaling fumes given off by the plant or by harvested hop cones. The sedative value of hops led to it being used as a cure for "uncontrolled sexual desires and a quarrelsome nature." As late as the 1980s, it was recommended that hop tea be taken "3–4 times a day for nervous heart conditions and excessive sexual urge." The tranquillizing effect commonly alleged in folklore for hops may have a logical basis in a sedative volatile alcohol which is present.
- It was customary in the thirteenth century to baptize babies with beer.
- In medieval Catholic Germany, anything liquid could be consumed during the Lenten fasting period. Sly monks brewed tasty *Starkbier*, a heavy, strong beer that considerably alleviated their pangs of hunger. Today, many breweries still produce this beer around the time of Lent (especially popular is Munich's Salvator Starkbier brewed at Nockherberg).
- Charles Darwin (1809–1882), the great student of evolution, entertained himself while sick in bed in 1882 by studying a hop plant growing on his windowsill. He noted that the tip of the stem completed a revolution in 2 hours. The hop had obviously been planted outside his window in accord with the tradition that it would help people in the bedroom to fall asleep.
- The patent office of the United States once granted a patent to a man who claimed to have "invented" the hop's habit of winding from left to right (i.e., circling clockwise, viewed so

that the twining stem is growing toward the observer). Many other kinds of vines, runner beans for example, climb anticlockwise.

- Ladies-in-waiting at the court of English King Henry VIII (1491–1547) were allowed 1 imperial gallon (4.5 L) of beer for breakfast.
- Each youngster in the Children's Hospital of Norwich, England in 1632 received 9 L (2 imperial gallons) of beer daily.
- "Ale conners" were officials in Old England who were responsible for overseeing ale quality (William Shakespeare's father was an ale conner). A quaint custom of testing the quality of brew involved pouring some on a bench and sitting in the puddle wearing leather breeches for about half an hour. If sugar remained in the ale (indicating incomplete fermentation), the breeches would stick to the bench.
- In the nineteenth century, hops was an ingredient in many patent medicines. One of these was hop bitters, comprised of hops in 30% alcohol. Its advertising slogan was "Take hop bitters three times a day, and you will have no doctor bills to pay."
- Before yeast cakes became widely available in stores, yeast for bread-making was prepared by culturing wild yeast in a decoction of hops and water. Some of this was mixed with bread dough, adding flavor to the bread, and apparently preventing the yeast from spoiling by virtue of the antiseptic properties. Prospectors in the Alaska Gold Rush were called sourdoughs because they kept a pot of the starter material with them to make bread. Sourdough starter has been called an eternal flame of cooking that should never be extinguished. San Francisco is renowned for its sourdough bread. It has been claimed that a crock of San Francisco sourdough, removed from its native city, loses its power to produce top-quality bread.
- Before modern film was developed, photographic plates were coated with a solution containing hop juice.
- Hop stems contain considerable fiber, and have been used in making paper and twine. The stems were also once used in basketry and wickerwork.
- Among the many curious Hoosier (Indiana) folk remedies are the following: for tuberculosis, take finely powdered hops in rye whisky—morning, noon, evening, and night. Prior to the use of antibiotics, hop essences were commonly used to treat tuberculosis patients, and certain components are, in fact, very effective against gram-positive bacteria. Another curious Hoosier hops remedy was for earache: tie a small muslin bag of warm hops over the ears at night.
- It has often been suggested that hops are psychoactive, like the related marijuana (*Cannabis sativa* L.). This may be partly due to the peddling by British merchants in the nineteenth century of a substance they called hopeine, alleged to be a narcotic derived from the finest wild American hops. In fact, it was a mixture of an aromatic oil and morphine.
- It is ironic that the closest relative of the genus *Humulus* is *Cannabis*, the marijuana plant. Both genera possess numerous small secretory glands producing a resin, but while the aliphatic acids of the hop plant resin provide flavor for legal intoxicants, the intoxicating tetrahydrocannabinol of the marijuana plant is illegal. Many have been tempted to smoke hop leaves, although they are devoid of the mood-altering chemicals found in marijuana.
- In parts of Europe, the presence of male plants in hop yards is forbidden by law.
- It is a popular misconception that American beers generally have much less alcohol than Canadian beers (in fact, the average Canadian beer is a mere 0.5% stronger than the average American beer). The error arises because Canadian alcohol content is reported by volume, whereas American is often reported by weight. A 5% Canadian beer is actually equal to a 4.5% American.
- A study of 30,000 men in Finland found that drinking a beer a day reduced the risk of kidney stone formation by 40%. It was suggested that the hops in beer might provide compounds that slow the release of calcium form bone, reducing calcium excretions that are the building blocks of kidney stones.

- Researchers have isolated beer compounds with powerful antioxidant effects. Antioxidants protect against substances called free radicals, which are thought to worsen a range of diseases, including heart disease, stroke, and certain cancers. The health-promoting antioxidants are called prenylated flavonoids, and are in beer because of the presence of hops. Unfortunately, to maximize health benefits, one would have to drink 450 L (about 1000 American pints) of beer a day.
- Beer has been classified as a staple food in Bavaria (southern Germany).
- In West Africa, Star beer is so popular that an empty Star beer bottle is often used in recipes as a unit of measurement (approximately equal to 2.8 cups).
- Some beer experts hold that a head of foam on beer enhances flavor by trapping the aroma of the hops, although probably the custom of producing a head on beer has more to do with tradition than with taste. If a beer will not produce a head, it has lost its carbonation and gone flat. To produce a head, tilt the glass on its side, fill about one-third, and then straighten and pour the beer straight in. Experienced beer drinkers buying draught beer in a bar do not want a head because they get less beer.
- The foam on beer in television commercials is usually made of detergent suds.
- In 1991, the British government promised beer drinkers a full pint of ale. The problem was that foam can take up a lot of room in the glass, depending on the skill of the bartender, and a 90% full pint was the minimum required to avoid prosecution. But in 2002, the government decided that a 95% full glass would be required. It was estimated that this would put an extra 60 million pints of beer into British bellies each year, about 4 pints per drinker. Britain managed to preserve the pint, one of the great measures of British accomplishment, from the European Union's requirement that its members adopt metric measures. (A British or Imperial pint is 20% larger than an American pint.)

## KEY INFORMATION SOURCES

Bassil, N.V., Gilmore, B., Oliphant, J.M., Hummer, K.E., and Henning, J.A. 2008. Genic SSRs for European and North American hop (*Humulus lupulus* L.) *Genet. Resourc. Crop Evol.* 55: 959–969.

Beach, D.R. 2000. *Homegrown hops: An illustrated how-to-do-it manual.* Second edition. D.R. Beach, Junction City, OR. pp. 110.

Brown, J.F. 1980. *Guinness and hops.* Guinness, London, U.K. pp. 264.

Burgess, A.A. 1964. *Hops: Botany, cultivation, and utilization.* Interscience Publishers, New York, NY. pp. 300.

Cantrell, P.A., II. 2000. Beer and ale. In *The Cambridge world history of food.* Edited by K.F. Kiple and K.C. Ornelas. Cambridge University Press, Cambridge, U.K. pp. 619–625.

Cronk, A. 1959. *English hops glossary: Over 300 terms used in hop-growing & the hop-trade.* St. Leonard's Press, Bradford, U.K. pp. 32.

DeLyser, D.Y., and Kasper, W.J. 1994. Hopped beer: The case for cultivation. *Econ. Bot.* 48: 166–170.

Eckel, A., and Fritz, D. 1990. Forcing of hop shoots as a vegetable. 1. Comparison of various hop cultivars. *Gartenbauwissenschaft* 55: 34–36.

Edwardson, J.R. 1952. Hops—Their botany, history, production and utilization. *Econ. Bot.* 6: 160–175.

European Brewing Convention Technology and Engineering Forum. 1997. *Hops and hop products: manual of good practice.* Getränke-Fachverlag Hans Carl, Nürnberg, Germany. pp. 185.

Faulkner, C. 1992. *Hops and hop-pickers.* Faulkner Publications, Quarry Bank, Brierley Hill, West Midlands, U.K. pp. 70.

Filmer, R. 1982. *Hops and hop picking.* Shire Publications, Aylesbury, Bucks, U.K. pp. 82.

Gimbel, L.S. 1973. *Steiner's guide to American hops.* Hopsteiner, New York. pp. 77. (cf. Steiner 1977, 1986, below)

Hampton, R., Small, E., and Haunold, A. 2001. Habitat and variability of *Humulus lupulus* var. *lupuloides* in upper Midwestern North America: A critical source of American hop germplasm. *J. Torrey Bot. Soc.* 128: 35–46.

Haunold, A. 1993. Agronomic and quality characteristics of native North American hops. *Amer. Soc. Brew. Chem. J.* 51: 133–137.

Howard, E.L., Whittock, S.P. Jakše, J., Carling, J., Matthews, P.D., Probasco, G., Henning, J.A., Darby, P., et al. 2011. High-throughput genotyping of hop (*Humulus lupulus* L.) utilising diversity arrays technology (DArT). *Theor. Appl. Genet.* 122: 1265–1280.

Lawrence, M. 1990. *The encircling hop: A history of hops and brewing*. SAWD, Sittingbourne, U.K. pp. 120.

Moir, M. 2000. Hops—a millennium review. *J. Am. Soc. Brew. Chem.* 58: 131–146.

Murakami, A., Darby, P., Javornik, B., Pais, M.S.S., Seigner, E., Lutz, A., and Svoboda, P. 2006. Microsatellite DNA analysis of wild hops, *Humulus lupulus* L. Genet. Resourc. *Crop Evol.* 53: 1553–1562.

Nelson, C.E., Early, R.E., and Mortensen, M.A. 1966. *Effects of plant spacing, fertility, and irrigation on yields and quality of Yakima Cluster hops*. Washington Agricultural Experiment Station, College of Agriculture, Washington State University, Pullman, WA. pp. 11.

Neve, R.A. 1991. *Hops*. Chapman and Hall, London, U.K. pp. 266.

Neve, R.A. 1995. Hops. In *Evolution of crop plants*. Second edition. Edited by J. Smartt and N.W. Simmonds. Longman Scientific & Technical, Burnt Mill, Harlow, Essex, U.K. pp. 33–35.

Reeves, P.A., and Richards, C.M. 2011. Species delimitatation under the general lineage concept: An empirical example using wild North American hops (Cannabaceae: *Humulus lupulus*). *Syst. Biol.* 60: 45–59.

Sharpe, F.R. 1978. *Chemistry and technology of the essential oil of hops*. Brewing Research Foundation, Nutfield, U.K. pp. 180.

Small, E. 1978. A numerical and nomenclatural analysis of morpho-geographic taxa of *Humulus*. *Syst. Bot.* 3: 37–76.

Small, E. 1980. The relationships of hop cultivars and wild variants of *Humulus lupulus*. *Can. J. Bot.* 58: 676–686.

Small, E. 1981. A numerical analysis of morpho-geographic groups of cultivars of *Humulus lupulus* L. based on samples of hops. *Can. J. Bot.* 59: 311–324.

Small, E. 1997. Cannabaceae. In *Flora of North America*. Vol. 3. Edited by Flora North America Editorial Committee. Oxford University Press, New York. pp. 381–387.

Srinivasan, V., Goldberg, D., and Haas, G.J. 2004. Contributions to the antimicrobial spectrum of hop constituents. *Econ. Bot.* 58(Suppl.): S230–S238.

Steiner Inc. 1977. *Steiner's guide to American hops, book II*. S.S. Steiner Inc. Hopsteiner, New York. pp. 74.

Steiner Inc. 1986. *Steiner's guide to American hops, book III*. S.S. Steiner Inc., New York. pp. 195.

Stevens, R. 1967. The chemistry of hop constituents. *Chem. Rev.* 67: 19–71.

Torrens, R.C. 1967. *The hops industry of the United Kingdom*. U.S. Dept. of Agriculture, Foreign Agricultural Service, Washington, DC. pp. 28.

United States Brewers Foundation. 1953. *Barley, hops and history*. United States Brewers Foundation, New York. pp. 28.

Verhagen, L.C. 1988. Hop analysis. In *Beer analysis*. Edited by H.F. Linskens, J.F. Jackson, and S. Aastrup. Mod. Methods Plant Anal. New Ser. 7. Springer-Verlag, Berlin, NY. pp. 67–87.

Zimmermann, C.E. 1981. *English and foreign publications on hops*. U.S. Dept. of Agriculture, Science and Education Administration, Washington, DC. pp. 89.

## Specialty Cookbooks

Fahy, C. 1978. *Cooking with beer*. Dover Publications, New York, NY. pp. 144.

Fisher, J., and Fisher, D. 1998. *The homebrewer's garden: How to easily grow, prepare, and use your own hops, brewing herbs, malts*. Storey Books, Pownal, VT. pp. 187.

Garetz, M. 1994. *Using hops: The complete guide to hops for the craft brewer*. HopTech, Danville, CA. pp. 222.

# 52 Huckleberries

Family: Ericaceae (heath family)

## NAMES

### Eastern Huckleberries

Scientific names: *Gaylussacia* species, especially the following:

- Common huckleberry—*G. baccata* (Wangenh.) K. Koch
- Dangleberry—*G. frondosa* (L.) Torr. & A. Gray
- Dwarf huckleberry—*G. dumosa* (Andrews) A. Gray
- The name "huckleberry" arose in America, a corruption of the English word hurtleberry (explained in Chapter 15 on Blueberries).
- In the United States, the name "huckleberry" is often used for blueberry (*Vaccinium*) species, especially in the west. The western blueberry species are sometimes called "western huckleberries" in contrast to the *Gaylussacia* species of eastern North America, which are distinguished as "eastern huckleberries." "Huckleberries" available in the marketplace could be blueberry (*Vaccinium*) species that are not usually considered to be true huckleberries. Huckleberries and blueberries often grow wild near each other, and many people pick huckleberries thinking that they are blueberries.
- Both blueberries and huckleberries are sometimes called "whortleberries."
- The "garden huckleberry" is an unrelated plant, *Solanum scabrum* Mill., which also produces edible berries. Unfortunately, some people use the name "garden huckleberry" for garden-planted *Gaylussacia* species.
- "Fool's huckleberry" (or false huckleberry) is *Rhododendron menziesii* Craven subsp. *menziesii* (*Menziesia ferruginea* Sm.), a shrub with inedible fruit, native to western North America.
- The common huckleberry (*G. baccata*) is also called black and high-bush huckleberry. Unfortunately, several western *Vaccinium* species are also called black huckleberry, including *V. membranceum*, featured below.
- Black snaps, crackerberry, and crackers are old names for the common huckleberry that were coined because the seeds crack or snap between the teeth.
- The dwarf huckleberry, *G. dumosa*, is also called bush huckleberry and bog huckleberry.
- The colorful name "dangleberry" is probably derived from "tangleberry," although it has also been claimed that dangleberries are so named because they dangle on long fruit stalks while on the plant. The species is also known as hairy dangleberry, blue huckleberry, and blue tangle. Unfortunately, it is also known as "dwarf huckleberry," the same name as applied to *G. dumosa*. However, the dangleberry is generally the tallest of the three species discussed here, and calling it "dwarf" is inappropriate. Like the common huckleberry, it too is sometimes called high-bush huckleberry.
- The genus name *Gaylussacia* commemorates French chemist Louis Joseph Gay-Lussac (1778–1850).
- *Baccata* in the scientific name *G. baccata* is Latin for berry-bearing.
- *Frondosa* in the scientific name *G. frondosa* is Latin for leafy, referring to the leafy flower stalks.
- *Dumosa* in the scientific name *G. dumosa* is Latin for bushy.

**FIGURE 52.1** Western huckleberries (*Vacciniuum* species). (a)–(d) Black huckleberry (*V. membranaceum*). (a) Berries. (Courtesy of Julie [CC By 2.0].) (b) Fruiting branches. (Courtesy of B. Bartlett, U.S. Department of Agriculture.) (c) Flowers. (Courtesy of Julie [CC By 2.0].) (d) Distribution map of black huckleberry. (e) Distribution map of Cascade huckleberry (*V. deliciosum*).

## Western Huckleberries

Scientific names: *Vaccinium* species, especially the following:

- Black huckleberry—*V. membranaceum* Douglas ex Torr. (*V. globulare* Rydb.)
- Cascade huckleberry—*V. deliciosum* Piper (*V. coccineum* Piper)
- The black huckleberry, *V. membranaceum* (which in fact usually has black fruit) is also known as big huckleberry, black blueberry, globe huckleberry, tall blueberry, and thin-leaved huckleberry (the leaves are indeed thin).
- The Cascade huckleberry, *V. deliciosum* (so-named because it grows in the Cascade mountain range from northern California into British Columbia), is also known as alpine

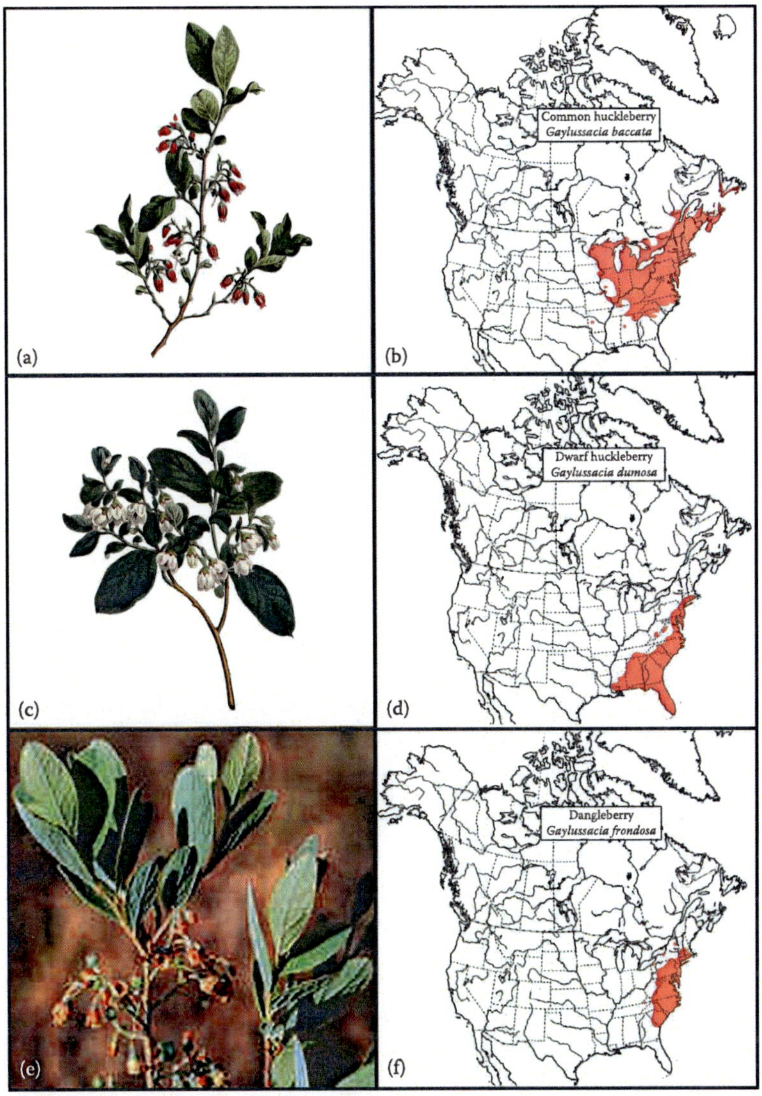


**FIGURE 52.2** Eastern huckleberries (*Gaylussacia* species). (a) Flowering branch of common huckleberry (*G. baccata*). (From Curtis, W., *Botanical Magazine*, Vol. 32, Plate 1288, 1805.) (b) Distribution map of common huckleberry. (c) Flowering branch of dwarf huckleberry (*G. dumosa*). (From Curtis, W., *Botanical Magazine*, Vol. 28, Plate 1106, 1808.) (d) Distribution map of dwarf huckleberry. (e) Flowering branch of dangleberry (*G. frondosa*). (Courtesy of USDA-NRCS PLANTS Database.) (f) Distribution map of dangleberry.

blueberry, blue huckleberry, blue-leaved huckleberry, Cascade blueberry, mountain bilberry, mountain huckleberry, and Rainier bilberry (Mount Rainier in the Cascade range of Washington state, and the allied Mount Rainier National Park, are prime sites for this huckleberry).

- *Membranaceum* in the name *V. membranaceum* is Latin for membranous, a reference to the leaves.
- *Deliciosum* in the name *V. deliciosum* is Latin for delicious, referring to the berries.
- See Chapter 15 on Blueberries for information on the genus name *Vaccinium*.

## GEOGRAPHY AND ECOLOGY OF WILD PLANTS

### Eastern Huckleberries

Common huckleberry (*G. baccata*) is native to southeastern Canada and the eastern United States. It occurs in sandy or rocky soil, in dry or moist woods, thickets, clearings, bogs, and coastal dunes. The grounds occupied are usually acidic, and often dry. The plant has moderate tolerance to shade, becoming dwarfed in deep shade and developing densely in canopy gaps and clearings. Leaf break and flowering occur in early spring, blooming continuing through early summer. Fruit maturation begins in late summer, with fruits often remaining on the plants in the fall.

The dangleberry (*G. frondosa*) is native to the eastern and southeastern United States. It occurs in moist woods, swamps, bogs, thickets, and woods. The very similar *G. bigeloviana* (Fernald) Sorrie & Weakley has often been included within *G. frondosa* (as *G. frondosa* var. *bigeloviana* Fernald); it overlaps much of the distribution of the dangleberry, but extends northward along the Atlantic coast into the maritime region of Canada.

Dwarf huckleberry (*G. dumosa*) is native to the eastern United States and eastern Canada. In the northern part of its range it occurs in sphagnum bogs in maritime regions, while in the southern part of its range it occupies dry barrens and pinelands, in sandy soils rather than in bogs.

### Western Huckleberries

The black huckleberry (*V. membranaceum*) grows at elevations between 600 and 3500 m above sea level (2000–11,500 feet a.s.l.) throughout forested areas of Idaho, western Montana, western Wyoming, Washington, British Columbia, and southwestern Northwest Territories, with small outcrops in Utah, California, Arizona, Michigan, and Ontario. It occupies coniferous woods, talus slopes, subalpine fir forests, and alpine heaths. This species has wide ecological amplitude, tolerating dry sites, but generally occurring on moist, somewhat deep, well-drained soils, and preferring a pH of about 5.5. The species is moderately shade tolerant, growing both under partially closed forest canopy and in sunny openings. Black huckleberry is often found in areas that have been disturbed. In the Pacific Northwest, Native Americans regularly burned vegetation to stimulate growth of black huckleberry.

The Cascade huckleberry (*V. deliciosum*) occurs at elevations of 580–2000 m (1900–6600 feet) from northern California to British Columbia, with a few collections reported from Idaho. It is found in alpine meadows, subalpine coniferous woods, and talus slopes. The species is relatively tolerant of waterlogged soils, and often grows near the edges of ponds.

## PLANT PORTRAIT

### Eastern Huckleberries

The genus *Gaylussacia* contains about 40 species of shrubs, mostly indigenous to South America, but several are native to North America, of which the species mentioned here bear the most edible fruit. Eastern huckleberries are quite similar in appearance and flavor to blueberries. However, blueberries have many, tiny, soft seeds that are too small to observe, while the seeds of huckleberries (up to 10 in a single berry) are hard and large enough to notice. Eastern huckleberries are harvested from wild plants, native to North America. The plants are deciduous (losing their leaves every fall), up to 2 m (6.6 feet) in height, and bear fruits in small clusters. The berries are blackish-blue or reddish-black, usually about 8 mm (0.3 inch) or less in diameter, sweet or slightly tart when ripe.

The common huckleberry (*G. baccata*) is the best known eastern huckleberry. It is 0.3–1 m (1–3 feet), rarely as tall as 2 m (6.6 feet) in height, producing pink or pale red flowers, and young growth that is resinous and sticky. The black, shiny berries are about 1 cm (0.4 inch) in diameter, ripen in early summer, and are said to be the sweetest of the huckleberries.

The dangleberry (*G. frondosa*) is taller than the common huckleberry, growing 1–2 m (3–6.6 feet) in height. The plant produces greenish flowers. The fruit is often as large as 1.3 cm (0.5 inch) in diameter, is dark blue with a whitish bloom, and is sweet and pleasant in taste.

The dwarf huckleberry (*G. dumosa*) is a low plant, 10–50 cm (4–20 inches) high. It produces white or pink flowers, and black fruit. Its fruits are inferior in taste to those of the above species.

### Western Huckleberries

Like blueberry species, western huckleberries belong to the genus *Vaccinium*. All shrubby species of *Vaccinium* growing in western North America have been called "huckleberries." However, the commercial blueberry species (see Chapter 15 on Blueberries) are in section *Cyanococcus* of the genus, while the most important (i.e. , harvested) western huckleberries are in a different group, section *Myrtillus*. Of the eight species of section *Myrtillus*, the two featured here have the largest and/or best tasting berries, and are the chief sources of collected wild huckleberries. These western huckleberries produce berries singly in the axils of leaves on new shoots (i.e., of the current year), whereas the commercial blueberries develop clusters of berries on 1-year-old wood, and produce larger amounts of berries. Attempts are underway in the United States to develop the two species as cultivated crops.

The black huckleberry (*V. membranceum*) ranges from 25–120 cm (1–4 feet) in height, and produces black or deep purple or red berries 9–11 mm (about 0.4 inch) in diameter occasionally as wide as 13 mm (0.5 inch).

The Cascade huckleberry (*V. deliciosum*) is a dwarf shrub, 7–28 cm (3–11 inches) high. It produces pink or creamy pink flowers and bluish (rarely black) berries of comparable size to the above species.

In addition to culinary uses, western huckleberries are employed to a small extent to provide aroma for soaps, lotions, shampoos, and candles. The leaves and berries of *V. membranaceum* are used for the production of over-the-counter herbal medicinal preparations, particularly in the form of capsules and extracts.

## CULINARY PORTRAIT

Huckleberries are almost entirely collected from the wild, and are sold locally. Huckleberries available in the marketplace are mostly western huckleberry (*Vaccinium*) species (huckleberry products coming from western North America are certain to be based on *Vaccinium* species). Fresh huckleberries are in season from late summer to late fall. The fruits are not edible until quite dark and soft. They are occasionally available fresh or frozen in supermarkets. Huckleberries are eaten like blueberries. The seeds of eastern huckleberries are especially large, and need to be removed. The seeds of western huckleberries are small, but often larger than in blueberries, producing a somewhat gritty texture. Western huckleberry fruits are difficult to remove from the plant undamaged, and are mostly frozen or processed. The texture of huckleberries is harmed by freezing, but their taste and color are undiminished. The fruit is consumed fresh, or made into preserves, syrups, jams, jellies, wine, and filling for pies, muffins, pancakes, and tarts. A variety of huckleberry-flavored goods such as tea, coffee, soft drinks, ales, salad dressings, honey, and candies are marketed, often produced as cottage industries (i.e., very small-scale). Upscale restaurants are increasingly offering gourmet meals with huckleberries, especially as sauces and syrups for desserts such as pastry, sorbets, and ice cream, but more and more as an accompaniment to savory foods, especially game meat. Because they are relatively rare and require effort to acquire, huckleberries generally are more expensive than blueberries. The flavor of huckleberries is more intense than that of blueberries, and while this is not to everyone's taste, many prefer the more pronounced flavor. Huckleberry flavor is strong, and when substituting huckleberries for blueberries it is often recommended that half of the amount of fruit be added.

### Culinary Vocabulary

- "Fly cake," "fly pie," and "roach cake" are old lunch counter terms for a raisin cake or huckleberry pie.

## PROSPECTS

Both eastern and western huckleberries are undomesticated wild plants, the berries (especially of the western species) collected from the wild except for plants grown in the occasional home garden. Managing wild stands has potential for increasing yields. It has been estimated that managed wild plants of the black huckleberry (*V. membranaceum*) can produce berries worth about $2000/ha. This can be a useful source of income for pickers: Western huckleberries typically wholesale at $4.00–7.00/L (about an American quart), and under ideal conditions pickers harvest 3.8 L/hour (subsequently the berries must be cleaned of debris). However, the volume of berries harvested is often quite variable from year to year and among sites, as is typical of wild fruit crops that are strongly influenced by weather conditions. Future development of the industry depends on the plants being cultivated and domesticated, with the selection of improved varieties, in the same manner as blueberry species. Indeed, current commercial blueberry crops provide the chief competition and deterrent to development of huckleberries as new crops. However, the widespread marketing of huge but low-flavor, watery blueberries suggests that the highly flavored huckleberries could become increasingly popular with consumers. There is much interest currently in the health benefits from antioxidants in berries, and it may be that huckleberries could also be desirable in this regard. Attempts are underway in the United States to develop the two western huckleberry species as managed, and possibly also as cultivated crops, notably at the Sandpoint Research Center of the University of Idaho. To facilitate this development, several publications on growing western huckleberries have been produced (see "Key Information Sources"). These include analyses of commercial prospects, as well as information on propagating, planting, pruning, harvesting, and marketing.

### Eastern Huckleberries (*Gaylussacia*)

At this time, there are no efforts underway to develop *Gaylussacia* species as either a managed wild crop, or a new, cultivated crop. Reasons that have been advanced to explain the lack of interest have focussed on the large, hard seeds. These are indeed an impediment to marketing eastern huckleberries as a fresh fruit crop but, as is evident for western huckleberries, there are numerous products that can be developed to use the pulp and juice. As the source of the best tasting fruit, *G. baccata* is the species of *Gaylussacia* most deserving of development. For this species to be developed into a commercial crop, it seems necessary to first select superior strains. An investment in research and development is required, and this, combined with a general reluctance to add another berry crop to the already competitive market, represents a substantial hurdle. One possibility, however, is to develop and manage the species in the manner of lowbush blueberry (*V. angustifolium*), which like *G. baccata* is native to eastern Canada and the northeastern United States, where most commercial production occurs. Lowbush blueberry stands are managed much like garden plants—they are fertilized, and pests and weeds are controlled, but in many cases, natural stands are simply adopted, rather than planted, and eastern huckleberries could be treated in the same way. The facts that *Gaylussacia* is adapted to habitats unsuitable for most agricultural endeavours, and that there are environmental advantages to developing native crops in native habitats suggests that eastern huckleberries represent a potentially profitable new crop initiative worth exploring.

### Western Huckleberries (*Vaccinium*)

The popularity of wild-collected western huckleberries as a gourmet item in upscale restaurants and occasionally in marketed, processed foods shows that there is a market niche for the fruit that can be expanded. In Canada, there is a small amount of production of various goods based on wild-harvested fruit. In the United States, there have been several studies of how to manage western huckleberries as wild-growing crops, and there is some consideration being given to development of cultivars. Western huckleberries grow slowly, and are notably less productive than the commercial blueberries, so that considerable investment in research may be required to raise their status to the level of the commercial blueberry species, limiting the probability of expanding the industry. However, their superior taste characteristics ensure that they will at least remain an ongoing gourmet treat harvested from the wild.

## CURIOSITIES OF SCIENCE AND TECHNOLOGY

- Languages evolve, and huckleberry is illustrative of this. Huckleberries were extremely common in pioneer days in the eastern United States, and came to be used in many expressions that are now obsolete. "As thick as huckleberries" meant very thick. "To get the huckleberry" was used like "getting a raspberry" (or getting razzed) is today, that is, to be laughed at or ridiculed. "To be a huckleberry to someone's persimmon" was to be nothing in comparison with something else, while "to be a huckleberry over someone's persimmon" meant to outrank someone.
- It has been claimed that a colony of the box huckleberry, *G. brachycera* (Michx.) A. Gray, a species native from Delaware and West Virginia south to Tennessee, holds the record for the greatest area covered by a single plant. The colony was found in 1920, at which time it covered an area of 40.5 ha (100 acres) near the Juniata River in Pennsylvania. It is thought to have started 13,000 years ago from a single plant, which multiplied vegetatively to cover a huge area.
- Many species occur separately in the mountains of western North America and in the eastern area of North America around the Great Lakes, usually in areas with rather similar ecological conditions. *Vaccinium membranaceum* is illustrative of this pattern. Other examples include giant rattlesnake plantain (*Goodyera oblongifolia* Raf.), thimbleberry (*Rubus parviflorus* Nutt.), devil's club (*Oplopanax horridus* (Sm.) Miq.), and oval-leaved bilberry (*Vaccinium ovalifolium* Sm.). The pattern suggests that in the past these species occupied a continent-wide area, that was split up by geological and climate changes.
- Huckleberries are a favorite food of bears, which is worth remembering when collecting this fruit from wild stands.
- In 2000, the "huckleberry" was declared to be the "official fruit" of Idaho, a measure that was initiated by a group of fourth graders. (Dr. Pam Brunsfeld, curator of the University of Idaho Stillinger Herbarium (ID), has noted (personal communication) that the fruit in question refers to species of *Vaccinium*, especially *V. membranaceum*.)

## KEY INFORMATION SOURCES

### Huckleberries in General

Small, E., and Catling, P.M. 2005. Poorly known economic plants of Canada—45. Eastern huckleberries (*Gaylussacia* spp.) and western huckleberries (*Vaccinium* spp.). *Bull. Can. Bot. Assoc.* 38(2): 15–23.

### Eastern Huckleberries

(*Note: G. frondosa* var. *bigeloviana* Fernald. is interpreted in this chapter as the species *G. bigeloviana.*)

Ballington, J.R., Ballinger, W.E., and Maness, E.P. 1988. Anthocyanin, aglycone and aglycone-sugar content of fruits of temperate North American species of *Gaylussacia* (Ericaceae). *Can. J. Plant Sci.* 68: 247–253.

Bonner, F.T, and Halls, L.K. 1974. *Gaylussacia baccata* (Wangh.) K. Koch—black huckleberry. *Agric. Handb. U.S. Dep. Agric.* 450: 427–428.
Camp, W.H. 1935. Studies in the Ericales. I. The genus *Gaylussacia* in North America north of Mexico. *Bull. Torrey Bot. Club* 62: 129–132.
Camp, W.H. 1941. Studies in the Ericales: a review of the North American Gaylussacieae; with remarks on the origin and migration of the group. *Bull. Torrey Bot. Club* 68: 531–551.
Duncan, W.H., and Brittain, N.E. 1966. The genus *Gaylussacia* (Ericaceae) in Georgia. *Bull. Ga. Acad. Sci.* 24: 13–26.
Fernald, M.L. 1911. The northern variety of *Gaylussacia dumosa*. *Rhodora* 13: 95–99.
Floyd, J.W. 2002. Phylogenetic and biogeographic patterns in *Gaylussacia* (Ericaceae) based on morphological, nuclear DNA, and chloroplast DNA variation. *Syst. Bot.* 27: 99–115.
Gauthier, R. 1983. *Gaylussacia dumosa* (Andr.) T. & G. var. *bigeloviana* Fern. new in the flora of Quebec. *Nat. Can.* 110: 411–420. [In French]
Gauthier, R., and Garneau, M. 1998. The situation of Bigelow's dwarf huckleberry (*Gaylussacia dumosa var. bigeloviana*) in Quebec. Gouvernement du Québec, ministère de l'Environnement et de la Faune, Direction de la conservation et du patrimoine écologique, Québec. pp. 36.
Harper, K.A. 1995. Effect of expanding clones of *Gaylussacia baccata* (black huckleberry) on species composition in sandplain grassland on Nantucket Island, Massachusetts. *Bull. Torrey Bot. Club* 122: 124–133.
Healy, W.M., and Robinette, S.L. 1974. Huckleberries. *USDA For. Serv. Gen. Tech. Rep. NE U.S. Dep. Agric. For. Serv. Northeast For. Exp. Stn.* 9: 89–92.
Hughes, R.H. 1977. Dwarf huckleberry/*Gaylussacia dumosa* (Andr.) T. and G. *For. Serv. Tech. Rep. So. U.S. South. For. Exp. Stn.* 16: 36–37.
Matlack, G.R. 1997. Resource allocation among clonal shoots of the fire-tolerant shrub *Gaylussacia baccata*. *Oikos* 80: 509–518.
Matlack, G.R., Gibson, D.J., and Good, R.E. 1993. Regeneration of the shrub *Gaylussacia baccata* and associated species after low-intensity fire in an Atlantic coastal plain forest. *Am. J. Bot.* 80: 119–126.
Sleumer, H. 1967. The genus *Gaylussacia*. H.B.K. Bot. *Jahrb.* 86: 309–384. (In German.)
Wherry, E.T. 1972. Box huckleberry as the oldest living protoplasm. *Castanea* 37: 94–95.

## Western Huckleberries

Barney, D.L. 1999. *Growing western huckleberries*. Cooperative Extension System, University of Idaho, Moscow, ID. pp. 22.
Barney, D.L. 2003. Effects of light, surface sterilization, and fungicides on the germination of black huckleberry seeds. *Small Fruits Rev.* 2(2): 73–80.
Barney, D.L. 2003. Prospects for domesticating western huckleberries. *Small Fruits Rev.* 2(1): 15–29.
Carroll, M.S., Blatner, K.A., and Cohn, P.J. 2003. Somewhere between: social embeddedness and the spectrum of wild edible huckleberry harvest and use. *Rural Sociology* 68: 319–342.
Kerns, B.K., Alexander, S.J., and Bailey, J.D. 2004. Huckleberry abundance, stand conditions, and use in Western Oregon: evaluating the role of forest management. *Econ. Bot.* 58: 668–678.
Minore, D. 1972. *The wild huckleberries of Oregon and Washington: A dwindling resource*. Pacific Northwest Forest and Range Experiment Station, U.S. Dept. of Agriculture, Portland, OR. pp. 20.
Minore, D., Smart, A.W., and Dubrasich, M.E. 1979. *Huckleberry ecology and management research in the Pacific Northwest*. U.S. Dept. of Agriculture, Forest Service, Pacific Northwest Forest and Range Experiment Station, Portland, OR. pp. 50.
Stark, N., and Baker, S. 1992. *The ecology and culture of Montana huckleberries: A guide for growers and researchers*. Montana Forest and Conservation Experiment Station, University of Montana, Missoula, MT. pp. 87.
Stark, N., Baker, S., and Essig, D. 1989. Allocation of nutrients in huckleberries. *J. Am. Soc. Hortic. Sci.* 114: 259–264.

## Specialty Cookbooks

Note that huckleberry recipes on the Web are usually based on the western species.

## Eastern Huckleberries

Hendrickson, R. 1981. Heavenly huckleberries. In *The berry book. The illustrated home gardener's guide to growing and using over 50 kinds and 500 varieties of berries*. Doubleday & Company, Garden City, NY. [chapter 5: "Heavenly Huckleberries," pp. 90–95.]

## Western Huckleberries

(*Note:* The recipe sources provided in the chapter on blueberry can also be used.)

Bowen, A. 1988. *The huckleberry book*. American Geographic Pub., Helena, MT. pp. 90.
Simonik, S. 1997. *Huckleberry haus cookbook: Over 170 delicious huckleberry recipes*. Stan Simonik, Priest Lake, ID. pp. 162.
Spokane Community Centers Foundation. 1988. *Spokane cooks! Northwest: All new restaurant recipes, featuring old Spokane favorites, huckleberries and favorite Washington wines*. Spokane Community Centers Foundation, Spokane Community Centers Foundation, Spokane, WA. pp. 220.
University of Alaska. 1960–1971? *Wild blueberries and huckleberries*. Publication #13. Cooperative Extension Service, University of Alaska, Division of Statewide Services, Fairbanks, AK. pp. 12. (Unpaginated; this publication deals with different species of *Vaccinium* than those mentioned here, but the recipes provided may be used.)

Krumm (1991) has 27 western huckleberry recipes, Freitus and Haberman (2005) have 7 recipes, Stanek and Butcher (2007) have 9, Watts and Watts (2007) have 4, and Williamson (1995) has 17. Additional recipes for western huckleberries include: "Mark's huckleberry crisp" (Stewart 2000), Huckleberry pie (Snell 2006), and Huckleberry wine (Marie 2008). Stewart (2002) has recipes entitled Huckleberry cordial and Huckleberry tea, using eastern huckleberries. (See the Appendix to this book for details of the publications cited.)

# 53 Jerusalem Artichoke

Family: Asteraceae (Compositae; sunflower family)

## NAMES

Scientific name: *Helianthus tuberosus* L.

- The French explorer Champlain called the Jerusalem artichoke *artichaut du Canada* (Canadian artichoke; American artichoke was a similar early name). This and the name *poires de terre* ("ground pears") were very early names used in France. Champlain used the name artichoke because he thought that the flavor of this potato-like vegetable resembled that of globe artichoke (*Cynara scolymus* L.). The globe artichoke, also known as French artichoke and green artichoke but often referred to simply as artichoke, is also a member of the sunflower family, but the portion harvested is the immature flower head, which is cooked as a vegetable. By contrast, the Jerusalem artichoke is occasionally called "root artichoke," although the edible tubers are stem tissue, not true roots. Still another "artichoke" is the Chinese artichoke (*Stachys affinis* Bunge; also known as Japanese artichoke and crosne), which also produces edible tubers. Curiously, in Burgundy, France, the potato (the vegetable most commonly compared to the Jerusalem artichoke) was once called "Indian artichoke."
- "Artichoke," a name originally used for the globe artichoke, is usually interpreted as a corruption of its Arabic name, *al'qarshuf*, taken up in English from the Italian dialect word *articiocco*.
- The "Jerusalem" in Jerusalem artichoke may be a corruption of Ter Neusen (a city in the Netherlands), the first source of the tubers transported to England. Another interpretation is that "Jerusalem" is derived from *girasole*, an Italian word for sunflower. It has also been suggested that the early North American pilgrims may have conceived of their new home as a "new Jerusalem" and applied the term Jerusalem to this vegetable, one of their new staple foods. It is also possible that Jerusalem was added to the name by Europeans simply to indicate that the plant originated in some distant, foreign country.
- The Jerusalem artichoke is neither an artichoke nor from Jerusalem.
- The sterile hybrid between the sunflower and the Jerusalem artichoke has been termed the sunchoke. This perennial produces abundant edible tubers and is much more vigorous than either of its parents. However, it has not been exploited as a cultivated crop with much success. The name sunchoke is now much more commonly used as an alternative commercial name for the Jerusalem artichoke. In fact, American Indians called the Jerusalem artichoke "sun root."
- The French (rarely also English) name for the Jerusalem artichoke, *topinambour*, was first used in France to arouse interest in the newly introduced vegetable, following the sensation that was generated by the exhibition in 1613 of six natives from the Topinambuous tribe of Brazil.
- For information on the genus name *Helianthus*, see Chapter 94 on Sunflower.
- *Tuberosus* in the scientific name *H. tuberosus* is Latin for tuberous, descriptive of the edible tuber.

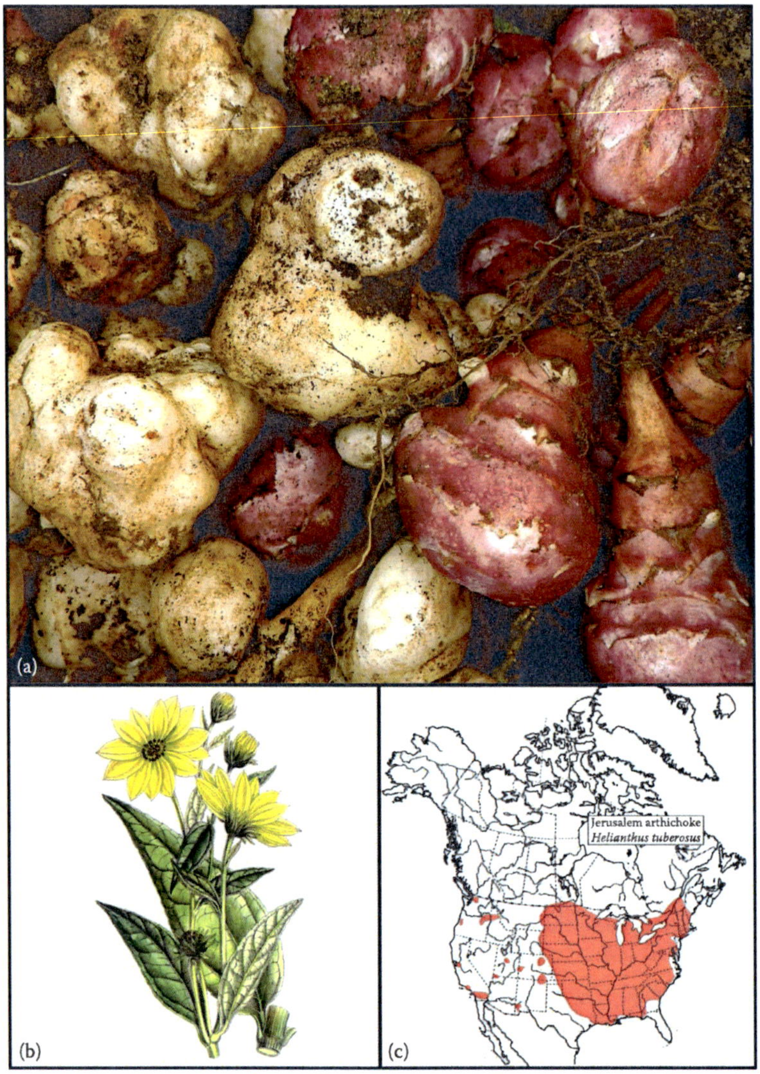


**FIGURE 53.1** Jerusalem artichoke (*Helianthus tuberosus*). (a) Tubers. (Courtesy of net_efekt [Flickr/CC-attribution].) (b) Flowering plant. (From Curtis, W., *Botanical Magazine*, Vol. 123, Plate 7545, 1897.) (c) Distribution map.

## GEOGRAPHY AND ECOLOGY OF WILD PLANTS

Wild Jerusalem artichoke is native to the eastern and southern United States and tropical America. The species was cultivated by Native Americans in pre-Columbian times, and this likely expanded its natural distribution area. The map shown here for this species (Figure 53.1c) indicates the main distribution area. Weedy and escaped forms have been collected from almost all states of the United States, and from all provinces of Canada except British Columbia, Alberta, and Newfoundland and Labrador. Outside of North America, *Helianthus tuberosus* is widely established as a weed in temperate areas of both the Northern and Southern Hemispheres. As a weed, it is found along roadsides and in fields and waste areas. It is capable of surviving as far north as Alaska. The species prefers a loose soil that is neither particularly acidic nor basic, and can tolerate poor soils. It prefers full sunlight, but can survive in light shade.

## PLANT PORTRAIT

Jerusalem artichoke is a stout herbaceous plant resembling a sunflower, growing 1–3.7 m (3–12 feet) tall. It is a perennial, overwintering by its tubers (swollen underground stems), but is commonly grown as an annual. The plant was reported in 1605 by French explorer Samuel de Champlain (1567–1635) to have been cultivated by Native Americans in what is now Cape Cod, Massachusetts. Small amounts of Jerusalem artichoke have been grown for many years in the United States, Canada, and Europe, as well as in Africa and Asia, but the market has remained very limited. In wild plants the tubers are slender, usually enlarged at their tips, and commonly red-skinned. In domesticated forms the tubers are notably thicker, crisp-fleshed (like an apple), whitish or yellowish, sometimes tinged with pink, and sometimes purple- or red-skinned. Shape varies from club-like to spherical, and short knobby branches are present. The Jerusalem artichoke has been described as resembling gingerroot or (more often) a knobby, new potato in appearance and size. The "knobs" are buds, comparable to the "eyes" of potatoes. In many of the cultivars, the tubers are 5–8 cm (2–3 inches) across and up to 10 cm (4 inches) long.

## CULINARY PORTRAIT

Jerusalem artichoke tubers are a specialty vegetable, consumed much like potatoes—with or without their skins, boiled, baked, roasted, fried, steamed, mashed, sautéed, and incorporated into soups and stews. They may be served raw in salads. Although quite different in taste from potatoes, and their distinctive flavor has often been acclaimed by gourmets, the Jerusalem artichoke is frequently negatively compared with the much more familiar potato, and this has contributed to its lack of popularity. The sugar inulin is the main storage carbohydrate in the tubers, not starch as in many other root crops. The lack of starch and the presence of inulin resulted in the recommendation, especially in the early twentieth century, that the Jerusalem artichoke is a suitable food for diabetics, which may be consumed in moderate amounts without serious increase in blood glucose. As noted below, however, some people cannot digest inulin and severe flatulence may result.

Jerusalem artichokes are typically sold in plastic packages in supermarkets. Because they are not in demand, they should be examined carefully. Avoid those that have a greenish tinge, or show evidence of sprouting, or are shriveled, wrinkled, dry, and light in weight, or moldy. After purchasing, the tubers should be consumed promptly, but they can be kept in plastic in a refrigerator for up to 2 weeks. The sweetness increases with refrigeration after harvesting, and it is advisable for those who grow their own Jerusalem artichokes to refrigerate them for 1–2 days before eating (or to simply leave the tubers in the ground until frost). They should be thoroughly scrubbed with a vegetable brush to remove any adhering sand or soil, but as much of their nutrients are stored just beneath the thin skin, it is best not to peel them (although cooking with the skins on may cause a darkening of the skins because of the high iron content). Do not use aluminum or iron cooking utensils, as these metals tend to oxidize on contact with the tubers. When cut, the tubers oxidize quickly. To prevent the resulting brown discoloration, immerse the tubers in acidulated water (with lemon or vinegar). For those who insist on removing the skin, this is best done immediately after the tubers have been cooked, before they cool and harden. Recommended times for cooking are as follows: 30–45 minutes for baking whole in an oven (as for potatoes), 10–15 minutes for steaming, and 5–7 minutes for sautéeing. Overcooking tends to turn the flesh into unappetizing mush. In Europe, Jerusalem artichokes are most frequently simply parboiled and sautéed in butter or served with a cream sauce or cheese. In California, they are often marinated in dressings, combined in salads, or substituted for water chestnuts in stir-fry dishes.

Inulin, the chief storage carbohydrate of Jerusalem artichoke, is indigestible, and is a chief cause of digestive difficulties in many people who consume the vegetable. (Just as flatulence is caused when beans are consumed because bacteria convert indigestibles that are present to carbon dioxide gas, the same thing occurs when inulin is consumed.) As noted above, prolonged cold makes

Jerusalem artichoke sweeter—in effect by converting some of the inulin to fructose sugar. Boiling the sliced tubers in a large volume of water significantly reduces the indigestibles. Cooking the whole tubers for 24 hours converts the majority of the indigestible substances in Jerusalem artichoke to sugar, not only making the vegetable digestible but also sweet. However, some people do not like the resulting taste.

Jerusalem artichoke can be used as a coffee substitute or coffee additive. Freshly dug tubers are washed, cut into small slices, and roasted in a slow oven at 120°C (248°F) for 1–2 hours, until they become dark brown and crisp. The pieces are ground and stored in a closed container. Because of the inulin content, the resultant drink has a sweeter flavor than regular coffee.

### Culinary Vocabulary

- In nineteenth-century England, a soup made from Jerusalem artichoke was once called "Palestine," under the assumption that the vegetable came from Jerusalem. Known for some 300 years, Palestine soup is all but forgotten today.
- In Europe, elegant Jerusalem artichoke meals are sometimes prepared by cutting the tubers to be cooked into small oval shapes called "pigeon eggs."

## PROSPECTS

Jerusalem artichoke is a significant vegetable and industrial crop. It has been thought to have considerable potential, and this has resulted in large attempts to expand its market over the last several decades. As a vegetable, it competes directly with potato, the world's fourth most important food plant, and it is unlikely to make much headway in this competition in the future. Jerusalem artichoke has been rather disappointing to date as an alcohol source, but given the current interest in non-fossil fuels, it deserves additional examination in this regard.

## CURIOSITIES OF SCIENCE AND TECHNOLOGY

- The name "inulin" was given to the storage molecule (chains of fructose) predominant in Jerusalem artichoke because the Scottish chemist Thomas Thomson (1773–1852) first extracted it from roots of *Inula helenium* L. (commonly known as elecampane).
- Pit baking is an ancient technique dating back to Neanderthal times 100,000 years ago, and commonly used by indigenous peoples in many parts of the world (see Chapter 22 on Camas for details). A large pit excavated in the ground is lined with wood, which is covered with rocks. Food is placed on the rocks, and covered over with more rocks and earth. The wood is set alight, and smoulders, often for days, heating the rocks and slowly cooking the food. The technique is particularly valuable for inulin-containing tubers and bulbs that have often been used as food by indigenous peoples. As noted above, prolonged slow cooking converts indigestible inulin to digestible sugars.
- Although Jerusalem artichoke has never become popular, there have been times of desperation when it was appreciated. During a famine that occurred throughout Europe in 1772, the Jerusalem artichoke proved that it could be quickly and easily grown to provide nourishment. About 1801, people were starving at a settlement near the site of Winnipeg in Canada, and survived mostly by eating Jerusalem artichoke. The Lewis and Clark explorers (see Chapter 71 on Pawpaw for detailed information on this 1804–1806 expedition of the western United States) collected wild Jerusalem artichokes when food was scarce. During World War II the vegetable was often consumed in several countries because the tubers could be bought without a ration card.
- A superstition developed in Europe that Jerusalem artichoke causes leprosy, based on the resemblance of the irregularly shaped roots to fingers deformed by the disease. This delayed its adoption as a food plant.

- Raw Jerusalem artichoke has a nutlike flavor. However, because of differences in body chemistry such as those that prevent some people from detecting bitterness in certain foods, not everyone can appreciate the nutty taste.
- It has been reported that growing Jerusalem artichoke on soils infested with nematodes (microscopic worm-like animals) can result in a reduction of 45% in the nematode population.
- Feeding Jerusalem artichoke to swine has been found to noticeably reduce the smell of their manure.
- In the early 1980s in Minnesota, Jerusalem artichoke was the basis of a get-rich-quick, pyramid scheme, promising up to $36,000 per hectare (about $15,000 per acre). Jerusalem artichoke seemed attractive, because it could be used as a fuel, a feed, a food, and a sugar crop. Unfortunately, there was no market for the new "wonder crop," and 450 farmers lost about $19 million. The mastermind behind the scheme was fined $20,000 and sentenced to 1 year in jail. (See Amato 1993 for additional information.)
- Because it is so tall, Jerusalem artichoke has sometimes been used to mark off property boundaries, for example, in the southern United States. In the Middle East, gardens are sometimes fenced off by hedges made of Jerusalem artichoke.

## KEY INFORMATION SOURCES

Amato, J.S. 1993. *The great Jerusalem artichoke circus: The buying and selling of the rural American dream.* University of Minnesota Press, Minneapolis, MN. pp. 244.

Baker, L., Thomassin, P.J., and Henning, J.C. 1990. The economic competitiveness of Jerusalem artichoke (*Helianthus tuberosus*) as an agricultural feedstock for ethanol production for transportation fuels. *Can. J. Agric. Econ.* 38: 981–990.

Boswell, V.R. 1936. *Studies of the culture and certain varieties of the Jerusalem artichoke.* U.S. Department of Agriculture, Washington, DC. pp. 70.

Denison, E. 1969. The Jerusalem artichoke, *Helianthus tuberosus. Mo. Bot. Gard. Bull.* 57(4): 4–6.

Fuchs, A. (Editor). 1993. *Inulin and inulin-containing crops.* Elsevier, Amsterdam, the Netherlands. pp. 417.

Gosse, G., and Grassi, G. (Editors). 1991. *Topinambour (Jerusalem artichoke).* In Proceedings of second workshop, Rennes, France, December 1988. Revised edition. Report No. CEC-EUR(13405. Commission of the European Communities, Directorate-General for Science, Research and Development, Brussels, Belgium. pp. 150. (In English, French, and Italian)

Grassi, G., and Gosse, G. (Editors). 1989. *Topinambour (Jerusalem artichoke).* In Proceedings of a workshop held in Madrid, Spain, September 30 and October 1, 1987. Report No. CEC-EUR(11855. Commission of the European Communities, Directorate-General Science, Research and Development, Brussels, Belgium. pp. 213. (In English, French, and Italian)

Kalloo, G. 1993. Jerusalem artichoke, *Helianthus tuberosus* L. In *Genetic improvement of vegetable crops.* Edited by G. Kalloo and B.O. Bergh. Pergamon Press, New York. pp. 747–750.

Kays, S.J., and Nottingham, S.F. 2007. *Biology and chemistry of Jerusalem artichoke:* Helianthus tuberosus *L.* CRC Press, Boca Raton, FL. pp. 478.

Kosaric, N., Cosentino, G.P., Wieczorek, A., and Duvnjak, Z. 1984. The Jerusalem artichoke as an agricultural crop. *Biomass* 5(1): 1–36.

Kosaric, N., Wieczorek, A., Cosentina, G.P., and Duvnjak, Z. 1985. Industrial processing and products from the Jerusalem artichoke. *Adv. Biochem. Eng. Biotechnol. (Berlin)* 32: 1–24.

Modler, H.W., Jones, J.D., and Mazza, G. 1993. Observations on long-term storage and processing of Jerusalem artichoke tubers (*Helianthus tuberosus*). *Food Chem.* 48: 279–284.

Pappalardo, J. 2008. *Sunflowers: the secret history: The unauthorized biography of the world's most beloved weed.* Overlook Press, Woodstock, NY. pp. 256.

Seiler, G.J. 1990. Protein and mineral concentrations in tubers of selected genotypes of wild and cultivated Jerusalem artichoke, *Helianthus tuberosus*, Asteraceae. *Econ. Bot.* 44: 322–335.

Seiler, G.J. 1993. Forage and tuber yields and digestibility of selected wild and cultivated genotypes of Jerusalem artichoke. *Agron. J.* 85: 29–33.

Swanton, C.J., and Hamill, A.S. 1983. *Jerusalem artichoke.* Factsheet. Ontario Ministry of Agriculture Food, Toronto, ON. pp. 3.

Routley, D.G. 1977. The versatile but unloved vegetable: Jerusalem artichoke. *Amer. Hort.* 56(4): 12–13.
Salaman, R. 1940. Why 'Jerusalem' artichoke? *J. Roy. Hort. Soc. N.S.* 65: 338–348, 376–383.
Schilling, E.E. 2006. *Helianthus*. In *Flora of North America north of Mexico*. Vol. 21. Edited by Flora of North America Editorial Committee. Oxford University Press, New York. pp. 141–169.
Shoemaker, D.N. 1927. *The Jerusalem artichoke as a crop plant*. U.S. Department of Agriculture, Washington, DC. pp. 32.
Stauffer, M.D., Chubey, B.B., and Dorrell, D.G. 1981. Growth, yield and compositional characteristics of Jerusalem artichoke as they relate to biomass production. In *Fuels from biomass and wastes*. Edited by D.L. Klass and G.H. Emert. Ann Arbor Science Publishers, Ann Arbor, MI. pp. 79–97.
Swanton, C.J., and Cavers, P.B. 1989. Biomass and nutrient allocation patterns in Jerusalem artichoke *Helianthus tuberosus*. *Can. J. Bot.* 67: 2880–2887.
Vervelde, G.J. 1996. *Helianthus tuberosus* L. In *Plant resources of South-East Asia, 9, plants yielding non-seed carbohydrates*. Edited by M. Flach and F. Rumawas. Backhuys Publishers, Leiden, the Netherlands. pp. 100–102.
Williams L.A., and Ziobro, G. 1982. Processing and fermentation of Jerusalem artichoke *Helianthus tuberosus* for ethanol production. *Biotech. Letters* 4(1): 45–50.
Wyse, D.L., and Wilfahrt, L. 1982. Today's weed: Jerusalem artichoke. *Weeds Today* 13: 14–16.

## Specialty Cookbooks

No cookbook dedicated to Jerusalem artichoke was located. Beatty (1987) has 2 Jerusalem artichoke recipes, Berglund and Bolsby (1971) have 8, Brill (2002) has 3, Marcus and Marcus (1982) have 8, Mogelon (2001) has 3, Morash (1982) has more than 20, Robe-Terry (1997) has 4, Stewart (2000) has 1, Tatum (1976) has 14, and Van Atta (1991) has 3. (See the Appendix to this book for details of the publications cited.) Numerous recipes are available on the Web, often under the name sunchoke. Recipes may also be found in cookbooks dedicated to diabetics.

# 54 Jojoba

Family: Simmondsiaceae (jojoba family; occasionally placed in the Buxaceae, the boxwood family)

## NAMES

Scientific name: *Simmondsia chinensis* (Link) C.K. Schneid.

- The word "jojoba" is of American Spanish origin, probably based on a term used by indigenous people in the native area of distribution of the species.
- Pronunciation: hoe-HOE-buh (based on the Spanish origin, the Js are pronounced as Hs).
- Jojoba is generally known as jojoba in the various languages of Europe.
- Jojoba is also known as coffee berry, coffee bush, goatnut, quinine nut, and quinine plant. Much less common old names include deer nut, gray box bush, and wild hazel.
- The name goatnut recognizes that goats commonly graze on the foliage and young twigs.
- The names coffee bush and coffee berry reflect use of jojoba berries to make a coffee-flavored drink.
- The genus name *Simmondsia* commemorates Thomas William Simmonds (1767–1804), an English botanist who died while exploring the island of Trinidad.
- *Chinensis* in the scientific name *S. chinensis* is Latin for "of China," a geographical adjective added by mistake by the botanist Johann H.F. Link in 1822. Plants that he had collected in China were confused with jojoba plants that had been collected in Baja California and, when he described the jojoba plants, Link thought that they had came from China.

## GEOGRAPHY AND ECOLOGY OF WILD PLANTS

Jojoba is native to the Sonoran desert (mostly southwest Arizona and California, northern Mexico, and Baja California), and has been collected at altitudes ranging from sea level to 1200 m (4000 feet). It is adapted to hot, dry desert conditions and full sun. Jojoba occurs naturally on well-drained soils that are near-neutral in pH.

## PLANT PORTRAIT

Jojoba is a perennial, evergreen shrub with leathery leaves. It grows to a height of 0.5–1 m (1.6–3.2 feet) in the wild, occasionally becoming a small tree as tall as 6 m (20 feet). There are separate male and female plants and, for both, small flowers are developed in the axils of leaves. Female flowers are small, pale green, and commonly solitary or in clusters. Male flowers are yellow, larger, and occur in clusters. The fruit is a dry, yellow-brown, acorn-like capsule with a pointed end, darkening as it ripens, about 2.5 cm (1 inch) long. The capsule contains one, sometimes two or three, dark brown peanut-sized seeds, 1.5–3.0 cm (0.6–1.2) inches long. The seeds are often called "nuts," and contain about 50% wax. The plants have a long life span (100–200 years). Jojoba is cultivated mainly in Arizona, northern Mexico, Argentina, Israel, South America, Australia, and India. Many countries with warm dry climates have initiated jojoba production.

Native Americans used jojoba seed and oil for cooking, hair care, and for treatments of many medical problems such as poison ivy, sores, wounds, colds, cancer, and kidney difficulties. The principal use today for jojoba is as the source of a multipurpose industrial oil. Jojoba is unique

**FIGURE 54.1** Jojoba (*Simmondsia chinensis*). (a) Branch with fruits. (Courtesy of Kenneth Bosma [CC By 2.0].) (b) Bottle of jojoba oil. (Courtesy of Jason Rogers [Flickr/CC-attribution].) (c) Wild shrub. (Courtesy of Homer Edward Price [Flickr/CC-attribution].) (d) Seeds. (Courtesy of Steve Hurst, U.S. Department of Agriculture.) (e) Cultivated shrub (note manicured appearance). (Courtesy of U.S. Department of Agriculture.) (f) Distribution map.

among plants in producing a seed oil that is a liquid wax. The most important uses are in the cosmetic industry, and as a high-temperature, high-pressure lubricant. Among the varied cosmetic products with jojoba are shampoos, hair conditioners, hair sprays, facial oils, body oils, bath oils, hand lotions, moisturizers, suntan lotions, make-up removers, shaving creams, lipsticks and lip glosses, vanishing creams, cleansing creams, and skin fresheners.

Jojoba oil is light gold, odorless, and resembles sperm whale oil both in composition and properties. It serves as a replacement for the whale oil, which is obtained from the blubber and the head cavity of the sperm whale. Until the early 1970s, sperm whale oil was a common ingredient in high-quality lubricants. It was used notably in vehicle differentials and transmissions, in hydraulic fluids that need a low coefficient of friction, and in cutting and drawing oils. Most automobile

**FIGURE 54.2** A sperm whale, the source of sperm whale oil before it was replaced by jojoba oil. (From Rice, D., *Animals, a picture sourcebook: over 700 copyright-free illustrations for direct copying and reference*, Van Nostrand Reinhold, New York, 1979.)

transmissions contained 5%–25% sperm whale oil. In 1971, the United States banned the import of sperm whale oil, and jojoba oil acquired importance as its substitute. During the 1970s, when saving whales became a particularly fashionable goal, jojoba was affectionately labeled "the desert whale" in recognition of its role in replacing the sperm whale as a source of oil.

In the late 1970s, the popular press labeled jojoba oil as the "golden wax," in response to an explosion of interest in investing in its development. The oil may be used to produce pharmaceuticals and commercial products such as lubricants, waxes, candles, and rubber compounds such as varnishes, rubber adhesives, and linoleum. Hundreds of jojoba products are marketed today. Jojoba is considered to be an excellent arid-land industrial crop, and starting in the 1970s there has been huge international interest in its cultivation and development. Unfortunately, there also has been bad planning and overinvestment in relation to the capacity of the marketplace, and there have been business failures in several countries.

## CULINARY PORTRAIT

Jojoba seeds were eaten raw or parched by Indians, and also ground into a powder that was used to make a beverage, boiled into an oily cake, and made into a preparation resembling peanut butter. Settlers and miners adopted several of these uses from the Indians, especially the preparation of a coffee-flavored drink. The seeds are still sometimes eaten raw, parched, or roasted and served with melted butter and salt. The ground seeds are also occasionally boiled and made into a coffee-like beverage in the Indian fashion. In Mexico, roasted, ground nuts are combined with the yolk of a hard-boiled egg and boiled in water with milk, sugar, and a vanilla bean, and the result is said to taste like chocolate. Jojoba seeds taste somewhat like hazelnut, but are slightly bitter until roasted. Humans cannot digest the waxes in jojoba seeds, which will act as a laxative if consumption is substantial.

Jojoba seeds contain about 15% protein, and after the oil is removed from the seed, the remaining meal has about 30% protein. However, the seed also contains about 11% antinutritional compounds, including toxic cyanide-containing compounds. The seeds are therefore slightly poisonous, and cannot be recommended as food. Methods are being explored to detoxify the protein meal from the seed, so that at least it can be used for livestock feed. Jojoba has been speculated to have some potential as a low-calorie edible oil, but this remains to be demonstrated. To date, experimental salad oil, vegetable oil, and shortening have been produced.

Jojoba oil is different from most, perhaps all other oils from plants, in being composed mostly of liquid waxes (esters). These may self-polymerize in the presence of sunlight, so jojoba oil should be stored in brown glass bottles, in the dark, or in closed metal cans.

### Culinary Vocabulary

- The expression "edible jojoba oil" is used for a number of commercially marketed products. As noted above, raw jojoba oil contains toxic constituents, and the degree to which these need to be removed before the oil can be safely consumed as an edible product is not yet clear.

## PROSPECTS

Jojoba is already a commercially successful crop as a source of industrial oil. The possibility exists of detoxifying the oil to make it usable for food purposes, and this would diversify and expand the value of jojoba. Given that the crop is adapted to hot, dry regions where few other crops can thrive, the possibility of developing jojoba into a food crop appears to be a worthwhile project.

## CURIOSITIES OF SCIENCE AND TECHNOLOGY

- Half of jojoba seedlings are males. In plantations established from seed, only 10% of the plants are required to be males, to pollinate the females for seed production. Removing the excess males is troublesome, since they must be grown for several years to determine their sex. Today, jojoba is often grown from vegetative cuttings instead of seeds. Examples of all-female varieties are 'Barindji', 'Wadi Wadi', and 'Waradgery'; some all-male varieties are 'Dadi Dadi' and 'Guyambal'. Generally, more than 10 plants of a female variety are grown for each male plant.
- Jojoba has several adaptations suiting it to its hot, arid environment. The leaves are vertically oriented on the plant to reduce exposure to the sun, which in turn reduces the extreme heat of the desert environment, and the associated loss of water. Another survival trick the plant uses to conserve water is to drop its leaves under conditions of extreme drought. Taproots have been observed at depths of 13 m (40 feet), one author stating a depth of 25 m (82 feet) is possible. As well, the plant may have several taproots seeking deeply located water in different locations.
- Jojoba relies on wind to transfer pollen from the male plants to the females. It has been shown that the vertically oriented leaves of the plant not only serve to reduce midday leaf temperatures during the summer but also cause wind eddies to develop, which direct the pollen toward the female flowers.
- In the 1970s, when extraction of jojoba oil produced large quantities of de-oiled seeds, the attempt was made to feed these to cows, as is commonly done with many other oilseeds after their oil has been removed. Rather than becoming fatter, however, the cows dramatically lost weight. This led to examination of the chemistry of jojoba seeds to try to find constituents that could be used to reduce the weight of humans. It was discovered that the meal left after oil has been extracted from jojoba contains an appetite suppressant (simmondsin). This was then shown to reduce weight in experimental animals but it has not been authorized for medical

use in humans. In the United States, substances are often legally sold as "dietary supplements" so long as they do not make health claims, and simmondsin is sold as part of dietary supplements claiming that weight loss will result. Jojoba oil may well be marketable as an edible oil for dieters, but much costly research will be needed to prove its effectiveness and safety.

## KEY INFORMATION SOURCES


Al-Ani, H.A., Strain, B.R., and Mooney, H.A. 1972. The physiological ecology of diverse populations of the desert shrub *Simmondsia chinensis*. *J. Ecol.* 60: 41–57.

Bailey, D.C. 1980. Anomalous growth and vegetative anatomy of *Simmondsia chinensis*. Am. *J. Bot.* 67: 147–162.

Baldwin, A.R. (Editor). 1988. *Proceedings 7th international conference on jojoba and its uses (Phoenix, AZ, 1988)*. American Oil Chemists' Society, Champaign, IL. pp. 453.

Benzioni, A. 1995. Jojoba domestication and commercialization in Israel. *Hort. Rev.* 17: 233–266.

Benzioni, A., and Forti, M. 1989. Jojoba. In *Oil crops of the world*. Edited by G. Röbbelen, R.K. Downey, and A. Ashri. McGraw-Hill, New York. pp. 448–461.

Buchmann, S.L. 1987. Floral biology of jojoba (*Simmondsia chinensis*), an anemophilous plant. *Desert Pl.* 8(3): 111–124.

Davidson, F.G., and Bernardi, M. 1985. *Miracle plants: Jojoba and yucca*. Century, London, U.K. pp. 64.

Dvoskin, D. 1987. *Potential utilization of agricultural resources: the case of jojoba*. U.S. Department of Agriculture, Economic Research Service, Resources and Technology Division, Washington, DC. pp. 33.

Elias-Cesnik, A. 1982. *Jojoba: guide to the literature*. Office of Arid Lands Studies, College of Agriculture, University of Arizona, Tucson, AZ. pp. 221.

Gentry, H.S. 1958. The natural history of jojoba (*Simmondsia chinensis*) and its cultural aspects. *Econ. Bot.* 12: 261–291.

Ismail, A.M.A. 1988. The ecological and agronomic role of seed polymorphism in *Simmondsia chinensis*. *J. Arid Environ.* 14: 35–42.

Kohorn, L.U. 1994. Shoot morphology and reproduction in jojoba: Advantages of sexual dimorphism. *Ecology* 75: 2384–2394.

Linam, D. 1981. *Jojoba fever: a survey of a new agricultural industry*. Burbank Books, Burbank, CA. pp. 94.

Linam, D. 1982. *Jojoba fever II: A continuing survey of a new agricultural industry*. Burbank Books, Burbank, CA. pp. 144.

Naqvi, H.H., and Ting, I.P. 1990. Jojoba: a unique liquid wax producer from the American desert. In *Advances in new crops*. Edited by J. Janick and J.E. Simon. Timber Press Portland, OR. pp. 247–251.

Niklas, K.J., and Buchmann, S.L. 1985. Aerodynamics of wind pollination in *Simmondsia chinensis* (Link) Schneider. *Am. J. Bot.* 72: 530–539.

Oyen, L.P.A. 2001. *Simmondsia chinensis* (Link) C.K. Schneider. In *Plant resources of South-East Asia, 14, vegetable oils and fats*. Edited by H.A.M. van der Vossen and B.E. Umali. Backhuys Publishers, Leiden, the Netherlands. pp. 134–138.

Princen, L.H., and Rossi, C. (Editors). 1996. *Proceedings of the 9th international conference on jojoba and its uses and of the third international conference on new industrial crops and products (Catamarca, Argentina)*. Association for the Advancement of Industrial Crops, Peoria, IL. pp. 483.

Rost, T.L., Simper, A.D. Schell, P., and Allen, S. 1977. Anatomy of jojoba (*Simmondsia chinensis*) seed and the utilization of liquid wax during germination. *Econ. Bot.* 31: 140–147.

Saguy, I.S., Shani, A., Weinberg, P., and Garti, N. 1996. Utilization of jojoba oil for deep-fat frying of foods. Lebensm. Wiss. Technol. London 29: 573–577.

Sherbrooke, W.C. 1978. *Jojoba: An annotated bibliographic update. Arid Lands Resour. Inf. Pap. 5(Suppl.)*. University of Arizona, Office of Arid Lands Studies, Tucson, AZ. pp. 80.

Sherbrooke, W.C. 1978. Jojoba—Sonoran desert shrub. *Pac. Discov.* 31(2): 22–29.

Sherbrooke, W.C., and Haase, E.F. 1974. *Jojoba: a wax-producing shrub of the Sonoran Desert: literature review and annotated bibliography*. University of Arizona, Office of Arid Lands Studies, Tucson, AZ. pp. 141.

Thomson, P.H. 1975. Jojoba horticulture. *California Rare Fruit Growers Yearbook* 7: 98–162.

Thomson, P.H. 1982. *Jojoba handbook*. Third edition. Bonsall Publications, Bonsall, CA. pp. 162.

United States National Research Council, Advisory Committee on Technology Innovation, Ad Hoc Panel. 1985. *Jojoba: New crop for arid lands, new raw material for industry*. National Academy Press, Washington, DC. pp. 102.

United States National Research Council, Committee on Jojoba Utilization, National Academy of Sciences. 1975. *Products from jojoba: A promising new crop for arid lands*. National Academy of Sciences, Washington, DC. pp. 30.

Wardlaw, I.F., Begg, J.E., Bagnall, D., Dunstone, R.L. 1983. Jojoba: temperature adaptation as expressed in growth and leaf function. *Aust. J. Plant Physiol.* 10: 299–312.

Waser, N.M. 1984. Sex ratio variation in populations of a dioecious desert perennial, *Simmondsia chinensis*. *Oikos* 42: 343–348.

Wisniak, J. 1987. *The chemistry and technology of jojoba oil*. American Oil Chemists' Society, Champaign, IL. pp. 272.

Wisniak, J. 1994. Potential uses of jojoba oil and meal—a review. *Ind. Crops Prod.* 3: 43–68.

Yermanos, D.M. 1974. Agronomic survey of jojoba in California. *Econ. Bot.* 28: 160–174.

## Specialty Cookbooks

Cookbooks for jojoba are not available, as the plant requires breeding to provide tasty, safe culinary material. Nevertheless, some jojoba cooking (as opposed to cosmetic) recipes are available on the Web and in wild-food books. Farnsworth (1999) has 2 jojoba recipes, Hall and Hall (1980) have 1, Niethammer (1999) has 2, and Williamson (1989) has 19. (See the Appendix to this book for details of the publications cited.)

# 55 Joshua Tree

Family: Agavaceae (agave family)

## NAMES

Scientific name: *Yucca brevifolia* Engelm. (*Y. arborescens* (Torr.) Trel.)

- The name "Joshua tree" has been attributed to Mormon pioneers in the eighteenth century traveling across the Mojave (sometimes spelled Mohave) Desert from Utah to the city of San Bernardino (which they founded in 1854), near Los Angeles. The seemingly beckoning branches of the tree were said to represent the biblical patriarch Joshua signaling the Promised Land to the Mormon settlers. The grotesquely extended branches of the Joshua tree were likened to the outstretched arms of Joshua as he pointed with his spear toward the city of Ai (Joshua 8:18, authorized version, Old Testament). The biblical Joshua appeared in Moses' lifetime in increasingly important positions—as a warrior, as the assistant to Moses, as one of the spies, and finally as Moses' designated successor as the leader of Israel. The Old Testament Book of Joshua explains the fate of Israel, covering among other topics the conquest of the Promised Land, the divine appointment of Joshua, the dry crossing of the Jordan River, the fall of Jericho, and the necessity of renouncing the worship of other gods and remaining faithful.
- The Joshua tree has also been called Joshua-tree yucca, tree yucca, and yucca-palm.
- The genus name *Yucca* is based on *yuca*, the name for cassava (*Manihot esculenta* Crantz) used by indigenous New World (Carib) people. However, the only points of similarity are that both plants occur in hot climates and in the Americas. John Gerard (1545–1612), English botanist and barber-surgeon, was responsible for this error, which was taken up by Linnaeus when he adopted the genus name *Yucca*.
- *Brevifolia* in the scientific name *Y. brevifolia* is based on the Latin *brevi*, short, + *folia*, leaves. Although not really short, the leaves are much shorter than those of some of the other species of the genus *Yucca*.
- Pronunciation of the scientific name: *Yucca* (YUK-uh) *brevifolia* (brev-ee-FOH-lee-uh).

## GEOGRAPHY AND ECOLOGY OF WILD PLANTS

The Joshua tree is one of the most characteristic plants of the Mojave Desert, and extends southward to the Sonoran Desert. The Joshua tree grows from southern California, Mexico, and western Arizona eastward into southern Nevada and southwestern Utah, at elevations of 450–2200 m (1500–7200 feet). It is most abundant in the vicinity of Joshua Tree National Park located near Palm Springs in California. Annual rainfall in many Joshua tree woodlands averages 13 cm (5 inches) or less, although some sites receive as much as 25 cm (10 inches). Most of the precipitation occurs in winter or early spring. Regrettably, extensive vandalism has occurred in many Joshua tree woodlands in California, with larger trees burned or defaced.

Joshua trees grow in areas periodically subjected to fire, and the plants have numerous specialized adaptations to survive fire. The flaky alligator-like bark of older trunks serves as insulation against fire. The height of older trees protects the growing points, which are high above the low vegetation that tends to burn. A protective sheath of thick, green leaves surrounds the buds and also serves to insulate the buds from fire. Vigorous root and stump sprouting after fire reproduces the plants quickly, and the seeds can remain viable in the soil for years.

**FIGURE 55.1** Joshua tree (*Yucca brevifolia*). (a) Tree. (Courtesy of Marshal Hedin [Flickr/CC-attribution].) (b) Fruits. (Courtesy of Stan Shebs [CC By 3.0].) (c) Distribution map.

The Joshua tree relies solely on the yucca moth (*Tegeticula synthetica*) for pollination. Flying at night, the moth is attracted to the white blossoms of the plant. She collects the sticky pollen from some flowers, then transfers the pollen to another flower, and injects her own eggs into that flower. The pollen fertilizes the ovules of the flower, which develop into seeds, some of which (generally about 7%) are eaten by the moth larvae, but many of the seeds are spared to disperse and produce new trees. Without this insect, there would be no seeds, and the species could become extinct, although at some sites the plant reproduces primarily by vegetative means, sprouting from the roots and underground stems (rhizomes).

## PLANT PORTRAIT

The Joshua "tree" is "arborescent," that is, tree-like, with a huge, reddish-brown to gray trunk-like stem (sometimes it has two or three stems) 40–80 cm (16–31 inches) in diameter, much thicker at the bottom where it tapers out. The plant is 5–13 m (16–42 feet) tall, rarely up to 20 m (66 feet) in height. A stout flowering stem is produced, bearing greenish-white, yellowish, cream, or gray, unpleasant-smelling flowers 4–7 cm (1.6–2.8 inches) long. The fruits are globose or egg-shaped, 6–10 cm (2.4–4.0 inches) long, and are generally produced only in wetter years. Joshua trees are evergreen, with numerous, stiff, dagger-like, sharp-pointed leaves 15–40 cm (6–16 inches) long at the ends of branches.

The unique appearance of the Joshua tree makes it a very desirable ornamental. Unfortunately, many trees have been dug up for planting in urban or residential areas, despite a very low rate of survival when transplanted. In cold climates, the Joshua tree can be grown in large pots, stored in a frost-free place during the winter. Seeds and plants are widely available from commercial sources.

## CULINARY PORTRAIT

The Joshua tree has been a minor wild food source since prehistorical times. Native Americans roasted and ate the flower buds. The young fruits were also roasted. Young seeds and seed pods were ground up to make a meal or mush that was eaten raw or cooked (and said to resemble banana in taste). An alcoholic beverage was made from the fermented buds and flowers. The roots have been eaten raw, boiled, or roasted. The most palatable part of the plant is the young flowering tissue. Modern wild food collectors sometimes consume the flower clusters, but before the buds open, as the open flowers are less palatable. These are best parboiled in salt water to remove the bitterness, drained, and cooked again, and served like cauliflower. Before being cooked, the flowers should be checked for presence of insects. Several other species of *Yucca* of the southwestern United States are similarly employed by wild food collectors, and indeed are more palatable, although less interesting.

An entrepreneur from England extracted the liquid found in the root system of the Joshua tree and created a drink similar to that of a pint of English bitter. Nobody could tolerate it, and the venture failed. The roots of *Yucca* species are rich in saponins—foaming substances that would give a pleasant appearance, if unpleasant taste, to beer.

### CULINARY VOCABULARY

- Although some species of *Yucca*, including *Y. brevifolia*, have edible albeit unpalatable roots, the expression “yucca root” is almost always a misspelling of “yuca root,” that is, the widely used tropical cassava (manioc), important for its starchy tubers. However, “yucca root powder” is a preparation made from the roots of various *Yucca* species, valued for the content of steroid-like saponins, claimed to be useful for reducing inflammation associated with arthritis.
- Although *Yucca* species were once fermented into alcoholic beverages by Native Americans, in modern times, a “yucca cocktail” is not made with *Yucca*, but rather from lemons, limes, sugar, vodka, and ice.
- “Yuccatash” is a variation, using *Yucca*, of succotash (a southern American dish of corn, lima beans, and sometimes sweet peppers).

## PROSPECTS

Joshua tree is scarcely edible by today’s standards, has virtually no prospects of becoming a commercial food crop, and is the lowest ranking of the food plants in this book. Its inclusion here is based on its uniqueness among North American plants that have been used as food. Aside from saguaro cactus (also discussed in this book), it is the only edible plant species in North America to have a national park named for it. Joshua tree is instantly recognizable and has been repeatedly viewed in motion pictures with a Western theme. Its austere appearance and meager nutritional value are reminders of the much more desirable food plants that nature provides. The uniqueness, mystique, and wide recognition of Joshua tree possibly could be exploited by putting extracts of it in food products, such as liqueurs and other alcoholic beverages. Since the plants are protected by legislation, this would probably require establishing orchards, which could be useful for protecting the species from becoming endangered.

## CURIOSITIES OF SCIENCE AND TECHNOLOGY

- The extinct ground sloth of the southwestern United States area fed almost entirely on *yucca* species, with a special fondness for the Joshua tree.
- Native Americans used the rootlets of the Joshua tree to make a red dye, and prepared rope, sandals, mats, and baskets from the leaf fibers.
- In the past, the wood of the Joshua tree was used in a variety of ways. Ancient Cliff Dwellers of the southwest often incorporated Joshua tree beams into their houses. Early ranchers used the trunks to build fences. The pulp was made into fine paper and several editions of *The London Telegraph* newspaper in England were printed on newsprint from Joshua tree pulp. The wood was used during war time for making splints for injured arms and legs. Joshua tree wood has a unique and attractive grain, and today it is used in novelty products such as picture frames, book covers, postcards, napkin rings, and boxes. Veneer from Joshua tree is sometimes available in southern California.
- The fascinating erratic branching of the Joshua tree is often partly because of damage from the yucca-boring weevil (*Scyphophorus yuccae*), which destroys the growing tips of branches, causing them to branch additionally. Flowering at the end of branches also stops growth of those branches and is also responsible for the peculiar branching.
- Joshua trees do not produce annual rings in the trunk, so their age cannot be easily determined. However, their overall shape provides clues to how old the plants are. Juvenile trees are generally unbranched, middle-aged trees are forked and dense, and the oldest trees generally have a single stem and an open crown. Joshua tree grows an average of 8 cm (about 3 inches) annually for the first 10 years, and about 1.3 cm (0.5 inch) afterward. Dividing the height of a Joshua tree by the average annual growth of 1.3 cm or 1/2 inch provides a rough estimate of age. The tallest Joshua tree in Joshua Tree National Park is 12 m (40 feet) high and has been estimated to be over 900 years old.
- In their natural habitat, Joshua trees are often the tallest plants, and so they are frequently struck by lightning.

## KEY INFORMATION SOURCES

Brittingham, S., and Walker, L.R. 2000. Facilitation of *Yucca brevifolia* recruitment by Mojave Desert shrubs. *West. North Am. Nat.* 60: 374–383.

Bronaugh, W. 1993. The biggest Joshua tree. *Am. Forests* 99(5/6): 36.

Cornett, J.W. 1999. *The Joshua tree*. Nature Trails Press, Palm Springs, CA. pp. 32. (A guide to Joshua Tree National Monument)

DeFalco, L.A., Esque, T.C., Scoles-Sciulla, S.J., and Rodgers, J. 2010. Desert wildfire and severe drought diminish survivorship of the long-lived Joshua tree (*Yucca brevifolia*; Agavaceae). *Am. J. Bot.* 97: 243–250.

Gilliland, K.D., Huntly, N.J., and Anderson, J.E. 2006. Age and population structure of Joshua trees (*Yucca brevifolia*) in the northwestern Mojave Desert. *West. North Am. Nat.* 66: 202–208.

Gossard, G.H. 1992. *The Joshua tree: A controversial, contradictory desert centenarian*. Yellow Rose Publications, Tehachapi, CA. pp. 112.

Hess, W.J., and Robbins, R.L. 2002. *Yucca*. In *Flora of North America north of Mexico*. Vol. 26. Edited by Flora of North America Editorial Committee. Oxford University Press, New York. pp. 423–439.

Kaiser, J. 2002. *Joshua tree: The complete guide*. Destination Press, Santa Monica, CA. pp. 189. (A guide to Joshua Tree National Monument)

Keith, S.L. 1982. A tree named Joshua. *Am. Forests* 88(7): 40–42.

Lenz, L.W. 2001. Seed dispersal in *Yucca brevifolia* (Agavaceae): Present and past, with consideration of the future of the species. *Aliso* 20: 61–74.
Lenz, L.W. 2007. Reassessment of *Yucca brevifolia* and recognition of *Y. jaegeriana* as a distinct species. *Aliso* 24: 97–104.
Maxwell, C.G. 1971. The tree that is not a tree. *Am. Forests* 77(3): 4–5.
Pellmyr, O., and Segraves, K.A. 2003. Pollinator divergence within an obligate mutualism: Two yucca moth species (Lepidoptera; Prodoxidae: Tegeticula) on the Joshua tree (*Yucca brevifolia*; Agavaceae). *Ann. Entomol. Soc.* America 96: 716–722.
Phillips, E.A., Page, K.K., and Knapp, S.D. 1980. Vegetational characteristics of two stands of Joshua tree woodland. Madroño 27: 43–47.
Preissel, H.G. 1978. Joshua trees, *Yucca brevifolia* from the Wild West Ornamentals. *Gartenpraxis (Germany)* 10: 518–520. (In German)
Rasmuson, K.E., Anderson, J.E., and Huntly, N. 1994. Coordination of branch orientation and photosynthetic physiology in the Joshua tree (*Yucca brevifolia*). *Great Basin Nat.* 54: 204–211.
Sanford, M.P., and Huntly, N. 2009. Selective herbivory by the desert woodrat (*Neotoma lepida*) on Joshua trees (*Yucca brevifolia*). *Western North Am. Nat.* 69: 165–170.
Simpson, P. 1975. *Anatomy and morphology of the Joshua tree (*Yucca brevifolia*): An arborescent monocotyledon*. PhD thesis. University of California, Santa Barbara, CA. 524 leaves.
Smith, S.D., Hartsock, T.L., and Nobel, P.S. 1983. Ecophysiology of *Yucca brevifolia*, an arborescent monocot of the Mojave Desert. *Oecologia* 60: 10–17.
Vander Wall, S.B., Esque, T., Haines, D. Garnett, M., and Waitman, B.A. 2006. Joshua tree (*Yucca brevifolia*) seeds are dispersed by seed-caching rodents. *Ecoscience* 13: 539–543.
Vuncannon, D.H. 1996. *Joshua Tree: The story behind the scenery*. KC Publications, Las Vegas, NV. pp. 48. (A guide to Joshua Tree National Monument)
Wimmer, A.T.T., and Merritt, R. 2008. Glucose-6-phosphate isomerase variation and genetic structure in *Yucca brevifolia* (Agavaceae). *Madroño* 55: 285–290.

## Specialty Cookbooks

Books on wild food plants of North America frequently include recipes for preparing flowers and fruits of *Yucca* species, notably for *Y. baccata* Torr. (Spanish bayonet, banana yucca, of the southwestern United States) and *Y. glauca* Nutt. (soapweed, of the central United States), and these recipes may be adopted for Joshua tree. For example, Niethammer (1999) has two recipes for *Yucca* flowers, and Marie (2008) has one, intended for other species of *Yucca* than Joshua tree, but adaptable to it. Niethammer (1999) also has a fruit recipe that can be adapted, but the result is likely to be much less tasty than for the fruits of recommended species. (See the Appendix to this book for details of the publications cited.)

# 56 Juniper

Family: Cupressaceae (cypress family; sometimes placed in Pinaceae, pine family)

## NAMES

Scientific name: *Juniperus communis* L.

- The name juniper and the genus name *Juniperus* are probably derived from the old Latin name for the plant, although it has been claimed that the root is the Celtic word, *jenupus*, variously interpreted as meaning bitter (a reference to the taste of juniper), or rough, rude, in allusion to the stiff habit of the shrubs.
- Among the colorful old names by which juniper is known are "bastard-killer" (for its wide use in producing abortions), "fairy circle," and "horse-saver."
- Occasionally, *Gaylussacia brachycera* (Michx.) A. Gray, the box huckleberry, a shrub native to the eastern United States, is called the juniperberry.
- *Communis* in the scientific name *J. communis* is the Latin word for common.

## GEOGRAPHY AND ECOLOGY OF WILD PLANTS

*Juniperus communis*, the most widespread of juniper species, is native to the colder regions of North America and Eurasia. In North America, it grows from Alaska south to Oregon in the west, and across the continent to northern Quebec and the northeastern United States. The species typically grows in rocky, infertile soils, in fields, meadows, pastures, open woods, and other settings, almost from sea level to alpine sites.

## PLANT PORTRAIT

The juniper is an evergreen with needlelike leaves, about 1 cm (0.4 inch) long, arranged in three's. Common juniper is a very variable species, found in several forms. A small tree of pyramidal or columnar form up to 10 m (33 feet), or rarely to 15 m (49 feet), commonly occurs in Europe. Some cultivated varieties are small trees. However, tree forms are rare in North America, where most wild plants are prostrate, mat-forming shrubs. Small cones, some male, others female, develop in early summer at the base of a few of the needlelike leaves. Male cones are yellow while female cones are hard, pungent, and bluish-green with a whitish, waxy coating. Female cones become fleshy berries, requiring 2 (sometimes 3) years to ripen, and turning gradually from green to blue-black. For the most part, plants produce either male cones or female cones, rarely both. The berries are about the size of a pea, dark blue, nearly globular, and smooth-skinned, with several bony seeds embedded in a brownish pulp. Both unripe green and ripe blue-black berries will be found on the plant. The green-blue bloom on the surface of the fresh berries tends to disappear after drying.

Juniper berries are harvested from wild trees in northern areas for local use, especially in southern Europe. Commercial crops have been harvested in Canada and the United States, but most berries in commerce come from eastern Europe. In addition to culinary uses, juniper oil is employed for fragrance in soaps, detergents, creams, lotions, shampoos, and perfumes. Junipers are popular ornamentals, and are widely planted in parks and around homes.

**FIGURE 56.1** Juniper (*Juniperus communis*). (a) Fruiting branch. (Courtesy of U.S. National Park Service.) (b) A branch with male flowers (upper left) and a branch with fruits (lower right). (From Köhler, H.A., *Köhler's Medizinal Pflanzen*, Verlag von F.E. Köhler, Germany. Vol. 2, 1887.) (c) Sprawling shrub. (Courtesy of Dave Powell, U.S. Department of Agriculture Forest Service, online at Bugwood.com [CC By 3.0].) (d) Distribution map.

The most widespread culinary use of juniper is as a flavorant of gin. Dr. Franciscus Sylvius (1614–1672), a Dutch medical professor, invented gin in 1650, working at the University of Leyden in Holland. It is believed that he was conducting research on a treatment for kidney stones and mixed two diuretics (substances that stimulate urination), juniper berries and grain alcohol, in the hope of flushing the stones from the system. William of Orange ascended to the English throne in 1689 and ruled jointly with Mary II. William came from Holland, famous for its gin, and to encourage its importation into England, he taxed other alcoholic beverages, which increased gin consumption. In time, the English began distilling their own gin, and by 1737, nearly 4.5 L (1.2 U.S. gallons) of gin was manufactured annually in England for every man, woman, and child. In 1727, London alone had 6187 gin shops. Tavern keepers offered to make one drunk for a penny, dead drunk for two pence. Concern about the "gin epidemic" led Parliament in 1751 to place a very high tax on gin and restrict its sale, and the population switched back to ale and beer. By 1782, gin drinking had fallen drastically, whereas beer consumption had risen to 155 L (41 U.S. gallons) annually for every person. Of the five principal spirits (brandy, whiskies, rum, vodka, and gin), gin has the most troubling reputation to live down. Over the years, it has been the curse of the urban poor, the ruin of young girls, the bathtub brew that rotted guts during the American Prohibition, and the first resort of the miserable unable to cope with life's difficulties.

## CULINARY PORTRAIT

The aroma of juniper is bittersweet, very reminiscent of gin, pine, and turpentine. The taste is sweet, pungent, slightly burning, with a bitter aftertaste. Juniper has been used to flavor beverages for centuries. A fermented brew made with the fruits was produced in early France. Boughs, with or without the fruit, were employed to make a beverage tea by various native groups in North America. Today, juniper fruits are commonly fermented to produce alcoholic beverages such as beer and brandy. In France, they are fermented with barley to make *genevrette*, a kind of beer. Essential oil is also extracted from juniper fruits and is used largely as a flavoring for gin, liqueurs, and such cordials as sloe gin. Gin has been one of the most popular alcoholic beverages of the Western world for well over 300 years. Ground, roasted berries are sometimes used as a substitute for coffee and as an alternative for pepper. The dried berries can be ground in a peppermill. Berries are made into a conserve in Sweden. The aromatic, resinous flavor goes well in marinades and with wild game, beef, pork, pickled fish, stuffings, pâtés, sauerkraut, gravies, and potatoes. The flavor also combines well with garlic and other aromatic herbs such as marjoram and rosemary. As the flavor is strong, juniper berries should be used very sparingly. Juniper berries are rarely used as a culinary ingredient in the kitchens of English-speaking countries, whereas cooks in the mountainous regions of Italy, France, and Germany commonly use juniper to flavor rich game dishes. Extracts and oils are used in soft drinks and other nonalcoholic beverages, frozen dairy desserts, candy, chewing gum, baked goods, gelatins, puddings, meat products, and condiments.

"True thrushes" are a group of small- to medium-sized songbirds such as the American robin and the European blackbird (the classification of the family, the Turdidae, is unsettled; one specialist recognizes 162 species in the world). Several of the species commonly consume juniper berries and, since classical Roman times, captive birds have been fattened on juniper berries and eaten. Thrushes fattened on juniper berries are still sold in street markets in Italy, leading one author to remark that Italy is the land "where songbirds are regarded as delicatessen." In Poland, snipe (a group of slender-billed game birds in the family Scolopacidae) that have fed on juniper berries acquire the flavor of gin and are often grilled over charcoal.

The gin martini may be the most popular form of consumption of juniper. Traditionally, an olive is added, but some bartenders recommend against this because the strong pickled taste can obscure the citrus aroma of many gins. Instead, a twist of lemon or orange may be substituted for the olive. Adding vermouth makes the beverage smoother. A recommended recipe: five parts of gin to one part of dry vermouth, with added dashes of orange bitters.

## WARNING

Consumption of juniper berries should be limited. Juniper and juniper extracts should not be eaten by expectant mothers because they increase intestinal movements and uterine contractions, and indeed in rare cases can cause abortion. Juniper should also not be consumed if suffering from kidney disease because it stimulates urination and may cause kidney irritation. Ingestion of berries or oil of evergreen species with needlelike foliage, other than those of *J. communis*, should not be attempted, as some may be toxic. The deadly berries of ornamental evergreen yew trees and shrubs (of the genus *Taxus*) could be mistaken for juniper. Only professionals should use juniper oil for culinary purposes—it is too concentrated and potentially poisonous. The medicinal use of small amounts of oil of some species of juniper has resulted in death on occasion.

### Culinary Vocabulary

- The word gin traces to the Latin *juniperus* through the Old French word for the plant, *genèvre.*
- *À la liégoise* refers to a classical French and Belgian method of cooking using juniper berries and alcohol, and applies particularly to veal kidneys braised in butter with juniper berries and flambéed with a little gin.
- *À la ardennaise* refers to game cooked with juniper berries or juniper-flavored alcohol.
- *Sauce moscovite* is a French sauce prepared with pepper and juniper berries, with a garnish of raisins and pine nuts, and served with game.
- A "Dog's Nose" is a cocktail made with hot beer, gin, and sugar.
- A "Tom Collins" is a tall iced drink made with gin, lemon or lime juice, soda water, and sugar, and commonly garnished with a maraschino cherry. It is frequently thought to commemorate a barman who invented it, but this is unclear. In nineteenth-century Australia, the beverage was known as "John Collins" (a name now used when the gin is Hollands gin). Another theory derives the name from "Old Tom," an eighteenth- and nineteenth-century British colloquialism for gin in general and gin strongly sweetened with sugar.
- "Borovicka" (pronounced bor-roh-vee-kah) is a Czech/Slovak juniper-flavored brandy reminiscent of gin.
- "Limerick hams" are world-famous cured hams from the Limerick region of Ireland. By tradition, these are smoked over juniper branches and berries.
- "Westphalian hams" (pronounced wehst-FAIL-ee-uhn) are German boneless hams cured in a salt mixture (often with juniper berries) and smoked over a fire of beechwood logs, juniper twigs, and juniper berries.
- *Pebronata* (pronounced peh-bro-nah-tah) is a Corsican dish prepared with beef or wild boar braised with juniper berries.

## PROSPECTS

Although commercial crops of juniper berries have been harvested in Canada and the United States, the major producing countries are Austria, France, Hungary, Italy, Germany, Poland, Spain, and the countries of the former Czechoslovakia, the U.S.S.R., and Yugoslavia. Supplies are especially common from eastern Europe. The well-established commercial crops of juniper in Europe make it difficult to grow it competitively for commercial extracts in North America. The plant is very common in the wild, and the wild supply is sufficient to supply the relatively small amounts that are needed. The value of the species is somewhat enhanced because the wood is useful, although produced in small quantities. Juniper is not a good candidate for investment in selection and development as a cultivated food crop, although ornamental cultivars are in demand.

## CURIOSITIES OF SCIENCE AND TECHNOLOGY

- Juniper has a history of medicinal use dating as far back as 1550 BC. A remedy to treat tapeworm was found on papyrus from ancient Egypt from that time.
- Juniper branches and berries were burned in ancient Egyptian temples as a part of purification ceremonies. Because juniper berries have a relatively high oil content, they tend to burn with considerable smoke, so they are ideal for using as a purification smudge. Many others after the Egyptians also burned juniper for ritualistic and health reasons. In North America, juniper was used by indigenous peoples as a fumigant, deodorizer, disinfectant, and protective agent against bad spirits, ghosts, and diseases. It was included in the wash water to cleanse bedding and personal effects of those who had just died. In Europe, juniper branches were often burned as a fumigant during plagues; for example, in French hospitals during the

smallpox epidemic of 1870. During the 1918 Spanish flu epidemic, which is believed to have killed some 20,000,000 people worldwide, several hospitals experimented with spraying vaporized essential oils into the atmosphere of flu wards in an attempt to prevent airborne infection spreading. Juniper was one of the oils that were found to be particularly effective.

- The Swiss city of Geneva is said to have been named for the numerous wild juniper plants in the surrounding area (through the Old French word for the plant, *genèvre*; the modern French for juniper is genièvre).
- The "Victorian language of flowers" was a secret coded language in Victorian times, with flowers and plants symbolic of certain messages; so, when the flower or plant was mentioned in a letter, those who knew the code could understand the hidden information. "Juniper" meant "succour" (to help someone in distress) and "protection."
- The dull red wood of eastern red cedar (*J. virginiana* L.), a species indigenous to eastern North America, has long been valued as a moth-repellent for lining wardrobe closets and chests, and is one of the main species of *Juniperus* used to produce "cedarwood oil." This species was named *bâton rouge* by French settlers in Canada, meaning "red stick," and after they encountered the same tree in Louisiana, they applied the name to the present capital city of that state, Baton Rouge. In the southern United States, the trees may exceed 18 m (about 60 feet) in height, but is smaller toward the north.
- One kilogram (2.2. pounds) of juniper berries is typically used to flavor over 400 L (about 106 U.S. gallons) of gin.
- Common juniper is possibly the most widely distributed tree in the world.
- Juniper may cause one's urine to smell like violets.
- As noted above, juniper plants are usually male (producing male flowers) or female (producing female flowers), with occasional trees producing both male and female plants. There are many planted trees that similarly produce both male and female plants (e.g., some oaks, aspens, poplars, honey locust, horse chestnut, yew, ash, and ginkgo), and because female plants often produce messy fruits, there has been a very strong tendency for horticulturists to have selected and planted only male trees for ornamental purposes. Although male trees do not produce fruits, most are wind pollinated, and this has increased the allergy problem for many people.
- Rabbits born to mothers fed juniper berries during their pregnancy have been found to prefer juniper-flavored food. This is one of several experimental observations that indicate that prenatal experience can alter taste development.

## KEY INFORMATION SOURCES


Adams, R.P. 1987. Investigation of *Juniperus* species of the United States for new sources of cedarwood oil. *Econ. Bot.* 41: 48–54.

Adams, R.P. 1993. *Juniperus*. In *Flora of North America north of Mexico*. Vol. 2. Edited by Flora of North America Editorial Committee. Oxford University Press, New York. pp. 412–320.

Adams, R.P. 2011. *The junipers of the world: The genus* Juniperus. Third edition. Trafford Publisher, Victoria, BC. pp. 428.

Adams, R.P., and Pandey, R.N. 2003. Analysis of *Juniperus communis* and its varieties based on DNA fingerprinting. *Biochem. Syst. Ecol.* 31: 1271–1278.

Adams, R.P., Pandey, R.N., Leverenz, J.W., Dignard, N., Hoegh, K., and Thorfinnsson, T. 2003. Pan-Arctic variation in *Juniperus communis*: historical biogeography based on DNA fingerprinting. *Biochem. Syst. Ecol.* 31: 181–192.

Ashworth, V.E.T.M., O'Brien, B.C., and Friar, E.A. 1999. Fingerprinting *Juniperus communis* L. cultivars using RAPD markers. *Madroño* 46: 134–141.

Ashworth, V.E.T.M., O'Brien, B.C., and Friar, E.A. 2001. Survey of *Juniperus communis* (Cupressaceae) L. varieties from the Western United States using RAPD fingerprints. *Madroño* 48: 172–176.

Chatzopoulou, P.S., and Katsiotis, S.T. 1993. Chemical investigation of the leaf oil of *Juniperus communis* L. *J. Essent. Oil Res.* 5: 603–607.

Diotte, M., and Bergeron, Y. 1989. Fire and the distribution of *Juniperus communis* L. in the boreal forest of Quebec, Canada. *J. Biogeography* 16: 91–96.
Dodd, R.S., and Poveda, M.M. 2003. Environmental gradients and population divergence contribute to variation in cuticular wax composition in *Juniperus communis*. *Biochem. Syst. Ecol.* 31: 1257–1270.
Garcia, D. 1998. Interaction between juniper *Juniperus communis* L. and its fruit pest insects: Pest abundance, fruit characteristics and seed viability. *Acta Oecol.* 19: 517–525.
Garcia, D., Zamora, R., Gomez, J.M., and Hodar, J.A. 2002. Annual variability in reproduction of *Juniperus communis* L. in a Mediterranean mountain: relationships to seed predation and weather. *Ecoscience* 9: 251–255.
Garcia, D., Zamora, R., Gomez, J.M., Jordano, P., and Hodar, J.A. 2000. Geographical variation in seed production, predation and abortion in *Juniperus communis* throughout its range in Europe. *J. Ecol.* 88: 436–446.
International Organization for Standardization. 1984. *Juniper berries (*Juniperus communis *Linnaeus).* International Organization for Standardization, Geneva, Switzerland. pp. 3.
International Organization for Standardization. 1991. *Oil of juniper berry (*Juniperus communis *Linnaeus).* International Organization for Standardization, Geneva, Switzerland. pp. 5.
Kallio, H., and Junger-Mannermaa, K. 1989. Maritime influence on the volatile terpenes in the berries of different ecotypes of juniper (*Juniperus communis* L.) in Finland. *J. Agric. Food Chem.* 37: 1013–1016.
Lawrence, B. M. 1987. Progress in essential oils. Juniperberry oil. *Perf. Flav.* 12(6): 59–61.
Miller, H.J. 1974. Anatomical studies of *Juniperus communis* L. ssp. *communis* and *Juniperus communis* L. ssp. *nana* Syme. *Acta Bot. Neerl.* 23: 91–98.
Ortiz, P.L., Arista, M., and Talavera, S. 2002. Sex ratio and reproductive effort in the dioecious *Juniperus communis* subsp. *alpina* (Suter) Celak. (Cupressaceae) along an altitudinal gradient. *Ann. Bot.* 89: 205–211.
Poveda, M.M., Souqual, M.C., Fauvel, M.T., Gamisans, J., and Gauquelin, T. 2002. Alkane composition diversity among populations of dwarf forms of *Juniperus communis* L.: comparison between western Europe and northern American populations. *Bot. J. Linn. Soc.* 140: 165–168.
Shahmir, F., Ahmadi, L., Mirza, M., and Korori, S.A.A. 2003. Secretory elements of needles and berries of *Juniperus communis* L. ssp. *communis* and its volatile constituents. *Flavour Fragr. J.* 18: 425–428.
Van Der Merwe, M., Winfield, M.O., Arnold, G.M., and Parker, J.S. 2000. Spatial and temporal aspects of the genetic structure of *Juniperus communis* populations. *Mol. Ecol.* 9: 379–386.
Ward, L.K. 1982. The conservation of juniper: Longevity and old age [*Juniperus communis communis*, England]. *J. Appl. Ecol.* 19: 917–928.

## Specialty Cookbooks

Aside from its commercial use in alcoholic beverages, juniper is a very minor culinary flavoring substance. Nevertheless, recipes are available in cookbooks, especially those concerned with wild plants or European cuisine. The following are examples. Leyel (1938) has recipes for Conserve of juniper and Rob of juniper. [A rob is "the juice of a fruit made so thick by the heat either of the sun or the fire that it is capable of being kept safe from putrefaction."] Brackett and Lash (1975) have recipes for Juniper and mushroom sauce for roast chicken, duck, or two game hens; Sauerkraut with juniper berries and wild carrot seed; Juniper ratafia; and Juniper marinade for wild game or beef. Turner and Szczawinski (1978) present recipes for Juniper-bough tea, Juniper-berry tea, Juniper coffee, and Juniper mocha. Rose (1979) gives recipes entitled Indian stew with lamb and Ollie's onions. Bremness (1988) has recipes for Lamb's kidneys sautéed with juniper, Pork chops marinated with juniper and garlic, Cabbage and juniper, and Mulled pears with juniper. Holt (1989) gives recipes for Game pâté with juniper berries, Wild duck with juniper berries, and Apple and herb [juniper] jelly. Norman (1990) presents recipes for Potatoes with bacon and juniper and Onion and juniper bread. Watts and Watts (2007) have recipes for Venison roast with juniper berry rub, and Juniper berry sauce. Other recipes are Juniper berry butter sauce (Bryan and Castle 1974); Game pâté with juniper berries (Boxer and Back 1980); Pork spare rib with juniper berry stuffing (Michael 1980); Juniper lamb stew (Hughes 1982); Juniper-berry tea (Marcin 1983); Juniper marinade (Schofield 1989); Fillet of beef with juniper berries and *crique ardechoise* (potato pancake) (Holt 1991); and Juniper butter (Gray 2011). Williamson (1995) has 12 recipes for Rocky Mountain juniper (*J. scopulorum* Sarg.), which may also be employed. Williamson (1989) has 21 recipes for several western North American juniper species. (See the Appendix to this book for details of the publications cited.)

# 57 Labrador Tea

Family: Ericaceae (heath family)

## NAMES

Scientific names: *Ledum* or *Rhododendron* species. Recent studies indicate that species of the genus *Ledum* should be transferred to the genus *Rhododendron*, but this is not yet widely accepted. The old (*Ledum*) species names are employed in this chapter.

- "Labrador" in the name "Labrador tea" reflects the northern distribution and the common use as tea by native North Americans.
- The name "crystal tea" for *L. palustre* subsp. *palustre* is also used in the illicit drug trade for LSD (d-lysergic acid diethylamide), whereas "crystal T" is PCP (phencyclidine, angel dust), another dangerous narcotic.
- The genus name *Ledum* is from the Greek *ledon*, referring to a species of *Cistus* (presumably *C. ladaniferus* L.) from which an aromatic resin (ladanum), which has a similar odor, was once obtained.
- The genus name *Rhododendron* is from the Greek *rhodos*, rose (for the showy flowers), and *dendron*, tree. Some species of the genus are trees, although the name rhododendron seems to have been applied by the classical Greeks to oleander (*Nerium oleander* L.).
- *Palustre* in the scientific name *L. palustre* is Latin for marshy, indicative of where the plants grow.
- *Glandulosum* in the scientific name *L. glandulosum* is Latin for gland-bearing. *Ledum* species have tiny epidermal glands on their leaves.

## GEOGRAPHY AND ECOLOGY OF WILD PLANTS

The principal North American Labrador tea, *L. palustre* subsp. *groenlandicum*, has a northern transcontinental distribution in North America, occurring from coast to coast in Canada, and occasionally north of the tree line. It has been locally naturalized in Germany and Great Britain. The plants reside in the mixed temperate and boreal vegetation zones of North America. They are found mostly in open areas of sphagnum bogs across Canada and the northern United States, but also in swamps, acid mountain meadows, and moist to wet forests. The plants develop into thick patches and are often dominant or codominant. They reproduce vegetatively by spread of rhizomes (underground stems) growing close to the soil surface, and reproduce to a much lesser extent by seeds. Typically, the soils occupied are very acidic and have limited nutrient availability.

Dwarf Labrador tea, *L. palustre* subsp. *decumbens*, is found in the Arctic from western Greenland and Labrador to Alaska, and in eastern Eurasia. It mostly inhabits tundra and muskeg in Canada, but also occurs in the drier areas of bogs. It may dominate or codominate dwarf shrub associations in bogs, muskegs, and open tundra.

Trapper's tea, *L. glandulosum*, extends northward from California to northern Wyoming, just reaching southern British Columbia and southwestern Alberta, and is especially in the western American Cordillera. It is found in moist to wet meadows and forests, extending from subalpine to montane zones. It often occurs in moist to moderately dry coniferous forest, occasionally in moist, somewhat boggy forest. Except near the tree line, it is rare in fully open areas.

**FIGURE 57.1** Labrador tea (*Ledum* species). (a) Twig of common Labrador tea (*L. palustre* subsp. *groenlandicum*). (Courtesy of Jeronimo 2412 [CC By 3.0].) (b) Cluster of flowers of common Labrador tea. (Courtesy of magnolia 1000 [Flickr/CC-attribution].) (c) Trapper's tea (*L. glandulosum*). (Courtesy of Dave Powell, U.S. Department of Agriculture Forest Service, online at Bugwood.com [CC By 3.0].) (d) Dwarf Labrador tea (*L. palustre* subsp. *decumbens*). (From Loddiges, C. & Sons, *The botanical cabinet*, J. & A. Arch, Cornhill, U.K. Vol. 11, Plate 1049, 1825.) (e) Distribution map of common Labrador tea. (f) Distribution map of dwarf Labrador tea (in red) and trapper's tea (in green).

## PLANT PORTRAIT

In Table 57.1, four different groups representing two species are called "Labrador tea." For practical purposes, the main group of interest is *L. palustre* L. subsp. *groenlandicum*, which is the dominant "Labrador tea" of North America and the one that is primarily used for culinary purposes. This is an evergreen erect shrub, generally 0.5–1 m (20 inches–3 feet) in height, with a sweet, spicy aroma. The leaves are 2.5–7.5 cm (1–3 inches) long, leathery in appearance, conspicuously rolled under at the margins, and densely wooly underneath, the wooly fuzz becoming a deep rust color as the leaves

**TABLE 57.1**
**Principal Species and Subspecies of Labrador Tea**

| Scientific Name When Considered to Be *Ledum* | Scientific Name When Considered to Be *Rhododendron* | Common Names | Distribution |
|---|---|---|---|
| *L. palustre* L. subsp. *groenlandicum* (Oeder) Hultén (*L. groenlandicum* Oeder) | *R. groenlandicum* (Oeder) Kron & Judd | Labrador tea, common Labrador tea, bog Labrador tea, bog tea, Hudson's Bay tea, James tea, St. James tea, Indian tea, wooly tea | Across Canada and the United States |
| *L. palustre* subsp. *decumbens* (Ait.) Hultén (*L. decumbens* (Ait.) Lodd. ex Steud.) | *R. tomentosum* Harmaja subsp. *subarcticum* (Harmaja) G.D. Wallace (*R. subarcticum* Harmaja) | Dwarf Labrador tea, northern Labrador tea | Arctic and subarctic plants of northeastern Asia and North America |
| *L. palustre* subsp. *palustre* | *R. tomentosum* subsp. *tomentosum* | Labrador tea, wild rosemary, swamp rosemary, swamp tea, marsh Labrador tea, marsh cistus, marsh tea, crystal tea | Eurasia |
| *L. glandulosum* Nutt. | *R. columbianum* (Piper) Harmaja (*R. neoglandulosum* Harmaja) | Trapper's tea, glandular Labrador tea, western Labrador tea | California and northern Wyoming north to southern British Columbia and Alberta |

mature. Dense, round-topped clusters of small white flowers are on the ends of some of the upper branches and ripen into small, brown, many-seeded fruits.

The European version of Labrador tea, marsh tea (*L. palustre* subsp. *palustre*), has been shown to be chemically different from the main North American Labrador tea, and indeed appears to have poisonous properties and to be capable of causing abortion. Although there is therefore very good reason to not use the European Labrador tea as a beverage, studies of the potential toxicity of North American Labrador tea are required to establish with certainty that it is completely safe, and until this is done, consumption should be limited.

Astringent (mouth-puckering) tannins are present in Labrador tea and, indeed, are also in the most common hot beverages: tea, coffee, and cocoa. They have been experimentally shown to interfere with digestibility of proteins and to have carcinogenic potential, but there is little evidence of adverse effect in humans. Concern has been expressed that long-term consumption of tannins may be harmful, and the recommendation has been made to reduce intake during pregnancy because of interference with iron absorption. Tannins are precipitated in beverages by the addition of milk, removing the potential for harm.

Native peoples of North America made very extensive use of Labrador tea for medicinal purposes, often as a general tonic. Illnesses and conditions treated included kidney problems, "weak blood," colds, coughs, tuberculosis, arthritis, heartburn, and headache. Externally, preparations were used to treat sores, burns, wounds, and skin problems. Similarly, *L. palustre* of Europe was used in folk medicine to treat a very wide variety of problems, including bronchitis, bugbites, fever, cold, cough, gout, inflammations, leprosy, malignancies, rheumatism, and sore throat. *Ledum* preparations are generally considered obsolete in Western medicine, but there is still some use in Europe and Asia.

## CULINARY PORTRAIT

Labrador tea has been the most popular indigenous beverage tea in North America for centuries, used not only by native peoples but also by explorers, settlers, traders, and trappers. Today, it is occasionally found as a specialty commercial tea preparation and continues to be used by indigenous groups.

There is not universal agreement on the best time to harvest Labrador tea. Some authorities recommend collecting the leaves in the spring, before flowering, others in late fall. It has been claimed that the flowers, if gathered at the proper time and carefully dried, make a far more pleasant beverage than the leaves. Native peoples of North America have used the tea as a spice for meat. In Finland, the European form of Labrador tea is used to flavor liqueurs. It has been said that subspecies *decumbens* produces a superior tea compared to subspecies *groenlandicum*, but given the extraordinary chemical variation known to occur, this conclusion seems premature. Trapper's tea (*L. glandulosum*) has been claimed to produce a bitter drink of inferior quality to that of Labrador tea.

It is often recommended that because of poisonous constituents in the leaves of Labrador tea, the beverage should not be too concentrated, ingested in large amounts, or consumed too often. For example, according to Heller (1981, listed in the Appendix to this book), "Oldtimers in Alaska advise that it not be used in too large quantities otherwise it may be cathartic and cause intestinal disturbances." Farnsworth (1999; see the Appendix to this book) warned "Labrador tea should be drunk in moderation by most people, and not consumed at all by people who have high blood pressure or a heart condition. The leaves contain a poisonous substance called ledol, which can cause cramping and paralysis." Also, it is often suggested that to limit possible extraction of toxins, the leaves should be steeped rather than maintained in boiling water (although some recommend boiling the leaves for up to 10 minutes). Nevertheless, as pointed out above, the North American plant has not been proven to be toxic, whereas the European plant is definitely poisonous.

Swamp laurel (*Kalmia polifolia* Wang.) and sheep laurel (*K. angustifolia* L.) very commonly grow in association with *Ledum*, and look rather similar. These shrubs are poisonous, and indeed have killed browsing livestock, so that collectors should know how to distinguish them from Labrador tea. The laurels normally have pink flowers and lack the characteristic lower leaf fuzz of Labrador tea.

### Culinary Vocabulary

- Rosemary (*Rosmarinus officinalis* L.) is a very common culinary herb. The name "wild rosemary" is occasionally encountered for Labrador tea, which could lead to confusion. The quite poisonous bog rosemary genus, *Andromeda*, might also be confused with Labrador tea.

## PROSPECTS

Labrador tea is not a cultivated food plant and is unlikely to become one. There are some ornamental cultivars, including the dwarf cultivars Compactus (assigned to subsp. *groenlandicum*) and Minus (assigned to subsp. *palustre*), but there has been no selection of cultivars for edible purposes. Considerable knowledge of chemotypes (chemical races) useful for developing medicinal cultivars is available, but there is very limited information on which plants are most suitable for use as a flavorful tea. In any event, wild plants are very widespread, so that the wild supply is more than adequate to meet demand. Extensive studies have been conducted in Europe on the sustainable management and harvest of natural stands of *L. palustre* for medicinal purposes (although very little has been published in English). By contrast, there does not appear to have been scientific research on *Ledum* for food purposes, that is, as a commercial tea. The reputation of Labrador tea as somewhat poisonous doubtlessly has limited its development. There seems to be widespread misunderstanding of the extent to which the plant is toxic. The most significant toxic compound, ledol, seems to be absent in Canadian plants, and there is limited evidence that poisonous concentrations of any other compounds are present. There is a need for basic scientific and management studies, including harvest, drying, and preservation methods. Herbal teas are very popular today, so there would seem to be market opportunities for Labrador tea. However, the rather inferior taste limits

its market potential. Systematic study of variation of the flavor constituents (presumably including the essential oils) could result in the identification of particularly desirable populations. There may be a market niche to be developed, based on finding a variant with proven negligible toxicity and improved flavor. Herbal teas offered in the trade today are very frequently blends of different species, and there may well be attractive blends with Labrador tea waiting to be developed. All things considered, however, Labrador tea is unlikely to become popular, and seems destined to remain of interest only for a niche market.

## CURIOSITIES OF SCIENCE AND TECHNOLOGY

- The Berserkers were ancient Viking warriors, reputedly invulnerable, enormously strong, and wildly frenzied in battle. In the Scandinavian sagas, Berserkers are often described as being fantastically ugly and guilty of monstrously depraved behavior, and their reputation terrified their opponents. The Danish historian Saxo Grammaticus (about 1150–1220) recorded the activities of a band of Berserkers in his *Gesta Danorum*: "The young warriors would harry and pillage the neighbourhood, and frequently spilled great quantities of blood. They considered it manly and proper to devastate homes, cut down cattle, rifle everything and take away vast hauls of booty, burn to the ground houses they had sacked, and butcher men and women indiscriminately.... they ravished other men's wives and daughters... scarcely any place in the land was free from the imprints of their lust." It has been suggested that intoxication from European Labrador tea contributed to the deranged but effective wildness of the Berserkers.
- In Scandinavia in past times, strong-smelling European Labrador tea branches were placed among harvested cereals stored in barns to drive away mice.
- In some parts of Germany, European Labrador tea branches were kept in bedrooms to discourage fleas from entering.
- Because of its alleged narcotic properties, European Labrador tea leaves were used in Germany and Poland to make beer more intoxicating. The Kwakiutl people of Vancouver Island and adjacent British Columbia also considered the leaves of Labrador tea to be narcotic. Women of some indigenous peoples used Labrador tea as a mild narcotic to reduce the pain of childbirth. Ledol, often present in European Labrador tea, has been characterized as a narcotic toxin.
- Labrador tea was often consumed during the American Revolution (1775–1783), when common (Chinese) tea was unavailable.
- When the Labrador Innu people traditionally followed the caribou, their children contributed to transporting the load by carrying dolls, each stuffed with approximately 1 kg (2.2 pounds) of Labrador tea. Labrador tea dolls are sold today as craft items.
- The English explorer Sir John Franklin (1776–1847) set out for the Canadian arctic coast in 1845 with two ships, one pessimistically named the Terror. In 1846, the ships became icebound in Victoria Strait, and all 105 men died of starvation and scurvy, although the mission had in fact succeeded in discovering the Northwest Passage. The crew tried to survive by eating boiled leather flavored with dwarf Labrador tea, which, Franklin recorded in his journal, tasted like rhubarb. Unknown to the expedition but well known to the native inhabitants was the fact that Labrador tea, which is rich in vitamin C, counteracts scurvy and is more than just a flavoring.
- European Labrador tea has been featured on postage stamps issued by Germany and Russia.
- For more than a century, ecologists were puzzled over the fact that *Ledum* and other woody plants found in bogs have features retarding water loss, such as thick waxy cuticle, inrolled leaf margins, and dense hairs, despite the fact that the bog substrate is soaked with water. The current explanation for drought-protective leaves in bog woody plants is that because

the leaves are perennial, the water-retaining features are necessary to prevent becoming dried out in winter, when water is unavailable. And why should the leaves be perennial? It has been shown that in soils that have very limited nutrient availability (especially nitrogen), such as sphagnum moss bogs, retention of nutrients in leaves for several years is an effective strategy for outcompeting deciduous plants that must re-obtain the scarce nutrients every year for new leaves.

## KEY INFORMATION SOURCES


Belleau, F., and Collin, G. 1993. Composition of the essential oil of *Ledum groenlandicum*. *Phytochemistry* 33: 117–121.

Dalgaard, V., and Fredskild, B. 1993. ×*Ledodendron vanhoeffeni* (syn.: *Rhododendron vanhoeffeni*) refound in Greenland. *Nord. J. Bot.* 13: 253–255.

Harmaja, H. 1991. Taxonomic notes on *Rhododendron* subsection *Ledum* Ericaceae with a key to its species. *Ann. Bot. Fenn.* 28: 171–173.

Hitchock, C.L. 1956. The *Ledum glandulosum* complex. *Leafl. West. Bot.* 8: 1–24.

Judd, W.S., and Kron, K.A. 2009. *Rhododendron*. In *Flora of North America*, Vol. 8. Edited by Flora of North America Editorial Committee. Oxford University Press, New York. pp. 455–473.

Karlin, E.F., and Bliss, L.C. 1983. Germination ecology of *Ledum groenlandicum* and *Ledum palustre* ssp. *decumbens*. *Arct. Alp. Res.* 15: 397–404.

Kron, K.A., and Judd, W.S. 1990. Phylogenetic relationships within the Rhodoreae (Ericaceae) with specific comments on the placement of *Ledum*. *Syst. Bot.* 15: 57–68.

Kudo, G. 1995. Leaf traits and shoot performance of an evergreen shrub, *Ledum palustre* ssp. *decumbens*, in accordance with latitudinal change. *Can. J. Bot.* 73: 1451–1456.

Kurashige, U., Mine, M., Kobayashi, N., Handa, T., Takayanagi, K., and Yukawa, T. 1998. Investigation of sectional relationships in the genus *Rhododendron* (Ericaceae) based on *matK* sequences. *J. Jpn. Bot.* 73: 143–154.

Lantai, K., and Kihlman, B. 1995. The chromosome numbers of *Ledum palustre* ssp. *decumbens* and of some related taxa. *Hereditas* 122: 181–184.

Mikhailova, N.S., and Rybalko, K.S. 1980. Chemical constitution of *Ledum palustre*. *Chem. Nat. Compd.* 16: 131–135.

Mikhailova, N.S., Rybalko, K.S., and Sheichenko, V.I. 1979. A study of the chemical composition of the essential oil of *Ledum palustre*. *Chem. Nat. Compd.* 15: 278–282.

Prudhomme, T.I. 1983. Carbon allocation to antiherbivore compounds in a deciduous [*Betula glandulosa*] and an evergreen [*Ledum groenlandicum*] subarctic shrub species. *Oikos* 40: 344–356.

Reichardt, P.B., Bryant, J.P., Anderson, B.J., Phillips, D., Clausen, T.P., Meyer, M., and Frisby, K. 1990. Germacrone defends Labrador tea from browsing by snowshoe hares. *J. Chem. Ecol.* 16: 1961–1976.

Riebesell, J.F. 1981. Photosynthetic adaptations in bog and alpine populations of *Ledum groenlandicum*. *Ecology* 62: 579–586.

Savile, D.B.O. 1969. Interrelationships of *Ledum* species and their rust parasites in Western Canada and Alaska. *Can J. Bot.* 47: 1085–1100.

Shaver, G.R. 1983. Mineral nutrition and leaf longevity in *Ledum palustre*: The role of individual nutrients and the timing of leaf mortality in an Alaskan evergreen shrub. *Oecologia* 56: 160–165.

Sifton, H.B. 1940. Lysigenous air spaces in the leaf of Labrador tea, *Ledum groenlandicum* Oeder. *New Phytol.* 39: 74–79.

Small, E., and Catling, P.M. 2000. Poorly known economic plants of Canada—26. Labrador tea, *Ledum palustre sensu lato* (*Rhododendron tomentosum*). *Bull. Can. Bot. Assoc.* 33(3): 31–36.

Smith, E.M., and Hadley, E.B. 1974. Phytosynthetic and respiratory acclimation to temperature in *Ledum groenlandicum* populations. *Arct. Alp. Res.* 6: 13–27.

Tattje, D.H.E., and Bos, R. 1981. Composition of essential oil of *Ledum palustre*. *Planta Medica* 41: 303–307.

Wallace, G.D. 1992. *Ledum* in the new Jepson manual and a new combination for *Ledum* in *Rhododendron* (Ericaceae). *Madroño* 39: 77.

## Specialty Cookbooks

Because it is rarely employed for anything but producing tea, there are few recipes available for Labrador tea. Turner and Szczawinski (1978) provided an excellent account of tea preparation. They recommended collecting the young leaves in early summer, drying them in paper bags in a warm place until they are completely dry and brittle, and storing them in tightly sealed containers, which will preserve them for years. (Alternatively, Web-based firms can be contacted for a supply.) Turner and Szczawinski suggested placing a generous handful of leaves in 1 L (four cups) of boiling water and simmering for about 5 minutes (a weaker preparation can be made by steeping the leaves in boiling water in a teapot, the way ordinary tea is made). A few drops of lemon juice and a little honey may enhance the flavor; sugar and cream may also be added, but end to mask the delicate flavor. Mogelon (2001) advised "Pour boiling water over tea leaves and let the mixture infuse for 5–10 minutes. (Don't boil the leaves in the water as it may release a harmful alkaloid.)" (The presence of harmful alkaloids has not been shown.) Additional information on preparing Labrador tea is provided by Farnsworth (1999) and Stewart (2002). (See the Appendix to this book for details of the publications cited.)

# 58 Lingonberry

Family: Ericaceae (heath family)

## NAMES

Scientific name: *Vaccinium vitis-idaea* L.

- "Lingonberry" (and variants such as lingenbery, lingberry, linberry, lingon, lingen) is derived from the Swedish word *lingon*, meaning lingonberry.
- The lingonberry is also known as alpine cranberry, cowberry, dry ground cranberry, foxberry, lowbush cranberry (in Alaska), moss cranberry, mountain cranberry, partridgeberry, red berry, red bilberry, red whortleberry, and rock cranberry. Most of these alternative names are also used for different species.
- "Partridgeberry" is the name applied to the lingonberry in Newfoundland, the chief area of collection in North America. The name "partridgeberry" may be based on a historical confusion of the lingonberry with the plant usually known as partridgeberry, *Mitchella repens* L. of the Rubiaceae (madder family).
- Names are very important in marketing fruits, and some plant names based on animals are undesirable while others are more acceptable (the "monkey peach" was a market failure until it became known as "kiwi," taken from the name of the New Zealand flightless bird). Calling the lingonberry "cowberry" or "foxberry" is undesirable for marketing purposes, although "partridgeberry" may be more attractive. Some marketers have suggested that the European name "lingonberry" is best. In Toronto, the product has been marketed as "mountain cranberry."
- See Chapter 15 on Blueberries for information on the genus name *Vaccinium*.
- *Vitis-idaea* in the scientific name *V. vitis-idaea* is Latin for grape of Mount Ida. There are two Mount Ida's of classical Greek mythology, one in Crete, the other in Asia Minor. A number of species are named after Mount Ida that do not actually grow in either location, and this includes the lingonberry. Not only is "grape of Mount Ida" a misnomer in this respect, but neither the fruit nor the plant is grape-like.

## GEOGRAPHY AND ECOLOGY OF WILD PLANTS

The lingonberry is circumpolar and circumboreal, occurring worldwide in northern temperate, boreal, and subarctic areas. The species is widely distributed in Eurasia. In North America, lingonberry ranges from above the Arctic circle southward to Minnesota, the mountains of Maine, New Hampshire, and Vermont, and along the eastern coast to Massachusetts. It is transcontinental, growing from eastern Newfoundland to Alaska and British Columbia.

Lingonberry grows in a great variety of situations, most often in infertile, acidic ground, including peat. Soils occupied are often rocky, but the plants are also common in sandy and clay loams. In North America, the species occurs in jack pine stands, spruce forests, raised bogs, muskegs, dry rocky barrens, lichen woodlands, and in a wide range of exposed habitats including heaths, high moors, headlands, tundra, cliffs, and mountain summits. In its southernmost range, it is found particularly in bogs, while in the north it occupies both wet and dry sites. The species is not a principal pioneer of newly opened areas, but nevertheless is often an early colonizer of burned forests and other open habitats.

**FIGURE 58.1** North American lingonberry (*Vaccinium vitis-idaea* var. *minus*). (a) Sprawling plants in fruit in Alaska. (Courtesy of Dawn Endico [CC By 3.0].) (b) Clusters of flowers. (Courtesy of magnolia 1000 [Flickr/CC-attribution].) (c) Distribution map.

## PLANT PORTRAIT

The lingonberry is a low, creeping shrub, with diffusely branched, slender stems growing horizontally on the surface of the ground. The plants typically produce dense colonies, often forming large mats. Two wild varieties have been recognized, var. *vitis-idaea* (= var. *majus* Lodd. et al.), mostly in Europe and Asia, and var. *minus* (Lodd. et al. (= *V. vitis-idaea* (Lodd. et al.) subsp. *minus* Hultén), mostly in North America, Greenland, and eastern Asia. The former is larger, growing 10–25 cm (4–10 inches), sometimes to 40 cm (16 inches) in height, while the latter grows 2–10 cm (less than an inch to about 4 inches). About two dozen cultivated varieties, derived from the wild Eurasian variety, are available. The leaves of the lingonberry plant are leathery and evergreen. The small flowers are pink or reddish in the North American wild variety and white or pink in the European wild variety. The small, dark red berries are usually 6–10 mm (0.2–0.4 inch) in diameter, with a very acid taste similar to that of cranberries. Commercially marketed berries are usually pea-sized,

**FIGURE 58.2** European lingonberry (*Vaccinium vitis-idaea* var. *vitis-idaea*) plants and culinary preparations. (a) Twig with fruit. (Courtesy of Jonas Bergsten.) (b) Cluster of flowers. (Courtesy of fi:Käyttäjä:kompak [CC By 3.0].) (c) Plants in fruit. (Courtesy of Jonas Bergsten.) (d) Fruiting and flowering branches. (From Hallier, E., *Flora von Deutschland* [edition 5 by D.F.L. von Schlechtendal et al.], F.E. Köhler, Gera-Untermhaus, Germany, Vol. 20, Plate 2038, 1885.) (e) A pot of cooked lingonberry jam. (Courtesy of Kimtaro/Kim Ahlström [Flickr/CC-attribution].) (f) Palt, a traditional Swedish dumpling, topped with lingonberry jam. (Courtesy of Mrs. Gemstone [CC By 2.0].)

somewhat smaller than cranberries. The seeds are extremely small and generally not noticeable in the fruit or processed products.

Lingonberry is a familiar wild fruit in northern Europe, where its harvest and processing are well established, and there is continuing breeding of improved cultivars. The species is normally harvested from the wild, but is sometimes grown, and is currently being developed as a crop in Poland, Germany, the Soviet Union, and the Pacific Northwest of the United States. Lingonberry is an important, mostly wild-collected berry crop in Russia, Scandinavia, the Baltic countries, Poland, and to a lesser extent in Japan and Germany. The Scandinavian countries harvest as much as 450 million kg (1 billion pounds) in some years. Perhaps Sweden has the greatest affection for the lingonberry. The berry is so valued in Sweden that it has been termed "red gold" (because of the fruit's red color), and it has been argued that the lingonberry should be declared to be the "national fruit" of Sweden. As late as the early twentieth century, Swedish immigrants to the midwest of the United States found that cranberries did not adequately replace the familiar lingonberry of their homeland, and imported the fruit from Sweden and Newfoundland. In North America, lingonberry is a minor, mostly wild-collected crop of unmanaged natural stands, especially of Newfoundland and Labrador (more than 100,000 kg/year, or more than 220,000 pounds/year are harvested) and Nova Scotia (about 5000 kg/year, or about 11,000 pounds/year). The province of Newfoundland and Labrador is the largest lingonberry-producing area of North America. There are small harvests along the eastern shore of Nova Scotia, and Cape Breton Island, and also in La Ronge, Saskatchewan. Small amounts are collected in the United States for commercial sale.

In addition to culinary use, the lingonberry plant is an attractive ornamental for cool climates, valued for its lustrous evergreen foliage, white flowers, and bright red berries. The species can serve as a groundcover, and also is suitable for borders, beds, and containers.

## CULINARY PORTRAIT

Lingonberry fruits have been described as "shaped like blueberries but acid like cranberries." They are pleasantly tart and aromatic, but may be bitter, especially when unripe. The berries have been said to taste best if they overwinter and are eaten during the melting of the snow. The seeds are very small and are not noticeable in the fruit or processed products. Lingonberries may be eaten raw, but most people think they are too bitter. Lingonberries are high in benzoic acid, resulting in a long shelf life—more than 8 weeks under normal refrigeration. They can be used in any recipe calling for cranberries, but are most often made into a sauce, which can be prepared as follows: combine 3 cups washed fruits, 1.3 cups sugar, and 1 cup water; boil for 10 minutes, skim, and cool. The berries are also used in preserves, jams, jellies, syrups, juice, candy, ice cream, pastries (especially bread and pies), condiments, and wines and liqueurs. The sharp, distinctive taste of lingonberry juice is often favored by bartenders for preparing cocktails. In the Scandinavian countries, lingonberry sauce is served with pancakes, omelets, and puddings, and lingonberry relish accompanies meat (especially venison) and poultry much like cranberries in North America. In Finland, a sweet fruit soup called mehukeitto is often made with lingonberries, and lingonberry soup is common in other European countries where the berries are available. Whipped lingonberry pudding served with cream is a traditional Finnish dessert. In Scandinavia and Russia, lingonberries are used extensively in purées and compôtes.

### CULINARY VOCABULARY

- In Europe, lingonberry sauce is often labeled for export as cranberry.
- "Angel Food" has been claimed (among other theories) to have evolved from the Swedish dessert *anglamat*, made with whipped cream, lingonberry jam, and cake crumbs.
- *Plättar* are small, Swedish pancakes that are traditionally eaten with lingonberries.
- *Palt* (see Figure 58.2f) is a meat-filled dumpling traditionally served in Sweden along with lingonbery preserves and butter (*Blodpalt* is a common variation made with the blood of cattle, swine, or reindeer).

## PROSPECTS

Lingonberry is an important crop in Europe, but the industry is in the early stages of development in North America. Newfoundland is the major center of harvest in North America, but the berries can be obtained or grown in many northern areas of the continent. The market for domestically produced lingonberries in North America is very small at present, but likely has very good potential. The fruit has an attractive appearance and interesting taste, and could become a gourmet item in North America. There is also good potential for exports to other areas of the world. The long shelf life and appealing appearance of lingonberries offer some prospect for fresh market sales, but because the fruit is much more palatable when cooked, lingonberry is especially suitable for processing as juices, preserves, liqueurs, and frozen products, and these are the consumer items that should be targeted. Research has been underway in recent years in Newfoundland and Nova Scotia, as well as in several locations in the northern United States. Because of its adaptation to northern areas where few other crops are available, the lingonberry is highly deserving of development both as a managed wild resource and as a cultivated crop.

## CURIOSITIES OF SCIENCE AND TECHNOLOGY

- Lingonberries have been used medicinally for centuries. In northern Europe, delivering mothers once drank a decoction of lingonberry to prevent or reduce bleeding. Alaskan Native Americans used lingonberries as remedies for headaches and sore throats. In modern times, the lingonberry has been occasionally used medicinally as a urinary antiseptic, although more commonly its relative the cranberry has been employed for this purpose.
- *The Kalevala* is a collection of records compiled by Elias Lönnrot (1802–1884) from ancient oral poetry from northeastern Finland and those parts of the Russian province of Archangel where Karelian (a language closely related to Finnish) was spoken. It is known as the Finnish national epic, and greatly increased the nationalism of the Finnish people. One of the myths concerned virgin childbirth to the maid Marjatta, who became pregnant upon eating a lingonberry (alternatively interpreted as the bilberry). The child was born in a forest and grew up to become a messianic hero and the new king. The obvious similarity with the Christian account of the virgin birth of Christ possibly illustrates how beliefs from one culture diffused into other cultures in past times.
- In the British Isles, the "whortleberry" (likely the lingonberry) serves as the badge of many clans, including Davidson, MacBean, MacDuff, MacGillivray, MacLeod, MacQueen, and Shaw.
- In Newfoundland, the leaves of lingonberry "are said to be excellent if dried to use as pipe tobacco" (Anonymous 1977, cited in the Appendix to this book).

## KEY INFORMATION SOURCES


Ballington, J.R. 2008. Lingonberry. In *The encyclopedia of fruit & nuts*. Edited by J. Janick and R.E. Paull. CABI, Wallingford, Oxfordshire, U.K. pp. 361–362.

Chester, A.L., and McGraw, J.B. 1983. Effects of nitrogen addition on the growth of *Vaccinium uliginosum* and *Vaccinium vitis-idaea*. *Can. J. Bot.* 61: 2316–2322.

Estabrooks, E.N. 1997. Native lingonberry (*Vaccinium vitis-idaea* var. *minus*) as a new crop in New Brunswick, Canada. *Acta Hortic.* 446: 125–127.

Freeland, K. 1991. *Vaccinium vitis-idaea* var. *minus*. *Am. Nurseryman* 173(8): 162.

Galetta, G.J., and Ballington, J.R. 1996. Blueberries, cranberries and lingonberries. In *Fruit breeding, Vol. II: Vine and small fruit crops*. Edited by J. Janick and J.N. Moore. Prentice Hall, New York. pp. 1–107.

Gustavsson, B.A. 1993. Lingonberry breeding and cultivation (*Vaccinium vitis-idaea* L.). *Acta Hortic.* 346: 311–314.

Gustavsson, B.A. 2001. Genetic variation in horticulturally important traits of fifteen wild lingonberry *Vaccinium vitis-idaea* L. populations. *Euphytica* 120: 173–182.

Hall, I.V., and Beil, C.E. 1970. Seed germination, pollination, and growth of *Vaccinium vitis-idaea* var. *minus* Lodd. *Can. J. Plant Sci.* 50: 731–732.

Hall, I.V., and Shay, J.M. 1981. The biological flora of Canada. 3. *Vaccinium vitis-idaea* L. var. *minus* Lodd. supplementary account. *Can. Field-Nat.* 95: 434–464.

Hjalmarsson, I. 1993. Domestication of wildgrowing lingonberries (*Vaccinium vitis-idaea*) in Sweden. *Acta Hortic.* 346: 305–310.

Hjalmarsson, I., and Ortiz, R. 1998. Effect of genotype and environment on vegetative and reproductive characteristics of lingonberry (*Vaccinium vitis-idaea* L.). *Acta Agricult. Scandinav. (Section B. Soil Plant Sci.)* 48: 255–262.

Hjalmarsson, I., and Ortiz, R. 2001. Lingonberry: Botany and horticulture. *Hortic. Rev.* 27: 79–123.

Holloway, P.S. 1984. Lingonberry cultivation. *Agroborealis (Alaska Agric. Exp. St., Fairbanks)* 16(2): 15–20.

Holloway, P.S., Stushnoff, C., and Wildung, D.K. 1983. Chilling and budbreak in lingonberries, *Vaccinium vitis-idaea* L. *J. Am. Soc. Hortic. Sci.* 108: 88–90.

Holloway, P.S., Veldhuizen, R.M. van Stushnoff, C., and Wildung, D.K. 1982. Effects of light intensity on vegetative growth of lingonberries. *Can. J. Plant Sci.* 62: 965–968.

Holloway, P.S., Veldhuizen, R.M. van Stushnoff, C., and Wildung, D.K. 1982. Vegetative growth and nutrient levels of lingonberries grown in four Alaskan substrates. *Can. J. Plant Sci.* 62: 969–977.

Kruger, E. 1985. Mineral nutrition of cultivated lingonberries (*Vaccinium vitis-idaea* L.). *Acta Hortic.* 165: 287–294.

Labokas, J., and Budriuniene, D. 1989. Vegetative propagation of lingonberry. *Acta Hortic.* 241: 270–272.

Landolt, E. 1996. *Vaccinium vitis-idaea* L. subsp. *minus* (G. Lodd.) Hulten (Ericaceae), an overlooked circumpolar-arctic taxon of the Alps. *Anales Jardin Bot. Madrid* 54: 277–284.

Lehmushovi, A. 1975. Methods of propagating the cowberry. *Annal. Agric. Fenniae (Finland)* 14: 325–333.

Lehmushovi, A. 1977. Some aspects of the cowberry trials in Finland. *Annal. Agric. Fenniae (Finland)* 16: 57–63.

Lehmushovi. A., and Säkö, J. 1975. Domestication of the cowberry (*Vaccinium vitis-idaea* L.) in Finland. *Annal. Agric. Fenniae (Finland)* 14: 227–230.

Nuortila, C., Tuomi, J., and Laine, K. 2002. Inter-parent distance affects reproductive success in two clonal dwarf shrubs, *Vaccinium myrtillus* and *Vaccinium vitis-idaea* (Ericaceae). *Can. J. Bot.* 80: 875–884.

Penney, B.G., Gallagher, C.A., Hendrickson, P.A., Churchill, R.A., and Butt, E. 1997. The wild partridgeberry (*Vaccinium vitis-idaea* L. var. *minus* Lodd.) industry in Newfoundland and Labrador and the potential for expansion utilizing European cultivars. *Acta Hortic.* 446: 139–142.

Penney, B.G., McRae, K.B., Hall, I.V., Morris, R.F., and Hendrickson, P.A. 1985. Effect of harvest date and location on the yield of *Vaccinium vitis-idaea* L. var. *minus* Lodd in eastern Newfoundland. *Crop Res. (Edinburgh)* 25(1): 21–26.

Persson, H.A., and Gustavsson, B.A. 2001. The extent of clonality and genetic diversity in lingonberry (*Vaccinium vitis-idaea* L.) revealed by RAPDs and leaf-shape analysis. *Mol. Ecol.* 10: 1385–1397.

Ritchie, J.C. 1955. Biological flora of the British Isles: *Vaccinium vitis-idaea* L. *J. Ecol.* 43: 701–708.

Saario, M. 2000. Fresh lingonberry quality as affected by storage conditions and packaging. *J. Food Qual.* 23: 453–463.

Small, E., Catling, P.M., and McKenzie, D.B. 2003. Poorly known economic plants of Canada—39. Lingonberry, *Vaccinium vitis-idaea* L. *Bull. Can. Bot. Assoc.* 36(4): 61–65.

St. Pierre, R. 1996. *The lingonberry. A versatile wild cranberry.* Third edition. Department of Horticulture Science, University of Saskatchewan, Saskatoon, SK. pp. 10.

Stang, E.J. 1994. Lingonberry cultivars: Building blocks for an industry? *Fruit Var. J.* 48(1): 3–6.

Stang, E.J., Anderson, M.D., Pan, S., and Klueh, J. 1993. Lingonberry cultural management research in Wisconsin, USA. *Acta Horic.* 346: 327–333.

Stang, E.J., Weis, G.G., and Klueh, J. 1990. Lingonberry: potential new fruit for the northern United States. In *Advances in new crops*. Edited by J. Janick and J.E. Simon. Timber Press, Portland, OR. pp. 321–323.

Taulavuori, K., Taulavuori, E., Niinimaa, A., and Laine, K. 2001. Acceleration of frost hardening in *Vaccinium vitis-idaea* by nitrogen fertilization. *Oecologia* 127: 321–323.

Torrey, G.S. 1914. *The partridge berry.* Newfoundland Department of Agriculture and Mines, St. John's, Newfoundland. pp. 12.

Williams, M.E. 1986. *Lingonberry and bake apple study in Cape Breton.* Nova Scotia Department of Agriculture and Marketing, Bras d'Or Institute, University College of Cape Breton, Nova Scotia. pp. 125.

## Specialty Cookbooks

University of Alaska. 2011. *Lowbush cranberries or lingonberries.* Cooperative Extension Service FNH-00110. University of Alaska, Fairbanks, AK. pp. 4.

Scandinavian cookbooks are a rich source of recipes for lingonberry (see Chapter 29 on Cloudberry for examples). So is the Web. Stanek and Butcher (2007) have 28 recipes. Additional excellent recipes in various cookbooks include "Partridgeberry bread," "Partridgeberry catsup," "Partridberry crisp," "Partridgeberry conserve," "Partridgeberry jam," "Partridgeberry jelly," "Partridberry muffins," "Partridgeberry pudding," "Partridberry punch," "Partridgebery sauce," "Partridgeberry salad," and "Partridgeberry and carrot jam" (Anonymous 1977); "Partridgeberry Tart" (Baird 1980); "Newfoundland berry pudding with dark rum 'n' butter sauce" (Stewart 2000); and Partridgeberry sauce" and "Partridgeberry tarts" (Scott 2010). (As noted in the early part of this chapter, in Newfoundland the lingonberry is called partridgeberry). (See the Appendix to this book for details of the publications cited.)

The following self-published book is difficult to obtain, but is dedicated to lingonberry recipes:

Bellingham Damekor (Norwegian Women's Chorus). *A lingonberry cookbook.* Bellingham Damekor, Bellingham, WA.

# 59 May-Apple

Family: Berberidaceae (barberry family; alternatively placed in Podophyllaceae, the May-apple family)

## NAMES

Scientific name: *Podophyllum peltatum* L.

- The name "May-apple" (May apple, Mayapple) reflects development of apple-like fruits in May.
- Many plant names with "May" are alternatively spelled with the m capitalized or not capitalized (see discussion in Chapter 46 on Hawthorn), and this is the case with May-apple.
- May-apple has also been called Devil's apple, hog-apple, Indian apple, and racoon berry—all names indicating that the fruit was not held in high esteem.
- May-apple is sometimes called American mandrake, mandrake, and wild mandrake. The European mandrake is *Mandragora officinarum* L., a plant used through history for medicines and potions. The transference of name to the American species is based on the common use of the roots of both plants for medicinal purposes.
- Another occasional name for May-apple is wild jalap. Jalap is *Ipomoea purga* (Wender.) Hayne (*Exogonium purga* (Wender.) Benth.), a Mexican plant with a root that is used medicinally. By extension, since May-apple's root is also used medicinally, it came to be called wild jalap.
- Still more names for May-apple are vegetable calomel and vegetable mercury. "Calomel" is mercurous chloride, which was used medicinally in early times; the names "vegetable calomel" and "vegetable mercury" reflect use of May-apple for the same medicinal purposes (*Acorus calamus* L. [Chapter 1, Figure 1.7b], a wild medicinal plant found in North America, has been called "sweet calomel," although it is more usually known as sweet flag).
- The yellow, lemonlike fruit of May-apple has given rise to the names ground-lemon (ground lemon), wild lemon, wild citron, and yellow berry.
- In addition, May-apple is called duck's foot and umbrella plant, both names referring to the appearance of the foliage.
- Occasionally in the southern United States, the Maypop, *Passiflora incarnata* L., has been called May-apple (see Chapter 60 on Maypop).
- The genus name *Podophyllum* is based on the Greek *podos*, a foot (referring to the leaf stalk), and *phyllon*, a leaf. This is descriptive of the leaves of the plant, which are made up of a prominent stalk or "foot" and a leaf blade.
- *Peltatum* in the scientific name *P. peltatum* is Latin for shield-shaped, descriptive of the leaf blade.

## GEOGRAPHY AND ECOLOGY OF WILD PLANTS

May-apple occurs throughout the eastern United States, from southern New England to southern Minnesota south to eastern Texas and northern Florida. In Canada, it is found in much of southern Ontario and Quebec (in Quebec it may not be collected or destroyed under penalty of substantial fines).

**FIGURE 59.1** May-apple (*Podophyllum peltatum*). (a) Parts of a May-apple plant. (From Barton, W.P.C., *Vegetable material medica of the United States*, M. Carey Son, Philadelphia, PA, Vol. 2, Plate 25, 1818; orange fruit at left from Millspaugh, C.F., *American medicinal plants*, Boericke & Tafel, New York, Vol. 1, Plate 17, 1887.) (b) Flowering plant, long section of fruit. (From Köhler, H.A., *Köhler's Medizinal Pflanzen*, Verlag von F.E. Köhler, Germany, Vol. 2, 1887.) (c) Distribution map.

The species is found in moist, shady, deciduous, rich woodlands, forest edges, thickets, and marshy meadows and ditches. North American Indians possibly cultivated and certainly dispersed May-apple.

## PLANT PORTRAIT

May-apple is a perennial herb, growing up to 60 cm (2 feet) tall, and producing one to three deeply lobed leaves. The foliage has the appearance of miniature umbrellas. The branched rhizomes (underground stems) are about 6 mm (0.24 inch) thick, and sometimes as long as a meter (about a yard). The solitary, waxy-white, nodding, flowers look much like a huge apple blossom, and are 1.5–4 cm (0.8–1.6 inches) across, 1.5–6 cm (0.8–2.4 inches) long. Flowering occurs in May in the north; and the yellow, oval or roundish, mucilaginous, pulpy fruit, 2–4 cm (0.8–1.6 inches) across and 3.5–5.5 cm (1.4–2.2 inches) long, ripens in July and August, turning from green to yellow (rarely orange or maroon). Plants with pink flowers and maroon or red fruits also occur rarely. North American Indians and early settlers made jam and preserves from the ripe fruits, and they cultivated the plant to a limited degree.

Native Americans used May-apple as a purifying medicine to expel parasitic worms, and in the treatment of certain cancers, such as tumorous skin growths. A brew of powdered May-apple was used as a laxative (the purgative action is very strong). The powder was also used as a poultice to get rid of warts. Settlers came to use May-apple for a variety of illnesses, including typhoid fever, cholera, dysentery, hepatitis, rheumatism, kidney, bladder and prostate problems, and venereal diseases. Of particular interest is the fact that both the North American and the Asian May-apple (Himalayan May-apple, *Podophyllum hexandrum* Royle [*P. emodi* Wall. ex Honigberger]) contain anticancer agents. At present, extracts of May-apple are used externally for genital warts and some skin cancers. The market for drugs from May-apple exceeds 100 million dollars in some years. The supply is obtained mostly from material collected in the wild, particularly in Indiana, Kentucky, North Carolina, Tennessee, and Virginia.

The roots as well as the unripe fruit, seeds, and leaves can be fatally poisonous. Grazing animals avoid the plant because of its bitterness, but pigs have died after eating May-apple shoots. May-apple should not be used medicinally during pregnancy, as it induces birth abnormalities and can cause abortion (even skin application has caused minor fetal anomalies). Extreme caution is recommended in the handling of May-apple. Fragments of root flicked into the eye during grinding can cause considerable swelling, internal bleeding, severe pain, and temporary loss of sight. Even handling the rhizomes can cause dermatitis, and there are recorded cases of death following external application. May-apple is too toxic to attempt self-medication. It is sold through mail-order nursery catalogues, and the potential danger of handling the plant is not always made clear. Of course, those who choose to collect the plant from the wild need to be especially aware of the risks from contact.

## CULINARY PORTRAIT

The fruit is disagreeably scented when immature but considered pleasantly fragrant when ripe by most people. It has an odd subacid, faintly strawberry or guava flavor. Beatty (1987, see the Appendix to this book) wrote, "The people I know who collect May-apples use them in jams and jellies. But if you do want to try your May-apples raw, peel them and discard the skin. The taste is like a cross between a strawberry and a grape, and the pulp and seeds are very grape-like in appearance, only larger." The fully ripe fruits make excellent jellies and marmalades. Drinks made by adding May-apple fruit to lemonade or Madeira wine are locally popular. Some people, however, find the ripe fruit nauseous, and it must be remembered that the unripe fruits are not only bitter, but poisonous. This wild fruit is available only by personal collection.

### CULINARY VOCABULARY

- Pinxter-flower or honeysuckle azalea, *Rhododendron periclymenoides* (Michx.) Shinners (*Azalea nudiflora* L.) is a deciduous shrub native to the eastern United States that is frequently grown as an ornamental. All parts of the plant are said to be toxic; nevertheless, irregular, juicy, crisp growths called "May apples" that replace flowers on the twigs are eaten as a thirst quencher, and used in salads and as a pickle.

## PROSPECTS

May-apple is basically a medicinal plant, not a food plant. The Eurasian and American May-apples are major sources of anticancer drugs. The North American supply comes primarily from material collected in the wild, mostly in Indiana, Kentucky, North Carolina, Tennessee, and Virginia.

Because the wild harvest is laborious, and the drug content of wild plants is variable, there is reason to select high-yielding clones for cultivation. This consideration is a key to the possible development of May-apple as a food plant. Once cultivation of the plant becomes established, it would be feasible to use the same (perennial) plants that are being cultivated as a fruit source. The plant, including the immature fruit, is quite toxic, so commercial development of May-apple fruits would require careful harvesting and handling. Nevertheless, the fruit is a decided novelty, and would likely find a market niche. A special problem with May-apple is that it is an alternate host of wheat rust, so the plants can be a source of infection for wheat. May-apple fruit may abort when the plant is infected.

## CURIOSITIES OF SCIENCE AND TECHNOLOGY

- In the Ozarks, May-apple was thought to be extremely valuable as an indicator of the probable presence of the more precious and rare ginseng. (The Ozark Mountains are eroded tablelands 450–750 m [1500–2500 feet] high, in the central United States, including northern Arkansas, southern Missouri, Northeastern Oklahoma, with an eastern extension in southern Illinois.)
- In the late nineteenth century, many North American physicians were using May-apple root in the treatment of complaints of the liver. This use continued into the twentieth century, for example, in the well-known Carter's liver pills.
- May-apple sometimes forms circles of plants ("fairy rings") as do some mushrooms and ferns. Such rings are always the result of outward growth of a colony, the central parts dying because nutrients in the soil have been used up.
- May-apple is one of several eastern North American plant species that have very close relatives in eastern Asia. The similarity between the deciduous forests of eastern Asia and eastern North America, along with other evidence, suggests that a temperate deciduous forest formed a continuous band around the northern hemisphere 15–20 million years ago. This band was later split by geological events, leaving two major fragments, the one surviving in eastern Asia and the other in eastern North America.

## KEY INFORMATION SOURCES

Bedir, E., Khan, I., and Moraes, R.M. 2002. Bioprospecting for podophyllotoxin. In *Trends in new crops and new uses*. Edited by J. Janick and A. Whipkey. ASHS Press, Alexandria, VA. pp. 545–549.

Bedows, E., and Hatfield, G.M. 1982. An investigation of the antiviral activity of *Podophyllum peltatum*. *J. Nat. Prod.* 45: 725–729.

Benner, B.L., and Watson, M.A. 1989. Developmental ecology of mayapple seasonal patterns of resource distribution in sexual and vegetative rhizome systems. *Funct. Ecol.* 3: 539–548.

Catling, P.M., and Small. E. 1994. Poorly known economic plants of Canada—2. May-apple, *Podophyllum peltatum* L. *Bull. Can. Bot. Assoc.* 27(2): 30–31.

Constable, J.V.H., Peffer, B.J., and DeNicola, D.M. 2007. Temporal and light-based changes in carbon uptake and storage in the spring ephemeral *Podophyllum peltatum* (Berberidaceae). *Environ. Exper. Bot.* 60: 112–120.

Couillard, L., and Forest, G. 1998. Endangered species of Quebec—May-apple. Gouvernement du Québec, Ministère de l'Environnement et de la Faune, Direction de la conservation et du patrimoine écologique, Quebec. pp. 4. (In French)

Cushman, K.E., Moraes, R.M., Gerard, P.D., Bedir, E., Silva, B., and Khan, I.A. 2006. Frequency and timing of leaf removal affect growth and podophyllotoxin content of *Podophyllum peltatum* in full sun. *Planta Medica* 72: 824–829.

DeMaggio, A.E., and Wilson, C.L. 1986. Floral structure and organogenesis in *Podophyllum peltatum* (Berberidaceae). *Am. J. Bot.* 73: 21–32.

Duke, J.A. 1983. The marvellous mayapple. *The Botanical Grower* [Newsletter] 1(1): 3–4.

Ernst, W.R. 1964. The genera of Berberidaceae, Lardizabalaceae, and Menispermaceae in the southeastern United States. *J. Arnold Arbor.* 45: 1–35.

Eyberger, A.L., Dondapati, R., and Porter, J.R. 2006. Endophyte fungal isolates from *Podophyllum peltatum* produce podophyllotoxin. *J. Nat. Prod.* 69: 1121–1124.

Fisher, A.A. 1981. Severe systemic and local reactions to topical podophyllum resin. *Cutis* 28: 233.

Geber, M.A., Kroon, H. de, and Watson, M.A. 1997. Organ preformation in mayapple as a mechanism for historical effects on demography. *J. Ecol.* 85: 211–223.

George, L.O. 1997. *Podophyllum.* In *Flora of North America*, Vol. 3. Edited by Flora of North America Editorial Committee. Oxford University Press, New York. pp. 287–288.

Graham, N.A., and Chandler, R.F. 1990. Herbal medicine—*Podophyllum. Can. Pharm. J.* 123: 330–333.

Krochmal, A., Wilkins, L., Van Lear, D., and Chien, M. 1974. Mayapple, *Podophyllum peltatum* L. USDA Forest Service Research Paper NE-296. U.S. Department of Agriculture, Upper Darby, PA. pp. 9.

Kroon, H. de, Whigham, D.F., and Watson, M.A. 1991. Developmental ecology of mayapple: effects of rhizome severing, fertilization and timing of shoot senescence. *Funct. Ecol.* 5: 360–368.

Lata, H., Moraes, R.M., Douglas, A., and Scheffler, B.E. 2002. Assessment of genetic diversity in *Podophyllum peltatum* by molecular markers. In *Trends in new crops and new uses.* Edited by J. Janick and A. Whipkey. ASHS Press, Alexandria, VA. pp. 537–544.

Laverty, T.L. 1992. Plant interactions for pollinator visits: a test of the magnet species effect. *Oecologia* 89: 502–508.

Laverty, T.M., and Plowright, R.C. 1988. Fruit and seed set in mayapple (*Podophyllum peltatum*): Influence of intraspecific factors and local enhancement near *Pedicularis canadensis. Can. J. Bot.* 66: 173–178.

Maqbool, M., Cushman, K.E., and Moraes, R.M. 2002. Propagule type and planting time for field-established mayapple. In *Trends in new crops and new uses.* Edited by J. Janick and A. Whipkey. ASHS Press, Alexandria, VA. pp. 533–536.

McFarland, M.F., III, and McFarland, J. 1981. Accidental ingestion of podophyllum *Podophyllum peltatum*, mayapple. *Clin. Toxicol.* 18: 973–977.

Meijer, W. 1974. *Podophyllum peltatum*—Mayapple a potential new cash-crop plant of eastern North America. *Econ. Bot.* 28: 68–72.

Miller, R. 1985. Podophyllin. *Int. J. Dermatol.* 24: 491–498.

Moraes, R.M., Bedir, E., Barrett, H., Burandt, C., Jr., Canel, C., and Khan, I.A. 2002. Evaluation of *Podophyllum peltatum* accessions for podophyllotoxin production. *Planta Med.* 68: 341–344.

Moraes, R.M., Burandt, C., Jr., Ganzera, M., Li, X., Khan, I., and Canel, C. 2000. The American mayapple revisited—*Podophyllum peltatum*—(still a potential cash crop? *Econ. Bot.* 54: 471–476.

Moraes, R.M., Lata, H., Bedir, E., Maqbool, M., and Cushman, K. 2002. The American mayapple and its potential for podophyllotoxin production. In *Trends in new crops and new uses.* Edited by J. Janick and A. Whipkey. ASHS Press, Alexandria, VA. pp. 527–532.

Parker, M. 1989. Disease impact and local genetic diversity in the clonal plant *Podophyllum peltatum. Evolution* 43: 540–547.

Policansky, D. 1983. Patches, clones, and self-fertility of mayapples (*Podophyllum peltatum* L.). *Rhodora* 85: 253–256.

Rosenstein, G., Rosenstein, H., Freeman, M., and Weston, N. 1976. Podophyllum—a dangerous laxative. *Pediatrics* 57: 419–421.

Rust, R.W., and Roth, R.R. 1981. Seed production and seedling establishment in the mayapple, *Podophyllum peltatum.* L. *Am. Midl. Nat.* 105: 51–60.

Sohn, J.J., and Policansky, D. 1977. The costs of reproduction in the mayapple *Podophyllum peltatum* (Berberidaceae). *Ecology* 58: 1366–1374.

Swanson, S., and Sohmer, S.H. 1976. Reproductive biology of *Podophyllum peltatum* (Berberidaceae): the comparative fertility of inter- and intra-populational crosses. *Transact. Wisconsin Acad. Sci. Arts Letters* 64: 109–114.

Swanson, S., and Sohmer, S.H. 1976. The biology of *Podophyllum peltatum* L. (Berberidaceae), the may apple, 2: the transfer of pollen and success of sexual reproduction. *Bull. Torrey Bot. Club* 103: 223–226.

Watson, M.A., and Lu, Y. 1999. Timing of shoot senescence and demographic expression in the clonal perennial *Podophyllum peltatum* (Berberidaceae). *Oikos* 86: 67–78.

Watson, M.A., Scott, K., Griffith, J., Dieter, S., Jones, C.S., and Nanda, S. 2001. The developmental ecology of mycorrhizal associations in mayapple, *Podophyllum peltatum*, Berberidaceae. *Evolutionary Ecol.* 15: 425–442.

Whisler, S.L., and Snow, A.A. 1992. Potential for the loss of self-incompatibility in pollen-limited populations of mayapple (*Podophyllum peltatum*). *Am. J. Bot.* 79: 1273–1278.

## Specialty Cookbooks

Beatty (1987) has three May-apple recipes, and Tatum (1976) has four. Brill (2002) and Boorman (1969) each have two, and Johnson (1989) and Kluger (1973) each have one. (See the Appendix to this book for details of the publications cited.)

# 60 Maypop

Family: Passifloraceae (passionflower family)

## NAMES

Scientific name: *Passiflora incarnata* L.

- The name "Maypop" has been explained on the basis that the plant seemingly pops out of the ground in May. Another (perhaps contrived) explanation is that if one squeezes the fruit, it "may pop" like a balloon. The name Maypop might be derived from a similar Native American word for *P. incarnata*, such as the Algonkian *maricock* or *maracock*, or the Powhatan *mahcawq*. It has been suggested that Native Americans acquired the word from the Tupi Indians of South America, who called the plant *maraca-cui-iba*, the rattle fruit, because seeds in the gourd-like fruits produce a rattling sound when the fruits are dried.
- Local names used historically in the southern United States for the Maypop include: apricot, apricot vine, field apricot, ground ivy, holy trinity flower, May-apple (likely to result in confusion with the true May-apple, *Podophyllum peltatum* L.; see Chapter 59), mollie cockle, molly-pop, old-field apricot, purple passionflower, wild apricot, and wild passion vine.
- "Maypop" and "maypop" are alternative spellings. The capitalization of the initial M in plant names beginning with "may" is inconsistent (cf. information for "mayhaw" in Chapter 46 on Hawthorns).
- *Incarnata* in the scientific name of the Maypop, *P. incarnata*, means flesh-colored, descriptive of the yellowish fruit.
- *Passiflora* species are known as passion flowers (alternatively passionflowers) or "flowers of passion." Spanish missionaries to South America, who encountered the plants in the jungles of Brazil in the sixteenth century, thought the flower represented some of the objects associated with the Crucifixion or suffering of Christ, and coined the name passionflower. The "passion" in the name is based on the Passion of Christ, the period between the Last Supper and his death. The 10 apostles who remained faithful to Jesus (leaving out Peter who denied him and Judas who betrayed him) throughout the Passion are symbolized by the five petals and the five petal-like sepals, while a fringe of colored, hairlike filaments above the petals (botanically known as the corona) was thought to represent the crown of thorns that Jesus wore (or by some, considered to symbolize a halo). The three spreading styles atop the stigma represented the three nails by which Christ was attached to the cross. The five hammer-like anthers on top of the stamens were thought by some to represent the hammer used to drive the nails; others interpreted the five stamens as the wounds of Christ (hence the name "flower of the five wounds" sometimes used in Peru, the West Indies, and elsewhere) and the ovary (unfertilized seed case) as the hammer.
- The genus name *Passiflora* has the same general basis as the "passion" in passionflower or passion fruit, noted above. The name *Passiflora* is specifically from the Latin *passio*, passion + *flos*, flower.

**FIGURE 60.1** Maypop (*Passiflora incarnata*). (a) Fruiting plant. (Courtesy of Pollinator [CC By 3.0].) (b) Potted plant in flower. (From Holtzbecker H.-S., *Gottorfer Codex*. 4 Vols. Produced in Gottorp Castle, Slesvig (modern Germany, 1649–1659.) (c) Flower. (Courtesy of Oliver P. Quillia [CC By 3.0].) (d) Cross section of a fruit. (From Edwards, S.T., *Botanical Register*, Vol. 2, Plate 152, 1816.) (e) Distribution map.

## GEOGRAPHY AND ECOLOGY OF WILD PLANTS

Maypop is one of the few species of *Passiflora* native to the United States, occurring from Iowa to New Jersey, south to Florida and Texas. It also is native to Bermuda. The Maypop is a common roadside weed throughout many areas of the southeastern United States, often growing in large masses in ditches and open fields. The species grows so aggressively in some areas that it is considered invasive; in Texas it has been declared to be a noxious weed. Maypop is intolerant of shade, and does not occur in shady forests. It grows well in infertile, sandy, acidic soils. Unlike most of the other 400 or so tender species of *Passiflora*, Maypop can overwinter in freezing climates, as far north as U.S. Department of Agriculture (USDA) climate zone 6 (it is said to be "semi-frost hardy"), although the vines die back to ground level. In warmer areas the foliage can be deciduous (not evergreen, as in most species of the genus).

## PLANT PORTRAIT

Most species of *Passiflora* are native to South America, but about 20 are found in Asia and Australia. Of the approximately 500 species, about 60 produce edible fruit (called passion fruit [less frequently passionfruit] and granadilla), of which only 10–12 have been commercially exploited for their fruits. The most widely marketed species, in decreasing order of importance, are *P. edulis* Sims, *P. quadrangularis* L., and *P. ligularis* Juss. The most common types have orange, purple, or yellow skin, and are egg- to apple-sized. The fruit has a hard, leathery shell, which encloses many small, dark, crunchy, edible seeds that are enveloped by juicy, tart, gelatinous flesh. The flesh may be yellow, orange, pinkish-green, white, or colorless. Passionfruit is grown widely throughout the tropics and subtropics. Commercial production occurs in South America, the Caribbean, Mexico, Australia, New Zealand, Southeast Asia, Taiwan, India, Africa, the Mediterranean area, Hawaii, and South Africa. There are small crops in California and Florida.

Maypop is a fast-growing, perennial vine with stems 3–6 m (10–20 feet) long, but only half this length in the northern areas of its range. It climbs over vegetation using tendrils for attachment, or simply trails on the ground. The intricate flowers are large, 5.7–7.6 cm (2–3 inches) in diameter. The fruit is oval, about the size of a hen's egg, green when immature, yellow, brown, or orange when ripe, with greyish flesh.

Maypop has not been developed as a commercial fruit, like the passion fruits mentioned above, but it hybridizes with *P. edulis*, the principal commercial passion fruit, and has been used to breed superior cultivars of the latter. Maypop was once a favorite wild fruit, harvested by Native Americans and settlers in the southern United States. It has been suggested that in the past, cultivated selections of superior fruit were made, but that these may have been lost. Today, Maypop is widely grown in tropical areas, but as an ornamental, not as a fruit source. It can be grown as far north as USDA climate zone 5, which includes parts of southern Canada. The plants are also excellent indoors in containers. Extracts from *P. incarnata* (also from *P. edulis*) are commonly used as sedatives in herbal medicine. Passionflower tea is consumed to some extent for its presumed medicinal value.

## CULINARY PORTRAIT

Maypop is not a commercial passion fruit, but its fruits can be used in the same way. The fruit usually has delicious pulp (although some plants produce unpleasant fruits), but is seedy and best made into jelly. Overripe fruit ferments into a foul paste.

Maypop fruits can be prepared in the same way as the fruit of commercial passion fruit species. The fruits can be cut lengthwise and the seedy pulp scooped out of the inedible shell with a spoon. The seeds of passion fruits are usually removed by squeezing the pulp through two thicknesses of cheesecloth or pressing the pulp through a strainer. Alternatively, the seeds may simply be eaten along with the pulp. Passion fruit can be prepared as a dessert with cream and sugar, or sprinkled with lemon or lime juice; or used in fruit salads and beverages; or made into jellies and jams. Passion fruit juice can be cloying alone, but is an excellent flavoring even when used in small quantities. It is used in sauces, gelatin desserts, candy, ice cream, yogurt, sherbet, icing, pies, soups, and cocktails. The fresh juice and concentrate are superb mixers with alcoholic beverages, such as gin, vodka, and rum. The juice is sometimes fermented to make alcoholic beverages. Most people are familiar, often unknowingly, with the flavor of passion fruit from having tasted Hawaiian Punch. It has been said that passionflower juice gives the punch to Hawaiian Punch.

### Culinary Vocabulary

- A "Hurricane" is a cocktail made with dark rum, passion fruit flavoring, and citrus juices, served in a hurricane glass (a footed glass that is bulbous at the bottom, tapering to a flaring cylinder at the top, used for blended or frozen tropical and specialty drinks). The origin of the beverage is obscure, but it has long been associated with Pat O'Brien's French Quarter Bar in New Orleans, Louisiana.
- A "Grand Passion" is a cocktail prepared with passion fruit liqueur, gin, and Angostura bitters.

## PROSPECTS

The Maypop is an attractive wild fruit, but it is not competitive with the passion fruit species currently grown as crops. Its chief virtue as a potential crop compared with the established *Passiflora* species is that it can be grown in colder, more northern areas. The species is very vigorous (in fact, weedy), produces quite tasty fruits, and indeed, cultivated fruit varieties appear to have been selected in the past by Native Americans. The Maypop, therefore, certainly has the potential to be developed into a new fruit. However, the fact that advanced cultivars of other passion fruit species are currently being grown in tropical areas is a major obstacle to development of the Maypop. Tropical areas are naturally more productive (because of the longer season and more intense sunlight) than temperate areas, so there is limited need to develop cold-hardy passion fruits. Nevertheless, there are advantages to growing plants in northern areas, particularly to supply local markets and avoid transportation, so the Maypop remains an interesting possibility as a future new crop. Moreover, the species can also be grown as a medicinal crop and as an ornamental, so there are multiple incentives to invest in its development.

## CURIOSITIES OF SCIENCE AND TECHNOLOGY

- Captain James Smith (who brought 105 people to Virginia and founded Jamestown in 1607, the first permanent English settlement in the New World) observed Native Americans cultivating the Maypop for its fruits.
- In 1865, Queen Victoria (1819–1901) had a wreath of passionflowers sent to the funeral of Abraham Lincoln (1809–1865, 16th president of the United States). The Queen declared that the passionflower was the ideal flower for funerals. Gravestones with elaborately carved passionflower vines became very popular for the wealthy in England.
- The Maypop was named the state wildflower of Tennessee in 1919, after a vote by the schoolchildren of the state. The Cherokee in the state called the plant *ocoee*, which gave rise to the names Ocoee River and Ocoee Valley. (In 1933, the iris was adopted as the state flower of Tennessee, the previous choice of the Maypop simply being ignored. In 1973, the state legislature voted to declare the Maypop as the "state wildflower" and the iris as the "state cultivated flower.")
- The caterpillars (larvae) of several species of zebra butterflies can only eat passionflower leaves, so if one wants to attract butterflies to a garden, growing the cold-hardy Maypop is a good idea. Several passionflower species have adaptations to prevent damage by the caterpillars. Some plants disguise themselves by producing leaves that do not look like typical passionflower leaves. Some species have nectar-secreting glands to attract wasp and ant parasites of the larvae. Some species produce chemicals that poison the larvae. And still other species produce tiny egg-like particles on the leaves that send the message to the female butterfly that the leaves are already occupied, so she should find another plant on which to lay her eggs.
- The Maypop, like some other species of *Passiflora*, has nectar-producing glands on the leaf stalks, which attract ants, which in turn have been experimentally shown to

protect the plant against attack from other insects. This type of cooperative arrangement whereby plants produce nectar on secretory glands outside of the flowers to attract ants for protective purposes is widespread. It has been shown that the Maypop and some other species of *Passiflora*, when their leaves have been damaged by insects, seem to have the ability to increase their production of secretory glands to attract more protective ants.

## KEY INFORMATION SOURCES


Abeysinghe, A. 1973. *Commercial passion fruit: Cultivation, processing, and marketing*. Ministry of Plantation Industries, Govt. of Sri Lanka, Colombo, Sri Lanka. pp. 29.

Anonymous. 1974. *Passion fruit culture in Hawaii*. University of Hawaii, Cooperative Extension Ser., Honolulu, HI. pp. 35.

Anonymous. 1978. *Passion fruit industry analysis*. Vol. 2. College of Tropical Agriculture, University of Hawaii, Manoa, HI.

Arjona, H.E., Matta, F.B., and Garner, J.O., Jr. 1991. Growth and composition of passion fruit (*Passiflora edulis*) and maypop (*P. incarnata*). *HortScience* 26: 921–923.

Bora, P.S., and Narain, N. 1997. Passion fruit. In *Postharvest physiology and storage of tropical and subtropical fruits*. Edited by S.K. Mitra. CAB International, Wallingford, U.K. pp. 375–386.

Buchbauer, G., and Jirovetz, L. 1992. Volatile constituents of the essential oil of *Passiflora incarnata* L. *J. Essent. Oil Res.* 4: 329–334.

Casimir, D.J., Keffod, J.F., and Whitfield, F.B. 1981. Technology and flavor chemistry of passion fruit juices and concentrates. In *Advances in food research*. Vol. 27. Edited by C.O. Chichester, E.M. Mrak, and G.F. Stewart. Academic Press, New York. pp. 243–295.

Chan, H.T., Jr. 1980. Passion fruit. In *Tropical and subtropical fruits: composition, properties and uses*. Edited by S. Nagy and R.E. Shaw. AVI Publishing Company, Westport, CT. pp. 300–315.

Chitra, V., Gowri, K., Tamilanban, T., and Soni, D. 2009. Neuropharmacological screening of aerial parts of *Passiflora incarnata* Linn (Passifloraaceae). *Arch. Appl. Sci. Res.* 1: 254–262.

Dhawan, K., Kumar, R., Kumar, S., and Sharma, A. 2001. Correct identification of *Passiflora incarnata* Linn., a promising herbal anxiolytic and sedative. *J. Med. Food.* 4: 137–144.

Fisher, A.A., Purcell, P., and Le Couteur, D.G. 2000. Toxicity of *Passiflora incarnata* L. *J. Toxicol. Clin. Toxicol.* 38: 63–66.

Fore, S.A., and Spira, T.P. 2002. Macrogeographic allozyme variation in *Passiflora incarnata*. *J. Torr. Bot. Soc.* 129: 326–330.

Glassman, S. 1978. North America's least known fruit? *North American Pomona* 11(3): 117–119.

Gremillion, K.J. 1989. The development of a mutualistic relationship between humans and Maypops (*Passiflora incarnata* L.) in the southeastern United States. *J. Ethnobiol.* 9: 135–138.

Hoch, J.H. 1934. The legend and history of *Passiflora*. *Am. J. Pharm.* 1934: 166–170.

Knight, R.J., Jr. 1969. Edible-fruited passionvines in Florida: The history and possibilities. *Proc. Trop. Reg., Amer. Soc. Hort. Sci.* 13: 265–274.

Martin, F.W., and Nakasone, H.Y. 1970. The edible species of *Passiflora*. *Econ. Bot.* 24: 333–343.

McGuire, C.1998. Field performance and phenotypic variation of *Passiflora incarnata* L. in New York State. *HortScience* 33: 240–241.

McGuire, C.M. 1999. *Passiflora incarnata* (Passifloraceae): A new fruit crop. *Econ. Bot.* 53: 161–176.

McKenzie, D.E., Chalker, F.C., and Buggie, G.J. 1981. *Passionfruit growing*. North Coast Region, Dept. of Agriculture, Coffs Harbour, N.S.W., Australia. pp. 24.

McLain, D.K. 1983. Ants, extrafloral nectaries and herbivory on the passion vine, *Passiflora incarnata*. *Am. Midl. Nat.* 110: 433–439.

Menzel, C.M., and Simpson, D.R. 1994. Passionfruit. In *Handbook of environmental physiology of fruit crops. Vol. 2, Subtropical and tropical crops*. Edited by B. Schaffer and P.C. Anderen. CRC Press, Boca Raton, FL. pp. 225–241.

Menzel, C.M., Winks, C.W., and Simpson, D.R. 1989. Passionfruit in Queensland. 3. Orchard management. *Queensland Agric. J.* 115: 155–164.

Morton, J. 1987. Passionfruit. In *Fruits of warm climates*. Edited by J.F. Morton. Creative Resource Systems, Winterville, NC. pp. 320–328.

Mott, J. 1969. *The market for passion fruit juice*. Tropical Products Institute, Ministry of Overseas Development, London, U.K. pp. 15.

Patel, S.S., Saleem, T.S.M., Ravi, V., Shrestha, B., Verma, N.K., and Gauthaman, K. 2009. *Passiflora incarnata* Linn: A phytopharmacological review. *Internat. J. Green Pharm.* 3: 277–280.

Patel, S.S., Verma, N.K., and Gauthaman, K. 2009. *Passiflora incarnata* Linn: A review on morphology, phytochemistry and pharmacological aspects. *Pharm. Rev.* 3: 175–181.

Patil, A.S. 2010. Exploring *Passiflora incarnata* (L.): a medicinal plant's secondary metabolites as antibacterial agent. *J. Med. Plants Res.* 4: 1496–1501.

Ulbricht, C., Basch, E., Boon, H., Karpa, K.D., Gianutsos, G., Nummy, K., Seamon, E., et al. 2008. An evidence-based systematic review of passion flower (*Passiflora incarnata* L.). *J. Diet. Suppl.* 5: 310–338.

Ulmer, T., and MacDougal, J.M. 2004. Passiflora: *passionflowers of the world.* Timber Press, Portland, OR. pp. 430.

Vanderplank, J. 1996. *Passionflowers and passionfruit.* Revised edition. Cassell, London, U.K. pp. 224.

Wehtjie, G., Reed, R.B., and Dute, R.R. 1985. Reproductive biology and herbicidal sensitivity of maypop passionflower (*Passiflora incarnata*). *Weed Sci.* 33: 484–490.

Whittaker, D.E. 1972. Passion fruit: agronomy, processing and marketing. *Trop. Sci.* 14: 59–77.

Williams, C.G. 1954. Preparing passion fruit for the fresh fruit market. *Queensland Agr. J.* 78(2): 81–88.

Winks, C.W., Menzel, C.M., and Simpson, D.R. 1988. Passionfruit in Queensland. 2. Botany and cultivars. *Queensland Agric. J.* 114: 217–224.

Wohlmuth, H., Penman, K.G., Pearson, T., and Lehmann, R.P. 2010. Pharmacognosy and chemotypes of passionflower (*Passiflora incarnata* L.). *Biol. & Pharm. Bull.* 33: 1015–1018.

## Specialty Cookbooks

Gebhard, M.L., and Butler, W.H. 1967. *Pineapples, passion fruit, and poi: Recipes from Hawaii.* Charles E. Tuttle Co., Rutland, VT. pp. 127.

Hawaiian Electric Company. 1956. *Reddy's passion fruit recipes.* Hawaiian Electric Company, Honolulu, HI. pp. 9.

Nabhan, G. (Editor). 2008. *Renewing America's food traditions: Saving and savoring the continent's most endangered foods.* Chelsea Green, White River Junction, VT. pp. 350. (Has recipe for "Maypop passionfruit jelly".)

Vinacke, W.R. 1955. *Passion fruit recipes for sweetened, frozen, pure passion fruit juice.* Food Processing Laboratory, University of Hawaii, Honolulu, HI. pp. 11.

Freitus (1980) has four Maypop recipes, Freitus and Haberman (2005) have seven, and Tatum (1976) has three. (See the Appendix to this book for details of the publications cited.)

# 61 Mesquite

Family: Fabaceae (Leguminosae; pea family)

## NAMES

Scientific names: *Prosopis* species:

- Honey mesquite—*P. glandulosa* Torr. [*P. juliflora* (Sw.) DC. var. *glandulosa* (Torr.) Cockerell; "*P. chilensis*" of authors; in most of the general literature *P. glandulosa* is called *P. juliflora*]
- Screwbean mesquite—*P. pubescens* Benth.
- Velvet mesquite—*P. velutina* Wooten [*P. juliflora* var. *velutina* (Wooten) Sarg.]
- The name "mesquite" (also spelled mesquit, mezquit, and mezquite) is derived from the Spanish *mezquite*, which in turn is based on the American Indian (Nahuatl) *mizquitl*, the indigenous people's name for mesquite. A "mesquital" is an area in which the vegetation is dominated by mesquite.
- Pronunciation: meh-SKEET OR MEH-skeet.
- The word algarroba (also spelled algaroba, algarobo, algarrobo, and algeroba) is used for several species of the legume family, usually the carob (*Ceratonia siliqua* L.), but also mesquite. Algarrobo is the Old World name for carob, and the Conquistadores applied it to *Prosopis* species because of their carob-like appearance. In South America, algarrobo is used in Spanish areas, and algaroba in Portuguese areas.
- Honey mesquite is named for its sweet pods. Honey mesquite is also called common mesquite, honey locust (a name usually used for other legume species), honey pod, ironwood (a name used for many unrelated species with hard wood), and western honey mesquite. *Glandulosa* in the scientific name of the species *P. glandulosa* refers to small nectar glands in the flowers.
- Screwbean mesquite is also called mescrew, screw bean, screw-pod mesquite, tornillo, and twisted bean. These names are based on the spiral, bean-like pods. The plant is also known as scrub mesquite. *Pubescens* in the scientific name of the species, *P. pubescens*, is Latin for hairy, referring to the pods that are densely covered with white hairs when young.
- *Velutina* in the scientific name of velvet mesquite, *P. velutina*, is Latin for velvety, and refers to the velvety hairs of the young pods and leaves.
- The genus name *Prosopis* is an old Greek name for a kind of prickly plant, usually identified as the burdock. The name was perhaps chosen in recognition of the abundant spines of the plants. Pronunciation: PROS-oh-pis.

## GEOGRAPHY AND ECOLOGY OF WILD PLANTS

Honey mesquite is found from southern California east to southern Kansas and Louisiana, and south to northern Mexico. Screwbean mesquite is found from western Texas to southern Nevada, and in Utah, Arizona, New Mexico, southern California, and northern Mexico. Velvet mesquite is distributed primarily in northern Mexico, and also occurs in the southwestern United States, mostly in Arizona.

Mesquite species are typically adapted to arid areas by possession of deep roots that can reach low water tables. However, some species also take advantage of rainfall by having lateral roots near

**FIGURE 61.1** Honey mesquite (*Prosopis glandulosa*). (a) Male catkins. (Courtesy of Joe Decruyenaere [CC By 2.0].) (b) Branches bearing pods. (Courtesy of Don A.W. Carlson [CC By 3.0].) (c) Tree. (Courtesy of Homer Edward Price [Flickr/CC-attribution].) (d) Distribution map.

the ground surface. Like numerous other members of the pea family, mesquite roots are associated with nitrogen-fixing bacteria that can considerably improve soil fertility by transforming inert nitrogen from the air into a form that plants can use for growth.

American cattlemen regard mesquite as a range weed and eradicate it when possible. Because of overgrazing, mesquite has invaded former grasslands where it did not grow a century ago. Mesquite drains soil of its water, and from an agricultural point of view represents competition for more desired rangeland and cultivated plants. The war against mesquite is being waged primarily with fire and herbicides. A more sensible approach that is underway is to harvest wild mesquite for honey, furniture, firewood, and charcoal.

Mesquite was introduced and escaped from cultivation in Australia, the Philippines, Sudan, and Hawaii, and has become a damaging weed in these countries. Nevertheless, it continues to be

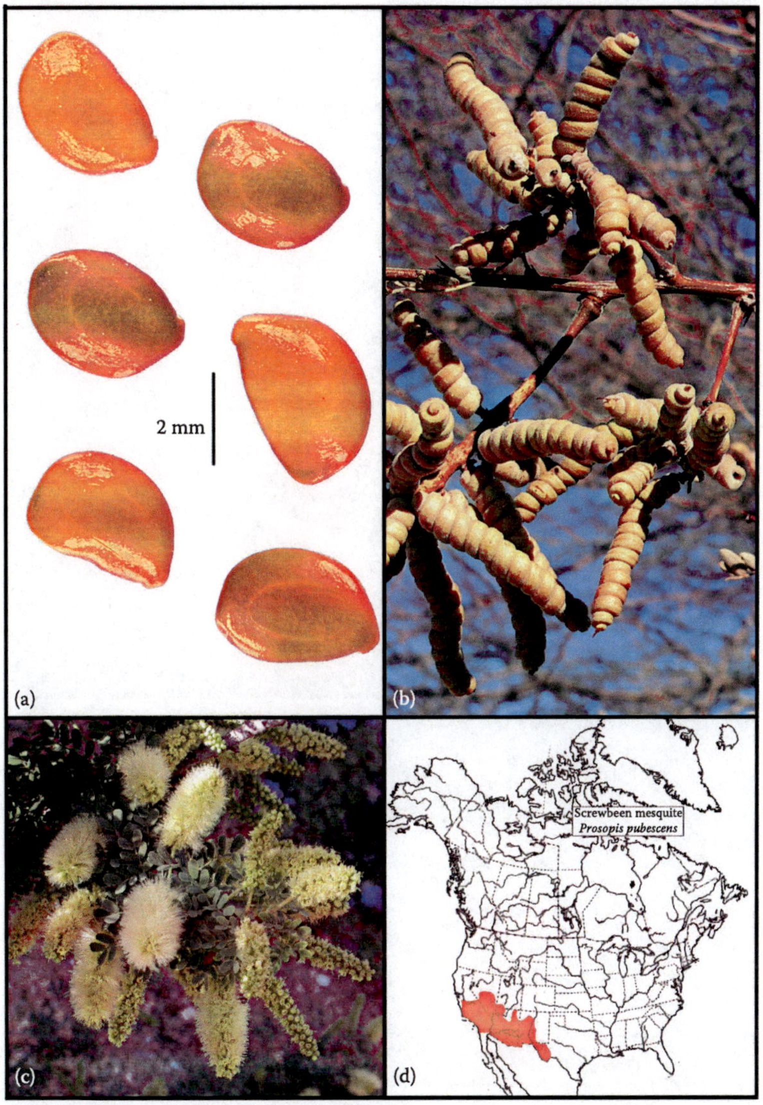


**FIGURE 61.2** Screwbean mesquite (*Prosopis pubescens*). (a) Seeds. (Courtesy of Steve Hurst, U.S. Department of Agriculture.) (b) Pods. (Courtesy of Stan Shebs [CC By 3.0].) (c) Flowering branches. (Courtesy of Curtis Clark [CC By 2.5].) (d) Distribution map.

introduced in arid regions (e.g., in India, China, Peru, Chile, Pakistan, and many African nations) to combat desertification—the gradual conversion of drylands supporting considerable vegetation to deserts that support very few plants. Introduced mesquite requires little water, and often serves to prevent erosion while providing food for man and beast.

## PLANT PORTRAIT

There are about 44 species of *Prosopis*, a genus mostly made up of trees and shrubs, occupying desert and semidesert regions of the Americas, southwestern Asia, and Africa. The plants are prominent in Mexico, Peru, Chile, Argentina, Brazil, Sahelian (sub-Saharan) Africa, Haiti, Pakistan, and

**FIGURE 61.3** Velvet mesquite (*Prosopis velutinosa*). (a) Male catkins. (Courtesy of Sue [CC By 3.0].) (b) Pods. (Courtesy of Sue.) (c) Tree. (Courtesy of Sue [CC By 3.0].) (d) Distribution map.

the arid parts of India. In these locations, mesquites are often used as firewood, livestock feed (the pods), and for soil improvement (because of the nitrogen-fixing bacteria in the root nodules). The South American algarrobo (*P. pallida* (Humb. & Bonpl. ex Willd.) Kunth) is considered the sweetest in flavor for humans; in Peru it is harvested and processed into flour for export to the United States, and in some areas of South America its sweet pods are made into a syrup used in various beverages. The discussion in this chapter concentrates on the species of the United States. Mesquite in other regions of the world is used similarly.

Mesquite is the most common shrub or small tree of the American southwestern desert region, occupying 30 million ha (74 million acres) in the United States. The three species featured in this chapter are the most common American mesquites. These have deciduous leaves that are made up of numerous small leaflets arranged on both sides of long midribs. The branches typically bear prominent spines.

Honey mesquite is a shrub or small tree, 1–12 m (3–40 feet) in height, with a trunk up to 30 cm (12 inches), or very rarely 1.2 m (4 feet) in diameter. It has clusters of small, fragrant, creamy yellow flowers, and flat, narrow, yellow-green, bean-like pods 10–23 cm (4–9 inches) in length.

Screwbean mesquite is a shrub or small tree 2–10 m (6.6–33 feet) in height, developing a trunk on larger trees 20–30 cm (8–12 inches) in diameter. The branches are twisted. Its light-brown to reddish, smooth bark separates into long, shaggy strips. The plant bears crowded clusters of small, light yellow flowers in cylindrical spikes approximately 5 cm (2 inches) long, and they develop fruits 2.5–5 cm (1–2 inches) long, which are hard, tightly spiraled, brown to yellow, with sweet pulp. The fruit, very tightly coiled into a spring-like cylinder, makes this species easy to identify.

Velvet mesquite is a large shrub or medium-size tree reaching a height of 15 m (49 feet) and developing a trunk up to 60 cm (2 feet) in diameter. Like the above species, the dark-brown, smooth bark separates into long, shaggy strips. The plant grows in dense thickets. The flowers are small, fragrant, greenish yellow, in slender, cylindrical spikes up to 10 cm (4 inches) long. The fruit is a slender, brown pod 10–22 cm (4–9 inches) in length. Velvet mesquite is difficult to distinguish from honey mesquite, and the two are sometimes placed in the same species.

Mesquite pods are among the earliest known foods of prehistoric mankind in the New World, fossils indicating that the pods were eaten more than 8000 years ago. Native Americans used the pods as food, and the bark for basketry, fabrics, and medicine. The women used mesquite bark fiber to make diapers, skirts, and other articles of clothing, as well as for weaving baskets, ropes, and twine. Mesquite bark was also used to make a poultice for treating wounds and illnesses. Mesquite pods were probably the most important wild plant food staple of Native Americans of the southwestern United States. The gum exuded from mesquite trunks was used as candy, as a glue for mending pottery, and as a black dye. Mesquite gum from the trunk of the plants resembles gum arabic (a gum from certain *Acacia* species). This gum can be obtained by making incisions into the trunk of the tree, but the yield is poor, and although such gum has occasionally been offered for sale in North America, it is not a significant item of commerce. A somewhat different mesquite gum with properties comparable to the gum from carob and guar [*Cyamopsis tetragonoloba* (L.) Taub] can be obtained from the ground seeds, and has some potential to be produced on a commercial scale.

Mesquite wood is very hard and has been made into fenceposts, railway sleepers, parquet flooring, tool handles, handicrafts, and such articles of furniture as cabinets, game tables, and desks. Early pioneers used mesquite wood for wagon wheels. Today, mesquite lumber is produced in Texas, New Mexico, Arizona, Argentina, and Mexico. Charcoal briquets for barbecuing are made out of mesquite wood in Texas, and recently, the industry has become very large in Mexico, which exports its briquets to the United States. Similar wood products made from honey mesquite include chips, chunks, nuggets, and sticks. "Liquid smoke," an extract or concentrate of mesquite wood, is another associated commercial product.

Although mesquite pods are important as a livestock feed, and considered to be excellent for cattle and horses, when cattle consume large amounts of beans continuously over a 2-month period, serious digestive disturbances or death may occur. This does not appear to be because of toxic constituents that directly affect the animals, but rather because of inhibition of the rumen bacteria that are essential to the animals to digest cellulose and synthesize vitamins. Mesquite pods do not seem to be toxic to humans, although some people develop an allergic dermatitis from contacting the plants.

## CULINARY PORTRAIT

Native Americans in the deserts of the southwestern United States and nearby Mexico relied on the mesquite pod as a dietary staple from which they made tea, syrup, and a ground meal called pinole. The meal was made into bread, pancakes, and a variety of other products. Water was added to form a sweet, nutritious mixture that was used as a nonalcoholic beverage, or fermented to produce a weak beer. Green mature pods were consumed and cooked like green beans. Flowers were eaten, raw or roasted. Today, flour products from mesquite pods remain popular, but are only sporadically prepared by native North Americans in southern California and Arizona, Sonora in Mexico, and parts of South America. Wild food enthusiasts also grind the seed pods, being careful to use tree-ripened beans, never the green seeds, which are not palatable. The yellow flour prepared from honey mesquite tastes sweet, and can be mixed with regular flour for use in baking. Toasted mesquite seeds are sometimes added to coffee. Mesquite pod flour and mesquite bean jelly are sometimes available in stores as specialty items. A sweet mesquite soft drink is also consumed in South America.

Mesquite is an ideal firewood because it burns slowly and is smokeless. It is used extensively to create aromatic charcoal for barbecuing, and "mesquite-grilled" food is a specialty of southwest of the United States.

Mesquite is extremely attractive to bees and the plant is famous for its fragrant honey, considered to be a delicacy.

### Culinary Vocabulary

- In some South American countries, an extract from the ripe pods is concentrated into a honeylike product called *algarrobina*, which is widely used and sold.
- *Mora* is a dried jalapeño chile pepper smoked over mesquite wood.

## PROSPECTS

Mesquite species have minor culinary uses, but are also useful for a variety of other products, particularly lumber and forage. Mesquites have considerable weed potential and can be very destructive invasive plants. However, they can grow in the extremely arid and hot conditions that are widespread in subtropical areas. Accordingly, they need to be considered as potential agricultural crops for a world that is running out of water and prime arable land. Mesquites are unlikely to become major food crops for humans, but they are multipurpose plants that can be developed profitably for a variety of products.

## CURIOSITIES OF SCIENCE AND TECHNOLOGY

- In the days of the Wild West, velvet mesquite wood was often used as a tombstone.
- Indigenous peoples played several athletic games with mesquite materials. The Papago held a footrace in which a ball the size of a croquet ball, usually made of mesquite wood or mesquite gum, was kicked. The Pima played a game that was similar to field hockey, using mesquite sticks. The Maricopa also played a game similar to field hockey, using a mesquite ball.
- Among the Yuma, a baby cradle constructed of mesquite slats was traditionally made by the father immediate after the birth of his child.
- The first cannon used by the Mexicans in their fight for independence from Spain was made by hollowing out a mesquite log. The cannon worked well, and is exhibited at the National Museum in Mexico City.
- In 1841, the Republic of Texas sent a 320 man expedition from Texas to Santa Fe to annex New Mexico. Crossing the Texas Panhandle in the summer, the men found themselves out of provisions, and turned to mesquite, eating immense quantities. Grateful for the presence of the trees, they called the beans "manna from heaven."
- In 1867, mesquite timbers were used to construct Fort Richardson near Jacksboro, Texas, the site originally called Mesquiteville. This was the northernmost of a line of federal forts established after the American Civil War.

- Mesquite wood was used as street and walkway paving in early Texas. In San Antonio, when the streets are excavated, the wooden paving blocks are commonly encountered on the undersurface of the roads.
- In the Pleistocene era (which ended 11,000 years ago), the megafauna (large animals, including mastodons and ground sloths) ate, passed, and dispersed mesquite seeds. The seeds are encased in hard coverings that permit them to survive the passage through an animal and to remain dormant but viable for years in the ground until conditions are suitable for germination. Today, cattle have taken over much of the role of seed dispersal. Because it takes days for seeds to pass through the digestive tracts of domestic animals, seeds are dispersed great distances. Passage through the digestive tract of an animal often seems to improve germination of the seeds, and is also of benefit to the plant in killing certain beetles (bruchids) that specialize on legume seeds. In southern Texas, honey mesquite seedlings have been observed in 75% of cattle dung piles. It has been estimated that in Texas a single cow chip may contain as many as 1000 viable mesquite seeds.
- Velvet mesquite roots penetrate to an astonishing depth of 49 m (160 feet) to reach deep, underground water supplies. Even more impressive, a taproot of honey mesquite 58 m (190 feet) deep has been observed. With such extensive root systems, there may be more wood below ground than above.
- It has been estimated that over 75% of a coyote's diet in late summer in the southwestern United States is mesquite beans.
- It has been claimed that mesquite wood is the most dimensionally stable of all woods in having equal shrinkage values in all directions of only 2%–3%, and in having total shrinkage values (in volume) of 4%–5%. Other fine woods such as oak, cherry, walnut, teak, mahogany, and Indian rosewood have volumetric shrinkage values in the range of 8%–15%. The result of the low shrinkage value is that furniture made from mesquite will have less expansion and contraction when ambient humidity levels change, such as from summer to winter in northern regions. Thus furniture will not crack and drawers will not stick with changing humidity levels.

## KEY INFORMATION SOURCES


Bell, W.H., and Castetter, E.F. 1937. *The utilization of mesquite and screwbean by the aborigines in the American southwest.* University of New Mexico Press, Albuquerque, NM. pp. 55.

Burkart, A. 1976. A monograph of the genus *Prosopis* (Leguminosae subfam. Mimosoideae). *J. Arnold Arb.* 57: 219–249, 450–525.

Cornejo-Oviedo, E., Meyer, J.E., and Felker, P. 1991. Thinning sapling high density mesquite (*Prosopis glandulosa* var. *glandulosa*) stands to maximize lumber production and pasture improvement. *For. Ecol. Manag.* 46: 189–200.

Del Valle, F.R., Escobedo, M., Munoz, M.J., Ortega, R., and Bourges, H. 1983. Chemical and nutritional studies on mesquite beans *Prosopis juliflora*. *J. Food Sci.* 48: 914–919.

Felker, P. 1979. Mesquite—an all-purpose leguminous arid land tree. In *New agricultural crops*. Edited by G.A. Ritchie. Westview Press for the American Association for the Advancement of Science, Washington, DC. pp. 89–132.

Felker, P. 1996. Commercializing mesquite, leucaena, and cactus in Texas. In *Progress in new crops*. Edited by J. Janick. ASHS Press, Alexandria, VA. pp. 133–137.

Felker, P., Meyer, J.M., and Gronski, S.J. 1988. Application of self-thinning in mesquite (*Prosopis glandulosa* var. *glandulosa*) to range management and lumber production. *For. Ecol. Manag.* 31: 225–232.

Folliott, P.F., and Thames, J.L. 1983. *Handbook on taxonomy of* Prosopis *in Mexico, Peru and Chile*. Food and Agriculture Organization, Rome, Italy. pp. 37.

Glendening, G.E., and Paulsen, H.A. 1955. *Reproduction and establishment of velvet mesquite as related to invasion of semidesert grasslands*. U.S. Depatment of Agriculture, Washington, DC. pp. 50.

Hilu, K.W., Boyd, S., and Felker, P. 1982. Morphological diversity and taxonomy of California mesquites (*Prosopis*, Leguminosae). *Madroño* 29: 237–254.

Meyer, R.E. 1971. *Morphology and anatomy of honey mesquite*. U.S. Department of Agriculture, Washington, DC. pp. 186.
Oduol, P.A., Felker, P., McKinley, C.R., and Meier, C.E. 1986. Variation among selected *Prosopis* families for pod sugar and pod protein contents. *For. Ecol. Manag*. 16: 423–431.
Parker, H.W. (Editor). 1982. *Mesquite utilization—papers presented at the symposium on mesquite utilization, Texas Tech University, Lubbock, Texas, October 29 & 30, 1982*. College of Agricultural Sciences, Texas Tech University, Lubbock, TX. pp. 252.
Pasiecznik, N.M. 2001. *The* Prosopis julifora–Prosopis pallida *complex: A monograph*. Henry Doubleday Research Association, Coventry, U.K. pp. 162.
Rogers, K. 2000. *The magnificent mesquite*. University of Texas Press, Austin, TX. pp. 159.
Sawal, R.K., Ratan, R., and Yadav, S.B.S. 2004. Mesquite (*Prosopis juliflora*) pods as a feed resource for livestock—a review. *Asian Australasian J. Animal Sci*. 17: 719–725.
Schuster, J.L. 1969. *Literature on the mesquite (*Prosopis *L.) of North America: An annotated bibliography*. Texas Tech University, Lubbock, TX. pp. 84.
Sharma, I.K. 1981. Ecological and economic importance of *Prosopis juliflora* in the Indian Thar Desert. *J. Econ. Taxon. Bot*. 2: 245–248.
Simpson, B.B. (Editor). 1977. *Mesquite, its biology in two desert scrub ecosystems*. Dowden, Hutchinson & Ross, Stroudsburg, PA. pp. 250.
Sosebee, R.E., and Dahl, B.E. 1980. *Effects of mesquite spraying on other rangeland resources*. Bureau of Land Management, Santa Fe, NM, and Denver, CO. pp. 387.
Texas Agricultural Experiment Station. 1973. *Mesquite: Growth and development, management, economics, control, uses*. Texas Agricultural Experiment Station, Texas A&M University, College Station, TX. pp. 84.
Ward, C.R. 1977. *Annotated checklist of New World insects associated with* Prosopis *(mesquite)*. Agricultural Research Service, U.S. Department of Agriculture, Washington, DC. pp. 115.
Whitson, R.E., and Scifres, C.J. 1980. *Economic comparisons of alternatives for improving honey mesquite-infested rangeland*. Texas Agricultural Experiment Station, Texas A&M University System, College Station, TX. pp. 185.
Wilson, T.B., Webb, R.H., and Thompson, T.L. 2001. *Mechanisms of range expansion and removal of mesquite in desert grasslands of the southwestern United States*. U.S. Department of Agriculture, Forest Service, Rocky Mountain Research Station, Ogden, UT. pp. 23.

## Specialty Cookbooks

Beinhorn, C., and Asher, G. 1986. *Beinhorn's mesquite cookery*. Texas Monthly Press, Austin, TX. pp. 137.
Hidalgo County Historical Museum. 1996. *Mesquite country: Tastes and traditions from the tip of Texas*. 1. Hidalgo County Historical Museum, Edinburg, TX. pp. 286.
Lee, J.W. 1986. *Mesquite: Its roots are deep in the heart of Texas*. Hummer Press, Texas, Reagan Wells/Uvalde, TX. pp. 115.
Powell, J. 1986. *Mesquite cookery*. McGraw-Hill, New York. pp. 148.

Freitus (1980) has 2 recipes based on mesquite, Hall and Hall (1980) have 2, Niethammer (1999) provides 10, and Williamson (1989) has 20. (See the Appendix to this book for details of the publications cited.)

# 62 Mexican Oregano

Family: Verbenaceae (vervain family)

## NAMES

Scientific names: *Lippia graveolens* Kunth

- The name "oregano" is derived from the Latin name for the European oregano plant, *origanum*. European oregano (usually just called oregano) is *Origanum vulgare* L.
- Mexican oregano is also known as Mexican marjoram, Mexican sage, Mexican wild sage, mintweed, redbrush lippia, scented lippia, scented matgrass, and té de pais. The name "wild marjoram" is occasionally used, but is likely to be confused with marjoram (*Origanum majorana* L.) as well as other species. The name "Puerto Rican oregano" is sometimes used, but this is also an ambiguous name, applied to other flavoring herbs such as *Plectrantus amboinicus* (Lour.) Spreng. (commonly known as country borage) and *Lippia micromera* Schauer (also known as Spanish thyme).
- The Mexican name té de pais is Spanish for "country tea," reflecting use as a herbal tea.
- Some other species are sometimes called Mexican oregano, and are used as culinary herbs in Mexico and sometimes in the southern United Sates. These include the following species of the mint family: *Monarda fistulosa* L. var. *menthifolia* (Graham) Fernald, *M. austromontana* Epling, and *Poliomintha longiflora* Gray. Additional species of *Lippia* sometimes harvested from the wild and used as oregano in Mexico and Central America include *L. affinis* Schauer, *L. cardiostegia* Benth., *L. formosa* T.S. Bandeg., *L. fragrans* Turcz., *L. origanoides* HBK, *L. palmeri* Wats., and *L. umbellata* Cav.
- So-called "Spanish oregano" oil is derived from *Coridothymus capitatus* (L.) Rchb. f. (*Thymus capitatus* (L.) Hoffm. & Link.). Spanish oregano oil is used in soaps, cosmetics, lotions, and perfumes; to flavor soft drinks, desserts, and candies; is employed as an antiseptic for pharmaceutical and dental products; and is used to produce an inferior oregano spice.
- The genus *Lippia* is named after Auguste (also Augustin and Augustus) Lippi (1678–1704), a Parisian botanist and physician who was killed in Abyssinia (Ethiopia).
- *Graveolens* in the scientific name *L. graveolens* is Latin for strong-smelling.

## GEOGRAPHY AND ECOLOGY OF WILD PLANTS

Mexican oregano is native to southern North America (reaching as far north as Texas and southern New Mexico, and to be expected in southern California) and occurs in Central America as far south as Nicaragua. It is adapted to scrublands, rocky slopes and hills, arroyos (steep-sided stream beds that are usually dry except after heavy rains), and arid conditions.

## PLANT PORTRAIT

At least 39 species of 16 genera from six plant families are colloquially called oregano, origanum, or some similar name. The name "oregano" is frequently used as a term for a certain spice flavor and aroma rather than for a particular plant species. The flavor is often determined by the relative

**FIGURE 62.1** Mexican oregano (*Lippia graveolens*). (a) Plant and twig with flowers. (Drawing courtesy of S. Rigby, colored by B. Flahey.) (b) Dish of Mexican oregano. (Courtesy of Leslie Seaton [Flickr/CC-attribution].) (c) Bottle of Mexican oregano. (d) Pizza, the predominant oregano dish. (Courtesy of Scott Bauer, U.S. Department of Agriculture.) (e) Distribution map.

concentrations of the chemicals carvacrol and thymol in the essential oil of the plants, the former primarily responsible for the distinctive "oregano" taste. About half of the "oregano" used in the United States is obtained from Mexico and is derived from the dried leaves and flowers of Mexican oregano. The rest generally comes from several species of *Origanum*.

Mexican oregano is a shrub or occasionally a small tree, 1–2.7 m (3.3–9 feet) tall. The small flowers are white or yellowish, often with a yellow eye, and tend to be present throughout the year, but especially after a period of rain. The leaves are up to 6 cm (2.5 inches) long, fragrant, evergreen in their natural habitats, and are borne on hairy or felty, slim branches. The plant was used in pre-Hispanic Mexico for medicinal rather than culinary purposes. Today, Mexican oregano is often sold as true oregano in Mexico and the southern United States. Mexico is the major producer, most of the supply harvested from the wild.

## CULINARY PORTRAIT

The leaves of Mexican oregano have an intense oregano aroma that is preferred by many over European oreganos (i.e., species of *Origanum*). The flavor of Mexican oregano has been described as sweeter and somewhat sharper and more pungent than that of oregano. Mexican oregano is employed in many Mexican-style foods (Mexican and Tex-Mex). It is also added to Italian-style foods in the western United States. However, on the east coast of the United States the European oreganos are usually used in Italian foods. Mexican oregano is added to flavor fish, sausages, meatballs, tomato sauces, and other dishes requiring a strong oregano flavor. The dried leaves are also brewed to make a herbal tea. Mexican oregano has been shown to have the ability to reduce sexual reproductive capacity. The chemical that might contribute to the antifertility, lapachenole, is carcinogenic. Nevertheless, little concern has been expressed to date about the use of Mexican oregano as food.

Oregano is the predominant flavorant of pizza, and indeed both pizza and the flavor of oregano were not popular in North America before World War II. Italian food remained of minor interest in North America until the return of American troops who had fought in Italy and become fond of pizza. Today, pizza is the most popular North American fast food.

In northern regions, Mexican oregano makes an excellent container plant, which can be overwintered indoors on a sunny windowsill. Having a plant available allows one not only to harvest the leaves for use as a culinary herb, but also to use the branches as flavor-enhancing skewers for shishkebabs and other foods, and to strew branches over charcoal to impart an interesting smoked taste to grilled foods.

### CULINARY VOCABULARY

- Among the most popular Mexican foods that are spiced with Mexican oregano are salsas (cold sauces prepared from certain vegetables, especially cilantro, chiles, onions, and tomatoes), rajas (roasted and seasoned strips of chile used as filling for tortillas and quesadillas and as a base for various dishes), pozole (a pork and hominy soup), and adobos (strongly flavored seasoning pastes).

## PROSPECTS

European oregano (*Origanum vulgare*) is the chief competitor of Mexican oregano, but both species have established market niches. Mexican oregano could be grown as a perennial plant in much of the southern United States. It has limited chance of outdoor commercial production in Canada and in the cooler, shorter-season areas of the United States since it is adapted to long-season, warm climates, and can be produced more efficiently in countries such as Mexico. It is hardy as far north as USDA climate zones 9, 10, and 11. It could be grown in northern greenhouses to supply fresh oregano to local specialty shops. A special problem for Mexican oregano is that its natural populations are being significantly overharvested and exterminated in Mexico, so valuable germplasm that may be useful for breeding improved cultivars is in danger of extermination.

## CURIOSITIES OF SCIENCE AND TECHNOLOGY

- In Mexico, folk healing practices are commonly carried out by curanderas (female shamans), and these traditions are also observed sometimes in the southwestern United States. Mexican tea is one of the very common herbs used in curandera, especially as a hot tea for coughs, and for indigestion and stomach gas.
- An ounce (28 g) of oregano can season 500 slices of pizza.
- A legend explains the origin of modern pizza. In 1889, King Umberto and his wife Margherita vacationed in Napoli (Naples), Italy. Pizza at that time was merely white bread

flavored with tomato paste, and was a popular food of the poor masses. To honor the queen, a local baker devised a richer kind of pizza, adding white mozzarella cheese and green basil leaves, thus reflecting the colors of the Italian flag. This became known as pizza Margherita and spread all over Italy and eventually the rest of the world.

- An analysis of foods that were liked and disliked during pregnancy showed a tendency to dislike sauces flavored with oregano; foods craved included ice cream, sweets (particularly chocolate), fruits, and fish (Reference: Hook, E.B. 1978. Dietary cravings and aversions during pregnanacy. *Am. J. Clin. Nutr.* 31: 1355–1362).
- The production of Mexican oregano in Mexico is believed to have increased after the nuclear explosion at Chernobyl of the former USSR (now in Ukraine) in 1986, because of reduced competition from oregano on the international market from Greece and Turkey, as their crops were thought to have been affected by radioactive fallout.

## KEY INFORMATION SOURCES

Borrego-Escalante, F., Bustamante-Garcia, L., Oyervides-Garcia, A., and Navarro-Guerrero, E. 1992. Germination and propagation of the oregano plant (*Lippia berlandieri*). In *Proceedings First International Conference on new industrial crops and products*. Edited by H.H. Naqvi, A. Estilai, and I.P. Ting. Association for the Advancement of Industrial Crops, Riverside, CA. pp. 211–213.

Calvo-Irabien, L.M., Yam-Puc, J.A., Dzib, G., Escalante-Erosa, F., and Pena-Rodriguez, L.M. 2009. Effect of postharvest drying on the composition of Mexican oregano (*Lippia graveolens*) essential oil. *J. Herbs, Spices Med. Plants* 15: 281–287.

Compadre, C.M., Hussain, R.A., Leon, I., and Enríquez, R.G. 1987. Volatile constituents of *Montanoa tomentosa* and *Lippia graveolens*. *Planta Med.* 53: 495–496.

Dominguez, X.A., Sanchez, V.H., Suarez, M., Baldas, J.H., and del Rosario-Gonzalez, M. 1989. Chemical constituents of *Lippia graveolens*. *Planta Med.* 55: 208–209.

Dunford, N.T., and Silva Vazquez, R. 2005. Effect of water stress on plant growth and thymol and carvacrol concentrations in Mexican oregano grown under controlled conditions. *J. Appl. Hort. (Lucknow)* 7: 1, 20–22.

Gonzalez, M.C., Gurrola, N., Mere, A., and Calderon, J. 2003. Production potential of oregano (*Lippia graveolens* H.B.K.) in the State of Durango. Revista Chapingo. *Serie Ingenieria Agropecuaria* 6(1/2): 155–158. (In Spanish.)

Kintzios, S.E. 2002. *Oregano: the genera* Origanum *and* Lippia. Taylor and Francis, New York. pp. 277.

Lin, L.Z., Mukhopadhyay, S., Robbins, R.J., and Harnly, J.M. 2007. Identification and quantification of flavonoids of Mexican oregano (*Lippia graveolens*) by LC-DAD-ESI/MS analysis. *J. Food Comp. Anal.* 20: 361–369.

Martínez-Natarén, D.A., Parra-Tabla, V., Dzib, G., and Calvo-Irabién, L.M. 2011. Morphology and density of glandular trichomes in populations of Mexican oregano (*Lippia graveolens* HBK, Verbenaceae), and the relationship between trichome density and climate. *J. Torrey Bot. Soc.* 138: 134–144.

Martínez-Rocha, A., Puga, R., Hernández-Sandoval, L., Loarca-Piña, G., and Mendoza, S. 2008. Antioxidant and antimutagenic activities of Mexican oregano (*Lippia graveolens* Kunth). *Plant Foods Hum. Nutr.* 63: 1–5.

Ocampo-Velázquez, R.V., Malda-Barrera, G.X., and Suárez-Ramos, G. 2009. Reproductive biology of Mexican oregano (*Lippia graveolens* Kunth) in three exploitation conditions. *Agrociencia* 43: 475–482. (In English and Spanish.)

Ong, H.C., and Brink, M. 1999. *Lippia graveolens* Kunth. In *Plant resources of south-east Asia, 13, spices*. Edited by C.C. de Guzman and J.S. Siemonsma. Backhuys, Leiden, the Netherlands. pp. 137–139.

Osorno-Sánchez, T., Flores-Jaramillo, D., Hernández-Sandoval, L., and Lindig-Cisneros, R. 2009. Management and extraction of *Lippia graveolens* in the arid lands of Queretaro, Mexico. *Econ. Bot.* 63: 314–318.

Pino, J.A., and Borges, P. 1999. Volatile compounds of spices: 9. Mexican oregano. *Alimentaria* 36(301): 83–85. (In Spanish.)

Pino, J.A., Hernandez, I.A., Alvarez, M., and Roncal, E. 1994. The effect of grinding on the quality of Mexican oregano. *Alimentaria* 31(249): 53–54. (In Spanish.)

Pino, J., Rosado, A., Baluja, R., and Borges, P. 1989. Analysis of the essential oil of Mexican oregano (*Lippia graveolens* HBK). *Die Nahrung* 33: 289–295.

Rivero-Cruz, I., Duarte, G., Navarrete, A., Bye, L., Linares, E., and Mata, R. 2011. Chemical composition and antimicrobial and spasmolytic properties of *Poliomintha longiflora* and *Lippia graveolens*. *J. Food Sci.* C309–C317.
Salgueiro, L.R., Cavaleiro, C., Goncalves, M.J., and Proenca da Cunha, A. 2003. Antimicrobial activity and chemical composition of the essential oil of *Lippia graveolens* from Guatemala. *Planta Med.* 69: 80–83.
Uribe-Hernández, C.J., Hurtado-Ramos, J.B., Olmedo-Arcega, E.R., and Matinez-Sosa, M.A. 1992. The essential oil of *Lippia graveolens* H.B.K. from Jalisco, Mexico. *J. Essent. Oil. Res.* 4: 647–649.
Vernin, G., Lageot, C., Gaydou, E.M., and Parkanyi, C. 2001. Analysis of the essential oil of *Lippia graveolens* HBK from El Salvador. *Flavour Fragrance J.* 16: 219–226.

## Specialty Cookbooks

(*Note:* Oregano and Mexican oregano are interchangeable in recipes.)

Belsinger, S., and Wilcox,T.M. 2004. Our favorite oregano in the kitchen. *The Herbarist* 70: 4–10.
Franks, J. 2012. *Oregano greats: delicious oregano recipes: The top 100 oregano recipes*. Emereo Pty Limited, Dayboro, Australia. pp. 118.
Pitzer, S. 1996. *Growing and using oregano*. Storey Publishing, Pownal, VT. pp. 32.

# 63 Mountain Mint

Family: Lamiaceae (Labiatae; mint family)

## NAMES

Scientific name: *Pycnanthemum pilosum* Nutt. (*Pycnanthemum verticillatum* (Michx.) Pers. var. *pilosum* (Nutt.) Cooperr.)

- True mints are species of the genus *Mentha*, most of which are native to Eurasia. The name "mountain mint" for *P. pilosum* is based on the minty smell of the species. Some, but by no means all of the species of *Pycnanthemum* are characteristic of mountainous areas.
- Mountain mint (mountain-mint) has also been called American mountain mint, hairy mountain mint, and pilose mountain mint. The name hoary mountain mint is sometimes used, but is better employed for *P. incanum*, mentioned below. When *P. pilosum* is considered to be a variety of *P. verticillatum* (i.e., *P. verticillatum* var. *pilosum*), it is also called whorled mountain mint.
- The name mountain mint is sometimes also applied to lesser calamint, *Calamintha nepeta* L. (*Clinopodium nepeta* (L.) Kuntze subsp. *nepeta*), a minor flavoring herb of Europe. *Monardella odoratissima* Benth. is also called mountain mint. It is native from Washington to New Mexico and is often grown as a fragrant ornamental.
- The genus name *Pycnanthemum* comes from the Greek *pycnos*, dense, and *anthemon*, blossom, referring to the compact flower heads.
- *Pilosum* in the scientific name *P. pilosum* is Latin for hairy.

## GEOGRAPHY AND ECOLOGY OF WILD PLANTS

*Pycnanthemum pilosum* is found in dry to moist woods, thickets, and clearings of eastern North America. In the north, it reaches Michigan, Ontario, and Maine; in the south, it extends from Arkansas to South Carolina. Mountain mint is considered to be endangered in several states. The plant has been introduced in regions neighboring the main distribution area, including Massachusetts, Connecticut, Pennsylvania, and eastern Quebec. The species is often found in the partial shade of wooded roadsides and in forest openings, and also occupies exposed hillsides. It frequently occurs on soils of low fertility.

## PLANT PORTRAIT

There are about 19 species of *Pycnanthemum*. They are perennial, aromatic herbs, distributed in eastern North America and California. Usually known as mountain mints, these produce a pungent, often mint-like odor when crushed. Several of the species have been used by North American Indians as medicinals and condiments. *Pycnanthemum incanum* Michx., hoary mountain-mint, was used extensively by Native Americans for medicinal purposes. It is sometimes used as a tea by wild food enthusiasts. *Pycnanthemum virginianum* (L.) T. Durand & B.D. Jacks. ex B.L. Rob. & Fernald, Virginia mountain mint or wild basil, grows from Quebec to North Dakota and south to Georgia and Oklahoma. Native North Americans used the buds and fragrant flowers to season soup and meat and the roots and leaves for medicine and to bait mink traps. This species is sometimes cultivated in gardens but is rarely used as a culinary herb. *Pycnanthemum pilosum* is the principal culinary species of the genus.

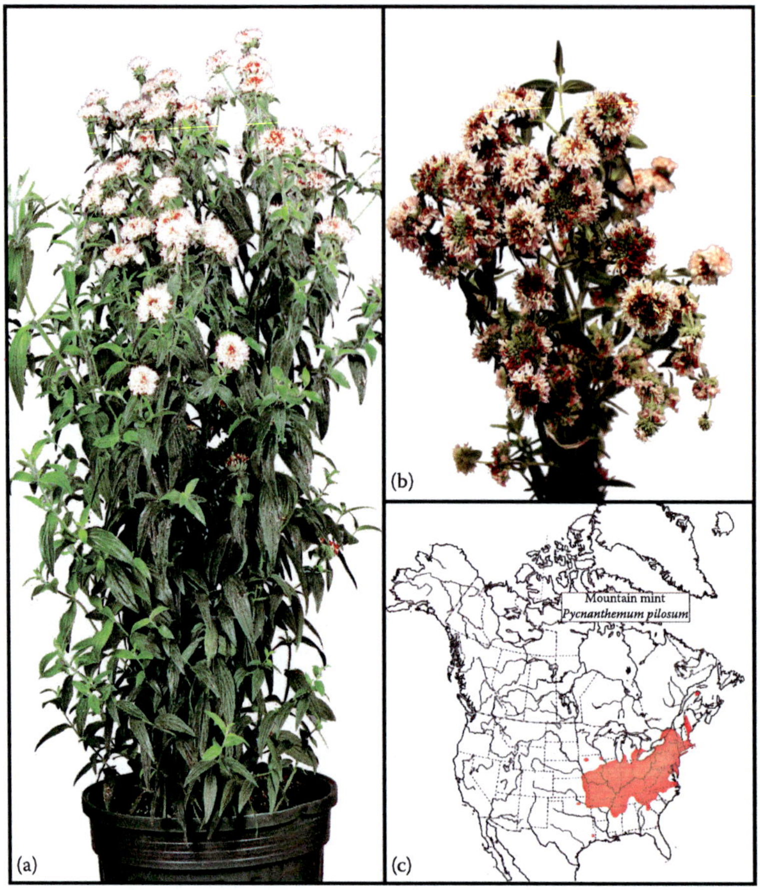


**FIGURE 63.1** Mountain mint (*Pycnanthemum pilosum*). (a) Plant in flower. (Courtesy of Eric Johnson.) (b) Plant in flower. (Courtesy of E. Small and B. Brookes.) (c) Distribution map.

*Pycnanthemum pilosum* is commonly 90–100 cm (about a yard) tall, but sometimes grows up to 150 cm (5 feet) in height. It develops clumps and produces branching stems, some with clusters of small pink flowers and leaves that are very aromatic when bruised. Mountain mint is grown as an ornamental in North America and is a fine nectar plant for honeybees.

## CULINARY PORTRAIT

The fresh and dried leaves are used for tea, the flavor described as mint-like and delicious. Mountain mint jelly as well as other products are occasionally produced as a cottage industry. The presence of (sometimes considerable) amounts of the toxic chemical pulegone in mountain mint (as well as in some other culinary herbs) suggests that large quantities should not be eaten, and the herb should be avoided during pregnancy.

### CULINARY VOCABULARY

- In the bar trade, "juleps" are cocktails made from gin, rum, or distilled spirit, and occasionally flavored with citrus juice. "Mint juleps" are made from bourbon, sugar, and mint, served with finely crushed ice. It has been claimed that American mint juleps were invented in the early nineteenth century. However, sweet herbal beverages called juleps are ancient beverages known in Iran and Arab countries. The word julep comes from the Persian *gulab*, from *gul* + *ab*, rose + water, but eventually julep came to mean any drink sweetened with an aromatic herb.

## PROSPECTS

Mint is one of the world's most important flavors after vanilla and citrus. However, spearmint (*Mentha spicata* L.), peppermint (*M.* ×*piperita* L.), and Scotch mint (*M.* ×*gracilis* Sole) are the predominant sources of mint flavor. Mountain mint is basically a wild plant of no significant commercial importance. Mountain mint jelly is occasionally sold as a cottage industry, as well as other products. The species does not seem to be cultivated commercially as a crop, although as detailed in Pellet (1950), a commercial demonstration of its potential as an oil plant was carried out in 1949, and it has potential as bee forage. The plant does seem well adapted to machine planting and harvesting. Because it is grown as an ornamental and is an indigenous North American plant, it is available for use as a culinary herb in much of North America. Although the commercial value of mountain mint is insignificant at present, it could be developed into a successful source of chemical extractives of value in the marketplace. Carr and Hunter (1973) stated that *P. pilosum* may not only be a valuable crop to bee keepers, but it could give returns comparable to agricultural crops if harvested for its essential oils.

## CURIOSITIES OF SCIENCE AND TECHNOLOGY

- The Mesquaki and Winnebago Indians employed the pungent-smelling mountain mint to scent traps used to catch mink. The rationale was either to serve as an attractant or to mask the odor of previous mink caught in the traps.
- The Mesquaki also used mountain mint as snake bait, chewing some of the leaves and spitting them onto the end of a stick, then holding the stick in front of a snake's mouth. This has been said to be as effective as catching birds by putting salt on their tail feathers, but there might be a rationale. Snakes have an acute sense of smell (the forked tongue picks up chemicals in the air and transfers these to a sensory apparatus, Jacobson's organ, in the roof of the mouth; the same organ is in the human nose). Snakes might be intrigued and distracted by the mint, making them easier to catch. One snake expert commented that the snakes might be just as intrigued by the spit.
- Still another use of mountain mint by the Mesquaki was to make a snuff of the dried leaves and use it for headache. It was often also applied to the nostrils of those about to die in hope of reviving them.
- The Koasati Indians stopped nosebleeds by placing soaked plant material of mountain mint in the nostrils.
- The Choctaw Indians blew a boiled preparation of mountain mint leaves over people who were sick with the expectation that this might cure them.
- The Cherokee Indians drank a boiled preparation of mountain mint while eating slightly immature corn, in order to prevent diarrhea.
- There are about 100,000 florets in one well-grown plant of mountain mint.
- Mountain mints were once prized by Native American tribes because the flower buds were considered effective tenderizers for buffalo meat.

## KEY INFORMATION SOURCES

Ayers, G.S. 1998. The qualities of mountain mint (*Pycnanthemum pilosum*) honey produced in a cage with no other floral source. *Am Bee J.* 138: 367–371.

Carr, R.K., and Hunter, G.E. 1973. Flavonoid studies of *Pycnanthemum*: Labiatae. *J. Tenn. Acad. Sci.* 48(1): 2–5.

Chambers, H.L. 1961. Chromosome numbers and breeding systems in *Pycnanthemum* (Labiatae). *Brittonia* 13: 116–128.
Chambers, H.L. 1993. Chromosome survey and analysis of artificial hybrids in *Pycnanthemum. Castanea* 58: 197–208.
Chambers, H.L., and Chambers, K.L. 1971. Artificial and natural hybrids in *Pycnanthemum* (Labiatae). *Brittonia* 23: 71–88.
Chambers, K.L., and Chambers, H.L. 2008. Infrageneric classification and nomenclatural notes for *Pycnanthemum* (Lamiaceae). *J. Bot. Res. Inst. Texas* 2: 193–199.
Chambers, H.L., and Hamer, J. 1992. More about picky pycnanthemums. *Tipularia* 7(1): 19–23.
Chilquist, V., Gilmour, R., and Jannke, P. 1949. Preliminary report on the volatile oil of *Pycnanthemum pilosum* Nutt. *J. Am. Pharm. Assoc.* 38: 204–206.
Crins, W.J. 1989. Status of the hoary mountain mint, *Pycnanthemum incanum* (Lamiaceae), in Canada. *Can. Field-Nat.* 103: 283–286.
Djao, E.H., and Schwartings, A.E. 1951. Chemical investigation of the volatile oil of *Pycnanthemum pilosum* Nutt. *J. Am. Pharm. Assoc.* 40: 101–103.
Grant, E., and Epling, C. 1943. A study of *Pycnanthemum* (Labiatae). *Univ. Calif. Publ. Bot.* 20: 195–240.
Lawrence, B.M. 1982. Addition to the chemical composition of *Pycnanthemum* essential oils. In *Cultivation and utilization of aromatic plants*. Edited by C.K. Atal and B.M. Kapur. Regional Research Laboratory, Council of Scientific & Industrial Research, Jammu-Tawi, India. pp. 677–680.
Pellett, F.C. 1947. Mountain mint. *Am. Bee J.* 87: 172–173.
Pellett, F.C. 1950. [Mountain] Mint, a new source of essential oil. *Am. Bee J.* 90: 66–67.
Ranger, L.S. 1990. Those picky pycnanthemums. *Tipularia* 5: 11–14.
Sørensen, P.D., and Matekaitis, P.A. 1981. A lemon-scented *Pycnanthemum* (Lamiaceae). *Rhodora* 83: 145–146. [*P. virginianum*]
Widrlechner, M. 1992. A field evaluation of native mint family plants as honeybee forage in Iowa. In *Proceedings of the twelfth North American prairie conference, a vanishing heritage, 5–9 Aug. 1990, Cedar Falls, Iowa.* Edited by D.D. Smith and C.A. Jacobs. University of Northern Iowa, Cedar Falls, IA. pp. 39–42.
Williams, R. 2005. The phylogeny of *Pycnanthemum*: Hybridization and polyploidy in a taxonomically confusing genus. *Herbarist* 71: 34–41.
Wroblewska, A., Ayers, G.S., and Hoopingarner, R.A. 1998. Nectar production dynamics and honey potential of mountain mint (*Pycnanthemum pilosum* Nutt.). *Am. Bee J.* 138: 195–200.

## Specialty Cookbooks

There do not appear to be any cookbooks dedicated to mountain mint. This herb may be used in recipes calling for mint flavor.

# 64 Nettle (Stinging Nettle)

Family: Urticaceae (nettle family)

## NAMES

Scientific name: *Urtica dioica* L.

- American stinging nettle—*U. dioica* subsp. *gracilis* (Aiton) Selander
- European stinging nettle—*U. dioica* subsp. *dioica*
- Hoary nettle—*U. dioica* subsp. *holosericea* (Nutt.) Thorne
- The modern English word "nettle" is derived from the Old English *netele*. The word is based on the Anglo-Saxon *noedl*, needle, referring to the sting of the plant or its use in making thread. The word may have come from older languages, such as from the Old High German *nazza* or the Greek *adikē*. A somewhat different claim is that the word nettle evolved through the German from the Latin *nassa*, fish net (nettle has been called net plant because nettle twine was long used for weaving nets), which in turn is similar to the ancient Sanskrit word *nahyati*, he binds.
- Nettle is also known as common nettle, common stinging nettle, great stinging nettle, and stinging nettle, and much less widely as hidgy pidgy, devils leaf, and net plant. The subspecies *gracilis* is called American stinging nettle, Indian spinach, Lyall's American stinging nettle, slender stinging nettle, slim American stinging nettle, tall nettle, and tall wild nettle. The subspecies *dioica* has been called devil's leaf, European stinging nettle, hidgy pidgy, net plant, and Swedish hemp.
- Some plants that are sometimes called nettles are unrelated to the nettle family. "Dead nettles" are species of the genus *Lamium*, which belongs to the mint family (Lamiaceae). "Hemp nettle," *Galeopsis tetrahit* L., is another member of the mint family. "Horse nettle," *Solanum carolinense* L., is a native herb and troublesome weed of eastern and central North America that belongs to the potato family (Solanaceae).
- In Florida, "stinging nettle" is an ambiguous name, referring not only to *U. dioica* but also to *Cnidoscolus stimulosus* (Michx.) Engelm. & A. Gray of the spurge family (Euphorbiaceae), a plant found in coastal plains from Virginia to southern Florida and west to Oklahoma, Texas, and Louisiana.
- "Sea nettles" are dangerous species of jellyfish, which have sometimes stung swimmers to death.
- The genus name *Urtica* is based on the Latin *urere*, to burn, and refers to the stinging caused by the hairs.
- *Dioica* in the scientific name *U. dioica* means "two houses," a botanical way of saying that male and female structures are separated (at least in Old World subspecies), with male flowers only on one kind of plant and female flowers only on another kind.

## GEOGRAPHY AND ECOLOGY OF WILD PLANTS

Common nettle is a very variable species, with some subspecies originating from Eurasia and others from North America (the classification of the species is controversial). One subspecies is native to and widespread in Eurasia, while two subspecies are native to North America. European stinging

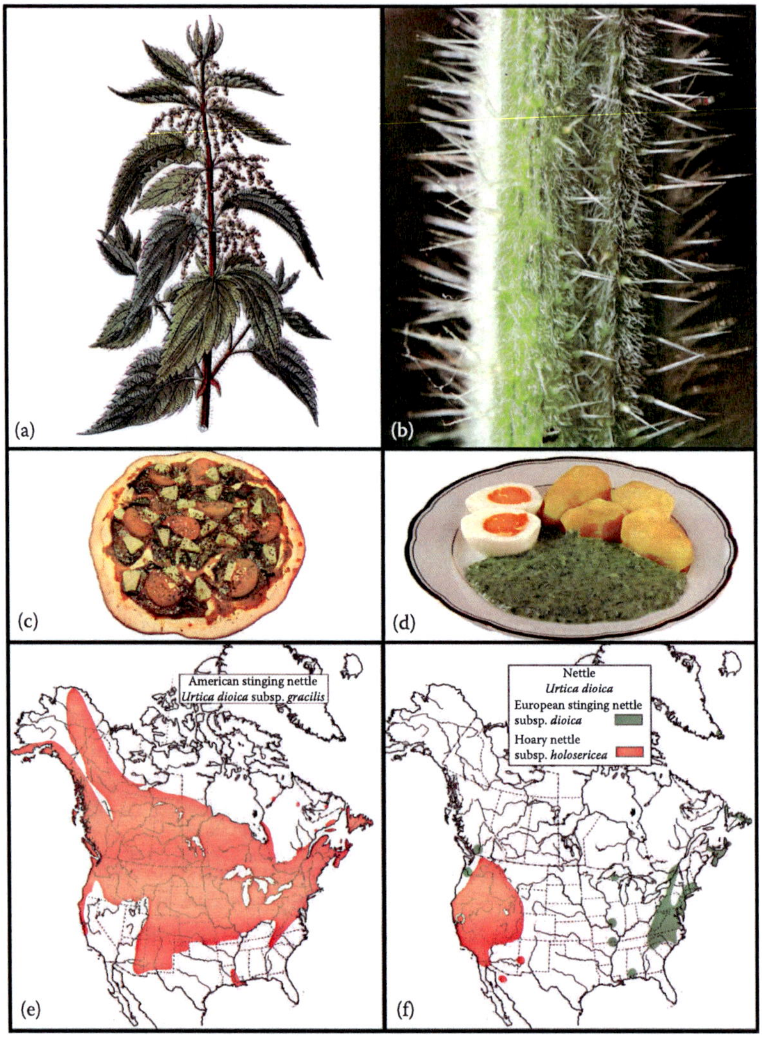


**FIGURE 64.1** Stinging nettle (*Urtica dioica*). (a) Introduced European stinging nettle (*U. dioica* subsp. *dioica*). (From Hallier, E., *Flora von Deutschland* (edition 5 by D.F.L. von Schlechtendal et al.), F.E. Köhler, Gera-Untermhaus, Germany. Vol. 9, Plate 903, 1882.) (b) Close-up of stinging hairs on stem of stinging nettle. (Courtesy of Jerry Kirkhart [Flickr/CC-attribution].) (c) Pizza with topping of stinging nettle. (Courtesy of Kess Vargavind [Flickr/CC-attribution].) (d) Food plate with nettle purée (in green). (Courtesy of Kobako [CC By 2.5].) (e) Distribution map of American stinging nettle (*U. dioica* subsp. *gracilis*). (f) Distribution map; hoary nettle (*U. dioica* subsp. *holosericea*) in red, European stinging nettle in green.

nettle (*U. dioica* subsp. *dioica*) is native to Eurasia, and beginning in the mid-1800s, it has become established along the Atlantic coast of North America. American stinging nettle (*U. dioica* subsp. *gracilis*) is the predominant nettle in North America. Hoary nettle (*U. dioica* subsp. *holosericea*) is native to the western United States, reaching close to the British Columbia border in Washington.

Nettle prefers moist, rich, circumneutral soil and full sun. Although best adapted to moist to wet, open sites, it also occurs in drier areas and in moderate shade such as found in thickets and forest edges. The species is a colonizer, growing well in disturbed areas such as woodland clearings, waste places, vacant lots, and pastures; along ditches, roadsides, and hedgerows; and in periodically flooded areas including shores and marshes. Nettle is listed in some jurisdictions as a noxious weed.

## PLANT PORTRAIT

True "nettles" in a broad sense are species of the nettle family, found chiefly in the tropics and subtropics. The plants of several genera of nettles are covered with bristly, small, stinging hairs that inject an irritant substance under the skin when touched. The stinging hairs are marvels of design. They consist of a long stiff, silicified (hardened with silica) shaft, narrowing toward the point and provided with a small bulbous tip. The shaft just under the tip is not hardened with silica, so a very slight touch breaks off the bulbous end. The tip breaks off at a slant, leaving a finely pointed hollow shaft, like the needle of a hypodermic syringe, which makes a puncture in the skin through which the secretion enters. This results in itchy, raised welts on the skin. On the positive side, the young foliage of many temperate nettles can be eaten as greens, much like spinach. In particular, the perennial herb *U. dioica*, commonly called stinging nettle or just nettle, has been used for food and is the subject of this chapter. Nettles are perennial herbs, growing up to a meter (about a yard) in height or even three times as high in rich soils. They have inconspicuous greenish flowers and coarsely toothed leaves. The plants overwinter and spread by rhizomes (Strictly, "rhizomes" are underground creeping stems, whereas aboveground creeping stems are "stolons." Nettle rhizomes grow shallowly underground but are also sometimes found aboveground.). A rhizome cutting has been observed to grow into a patch covering 3 m (10 feet) in diameter in a single season.

Nettle stems were one of the earliest sources of fiber in Europe, employed to make cloth and occasionally also paper, although such usage was generally replaced with flax and hemp. Nettle fabric has been found in burial sites dating back to the Bronze Age, including a Danish grave with nettle cloth wrapped around cremated bones. Independently of the European use of stinging nettle as a source of fiber, indigenous people of North America also used fiber from the plant for preparing twine, rope, and fishing nets. For centuries, fiber from nettle was frequently used to weave tablecloths and sheets in Scotland, notably fine sheets for Mary, Queen of Scots (1542–1587). The term "nettlecloth" came to be used for all manner of fine fabric in Scotland, whether made from nettle or not. Napoleon's troops were clad in part with nettle fiber. Nettle fiber was used extensively in Germany, Austria, Scotland, and Norway as late as World War I, for cord, fishing line, and sailcloth. During World War I when cotton was in short supply, the uniforms of German soldiers were partly manufactured from nettles (by 1918, over 23,000 ha or 50,000 acres were cultivated in Germany). In modern times, there is interest in resurrecting the fiber use of nettles because the plant can produce large amounts of material on an acreage basis.

Nettle has a long history of use for medicinal purposes, both in the Old and New Worlds. The conditions treated have been quite varied in Europe, including anemia, skin disorders, and allergies, and nettle has also been used as a "blood purifier." According to old folklore, nettles have the ability to prevent baldness, a consequence of the old "Doctrine of Signatures" idea that plants have characteristics that signal their usefulness (in this case, the profusion of hairs on the plant suggested that it could stimulate hair growth). North American indigenous people used nettle to treat acne, eczema, diarrhea, intestinal worms, and urinary tract infections.

Dried nettles completely lose their stinging ability and can be fed to livestock as fodder. In Sweden, Russia, and elsewhere, nettles are in fact used as livestock feed. However, few animals will eat fresh nettle.

Nettles have also been used as human food wherever they grow, but for the most part only to a limited extent. Nettle-eating contests in the United Kingdom (based on who can eat the longest total length of the plants) have given the plant some notoriety.

## CULINARY PORTRAIT

*To nettle* means to aggravate, annoy, bother, bug, chafe, disturb, exasperate, fret, gall, irk, irritate, peeve, provoke, put out, rile, ruffle, vex, get in one's hair, get on one's nerves, and get under one's skin. It is not surprising, therefore, that the nettle has an image problem that discourages its use as food. Nevertheless, nettle does have culinary virtues. Nettle is one of the traditional "five bitter herbes" (the others are horseradish, coriander, horehound, and lettuce) eaten by Jews during the 8 day Passover to symbolize the bitterness of their enslavement by the ancient Egyptians. Young, fresh leaves are cooked like spinach. Leaves are best picked (wearing leather, rubber, or plastic-coated gloves, long sleeves, and long pants) before the plants begin to flower, as they become progressively coarse (it is recommended that shoots be harvested when less than 30 cm or 1 foot in length). Boiling removes the irritation ability of the plant, as indeed does drying, although dried plants are not nearly as tasty. The young stalks are nutritious, high in vitamins A and C, and protein, and they have more iron than spinach, with a very similar taste. Nettle can be used in quiches, stir fries, and soups. Nettle soup, made of some combination of nettle with potatoes, leeks, watercress, cabbage, or beans, is particularly recommended. Nettle also goes well braised and seasoned with onions, garlic, and nutmeg. Nettle beer and pudding are occasionally made. Wild-growing plants can be harvested (one should be certain of the identification before consuming), or plants can be grown in home gardens from seeds obtained from seed supply houses. Nettle has been a relatively popular food plant in England but not in North America. However, stinging nettles have turned up on the menus of some American gourmet restaurants. It should be kept in mind that nettles are used in herbal medicine for its diuretic (urine-stimulating) property, and consuming nettles is likely to result in exercising one's kidneys.

### Culinary Vocabulary

- *Satuk* is a Himalayan nettle leaf soup, typically flavored with black pepper and ginger, and perhaps with other ingredients such as stock, honey, and milk.

## PROSPECTS

There is some interest in Europe and North America in developing a commercial crop of nettles. In fact, nettle is sometimes commercially cultivated for medicinal and cosmetic extracts, although it is mostly collected from the wild. Recent scientific studies have suggested that nettle "root" (i.e., rhizome) extracts may be useful for relief of benign prostatic hyperplasia (enlargement of the prostate gland not due to cancer). It has been shown that constituents in the root affect sex hormones, and this may be related to the effect on the prostate. Commercial preparations of nettle tea have gained popularity for weight loss and maintenance. However, the principal commercial development of the plant has been as a fiber source.

Cultivation of European stinging nettle (subspecies *dioica*) has a long history in Europe for fiber production. During the first half of the twentieth century, the German fiber plant specialist Gustav Bredemann, working in Hamburg, selected forms of stinging nettle with greatly increased content of fiber (up to 15% of the dry stem weight, it has been claimed, although a trial in 1995 using 30 Bredemann selections achieved less than 13% for the best selection). Wild plants, by comparison, contain only about 5% fiber. Male plants have been found to be slightly greater in productivity than female plants, an important consideration since clones (cuttings) rather than seeds are often being used currently to establish the crop. The plants are moderately resistant to insects and diseases (not surprising since the most productive forms are little changed from the wild) but require high amounts of nitrogen and phosphorus to achieve good yield. Modern commercial methods for extracting and preparing the fiber are still at the experimental stage, and stinging nettle is not yet a profitable fiber crop. The textile potential of the plant is currently receiving considerable attention, particularly in Austria, Germany, Finland, and the United Kingdom. The Italian fashion house Corpo Nove has marketed jeans and jackets made from nettle. In Germany and central Europe, the species is considered to be a promising candidate for sustainable production of natural fiber.

Nettle is high in protein (crude contents of 20%–25% have been reported). Accordingly, it has potential for industrial and livestock feed applications. Although nettle has quite limited prospects of being developed into a significant salad food, the fact that there are good reasons to domesticate it for textiles and fodder means that the species also has some prospect for human food use.

## CURIOSITIES OF SCIENCE AND TECHNOLOGY

- Roman soldiers are reported to have carried nettle in the cold (by Roman standards) climate of Britain as an irritant to keep their legs warm, flogging their legs and arms with the foliage to keep their circulation going. These reports are of uncertain validity.
- Corinthian architecture, considered the most ornate of classical Greek architectural styles, is characterized especially by bell-shaped capitals (i.e., the heads or tops of the supporting columns). It has been claimed that nettle leaves were the model for the Corinthian capitals.
- In Daniel Moerman's monumental book detailing uses of plants by North American Indians (*Native American Ethnobotany*, 1998, Timber Press), stinging nettle is listed as one of the ten plants with the greatest number of uses. Among the uses were the following: The Iroquois combined stinging nettle and dried snake's blood as a "witching medicine." The Makah rubbed stinging nettle on their bodies for several purposes: to give them strength to hunt whales and seals and as a purification ritual after handling a corpse; they also rubbed the plants on their fishing lines to improve their catch (it is possible these practices were beneficial by lessening human odors). The Thompson applied a decoction of stinging nettle root to treat hemorrhoids (which may seem like a good way to make a painful condition even more painful but has some logic in the counterirritant value of the plant).
- A common remedy dating to the Middle Ages in Europe, also used by Native American healers, and still occasionally practiced, is urtification: flogging with nettles, particularly for joint problems. Modern research has shown that the nettle irritant increases blood flow and stimulates the production of cortisone, which may counteract the inflammation associated with diseases such as rheumatism.
- Young nettle shoots were once consumed as a remedy for scurvy (caused by a deficiency of vitamin C), a reasonable use since the plants are high in vitamin C.
- It was once thought that apple trees that grow near nettles produce bigger apples. Because nettles often grow in rich soil, this may actually be true.
- Nettle juice was once used to curdle milk.
- In 1942 in England, 90 tons of dried nettle were used for military purposes: to manufacture a green dye for camouflage as well as a green tonic. The green color is due to nettle's high content of chlorophyll, which is evident when nettle is cooked, the green color of the leaves persisting for some time. Nettle has been used as a commercial source of chlorophyll, although there are numerous other species employed for this purpose today.
- Male gorillas are the largest of the apes, reaching a height of 1.5–2 m (5–6.6 feet) and an arm spread of 2.7 m (9 feet), and weighing about 200 kg (450 pounds) in the wild and sometimes more than 270 kg (600 pounds) in zoos when they become obese. (Notwithstanding the expression "800 pound gorilla", animals this large have not been recorded.) Despite their fearsome size and appearance, they are vegetarians, living on a variety of plants, including nettles.
- "Urticaria" (also known as "nettle rash") is a term used in medicine to indicate reddish, itchy swelling on the skin similar to the result of contact with stinging nettle, regardless of the cause (most of the time the cause is not exposure to nettles, but to other allergens). A synonym is "hives." Surprisingly, nettle has been used to treat nettle rash! (The herb has also been used to treat hayfever, in part caused by nettle pollen!) Several other words bear testimony to the stinging power of the genus *Urtica*: urticate (to sting or produce a rash), urticant (producing itching or stinging), and urtication (an itching and stinging sensation). "Urticose" is descriptive of an area with abundant nettles.
- In Berkshire, England, the following verse was recommended to relieve the sting of nettles:
  Out nettle, In dock;
  Dock shall hev, A new smock
  Nettle shan't, Ha'narrun.

In fact, the juice of dock plants (species of *Rumex*) is used as an outdoor, first-aid antidote to the sting of nettles. Docks tend to grow in the same places as nettles, and as dock juice is basic, it will relieve the stinging (as noted below, liquid that is basic [i.e., with a pH > 7] will neutralize the formic acid responsible for the painful reaction to nettle).

- The stinging sensation from contact with stinging nettle hairs has been attributed primarily to the presence of formic acid. However, while formic acid is present, recent research has indicated that the principal stinging constituents identified to date are chemicals called histamine, acetylcholine, and 5-hydroxytryptamine (serotonin), stored at the base of the hairs. Formic acid is the chemical warfare weapon that ants have in their saliva glands and bees have in their stingers, while the other chemicals are in some snake venoms. The degree and length of itchiness depends on an individual's sensitivity, some suffering for as long as 24 hours, others for less than an hour. Formic acid can be neutralized by mixing it with a base, such as a paste of baking soda made with a little water. Human spit tends to be slightly basic and may help. The juice of nettle is also the antidote for its own sting, although care is required to obtain it. Other remedies include many common toothpastes, dilute alcohol, and water.
- The Clairol Company has used more than 40 tons of nettles in some years in hair conditioners. Nettles have a strong reputation for making hair soft and shiny and for eliminating dandruff.
- Nettles are a basic food of several attractive butterflies, both in Europe and the Americas. Some knowledgeable gardeners keep native plants such as nettles in a corner of their garden to attract butterflies and other pleasant animals.

## KEY INFORMATION SOURCES


Bassett, I.J., Crompton, C.W., and Woodland, D.W. 1977. The biology of Canadian weeds. 21. *Urtica dioica* L. *Can. J. Plant Sci.* 57: 491–498.

Bombardelli, E., and Morazzoni, P. 1997. *Urtica dioica* L. *Fitoterapia* 68: 387–402.

Boot, R., Raynal, D.J., and Grime, J.P. 1986. A comparative study of the influence of drought stress on flowering in *Urtica dioica* and *Urtica urens*. *J. Ecol.* 74: 485–495.

Boufford, D.E. 1997. *Urticaceae Jussieu,* Nettle family. In *Flora of North America North of Mexico*. Volume 3. Edited by Flora of North America Editorial Committee. Oxford University Press, Oxford, UK. pp. 400–413.

Geltman, D.V. 1986. Systematic and ecological-geographic characteristics of species from those related to *Urtica dioica* (Urticaceae) in the flora of the USSR. *Bot. Zh. (Leningrad)* 71: 1480–1490. (In Russian, English summary)

Great Britain Ministry of Agriculture and Fisheries. 1953. *Stinging nettles*. Revised edition. H.M.S.O, London, UK. pp. 5.

Guil-Guerrero, J.L., Rebolloso-Fuentes, M.M., and Torija-Isasa, M.E. 2003. Fatty acids and carotenoids from stinging nettle (*Urtica dioica* L.). *J. Food Compos. Anal.* 16: 111–119.

Gulcin, I., Kufrevioglu, O.I., Oktay, M., and Buyukokuroglu, M.E. 2004. Antioxidant, antimicrobial, antiulcer and analgesic activities of nettle (*Urtica dioica* L.). *J. Ethnopharmacol.* 90: 205–215.

Hara, T., and Srutek, M. 1995. Shoot growth and mortality patterns of *Urtica dioica*, a clonal forb. *Ann. Bot.* 76: 235–243.

Holter, V. 1979. Distribution of *Urtica dioica* L. in relation to edaphic factors. *Plant Soil* 52: 599–603.

Kavalali, G.M. 2003. Urtica: *Therapeutic and nutritional aspects of stinging nettles*. Taylor & Francis, London, UK. pp. 83.

Koch, E. 2001. Extracts from fruits of saw palmetto (*Sabal serrulata*) and roots of stinging nettle (*Urtica dioica*): viable alternatives in the medical treatment of benign prostatic hyperplasia and associated lower urinary tracts symptoms. *Planta Med.* 67: 489–500.

Luna, T. 2001. Propagation protocol for stinging nettle (*Urtica diocica*). *Native Plants J.* 2: 110–111.

Manandhar, N.P. 1999. Botany and ethnobotany of *Urtica dioica* L. (stinging nettle). *J. Econ. Taxon. Bot.* 23: 38–40.

Miller, N.G. 1971. The genera of the Urticaceae in the Southeastern United States. *J. Arnold Arb.* 52: 40–68.

Mitich, L.W. 1992. The nettles. *Weed Technol.* 6: 1039–1041.
Mutikainen, P., and Walls, M. 1995. Growth, reproduction and defence in nettles: Responses to herbivory modified by competition and fertilization. *Oecologia* 104: 487–495.
Pollard, A.J., and Briggs, D. 1982. Genecological studies of *Urtica dioica* L. I. The nature of intraspecific variation in *Urtica dioica. New Phytol.* 92: 453–470.
Pollard, A.J., and Briggs, D. 1984. Genecological studies of *Urtica dioica* L. III. Stinging hairs and plant-herbivore interactions. *New Phytol.* 97: 507–522.
Pullin, A.S., and Gilbert, J.E. 1989. The stinging nettle, *Urtica dioica*, increases trichome density after herbivore and mechanical damage. *Oikos* 54: 275–280.
Schofield, J.J. 1998. *Nettles: Both potent herb and delectable food, this common "weed" can ease allergies, prostate enlargement and more.* Keats Pub, New Canaan, CT. pp. 89.
Small, E., and Catling, P.M. 2003. Poorly known economic plants of Canada—38. Stinging nettle, *Urtica dioica* L. *Bull. Can. Bot. Assoc.* 36(3): 34–39.
Srutek, M. 1995. Growth responses of *Urtica dioica* to nutrient supply. *Can. J. Bot.* 73: 843–851.
Srutek, M. 1997. Growth responses of *Urtica dioica* L. to different water table depth. *Plant Ecol.* 130: 163–169.
Srutek, M., and Teckelmann, M. 1998. Review of biology and ecology of *Urtica dioica. Preslia* 70: 1–19.
Swan, C. 1980. *Nettles: Healers of the wild: Their therapeutic, cosmetic and culinary use.* Thorsons, Willingborough, UK. pp. 64.
Thurston, E.L. 1974. Morphology, fine structure, and ontogeny of the stinging emergence of *Urtica dioica. Am. J. Bot.* 61: 809–817.
Tuberville, T.D., Dudley, P.G., and Pollard, A.J. 1996. Responses of invertebrate herbivores to stinging trichomes of *Urtica dioica* and *Laportea canadensis. Oikos* 75: 83–88.
Vassilyev, A.E. 1994. Developmental and comparative ultrastructure of glandular hairs in the two Urticaceae, 1: *Urtica dioica. Nordic J. Bot.* 14: 531–545.
Weib, F. 1993. Effects of varied nitrogen fertilization and cutting treatments on the development and yield components of cultivated stinging nettles. *Acta Hortic.* 331: 137–144.
Woodland, D.W. 1982. Biosystematics of the perennial North American taxa of *Urtica*. II. Taxonomy. *Syst. Bot.* 7: 282–290.
Woodland, D.W., Bassett, I.J., and Crompton, C.W. 1976. The annual species of stinging nettle (*Hesperocnide* and *Urtica*) in North America. *Can. J. Bot.* 34: 374–383.
Woodland, D.W., Bassett, I.J., Crompton, C., and Forget, S. 1982. Biosystematics of the perennial North American taxa of *Urtica*. I. Chromosome number, hybridization, and palynology. *Syst. Bot.* 7: 269–281.

## Specialty Cookbooks

Jumbalaya, J. 2003. *The essential nettle, dandelion, chickweed & thistle cookbook.* Johnny Jumbalaya, UK. pp. 100.

Berglund and Bolsby (1971) have four nettle recipes, Brill (2002) has 11, Freitus (1980) has five, Gray (2011) has one, Hahn (2010) has three, Hunt (1977) has two, Knutsen (1975) has three, Schufer (2011) has one, Szczawinski and Turner (1978) have five, and Williamson (1995) has 12. Also see Kavali (2003), Schofield (1998), and Swan (1980), cited above and Stewart (2002) for information on the culinary preparation of nettles. (See the Appendix to this book for details of the publications cited.)

# 65 Nodding Onion

Family: Liliaceae (lily family; sometimes placed in Alliaceae, the onion family)

## NAMES

Scientific name: *Allium cernuum* Roth

- The name "nodding onion" is based on the onion-like bulb and the drooping top of the flower stalk.
- Nodding onion is also known as lady's leek, nodding pink onion, and nodding wild onion, and also by the name wild onion, used for several other species of *Allium*.
- The genus name *Allium* is the classical Latin name for garlic (*A. sativum* L.).
- *Cernuum* in the scientific name is Latin for nodding.

## GEOGRAPHY AND ECOLOGY OF WILD PLANTS

*Allium cernuum* is the most widely distributed of the almost 100 species of *Allium* in North America and north of Mexico. In Canada, it is found in the southern half of British Columbia; southern and western Alberta; and rarely in Saskatchewan, Manitoba, and southwestern Ontario. In the United States, it is widespread, except in the southwest, southeast, and central tier of states. It also extends into northern Mexico. The species grows from coastal areas up to subalpine locations, in a variety of habitats, including woods, rocky outcrops, slopes, cliffs, open thickets, and prairies. It grows well in dry soils but can also be found in moist ground.

## PLANT PORTRAIT

Nodding onion is a perennial, onion-smelling herb made up of a rosette of grass-like, slim, elongated leaves arising near the ground from a slender, conical, grayish or brownish bulb covered with a membranous layer. Typically, a cluster of bulbs occurs at the base of a plant, the bulbs 1–2 cm (0.4–0.8 inch) in diameter. A clump of plants is frequently developed from the cluster of bulbs at the base of a given plant. Each bulb produces a flowering stalk 10–50 cm (4–20 inches) tall, terminating in an umbrella-like group of bell-shaped white or rose flowers, the stalk curving backward at the top so the bunch of flowers faces the ground. The flowers transform into papery, black capsules about 4 mm (0.16 inch) long, with small, dull-black seeds. Ornamental garden cultivars of nodding onion are available, typically with rose-pink flowers. Like most other *Alliums*, nodding onion provides a good display in beds, borders, and rock garden.

## CULINARY PORTRAIT

The common or true onion, *A. cepa* L., is employed to prepare countless dishes, both hot and cold, and provides a model for ways to utilize nodding onion. Common onion bulbs as well as the chopped leaves are used raw or cooked in many vegetable dishes to which they impart an added zest. Onions are boiled, fried, stewed, baked, creamed, roasted, and pickled. They are deep-fried as onion rings. Onions are both a vegetable (i.e., eaten in considerable quantity) and a condiment or spice (i.e., used in small quantities as a flavorant).

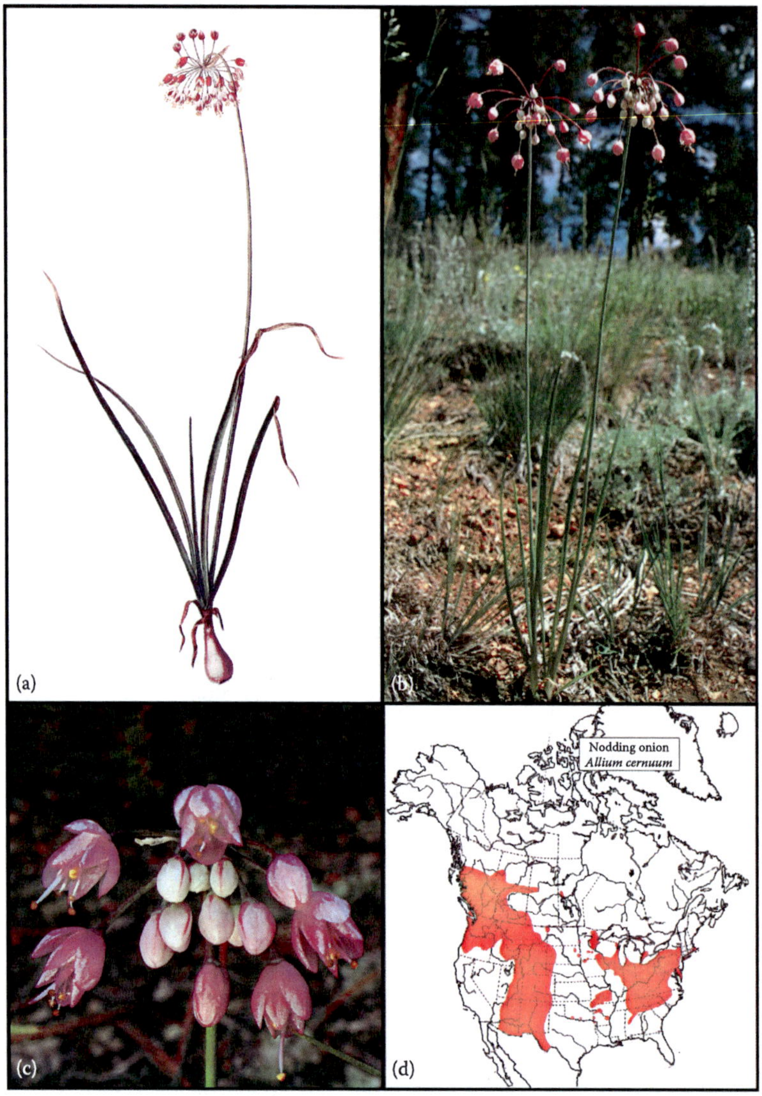


**FIGURE 65.1** Nodding onion (*Allium cernuum*). (a) Painting. (From Redouté, P.J., *Les liliacée*. Redouté, Paris, France. Vol. 6, Plate 345, 1811–1812.) (b) Plants growing in woods. (Courtesy of Dave Powell, U.S. Department of Agriculture Forest Service, online at Bugwood.org [CC By 3.0].) (c) Flowering head. (Courtesy of Alan Cressler, reproduced with permission.) (d) Distribution map.

Raw, cooked, and dried nodding onion bulbs were widely consumed by North American indigenous people and European settlers. The edible bulb has a strong onion taste and is often used raw to add taste to prepared foods, especially salads. The edible flowers are strong-tasting and also make a good addition to salads. The bulbs make an excellent pickle. Cooking reduces the taste and odor of the bulbs. Inulin, which is difficult to digest, is a major storage carbohydrate of nodding onion, and cooking transforms much of it to sweet, much more digestible fructose.

*Allium* species release irritating, tear-producing vapors when cut, and this can be significant when large amounts are being prepared. The tears are due to the rupture of tissues that release sulfurous molecules which react with air to create the irritant chemical allyl sulfate. Using a sharp knife and keeping the face as far away as possible, or wearing goggles or eyeglasses can be helpful. Some cooks cut

*Allium* bulbs under a stream of cold water or cool them for an hour in the refrigerator or 15 minutes in the freezer before cutting. Blanching for several minutes in boiling water, followed by soaking in cold water immediately afterwards to stop the cooking process, also moderates the sharpness of raw bulbs.

There are other cookery precautions that should be kept in mind for *Allium* bulbs. Juice may be released when they are cut and could be absorbed by countertops and wooden cutting surfaces, as well as by hands. Lemon juice or vinegar is sometimes used to remove this juice. The odor of *Allium* bulbs stored in a refrigerator tends to be absorbed by other foods.

## WARNING

See the warnings in Chapter 23 on Canada Garlic regarding hazards associated with *Allium* species.

### Culinary Vocabulary

- A "rope of onions" is a customary way of referring to a group of onions (similarly, one refers to a rope or string of pearls).
- "Skunk egg" is an obsolete cowboy name for an onion.

## PROSPECTS

Nodding onion has not been domesticated, is not yet significantly cultivated for food, and faces considerable established competition in the marketplace. There are numerous commercial *Allium* species (onion, garlic, leek, etc.) and hundreds of species of the genus around the world with edible potential. These considerations suggest the chances of nodding onion becoming a commercially significant new crop are very limited. Nevertheless, the species has exceptionally good taste, and the fact that it is the most widespread of the almost 100 North American wild *Allium* species indicates that it has the kind of natural vigor that makes plants suitable for domestication. Because the genus *Allium* is so important commercially, there are numerous researchers with expertise on its species and a considerable body of knowledge already available that can facilitate development of nodding onion. All factors considered, developing nodding onion for the marketplace is a very good idea.

## CURIOSITIES OF SCIENCE AND TECHNOLOGY

- The strong odor of nodding onion determined the name given to it by several native peoples of North America. The Gitksan name translates as "Raven's underarm odor" and the Carrier name was "stink-grass."
- In 1778, Captain James Cook sailed into Nootka Sound on the southwestern coast of Vancouver Island. His vessel *Resolution* was the first British ship to visit the region. The crew found "wild garlick" there and used it for food. The species was undoubtedly nodding onion.
- The juice of nodding onion has been slathered on skin to repel mosquitoes and other biting insects.
- Jainism is a religious sect that originated in India in the sixth century BC. There are more than 2 million Jains, mostly in India. The vegetarian dietary code is demanding—no eating at night (animals must be respected, and tiny flying creatures could be inadvertently eaten), no consumption of parts of plants that are essential to their lives (such as roots), and no eating of any member of the onion family.

## KEY INFORMATION SOURCES

(For additional information, see Chapter 23 on Canada Garlic and Chapter 97 on Wild Leek.)

Asarkova, A.F., Stikhin, V.A., Cherkasov, O.A., and Miasuradze, N.I. 1983. Diosgenin from *Allium nutans* and *Allium cernuum*. *Chem. Nat. Compounds* 19: 621.

Chester, M., Sykorova, E., Fajkus, J., and Leitch, A.R. 2010. Single integration and spread of a *Copia*-like sequence nested in rDNA intergenic spacers of *Allium cernuum* (Alliaceae). *Cytogenet. Genome Res.* 129(1/3): 35–36.
Choi, H.J., and Cota-Sánchez, J.H. 2009. Nodding onion (*Allium cernuum*), a rare plant in Saskatchewan. *Blue Jay* 67: 249–252.
Dubouzet, J.G., and Shinoda, K. 1998. Phylogeny of *Allium* L. subg. *Melanocrommyum* (Webb et Berth.) Rouy based on DNA sequence analysis of the internal transcribed spacer region of nrDNA. *Theor. Appl. Genet.* 97: 541–549.
Friebe, B. 1989. Nucleolar activity of B-chromosomes in *Allium cernuum* (Alliaceae). *Plant Syst. Evol.* 163: 87–92.
Grun, P. 1958. Cytological instability of accessory chromosomes from native populations of *Allium cernuum*. In *Proceedings of the X International Congress of Genetics, McGill University, Montreal Canada, Aug. 20–27, 1958, Vol. 2.* Southam Printing Company, Montreal, QC. pp. 106–107.
Grun, P. 1959. Variability of accessory chromosomes in native populations of *Allium cernuum*. *Am. J. Bot.* 46: 218–224.
Jacobsen, T.D. 1979. Numerical analysis of variation between *Allium cernuum* and *Allium stellatum* (Liliaceae). *Taxon* 28: 517–523.
Jacobsen, T.D. 1980. The neotypification of *Allium cernuum* Roth (Liliaceae). *Taxon* 29: 150–151.
Murphy, J.B. 1949. Flower development of *Allium cernuum* Roth. *Am. J. Bot.* 36: 800–801.
Ricroch, A., Yockteng, R., Brown, S.C., and Nadot, S. 2005. Evolution of genome size across some cultivated *Allium* species. *Genome* 48: 511–520.
Saghir, A.R.B., Mann, L.K., Ownbey, M., and Berg, R.Y. 1965. Composition of volatiles in relation to taxonomy of American alliums, Liliaceae. *Am. J. Bot.* 53: 477–484.
Schuett, E.M., and Vamosi, J.C. 2010. Phylogenetic community context influences pollen delivery to *Allium cernuum*. *Evol. Biol.* 37: 19–28.

## Specialty Cookbooks

(*Note:* there are numerous cookbooks featuring onions; the following provide numerous recipes. One may experimentally substitute nodding onions for regular onion in recipes combining various ingredients, but the amount should be reduced.)

Bailey, L. 1995. *Lee Bailey's onions*. Clarkson Potter, New York, NY. pp. 80.
Bothwell, J. 1976. *The onion cookbook*. Dover Publications, New York, NY. pp. 166.
Cavage, B. 1987. *The elegant onion: The art of allium cookery*. Storey Communications, Pownal, VT. pp. 156.
Ciletti, B.J. 1998. *The onion harvest cookbook*. Taunton Press, Newtown, CT. pp. 169.
Cool, J.Z. 1995. *Onions*. Collins Publishers San Francisco, CA. pp. 95.
Dille, C., and Belsinger, S. 1996. *The onion book: A bounty of culture, cultivation, and cuisine*. Interweave Press, Loveland, CO. pp. 96.
Glover, B. 2002. *Know your onions*. Southwater, London, U.K. pp. 128.
Greig, D. 1990. *Oh, for an onion!* Kangaroo Press, Kenthurst, NSW, Australia. pp. 80.
Griffith, L., and Griffith, F. 1994. *Onions, onions, onions*. Chapters Publication, Shelburne, VT. pp. 384.
Mendelsohn, O.A. 1966. *A salute to onions; some reflections on cookery and cooks*. Hawthorne Books, New York, NY. pp. 190.
Moon, R. 2000. *Onions, onions, onions*. Firefly Books, Willowdale, ON. pp. 144.
Roberts-Dominguez, J. 1996. *The onion book*. Doubleday, New York, NY. pp. 339.
Rogers, M.R. 1995. *Onions: A celebration of the onion through recipes, lore, and history*. Addison-Wesley Pub. Co., Reading, MA. pp. 193.

Niethammer (1999) has three recipes specifically for nodding onion. Berglund and Bolsby (1971) have 8 recipes for preparing wild onions, and Freitus and Haberman (2005) have 10. Angier (1969) provides general advice on the culinary preparation of wild *Allium* species. (See the Appendix to this book for details of the publications cited.)

# 66 Northern Gooseberry

Family: Grossulariaceae (currant family; traditionally placed in the Saxifragaceae, the saxifrage family, but recently the genus *Ribes* has been segregated as the only genus in the Grossulariaceae)

## NAMES

Scientific name: *Ribes hirtellum* Michx.

- The northern gooseberry is also called hairystem gooseberry, hairy gooseberry, or smooth gooseberry, depending on the degree of hairiness of the variety.
- Other names for the northern gooseberry include American gooseberry, Canada gooseberry, swamp gooseberry, and wild gooseberry.
- The name "gooseberry" is of obscure origin and has been variously interpreted. It may be a corruption of the German *Krausbeere*, crisped (wrinkled) berry. Another view is that the term is a corruption of the German *Jansbeeren*, John's berry, so named because it ripens during the Feast of St. John. *Jansbeeren* is said to have changed to the German *Gansbeeren*, which was translated into English as "gooseberry" because *gans* means "goose" in German. It might seem that the word could have arisen because geese eat gooseberries, but the birds are not fond of the fruit. The name has also been said to simply reflect use of the berries as a seasoning for geese, although this is not likely the actual basis for the word. However, the sixteenth century English names *grosiers* and *groser bush* may be derived from the French *groseille*, or the sixteenth century botanical Latin *grossula*, used for gooseberries or currants, and one of these names may simply have evolved into "gooseberry."
- In Great Britain, the gooseberry was called the "squinancy berry" (quinsyberry) because of its common use, as a jelly, in treating quinsy (peritonsillar abscess; quinsy is an obsolete medical term, derived from the Latin *cynanche*, sore throat).
- An early name for the gooseberry, used through the Middle Ages in some regions of England, was "feaberry" or "feverberry." The fruit had a reputation for its cooling property and an ability to control fevers. Remarkably, the same type of application, using American rather than European gooseberries, developed in America. North American Indians in the Rocky Mountains ate what they called porridge of cooked gooseberries for fever and for chills.
- So common was the use of gooseberry for brewing in Scotland that the fruit was called "the Scottish hairy grape" (European gooseberries tend to have a hairy or fuzzy fruit surface.)
- In the eighteenth century in Europe, gooseberry and currant shrubs were often referred to as "bush trees" or frequently simply as trees. This is because the plants were commonly pruned in a standard (tree-like) shape on a short trunk.
- Several quite unrelated species (i.e., plants that are not related to *Ribes*) have "gooseberry" in their names. For example, "Chinese gooseberry" is kiwi (*Actinidia deliciosa* (A. Chev.) C.F. Liang & A.R. Ferguson); "gooseberry gourd" is the West Indian gherkin (*Cucumis anguria* L.); cape gooseberry and dwarf cape gooseberry are species of *Physalis*; the Indian gooseberry is *Phyllanthus emblica* L. (a tropical fruit plant better known as emblic); and the Otaheite gooseberry is *Phyllanthus acidus* (L.) Skeels, another tropical fruit.
- "Jostaberries" are hybrids of the Eurasian black currant (*R. nigrum* L.) and the American gooseberry, *R. hirtellum*, produced in Germany during the1930s to the 1950s. They have

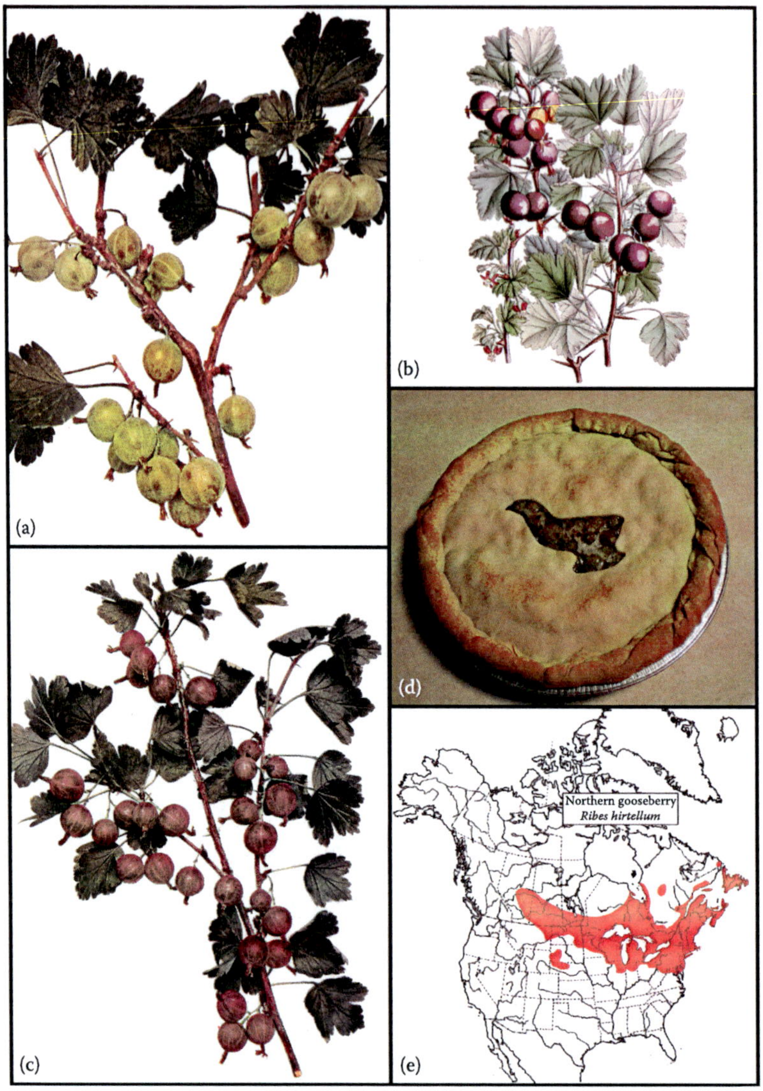


**FIGURE 66.1** Northern gooseberry (*Ribes hirtellum*). (a) Cultivar 'Downing'. ([a] and [c] are hybrids of northern gooseberry and the European gooseberry, *R. uva-crispa*.) (From Hedrick, U.P., *The small fruits of New York*, Department of Farms and Markets, New York, 1925.) (b) Branches with fruit. (From Curtis, W., *Botanical Magazine,* vol. 112, Plate 6892, 1886.) (c) Cultivar 'Hoenings Earliest'. (d) Gooseberry pie. (Courtesy of Celeste Lindell [Flickr/CC-attribution].) (e) Distribution map.

been commercially promoted in recent years in the United States but have been disappointing in flavor to date.

- The genus name *Ribes* is said to derive from the Arabic *ribas*, rhubarb, presumably applied to the gooseberry because of a supposed similarity in taste that was noted by Arabs while they expanded into Spain centuries ago. Alternatively, the name has been interpreted as originating from the Old Danish colloquial *ribs*, red currant, or as the Latinized form of *riebs*, an old German word for currant.
- *Hirtellum* in the scientific name *R. hirtellum* is Latin for bristly, a reference to the prickles of the plant.

## GEOGRAPHY AND ECOLOGY OF WILD PLANTS

The northern gooseberry is native to the southern portion of central and eastern Canada (from Alberta to Newfoundland) and the adjacent areas of the United States (from the Dakotas to Virginia). The species has been recorded in various habitats, including rocky woods and cliffs, coastal thickets and heathlands, river banks, bogs, clearings, cedar and tamarack swamps, and in wet woods and forests.

## PLANT PORTRAIT

About 50 species of *Ribes* are called "gooseberries," the remaining 100 or so are "currants." Two species of gooseberry are important commercially: the northern gooseberry discussed in this chapter, and the European (or English gooseberry, *R. uva-crispa* L. (*Grossularia reclinata* (L.) Mill.), which is native to the Atlas Mountains of North Africa and to Europe. The berries of a number of wild North American gooseberry species have been collected for food, but all but the one highlighted in this chapter have been considered to be too low in quality to merit cultivation. Cultivated gooseberry species are shrubs, typically about a meter (about a yard) or more high. European varieties are derived primarily from the European gooseberry, while American varieties usually are hybrids of this species and the northern gooseberry. American varieties have smaller berries which taste somewhat different from the European kinds. Gooseberries can be green, white, yellow, or shades of red from pink to purple and almost black.

The northern gooseberry is a straggling to erect deciduous shrub, 0.5–1.5 m (1.6–5 feet) in height. The twigs are smooth or have scattered, bristle-like prickles, which fall off with the bark after the first year of growth. Spines are present at the nodes of the twigs (where the leaves arise). The flowers are produced in the spring. They are about 1 cm (0.4 inch) in length and are greenish-yellow to purplish. In the summer, the shrubs produces palatable, globose fruits that are greenish, purplish, or purple-black, 6–12 mm (0.24–0.5 inch) in diameter.

Gooseberries seem not to have been mentioned by early Greek and Roman writers. These fruits were first recorded in a thirteenth century document that noted they were cultivated in Europe. By the seventeenth century, red-, green-, and dark-fruited varieties were described. In the eighteenth century, "gooseberry clubs" were established in Britain with the goal of giving prizes for the heaviest fruit. This resulted in the creation of hundreds of new varieties and stimulated the industry in nineteenth century Britain. Subsequently, popularity of gooseberries declined in Britain, as the sourish berries became less attractive than imported, luscious exotic fruits. European gooseberries brought to the New World proved highly susceptible to American gooseberry mildew and somewhat unsuitable to the climate. Wild gooseberries were very popular in North America among settlers and native peoples, but as civilization developed, the shrubs were rarely cultivated. Nevertheless before the 1850s, some small-fruited varieties of the American *R. hirtellum* were grown in North America, as well as hybrids of the American and European species. Mildew-resistant varieties and hybrids were selected so that gooseberries could be grown well in North America. However, the introduction of the fungus known as white pine blister rust drastically curtailed gooseberry cultivation in North America, and today gooseberries are a very minor crop in the New World. To prevent white pine trees (and other susceptible pine species) from being infected by spores produced by a fungus that is resident on gooseberries, many jurisdictions do not allow their cultivation. Gooseberries are grown in substantial amounts in Germany, Poland, Britain, and other European countries. The bulk of the crop is processed, much of the harvest picked when green and hard. The North American crop is similarly treated, but the quantities raised are relatively small.

## CULINARY PORTRAIT

Gooseberries can be eaten fresh—with sugar, or added to fruit salads. They make very fine jellies, jams, preserves, sauces, syrups, puddings, chutneys, and sorbets. They also serve as an excellent tart accompaniment to meat, fish, and fowl, and an interesting component of baked goods, especially pies. Gooseberries are sometimes picked slightly underripe for cooking. In Germany, they have been dried and used like raisins. An old rule holds that green varieties taste better than red, and hairy ones better than smooth, but this is now thought to be only partly correct. Dessert varieties have larger, sweeter berries than varieties used for cooking purposes, but the latter tend to produce more flavor when cooked.

### Culinary Vocabulary

- A "fool" is a British dessert prepared by folding purе́ed fruit (traditionally gooseberry) into whipped cream.
- In France and England, gooseberry is a characteristic accompaniment to mackerel (hence the French culinary phrase *groseille à maquereau*, "mackerel with gooseberries").

## PROSPECTS

Gooseberries are an old crop that has diminished in popularity in recent times. The fruit has limited market and research interest in North America, and the North American wild species are considered inferior except for *R. hirtellum*, the northern gooseberry, which is mainly significant commercially in the form of hybrids with the much more popular European gooseberry. The disease white pine blister rust, associated with cultivation of *Ribes* species, complicates efforts to breed improved varieties. Given the increasing competition in the fruit market resulting from importation of numerous, delicious tropical fruits to more northern areas, the relatively uninteresting gooseberry seems to have limited prospects for development in North America.

## CURIOSITIES OF SCIENCE AND TECHNOLOGY

- Gooseberries have been thought to have curative properties. In the sixteenth century, they were recommended in the treatment of the plague. A stye (an inflamed swelling of a sebaceous gland at the margin of an eyelid) was once cured by lancing it with a gooseberry thorn, after passing it through a wedding ring.
- Nehemiah Grew (1641–1712), was the founder of plant anatomy, the first to publish the idea that there is sex in plants, and a pioneer in the use of the microscope. He chose the gooseberry as one of the first plants to be studied with the microscope, and he published a drawing of the plant's anatomy in his 1672 book *The Anatomy of Vegetables Begun*.
- The gooseberry has been used to produce alcoholic beverages at least since medieval times. In seventeenth century Scotland, monks made a strong wine-like ale from unmalted wheat, herbs, and gooseberries. Scottish women brewers ("ale wives") of the seventeenth and eighteenth centuries, known as the "luckies," were particularly responsible for gooseberry brews.
- "Gooseberry" was used in several English phrases, now obsolete. "Great (or big) gooseberry season" was a term used by journalists to refer to a period of limited interesting material, when it was necessary to include trivial material such as record-breaking gooseberries to fill up a newspaper. "Playing gooseberry" meant to be the odd person in a trio, such as a chaperone. "Playing old gooseberry" meant causing chaos or destruction. "Going gooseberrying" referred to stealing clothing hung out to dry. "Not worth a gooseberry" meant of little value. "Gooseberry wig" was a frizzy hairpiece so poorly made that it resembled a gooseberry bush.
- Hundreds of plants were assigned special meanings in the "Victorian Language of Flowers," popular in Victorian times, and many delighted in sending coded messages by this means, especially for romantic purposes. Gooseberry meant "anticipation."

- Because the best gooseberry cultivars were selected in England, most bear English names, including those of English politicians, such as Lords Brougham, Derby, and Eldon. In a subtle jibe at political practices, two varieties were named Ranter and Bribery.
- In 1852, a wining gooseberry at a competition was weighed at just below 57 g (2 ounces). This caught the attention of Charles Darwin (1809–1882), the father of evolution, who noted the extremely rapid selection that had turned the wild fruit, weighing about 7 g (1/4 ounce), from the size of a small pea to that of a small apple.
- Garnets, which are semiprecious gems, occur in shades of all colors except for blue. The "gooseberry garnet" is a light, greenish-brown garnet. A grape garnet is purplish-red to violet. A raspberry garnet is red. A cinnamon stone (garnet) is orange to orange-brown.

## KEY INFORMATION SOURCES


Allott, D.J., Robinson, D.W., and Uprichard, S.D. 1971. The response of gooseberries to non-tillage systems of management. *Hortic. Res.* 11: 166–176.

Arbury, J. 1996. Getting to grips with gooseberries. *Garden (London)* 121: 482–485.

Barney, D.L., and Hummer, K.E. 2005. *Currants, gooseberries, and jostaberries. A guide for growers, marketers, and researchers in North America.* CRC Press, Boca Raton, FL. pp. 266.

Beach, S.A. 1897. *Gooseberries.* New York Agricultural Experiment Station, Geneva, NY. pp. 48.

Berger, A. 1924. *A taxonomic review of currants and gooseberries.* New York State Agricultural Experiment Station, Geneva, NY. pp. 118.

Brennan, R.M. 1990. Currants and gooseberries (*Ribes*). In *Genetic resources of temperate fruit and nut crops.* Edited by J.N. Moore and J.R. Ballington, Jr. International Society for Horticultural Science, Wageningen, the Netherlands. pp. 459–388.

Crawford, M. 1997. *Currants and gooseberries: Production and culture.* Agroforestry Research Trust, Totnes, U.K. pp. 48.

Crowther, D.S. 1956. *Currants and gooseberries.* W.H. & L. Collingridge, Ltd., London, U.K. pp. 92.

Darrow, G.M. 1919. *Currants and gooseberries.* U.S. Department of Agriculture, Washington, D.C. pp. 40.

Darrow, G.M., and Detwiler, S.B. 1946. *Currants and gooseberries: Their culture and relation to white-pine blister rust.* Revised edition. U.S. Department of Agriculture, Washington, D.C. pp. 36.

Harb, J., and Streif, J. 2004. Quality and consumer acceptability of gooseberry fruits (*Ribes uva-crispa*) following CA [controlled atmosphere] and air storage. *J. Hortic. Sci. Biotech.* 79: 329–334.

Harmat, L., Porpaczy, A., Himelrick, D.G., and Galletta, G.J. 1990. Currant and gooseberry management. In *Small fruit crop management.* Edited by G.J. Galletta and D.G. Himelrick. Prentice Hall, Englewood Cliffs, NJ. pp. 245–272.

Hughes, H.M. 1972. Soil and bush management studies on gooseberries. *Exp. Hortic.* 24: 43–49.

Jefferies, C.J., Atwood, J.G., and Williams, R.R. 1982. Crop failure in gooseberry due to poor pollination. *Scientia Hortic.* 16: 147–153.

Kaack, K. 1992. Ripening of gooseberry (*Ribes uva-crispa*). *Tidsskrift Planteavl* 96: 271–278.

Keep, E.1975. Currants and gooseberries. In *Advances in fruit breeding.* Edited by J. Janick and J.N. Moore. Purdue University Press, West Lafayette, IN. pp. 197–268.

Kenny, A. 1971a. Storage of gooseberries in plastic bags. *J. Food Technol.* 6: 403–408.

Kenny, A. 1971b. Looking for a cheap method of extending the shelf-life of gooseberries. *Farm Food Res.* 2(1): 8–9.

Lanham, P.G., and Brennan, R.M. 1999. Genetic characterization of gooseberry (*Ribes grossularia* subgenus *Grossularia*) germplasm using RAPD, ISSR and AFLP markers. *J. Hortic. Sci. Biotech.* 74: 361–366.

MacNeill, M.M. 1969. Propagation of gooseberries by softwood cuttings under mist in a glasshouse. *East Malling Res. Sta. Annu. Rep.* 1968: 181–182.

Mesler, M.R., Cole, R.J., and Wilson, P. 1991. Natural hybridization in western gooseberries (*Ribes* subgenus *Grossularia*: Grossulariaceae). *Madroño* 38: 115–129.

Morin, N. 2009. *Ribes.* In *Flora of North America north of Mexico.* Vol. 8. Edited by Flora of North America Editorial Committee. Oxford University Press, New York, NY. pp. 9–42.

Rake, B.A. 1958. The history of gooseberries in England. *Fruit Yearb.* 10: 84–87.

Reich, L. 1986. Gooseberry revival. *Org. Garden.* 33(6): 62–67.

Schultheis, L.M., and Donoghue, M.J. 2004. Molecular phylogeny and biogeography of *Ribes* (Grossulariaceae), with an emphasis on gooseberries (subg. *Grossularia*). *Syst. Bot.* 29: 77–96.

Sinnott, Q.P. 1985. A revision of *Ribes* L. subg. *Grossularia* (Mill.) Pers. sect. *Grossularia* (Mill.) Nutt. (Grossulariaceae) in North America. *Rhodora* 87: 189–286.
Smolarz, K. 1993. Growth and yielding of gooseberry in response to long term (since 1923) differential fertilization. *Acta Hortic.* 352: 123–128.
Stewart, C.C. 1976. Gooseberries for hot-weather gardens [Missouri]. *Org. Gard. Farm.* 23(6): 62–64.
Suit, R.F., and Palmiter, D.H. 1945. *Control of gooseberry diseases.* New York State Agricultural Experiment Station, Geneva, NY. pp. 22.
Thornber, W.S. 1909. *Gooseberries for the home garden or commercial orchard.* Washington Agricultural Experiment Station, Pullman, WA. pp. 4.
Uhlarz, H. 1982. Ontogeny of the subfoliar prickles of *Ribes uva-crispa* L. (Grossulariaceae). Morphology and histology. *Beitr. Biol. Pflanz. (Berlin)* 57: 245–255. (In German, English summary)
United Kingdom Ministry of Agriculture, Fisheries and Food. 1979. *Gooseberries. Advisory Leaflet 215.* Revised edition. Agricultural Development and Advisory Service, Ministry of Agriculture, Fisheries and Food, Pinner, U.K. pp. 8.
Vaverka, M., and Vaverka, S. 2002. Evaluation of gooseberry cultivars resistance to the American gooseberry powdery mildew *Sphaerotheca morsuvae* Schwein and its chemical control. *Plant Protect. Sci.* 38(2): 583–587.
Wainwright, H., and Flegmann, A.W. 1985. The micropropagation of gooseberry (*Ribes uva-crispa* L.). I. Establishment in vitro. *J. Hortic. Sci.* 60: 215–221.
Yeager, A.F., and Latzke, E. 1933. *Gooseberries: Varieties, breeding, culture and use.* Agricultural Experiment Station, North Dakota Agricultural College, Fargo, ND. pp. 19.

## Specialty Cookbooks

Baker, J. 1996. *Kettle broth to gooseberry fool: A book of simple English food.* Faber and Faber, London, U.K. pp. 270.
Hendrickson, R. 1981. *The berry book. The illustrated home gardener's guide to growing and using over 50 kinds and 500 varieties of berries.* Doubleday & Company, Garden City, NY. [Chapter 11: A gander at gooseberries. pp. 168–183.]
McVicker, E. 1994. *Gooseberries.* E. McVicker, place of publication not stated. pp. 59.

Baird (1980) presents nine gooseberry recipes, Boorman (1969) has six, and Krumm (1991) has eight. Snell (2006) presents a recipe entitled Goosebery pudding, and Marie (2008) has a recipe for wild gooseberry wine. Marrone (2009) has 17 wild gooseberry recipes and Williamson (1995) has 10. (See the Appendix to this book for details of the publications cited.)

# 67 Ogechee Lime

Family: Cornaceae (dogwood family; *Nyssa* is sometimes split off into its own family, Nyssaceae)

## NAMES

Scientific name: *Nyssa ogeche* W. Bartram ex Marsh all

- Ogeechee in the vernacular name and *ogeche* in the scientific name reflect the discovery of the species in 1765 near the Ogeechee River in Georgia by early American explorer William Bartram (1739–1823), the "W. Bartram" in the authority for the scientific name. The "lime" in the name is based on the use of the juice as a substitute for lime juice (Ogeechee lime is not related to the true lime, *Citrus aurantiifolia* (Christm.) Swingle).
- Ogeechee lime has also been called bee tupelo, gopher plum, limetree, lone tupelo, Ogeechee plum, Ogeechee tupelo, sour tupelo, sour tupelo gum, white tupelo, and wild limetree.
- "Ogeechee" should not be confused with "Okeechobee," the name of a lake in southeast Florida.
- The word "tupelo" applied to *Nyssa* species is apparently from the Cree (Algonquian) *ito opilwa* meaning "swamp tree."
- The genus name *Nyssa* is based on Nyssa or Nysa, a water nymph of Greek legends. The choice of name reflects the growth of most *Nyssa* species in waterlogged soils.

## GEOGRAPHY AND ECOLOGY OF WILD PLANTS

Ogeechee lime is native to southwestern Alabama, northern Florida, southern and eastern Georgia, and southwestern South Carolina. It occurs almost entirely in southeastern Georgia and northern Florida, where it is extremely common. The species occurs in coastal plains and along the borders of rivers, swamps, ponds, and lakes in continuously wet, often flooded soils. It requires very moist habitats and is intolerant of shade, but is sometimes in the partial shade of wet woods.

## PLANT PORTRAIT

Ogeechee lime often develops into a small tree 7–11 m (23–36 feet) tall, sometimes to 15 m (49 feet) and rarely to 20 m (66 feet). The trees usually have numerous irregular branches, and when mature they may develop buttresses at their base. The plant is also often simply a much-branched shrub. The species has deciduous leaves. The trunks rarely exceed 60 cm (2 feet) in diameter but are occasionally up to 150 cm (5 feet). When a plant weakens or starts to die, sprouts frequently develop from the base and grow into independent trees, and this can result in a dense clump. Male trees bear only male flowers, whereas fruiting trees have female flowers and flowers with both male and female organs. The blooms are tiny, the male flowers in hanging clusters, and the female flowers solitary. Flowering occurs from March to May in the native area. The fruit is oblong, dull yellow, olive brown, orange, or red, 2–4 cm (0.8–1.5 inches) long, with acidic flesh. The fruits have one, rarely two, deeply grooved seeds up to 3 cm (1.2 inches) long, covered by

**FIGURE 67.1** Ogeechee lime (*Nyssa ogeche*). (a) Branch bearing fruit. (Painting courtesy of B. Flahey.) (b) Honey, the main economic product from Ogeechee lime. (Photo at left courtesy of Fifth World Art/Robert Neff [Flickr/CC-attribution]; photo at right courtesy of J.J. Harrison [CC By 3.0].) (c) Distribution map.

a pale, papery, seed coat. The fruits mature in July and August and persist until November and December after the leaves have dropped. Most people consider the eating quality of the fruits to be poor, but they are extremely attractive to ducks, otters, raccoons, opossums, squirrels, deer, and bears.

Ogeechee lime wood is in short supply, and so is of limited value. However, the wood has been used for specialty items, especially wood carving. Ogechee lime produces very colorful vivid yellow to deep purple foliage in the fall. Female trees are not often cultivated as ornamentals because the fruits falling in the autumn are very messy and the moisture requirement is demanding. However, male plants, which can be identified early in their development, are suitable for landscape use, provided the sites are moist.

**FIGURE 67.2** Ogeechee lime (*Nyssa ogeche*). Trees growing on river bank. (Courtesy of Gordon Landon, U.S. Forest Service.)

## CULINARY PORTRAIT

Ogechee lime fruits have a pleasantly subacidic, sourish flavor, and are used to make beverages, preserves, marmalade, and sauces. In Georgia, the fruit has been employed commercially to produce a chutney-like relish. Ogeechee lime is sometimes used as a substitute for limes and other sour citrus.

The species is extremely valuable as a source of nectar for honeybees, and thousands of hectares have been planted to provide honeybees with nectar for the production of "tupelo honey." This specialty honey is a light golden amber with a greenish tint, has a mild, pleasant taste, and will not granulate (crystallize) because of its high fructose to glucose ratio. Male trees are much more attractive to bees, but of course female trees are needed for fruit production.

Five of the approximately 10 species of *Nyssa* occur in eastern North America, and three in the southeastern United States, including Ogechee lime. The southeastern trees are employed for production of tupelo honey. The beekeepers place honeybee hives on platforms on the border of swamps and even on floats. The Apalachicola River of the Florida Panhandle is the center for certified tupelo honey production, and in some years the tupelo honey produced has a value approaching $1 million.

### CULINARY VOCABULARY

- The phrase "Ogeechee limes" has been used for preserves of the fruit of Ogeechee lime.
- "Monofloral honey" is honey which has a high value in the marketplace since it has a distinctive flavor or other qualities because it is predominantly from the nectar of one plant species. Tupelo honey is one of the most valuable of monofloral honeys.
- The word "lime" also means "calcium oxide," a building material produced from limestone (calcium carbonate). Surprisingly, limestone is occasionally used for culinary purposes: in Mexican masa harina (tortilla flour) and to enhance release of alkaloids in several narcotic plants consumed by mouth (e.g., betel nut, *Areca catechu* L.).

## PROSPECTS

As a fruit, Ogeechee lime has achieved only very local cottage industry usage for jam and preserves. However, as a source of special honey (tupelo honey), the species has become a significant marketing success, with a steady demand. Indeed, it seems clear that demand for the gourmet product will exceed supply for the foreseeable future. Numerous trees have been planted specifically for the production of honey. The tree has not been domesticated and is being maintained in its natural wild habitat; as this suffices for honey production, it is unlikely that the fruit will become significant. Nevertheless, given the need to maintain the trees for honey production, there would seem to be a market niche for the fruit.

## CURIOSITIES OF SCIENCE AND TECHNOLOGY

- Before the use of modern bee hives, honey bees were housed in sections of trees that had become naturally hollowed out. In the eastern United States, notably in the southeast, such tree sections were called "gums" because they were often cut from gum (*Nyssa*) trees.
- The city of Tupelo in Mississippi (the birthplace of Elvis Presley) was named for tupelo (*Nyssa*) trees.
- Thomas Earl "Tom" Petty was born in 1950 in Gainesville, Florida, and has become a very popular singer–songwriter and multi-instrumentalist. In his early years, he worked on the grounds of the University of Florida at Gainesville, where he planted an Ogeechee lime tree that has become known as "the Tom Petty tree."

## KEY INFORMATION SOURCES

Bonner, F.T. 1974. *Nyssa* L., tupelo. In *Seeds of woody plants in the United States*. Agriculture Handbook 450. Technically coordinated by C.S. Schopmeyer. U.S. Department of Agriculture Forest Service, Washington, DC. pp. 554–557.

Burckhalter, R. 1992. The genus *Nyssa* (Cornaceae) in North America: a revision. *Sida* 15: 323–342.

Darwin, W.N., Jr. 1972. *Tupelo (*Nyssa *spp.)*. American Woods FS-269. U.S. Department of Agriculture, Forest Service, U.S. Government Printing Office, Washington, DC. pp. 8.

Eyde, R.H. 1959. The discovery and naming of the genus *Nyssa*. *Rhodora* 61: 209–218.

Eyde, R.H. 1966. The Nyssaceae in the southeastern United States. *J. Arnold Arb.* 47: 117–125.

Eyde, R.H. 1997. Fossil record and ecology of *Nyssa* (Cornaceae). *Bot. Rev.* 63: 97–123.

Ganzera, M., Orsi, H., Zidorn, C., Khan, I.A., and Stuppner, H. 2005. Occurrence of grandifloroside-11-methyl ester in *Nyssa ogeche* (Nyssaceae). *Biochem. Syst. Ecol.* 33: 327–329.

Gilman, E.F., and Watson, D.G. 1993. Nyssa ogeche, *Ogechee lime*. Environmental Horticulture Department document ENH-579, University of Florida, Gainesville, FL. pp. 3.

Holroyd, H.B. 1906. *The utilization of tupelo*. U.S. Department of Agriculture Forest Service Circular 40. United States Department of Agriculture, Washington, DC. pp. 16.

Keyarts, E. 1964. Tupelo honey. *Am. Bee J.* 104: 90–91. [Source incorrectly identified as *N. sylvatica*.]

Kossuth, S.V., and Scheer, R.L. 1990. *Nyssa ogeche* Bartr. ex Marsh., Ogeechee tupelo. In *Silvics of North America*. Volume 2, Hardwoods. Agriculture Handbook 654. Technically coordinated by R.M. Burns and B.H. Honkala. U.S. Department of Agriculture Forest Service, Washington, DC. pp. 479–481.

Laird, C.R. 1963. Tupelo honey, a $75,000 industry in Northwest Florida. *Glean. Bee Cult.* 91: 678–681.

McGee, C.E., and Outcalt, K.W. 1990. *Nyssa sylvatica* Marsh., black tupelo. *N. sylvatica* Marsh. var. *sylvatica*; black tupelo (typical), *N. sylvatica* var. *biflora* (Walt.) Sarg., swamp tupelo. In *Silvics of North America*. Volume 2, Hardwoods. Agriculture Handbook. 654, Technically coordinated by R.M. Burns and B.H. Honkala. U.S. Department of Agriculture Forest Service, Washington, DC. pp. 482–489.

Rickett, H.W. 1945. Nyssaceae. *N. Am. Fl.* 28B: 313–316.

Sargent, C.S. 1893. *Nyssa*. *Silva N. Am.* 5: 73–84 + plates 217–220.

Saucier, J.R. 1982. *Tupelo (*Nyssa *spp.)*. American Woods FS-269. U.S. Department of Agriculture Forest Service, Southeastern Forest Expt. Station, Athens, GA. pp. 6.

Titman, P.W. 1949. Studies in the woody anatomy of the family Nyssaceae. *J. Elisha Mitchell Sci. Soc.* 65: 245–261.

Wheeler, D.H. 1937. The non-volatile acids of the fruit of *Nyssa ogeche*. *Science* 86: 588.
Xiang, Q.-Y., Soltis, D.E., and Soltis, P.S. 1998. Phylogenetic relationships of Cornaceae and close relatives inferred from *matK* and rbcL sequences. *Am. J. Bot.* 85: 285–297.

## Specialty Cookbooks

Freitus and Haberman (2005; see the Appendix to this book) have six recipes for black gum (*Nyssa sylvatica* Marsh.), which can be applied to the related Ogeechee lime.

# 68 Oregon Grape

Family: Berberidaceae (barberry family)

## NAMES

Scientific name: *Mahonia aquifolium* (Pursh) Nutt. (*Berberis aquifolium* Pursh)

The genera *Mahonia* and *Berberis* together include perhaps 500 species, 22 of which are in North America north of Mexico. They have been combined (as *Berberis*) or separated in different ways by various authors in recent times. Some species are treated as *Berberis* by some authors but as *Mahonia* by others, and this includes Oregon grape, the subject of this chapter. The name *M. aquifolium* is often used in horticultural articles, and in recent years it has come to be adopted by major online botanical databases, although the name *B. aquifolium* is also frequently encountered.

- Oregon grape (Oregon-grape) acquired its name from the grape-like appearance of the clusters of purplish fruit of the species and its common presence in Oregon.
- Oregon grape has also been called blue barberry, California barberry, holly barberry, holly-leaf barberry, holly-leaved barberry, holly grape, holly mahonia, mountain grape, Oregon holly grape, Rocky Mountain grape, and tall Oregon grape.
- The name Oregon grape is also applied to *M. nervosa* (Pursh) Nutt. (*B. nervosa* Pursh), a smaller but similar species with a comparable North American geographical distribution to that of *M. aquifolium* and fruits that are also palatable.
- The genus name *Mahonia* is named for the Irish-born American horticulturist Bernard M'Mahon (McMahon, about 1775–1816).
- The genus name *Berberis* is based on the Medieval Latin *barbaris*, which in turn is derived from an Arabian name for the barberry fruit.
- *Aquifolium* in the scientific name is based on Latin meaning "spiny-leaved" (similarly, English holly is *Ilex aquifolium* L.).

## GEOGRAPHY AND ECOLOGY OF WILD PLANTS

Oregon grape is native to western North America, from southern British Columbia to northern California, and in northern Idaho and western Montana. It has become naturalized in other areas of North America and in Australia, New Zealand, and Europe. In some locations, the plant is considered to be an invasive exotic weed.

The species occurs in the understory of open coniferous forests, in brushlands, open oak woods, canyons, valleys, and slopes. Although adapted to sunny, dry conditions, it also grows well in shade and moist soils. It tolerates poor soils, but often occurs in rocky and gravely areas. Oregon grape grows best in well-drained, moist, acidic soil and in shady locations.

## PLANT PORTRAIT

Oregon grape is a broad-leaved evergreen shrub, 0.3–3 m (1–10 feet) or rarely up to 4.5 m (15 feet) in height. The plant spreads laterally by roots and can form colonies by vegetative reproduction. The flowers are bright yellow. The leaves are compound, each leaf made up of 5–9 pairs of opposite leaflets and one unpaired leaflet at the tip. The leaflets (usually referred to as leaves by nonbotanists)

**FIGURE 68.1** Oregon grape (*Mahonia aquifolium*). (a) Branch bearing fruit. (Courtesy Louis.-M. Landry.) (b) Flowering branch. (From Loddiges, C. & Sons, *The botanical cabinet*. J. & A. Arch, Cornhill, UK, Vol. 18, Plate 1718, 1831.) (c) Distribution map.

are stiff, leathery, waxy, and prickly on the edges, making it advisable to wear gloves while harvesting the fruit, which is produced abundantly. The plant produces berries that are blue or purple and covered with a waxy bloom which produces a blue-gray sheen. The fruits are egg shaped or oblong, 6–12 mm (0.2–0.5 inch) long, and borne in clusters of a few to many. The seeds take up much of the volume of the fruit.

*Mahonia aquifolium* is widely grown as a deer-resistant ornamental, ground cover, hedge, and barrier. There are many garden cultivars and hybrids. Popular cultivars include 'Orange Flame', 'Mayhan', and 'Compacta'. The foliage has an interesting bronze or copper color when developing in the spring and turns an attractive red during the winter. The holly-like foliage is gathered from the wild for use as greenery by florists. The root system binds soil very well, and so the plant is used to stabilize slopes.

Oregon grape is resistant to infection by the fungus *Puccinia graminis* Pers., which affects many species of *Berberis*. Because the fungus is subsequently easily spread to wheat, causing the disease stem rust, there have been campaigns in North America for many years to eradicate barberry species, especially the introduced weedy European barberry, *B. vulgaris* L. It is illegal in the United States and Canada to sell or transport *Berberis* species or hybrids that are susceptible to the fungus or that have not been tested for susceptibility.

Native Americans used roots and leaves of Oregon grape to treat various medical problems, including arthritis, diarrhea, jaundice, and fever. The roots contain several alkaloids, particularly berberine, known to be antibacterial and anti-inflammatory and used in modern medicine to treat infection and inflammatory diseases such as eczema and psoriasis. The medicinal root bark is collected both from wild and cultivated plants. Goldenseal (*Hydrastis canadensis* L.) is another North American species with considerable amounts of berberine. Because Oregon grape also has this alkaloid, it has some of the same medicinal properties of goldenseal. Goldenseal has been overcollected from the wild, and to compensate for the diminishing supply, more wild Oregon grape is being harvested, leading to concern that it too could become endangered.

Berberine, the chief alkaloid of *M. aquifolium*, is a natural dye that is used to stain materials yellow. It is not authorized for use in food. Several plant species containing berberine have been used in different areas of the world as sources of yellow colorant. Native Americans used the yellow dye from Oregon grape particularly to color woven objects such as mats and baskets, and because of its popularity the dye was often traded and bartered.

An American species of *Mahonia*, *M. swaseyi* (Buckley ex Young) Fedde (*B. swaseyi* Buckley ex Young), which grows only in Texas, has been recommended for development as a new fruit species (Durand 1972). Unfortunately it is susceptible to *Puccinia graminis*, and such plant species that are extremely narrowly adapted to a small geographical region rarely have the characteristics that preadapt them to becoming commercial food crops.

## CULINARY PORTRAIT

The fruit of Oregon grape was eaten fresh in colonial times, and also made into preserves or simply dried. Today, the acidic, tart berries are mainly of interest to wild food collectors, who use them in pies, jams, jellies, beverages, and confectionary. The fruit is also sometimes fermented into wine. Although concern has been expressed that there might be undesirable levels of berberine, it has been reported that very little of the alkaloids present in other parts of the plant occur in the fruit. The flowers are also sometimes eaten or made into a lemonade-like drink.

Wild food collectors have offered the following tips for collecting Oregon grape. The fruit hanging at the tips of the branches and those maturing in the sun ripen first. The berries should be picked when they are soft and plump; wearing latex surgical gloves for picking will protect hands against the spines while allowing one to judge fruit softness. Dark blue berries are best; green (immature) and shriveled (overripe) fruit should be avoided or removed (along with insects) after collection. The small, hard seeds can be bitter and are best strained off after the berries are pulped. For wine, it should be noted that the berries contain little sugar, so more is necessary than for most wild-fruit wines.

### CULINARY VOCABULARY

- "Oregon grape" or "Oregon grapes" can be an ambiguous term, since Oregon is a significant grape-growing and wine-producing state (the commercial wine grape is *Vitis vinifera* L.; hybrids of it and other species are grown for table grapes). Some have proposed to use the hyphenated name "Oregon-grape" for M. *aquifolium* to avoid confusion.

## PROSPECTS

Compared to most other species harvested for their fruit, Oregon grape is not attractive. Its fruits are sour (tongue in cheek, it has been described as a "sour grape"), and the insipid fruit alone would not be a sufficient basis to invest in developing the species as a new commercial fruit (some species of *Berberis* native outside of North America have superior taste). However, its potential as a crop is suggested by the facts that (1) it is commonly grown as an ornamental so cultivation techniques are well studied, cultivars are available for additional selection, and other species could be used for hybridization and gene transfer; (2) Oregon grape is harvested from the wild in North America and also cultivated internationally for its medicinal properties, considerable research has been carried out on its pharmaceutical values, and the plant has several accepted medical applications, so it could be grown for more than one product, a distinct advantage for any crop; (3) the species has invasive tendencies in some locations, a characteristic that is shared by numerous major crops (aggressively colonizing; weedy plants are competitive, giving them an advantage as crops; shrinking violets, on the other hand, are less likely to be suitable for large-scale field cultivation); (4) Oregon grape could be grown under the pampered conditions of field cultivation provided for major crops, but more critically it is a tough plant that can be grown in poor soils, with very limited provision of irrigation and with minimal care so that it could be ideal for agriculturally marginal lands. Although the berries are small and seedy, they are produced in appreciable quantities and are the basis of a small cottage industry for production of jam, jelly, and wine. The key to developing Oregon grape as a fruit is selection of pomological varieties. The market for small berries is competitive and dominated by blueberry, raspberry, and strawberry, but most species are in the marketplace only at particular times and so it is possible for new fruits to be successful. Oregon grape has limited market prospects but is an interesting possibility.

## CURIOSITIES OF SCIENCE AND TECHNOLOGY

- In 1823, explorer David Douglas (1799–1834) sent some plants of Oregon grape back to England, where they were reproduced by cuttings and selling for the then astronomical price of about $50.00 for each shrub. However, in 1825, he collected and sent back seeds of the species, which allowed great numbers to be produced rapidly and greatly lowered the price.
- Oregon grape was officially recognized as the "state flower" of Oregon in 1899. As has frequently been the case when plants have been recognized as official symbols, considerable lobbying occurred prior to legislative recognition. In Oregon, the Oregon Horticultural Society and the Oregon Federation of Women's Clubs played important roles in the 1890s in persuading politicians to adopt Oregon grape as Oregon's official flower.
- The official governor's mansion in Salem, Oregon, was purchased in 1988 and subsequently was named "Mahonia Hall" after the state flower.
- Prickly leaved evergreen shrubs such as *Mahonia* and *Berberis* are among those often recommended for crime prevention through "access control." Spiny, thorny hedges act as barriers discouraging intruders from entering vulnerable locations.

## KEY INFORMATION SOURCES

Abrams, L. 1934. The mahonias of the Pacific states. *Phytologia* 1: 89–94.

Ahrendt, L.W.A. 1961. *Berberis* and *Mahonia*: A taxonomic revision. *J. Linn. Soc. Bot.* 57: 1–410.

Anonymous. 2003. Portrait exotica: *Mahonia aquifolium* (compared with other exotic and Indian *Mahonia* species.). *MFP News* 13: 4, 17–19.

Auge, H., and Brandl, R. 1997. Seedling recruitment in the invasive clonal shrub, *Mahonia aquifolium* Pursh (Nutt.). *Oecologia* 110: 205–211.

Berg, V.S. 1985. Physical, chemical and spectral properties of leaf surfaces of *Berberis aquifolium*. *Plant Cell Environ.* 8: 631–638.

Cernakova, M., and Kostalova, D. 2002. Antimicrobial activity of the protoberberine alkaloid, berberine, isolated from *Mahonia aquifolium*. *Folia Microbiol.* 47: 375–378.

Durand, H.K. 1972. Texas mahonia: A neglected economic plant. *Econ. Bot.* 26: 319–325.

Finn, C.E. 2008. Berberidaceae. *Berberis* spp. barberry; *Mahonia* spp. Oregon grape holly. In *The encyclopedia of fruit & nuts*. Edited by J. Janick and R.E. Paull. CABI, Wallingford, Oxfordshire, UK. pp. 161.

Gieler, U., Weth, A. von der, and Heger, M. 1995. *Mahonia aquifolium*: A new type of topical treatment for psoriasis. *J. Derm. Treat.* 6: 31–34.

Hudek, C., Melanie, B., and Rey, F. 2010. Root system traits in *Mahonia aquifolium* and its potential use in soil reinforcement in mountain horticultural practices. *Scientia Hort.* 125: 504–511.

Kim, Y., Kim, S., and Landrum, L.R. 2004. Taxonomic and phytogeographic implications from ITS phylogeny in *Berberis* (Berberidaceae). *J. Plant Res.* 117: 175–182.

Landolt, P.J., and Smithhisler, C.L. 2003. Characterization of the floral odor of Oregon grape: Possible feeding attractants for moths. *Northwest Sci.* 77: 81–86.

Lukic, P., Nikolic, R., and Gorunovic, M. 1972. Contribution to the comparative pharmacognostic study of *Berberis vulgaris* and *Berberis aquifolium*. *Acta Pharm. Jugosl.* 22: 19–24.

McCain, J.W., and Hennen, J.F. 1982. Is the taxonomy of *Berberis* and *Mahonia* (Berberidaceae) supported by their rust pathogens *Cumminsiella santa* sp. nov. and other *Cumminsiella* species (Uredinales)? *Syst. Bot.* 7: 48–59.

Moran, R.V. 1982. *Berberis claireae*, a new species from Baja California, and why not *Mahonia*. *Phytologia* 52: 221–226.

Neugebauer, H., and Brunner, K. 1939. The alkaloid content of *Berberis aquifolium* in the course of the year. *Arch. Pharm. Zentral. Deutsch.* 80: 241–245. (In German)

Ross, C., and Durka, W. 2006. Isolation and characterization of microsatellite markers in the invasive shrub *Mahonia aquifolium* (Berberidacaeae) and their applicability in related species. *Mol. Ecol. Notes* 6: 948–950.

Ross, C.A., Auge, H., and Durka, W. 2008. Genetic relationships among three native North American *Mahonia* species, invasive *Mahonia* populations from Europe, and commercial cultivars. *Plant Syst. Evol.* 275: 219–229.

Seidemann, J. 1998. The holly-leaved barberry plant (*Mahonia aquifolium* Pursh/Nutt.), a useful fruit plant? *Deutsche Lebensmittel-Rundschau.* 94: 217–220. (In German, English summary)

Slobodníková, L. Kost'álová, D., Labudová, D., Kotulová, D., and Kettmann, V. 2004. Antimicrobial activity of *Mahonia aquifolium* crude extract and its major isolated alkaloids. *Phytother. Res.* 18: 674–676.

Whittemore, A.T. 1997. *Berberis*. In *Flora of North America North of Mexico*, Vol. 3. Edited by Flora of North America Editorial Committee. Oxford University Press, Oxford, UK. pp. 276–286.

## Specialty Cookbooks

Krumm (1991) has two recipes for Oregon grape jelly, and Latorre (1977) has one. Freitus and Haberman (2005) have 8 recipes for Oregon grape and 10 barberry recipes, which are applicable to Oregon grape. Marrone (2009) has eight Oregon grape recipes, and Williamson (1995) has nine. Marie (2008) has recipes for Oregon grape wine and Oregon grape flower wine. Stewart (2002) provides information on the culinary preparation of Oregon grape. (See the Appendix to this book for details of the publications cited.)

# 69 Paper Birch

Family: Betulaceae (birch family)

## NAMES

Scientific name: *Betula papyrifera* Marsh all

- Paper birch is also known as American white birch, canoe birch, silver birch, and white birch. The names canoe birch and paper birch reflect the historical use of the bark by Native Americans for canoes and writing parchment, respectively. The names white birch and silver birch are based on the bark color. "White birch" is often used for white-barked *Betula* species other than *B. papyrifera* (especially Eurasian species).
- The genus name *Betula* is based on *betulla*, the Latin name for birch (some authorities ascribe *Betula* to the Celtic *betu*, meaning tree).
- *Papyrifera* in the scientific name *B. papyrifera* is from the Greek word *papurus*, meaning papyrus or paper, referring to the paper-thin bark that has actually been used as paper, + the Greek *fero*, or Latin *ferre*, to bear, carry, or bring, that is, "paper bearing."

## GEOGRAPHY AND ECOLOGY OF WILD PLANTS

*Betula papyrifera* occurs across northern North America, from Newfoundland to northwestern Alaska, from the tree line in the north southward to the northern United States, from Washington State to New York, with scattered populations in Iowa, Nebraska, the Dakotas, and on a few high mountains in North Carolina. The tree occurs in all provinces and territories of Canada, except Nunavut.

Paper birch occurs in moist, more or less open, upland forests, often on rocky slopes, and sometimes in swampy woods. It is common on the margins of forests, in clearings in closed forests, on rocky cliffs, in ravines, and along shores. The tree occupies a wide range of soil types, including peat, but does best on deep, well-drained, fertile soils. It is intolerant of shade and prefers cool, moist sites. Paper birch is adapted to colonizing areas cleared by fire, often regenerating by stump sprouts, and frequently appearing following logging and abandonment of cleared land. The thin, inflammable bark makes the tree susceptible to fire. The species commonly grows in pure stands or in associations with conifers.

## PLANT PORTRAIT

There are about 35 species of *Betula*, distributed in north temperate regions of the world. About 18 are native to North America north of Mexico. In North America, paper birch is the main species used for the production of birch syrup and accordingly is highlighted in this chapter. In addition to paper birch, *B. neoalaska* Sarg., known as Alaska birch, Alaska paper birch, and resin birch, is also often used as a source of birch syrup. This is native to Alaska and northern Canada from the Yukon to northwestern Ontario. Belarus, Russia, Scandinavia, and Ukraine use other species, native to their regions, for birch syrup production.

*Betula papyrifera* grows occasionally as a shrub but is usually a deciduous tree up to 35 m (115 feet) in height and 80 cm (32 inches) in diameter at breast height. There may be a single or

**FIGURE 69.1** Paper birch (*Betula papyrifera*). (a) Stand of young trees in winter. (Photos [a] and [c] are courtesy of Joseph O'Brien, U.S. Department of Agriculture Forest Service, online at Bugwood.org [CC By 3.0]). (b) Tree. (From Herman, D.E., et al., *North Dakota tree handbook*, U.S. Department of Agriculture, Bismarck, ND, 1996.) (c) Peeling bark. (d) Birch syrup. (Courtesy of Kahiltna Birchworks. The copyright holder of this file allows anyone to use it for any purpose, provided that Kahiltna Birchworks and http://www.alaskabirchsyrup.com/[Kahiltna Birchworks official website] are credited.) (e) Distribution map.

several stems (trunks), and often the presence of multiple trunks is the result of damage to a young plant by browsing animals. The tree is shallow-rooted, and most of the roots are found in the top 60 cm (2 feet) of soil. The bark of young trunks and branches is dark reddish-brown and smooth, and in mature trunks it is creamy to chalky white or pale (infrequently dark brown) with small black marks and scars and elongated horizontal lenticels (natural openings in the bark that allow for gaseous exchange). The chalky white covering on the bark can rub off onto clothing. The inner bark is often reddish-orange, turning black with age. The outer bark is easily peeled off in sheets; however, doing so results in the reddish-orange bark dying and turning an ugly black. The papery white outer

**FIGURE 69.2** Canoe building on the Ottawa River, Ontario, by British Artist William Henry Bartlett (1809–1854). (From Bartlett, W.H., *Canadian scenery illustrated,* G. Virtue, London, UK, 1840.)

bark never grows back, although a new bark develops if removal of bark is not so extensive that it kills the tree. The flowers appear in male and female catkins. The male catkins are 4–10 cm (1.5–4 inches) long at maturity in the spring. They hang down, singly or up to five in a cluster, from the ends of branches. The female catkins, up to 5 cm (2 inches) long, are upright and somewhat back from the ends of branches where the male catkins are located. The fruiting cones hang down and are cylindrical, brownish, and 2.5–5 cm (1–2 inches) long. The tiny, papery "seeds" (technically fruits or nutlets) are two-winged. Paper birch is considered to be a short-lived species; few trees live longer than 140 years (some for up to 200 years), and most survive for no longer than 70 years.

Historically, North American Indians used paper birch for a wide range of medicinal and construction purposes. The bark is waterproof because of its high oil content. It was used for many items, including canoes, baskets, containers for food and liquids, shelters, bathtubs, clothing, kindling, torches, drums, moose callers, paper, playing cards, and even as a cast for broken limbs. Canoes were made by stretching stripped bark over frames of white cedar, sewing the pieces of bark together with tamarack roots, and caulking the seams with pine or balsam fir resin. The bark has been used to make "sunglasses" to prevent damage from snow glare; this was simply a strip of bark about 5 cm (2 inches) wide in which lenticels functioned as eye apertures. Native Americans used the strong, flexible wood to construct such items as spears, bows, arrows, snowshoes, and sleds.

Today, paper birch wood is mostly valuable as lumber, plywood, veneer, fuel, and pulpwood. It is used for furniture and cabinets, flooring, and for specialty wood products including clothespins, spools, pool cues, golf tees, broom handles, and toys. The odorless and tasteless qualities of the wood, as well as its uniform texture and white appearance, make it useful for ice cream sticks, toothpicks, and tongue depressors. Although burned as firewood, giving off considerable heat even when green, birch tends to coat chimneys with a layer of creosote, posing a fire danger.

Among the other minor uses for the paper birch are the following. The species is a popular ornamental or landscape tree because of its graceful shape (often forming clumps) and attractive bark. It is also used for revegetation and soil stabilization, including severely disturbed sites such as mine spoils. Occasionally, paper birch is used as a source of medicinal extracts. Oil of wintergreen (methyl salicylate) was once obtained from wintergreen (*Gaultheria procumbens* L.), later from birch species, but today, this volatile oil is usually produced synthetically. Oil of wintergreen, toxic in the pure form, has medicinal applications and is used as a flavorant to impart a minty taste to candy, chewing gum, root beer, and other products.

The culinary uses of paper birch are limited. Occasionally, indigenous people ate fresh cambium (the growing portion of the trunk, between the wood and the bark). Birch leaves and twigs have been used to make tea. In Scandinavia, Russia, and the former European nations of the USSR, birch sap (not concentrated as in birch syrup, but typically pasteurized) is drunk as a tonic beverage. Birch syrup, produced in the manner of maple syrup, is a cottage industry in North America, especially in Alaska, but to a minor extent also in Canada. Alaska currently produces about 4000 L (about 1000 American gallons) annually. Birch trees are occasionally used to produce alcoholic or nonalcoholic birch beer in Europe and North America.

Birch syrup production is much like maple sugar production (see Chapter 93 on Sugar Maple). However, fructose is the primary sugar in birch syrup (sucrose is the main sugar in maple syrup), and because fructose burns at a lower temperature than sucrose, boiling of the syrup to evaporate most of the water needs to be done more carefully to avoid a scorched taste. Paper birch in northern areas of the range have a shorter harvest period than maple. The pipeline method of maple sugar collection is harder to use with birch, because birch has lower root and trunk sap pressure than maple. Fructose has a lower tendency to crystallize compared to sucrose, so birch syrup is thinner (less viscous at the same sugar concentration) than maple syrup. Birch syrup harvested later in the season tends to be darker in color and more pronounced in taste, and more advisedly used to flavor food preparations rather than as a syrup. Birch syrup is about half as sweet as maple syrup.

## CULINARY PORTRAIT

Birch trees have several very minor culinary uses. The sap can be tapped to yield a sugar or syrup but, compared to sugar maple, the amount produced is much less and it is more expensive. Between 80 and 110 L of birch sap are required (sometimes more depending on birch variety) to make just 1 L of syrup (30–40 L are needed for maple syrup), which sells for about $80.00 wholesale. Birch syrup is a very minor commodity in Europe, Alaska, and Canada. In North America, paper birch is the main source of birch syrup. Birch syrup is darker and slightly more bitter than maple syrup and is sometimes encountered as a glaze or flavoring used on very fancy meals in high-end restaurants. "Birch beer" is usually a carbonated soft drink made with birch twigs or bark from various *Betula* species, but may also be alcoholic, prepared by fermenting the sap or adding honey or sugar to promote fermentation. In North America, more than a dozen brands of birch beer are available. These are usually nonalcoholic, carbonated beverages, rather similar in taste to root beer, and often made by using extracts from the black birch (*B. lenta* L., also known as sweet birch and cherry birch). Birch wine is occasionally available commercially.

Birch syrup is used like maple syrup, particularly on pancakes and waffles, but sometimes also to impart flavor to sauces, glazes, dressings, soft drinks, and alcoholic beverages. The distinctive flavor has been described as semisweet, combining the tastes of honey, caramel, licorice, and molasses, with a spicy, balsamic aftertaste.

Xylitol is a natural sugar that can be obtained from many plants, but is frequently extracted commercially from the bark of paper birch. It is about as sweet as sucrose (table sugar) with about two-thirds the calories and is used as a sweetener in numerous food products. Xylitol is particularly used by diabetics because it has limited effect on insulin levels. It tends to reduce dental decay and so is often used in tooth products like chewing gum and toothpaste.

### Culinary Vocabulary

- In Pennsylvania, a "Red Bull" is an ice cream soda made with vanilla ice cream and birch beer, whereas a "Black Cow" is an ice cream soda made with chocolate ice cream and root beer.

## PROSPECTS

The birch syrup industry is very comparable to the maple syrup industry and shares the advantage of regular demand, but on a much smaller scale. Birch syrup is a rare, expensive commodity with appeal primarily to a gourmet, connoisseur market. Nevertheless, there is an appreciable market for birch syrup, particularly including upscale restaurants and individuals who are not deterred by the high cost, which is up to five times as much as maple syrup. Paper birch trees are common in the colder areas of North America, so that the natural wild supply is not limited. Like the maple syrup industry, the work is very labor-intensive and seasonal, which is often quite advantageous during the later winter and early spring period in northern North America when workers are available. Although the growth potential seems limited, birch syrup is an interesting possibility for entrepreneurs with a sense of adventure. Birch trees are a natural resource that is present in considerable quantities in some states and provinces, and the production of birch syrup represents the kind of enterprise for which governmental subsidization may be available.

## CURIOSITIES OF SCIENCE AND TECHNOLOGY

- The ancient "Irish Tree Alphabet" begins with the letter B (beth), meaning "birch." The system was created by the ancient Druids and used just for spiritual purposes. The alphabet consists of 5 vowels and 13 consonants, each letter named after a tree or shrub. Originally, each plant had particular historical importance, and the birch had considerable significance. It was used in ceremonies installing Roman consuls (the highest office of the ancient Roman Republic). Later in Europe, the tradition arose of using birch twigs to flog delinquents and expel evil spirits from lunatics. Birch rods were also used in rituals for driving out the spirit of the old year, because the tree produced its leaves very early in the year. Indeed, in Scandinavia, the production of birch leaves marked the beginning of the agricultural year, and farmers used the event as a signal to sow their spring wheat.
- In ancient Rome, birch was a symbol of power and authority. Birch sticks tied around an axe handle with the blade threateningly revealed were prominently displayed in front of Roman processions for dignitaries. These birch-clad axes were called "fasces," and came to symbolize the power of the Roman Republic. Fasces subsequently became the symbol of many legitimate political and labor movements, but when Benito Mussolini's Fascist Party took up the symbol, "fascism" and "fascists" were stigmatized and the words became pejorative terms.
- Paper birch was a sacred tree for the Ojibwa Indians, who placed bark on the coffins when burying their dead. While gathering material from the trees, offerings of tobacco were made to the Great Spirit, Winabojo, and to Grandmother Earth.
- Ojibwas claimed that the birch was never struck by lightning and offered a safe place during thunderstorms. Curiously reminiscent, European farmers sometimes planted birch trees around their houses in the belief that this would protect them against lightning. (Since birches are generally not the tallest trees in a region, perhaps they are less likely to be struck by lightning and this gave rise to the impression that they are immune.)
- Witches are said to have preferred birch twigs to make the ends of their broomsticks.
- Cradles were once made of birch wood in the belief that this would protect babies.

- Birch twigs bound in a bundle (called a birch) were once widely used as a whip to "birch" (punish) people. Sapling branches from birch trees were a favorite disciplinary tool of stern schoolmasters in early colonial North America.
- In Finland, Scandinavia, Estonia, Latvia, and Russia, birch twigs with the foliage left on are traditionally used to slap one's body in the sauna as a form of massage and to open the pores and increase blood circulation.
- The Ukrainian word for the month of March is *berezen*, meaning "time when the birch trees flower."
- Native Americans produced a beautiful art form called birch bark biting, by which marks were made by teeth to create intricate designs. Birch bark biting, a tradition dating back centuries, was usually carried out by women as a social activity at gatherings.
- Birch trees have been recognized as official emblems of several political regions. The paper birch was designated the New Hampshire state tree in 1947 and the provincial tree of Saskatchewan in 1988. Yellow birch (*B. alleghaniensis* Britton) was adopted in 1993 as the provincial tree of Quebec. The silver birch (*B. pendula* Roth) is the national tree of Finland. The "birch" is the national tree of Russia.
- Yellow-bellied sapsuckers (which are woodpeckers) often bore holes into birch trees to feed on the sap, and when the sweet sap oozing out of the holes attracts insects, they also eat the insects.

## KEY INFORMATION SOURCES

Ahtonen, S., and Kallio, H. 1989. Identification and seasonal variations of amino acids in birch sap used for syrup production. *J. Food Chem.* 33: 125–32.

Barnes, B.V., Dancik, B.P., and Sharik, T.L. 1974. Natural hybridization of yellow birch and paper birch. *Forest Sci.* 20: 215–221.

Beveridge, T., Bruce, K., and Kok, R. 1978. Carbohydrate and mineral composition of gray birch syrup. *Can. Inst. Food Sci. Tech. J.* 11: 28–30.

Cameron, M. 2005. Tapping into birch syrup opportunities. In *Proceedings: Linking healthy forests and communities—successful strategies and future directions*. Edited by D.L. Nicholls. General Technical Report PNW-GTR-631, Pacific Northwest Research Station, USDA Forest Service, Portland, OR. pp. 23–24.

Clennett, C., and Sanderson, H. 2002. *Betula papyrifera. Curtis's Botanical Magazine* 19: 40–48.

Dugle, J.R. 1966. A taxonomic study of western Canadian species in the genus *Betula*. *Can. J. Bot.* 44: 929–1007.

Furlow, J.J. 1997. *Betula*. In *Flora of North America North of Mexico*. Edited by Flora of North America Editorial Committee. Oxford University press, Oxford, UK, Vol. 3, 516–530.

Haack, R.A. 1996. Will global warming alter paper birch susceptibility to bronze birch borer attack? In *Dynamics of forest herbivory: Quest for pattern and principle*. Edited by W.J. Mattson, P. Niemelä, and M. Rousi. U.S. Dept. Agriculture Forest Service General Technical Report NC 183, St. Paul, MN. pp. 234–247.

Hutnik, R.J., and Cunningham, F.E. 1961. *Silvical characteristics of paper birch*. U.S. Dept. Agriculture Forest Service, Northeastern Forest Experiment Station, Upper Danby, PA, Station Paper No. 141. pp. 24.

Hyvarinen, M.J. 1968. *Paper birch, its characteristics, properties, and uses: A review of recent literature*. North Central Forest Experiment Station, Forest Service, U.S. Dept. Agriculture Forest Service General Technical Report NC 22, Saint Paul, MN. pp. 12.

Kallio, H. 1989. Aroma of birch syrup. *J. Agric. Food Chem.* 37: 1367–1371.

Kallio, H., Rine, S., Pangborn, R.M., and Jennings, W. 1987. Effect of heating on the headspace volatiles of Finnish birch syrup. *Food Chem.* 24: 287–300.

Kallio, H., Teerinen, T., Ahtonen, S., Suihko, M., and Linko, R.R. 1989. Composition and properties of birch syrup (*Betula pubescens*). *J. Agric. Food Chem.* 37: 51–54.

Kok, R., Norris, E. R., and Beveridge, T. 1978. Production and properties of birch syrup (*Betula populifolia*). *Can. Agric. Engineer.* 20: 5–9.

Larson, E.H. (Editor). 1969. *The birch symposium*. Northeastern Forest Experiment Station, Forest Service, U.S. Dept. Agriculture, Upper Darby, PA. pp. 183.

Oldenmeyer, J.L. 1982. Estimating production of paper birch and utilization by browsers. *Can. J. Forest Res.* 12: 52–57.

Peterson, E.B. 1997. *Paper birch managers' handbook for British Columbia.* B.C. Ministry of Forests, Victoria, British Columbia, and Forestry Canada. pp. 133.

Safford, L.O. 1983. *Silvicultural guide for paper birch in the Northeast* (revised). U.S. Dept. Agriculture Forest Service, Northeastern Forest Exp. Station, Broomall, PA, Research Paper NE 535. pp. 29.

Wilder, N., Lousier, J.D., and Hawkins, C. (Editors). 2002. *Proceedings of the Paper Birch Workshop and Conference, September 19–22, 2001, Prince George, British Columbia, Canada.* University of Northern British Columbia, Prince George, BC. pp. 166.

## Specialty Cookbooks

Dixon-Warren, H.D. 2007. *The birch syrup production manual.* Quesnel Community and Economic Development Corporation, Quesnel, British Columbia. pp. 103. (Presents more than 25 recipes.)

North House Folk School. 2007. *Celebrating birch: The lore, art, and craft of an ancient tree.* Fox Chapel Publishing, Lancaster, PA. pp. 192. (Has recipe for birch root beer.)

Additionally, Genest (2010) has four recipes using birch syrup and Hahn (2010) has two. Freitus (1980) and Johnson (1989) have recipes for preparing birch beer. Freitus (1980) has advice on preparing birch syrup and presents a recipe for making a flavoring by simply steeping birch twigs in hot water. (See the Appendix to this book for details of the publications cited.)

# 70 Paradise Tree

Family: Simaroubaceae (quassia family; also called simarouba family and tree of heaven family)

## NAMES

Scientific name: *Simarouba glauca* DC.

- The name "paradise tree" is based on the idea that the tree is extremely useful, providing everything wanted.
- Paradise tree is also known as bitterwood (a name used for several species of *Simarouba* with bitter-tasting wood and bark) and dysentery bark (in 1713, the bark was imported from Guiana to France as a remedy for dysentery). Other similar names include bitter ash, bitter damson, and Florida bitterwood. There are also numerous names in non-English languages.
- Paradise tree has also been called princess tree and olivo, a name used in parts of Central America.
- The name "tree of heaven" is sometimes used for paradise tree, but is more appropriately reserved for *Ailanthus altissima* (Mill.) Swingle, another tree of the Simaroubaceae. This was once planted as an ornamental, but has become an invasive weed in many areas of the world.
- The name paradise tree is sometimes used for *Melia azedarach* L., better known as Chinaberry. This Asian tree is widely planted as an ornamental, but is invasive in some areas.
- The expression "paradise tree" has been used in Europe to designate Christmas trees. In past centuries, the paradise tree was often an evergreen decorated with apples. This is an allusion to the apple tree in Paradise that led to the downfall of Adam and Eve; in the Middle Ages, such a tree was associated with the Feast of Adam and Eve, held on December 24.
- The genus name *Simarouba* is based on a native vernacular name in Guiana for *S. amara* Aubl. The vernacular (i.e., non-Latinized) name simarouba is sometimes used for *S. glauca*.
- *Glauca* in the scientific name is Latin for bluish-gray, generally caused by a "bloom" or whitish, silvery gray, or bluish-green covering. The word usually refers to the appearance of foliage due to a waxy surface.

## GEOGRAPHY AND ECOLOGY OF WILD PLANTS

Paradise tree is native to southern Florida, Mexico, Central America, and the Caribbean region (including Bahamas and Greater Antilles). It has been introduced in India. The species is adapted to warm, humid, tropical regions and occurs in a wide range of habitats. It is quite salt tolerant, and in southern Florida, it occurs in coastal hammocks.

## PLANT PORTRAIT

Paradise tree is a large shrub or a small- or medium-sized tree up to 20 m (65 feet) in height, with a maximum trunk diameter of about 80 cm (32 inches), but usually much smaller. The leaves are evergreen, large, compound (with 13–23 leaflets), leathery, and glossy. The flowers are small,

**FIGURE 70.1** Paradise tree (*Simarouba glauca*). (a) Branches bearing fruit. (Courtesy of U.S. Department of Agriculture.) (b) Foliage. (Courtesy of Forest & Kim Starr [CC By 3.0].) (c) Distribution map.

yellow or yellow-green and occur in long clusters. About 5% of trees produce male flowers only, whereas many trees have mainly male with a few bisexual flowers. Many trees have female flowers only and are used for fruit harvest. In plantations, about one of the mostly male plants is recommended for about 20 female plants. The fruits are oval-elongated, greenish-yellow, purple or black, and reminiscent of olives. Three layers need to be distinguished: a thin outer fleshy layer 2–3 mm (0.08–0.12 inch) thick, the hard outer shell (hull) of the seed (constituting 30% to 45% of the seed weight), and the "kernel" inside the seed hull (constituting about 55% to 70% of the seed). Up to two-thirds of the seed kernel is edible fat. When the seed oil is extracted, the remaining cake is rich in proteins (about 50%) but also contains toxic, bitter constituents. Detoxified seed cake is often fed to livestock. The tree is widely grown in South and Central American countries and in India since the 1960s. In addition to food use, the oil is also used in industrial manufacture of lubricants, soap, polishes, paints, and pharmaceuticals. Considerable research has been carried out attempting to develop paradise tree oil into biofuel. Lumber from the trees is widely used, and the seed hulls have been used to manufacture synthetic boards. The bark and leaves contains chemicals (triterpenes) that are employed medically to treat malaria, diarrhea, and amoeba-caused diseases. Additional extracts from the plant have been used to treat skin conditions.

## CULINARY PORTRAIT

The fat from paradise tree is odorless, greenish-yellow, and melts at 26.4°C (79.5°F). Major components are oleic acid (about 53%), stearic acid (about 30%), and palmitic acid (about 12%). Oleic acid is an unsaturated fatty acid, considered more nutritionally desirable than stearic and palmitic acids, which are saturated fatty acids (olive oil, which is a very healthy oil, is about 75% oleic acid). In Central and South America and in India, the seed oil is widely used as cooking oil and consumed as vegetable fat preparations such as margarine. The oil is said to be suitable for use as cocoa butter extender in chocolate products and in confectionary and bakery products. Fractions of the oil are also used in ice cream and mayonnaise.

The thin pulp of the fruit is astringent but semisweet, with about 12% sugars. It is consumed locally (the fruits are rather popular in El Salvador), and the flesh is sometimes processed into jam and nonalcoholic beverages or fermented into alcoholic beverages. The fruit is quite perishable and does not ship well, but has potential for incorporation into commercially prepared products.

### Culinary Vocabulary

- What is the difference between "fats" and "oils"? The word fat is often used to include oil ("lipids" is a still broader term that includes substances that are called fats and oils, but also unrelated compounds). However, in the context of culinary usage, the term "oils" is usually restricted to fats that are liquids at normal room temperature, while "fats" is usually used to refer to fats that are solids at normal room temperature.

## PROSPECTS

Paradise tree has proven to be an extremely useful source of edible oil, but is a crop that is best grown in tropical areas. It has achieved good market penetration and has potential to be used additionally for biofuel and other industrial oil applications. It has some medical value, and research may extend the market for this as well. The usefulness of paradise tree as a crop in the United States and Canada is limited because of its lack of adaptation to most of North America, although climate change may alter this situation in the future.

## CURIOSITIES OF SCIENCE AND TECHNOLOGY

- "DC." in the scientific name *Simarouba glauca* DC. is the standard author abbreviation for the famous Swiss botanist Augustin Pyramus de Candolle (1778–1841). The only author honored with a shorter abbreviation (L.) is Carl Linnaeus (1707–1778), the founder of scientific biological nomenclature, who died in the same year that de Candolle was born.
- The medicinal value of paradise tree bark for treating dysentery and other gastric conditions, amoeba-borne diseases, and malaria was known to indigenous peoples of Central America centuries ago. Modern science has confirmed the usefulness of paradise tree for treating these conditions, a tribute to the remarkable ability of people with extremely limited scientific knowledge to make important scientific discoveries.
- The use of religious terms in names of species reflects the values that people have acquired. "Paradise" of course is suggestive of extraordinary appearance or value. Thus, the paradise duck and the paradise fish are brilliantly colored and the paradise tree discussed in this

chapter is exceptionally useful. As noted above, the name "tree of heaven" is sometimes applied to the paradise tree, but more usually to *Ailanthus altissima*. However, this tree that was once considered to be a most desirable ornamental has turned out to be a horrible, stinking, difficult-to-control weed that exudes chemicals in the soil, which repress the growth of other plants; it is now sometimes called "tree from hell."

- The edible fruits developed on female trees of paradise trees attract birds, which can scatter the fruits. Unfortunately, the fruits can stain concrete and decks. As with several other trees grown for ornament that produce messy fruits, the problem can be avoided by planting male trees.
- Under the name *aceituno*, paradise tree fruits are one of the fruits used in El Salvador homes to decorate the Holy Cross on *Dia de la Cruz* (Day of the Holy Cross on May 3).
- Paradise tree, like many other plants, has "extrafloral nectaries"—small glands that are not associated with the flowers and that secrete nutritive (often sugary) sap that attracts insects. These glands are on the adaxial (upper) leaf surfaces and are thought to attract ants which, in return for receiving food from the plants, serve to dissuade harmful insects from attacking the plants (Koptur et al. 2010).

## KEY INFORMATION SOURCES

Armour, R.P. 1959. Investigations on *Simarouba glauca* DC. in El Salvador. *Econ. Bot.* 13: 41–66.

Bangar, S.S., Deshmukh, A.G., Dudhare, M.S., and Moharil, M.P. 2009. Antifungal and antibacterial activities of *Simarouba glauca* leaf and bark extracts. *J. Plant Disease Sci.* 4: 237–238.

Bawa, K.S., and Opler, P.A. 1978. Why are pistillate inflorescences of *Simarouba glauca* eaten less than staminate inflorescences? *Evolution* 32: 673–676.

Behura, N.C., Dehuri, P.K., Mishra, S.K., and Das, S.K. 2010. Feeding value of detoxified (ammonia treated and roasted) simarouba (*Simarouba glauca*) oilcake in broilers. *Indian J. Anim. Sci.* 80: 916–920.

Bhatnagar, S., Polonsky, J., Prange, T., and Pascard, C. 1984. New toxic quassinoid glucosides from *Simarouba glauca* (X-ray analysis). *Tetrahedron Letters* 25: 299–302.

Brizicky, G.K. 1962. The genera of Simaroubaceae and Burseraceae in the southeastern United States. *J. Arnold Arbor.* 43: 173–186.

Chevelas Polito, J. 1982. *Simarouba glauca*, a native species with multiple uses. *Ciencia Forestal* 6: 3–16. (In Spanish)

Chikara, J., Shethia, B.D., Rathod, M., and Pandya, J.B., 1998. *Simarouba glauca*—A rich source of oleic acid for rehabilitation of marginal land of the country. *J. Oil Tech. Assoc. India* 30: 177–179.

Clayton, J.W., Fernando, E.S., Soltis, P.S., and Soltis, D.E. 2007. Molecular phylogeny of the tree-of-heaven family (Simaroubaceae) based on chloroplast and nuclear markers. *Int. J. Plant Sci.* 168: 1325–1339.

Cronquist, A. 1944. Studies in the Simaroubaceae II. The genus *Simarouba*. *Bull. Torrey Bot. Club* 71: 226–234.

Fattepurkar, S.C., Patil, M.S., Gawai, D.C., Zende, M.K., and Londhe, J.N. 2009. Phenolic compounds in *Simarouba glauca*. *Asian J. Exp. Chem.* 4: 101–103.

Fernando, E.S., Gadek, P.A., and Quinn, C.J. 1995. Simaroubaceae, an artificial construct, evidence from rbcL sequence variation. *Am. J. Bot.* 82: 92–103.

Govindaraju, K., Darukeshawara, J., and Srivastava, A.K. 2009. Studies on protein characteristics and toxic constituents of *Simarouba glauca* oilseed meal. *Food Chem. Toxicol.* 47: 1327–1332.

Jeyarani, T., and Yella Reddy, S.Y. 2001. Cocoa butter extender from *Simarouba glauca* fat. *J. Am. Oil Chem. Soc.* 78: 271–276.

Koptur, S., Pascale, W., and Olive, Z. 2010. Ants and plants with extraflora nectaries in fire successional habitats on Andros (Bahamas). *Florida Entomol.* 93: 89–98.

Kureel, R.S., Gupta, A.K., and Pandey, A. 2009. *Simarouba—A potential tree borne oilseed for edible oil*. National Oilseeds & Vegetable Board, Ministry of Agriculture, Government of India, Gurgaon, India. pp. 15.

Lognay, G., Wathelet, J.P., Severin, M., and Ergo, A. 1981. Chemical composition of the *Simarouba glauca* DC. endocarp. Laboratory study. *Rev. Fr. Corps Gras* 28: 319–321. (In French.)

Pradhan, K. 2000. *Simarouba glauca*—A promising species for farm and forest. *Advances Forest. Res. India* 23: 106–117.

Raghunandan, P., Reddy, P.N., and Jayaveera, K.N. 2002. Characteristics and composition of *Simarouba glauca* DC. seeds and fat. *J. Oil Tech. Assoc. India* 34: 63–65.

Ramajayam, D., Joshi, S., Joshi, S., and Mishra, P.K. 2007. Simarouba (*Simarouba glauca* DC.). In *Underutilized and underexploited horticultural crops*. Edited by K.V. Peter. New India Publishing Agency, New Delhi, India. pp. 345–368.

Rao, Y.R., Jena, K.S., Sahoo, D., and Rout, P.K. 2002. Characteristics of *Simarouba glauca* seeds and fat of Orissa origin. *J. Food Sci. Tech.* 39: 552–555.

Rao, K.V.S.A., and Lakshminarayana, G. 1983. Characteristics and composition of *Simarouba glauca* D.C. seed and fat. *J. Food Sci. Tech.* 20: 176–177.

Rivero-Cruz, J.F., Lezutekong, R., Lobo-Echeverri, T., Ito, A., Mi, Q., Chai, H.B., Soejarto, D.D., Cordell, G.A., Pezzuto, J.M., and Swanson, S.M. 2005. Cytotoxic constituents of the twigs of *Simarouba glauca* collected from a plot in southern Florida. *Phytother. Res.* 19: 136–140.

Satpathi, B. 1984. *Simarouba glauca*: An oil-yielding tree. *Indian Farm.* 34: 22–23.

Yamamoto, D., and Shepherd, G.J. 1999. Distinctions among three *Simarouba* species. *Syst. Bot.* 23: 479–478.

## Specialty Cookbooks

Recipe sources for paradise tree were not located in the literature or on the Web. The flesh of the fruit can be prepared in the same ways that most sour fruits can be processed.

# 71 Pawpaw

Family: Annonaceae (custard apple family)

## NAMES

Scientific name: *Asimina triloba* (L.) Dunal

- The origin of the name "pawpaw" is uncertain. The most common explanations are (1) that the word is based on a Native American word and (2) that the name was transferred from the papaya, which is also called pawpaw.
- In parts of the United States, father and mother are, respectively, called "pawpaw" and "mawmaw" (in the manner of "papa" and "mama").
- Pawpaw (paw paw, papaw) has also been called American custard apple, American pawpaw, Arkansas banana, banana pawpaw, common pawpaw, custard apple, custard tree, dog banana, false banana, fetid shrub, frost banana, Hoosier banana, Indian banana, Indiana banana, jasmine, Kentucky banana, Michigan banana, monim, North American pawpaw, northern pawpaw, pawpaw apple, poor man's banana, Virginia papwaw, West Virginia banana, and wild banana.
- The fruit often looks somewhat like a thick, short banana, and the taste is somewhat reminiscent, hence "banana" is in many of the common names.
- The names "Indiana banana" and "Hoosier banana" are the most frequent of the "banana names," reflecting the common occurrence of the species in Indiana. (Hoosiers come from Indiana, the "Hoosier state." The origin of the name Hoosier is obscure. It has been suggested that the word probably comes from the "mountaineer" meaning [i.e., rustic people living in mountains] of the word hoosier, based on the fact that the south of Indiana was mainly settled by Kentucky mountaineers.)
- The pawpaw has been recorded in more than half of the U.S. states, all of which can claim it as their own, leading to the facetious name "whatever-state-the-pawpaw-happens-to-grow-in banana."
- In Australia and sometimes in other countries, the tropical papaya, *Carica papaya* L., is also known as pawpaw.
- The genus name *Asimina* is based on one of the Indian names *assimin*, *arsimin*, or *rassimin*, taken up by French colonists as *asiminier.*
- *Triloba* in the scientific name *A. triloba* is Latin for three-lobed, in respect to the three-parted structure of the flower (which has three sepals, three small petals, and three large petals).

## GEOGRAPHY AND ECOLOGY OF WILD PLANTS

Pawpaw grows wild in the hardwood forests of 26 states in the eastern United States, and ranges from northern Florida to southern Ontario in Canada and as far west as eastern Nebraska. The trees are most common in Indiana, Ohio, Illinois, and Michigan. The Canadian plants occur in two separate regions in extreme southwestern Ontario, where the lower Great Lakes moderate the temperature (pawpaw has been observed to survive to –35°C or –35°F). Fossils show that during the Pleistocene era, the pawpaw grew as far north as Toronto. Native Americans made considerable use of pawpaws and may have extended the range of the species far beyond its natural growing area.

**FIGURE 71.1** Pawpaw (*Asimina triloba*). (a) Young fruits. (Courtesy of Scott Bauer, U.S. Department of Agriculture.) (b) Yellow-billed cuckoo in fruiting branches. (From Audubon, J.J., *Birds of America*, R. Havell Jr., London, 4 vols, Plate 2, 1827–1838.) (c) Flower. (Courtesy of Kerry Wixted [Flickr/CC-attribution].) (d) Flowering twig. (From Michaux, F.A., *Histoire des arbres forestiers de l'Amérique septentrionale*. L. Haussman, Paris, France, Vol. 3, 1813.) (e) Distribution map.

Papaws occur in deciduous forests, usually in deep, moist, rich soil along streams, rivers, and especially river flats. The plants are tolerant of shade and characteristically constitute part of the undergrowth in forests. In the main part of its range in the United States, the pawpaw is also found in unshaded, relatively dry sites. It has been suggested that the frequent proximity to water is related to the fact that the fruits are buoyant and the seeds are distributed by floating fruits. An equally plausible explanation is that native people traveling by canoe along watercourses distributed the seeds. Because of the conversion of forest to agricultural and urban landscapes, pawpaws have notably decreased in North America in the last 200 years, and native stands are now small and isolated over much of the range. Although much reduced and locally extirpated, pawpaw still has an extensive geographic distribution.

## PLANT PORTRAIT

The pawpaw is a small, deciduous tree, 5–12 m (16–40 feet) in height (a record tree in Mississippi was 18 m [60 feet] high). The plants tend to sucker extensively at their bases, producing thickets, sometimes affectionately referred to as "paw paw patches," as highlighted in the well-known children's song: "Where, oh where, is dear little Nellie? Way down yonder in the pawpaw patch." Very rarely, trees reach a trunk diameter of 1.5 m (5 feet). The trees are short-lived, but occasionally survive for more than 50 years. Pioneers made bread and pudding from the fruit. The pawpaw was popular in the United States early in the twentieth century, but became less significant subsequently. About a hundred varieties have been selected, although perhaps only 50 are available today. Attempts are underway to improve the fruit and to make it more marketable.

The fruits are typically 3–15 cm (1–6 inches) long, oblong to banana-shaped. Larger fruits are often plump, like a mango. The fruits are borne singly or in clusters, which resemble the "hands" of a banana plant. Occasional fruits reach 1 kg (2.2 pounds) in weight, but typically weigh 140–450 g (5–16 ounces). The skin is generally smooth and thin, varying from green to bright yellow at maturity and turning brown or black after a frost. When ripe, the custard-like flesh is typically orange, often yellow, and sometimes white. The fruit contains two rows of large, brown to blackish, bean-shaped, laterally compressed seeds that are 1.3–3.0 cm (0.5–1.3 inches) long. Large fruits usually have 10–15 seeds.

## CULINARY PORTRAIT

Pawpaw fruit is very sweet, pleasant, unique, and attractive, and indeed has been declared to be one of the five most delicious fruits in the world (Cheatham et al. 2000, *The Useful Wild Plants of Texas*, Vol. 2). The taste is reminiscent of papaya with pineapple overtones and hints of banana and mango. When cut open, the aroma from the fruits fills a whole room. The fruit may be eaten when it becomes soft although some prefer to wait until after the skin has darkened. Pawpaw fruits are best consumed fresh when fully ripe (but almost-ripe fruit will ripen well in storage). The thin, leathery skin and the seeds should not be eaten. Tree-ripened fruit stored at room temperature lasts only 2 or 3 days, but with refrigeration can be held up to 3 weeks. The fruits bruise easily and are extremely difficult to get to market in a usable form. Accordingly, pawpaws are mainly a local fruit. Specialty processed products, such as pawpaw ice cream, are occasionally available outside of the natural range of the species. Pawpaws are delicious in preserves, puddings, cookies, pies, cakes, and breads. They are particularly suitable for combining in desserts that have a custard base and are an excellent substitute in recipes calling for bananas. When eaten raw, some people find the taste of pawpaw cloyingly sweet.

Some individuals are allergic to the fruits, getting a skin rash from handling them or experiencing serious nausea, vomiting, or diarrhea from eating them. The seeds and unripe fruits contain chemicals that induce vomiting. The seeds can easily be removed and discarded after cutting the fruits in half like an avocado. Recipes that call for "pawpaw seeds" are referring to papaya, which as noted earlier is often called pawpaw outside of North America. Because some people will throw up from eating pawpaw, only a limited amount should be consumed the first time it is eaten.

### Culinary Vocabulary

- "Pawpaw butter," made by cooking pawpaw pulp with pectin and sugar, is a popular preparation of home cooks in the Ozarks.

## PROSPECTS

The pawpaw is an extraordinarily attractive fruit, but one that requires considerable breeding and research before it can be produced and marketed profitably. The areas needing attention are harvest methodology; shelf life; development of processed products; development of markets;

increased yield, particularly through pollination research; and breeding for desirable characteristics (reduced seed size and number, skin color, pulp color and texture, reduced toxins and allergens, improved flavor, and fruit uniformity). Researchers have recently highlighted the following advantages of pawpaw as a crop for eastern North America: (1) adaptation to existing climate and soil; (2) nutritional/cosmetic value of the fruit; (3) presence of valuable natural compounds; (4) established nursery wholesale and retail production; (5) relatively high level of resistance to pests and pathogens (attributed to natural defense compounds); (6) considerable information available on crop culture (including greenhouse seedling production); and (7) increasing popularity as evidenced by numerous recent articles with titles such as "The return of the pawpaw," "Pawpaws making a comeback," "Spotlight on pawpaw," "A tropical tree for the temperate zone," and many others. Prospects for commercialization appear to be improving. The pawpaw has been found to develop superbly in many areas well outside of its native range, so it should be possible to establish orchards outside of eastern North America. Pawpaw has recently been strongly promoted as a diversification crop, with considerable research carried out at Kentucky State University with the goal of making it into a new high-value fruit crop.

## CURIOSITIES OF SCIENCE AND TECHNOLOGY

- During the Pleistocene era (1.8 million to 10,000 years ago), large mammals ruled the Earth. At the end of the period, it is thought that the pawpaw was in serious danger of becoming extinct, because of the disappearance of the giant fruit-eating animals, such as the mastodon, which likely ate and distributed the seeds. It is believed that people migrated to the Americas from the Old World just in time, the first Native Americans finding the fruits so delicious that they took over the role of being the main distributors of the seeds. Some of the mammals on which the pawpaw depended may well have been hunted to extinction by humans, but at least people seem to have played a role in preserving the species.
- *Pawpawsaurus* is a genus of dinosaurs, which lived about 100 million years ago. The genus was first recognized when *P. campbelli* Lee was discovered in Texas in 1989 by 12-year-old Johnny Maurice. This species was named in 1996 after the Paw Paw formation (a Cretaceous Period shale in eastern Texas). The herbivorous animal is estimated to have been about 1.8 m (6 feet) high and 6 m (20 feet) long. It weighed about 1800 kg (2 tons), resembled an armadillo, and had bony armor to protect it from predatory dinosaurs.
- Thomas Jefferson (1743–1826, third president of the United States) authorized an expedition in 1803 to explore the country west of the Mississippi, especially the "Louisiana Purchase," an immense tract of territory obtained from Napoleon (1769–1821) for $15 million, doubling the size of the United States at a cost of 7 cents/ha (3 cents/acre). The expedition was led by his young secretary, Captain Meriwether Lewis (1774–1809), and Lewis's friend, Lieutenant William Clark (1770–1838). Both were familiar with the frontier and with Indians through their service in the army. The two men were accompanied by 14 soldiers, nine frontiersmen from Kentucky, two French boatmen, and Clark's slave, York. By 1805, after an adventurous journey of over 18 months, they reached the Pacific Ocean. The party returned to St. Louis in 1806. Their journey lasted 2 years and 4 months, during which they traveled about 9650 km (6000 miles). They brought back much new material for map makers and specimens of previously unknown wildlife (including 178 plants and 122 animals previously not recorded), and they established travel routes for American settlers and traders. Although an obvious success in hindsight, the expedition failed in its main mission, that is, to find the water passage Jefferson mistakenly believed linked the Atlantic and Pacific oceans. During the return trip, The Lewis and Clark expedition found itself running very low on rations, and with 240 km (150 miles) to go, had less than one biscuit for each man. The journal of the expedition recorded that they feasted contentedly on almost nothing but pawpaws until they reached St. Louis.

- During the famous feud between the Hatfields and the McCoys along the Kentucky–West Virginia border on August 9, 1882, three sons of clan leader Randolph McCoy were tied to pawpaws and executed by the Hatfields.
- As late as the early 1900s, fishermen in the Ohio valley used strips of the inner bark of pawpaw trees for stringing fish. They likely learned how to do this from Native Americans, who also used the bark strips to make fabric and nets.
- Except for the pawpaw genus *Asimina*, which grows in temperate regions, the custard apple family is entirely tropical.
- The importance of a plant is often indicated by its use as a geographical name. The states of Kansas, Kentucky, Illinois, Indiana, Michigan, Missouri, Oklahoma, and West Virginia have towns named Pawpaw or some variation of the name.
- The pawpaw fruit is the largest of any tree fruit native to the United States and Canada.
- Pollination of pawpaw is by flies and beetles. To attract these insects, the flowers have dark, meat-colored petals and a fetid aroma. To insure pollination so that fruits will be produced, roadkill is sometimes collected and hung in trees of commercial plantings to attract flies!
- Racoons are quite fond of pawpaw fruit. In the United States, "coon" hunters often start their dogs in or near a pawpaw thicket in hopes of finding a fresh trail.
- Pawpaw fruit is about 75% water.
- Pawpaw plants produce annonaceous acetogenins and natural pesticide compounds, which have highly antitumor properties, currently under investigation for possible anticancer applications.

## KEY INFORMATION SOURCES


Archbold, D.D., Koslanund, R., and Pomper, K.W. 2003. Ripening and postharvest storage of pawpaw. *HortTechnology* 13: 439–441.

Archbold, D.D., and Pomper, K.W. 2003. Ripening pawpaw fruit exhibit respiratory and ethylene climacterics. *Postharvest Biol. Technol.* 30: 99–103.

Barrett, T. 1994. The pawpaw, a checkered past and promising future. *Small Farm Today* 11(3): 55–56.

Blossom, M. 1993. The pawpaw project. *Annu. Rep. North Nut. Grow. Assoc.* 84: 197–199.

Bowden, W.M., and Miller, B. 1951. Distribution of the papaw, *Asimina triloba* (L.) Dunal, in southern Ontario. *Can. Field-Nat.* 65: 27–31.

Callaway, M.B. 1990. *The pawpaw,* Asimina triloba. Kentucky State University, Frankfort, KY. pp. 22.

Callaway, M.B. 1992. The pawpaw. *Annu. Rep. North Nut. Grow. Assoc.* 83: 106–119.

Callaway, M.B. 1993. Pawpaw (*Asimina triloba*): A "tropical" fruit for temperate climates. In *New crops*. Edited by J. Janick and J.E. Simon. Wiley, New York, NY. pp. 505–515.

Callaway, M.B., and Callaway, D.J. 1992. Our native pawpaw: The next new commercial fruit? *Arnoldia* 52: 20–29.

Finneseth, C.H., Layne, D.R., and Geneve, R.L. 1998a. Morphological development of the North American pawpaw during germination and seedling emergence. *HortScience* 33: 802–805.

Finneseth, C.H., Layne, D.R., and Geneve, R.L. 1998b. Requirements for seed germination in North American pawpaw (*Asimina triloba* (L.) Dunal). *Seed Sci. Technol.* 26: 471–480.

Geneve, R.L., Pomper, K.W., Kester, S.T., Egilla, J.N., Finneseth, C.L.H., Crabtree, S.B., and Layne, D.R. 2003. Propagation of pawpaw—A review. *HortTechnology* 13: 428–433.

Greer, L., and Ames, G.K. 1997. *Pawpaw production*. Appropriate Technology Transfer for Rural Areas [Organization], Fayetteville, AR. pp. 5.

Huang, H., Layne, D.R., and Kubisiak, T.L. 2003. Molecular characterization of cultivated pawpaw (*Asimina triloba*) using RAPD markers. *J. Am. Soc. Hortic. Sci.* 128: 85–93.

Huang, H., Layne, D.R., and Riemenschneider, D.E. 1998. Genetic diversity and geographic differentiation in pawpaw populations from nine states as revealed by allozyme analysis. *Am. Soc. Hortic. Sci.* 123: 635–641.

Keener, C., and Kuhns, E. 1997. The impact of Iroquoian populations on the northern distribution of pawpaws in the northeast. *North Am. Archaeol.* 18: 327–342.

Kral, R. 1960. A revision of *Asimina* and *Deeringothamnus* (Annonaceae). *Brittonia* 12: 233–278.

Kral, R. 1997. *Asimina*. In *Flora of North America north of Mexico*. Vol. 3. Edited by Flora of North America Editorial Committee. Oxford University Press, Oxford, UK. pp. 14–19.

LaGrange, R.L., and Tramer, E.J. 1985. Geographic variation in size and reproductive success in the paw paw *Asimina triloba*. *Ohio J. Sci.* 85: 40–45.
Layne, D.R. 1996. The pawpaw: A new fruit crop for Kentucky and the United States. *HortScience* 31: 777–784.
McCage, C.M., Ward, S.M., Paling, C.A., Fisher, D.A., Flynn, P.J., and McLaughlin, J.L. 2002. Development of a paw paw herbal shampoo for the removal of head lice. *Phytomedicine* 9: 743–748.
McGrath, M.J., and Karahadian, C. 1994a. Evaluation of headspace volatiles and sensory characteristics of ripe pawpaws (*Asimina triloba*) from selected cultivars. *Food Chem.* 51: 255–262.
McGrath, M.J., and Karahadian, C. 1994b. Evaluation of physical, chemical, and sensory properties of pawpaw fruit (*Asimina triloba*) as indicators of ripeness. *J. Agric. Food Chem.* 42: 968–974.
McLaughlin, J.L. 1997. Anticancer and pesticidal components of pawpaw (*Asimina triloba*). *Ann. Rpt. North Nut. Grow. Assoc.* 88: 49–57.
Mohana Rao, P.R. 1982. Seed and fruit anatomy in *Asimina triloba*, with a discussion on the affinities of Annonaceae. *Bot. Jahrb. Syst. Pflanzengesch. Pflanzengeogr.* 103: 47–57.
Peterson, R.N. 1990. Pawpaw (*Asimina*). In *Genetic resources of temperate fruit and nut crops*. Edited by J.N. Moore and J.R. Ballington Jr. Int. Soc. Hort. Sci., Wageningen, the Netherlands. pp. 569–600.
Peterson, R.N. 2003. Pawpaw variety development: A history and future prospects. *HortTechnology* 13: 449–454.
Peterson, R.N., Cherry, J.P., and Simmons, J.G. 1982. Composition of pawpaw (*Asimina triloba*) fruit. *Annu. Rep. North Nut. Grow. Assoc.* 73: 97–107.
Pomper, K.W., Crabtree, S.B., Brown, S.P., Jones, S.C., Bonney, T.M., and Layne, D.R. 2003. Assessment of genetic diversity of pawpaw (*Asimina triloba*) cultivars with intersimple sequence repeat markers. *J. Am. Soc. Hortic. Sci.* 128: 521–525.
Pomper, K.W., and Layne, D.R. 2000. The Kentucky State University pawpaw story. *Pomona* 33: 13–19.
Pomper, K.W., and Layne, D.R. 2005. The North American pawpaw: Botany and horticulture. *Hort. Rev.* 31: 351–384.
Pomper, K.W., Layne, D.R., and Jones, S.C. 2003. Container production of pawpaw seedlings. *HortTechnology* 13: 434–438.
Ratnayake, S., Rupprecht, J.K., Potter, W.M., and McLaughlin, J.L. 1992. Evaluation of various parts of the paw paw tree, *Asimina triloba* (Annonaceae), as commercial sources of the pesticidal annonaceous acetogenins. *J. Econ. Entomol.* 85: 2353–2356.
Rexwinkle, E. 2000. The pawpaw patch—New promises from an old fruit. *AgVentures* 4(5): 10–14.
Rogstad, S.H., Wolff, K., and Schaal, B.A. 1991. Geographical variation in *Asimina triloba* Dunal (Annonaceae) revealed by the M13 "DNA Fingerprinting" probe. *Am. J. Bot.* 78: 1391–1396.
Shiota, H. 1991. Volatile components of pawpaw fruit (*Asimina triloba* Dunal.). *J. Agric. Food Chem.* 39: 1631–1635.
Small, E., and Catling, P.M. 2002. Poorly known economic plants of Canada—34. Pawpaw (Canada banana), *Asimina triloba* (L.) Dunal. *Can. Bot. Assoc. Bull.* 35(3): 28–33.
Stevenson, P. 1990. A passion for the pawpaw. *Am. Forests* 96(3/4): 40–49.
Templeton, S.B., Marlette, M., Pomper, K.W., and Jones, S.C. 2003. Favorable taste ratings for several pawpaw products. *HortTechnology* 13: 445–448.
Thomson, P.H. 1982. The paw paw. *California Rare Fruit Growers Yearb.* 14: 5–31.
Willson, M.F., and Schemske, D.W. 1980. Pollinator limitation, fruit production, and floral display in pawpaw (*Asimina triloba*). *Bull. Torrey Bot. Club* 107: 401–408.
Wood, R., and Peterson, S. 1999. Lipids of the pawpaw fruit: *Asimina triloba*. *Lipids* 34: 1099–1106.
Young, D.R. 1987. Daily and seasonal variations in the water relations of the understory tree *Asimina triloba*. *Acta Oecologica* 8: 59–68.
Zhao, G.X., Rieser, M.J., Hui, Y.H., Miesbauer, L.R., Smith, D.L., and McLaughlin, J.L. 1993. Biologically active acetogenins from stem bark of *Asimina triloba*. *Phytochemistry* 33: 1065–1073.

## Specialty Cookbooks

(*Note:* Recipes for using "pawpaw" are widely available on the Web, but are often for papaya, not pawpaw.)

Jones, S.C., and Layne, D.R. 1997. *Cooking with pawpaws*. Kentucky State University, Frankfort, KY. pp. 12.
Kluger, M. 1990. *The wild flavor. Delectable wild foods to be found in field and forest and cooked in country kitchen*. Henry Holt & Co., New York, NY. pp. 285.

Beatty presents six recipes featuring pawpaw, Brill (2002) has three, Freitus (1980) has three, Freitus and Haberman (2005) have six, Kluger (1973) has eight, Robe-Terry (1997) has five, and Schufer (2011) has two. (See the Appendix to this book for details of the publications cited.)

# 72 Pecan

Family: Juglandaceae (walnut family)

## NAMES

Scientific name: *Carya illinoinensis* (Wangenh.) C. Koch (commonly misspelled "*illinoensis*"; *C. oliviformis* (Michx.) Nutt., *C. pecan* (Marshall) Engl. & Gräbn.)

- "Pecan" is based on the Algonquin Indian *paccan*, meaning "hard-shelled nut" or "all nuts requiring a stone to crack." This originally included pecans, walnuts, and hickories. (A less likely explanation of the word pecan is that it is based on the Algonquin chief Peccan.) The French settlers in Louisiana adopted the name "pacanes" for the pecan, and the word evolved in English to pecan.
- Pecan is variously pronounced: pih-KAHN, pih-KAN, and PEE-kan.
- The pecan is also known as the sweet pecan.
- The genus name *Carya* is the classical Latin word for walnut, and was adopted by the botanist Thomas Nuttall in 1818 when he named the genus. The Latin name came from the Greek word for the walnut tree, *karua*. According to Greek mythology, Karua, that is, Carya, was the daughter of the king of Laconia (a region of far southern Greece, where supposedly the entrance to Hades was located), and she was changed into a walnut tree by Dionysus, the god of wine, for obscure reasons.
- *Illinoinensis* in the scientific name *C. illinoinensis* refers to the territory of Illinois, where the species was said to originate when it was first described (an alternative explanation holds that the word refers to the Illinois tribe on whose land the species grew).

## GEOGRAPHY AND ECOLOGY OF WILD PLANTS

The pecan is native to the valley of the Mississippi River, reaching north to southern Indiana, Illinois, western Kentucky, and Tennessee, ranging south to Mississippi and Louisiana, and west to Texas. It is also found in the mountains of northern Mexico and has been introduced in the southwestern United States. The pecan occurs in floodplains, bottomlands, and open woods. The distribution of pecan is associated with the Mississippi River drainage system of the United States. In this area, the climate ranges from humid to semiarid, with mild to harsh winters. Its very deep rooting system allows pecan to survive periods of drought. The pecan does best on loamy, well-drained land with a relatively high water table and soil that is more or less neutral in pH. The species is shade intolerant.

## PLANT PORTRAIT

The pecan is the most important native nut tree of North America. This deciduous tree occasionally grows to over 61 m (200 feet) in height, but usually is less than 30 m (100 feet). Some trees with massive trunks more than 2 m (about 7 feet) are known. There are reports of specimens having reached trunk diameters of 3.5 m (about 11 feet), but no known living tree is this large. The largest trees develop in southern Arkansas and eastern Texas. Pecan is long-lived, and some trees in the southeast of the United States are more than 1000 years old. Large trees can yield 181 kg (400 pounds) of

**FIGURE 72.1** Pecan (*Carya illinoinensis*). (a) Harvesting nuts with long poles in an orchard. ([a] and [b] Courtesy of Jerry A. Payne, U.S. Department of Agriculture, online at Bugwood.org [CC By 3.0].) (b) Fruits splitting open on branch. (c) Pecan pie. (Courtesy of Jonathunder [CC By 3.0].) (d) Kernels. (Courtesy of J.E. Theriot [Flickr/CC-attribution].) (e) Distribution map.

nuts in a single year. Pecans are cultivated mostly in the southeastern United States. Some connoisseurs maintain that pecans grown in the northern part of the range produce the best nuts.

Pecan nuts resemble walnuts, but are more elongated, with a smoother shell and a greater proportion of kernel in the shell. Pecan kernels look like slightly elongated and flattened walnut kernels, but have a somewhat sweeter, distinctive flavor and richer texture. The nuts of wild trees tend to be tart because of the presence of tannin in the thin covering skin. Unlike the walnut, when mature the husk of the pecan splits open into four segments, the husk splitting to below the middle. The unhusked nuts are 3.5–8 cm (1.4–3.1 inches) long, while husked nuts range from 2.5 to 7 cm (1–2.8 inches) in length. The husked nuts are smooth, light, or reddish brown.

Native American tribes in the south central region of the United States, especially in the Mississippi Valley, relied on the pecan as a dietary staple long before the arrival of Europeans.

Much of the early improvement of the pecan was carried out by Native Americans. Native peoples planted nuts in the vicinity of their campsites to provide food for their descendants. The largest, tastiest, and thinnest shelled selections were preferred, and thus were selected. More than 500 varieties of pecans exist today. "Paper shell" varieties are simply thin-shelled cultivars. The pecan industry in the United States largely developed during the twentieth century. More than 900 metric tons (about 200 million pounds) are produced annually. The most important commercial pecan-growing states are Alabama, Arizona, Arkansas, California, Florida, Georgia, Louisiana, Mississippi, New Mexico, North Carolina, Oklahoma, South Carolina, and Texas. Kansas, Kentucky, Maryland, Missouri, Tennessee, and Virginia also grow some pecan trees. Georgia is usually the largest producer, followed by Texas, Alabama, New Mexico, and Louisiana. The United States grows more than 80% of the world's pecans, and Mexico is the world's second largest producer. Outside of the United States, pecan planting has been carried out mainly in Mexico, Australia, Brazil, South Africa, Israel, Peru, and several other countries.

Pecan is an excellent hardwood, and pecan veneer and lumber are in demand for decorative paneling, flooring, and fine furniture. In addition to the culinary uses mentioned later, pecan oil extracted from the nuts is used to a limited extent in cosmetics and other special preparations. A minor problem with the pecan in its native area is its very large production of pollen in April and May, known to cause a severe form of hay fever in susceptible individuals.

## CULINARY PORTRAIT

Pecans are eaten raw, roasted, or salted, and used in confections, mixed nuts, cakes and other pastries, nut bread, and used for flavoring in baking and cooking. Particularly popular foods are pralines, pecan brittle, chocolate-covered pecan halves, pecan candies and fudge, and pecan ice cream. Pecan pie, made principally of pecans, molasses, and maple syrup, is a favorite dessert of the southern United States. Pecan nuts are a gourmet's delight, being relatively soft-textured and mild in flavor. The kernels are also used in home cooking, especially as meat substitutes, and are popular in salads, puddings and other desserts, stuffings (especially for poultry), creamed chicken, croquettes, and fish, and as a sandwich filling mixed with dates or raisins.

Pecans are among the most popular table nuts for a Thanksgiving or Christmas display because of their flavor and the attractive appearance of the shells. Manufacturers have found that the nuts sell better after the shells are washed, sanded, waxed, and polished. Heavily streaked shells are commonly bleached and the reddish brown color is enhanced by dyeing. The best varieties crack out in perfect wholes or halves, after one has figured out how to get the slippery shells to stay in the groove of the nutcracker. Pecan nuts absorb moisture and become stale or rancid more quickly than most other nuts. Therefore, they require a little more care in storing and processing to ensure freshness and crunchiness after shelling. It is difficult to purchase high-quality shelled pecans in bulk, and such nuts often are rancid or lacking in flavor. Buying nuts in their shells or shelled nuts in vacuum-sealed jars or cans is preferable. When purchasing fresh pecans in their shell, the best nuts will not rattle when shaken, will not be cracked or have holes, are heavy, and have not been stained artificially. Most nuts that reach consumers will have been carefully dried or toasted, stored at low temperature, and often are coated with antioxidants to prevent moisture absorption or staleness in candy, baked goods, or ice cream. The tendency to lose freshness quickly somewhat limits the use of pecans in granola-type cereals and nut mixes. Salted nuts are packaged in small quantities to ensure freshness, but as they are addictive, they seldom last long enough to get stale. For cooking, pecans rival the Persian (English) walnut in popularity within North America, but not in overseas markets where the Persian walnut has been in use for at least 2000 years. In the southern states the pecan has no serious rival. Specialties such as a Texan fruitcake, loaded with pecans and packaged in ornamental tins, have been shipped all over the continent. It has been estimated that over a quarter of the annual pecan production goes into bakery products, of which more than 4 million kg (8.8 million pounds) are used in fruitcakes alone. Although pecans are used in the same ways as walnuts, and walnuts are widely substituted for them both commercially and by home cooks, there are a few specialties such as pecan pie, pecan ice cream, and pralines in which only pecans can produce the desired sweet, devastatingly rich effect. Pecans are high in oil (65%–75%), and pecan oil is used in limited quantities for cooking. Because of the high oil and low water content, the food value of pecan kernels is very high—about 8000 calories/kg—more than four times that of most meats.

### Culinary Vocabulary

- "King cake" has been prepared in France as far back as the twelfth century, when such cakes were used to celebrate the coming of the three wise men to honor the birth of Christ. When French settlers came to New Orleans about 1870, they brought with them the tradition of preparing king cakes. At that time, it became customary to hide pecans (or other objects such as coins, beans, or peas) inside of king cakes. Wealthy Louisiana plantation owners sometimes placed precious stones or jewels in the cakes. In the mid-1900s, a small plastic baby representing Christ was placed in a cake served at a party, and anyone finding it was crowned King or Queen for the day, but had the obligation of hosting the following year's party.
- A praline is a confection made with almonds coated with caramelized sugar and nuts, usually almonds or hazelnuts (see Chapter 47 on Hazelnuts for additional information). In Texas and other areas of the United States, pralines are typically prepared with pecans.

## PROSPECTS

The pecan is one of the world's major nut crops and is produced mainly in the United States. The demand for it is high and stable, and the main competition comes from the world's other major nut species. The pecan is an important crop in many of the U.S. states, and receives considerable research and development support that ensures its future. Hurricanes, insects, and diseases are continuing concerns for the industry, and climate change may also pose a threat.

## CURIOSITIES OF SCIENCE AND TECHNOLOGY

- Alvar Núñez Cabeza de Vaca, a Spanish explorer, was captured by Indians in 1528 on an island near the Texas coast, and imprisoned for several years. He recorded that his captors lived on nothing but pecans for 2 months of the year.
- In 1847, the slave gardener "Antoine" (the only name by which he is remembered), on Oak Alley Plantation (now a U.S. National Historic Landmark) in southern Louisiana, perfected a technique of grafting a new type of pecan onto hardier pecan trees. This was a milestone in the development of the pecan industry, and in 1876 Antoine's new variety was given the name Centennial to commemorate the hundredth anniversary of the United States.
- An old law in Seminole County, Oklahoma, called for a fine of 5 dollars for anyone mutilating a pecan tree. The law was intended to discourage Native Americans from the traditional practice of chopping off the limbs of trees to facilitate gathering the nuts on private land.
- In 1906, James Stephen Hogg, the first native-born governor of Texas, declared on his death bed: "I want no monument of stone or marble, but plant at my head a pecan tree and at my feet an old-fashioned walnut... and when these trees shall bear, let the pecans and walnuts be given out among the plain people of Texas so that they may plant them and make Texas a land of trees." Following his wishes, trees were planted at Hogg's Austin grave, and their nuts were distributed for years, giving a start to many a Texas pecan grower.

- Texas adopted the pecan as its official state tree in 1919, becoming the first U.S. state to select a state tree (all of the other states have since chosen official trees). It has been estimated that there are 75 million wild pecan trees in the state. In 2001, Texas declared the pecan to be its official state "Health Nut."
- Alabama selected the pecan as its official "state nut" in 1982.
- Oklahoma declared pecan pie to be its official "state pie" in 1988.
- In 1988, at the Pecan Festival at Okmulgee, Oklahoma, the "world's largest pecan pie" was prepared. It was 12 m (40 feet) in diameter, and weighed 15 metric tons (16.5 tons).
- Sixty pecan trees were used to produce handles for the 10,000-plus Olympic torches carried in the 24,000 km (15,000 miles) relay for the 1996 Summer Olympics that culminated in the lighting of the Olympic flame in Atlanta, Georgia.
- The largest pecan orchard in the world is in the desert south of Tucson Arizona. This irrigated farm has 2400 ha (6000 acres) with more than 300,000 pecan trees.
- Pecan nuts were taken to the moon on the Apollo 13 and 14 space missions.
- Pecan pies use on average 227–340 g (8–12 ounces) of pecans (there are about 78 pecans used in every pecan pie). In 2010, the states of Georgia and Texas harvested about 320,000 metric tonnes (70 million pounds) of pecans, enough to make 140 million pecan pies, more than required to circle the earth [97,812,000 23 cm (9 inch) pies would be needed].
- Stacked end to end, it would take 11,624 pecans to reach the top of New York's Empire State Building. To reach the moon would require more than 10 billion pecans.
- Occasional large pecan trees have been known to produce more than 400 kg (more than 1000 pounds) in some years.
- More butter pecan ice cream is consumed in Texas than in any other state.
- In 1664, the Abbot de Rancé (1626–1700) at the monastery of La Trappe in Normandy, France, established the Trappists, a reformed branch of the Roman Catholic Cistercian Order. Trappist monasteries have been founded in many countries, and are required to be self-supporting by the labor of the monks. In the United States, making fruitcakes is one of the successful activities of the Trappists, especially at Assumption Abbey, located in the foothills of the Missouri Ozarks. The finished cakes are decorated with four pecan halves—carefully placed to form a cross.

## KEY INFORMATION SOURCES


Brison, F. 1986. *Pecan culture*. Texas Pecan Growers Association, College Station, TX. pp. 297.

Burkett, J.H. 1924. *The pecan in Texas, the state tree. The pecan: Its history, importance, economic value. Pecan streams, soils, orchards, production, insects, diseases, grades, propagation, etc.* Texas Department of Agriculture, Austin, TX. pp. 177.

Burkett, J.H. 1932. *The pecan in Texas: Pecan soils, districts, streams, orchards, groves, care, propagation, cultivation, covercrops, irrigation, production, insects, diseases, etc.* Texas Department of Agriculture, Austin, TX. pp. 239.

Campbell, R.D. 1978. Pecans for the true north. *Annu. Rep. Northern Nut Growers Assoc.* 69: 55–59.

Goff, W.D. 1989. *Pecan production in the southeast: A guide for growers.* Alabama Cooperative Extension Service in cooperation with the U.S. Department of Agriculture, Auburn University, AL. pp. 222.

Heaton, E.K. 1977. *Pecans: Handling, storage, processing & utilization.* Agricultural, Experiment Stations, University of Georgia, Athens, GA. pp. 79.

Herrera, E. 2000. *Pecan orchard management.* Cooperative Extension Service, New Mexico State University, Las Cruces, NM. pp. 60.

Hinrichs, H.A., and Ramming, D.W. 1973. Flowering and nut development of pecans. *Proc. Okla. Pecan Growers Assoc.* 43: 21–28.

Logsdon, G. 1979. Pecans grow surprisingly far north. *Org. Gard.* 26(11): 71–76.

Madden, G. 1976. Hardiness and fruiting of pecans in the north. *Annu. Rep. Northern Nut Growers Assoc.* 67: 32–35.

Madden, G.D., and Malstrom, H.L. 1975. Pecans and hickories. In *Advances in fruit breeding*. Edited by J. Janick and J.N. Moore. Purdue University Press, West Lafayette, IN. pp. 420–438.
Manaster, J. 1994. *The pecan tree*. University of Texas Press, Austin, TX. pp. 109.
McEachern, G.R., and Stein, L.A. 1986. *Planting and establishing pecan trees*. Texas Agricultural Extension Service, Texas A & M University System, College Station, TX. pp. 8.
Miller, G. 1977. The origin and development of pecan varieties. *Fruit Var. J.* 31: 16–21.
Millikan, D.F., Hibbard, A.D., and Guengerich, H.W. 1968. Planting problems with pecans. *Fruit. Var. Hort. Digest* 22: 2.
Nakayama, R.M. 1967. *Pecan variety characteristics*. Agricultural Experiment Station, New Mexico State University, Las Cruces, NM. pp. 13.
Osburn, M.R. 1966. *Controlling insects and diseases of the pecan*. Revised edition. U.S. Department of Agriculture, Washington, DC. pp. 55.
Payne, J.A., Malstrom, H.L., and KenKnight, G. 1980. *Insect pests and diseases of the pecan*. U.S. Department of Agriculture, Southern Fruit and Tree Nut Research Laboratory, New Orleans, LA. pp. 43.
Ree, B., and Knutson, A. 1997. *Field guide to the insects and mites associated with pecan*. Texas Agricultural Extension Service, Texas A & M University System, College Station, TX. pp. 127.
Rice, G.W. 1994. *Pecans: A grower's perspective*. PecanQuest Publications, Ponca City, OK. pp. 198.
Ritchie, J.D. 1979. What do you get when you cross a hickory with a pecan? You get a great-tasting nut called a hican. *Org. Gard.* 26(11): 77–79.
Santerre, C.R. 1994. *Pecan technology*. Chapman & Hall, New York. pp. 164.
Shuhart, DV. 1932. Morphology and anatomy of the fruit of *Hicoria pecan*. *Bot. Gaz.* 93: 1–20.
Sparks, D. 1974. The alternating fruit bearing problem in pecans. *Ann. Rep. North. Nut Growers Assoc.* 65: 145–158.
Sparks, D. 1986. Pecan. In *CRC handbook of fruit set and development*. Edited by S.P. Monselise. CRC Press, Boca Raton, FL. pp. 323–339.
Sparks, D. 1992. *Pecan cultivars: The orchard's foundation*. Pecan Production Innovations, Watkinsville, GA. pp. 446.
Sparks, D. 2005. Adaptability of pecan as a species. *HortScience* 40: 1175–1189.
Thompson, T.E., and Grauke, L.J. 1990. Pecans and other hickories (*Carya*). In *Genetic resources of temperate fruit and nut crops*. Edited by J.N. Moore and J.R. Ballington, Jr. International Society for Horticultural Science, Wageningen, the Netherlands. pp. 839–904.
Thompson, T.E., and Young, F. 1985. *Pecan cultivars: Past and present*. Texas Pecan Growers Association, College Station, TX. pp. 265.
True, R.H. 1919. *Notes on the early history of the pecan in America*. Smithsonian Institution Annual Report 1917. Washington, DC. pp. 435–448.
Wolstenholme, B.M. 1979a. The ecology of pecan trees. Part 1. Characteristics of the native habitat. *Pecan Quarterly* 13(2): 32–35.
Wolstenholme, B.M. 1979b. The ecology of pecan trees. Part 2. Climatic aspects of grown pecans. *Pecan Quarterly* 13(3): 14–19.
Woodroff, J.G., and Woodroff, N.C. 1927. The development of the pecan nut (*Hicoria pecan*) from flower to maturity. *J. Agric. Res.* 34: 1049–1063.

## Specialty Cookbooks

Altrusa Club of Tallahassee. 1978. *Pecan recipes*. Altrusa Club of Tallahassee, Tallahassee, FL. pp. 57.
Anonymous. 1925. *Eight hundred proved pecan recipes, their place in the menu, by 5083 housewives*. Keystone Pecan Research Laboratory, Manheim, PA. pp. 393.
Bastrop County Pecan Growers' Auxiliary. 2000. *Texas pecan cookbook*. G & R Publishing, Waverly, IA. pp. 277.
Bertling, T.S. 1986. *Cooking with pecans: Texas in a nutshell*. Eakin Press, Austin, TX. pp. 142.
Blazek, M. 1986. *Pecan lovers' cook book*. Golden West Publishers, Phoenix, AZ. pp. 120.
Courrégé, K. 1984. *Pecans, from soup to nuts!* Cane River Pecan Company, Natchitoches, LA. pp. 54.
Heyl, A. 1994. *Nuttin' but pecans*. Eastern Virginia Medical School Women, Norfolk, VA. pp. 256.
Johnson, R.H., and Rousseau, S.K. 1984. *The Great American cookbook: Featuring pecans—America's favorite nut*. National Pecan Marketing Council, Atlanta, GA. pp. 158.
Keystone Pecan Research Laboratory. 1925. *800 proved pecan recipes: Their place in the menu*. Macrae Smith Company, Philadelphia, PA. pp. 393.
Miller, C.L. 1980. *The pecan lovers' cookbook*. Myrtle Bank Publishers, Natchez, MS. pp. 78.

Snyder, R.M. 1994. *Pecans, past and repast*. Morris Publishing, Kearney, NE. pp. 68.
Texas Department of Agriculture. 1979. *Texas pecan recipes*. Revised edition. Department of Agriculture, Austin, TX. pp. 15.
Thames Pecan Co. 1987. *Pecan cookery*. Fourth edition. H.M. Thames Pecan Company, Mobile, AL. pp. 150.
Trull, E. 1993. *Nuts about pecans: recipes*. McClanahan Publishing House, Kuttawa, KY. pp. 128.
Wallace, C.E., and Harris, T. 1990. *Nuttin' 'cept pecan pies: Sixty-nine perfectly provocative pecan pie recipes from the traditional to the bizarre*. C.E. Wallace, Austin, TX. pp. 58.
Zehr, F.J. 1967. *The pecan cookbook*. Koinonia Publication, Americus, GA. pp. 202.
Zonta Club of Dallas. 1982. *The Dallas pecan cookbook: A collection of recipes used for gracious living and entertaining*. Cookbook Publishers, Lenexa, KS. pp. 130.

# 73 Piñon Pine

Family: Pinaceae (pine family)

## NAMES

Scientific name: *Pinus edulis* Engelm. (*P. cembroides* Zucc. var. *edulis* (Engelm.) Voss)

- Predominantly in English today, the word piñon, with the Spanish character ñ (an n with a tilde above it), is transliterated in English (especially American English) as pinon, a practice that is not followed in this chapter.
- The piñon pine is also known as American piñon, Colorado piñon, common piñon, common piñon pine, New Mexican piñon, nut pine, piñon, Rocky Mountain nut pine, Rocky Mountain piñon pine, true piñon, two-leaf nut pine, two-leaf piñon, two-leaved pine, and two-needle piñon.
- The genus name *Pinus* is based on the Latin *pinus* which came from the Greek *pitys*, pine or fir tree. The English word "pine" is derived from the Latin *pinus*.

## GEOGRAPHY AND ECOLOGY OF WILD PLANTS

The most important American pine used as a source of nuts, piñon pine (*P. edulis*), grows wild mostly in the southwestern Rocky Mountain region, including southwest Wyoming, Utah, western Colorado, the extreme western tip of Oklahoma, western Texas, New Mexico, adjacent Chihuahua (northwestern Mexico), and eastern Arizona. This species typically grows in pure stands or with juniper (species of *Juniperus*); together these species define an ecological zone, the piñon-juniper life zone of the intermountain region of western North America, a semidesert area covering 25,000,000 ha (61,000,000 acres). The less important single-leaf piñon, *P. monophylla* Torr. & Frém. is found in mountainous regions of Arizona, Utah, Nevada, California, and southward to Baja California. Another pine used for nuts (although the shells are very thick), Mexican piñon (*P. cembroides*), occurs in the Mexican states of Sonora, Chihuahua, and Puebla. Hybridization is known to occur between *P. edulis* and the other two species. Still another pine, the digger pine (*P. sabiniana* Douglas), has been proposed as meriting development as a new nut crop (Farris 1983).

The piñon pine occurs in woodlands, high plains, canyons, plateaus, lower mountain slopes, and mesas (small, high plateaus or flat tableland with steep sides, especially in the southwestern United States) at elevations of 1200–2750 m (4000–9000 feet). The species is adapted to a wide range of soils, moisture conditions, and temperature. It typically occupies dry, rocky soils and is very drought-resistant. According to Williamson (1995, cited in the Appendix to this book), "Unfortunately many of the stands that recently existed on public lands have been cut down by the Forest Service to encourage the growth of grasses to provide more grazing for cattle."

## PLANT PORTRAIT

"True pines" are species of the genus *Pinus*. There are about 100 species, almost all in north temperate areas of the world. About a dozen produce seeds that are sufficiently large, common, and tasty to be collected and sold commercially as "pine nuts," but for the most part these are marketed and consumed locally. Three species are most likely to appear as pine nuts in supermarkets. These are

**FIGURE 73.1** Piñon pine (*Pinus edulis*). (a) An old tree on the North Rim of the Grand Canyon, Arizona. (Courtesy of bloodredrapture [CC by 2.0].) (b) Seeds. (Courtesy of Steve Hurst, U.S. Department of Agriculture.) (c) Seeds emerging from cone. (Courtesy of Curtis Clark [CC By 2.5].) (d) Distribution map.

the piñon pine featured in this chapter (the nuts are known as piñon nuts or pinyon nuts); the Italian stone pine (pignolia pine, pignolia nut pine), *P. pinea* L. (the nuts are knows as pignolia nuts); and the Korean pine (Chinese pine), *P. koraiensis* Siebold & Zucc. (the nuts are known as Chinese or Korean pine nuts). Piñon nuts are harvested mainly from the Colorado piñon pine, *P. edulis*. This is a small, bushy (scrubby) tree growing occasionally to 15 m (49 feet), but usually less than half this height. The tallest recorded height of the species is 21 m (69 feet). Piñon pine does not start to bear nuts until the age of 25 years, and does not reach full production until the age of 75. The trees are slow-growing; it can take hundreds of years for the trunk to reach a diameter of 10 cm or 4 inches, and not surprisingly the species is used to produce dwarf bonsai plants. However, piñon pine can live for a thousand years, and specimens more than 700 years of age are not uncommon. The cones contain 15 to 30 small seeds, 10–15 mm (0.4–0.6 inch) in length. A large tree in a good year (which may be only once in 7 or 8 years) can yield over 9 kg (20 pounds) of seed. The seeds are harvested

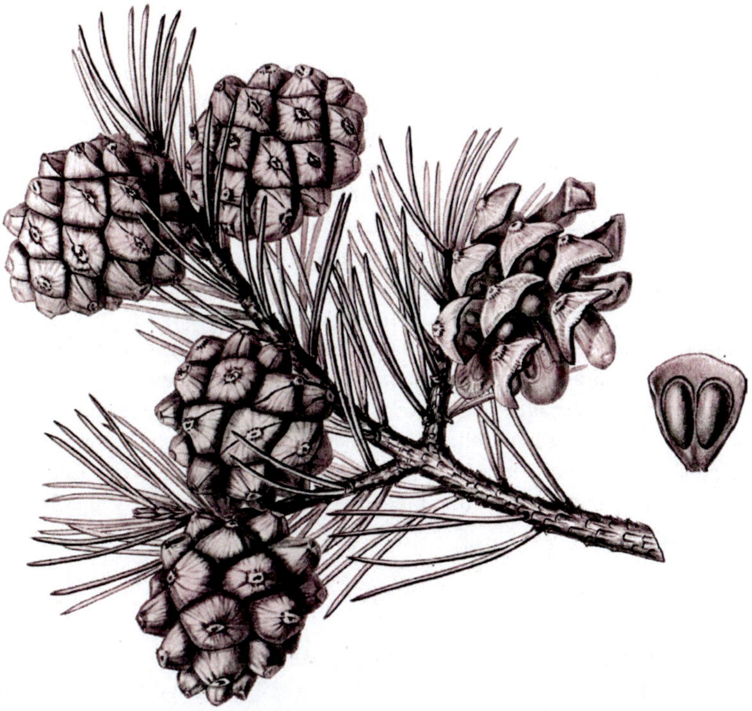

FIGURE 73.2 Piñon pine (*Pinus edulis*). Branch with cones, and (at right) cone scale with two seeds. (From Sargent, C.S., *The silva of North America*, Houghton, Mifflin and Company, Boston, MA, Vol. 11, Plate 552, 1897.)

entirely from wild trees; the species is not cultivated for seeds, although there are a few ornamental cultivars. American piñon pines are mostly stunted wild trees of the southwestern United States, and provide a much more modestly sized annual crop than do the more robust and commonly planted stone pine trees of Europe. Piñon pine cones are roasted to facilitate removal of the seeds (also known as nuts), a labor-intensive process that adds to their cost. North American pine nuts were a food source for indigenous peoples, especially of the southwestern United States, where there is evidence that they were consumed at least 6000 years ago. Today, seed harvest rights in many parts of the tree's range belong to Native American tribes. Reliable statistics are not available, but it is believed that more than 1 million kg (more than 2 million pounds) of piñon nuts are collected annually in the United States, and that collection has been declining since World War II, because of a shortage of labor and the cheapness of less expensive nuts, especially peanuts. Occasionally, irresponsible harvesters have cut down whole trees to facilitate collection of the cones.

## CULINARY PORTRAIT

As noted earlier, two other pine nuts are more important commercially than piñon nuts. The flavor of pignolia nuts (from the Italian pine) is light and delicate, and is considered the most desirable of the pine nuts. Pignolia nuts are eaten out of hand, raw, or roasted and salted, and are the predominant pine nut used in cooking in Europe. Chinese nuts are strong-flavored, with a pungent pine or turpentine aroma that can easily overpower some foods. Chinese pine nuts are more likely to be available in Asian markets than the other types, but increasingly they are the type most sold in all supermarkets. The resinous flavor is mostly neutralized during roasting.

Whatever the source, whole or ground pine nuts can be used in a variety of sweet and savory dishes. They have been combined with lamb, pork, veal, chicken, fish, duck, and game birds; added to stuffings, vegetables, sauces, soups, salads, sweetmeats, stews, cakes, cookies, puddings, marzipan, and other confectioneries; and coated with chocolates and eaten as candy. Pine nuts also serve as an attractive garnish.

Pine nuts are well known as a flavorful addition to the classic Italian pesto. Traditionally, pesto (pronounced PEH-stoh) is a thick sauce prepared cold with basil, garlic, olive oil, pine nuts, Parmesan (or some other sharp, hard cheese), and salt, with basil as the dominant constituent, and most often served with pasta or fish. In North America, a very wide variety of pesto pastes have become popular, often featuring some other herb than basil, as well as other or alternative ingredients, and served with such foods as poultry, meat, and vegetables.

Because of their high fat content, pine nuts turn rancid quickly (within 3 to 6 months of harvest), and should be purchased from a store with a rapid turnover. They may be stored airtight in a refrigerator for up to 3 months, or frozen for up to 9 months.

Only the nuts of pine trees have commercial value as food, but it is of interest that other parts of the tree have been consumed. The inner bark of many pine species is edible, and indeed some Native Americans cut it into spaghetti-like strips and cooked it with vegetables and meat. Several tribes and North American settlers pounded the bark into cakes, which were wrapped in moist leaves and baked slowly. Tea was frequently prepared from the needles. New England Shakers (the religious sect) prepared candy from young pine shoots boiled in maple syrup. Pine gum was frequently used like chewing gum.

### Culinary Vocabulary

- Piñon nuts are sometimes sold as "Indian nuts," referring to their use by North American Indians (although nuts of different species were once imported to North America from Asia under the name "Indian nuts").
- In the piñon nut industry, piñon pine (*P. edulis*) is often called "hard shell" while the single-leaf piñon (*P. monophylla*) is known as "soft shell." "Papershell piñons" is a phrase encountered for the very thin-shelled nuts of the papershell or Texas piñon, *P. remota* (Little) D.K. Bailey & Hawks., which occurs in Texas and Mexico.

## PROSPECTS

Piñon nuts are collected from the wild, a very laborious process that adds to their cost, making it difficult to compete in the marketplace against cheaper imports of European and Korean pine nuts. The wild supply of piñon nuts is variable from year to year (commercial supplies in some years are very low), which also lessens competitiveness. Human collectors must also compete against wildlife, most of the nuts being eaten by piñon jays, wild turkeys, packrats, bears, deer, and other animals. The trees have been subjected to overcollecting in the past, not for nuts but for firewood and charcoal, and are being harvested as Christmas trees and subjected to damage from cattle grazing, all of these activities resulting in a reduced supply. Moreover, climate change may put additional stresses on piñon trees in the future. Given these circumstances, the future of the piñon nut industry seems uncertain.

## CURIOSITIES OF SCIENCE AND TECHNOLOGY

- The Iroquois used the bark of rotten pine or chestnut to prepare a powder, which was dusted on the skin to absorb grease and perspiration.
- The piñon pine was an important part of the spiritual life of many Native Americans of the southwestern United States. The Navaho smeared pitch from the trees on corpses before burial. In December, before going outside, the Hopi applied a dab of piñon pitch to their foreheads as protection against sorcerers. Piñon gum was burned as incense in Navaho ceremonies, and selected branches from the piñon were used as ritual wands.

- Fumes from the pitch of piñon trees were inhaled by Native Americans of the Southwest as a treatment for head colds, coughs, and earache.
- One kilogram of the nuts of piñon pine has about 6500 calories (1 pound has about 3000 calories). A person should consume 10–18 calories per pound of body weight according to the U.S. National Institute of Health, so 1 pound of nuts is sufficient to maintain a 215 pound man for 1 day. According to Williamson (1995, cited in the Appendix to this book), "Male members of some Native American tribes would not allow their pregnant wives to eat the nuts since the weight gained often made delivery more difficult."
- Piñon pine nuts require 3 years of development on the trees before they are mature.
- Wood rats of the southwestern United States are notorious for gathering and hoarding piñon pine nuts. Such concealed caches were often plundered by indigenous peoples. Native Americans and settlers who became lost in the mountains often survived because they knew enough to hunt out and rob the piñon pine nuts stored in the nests of ground squirrels and pack rats. Today, people are often requested to replace any nuts they remove with pinto beans, to protect the animal's food supply.
- Piñon jays and several other bird species (including Clark's nutcracker, Steller's jays, and scrub-jays) also collect and cache piñon nuts, and are the principal dispersal agents of the seeds (which lack wings like many other pines, and therefore are not distributed by the wind). Piñon jays transport many more seeds than the other birds because they have a special adaptation: an expandable esophagus which carries more than 50 seeds.
- The piñon pine (*P. edulis*) was designated as the state tree of New Mexico in 1949, after being nominated by the New Mexico Federation of Women's Clubs.
- The single-leaf piñon (*P. monophylla*), one of the minor sources of edible piñon nuts, was designated as the state tree of Nevada in 1953 (the bristlecone pine [*P. aristata* Engelm.] was added as a second state tree in 1987).

## KEY INFORMATION SOURCES


Barger, N.N., Adams, H.D., Woodhouse, C., Neff, J.C., and Asner, G.P. 2009. Influence of livestock grazing and climate on pinyon pine (*Pinus edulis*) dynamics. *Rangeland Ecol. Manage.* 62: 531–539.

Botkin, C.W., and Shires, L.B. 1948. *The composition and value of pinon nuts*. Agricultural Experiment Station, New Mexico College of Agriculture and Mechanic Arts, Las Cruces, NM. pp. 14.

Critchfield, W.B., and Little, E.L. 1966. *Geographic distribution of the* pines *of the world*. U.S. Department of Agriculture, Forest Service, Washington, DC. pp. 97.

Cobb, N.S., Mitton, J.B., and Whitham, T.G. 1994. Genetic variation associated with chronic water and nutrient stress in pinyon pine. *Am. J. Bot.* 81: 936–940.

Farjon, A. 1984. *Pines, drawings and descriptions of the genus* Pinus. Backhuys, Leiden, the Netherlands. pp. 220.

Farjon, A., and Styles, B.T. 1967. Pinus *(Pinaceae)*. New York Botanical Garden, New York. pp. 291.

Farris, G.J. 1983. California pignolia: seeds of *Pinus sabiniana*. *Econ. Bot.* 37: 201–206.

Fresquez, P.R., Huchton, J.D., Mullen, M.A., and Naranjo, L., Jr. 2000. Radionuclides in pinon pine (*Pinus edulis*) nuts from Los Alamos National Laboratory lands and the dose from consumption. *J. Environ. Sci. Health, Part B*, B35(5): 611–622.

Garrett, P.W. 1979. *Species hybridization in the genus* Pinus. Forest Service, U.S. Department of Agriculture, Northeastern Forest Experiment Station, Broomall, PA. pp. 17.

Kral, R. 1993. *Pinus*. In *Flora of North America North of Mexico*. Vol. 2. Edited by Flora of North America Editorial Committee. Oxford University press, Oxford, U.K. pp. 373–398.

Little, E.L., Jr. 1987. Pinyon pine management for pinyon nuts. *Gen Tech. Rep.* 150: 45–48, 251. RM Rocky Mt. For. Range Exp. Stn. U.S. Department Agriculture Forest Service, Fort Collins, CO. (In Spanish, English summary)

Little, E.L., and Critchfield, W.B. 1969. *Subdivisions of the genus* Pinus *(pines)*. U.S. Forest Service, Washington, DC. pp. 51.

Matuska, J. 1983. Pinyon pine nuts—Nevada's new cash crop. *Our Public Lands, Washington, DC.* 33(4): 7–11.

Mirov, N.T. 1967. *The genus* Pinus. Ronald Press Company, New York. pp. 602.

Mirov, N.T., and Hasbrouck, J. 1976. *The story of pines*. Indiana University Press, Bloomington, IN. pp. 148.

Mitton, J.B., and Duran, K.L. 2004. Genetic variation in pinon pine, *Pinus edulis*, associated with summer precipitation. *Molec. Ecol.* 13: 1259–1264.
Perry, J.P. 1991. *The pines of Mexico and Central America*. Timber Press, Portland, OR. pp. 231.
Richardson, D.M. (Editor). 1998. *Ecology and biogeography of* Pinus. Cambridge University Press, Cambridge, UK. pp. 527.
Shreve, L.W. 1991. English walnut and pinyon pine nut potential in Texas. *Proc. Annu. Conf. Texas Pecan Growers Assoc.* 65(67/69): 51–53.
Skelly, J., and Christopherson, J. 2003. *Pinyon pine*. Nevada Cooperative Extension, Nevada. Division of Forestry, U.S. Forest Service, Reno. NV. pp. 37.
Weidenborner, M. 2001. Pine nuts: the mycobiota and potential mycotoxins. *Can. J. Microbiol.* 47: 460–463.

## Reports on Allergies

Beyer, A.V., Gall, H., and Peter, R.U. 1998. Anaphylaxis to pine nuts. *Allergy* 53: 1227–228.
De Las Marinas, D., Vila, L., and Sanz, M.L. 1998. Allergy to pine nuts. *Allergy* 53: 220–222.
Falliers, C.J. 1989. Pine nut allergy in perspective. *Annals Allergy* 62: 186–189.
Nielsen, N.H. 1990. Systemic allergic reaction to pine nuts. *Annals Allergy* 64(2 Part 1): 132–133.
Roux, N., Hogendijk, S., and Hauser, C. 1998. Severe anaphylaxis to pine nuts. *Allergy* 53: 213–214.
Senna, G., Roncarolo, D., Dama, A., and Mistrello, G. Anaphylaxis to pine nuts and immunological cross-reactivity with pine pollen proteins. *J. Investigat. Allerg. Clin. Immunol.* 10: 44–46.

## Specialty Cookbooks

Castellon, H. 1991. Favorite formula: pignoli (pine-nut cookies). *Bakery Production Marketing* 26(3): 147.
Lanner, R.M. 1981. *The piñon pine: a natural and cultural history*. University of Nevada Press, Reno, NV. pp. 208.

Freitus (1980) has seven recipes specifically using piñon pine, Farnsworth (1999) has 2, Niethammer (1999) has 4, Williamson (1989) has 19, and Williamson (1995) has 13. Brill (2002) has 3 "wild pine nut" recipes. Watts and Watts (2007) and Schufer (2011) each have a recipe entitled "Pine needle tea," which true to its name uses pine needles to prepare tea. Angier (1969), Freitus and Haberman (2005), and Hall and Hall (1980) provide advice on preparing piñon pine seeds. (See the Appendix to this book for details of the publications cited.)

# 74 Plums
## *North American Species*

Family: Rosaceae (rose family)

## NAMES

Scientific name: *Prunus* species. The most important species used for food are listed in Table 74.1.

- The English word "plum" is derived from the Latin word for the tree, *prunus*, as noted in the third bulleted entry, below.
- As noted in Table 74.1, "hog plum," "sloe" (or sloe plum), and "wild-goose plum" are names that have been applied to more than one species. The ambiguous name "wild plum" has been used for most of the American species, and several of them have also been called "red plum." The name "sloe" is best reserved for the European *P. spinosa* L.
- The genus name *Prunus* is the classical Latin name of the plum tree, from the Greek *prunos*.
- The Chickasaw plum is a reference to the Chickasaws, Native Americans of Mississippi and Alabama known to have harvested and spread the species.
- The scientific names of *Prunus* species in Table 74.1 have the following derivations: *alleghaniensis* (Latin for of the Allegheny region; note the retention of the old spelling), *americana* (Latin form of American), *angustifolia* (Latin for narrow leaves), *gracilis* (Latin for slender), *hortulana* (Latin for "of the garden," so named because the species became noticed through the work of horticulturists), *maritima* (Latin for "of or belonging to the sea," reflecting the beach habitat of the species), *mexicana* (Latinized form for "of Mexico"), *munsoniana* (for Thomas Volney Munson, 1843–1913, U.S. nurseryman and specialist on grape culture), *nigra* (Latin for black, referring to the dark branches), *subcordata* (Latin meaning somewhat cordate or heart-shaped, referring to the leaves), *umbellata* (Latin, meaning that the flowers are arranged on stalks that arise from one point).

## GEOGRAPHY AND ECOLOGY OF WILD PLANTS

The genus *Prunus* is widely distributed in North America, and the plum species addressed in this chapter differ in degrees of adaptation to temperature. Many of the native American plums (especially the northern species) are hardy, some tolerating a temperature of –58°C (–50°F). Hybrids with these are also often quite cold-hardy. Depending on species, the native American plums grow in a variety of substrates, from sandy soils that are well drained to heavy lands that have poor drainage.

## PLANT PORTRAIT

The genus *Prunus* is large, diverse, and economically important (for more information, see Chapter 25 on Cherries). The plums and apricots are in subgenus *Prunus* (often called subgenus *Prunophora* in the past). Of the commercial kinds of fruit in *Prunus* (almonds, apricots, cherries, peaches, etc.), the plums are the largest group. Asian and European plums have been placed in one section of the subgenus, section *Prunus* (often called section *Euprunus* in the past). American plums

## TABLE 74.1
## Most Important North American Plum (*Prunus*) Species for Breeding for Fruit, According to Various Authorities

| Where Listed | Common Name(s) | Scientific Name | Distribution | Description |
|---|---|---|---|---|
| a, b | Allegheny plum, American sloe, Porter's plum | *P. alleghaniensis* Porter | North central United States | Straggling shrub or small tree. Fruit is dark blue or purple, with a clinging pit. |
| a–e | American plum, American red plum, American wild plum, August plum, goose plum, hog plum, sloe, wild-goose plum | *P. americana* Marshall | Widespread in Canada & the United States, northern Mexico | Small tree. The most widely distributed wild plum in North America. Fruit mostly red, with cling or free pit. Skin is astringent. Many fruit cultivars. The most important horticultural North American plum. |
| a–c, e | Beach plum, shore plum | *P. maritima* Marshall | Eastern Canada, Eastern United States | Straggling shrub. Fruit usually dark purple, sometimes red, infrequently yellow; flesh medium firm, pit free, or clinging; taste ranges from inedible to excellent. Several cultivars. Also planted as an ornamental and soil binder. |
| a–e | Black plum, Canada plum, Canada black plum, horse plum | *P. nigra* Aiton | Eastern Canada, northeastern and north central United States | Small tree. Fruit red, orange, or yellowish, with astringent skin. Many northern cultivars. |
| b–e | Chickasaw plum, Cherokee plum, Florida sand plum, sand plum, mountain cherry, sandhill plum | *P. angustifolia* Marshall | Widespread in the United States | Shrub or small tree. Fruit bright red, sometimes yellow. Several cultivars. |
| b, c | Hog plum, flatwood plum, sloe plum | *P. umbellata* Elliot | Southeastern United States | Small tree. Fruit usually dark purple with bloom, sometimes red or yellow, with tough skin and very bitter flesh. |
| b–e | Hortulan plum, Miner plum, wild-goose plum | *P. hortulana* L.H. Bailey | North central and eastern United States | Small or medium tree. Fruit red or yellow, flesh firm, acid, with tough skin and clinging pit. Several cultivars. |
| b, c | Mexican plum, big tree plum | *P. mexicana* S. Watson | Central and eastern United States, northern Mexico | Small or medium tree, despite name "big tree." Fruit dark purplish red with bloom. |
| b–e | Munson plum, wild-goose plum | *P. munsoniana* W. Wight & Hedrick | Eastern and central United States | Small or medium tree. Fruit red, less commonly yellow, flesh yellow and juicy. Several cultivars. |
| c | Oklahoma plum, prairie cherry | *P. gracilis* Engelm. & A. Gray | Central United States | Straggling shrub. Fruit usually reddish. Resembles *P. maritima*. |

**TABLE 74.1 (*Continued*)**
**Most Important North American Plum (*Prunus*) Species for Breeding for Fruit, According to Various Authorities**

| Where Listed | Common Name(s) | Scientific Name | Distribution | Description |
|---|---|---|---|---|
| b, c | Pacific plum, Klamath plum, Pacific plum, Sierra plum, western plum | *P. subcordata* Benth. | California, Oregon | Shrub or sometimes small tree. Fruit dark red or purplish, sometimes yellow; flesh sub-acid; pit clinging (freestone selections available). Several cultivars. |

[a] Darrow, G.M. 1975. Minor temperate fruits. In *Advances in fruit breeding.* Edited by J. Janick and J.N. Moore. Purdue University Press, West Lafayette, IN. pp. 269–284.
[b] Weinberger, J.H. 1975. Plums. In *Advances in fruit breeding*. Edited by J. Janick and J.N. Moore. Purdue University Press, West Lafayette, IN. pp. 336–347.
[c] Ramming, D.W., and Cociu, V. 1990. Plums (*Prunus*). In *Genetic resources of temperate fruit and nut crops.* Edited by J.N. Moore and J.R. Ballington, Jr. International Society for Horticultural Science, Wageningen, the Netherlands. pp. 235–287.
[d] Facciola, S. 1998. *Cornucopia II. A source book of edible plants.* Kampong Publications, Vista, CA. pp. 713.
[e] Okie, W.R. 2008. *Prunus domestica*, European plum; *Prunus salicina*, Japanese plum. In *The encyclopedia of fruit & nuts*. Edited by J. Janick and R.E. Paul. CABI, Wallingford, Oxfordshire, U.K. pp. 694–705.

are in a different section, section *Prunocerasus*. There are 15–17 species of North American plums, the exact number recognized differing somewhat among specialists.

Plums are small- to medium-sized trees, generally growing up to 10 m (33 feet) in height, typically with rounded crowns when grown in the open (not in forests). The floral display is impressive, generally with numerous white-petaled flowers. The fruit skin is usually covered with a waxy layer or bloom, which can be buffed away, leaving a shiny surface.

Plums are a valuable north temperate fruit crop. The three main species (which are not native to North America), include the common plum (*P. domestica* L.), damson plum (*P. domestica* subsp. *insititia* (L.) C.K. Schneid., and Japanese plum (*P. salicinia* Lindl.). These produce the best fruit, but they require milder climates than many of the other plum species. Major producers of plums include Russia, China, the United States, and Romania. California produces about 90% of the U.S. crop.

Well over a dozen species of plum are cultivated for fruit. Plums were independently domesticated in Europe, Asia, and North America. The common plum originated in Europe, the damson plum in western Asia, the cherry plum (*P. cerasifera* Ehrh., also known as myrobalan plum) in western and central Asia, the Japanese plum and apricot plum (*P. simonii* Carrière) in China, and the American plum and other North American plums with domesticated forms come from North America. Modern cultivars are often the result of hybridization. The following North American species have often been used to generate hybrids: *P. americana*, *P. angustifolia*, *P. besseyi*, *P. hortulana*, *P. munsoniana*, and *P. nigra*. The North American species have frequently been crossed with the Japanese plum.

Hundreds of domesticated varieties have been generated from the North American plums. Of the North American plums, cultivars of American plum and black plum (and hybrids of these with other plums) are the most frequently grown. The American plum is a relatively small tree with small, tough-skinned fruit, but it has greater winter hardiness than the European plums. Several hundred cultivars of American plum have been created, although only a few are grown commercially. The

**FIGURE 74.1** Notably edible native plums of North America (*Prunus* species). (a) American plum (*P. americana*); fruiting branch of cultivar 'Wolf'. (From Hedrick, U.P., *The plums of New York*, Vol. 2. New York Department of Agriculture, Albany, NY, 1911.) (b) Distribution map of American plum. (c) Beach plum (*P. maritima*); fruiting branches. (Courtesy of Aznaturalist [CC By 3.0].) (d) Distribution of beach plum. (e) Black plum (*Prunus nigra*); fruiting branch of cultivar 'Cheney'. (From Hedrick, *The plums of New York,* Vol. 2. New York Department of Agriculture, Albany, NY, 1911.) (f) Distribution map of black plum.

Chickasaw plum, the Munson plum, and the hortulan plum are common in the southeastern and southcentral United States, and have been the basis of cultivars adapted to these regions.

The plum, both fresh and in the form of prunes, was one of the staples of prehistoric peoples. Plums were grown by the Romans, and ever since the fruit has been produced in Europe. Plums were brought to America with the first colonists, but were not nearly as widely planted as apples, pears, and cherries, and did not achieve notable popularity for some two centuries. It is believed that Native Americans may have carried out minor selection of native American plums, but only in the past two centuries have concerted efforts been made to select varieties of North American species.

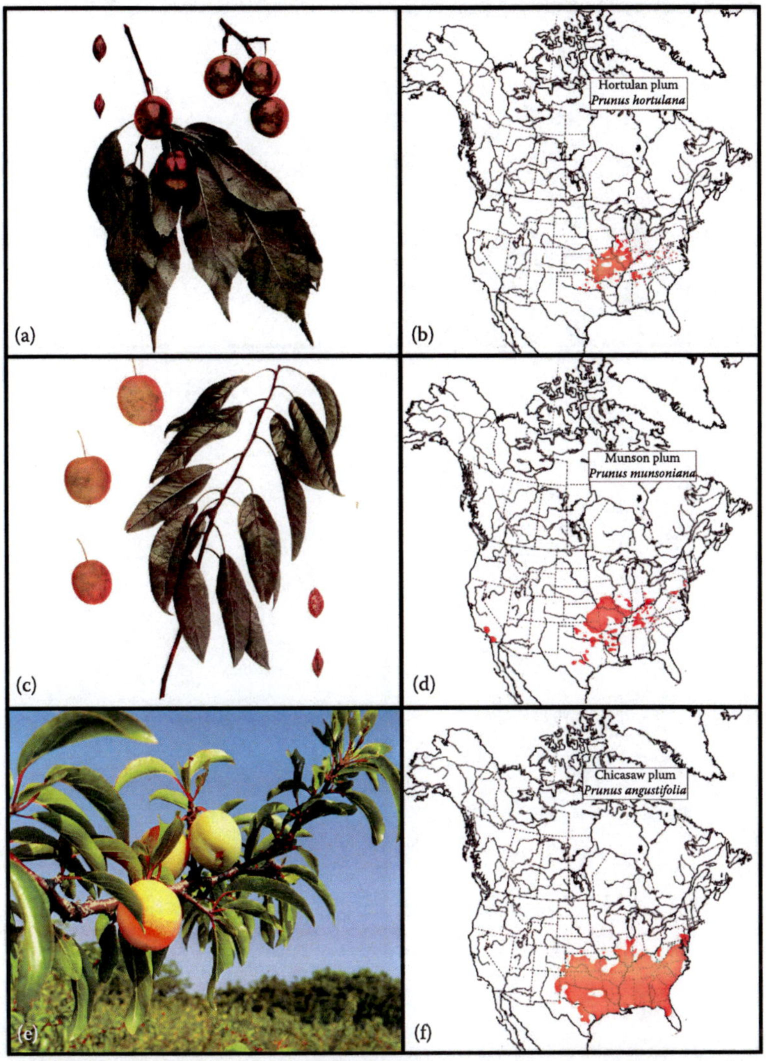


**FIGURE 74.2** More notably edible native plums of North America (*Prunus* species). (a) Hortulan plum (*P. hortulana*); fruiting branch of cultivar 'Wayland'. (From Hedrick, *The plums of New York,* Vol. 2. New York Department of Agriculture, Albany, NY, 1911.) (b) Distribution map of hortulan plum. (c) Munson plum (*P. munsoniana*); fruiting branch of cultivar 'Pottawattamie'. (From Hedrick, *The plums of New York,* Vol. 2. New York Department of Agriculture, Albany, NY, 1911.) (d) Distribution map of Munson plum. (e) Chicasaw plum (*P. angustifolia*). (Courtesy of Lm13700 [CC By 3.0].) (f) Distribution map of Chicasaw plum.

## CULINARY PORTRAIT

Plums are eaten fresh as dessert fruits or dried as prunes. They can usually be consumed with their skin, but if peeling is desired, the fruit can be immersed in boiling water for 30 seconds, then rinsed immediately afterward in cold water. Plums come in both freestone (the pit easily separating from the flesh) and clingstone (the pit clinging to the flesh) varieties, and the former is obviously easier to pit. Plums are made into jelly, jam, juice, liquor, brandy, cognac, and cordials, and are also used in baking and in confections. The fruit is an excellent accompaniment to pork and poultry, and goes well with pies, cakes, muffins, puddings, and ice cream. Plums are commonly used in sweet and sour sauce, and often candied or preserved in vinegar. Overcooking plums turns them into a purée,

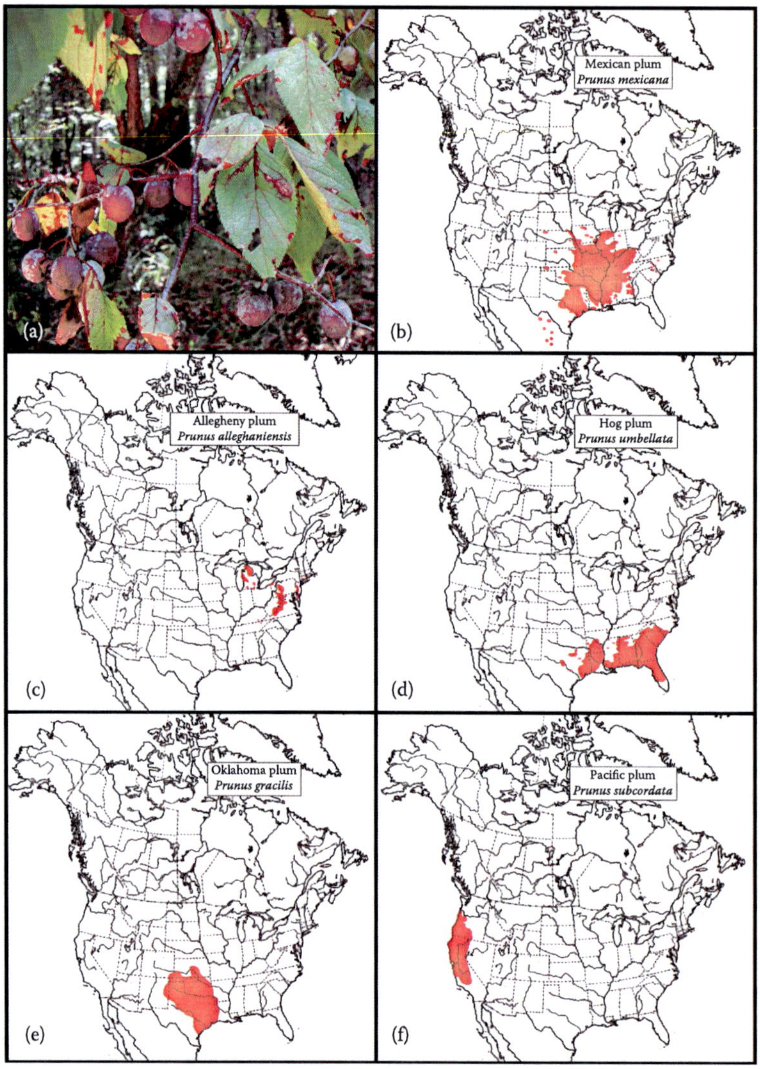


**FIGURE 74.3** Still more notably edible native plums of North America (*Prunus* species). (a) Hog plum (*P. mexicana*). (Courtesy of Jim Conrad.) (b) Distribution map of hog plum. (c)–(f) Distribution maps of additional plum species.

although stewed plums are popular. Plums are moderately perishable. They can be stored at room temperature to ripen, and ripe plums will keep in a refrigerator for several days. Plums and prunes are well known for their laxative effect.

Fruit from wild trees of most of the North American plum species is commonly collected and used for jams, jellies, pies, sauces, and preserves. The quantity and quality are generally not up to the standards of commercially marketed plums. Cultivars of the North American plum species are often excellent. Nevertheless, the North American species have contributed to the plum industry mostly in hybrids with Eurasian species, as sources of breeding stock for improving Eurasian species, and as rootstocks upon which Eurasian species can be grown for purposes of dwarfing or increasing hardiness in stressful environments.

**Note:** See the section "Warning" in Chapter 25 on Cherries, concerning the poisonous potential of the pits and foliage of species of *Prunus*.

## Culinary Vocabulary

- A "shrub" is a beverage, made by simply adding acidulated fruit juice to water, or made of rum or brandy, sugar, and fruit juice. Beach plums have traditionally been the basis of "beach plum shrub," a concoction prepared by mixing cider vinegar, cloves, sugar, and plum purée.
- "Plum pudding" is a steamed or boiled pudding frequently served at holiday times. Plum pudding typically does not contain plums. Before the seventeenth century in England, the term plum always meant plum. However, by the seventeenth century dried plums (i.e., prunes) began to be replaced by raisins in cooking. The dishes once made with plums, despite being now made with raisins, retained the term plum. Thus not only plum pudding is no longer made with plums, neither is plum cake, plum duff, etc. A "sugar plum" is a small candy in a ball or disk, or a sweetmeat, once again not made with plum.
- In North America, "prunes" are simply dried plums. However, in some countries the term prune is also used for the fresh fruit, and in Europe the word is also used to designate a distinct group of plums in which the fruit is usually reddish or blue, elongated, high in sugar content, and firm. Plums usually need to have firm flesh and high sugar content to be dried successfully without removal of the stone. Fermentation would occur at the pit were it not for the high concentration of sugars. All prune varieties belong to *P. domestica*. The kernels in the pits contain toxic hydrocyanic acid, and should not be consumed.

## PROSPECTS

The Eurasian plums of section *Prunus* (common plum, damson plum, Japanese plum) are the most important plums in agriculture and can be expected to remain so. However, these species have proven difficult to grow in many parts of North America because they are often poorly adapted to drought (especially in the prairies of North America), frost (particularly in northern regions), and fungal, bacterial, and viral diseases (notably in the eastern and southern states). The North American species discussed in this account have proven to be essential for generating superior hybrids with the Eurasian species. Some cultivars from the North American species are very good, but the fruit is generally not as desirable as from hybrids and the Eurasian plums. The likelihood of generating superior cultivars only from the American species is relatively limited, but remains significant. The chief importance of the North American plums will probably remain as contributors of genes for extending the adaptation of Eurasian plums so that they can withstand stressful climates and diseases and grow in areas in which they cannot grow well or at all at present.

## CURIOSITIES OF SCIENCE AND TECHNOLOGY

- The Chinese considered the plum to be one of the "five renowned fruits of antiquity." The others are peach, apricot, jujube, and chestnut.
- The Pilgrims had wild plums for dessert during their first Thanksgiving in 1621.
- Species of *Prunus*, in common with many tree species, often have lighter colored bark in warmer and dryer climates. The bark color often becomes lighter toward the south in North America.
- As with many trees, a heavy fruit yield one year may cause plums to "take a rest" the following year, developing a smaller crop (good years are termed "mast years").
- Potatoes should not be planted near plum trees, because an aphid which sometimes overwinters in the tree may carry a potato disease.

## KEY INFORMATION SOURCES

Ashton, R.W. 2008. *Plums of North America*. Third Millennium Publishing, Tempe, AZ. pp. 558.
Bailey, L.H. 1892. *The cultivated native plums and cherries*. Cornell University, Ithaca, NY. pp. 73.

Blair, D.S. 1954. *Plum culture in the less favourable parts of eastern Canada* (reprinted 1965, unaltered except title: Plums for cold areas of eastern Canada). Agriculture Canada Public, 849, Ottawa, ON. pp. 14.
Fogle, H.W. 1978. Plum improvement in the United States. *Acta Hortic.* 74: 35–40.
Groh, H., and Senn, H.A. 1940. *Prunus* in eastern Canada. *Can. J. Res. Sect. C, Bot. Sci.* 18: 318–346.
Hedrick, U.P., and Wellington, R. 1911. *The plums of New York*. J.B. Lyon Company, State Printers, Albany, NY. pp. 616.
Hummel, R. 1977. A classification of hardy North American *Prunus* cultivars and native species based on hardiness zones. *Fruit Var. J.* 31(3): 62–70.
Okie, W.R. 2008. *Prunus domestica*, European plum; *Prunus salicina*, Japanese plum. In *The encyclopedia of fruit & nuts*. Edited by J. Janick and R.E. Paul. CABI, Wallingford, Oxfordshire, U.K. pp. 694–705.
Ramming, D.W., and Cociu, V. 1990. Plums (*Prunus*). In *Genetic resources of temperate fruit and nut crops*. Edited by J.N. Moore and J.R. Ballington, Jr. International Society for Horticultural Science, Wageningen, the Netherlands. pp. 235–287.
Reid, W., and Gast, K.L.B. 1993. The potential for domestication and utilization of native plums in Kansas. In *New crops*. Edited by J. Janick and J.E. Simon. Wiley, New York. pp. 520–523.
Robertson, K.R. 1974. The genera of Rosaceae in the southeastern United States. *Prunus*. *J. Arnold Arb.* 55: 654–662.
Rohrer, J.R., Brien, M.A., and Anderson, J.A. 2008. Phylogenetic analysis of North American plums (*Prunus* sect. *Prunocerasus*: Rosaceae) based on nuclear leafy and *s6pdh* sequences. *J. Bot. Res. Inst. Texas* 2: 401–414.
Shaw, J., and Small, R.L. 2005. Chloroplast DNA phylogeny and phylogeography of the North American plums (*Prunus* subgenus *Prunus* section *Prunocerasus*, Rosaceae). *Am. J. Bot.* 92: 2011–2030.
Waugh, F.A. 1901. *Plums and plum culture: A monograph of the plums cultivated and indigenous in North America, with a complete account of their propagation, cultivation and utilization*. Orange Judd Company, New York. pp. 371.
Weinberger, J.H. 1975. Plums. In *Advances in fruit breeding*. Edited by J. Janick and J.N. Moore. Purdue University Press, West Lafayette, IN. pp. 336–347.
Wight, W.F. 1915. *Native American species of Prunus*. U.S. Department of Agriculture, Washington, DC. pp. 75.
Wight, W.F. 1915. *The varieties of plums derived from native American species*. U.S. Department of Agriculture, Washington, DC. pp. 44.

## Specialty Cookbooks

Bestwick, J. 2001. *Life's little peaches, pears, plums & prunes cookbook: 101 fruit recipes*. Avery Color Studios, Gwinn, MI. pp. 132.
Carver, G.W. 1917. *43 ways to save the wild plum crop*. Printed by Tuskegee students from funds supplied by the Ash Fund, Tuskegee, AL. pp. 12.
Eliot, E. 2001. *Peaches, pears & plums*. World Leisure Marketing, Halifax, NS. pp. 64.
Lawless, D. 1992. *Prunes: more than wrinkles*. N. Bowyer, Dayton, OR. pp. 53.
Nims, C.C. 2003. *Stone fruit: cherries, nectarines, apricots, plums, peaches*. WestWinds Press, Portland, OR. pp. 96.
Sunsweet Growers, Inc. 1959. *Visions of sugarplums: 29 delightful ways to serve Sunsweet prunes*. Sunsweet Growers, Inc., San Jose, CA. pp. 12.
United Prune Growers of California. 1933. *Prunes for epicures: 35 intriguing recipes*. United Prune Growers of California, San Francisco, CA. pp. 22.
Woodier, O. 2004. *Peaches and other juicy fruits: From sweet to savory—150 recipes for peaches, plums, nectarines, and apricots*. Storey Publishing, North Adams, MA. pp. 186.

Krumm (1991) presents 14 wild plum recipes, Freitus and Haberman (2005) have 10, Snell has 4, Marrone (2009) has 15, and Marie (2008) has 1 recipe for wild plum wine. Williamson (1995) has 12 recipes specifically for American plum (*P. americana*). (See the Appendix to this book for details of the publications cited.)

# 75 Pokeweed

Family: Phytolaccaceae (pokeweed family)

## NAMES

Scientific name: *Phytolacca americana* L. (*P. decandra* L.)

- The "poke" in pokeweed is based on the Virginia Algonquian Indian word for the plant, *pocan*, based on *pak*, blood, in reference to the red juice of the berries. The "weed" in pokeweed reflects the fact that the plant is indeed a weed.
- Pokeweed is also known as American cancer, American nightshade, American pokeweed, American spinach, bear's grape, cancer japap, cancer root, coakum, crowberry, garget, Indian poke, inkberry, inkweed, pigeonberry (a name also used for saskatoon; see Chapter 84). pocan, pocan bush, poke, pokeberry, pokeroot, red inkberry, red-ink plant, red weed, scoke, skoke berry, and Virginia poke.
- During the American Civil War (1861–1865), soldiers used the purplish-crimson juice from pokeweed berries as ink to write letters home, hence the name "inkberry."
- The extinct passenger pigeon loved the berries, which led to the old name "pigeonberry."
- The genus name *Phytolacca* is based on the Greek *phyton*, plant + French *lac*, lake, the word referring not to a body of water but a red pigment from the crimson juice of the berries.
- The word *americana* in the scientific name *P. americana* of course means American.

## GEOGRAPHY AND ECOLOGY OF WILD PLANTS

Pokeweed grows as a native plant from southern Ontario, Quebec, and New Brunswick in Canada to Minnesota and south to Florida, Texas, and northeastern Mexico. It has been introduced elsewhere and now grows as a weed in most of the United States, as well as in Europe and Africa. This species is common in rich, moist soil along fence rows, railroads, and roadsides, and in fields, waste ground, disturbed sites, open woods, pastures, prairies, and uncultivated land.

## PLANT PORTRAIT

Pokeweed is a foul-smelling perennial, bushy herb, often resembling a small tree. The plant grows 1–3.5 m (3–11 feet) in height, developing (in perennial forms) from an enormous, fleshy taproot 10–15 cm (4–6 inches) in diameter. The aboveground parts die during the cold season, the plant regrowing from the root. In the northern portion of its range, pokeweed grows as an annual. The stems are succulent, often reddish or purplish, usually hollow, and sometimes as large as 10 cm (4 inches) in diameter. The erect clusters of small, white or whitish-green flowers mature into drooping clusters of berries. The fruit is an 8–10-seeded, shiny berry, about 6 mm (0.24 inch) in diameter, green when young, maturing to purple-black, with inky reddish juice. A somewhat distinctive group of pokeweed plants, *P. americana* var. *rigida* (Small) Caulkins & Wyatt, is restricted to Florida and the Atlantic and Gulf coasts.

**FIGURE 75.1** Pokeweed (*Phytolacca americana*). (a) Branches bearing flowers and fruits. (Courtesy of Huwmanbeing.) (b) Worm-eating warblers on branches. (From Audubon, J.J., *The birds of America*, R. Havell Jr., London, 4 vols, Plate 34, 1827–1838.) (c) Young shrub. (Courtesy of H. Zell [CC By 3.0].) (d) Boy eating poisonous berries. (Courtesy of USDA APHIS PPQ Archive, online at Bugwood.com [CC By 3.0].) (e) Flowering branch. (Courtesy of Rabe! [CC By 3.0].) (f) Distribution map.

Pokeweed was used medicinally by Native Americans and early settlers to treat a variety of conditions, from hemorrhoids to headaches. It was also a popular constituent of medicines in the nineteenth century. American Indians and European colonists also used the berries, leaves, and young spring shoots as food. The Native Americans were aware of pokeweed's toxic properties, occasionally using it as a poison. In recent times in Appalachia and the southeastern United States, the spring vegetable is boiled in three or four changes of water, in apparent recognition of the toxic potential of the plant. For much of the last century, pokeweed has been mostly a poverty food of the rural southern United States. Pokeweed is grown as an ornamental in temperate region gardens, and cultivars are available for this purpose. Other species of *Phytolacca* are used as minor leaf vegetables in tropical America, China, India, and Africa.

## CULINARY PORTRAIT

Pokeweed is a wild (undomesticated) poisonous plant, and some people are known to have died from eating it. The root is the most poisonous part of the plant and occasionally has been known to kill pigs that dig up the roots and eat them. However, this is a rare occurrence because the plants are not attractive as food, either to wild or domesticated animals, or to humans. The mature leaves and fruits (and the seeds) are also quite toxic. Research with humans has shown that pokeweed can cause mutations (possibly leading to cancer) and birth defects. Since the toxic juice of pokeweed can be absorbed through the skin (particularly through abrasions), contact of plant parts with bare skin should be avoided, and the plant should be handled with gloves. Pokeweed cannot be recommended as food, and indeed authorities commonly advise that it should never be consumed. It is widely thought that eating the emerging spring shoots after boiling is safe if the cooking water is discarded, and it does seem that proper cooking destroy the main group of toxic chemicals. However, there also appear to be additional toxins in the plant that are not destroyed by cooking. Symptoms of poisoning from pokeweed include a burning sensation in the mouth, salivation, gastrointestinal cramps, vomiting, and bloody diarrhea. Consumption of large quantities can lead to anemia, altered heart rate and respiration, convulsions, and death from respiratory failure. The primary toxic compounds are thought to be oxalic acid, saponins (phytolaccotoxin and phytolaccigenin), and an alkaloid (phytolaccin).

Despite the toxicity of the plant, the young, green, asparagus-like shoots emerging in the spring are used as a boiled vegetable when less than 15 cm (6 inches) long and indeed have long been a favorite potherb in eastern North America, with taste and texture that have been compared to spinach and asparagus. There has even been some commercial canning of pokeweed. The berries are sometimes used to make pies and wine but are not nearly as commonly used as food as the young shoots. Few people seem concerned about the toxicity of pokeweed greens, provided that they are boiled at least twice. Pokeweed consumption, although less popular than in the past, remains a tradition in the Deep South of the United States. Typically "poke" is boiled for 20–30 minute intervals in at least two and often three water changes and the green water poured down the sink. Sometimes, baking soda is added to the first boil. The greens are often cooked with salt pork, or fried in bacon grease, and mixed with scrambled eggs, hard-boiled eggs, onions, and/or corn bread. The greens are also consumed like spinach and used in similar dishes. Poke stalks are also peeled and fried after rolling them in flour or corn meal. Jelly has been made from the berries, a dish that is particularly not recommended.

### Culinary Vocabulary

- "Poke" is also used as the name of a Hawaiian dish, which consists of diced, seasoned raw or marinated fish and accompanying condiments. "Korean-style" poke comes with chile.
- "Poke sallet" (poke salit, poke salet) is a southern U.S. phrase for poke greens used as a salad. The phrase has been claimed to be a corruption of "poke salad," but "sallet" is simply a Middle English term for cooked greens.

## PROSPECTS

Given that pokeweed is poisonous and authorities have strongly recommended it not be used as food, it may seem peculiar to suggest that it has potential as a commercial food plant. However, some toxic food plants can be made edible by appropriate processing (in fact, this seems to be the case for those who customarily consume pokeweed). More significant is the fact that edible varieties of many present-day food plants have been selected over the last several thousand years, beginning with wild forms that were quite toxic. Of course, since quite edible vegetables are now available, there would seem to be limited reason to domesticate another one. The key consideration today, however, is that modern genetic engineering has the potential to identify and turn off genes controlling the production of poisonous compounds, and should this prove to be the case with pokeweed, it would almost certainly achieve market success.

## CURIOSITIES OF SCIENCE AND TECHNOLOGY

- The Pawnee Indians used pokeweed berries as a source of red stain to paint horses and jewelry.
- Kiowa Indian girls fashioned necklaces out of dry pokeweed berries.
- It is widely claimed that the United States Declaration of Independence of 1776 was written in ink made from fermented pokeberry juice.
- In the eighteenth century, pokeweed berries in Portugal were added to low-grade wine to increase the red color. Although the taste was poorer, the appearance was better. To prevent this deceptive practice, the king ordered all pokeweed plant to be cut down before they produced berries. Nevertheless, the plant was widely cultivated in wine areas of Portugal, Spain, France, and Italy to adulterate wine.
- Supporters of James Knox Polk (1795–1849, eleventh president of the United States) wore pokeweed twigs instead of the customary campaign buttons during his 1845 presidential bid.
- Robins, towhees, mockingbirds, mourning doves, catbirds, bluebirds, and other migrating songbirds are very fond of poke berries, unfortunately often converting them to unwanted purple splotches. The berries of pokeweed are eaten without apparent toxic effects by wild birds, and feeding the berries to domestic fowl has also appeared to be harmless. However, the fruit is nevertheless toxic to humans, and children are believed to have died from eating the berries. The old idea that it is safe to eat fruits if wild birds are seen eating them is quite false. Poke plants should not be planted where young children are likely to be attracted to the berries.
- Pokeweed seeds can remain viable in the soil for at least 40 years.
- A protein (pokeweed antiviral protein) isolated from pokeweed is being tested for its ability to inhibit the replication of the HIV virus in human cells. The presence in some plants of proteins that destroy viruses and virus-infected cells is thought to protect the plants from disease. Curiously, pokeweed was considered to be a cure for other (venereal) sexually transmitted diseases in the 1800s.

## KEY INFORMATION SOURCES

Armesto, J.J., Cheplick, G.P., and McDonnell, M.J. 1983. Observations on the reproductive biology of *Phytolacca americana* (Phytolaccaceae). *Bull. Torrey Bot. Club* 110: 380–383.

Armstrong, W.P. 1989. A peek at pokeweed. *Pacific Hortic.* 50(4): 27–31.

Barnett, B.D. 1975. Toxicity of pokeberries (fruit of *Phytolacca americana*) for turkey poults. *Poultry Sci.* 54: 1215–1217.

Caulkins, D.B., and Wyatt, R. 1990. Variation and taxonomy of *Phytolacca americana* and *P. rigida* in the southeastern United States. *Bull. Torrey Bot. Club* 117: 357–367.

Driver, M.G., and Francis, F.J. 1944. Purification of phytolaccanin (betanin) by removal of phytolaccatoxin from *Phytolacca americana*. *J. Food Sci.* 44: 521–523.

Edwards, M.E., Harris, E.M., Wagner, F.H., Cross, M.C., and Miller, G.S. 1988. Seed germination of American pokeweed (*Phytolacca americana*). I. Laboratory techniques and autotoxicity. *Am. J. Bot.* 75: 1794–1802.

Farmer, R.E., and Hall, G.C. 1970. Pokeweed seed germination: effects of environment, stratification, and chemical growth regulators. *Ecology* 51: 894–898.

Fukuyama, Y., Hasegawa, T., Toda, M., Kodama, M., and Okazaki, H. 1992. Structures of americanol A and isoamericanol A having neurotrophic properties from the seeds of *Phytolacca americana*. *Chem. Pharm. Bull.* 40: 252–254.

Hamilton, R.J., Shih, R.D., and Hoffman, R.S. 1995. Mobitz type I heart block after pokeweed ingestion. *Vet. Human Toxicol.* 37: 66–67.

Hyatt, L.A. 1998. Spatial patterns and causes of overwinter seed mortality in *Phytolacca americana*. *Can. J. Bot.* 76: 197–203.

Jaeckle, K.A., and Freemon, F.R. 1981. Pokeweed *Phytolacca americana* poisoning. *Southern Medical J.* 74: 639–640.

Knight-Trent, A.H., and Cupp, M.J. 2000. Pokeweed. Toxicol. *Clin. Pharm. Herbal Products* 2000: 237–243.

Krochmal, A., and Le Quesne, P.W. 1970. *Pokeweed* (Phytolacca americana): *possible source of a molluscicide*. Northeastern Forest Experiment Station, Forest Service, U.S. Department of Agriculture, Upper Darby, PA. pp. 8.

Lewis, W.H., and Smith, P.R. 1979. Poke root herbal tea poisoning. *J. Am. Med. Assoc.* 242: 2759–2760.

McDonnell, M.J., Stiles, E.W., Chepick, G.P., and Armesto, J.J. 1984. Bird-dispersal of *Phytolacca americana* L. and the influence of fruit removal on subsequent fruit development. *Am. J. Bot.* 71: 895–901.

Mikesell, J. 1988. Comparative development of viable and aborted ovules in *Phytolacca americana* L. (Phytolaccaceae). *Bot. Gaz.* 149: 196–202.

Mikesell, J. 1996. Correlation of microsporangial and megasporangial differentiation along racemes of *Phytolacca americana* L. (Phytolaccaceae). *Bot. J. Linn. Soc.* 122: 253–267.

Mikesell, J.E., and Schroeder, A.C. 1980. Development of chambered pith in stems of *Phytolacca americana* L. (Phytolaccaceae). *Am. J. Bot.* 67: 111–118.

Mitich, L.W. 1994. Common pokeweed. *Weed Technol.* 8: 887–890.

Nienaber, M.A., and Thieret, J.W. 2003. *Phytolacca*. In *Flora of North America north of Mexico*. Vol. 4. Edited by Flora of North America Editorial Committee. Oxford University Press, New York, NY. pp. 6–9.

Nowicke, J.W. 1969. Palynotaxonomic study of the Phytolaccaceae. *Annals Missouri Bot. Gard.* 55: 294–364.

Owen, M.D.K. 1988. *Pokeweed (Phytolacca americana L.)*. Revised edition. Iowa State University Extension, Ames, IA. pp. 2.

Sauer, J.D. 1952. A geography of pokeweed. *Ann. Missouri Bot. Gard.* 39: 113–125.

Scanlon, D.H. III. 1979. Direct seeding of pokeberry on mine soils. *Reclamation Rev.* 2(1): 17–21.

Thomas, F. 1973. The versatile poke plant. *Ozark Highw.* 1973(2): 22–23.

Woo, W.S., Kang, S.S., Wagner, H., Seligmann, O., and Chari, V.M. 1978. Triterpenoid saponins from roots of *Phytolacca americana*. *Planta Medica* 34: 87–92.

Yusuf, R. 2001. *Phytolacca* L. In *Plant resources of South-East Asia, 12(1), medicinal and poisonous plants 1*. Edited by L.S. de Padua, N. Bunyapraphatsara and R.H.M.J. Lemmens. Backhuys Publishers, Leiden, the Netherlands. pp. 392–397.

## Specialty Cookbooks

Recipes for pokeweed are available in wild food books. Beatty (1987) presents four recipes, Brill (2002) has six, Farnsworth (1999) has one, Johnson (1989) has five, Kluger (1973) has two, Robe-Terry (1997) has four, Tatum (1976) has seven, and Schufer (2011) has three. Angier (1969) provides general advice on the culinary preparation of pokeweed. (See the Appendix to this book for details of the publications cited.) Recipes are also available on the Web. The potentially poisonous nature of pokeweed should be kept in mind and, given the danger, pokeweed is a particularly hazardous food for those who are not familiar with its preparation.

# 76 Prairie Turnip

Family: Fabaceae (Leguminosae; pea family)

## NAMES

Scientific name: *Pediomelum esculentum* (Pursh) Rydb. (usually found under the older name *Psoralea esculenta* Pursh).

- Prairie turnip has also been called breadroot, breadroot scurf pea, Cree turnip, Dakota turnip, ground apple, Indian breadroot, Indian potato, Indian turnip, large Indian breadroot, prairie apple, prairie potato, prairie wild turnip, scurfpea, shaggy prairie turnip, tipsin, tipsinna, and white apple. Several of these names are used frequently for other species. The words apple, potato, and turnip are misleading, and the adjective "Indian" is unnecessary.
- The French names pomme blanche (white apple) and pomme de prairie (prairie apple) have also been used, dating from the early French Canadian fur traders and voyageurs.
- The name "breadroot" is based on the farinaceous (floury) properties of the tubers (some other species of *Pediomelum* are also sometimes called breadroot).
- The genus name *Pediomelum* is based on Greek *pedion*, plain or field + *melon*, apple or fruit.
- *Esculentum* in the scientific name is Latin for edible.

## GEOGRAPHY AND ECOLOGY OF WILD PLANTS

Prairie turnip is native to central North America, including the southern prairie provinces of Canada (Alberta, Saskatchewan, and Manitoba), and much of the westcentral United States. The species is most common on dry prairies and foothills. It occurs on hillsides, plains, bluffs, stream valleys, grasslands, open woodlands and pine forests, and sometimes along roadsides. Prairie turnip grows best in deep, fertile, well-drained soils and prefers those with limestone. It is quite drought resistant. With the spread of cattle grazing and the establishment of crops on the prairies, the availability of prairie turnip has decreased significantly.

## PLANT PORTRAIT

Prairie turnip is a perennial herb that dies back to the ground before winter. The leaves are divided into five leaflets 3–6 cm (1.2–2.4 inches) long. A flowering stem 10–50 cm (4–20 inches) tall develops, bearing blue or purple flowers 12–20 mm (0.5–0.8 inch) in length, borne in conical clusters 2–8 cm (0.8–3 inches) long. The flowers often fade to a yellowish color as they age. After flowering, one-seeded, egg-shaped pods 5–7 mm (about ¼ inch) long are produced, with slender tips 1–2 cm (0.4–0.8 inch) long. The flowering stem and leaf stalks are densely covered with long, white hairs. In the fall, the stem with its fruits breaks off at ground level and becomes a tumbleweed, blowing across the prairie while scattering its seeds. From the taproot and sturdier, branch roots, tubers 4–15 cm (1.6–6 inches) long and 3–8 cm (1.2–3.1 inches) wide are produced, which are about 4–10 cm (1.6–4 inches) below ground level. The tubers have a dark, thick, rind-like covering and are whitish on the inside.

Prairie turnip was extensively harvested by Native Americans indigenous to the Great Plains and represented a dietary staple for the nomadic buffalo-hunting tribes of the northern Great Plains.

**FIGURE 76.1** Prairie turnip (*Pediomelum esculentum*). (a) Flowering branch and tuber. (From Goodale, G.L., *The wild flowers of America,* Bradlee Whidden, Boston, MA, 1886.) (b) Flower cluster. (Courtesy of U.S. National Park Service.) (c) Vegetative plants. (Courtesy of John McDaris.) (d) Distribution map.

The plant is still harvested in traditional fashion to a minor extent. In the early nineteenth century, trials were conducted in France to examine the possibility of developing the species into a substitute for potato, which at the time was experiencing problems with disease. In recent years, recommendations have been made to develop prairie turnip into a modern crop.

## CULINARY PORTRAIT

Prairie turnip was the most important staple plant food gathered by indigenous people of the Great Plains, valued for its nutritional quality and abundance. (Bison was the staple meat.) The tubers were peeled and prepared by roasting, frying, boiling, or mashing. They were also frequently dried for winter use. The tubers have a hard, dark, fibrous outer layer that facilitates storage. This outer layer is removed before eating, either raw or cooked. The dried root was often ground into a flour, which was commonly used in pemmican. The tubers are still gathered for food to a small extent by native people and wild food enthusiasts. In addition to cooking the starchy, tuberous root, it has also been consumed as a raw vegetable. The raw root is somewhat sweet, especially the younger tubers. The relatively bland and unique taste has been compared to turnip, raw green beans, unroasted peanuts, and fresh field corn.

### Warnings

Prairie turnip has been reported to contain furanocoumarins, chemicals that can cause photosensitivity (a rash that develops when skin is exposed to sunlight) in some people. Several plant species similar to prairie turnip are toxic, and it is essential that collected material be accurately identified before consumption. Digestive problems (flatulence and constipation) have sometimes been reported after eating prairie turnip.

### Culinary Vocabulary

- Wojapi is a classic fruit pudding made by Dakota and Lakota people and other Plains Indians, by combining berries with other ingredients including sugar. It can be thickened in traditional fashion with ground prairie turnip, instead of flour, as is usually done today.
- The expression "iron ration" means an emergency food supply (often supplied to soldiers), and prairie turnip sometimes met this definition. In the past, prairie turnip has been used as a last-resort food (especially by European settlers), but Great Plains Native Americans frequently relied on prairie turnip as a staple.

## PROSPECTS

Prairie turnip is an undomesticated "root crop" (the tubers are actually stem tissue) once of great importance to indigenous peoples of the Great Plains. In the modern marketplace, it would have to compete directly with the major popular Western root crops, notably potatoes, turnips, rutabagas, and carrots. The species has caught the attention of some scientists, who have recommended its development as a cultivated food plant. Of the dozens of root crops that were regularly consumed by Native Americans, prairie turnip is one of the most attractive possibilities. It would require considerable investment in breeding and management to raise its taste and productivity up to a standard that would make it sufficiently attractive to be profitable. Prairie turnip is naturally adapted to a set of soil and climate conditions that once made it a predominant wild food in the Great Plains. It is possible that outside North America, there may be similar conditions in which the plant would find circumstances to its liking, and it could be managed as a low-maintenance crop capable of producing substantial edible material.

## CURIOSITIES OF SCIENCE AND TECHNOLOGY

- There is often remarkable parallelism among the stories of different religions, likely reflecting a shared underlying morality. According to a widespread Native American myth (see Cowen 1991), the morning star (the planet Venus) descended from the sky, married a Blackfoot maiden, and took her to the heavens. The Moon (the mother of the morning star) gave her new daughter-in-law a special gift: a root digger, but ordered her not to harvest a certain sacred root plant, the prairie turnip, which could cause evil if touched. But like Adam and Eve, warned to avoid the apple tree, she succumbed to temptation and uprooted the turnip. And also like the Christian story of expulsion from Paradise, she was banished from the heavens and forced to live on the harsh prairies of Earth.

- Some anthropologists have concluded that in the nineteenth century, the equestrian tribes of the Great Plains, particularly the Lakota, Cheyenne, Comanche, and Blackfoot, were exceptionally healthy. One measure of this was that nomadic Plains Indians were taller on average than settlers in the region. The superior health despite hardships, wars, and the constant threats of smallpox and other epidemics has been credited to excellent nutrition, notably including consumption of prairie turnip.
- The Battle of the Little Bighorn (also known as Custer's Last Stand) took place in June of 1876 in what is now Montana. A combination of Native American tribes defeated a force of 700 men led by George Armstrong Custer. Because of its excellent nutritional qualities and its use as a staple food by the Indians, the prairie turnip has been characterized as the vegetable that helped them defeat Custer.
- Blackfoot Indians used prairie turnip for several medicinal uses, including gastrointestinal complaints. To treat colic, chewed roots were blown into a baby's rectum. Children cutting teeth were sometimes given pieces of prairie turnip to chew.
- Prairie turnip is thought to have been a favorite food of the plains grizzly bear, which became extinct in the early twentieth century.
- Topeka, Kansas, was named in 1855. According to one theory, the name traces to Native American words meaning "a good place to dig potatoes," and the "potatoes" were in fact prairie turnip ("Topeka" is said to be comprised of three words common in the languages of the Iowa, Omaha, and Kansa Indians: "To" meaning potato, "pe" meaning good, and "okae" meaning to dig.).
- The Pomme de Terre River in western Minnesota was named for prairie turnip. The French *pomme de terre* normally is translated as "potato," but the voyageurs (French explorer–traders) responsible for the name had the prairie turnip in mind.
- Prairie turnip tubers weighing 6.8 kg (15 pounds) have been collected.

## KEY INFORMATION SOURCES


Cowen, R. 1991. The sacred turnip. Dietary clues gleaned from tuber traditions. *Science News* 139: 316–317.

Grimes, J.W. 1990. A revision of the New World species of *Psoraleeae* (Leguminosae: Papilionoideae). *Mem. New York Bot. Gard.* 61: 1–114.

Kaldy, M.S., Johnston, A., and Wilson, D.B. 1980. Nutritive value of Indian bread-root, squaw-root, and Jerusalem artichoke. *Econ. Bot.* 34: 352–357.

Kaye, B., and Moodie, D.E. 1978. The *Psoralea* food resource of the Northern Plains. *Plains Anthropol.* 23: 329–336.

Kaye, B., and Moodie, D.W. 1981. Rooting for the truth: a reply to Reid on the importance and distribution of *Psoralea esculenta*. *Plains Anthropol.* 26: 81–83.

Kotiyal, J.P., and Sharma, D.P. 1992. Phytochemical studies of *Psoralea* species. *Bull. Medico-Ethno-Bot. Res.* 13: 209–223.

Maisch, J.J. 1889. Useful plants of the genus *Psoralea*. *Am. J. Pharm.* 61: 345–352.

Ockendon, D.J. 1965. A taxonomic study of *Psoralea* subgenus *Pediomelum* (Leguminosae). *Southw. Nat.* 10: 81–124.

Parker, J.L. 1968. Notes on Manitoba Canada plants, II. *Psoralea esculenta, Petalostemum candidum, Asparagus officinalis, Lotus corniculatus*. *Can. Field-Nat.* 82: 50–51.

Perera, G., and Reese, R.N. 2002. Nutritional values of *Psoralea esculenta*. *Proc. South Dakota Acad. Sci.* 81: 271.

Reid, K.C. 1977. *Psoralea esculenta* as a prairie resource: An ethnographic appraisal. *Plains Anthropol.* 22: 321–327.

Spessard, L.L. 1988. Seed-germination studies of *Psoralea esculenta* Pursh (Indian turnip) and *Psoralea argophylla* Pursh (silver scurfpea). *Trans. Nebraska Acad. Sci. Affil. Soc.* 16: 123–126.

Stahnke, A., Hayes, M., Meyer, K., Witt, K., Weideman, J., Fernando, A.P., Burrows, R., and Reese, R.N. 2008. Prairie turnip *Pediomelum esculentum* (Pursh) Rydb.: Historical and modern use, propagation, and management of a new crop. *Native Plants J.* 9: 46–58.

Stahnke, A.L., and Reese, R.N. 2003. Morphological examination of prairie turnip (*Psoralea esculenta* Pursh) roots. *Proc. South Dakota Acad. Sci.* 82: 113–117.
Warner, M.F. 1947. Lamare-Picquot and the breadroot. *Agric. Hist.* 20: 23–26.
Wedel, W.R. 1978. Notes on the prairie turnip (*Psoralea esculenta*) among the Plains Indians. *Nebraska Hist.* 59: 155–179.
Wiedeman, V.E. 1958. The *Psoralea esculenta* series of Oklahoma. *Proc. Oklahoma Acad. Sci.* 38: 7–9.

## Specialty Cookbooks

An especially delightful account of harvesting prairie turnip, as well as several recipes, is in Snell (2006). Her prairie turnip recipes are entitled Wild turnip flour and turnip meal; Wild turnip porridge; and Wild turnip bread. Freitus (1980) presents three prairie turnip recipes. Angier (1969) provides general advice on the culinary preparation of prairie turnip. (See the Appendix to this book for details of the publications cited.)

# 77 Raspberries

Family: Rosaceae (rose family)

## NAMES

Scientific names: *Rubus* species and hybrids

- Red raspberry—*R. idaeus* L.
- American red raspberry—*R. idaeus* var. *strigosus* (Michx.) Maxim. (*R. strigosus* Michx.)
- Black raspberry—*R. occidentalis* L.
- Purple (cane) raspberry—Crosses of red and black raspberries (*R. idaeus* × *R. occidentalis*).
- The name "raspberry" is claimed to be based on *rasp* (from English *raspen*, rough) + berry, which has been interpreted as an allusion to the thorny stems bearing the fruit. Another explanation is that the word owes its origin to a sweet, dark red wine that was imported into England from France starting in the fifteenth century. The wine was known in England as "respyce" or "raspis." The color or taste may have been like that of the raspberry, resulting in the raspberry acquiring its name.
- In common language, a "berry" is any small pulpy or juicy fruit, and in this broad sense, a raspberry is certainly a berry. Botanically, a berry is more narrowly defined as a fleshy fruit that is developed from a single flower. A raspberry is technically not a berry because it is developed from an aggregate of flowers.
- The black raspberry is also known as the blackcap and thimbleberry. In the mid-Atlantic United States, the black raspberry is often called blackberry, which leads to confusion with true blackberries (see Chapter 12 on Blackberries and Dewberries).
- The genus name *Rubus* is from the Latin *rubus*, bramble, bush, and is based on *ruber*, red, an allusion to the red dye obtained from fruits of some of the species.
- *Idaeus* in the scientific name *R. idaeus* is based on Mount Ida in Asia Minor (Turkey), which according to classical literature was heavily populated with wild raspberry.
- *Occidentalis* in the scientific name *R. occidentalis* is Latin for western.

## GEOGRAPHY AND ECOLOGY OF WILD PLANTS

The red raspberry grows wild in northern Europe, Asia, and North America. It is found in woods, in sunny or shady, dry or moist habitats with poor or good soils of a variety of textures, and it tolerates a fairly wide range of soil pH. The American form of the red raspberry has been interpreted differently by various authorities. According to one view of the American red raspberry, it is found from Alaska and subarctic Canada, southwards through all of the provinces of Canada and more than half of the states of the United States, to northern Mexico. Some authorities have concluded that it is also present in eastern Asia. A related species, *R. leudodermis* Douglas ex Torr. & A. Gray (whitebark raspberry), a native of western North America, has sometimes been considered to be a variety of the red raspberry.

The black raspberry grows wild in much of eastern Canada (New Brunswick, Ontario, and Quebec) and the eastern United States (Maine to North Dakota and Colorado; south to Georgia, Louisiana, Oklahoma, and Kansas). The species grows in both moist and dry situations, in both

FIGURE 77.1 American red raspberry (*Rubus idaeus* var. *strigosus*). (a) Red berries. (Courtesy of K. Kendall [Flickr/CC-attribution].) (b) Golden berries. (Courtesy of Amy Loves Yah [Flickr/CC-attribution].) (c) Cultivar 'Cuthbert'. ([c] and [d] From Hedrick, U.P., *The small fruits of New York,* Department of Farms and Markets, New York, 1925.) (d) Cultivar 'Golden Queen'. (e) Fruiting branches. (Courtesy of Alan Cressler, reproduced with permission.) (f) Distribution map.

sun and shade, and prefers soils that are gravelly or sandy, and slightly acidic. It occurs in thickets, ravines, bluffs, open woods, wet meadows, and pastures. It is a rather weedy plant in some places, such as along roadsides.

## PLANT PORTRAIT

Raspberries are perennial shrubs. Various species of *Rubus* are called raspberries and are often collected from the wild on all continents except Antarctica. However, the species highlighted here are predominantly the ones that are cultivated and found in commerce. The red raspberry has fruits that are dark or light red, never black, but sometimes yellow or white. The fruits are usually juicy

**FIGURE 77.2** Black raspberry (*Rubus occidentalis*). (a) Branches with fruit. (Courtesy of Ken Goulding.) (b) Hoosier black raspberry. (Courtesy of the U.S. Department of Agriculture artist E.I. Schutt; painting owned by the USDA, online at Bugwood.org [CC By 3.0].) (c) Cultivar 'Plum Farmer'. (From Hedrick, U.P., *The small fruits of New York,* Department of Farms and Markets, New York, 1925.) (d) Distribution map.

at maturity, and the canes are bristly with small prickles. By contrast, the black raspberry has fruits that are purple-black, never red at maturity, and rarely yellow or white. The mature fruits are firm, and the canes (shoots) are not bristly (although hooked spines are present). The European red raspberry (*R. idaeus* var. *idaeus*) and the American red raspberry (*R. idaeus* var. *strigosus*) are the ancestors of modern varieties of red raspberries. The nearly thornless canes of raspberries grow from the crown in 1 year, produce fruit the following year, then die. Some kinds produce a fall crop at the end of the canes of the current season. Plants that produce a normal crop during the summer and another crop in the fall are called "everbearing." The canes are stiff and may attain heights of over 3 m (10 feet) in the red and purple raspberries, but are smaller, 1.2–1.5 m (4–5 feet) in the black raspberry. Commercially, the canes are usually supported by parallel wires along each side of the row and not over 2 m (about 6 feet) high. The berries vary in diameter from 1.3–2.5 cm (0.5–1 inch). Red raspberries are much more popular than black raspberries, but the latter are also popular, mostly in regions of the eastern United States. In North America, most raspberries are cultivated on the West coast (including British Columbia, Washington, and Oregon).

## CULINARY PORTRAIT

Raspberries are widely eaten fresh as a dessert fruit out of hand, with sugar and whipped cream or with ice cream. They are preserved as juice and by canning and freezing. They may also be made into syrup for various flavoring purposes. Cooked raspberries are a basis for many jams and jellies, tarts, pies, puddings, and sauces. Raspberries are used to make specialty wines, liqueurs, beers, and vinegars. The chief problems with raspberries are their natural seediness and their high perishability.

### CULINARY VOCABULARY

- Chef Bernard Loiseau, owner and chef of the Hôtel de la Côte d'Or in Burgundy, France, was known for a unique raspberry dish—*millefeuille de framboises* with iced yogurt, a sensational preparation of flaky, crisp, barely warm pastry with perfect raspberries. (Warned in the Michelin 2003 restaurant guide that his menu was losing favor, and downrated by the GaultMillau restaurant guide, the humiliated chef shot himself to death in 2003.)

## PROSPECTS

Raspberries are a favorite garden fruit of the north temperate world. They are very perishable, and few fresh raspberries are transported long distances. Demand for berries is usually unlimited, and prices are high. Production costs are also high, and pick-your-own operations are popular. Considerable red raspberries are grown for the local fresh-market trade, but the market for processed products is also large. The cultivation and marketing of raspberries are very likely to remain stable for the foreseeable future.

## CURIOSITIES OF SCIENCE AND TECHNOLOGY

- The "Victorian language of flowers" was a secret coded language in Victorian times, with flowers and plants symbolic of certain messages; so when the flower or plant was mentioned in a letter, those who knew the code could understand the hidden information. Raspberry meant "remorse."
- Raspberry fruits were once used to dye hair.
- Professional food photography for advertisements has become an art, employing stylists and prop specialists in addition to photographers. The seductive images of luscious desserts that appear in magazines and cookbooks look better than they do in real life because they have been created by masters of illusion. Ice cream, for example, melts too rapidly under bright photography lights, so a combination of Crisco shortening, icing sugar, and jam is typically substituted. Raspberries pose a particular problem, because with magnification the tiny wisps of hair-like fiber that protrudes from the tiny drupelets making up the fruit can look unkempt, requiring careful manicuring, sometimes for hours.
- One curious definition of raspberry is that it is a derisive or contemptuous spluttering noise made by vibrating the extended tongue and the lips while exhaling (the negativity is reflected in the phrase "blow a raspberry"). This odd meaning is thought to be based on the obsolete expression "raspberry tart," a euphemism for fart, the spluttering noise said to imitate the passing of flatulence. The "Golden Raspberry Award Foundation"

presents annual prizes (razzies) similar to the Oscars, but for the worst achievements in movies. The award is a golf-ball-sized raspberry, spray-painted gold. These quite negative associations have nevertheless not prevented the word raspberry from being trademarked for computer devices and software, in the same manner as for apple and blackberry.

## KEY INFORMATION SOURCES

(For additional references, see Chapter 12 on Blackberries and Dewberries.)


Bucks, C. 1988. Raves for raspberries! *Org. Garden.* 43(8): 44–49.

Craig, D.L., Coulombe, L.J., and Rivard, I. 1984. *Cultivating red raspberries in eastern Canada.* Agriculture Canada, Ottawa, ON. pp. 30.

Crandall, P.C., and Daubeny, H.A. 1990. Raspberry management. In *Small fruit crop management.* Edited by G.J. Galletta and D.G. Himelrick. Prentice Hall, Englewood Cliffs, NJ. pp. 157–213.

Darrow, G.M. 1967. The cultivated raspberry and blackberry in North America—breeding and improvement. *Amer. Hort. Mag.* 46: 203–218.

Dickerson, G.W. 2004. *Raspberries for the home garden.* Revised edition. Cooperative Extension Service, College of Agriculture and Home Economics, New Mexico State University, Las Cruces, NM. pp. 4.

Ellis, M.A. 1991. *Compendium of raspberry and blackberry diseases and insects.* American Phytopathological Society Press, St. Paul, MN. pp. 100.

Hall, H.K., and Stephens, M.J. 1999. Fruit firmness in raspberries. *Acta Hortic.* 505: 93–99.

Henderson, D.E.H., and Raworth, D.A. 1991. *Beneficial insects and common pests on strawberry and raspberry crops.* Agriculture Canada, Ottawa, ON. pp. 33.

Jennings, D.L. 1995. Raspberries and blackberries. In *Evolution of crop plants.* Second edition. Edited by J. Smartt and N.W. Simmonds. Longman Scientific & Technical, Burnt Mill, Harlow, Essex, UK. pp. 429–434.

Keep, E. 1988. Primocane (autumn)-fruiting raspberries: A review with particular reference to progress in breeding. *J. Hortic. Sci.* 63: 1–18.

Klesk, K., Qian, M., and Martin, R.R. 2004. Aroma extract dilution analysis of cv. Meeker (*Rubus idaeus* L.) red raspberries from Oregon and Washington. *J. Agric. Food Chem.* 52: 5155–5161.

Lamonte, E.R., and O'Rourke, A.D. 1985. *The red raspberry industry in the Pacific Northwest of the United States and Canada.* College of Agriculture & Home Economics, Washington State University, Pullman, WA. pp. 55.

Lawrence, F.J. 1979. *Growing raspberries.* Revised edition. U.S. Department of Agriculture, Washington, DC. pp. 20.

Leemans, J.A. 1957. *A morphological classification of raspberry varieties.* Instituut voor de Veredeling van Tuinbouwgewassen, Wageningen, the Netherlands. pp. 140.

Logsdon, G. 1984. How to win at raspberries. *Org. Garden.* 31(12): 54–58.

Oldeheuvel, L. 1982. *A brief analysis of the international market for raspberries.* Province of British Columbia, Ministry of Agriculture and Food, Market Research and Analysis, Surrey, BC. pp. 73.

Oregon State University. 1983. *Raspberries: insect, disease, and weed control guide for commercial growers.* Pacific Northwest Extension Publication no. 233. Oregon State University, Extension Service, Corvallis, OR. pp. 16.

Pritts, M.P. 1996. Raspberries. *J. Small Fruit Viticult.* 4: 189–225.

Scheer, W.P.A. 1981.*Commercial red raspberry production.* Washington State University, Cooperative Extension Service, Pullman, WA. pp. 31.

Taylor, C.E. 1979. *Raspberries.* Grower Books, London, UK. pp. 39.

Whitney, G.G. 1984. The reproductive biology of raspberries and plant-pollinator community structure. *Am. J. Bot.* 71: 887–894.


### SPECIALTY COOKBOOKS


Beach, D.R. 1990. *Let's grow raspberries!* D. Beach, Junction City, OR. pp. 87.

Brock, V. 1997. *Raspberry story: a guide to cultivating & enjoying raspberries; includes 140 recipes, hot tips & sound advice.* Vivian Brock, Englewood, CO. pp. 90.

Gahaghan, L. 1994. *It's raspberry tyme: recipe book.* Third revised edition. L. Gahaghan, Kingsburg, CA. pp. 189.

Hériteau, J. 1984. *Mad about raspberries & strawberries*. Putnam Publishing Group, New York, NY. pp. 62.
Kile, S. 1984. *Remarkable red raspberry recipes*. Washington Red Raspberry Commission, Lacey, WA. pp. 170.
Olson, C.K. 1985. *Berries beautiful: An all-season cookbook for lovers of strawberries and raspberries*. C. Olson Publishing, Spy Hill, SK. pp. 110.
Peck, D., and Rosenstein, B. 1993. *The very special raspberry cookbook*. The Very Special Raspberry Committee, printed by Jumbo Jack's Cookbooks, Albuquerque, NM. pp. 272.
Rothschild, S. 1991. *Red raspberry recipes: Featuring the Robert Rothschild culinary collection*. Robert Rothschild Berry Farm in association with the American Cooking Guild, Urbana, OH. pp. 64.
Thompson Berry Farms. 1985. *The raspberry recipe book*. Thompson Berry Farms, Duluth, MN. pp. 41.

Krumm (1991) has ten wild raspberry recipes, Freitus and Haberman (2005) have 14, and Stanek and Butcher (2007) have 12. Marie (2008) has a recipe for wild raspberry jam. (See the Appendix to this book for details of the publications cited.)

# 78 Red Mulberry

Family: Moraceae (mulberry family)

## NAMES

Scientific name: *Morus rubra* L.

- The name "mulberry" is derived from Old English *morberie*, berry, + *mor*, from the Latin *morum*, mulberry.
- "Black mulberry" is occasionally applied to the red mulberry, which is quite inadvisable since it normally refers to another species (discussed below).
- The genus name *Morus* is derived from the ancient Latin *morus*, and the Greek *moron*, names for the mulberry. It has also been suggested that the Latin name was derived from the Latin *mora*, delay, a reference to the tardy expansion of the buds in the spring. Still another interpretation is that the name comes from the Celtic *mor*, black, in allusion to the color of the fruits of some species, probably the black mulberry.
- The mulberry genus *Morus* should not be confused with the paper mulberry, *Broussonetia papyrifera* (L.) Vent., a small tree native to China and Japan, which is widely planted in the Far East for its bark, used to manufacture paper, rope, and clothing.
- *Rubra* in the scientific name of the red mulberry is Latin for red, descriptive of the color of the fruit.

## GEOGRAPHY AND ECOLOGY OF WILD PLANTS

The red mulberry is a native of the eastern and Midwestern United States, reaching northward to southern Ontario and Vermont, west to South Dakota, and south to Florida and Texas. It is also found in Bermuda. Habitats include moist deciduous forests, forest edges, valleys, floodplains, and low, moist hillsides. In the southern part of its range, red mulberry is often found in pastures and along borders of fields. The species occurs as an understory tree or shrub but grows best in open situations. It is found in various soil types, with a preference for moist but well-drained soils. Although widespread, the tree is rarely abundant in locations where it occurs.

## PLANT PORTRAIT

There are about 10 species of *Morus*, which are deciduous trees and shrubs, all with more or less edible fruit that resembles a blackberry. Mulberry fruits have never become as desired in the New World as in the Old and have lost popularity over the centuries. Generally, mulberries are eaten locally and are not sold in commerce, because they are highly perishable with a short shelf life and the taste is not appealing to most people. The ripe fruit falls off easily and can be harvested by spreading sheets beneath a tree and shaking the branches. Mulberries should not be planted near patios or other such locations where staining from the fruits should be avoided (Only male trees, sometimes called "fruitless mulberries," are often planted for ornament.).

Mature red mulberry trees are 5–21 m (15–70 feet) high. The trunk diameter is rarely wider than 60 cm (2 feet). The leaves may be unlobed or may have two to five major lobes. Considerable variation in extent of lobing occurs among trees, indeed even among the leaves of a given branch.

**FIGURE 78.1** Red mulberry (*Morus rubra*). (a) Fruit. (Courtesy of Kmakinemi [CC By 3.0].) (b) Trees. (Courtesy of Richard Webb, self-employed horticulturist, Bugwood.org [CC By 3.0].) (c) Fruits at different stages of maturation. (Courtesy of Steven J. Baskauf [http://bioimages.vanderbilt.edu].) (d) Fruiting branches. (Courtesy of Franklin Bonner, USFS [ret.], Bugwood.org [CC By 3.0].) (e) Distribution map.

Moreover, the leaves produced by a given tree in one year often differ dramatically from the leaves produced in the next year. Both leaves and twigs contain a milky sap. Red mulberry plants can have male flowers only, female flowers only, or (less commonly) both kinds of flower. The flowers are yellowish green or reddish green and are arranged in catkins 1–2 cm (0.4–0.8 inch) long. Although male and female plants often need to be cultivated together to ensure sufficient fruit production, some plants are bisexual, and some cultivars (notably 'Tufts' and 'Hicks') are parthenocarpic (from the Greek words *parthenos* meaning virgin, and *carpic* meaning fruit), producing fruit without fertilization. The fruits are elongate, cylindrical, about 1 cm (0.4 inch) wide and 2.5–4 cm (1–1.6 inches) long. They are initially green, turning red, and becoming black or deep purple when fully ripe. There are several red mulberry cultivars used for fruit, but the species is not as well developed as the black and white mulberries (discussed below). Its fruits are tastier than those of the white mulberry, but not as good as those of the black mulberry. In colonial times in North America, settlers sometimes used the native red mulberry as food for silkworms.

Botanically, a "berry" is a fleshy fruit produced from a single ovary, although in common language the word usually means any small fleshy fruit. However, berries in the strict sense are not necessarily small (watermelons and pumpkins are berries). Although "berries" is part of the name of "mulberries," the fruit is not technically a berry—it is a "multiple fruit," developed from a cluster of very small flowers, the individual fruits of which coalesce together during development (figs and pineapples are other examples of multiple fruits).

The red mulberry is common (but apparently decreasing) in much of its range in the United States. In Canada, it is the tree species most threatened with extinction. There are only about 300 wild trees in the entire country. Habitat destruction, caused by land clearing and urbanization, is the most important cause, but the possibility of extinction is increased because of hybridization with the introduced white mulberry, which is eliminating the pure form of the species. Hybrids of red and white mulberry such as 'Illinois Everbearing' and 'Collier' have been promoted for cultivation because they produce large, flavorful, nearly seedless fruits and are hardy and vigorous. However, these cultivars may be invasive, escaping to natural habitats, and displacing the native red mulberry.

The black mulberry (*M. nigra* L.) is a small, picturesque spreading tree to 10 m (33 feet) in height. A native of central Asia (Iran, Asia Minor), it is widely cultivated for its black fruit throughout southern Eurasia and at higher altitudes in the tropics. The species has been grown since antiquity in the Old World as a source of fruit. It is referred to in the Bible, and by classical Greek and Roman writers. It was introduced into North America soon after the Discovery and has escaped from cultivation in many parts of the southern United States. The black mulberry is less hardy than the peach, and may be raised with protection as far north as New York and New England in the United States, and southern Ontario and southern British Columbia in Canada. There are numerous cultivars. The fruits are the largest, juiciest, and finest of any species in the genus. Black mulberry trees are long-lived, and specimens hundreds of years old have been recorded.

The white or silkworm mulberry (*M. alba* L.) is the most discussed tree in literature. It generally grows no higher than 15 m (49 feet), but sometimes attains 25 m (82 feet) in height, with spreading branches and a round-topped habitat. The fruits are white or pinkish, or sometimes dark purple (in which case the plant is often mistakenly identified as the black mulberry). The species is a native of China, cultivated there and elsewhere primarily for its leaves, which form the food of the silkworm, and also for its sometimes palatable (although generally insipid) fruits. The white mulberry is also grown for fiber in the stems used for papermaking, and is also frequently encountered as an ornamental. The tree is sometimes found in North America as an escape from cultivation.

People expect white mulberry to have only white fruits, red mulberry to have only red fruits, and black mulberry to have only black fruits. However, the fruits of white mulberry are red when immature, later becoming black, purple, or nearly white. The fruit of red mulberry is red when young, but becomes black or dark purple at maturity. The fruits of black mulberry are usually black.

## CULINARY PORTRAIT

In general, mulberries are basically a homegrown fruit, which is not marketed because the fruits bruise easily during handling and do not store well, although they can be frozen. Red mulberries are not available commercially and need to be collected from the wild or from the occasional cultivated plant. The fruits are conveniently harvested by spreading sheets under the tree and shaking the branches (but the sheets will become purple). Mulberry fruit gradually darkens and sweetens as maturity is approached and should not be picked until completely ripe. At full ripeness, mulberries are juicy with a musky, slightly acid, distinctive flavor. Red mulberry fruits are juicy and sweet. They are eaten raw as a dessert by some who appreciate the fresh flavor, or are stewed, made into syrup, preserves, jelly, jam, wine, liqueur, or incorporated into pies, tarts, bread, muffins, and cakes. In some regions of Afghanistan, dried mulberries, frequently mixed with ground almonds, are a dietary staple. Mulberry gin, made from black or red mulberries, is said to be better than sloe gin. To make a purple mulberry milkshake, add 2 cups vanilla ice cream, ½ cup milk, 1 cup mulberries, and blend. Many consider the highly saccharine flavor of fresh mulberries to be excessively sweet. When eaten fresh, it is often difficult to avoid staining hands, lips, and clothing with the red to black juice. Mulberry fruits are rather seedy (except in seedless varieties), and the seeds are best strained away.

Although its fruits are tastier than those of the white mulberry, some consider the fruit of red mulberry inferior to that of black mulberry. Others consider the fruit to be just as good. There are several red mulberry cultivars used for fruit, but the species is not as well developed as a food source as the black mulberry, which has been considered a delicacy since Roman times.

### Warnings

Mulberries stain clothes and skin purple. Children who like red mulberries are purple much of the time! The unripe fruits and milky sap in the leaves and young stems are toxic and have acquired a reputation for use as a mild hallucinogen. Touching the leaves and stems can produce dermatitis in susceptible people. The pollen from male plants can cause hay fever.

### Culinary Vocabulary

- "Rosolio" (pronounced roh-so-leh-o) is an Italian liqueur made with mulberry or cherry juice. (An alternative meaning is that it is a flavoring liquid made from rose petals and honey in Italy's Calabria and Sicily regions.).

## PROSPECTS

The potential economic value of native red mulberry has been largely overlooked. It is another example of the neglect of promising native species, while looking to familiar Old World plants as starting material for the development of North American agriculture. This relates to the old pioneer notion that North America was a hostile wilderness where successful agriculture could only be ensured by transplanting European crops. However, red mulberry has desirable agricultural characteristics. It is generally resistant to pests and grows rapidly. Some of the limitations to address in developing it as a crop include the fact that the fruits fall off and become soft soon after ripening and they also produce a relatively persistent stain. Although mulberries do not ship well when fresh, freezing and drying could extend their use.

## CURIOSITIES OF SCIENCE AND TECHNOLOGY

- In pre-Columbian times, the Natchez Indians of Louisiana used the red mulberry as the name of one of their 13 months.
- King James I (1566–1625) made considerable efforts to establish a silk industry in Britain: in 1606, a million mulberry trees were imported to England; in 1609, he ordered that 162 ha (400 acres) of mulberries be planted in St. James Park in London. All of this work came to nothing. Attempts to raise silk in the New World were even more disastrous. The king tried to compel the London Company to raise silk in Virginia, which in 1623 imposed a fine of 4.5 kg (10 pounds) of tobacco on every planter who did not cultivate at least 10 mulberry trees for every 40 ha (100 acres) of estate, and presented a reward of 50 times as much tobacco as the weight in silk produced. Similar attempts were made in Carolina and Georgia after they were founded. As colorfully described by one Georgia historian (cited in Bailey, L.H. 1898. The evolution of our native fruits, Macmillan), the trustees of Georgia had three goals: "to provide an asylum for the poor debtor and persecuted Protestant; to erect a silk, wine and drug-growing colony; and to relieve the mother country

of an overburdened population." For several years, the New World produced hundreds of kilograms of silk annually, and in 1759, over 4500 kg (over 5 tons) was exported from Georgia. Silkworm production waxed and waned, but in 1826 an astonishing resurrection of interest occurred, following a congressional report that recommended the preparation of a practical manual on the industry. The manual was produced in 1829 and resulted in a contagious desire to produce silk. This was heightened with the introduction of the *multicaulis* variant of the white mulberry (*M. alba* var. *muticaulis* (Perrott.) Loud., sometimes called the Philippine mulberry), thought to be the source of the best Chinese silk. Multicaulis mulberries reached America about 1826, and over the next 10 years were marketed with incredible success. However, the industry quickly failed, with the loss of fortunes and the end of commercial silkworm culture in North America. Before the American mulberry industry collapsed, enough silk was manufactured to weave the coronation robe of English King Charles II (1630–1685), worn in 1660. Because of widespread attempts in the United States in the past to grow white mulberry for silk, numerous streets are named "Mulberry Street." Since mulberry plants can live for over 200 years, these same streets may have several mulberry trees growing vigorously.

- Hundreds of plants were assigned special meanings in the "Victorian Language of Flowers," popular in Victorian times, and many delighted in sending coded messages by this means, especially for romantic purposes. The black mulberry translated as "I shall not survive you" or "sadness"; and the white mulberry indicated "wisdom."
- In the Cotton Belt and southern Corn Belt areas of the United States, farmers used to plant a row of mulberry trees along the south side of their hog lots to provide food for the animals. Fruits falling from the trees were gobbled up by the pigs, and also by flock of birds which bombed cars and drying laundry with their inky droppings. The bird nuisance gradually led to the abandonment of the system.
- During World War II, "mulberry" was a code name for prefabricated ports that were towed across from England to the Normandy coast to supply the Allied armies in France in 1944. Two such ports were installed.
- Dispersal of seeds by fruit-eating fish is well documented in some tropical regions, but it was unknown in North American until 2003 when it was reported that channel catfish (*Ictalurus punctatus*) consume red mulberry fruits in the floodplain forests of the Mississippi River (Chick et al. 2003). The catfish move into periodically flooded habitats and are believed to defecate seeds in habitats suitable for germination when flooding subsides.
- A "mulberry molar" is a tooth with more than the usual four cusps.
- Woodchucks are reported to climb trees occasionally to avoid predators or to rest and sun (I observed a young woodchuck climb 5 m [26 feet] up a white spruce tree to escape my dog). Some woodchucks are thought to climb mulberry regularly to browse on the highly palatable leaves (Swihart and Picone 1990).

## KEY INFORMATION SOURCES


Alonzo, D.S. 2001. *Morus* L. In *Plant resources of South-East Asia, 12(1), medicinal and poisonous plants 1*. Edited by L.S. de Padua, N. Bunyapraphatsara, and R.H.M.J. Lemmens. Backhuys Publishers, Leiden, the Netherlands. pp. 359–364.

Ambrose, J.D. 1990. *Status report on the red mulberry,* Morus rubra, *in Canada*. Committee on the Status of Endangered Wildlife in Canada, Ottawa, ON. pp. 17.

Ambrose, J. D. 1999. *Update COSEWIC status report on the red mulberry* Morus rubra *in Canada*. Committee on the Status of Endangered Wildlife in Canada. Ottawa, ON. pp. 11.

Bailey, L.H. 1892. Mulberries. *Bull. Cornell Univ. Agric. Exp. Sta.* 46: 223–243.

Burgess, K.S., and Husband, B.C. 2006. Habitat differentiation and the ecological costs of hybridization: the effects of introduced mulberry (*Morus alba*) on a native congener (*M. rubra*). *J. Ecol.* 94: 1061–1069.

Burgess, K.S., Morgan, M., Deverno, L., and Husband, B.C. 2005. Asymmetrical introgression between two *Morus* species (*M. alba, M. rubra*) that differ in abundance. *Molec. Ecol.* 14: 3471–3483.
Burgess, K.S., Morgan, M., and Husband, B.C. 2008. Interspecific seed discounting and the fertility cost of hybridization in an endangered species. *New Phytol.* 177: 276–284.
Catling, P.M., and Small, E. 2005. Poorly known economic plants of Canada—44. Red mulberry (*Morus rubra* L.). *Bull. Can. Bot. Assoc.* 38(1): 50–11.
Chick, J.H., Cosgriff, R.J., and Gittinger, L.S. 2003. Fish as potential dispersal agents for floodplain plants: first evidence in North America. *Can. J. Fish. Aquat. Sci.* 60: 1437–1439.
Core, E.L. 1974. Red mulberry. In *Shrubs and vines for northeastern wildlife*. Edited by J.D. Gill and W.M. Healy. U.S. Dept. Agriculture Forest Service General Technical Report NE-9. pp. 106–107.
Ercisli, S., and Orhan, E. 2007. Chemical composition of white (*Morus alba*), red (*Morus rubra*) and black (*Morus nigra*) mulberry fruits. *Food Chem.* 103: 1380–1384.
Griggs, W.H., and Iwakiri, B.T. 1973. Development of seeded and parthenocarpic fruits in mulberry (*Morus rubra* L.). *J. Hortic. Sci.* 48: 83–97.
Hotta, T. 1954. Fundamentals of *Morus* plants classification. *Kinugasa Sanpo* 390: 13–21.
Minamizawa, K. 1997. *Moriculture: science of mulberry cultivation.* Second edition. A.A. Balkema, Rotterdam, the Netherlands. pp. 431.
Monson, C.S. 1996. Mulberry trees: the basis and remnant of the Utah silk industry. *Econ. Bot.* 50: 130–138.
Moore, D.M., and Thomas, W.P. 1977. Red mulberry/*Morus rubra* L. *Forest Service Tech. Report (USA)* 16: 55–56.
Rangaswami, G. 1978. *Mulberry cultivation.* Food and Agriculture Organization of the United Nations, Rome, Italy. pp. 150. [Cultivation of white mulberry]
Sanchez, M.D. 1999. Mulberry: an exceptional forage available almost worldwide. *World Animal Review (FAO)* 93: 36–46.
Sanchez, M.D. 2002. *Mulberry for animal production: Proceedings of an electronic conference carried out between May and August 2000.* Food and Agriculture Organization of the United Nations, Rome, Italy. pp. 331. [Mulberry as forage]
Sandhya [Librarian]. 1988. *Bibliography on mulberry, 1900–1984.* Central Sericultural Research and Training Institute, Mysore, India. pp. 321.
Swihart, R.K., and Picone, P.M. 1990. Arboreal foraging and palatability of tree leaves to woodchucks. *Am. Midl. Nat.* 125: 372–374.
Tikader, A., and Ananda-Rao, A. 2001. Analysis of divergence in mulberry species through different clustering techniques. *Sericologia (France)* 41: 109–120. (In English and French)
Wunderlin, R.P. 1997. *Morus.* In *Flora of North America north of Mexico*, Vol. 3. Edited by Flora of North America Editorial Committee. Oxford University Press, New York. pp. 390–392.
Zhao, W.G., Zhang, Z.F., and Pan, Y.L. 2003. Molecular phylogeny of genus *Morus* based on cpDNA trnL-trnF sequence data. *Sericologia (France)* 43: 359–367. (In English and French)

## Recipe Sources

Freitus (1980) has 3 mulberry recipes, Freitus and Haberman (2005) have 8, Johnson (1989) has 4, Kluger (1973) has 8, Marie (2008) has 1, Marrone (2009) has 14, Russell (1975) has 7, Robe-Terry (1997) has 1, Tatum (1976) has 2, and Van Ata (1991) has 2. Williamson (1989) has 21 recipes for the Texas mulberry (*M. microphylla* Buckley), which occurs in Arizona, New Mexico, Texas, and Oklahoma. (See the Appendix to this book for details of the publications cited.)

# 79 Reed (Common)

Family: Poaceae (Gramineae; grass family)

## NAMES

Scientific name: *Phragmites australis* (Cav.) Trin. ex Steud. (*P. communis* Trin.)

- The English word "reed" traces to the Old English *hreod*, and seems to be related to the Old High German *hriot*, meaning reed.
- Common reed is also known as common reed grass, ditch reed, giant reed ("giant" is better reserved for the taller true giant reed, as noted below), giant reed grass, and reed grass (also spelled reed-grass and reedgrass).
- Common reed is sometimes misidentified as the true "giant reed," *Arundo donax* L., an exotic Asian introduction to North America, where it is a troublesome aquatic weed in the southern United States.
- The names cane grass and quilrod are used in Australia.
- Names like arrow grass, bamboo grass, and common rush are used locally, but lead to confusion with unrelated species.
- The genus name *Phragmites* is derived from the Greek word for fence, *phragma*, a reference to its fence-like growth along bodies of water.
- *Australis* in the scientific name *P. australis* is Latin for southern.

## GEOGRAPHY AND ECOLOGY OF WILD PLANTS

Common reed occurs on every continent except Antarctica, and is especially widespread in the temperate zones. It is present in almost all states of the United States, and is found in all provinces of Canada. The species is a semiaquatic, usually growing with its roots in water or at least in soil that is periodically wet. It inhabits fresh water, brackish (slightly saline) and alkaline wetlands, typically in marshes, swamps, wet meadows, and fens. It has been recorded in standing water up to 2 m (6.5 feet) deep, but generally occupies much shallower water. The species roots in a wide variety of soil types. Common reed is usually the dominant plant where it occurs, often forming very large, continuous patches. The Danube Delta (mostly in Romania), probably has the world's greatest stretch of reed beds (2700 $km^2$ or 1042 square miles). In European lakes, extremely large floating or loosely rooted mats of vegetation are often developed, dominated by common reed.

*Phragmites australis* reproduces very vigorously by vegetative spread. It grows so tall that few plants can outshade it. Although seed reproduction is comparatively rare, the seeds are sometimes numerous. They are normally dispersed by the wind, but sometimes by birds. Seeds will germinate in water depths up to 5 cm (6 inches), and salinities up to 2% (but germination is reduced in water with over 1% salt). Perhaps more important than reproduction by seed is reproduction by rhizome (underground stem) fragments. Such fragments can be carried to newly opened sites by machinery, floodwaters, birds gathering nesting material, and even by wind. Rhizome fragments are generated not only by human activity (e.g., plowing and cleaning out ditches) but also naturally by rodents. Common reed can also develop stolons (aboveground stems) that trail along the ground for as long as 15 m (49 feet). These very long runners typically develop out of water, serve to rapidly spread the plant, and produce shoots with abnormally small leaves.

**FIGURE 79.1** Reed (*Phragmites australis*). (a) Invasive plants in Louisiana. (Courtesy of Steve Hillebrand, U.S. Fish and Wildlife Service.) (b) Invasive plants in Maryland. (Courtesy of Mary Hollinger, U.S. National Oceanic and Atmospheric Administration.) (c) Rhizome with young stems. (Courtesy of Kenraiz [CC By 3.0].) (d) Flowering branch. (From Syme, J.T.B., *English botany*. Vol. 11, Plate 1727, 1873.) (e) Bundles of harvested stems for thatch roofing. (Courtesy of nl. Riet [dakdekkersriet] [CC By 3.0].) (f) Distribution map.

Common reed has been listed as one of the "world's worst weeds," and is legislated as a noxious weed in several jurisdictions in North America. It is a major pest of irrigation and flood control channels, where it invades the wet banks and sometimes extends into water 2 m (6.6 feet) deep, reducing flow. In addition, it causes water loss and stagnation in reservoirs and impedes navigation. Common reed is a particularly troublesome weed in marshes that have been disturbed by excess sediment from flooding, pollution, and cultivation. It does well in disturbed areas, especially along roadside ditches, railroad tracks, and on dredged soils and mine spoils. Common reed often becomes very aggressive in aquatic areas where the water table is lowered.

Invasive races have been introduced to North America from Eurasia, and are of considerable concern because they are displacing native forms of the species as well as other aquatic species. Invasive reed is particularly common in eastern North America, whereas native North American kinds of reed are relatively predominant in western North America. Native North American plants can frequently be distinguished by appearance from the Eurasian races, but botanical expertise is required as the differences are difficult to discern. Most native North American plants have been placed in *P. australis* subsp. *americanus* Saltonstall; most invasive common reed plants have been placed in *P. australis* subsp. *australis*.

## PLANT PORTRAIT

Common reed is a tall, robust, perennial grass, usually 1–4 m (3–13 feet), sometimes as high as 7 m (23 feet), with stout, jointed stems. The flat, pennant-shaped, gray-green leaves are 20–50 cm (8–20 inches) long. The plant produces a large plume-like, whitish (often purplish when young) flowering head 20–40 cm (8–16 inches) long, which appears to be feathery because of the presence of long, silky hairs. The edible grains are small, but the seed head may contain as many as a thousand seeds. In addition to the wild plants, there are variegated ornamental cultivars, often with thin gold and green stripes, that are grown as garden plants.

For the most part, common reed has not been developed as a commercial crop, except at the cottage industry level. The plant has often been used for mats, thatching, cordage, nets, roofing, and construction of huts. Pens for writing on parchment were once cut and fashioned from common reed. Other miscellaneous uses include animal bedding, upholstery stuffing, insulation, fencing, musical instruments, fishing rods, and basket weaving. North American Indians used the stems for a very wide variety of purposes, including shafts of arrows, prayer sticks, weaving rods, and pipestems. The plant has also been employed as a source of cellulose for making paper and construction board. Because of its short fibers, common reed is unsuitable for high-quality paper, but up to 20% can be added to paper products. In Europe, reeds are harvested to manufacture paper and cardboard. The rhizomes have served as raw material for manufacturing alcohol. The flowers heads are often a component of dried flower arrangements. The species is occasionally planted to dry out excessively wet areas, especially near water courses. The cultivar ‘Shoreline’ is used as a land stabilizer in the southern central Great Plains of North America.

## CULINARY PORTRAIT

Common reed is considered to be an attractive wild food for humans. The young emergent shoots, especially while still white, can be eaten like asparagus, or pickled. Very young leaves (before they are completely unfolded) may be cooked as a potherb. The young, green stems (while still fleshy) can be dried and pounded into a fine powder, which can be moistened and roasted like marshmallows. The rhizomes (underground stems) are full of starch in the fall and spring, and can be consumed like potatoes. The small seeds are firmly attached to their hulls, but can be ground into flour and consumed as a gruel. The rhizome has been used to produce a brown beer.

### Warning

Common reeds should not be collected as food unless it is certain that the water it is occupying is not polluted.

### Culinary Vocabulary

- Reeds (species of the genus *Phragmites*) are often confused with cattails (of the genus *Typha*), both growing in shallow water. In Europe, cattail species are often called reedmace (reed mace, reedmace), and so European wild food guides need to be used carefully to ensure that the identification of the “reed” in question is clear.

## PROSPECTS

The seeds of common reed are very small, limiting the possibility of developing the plant as a cereal. However, the species produces prodigious amounts of starchy rhizomes, and this source of edible and industrial materials has been largely ignored to date. Appropriate harvesting and processing technology is lacking, but given that there is a widespread desire to have reeds harvested for control purposes, it would be clever to develop means of exploiting this natural resource. There is also considerable potential for common reed as a source of biomass, principally for production of building materials such as fiberboard, insulation, and similar products. As well, biomass could be useful as fuel, fodder, and agricultural compost. Common reed occupies wetland habitats that otherwise are largely unsuitable for growing crops, so it offers the possibility of utilizing land currently not being used for agricultural purposes. Although developing reed into a crop is challenging, it represents the kind of innovation that is needed to revitalize agriculture today.

## CURIOSITIES OF SCIENCE AND TECHNOLOGY

- A "reed" was an ancient Hebrew unit of length, equal to 6 cubits. A cubit was the length of the forearm from the elbow to the tip of the middle finger, usually about 45 cm (18 inches). The use of reed units is mentioned in the bible, for example: "He measured the east side with the measuring reed, five hundred reeds, with the measuring reed round about" (Ezekiel 42:16).
- The word "reed," of enormous significance in music, traces to the use of the hollow stem of plants known as reeds. The giant reed (*Arundo donax*) noted earlier is often confused with common reed. It is probably the most important of the hollow reeds used for the last 5000 years to construct woodwind musical instruments. Today, reeds for musical instruments are still made to some extent from giant reed. Fundamentally, woodwind instruments consist of one or more tubes that function to resonate sound waves, usually generated by a stream of air. Modern woodwind instruments include the flute, pan pipes, clarinet, bagpipe, certain mouth organs, reed organs, and concertinas. Reed instruments also evolved into other musical instruments, such as the oboe, English horn, and bassoon.
- In the Exodus story, the Hebrews were in the desert, near starvation, and complained to Moses who appealed to God. The next day they were astounded as the dew dried into edible flakes and tasted like honey on the desert floor. It was the manna that was to sustain them through 40 years of wandering. But this manna could not be stored. Some people tried gathering more than they needed, but by the next day their surplus stunk and was full of maggots. And so for 40 years they lived day in and day out dependent on the grace of God. The story of manna teaches that the gifts of God are not to be hoarded. Early missionaries in California observed that a kind of sweet sugar solidifies and hardens on the leaves of reeds from which it is collected by the Indians. The gathering of the honeydew seems to have been one of the annual seasonal rounds of activity of the Native Americans of the Great Basin. The honey dew was rolled into balls the size of one's fist, wrapped in leaves, and stored in baskets until needed. Some of the Holy Fathers relished this Indian food and considered it "manna from Heaven," but lost their appetites when they observed that the main ingredients of the pressed mass were insects. The predominant insect is *Hyalopterus pruni*, called the mealy plum aphid because it spends its winter phase on plum trees. In the spring and early summer it migrates to summer hosts, including common reed, where it produces the honeydew.
- The exodus of the Israelites from Egypt is through the "Red Sea," as traditionally translated into English by the King James version of the Bible. This has given rise to an extended debate about the proper translation from the original Hebrew, biblical experts contending that the correct translation is "Sea of Reeds."

- In Australia, aboriginal peoples constructed large rectangular rafts of bound common reed, and used these to collect mussels from lakes. North American Indians also used common reed for rafts.
- Several North American indigenous groups used to fill reed stems with tobacco and smoke these like cigarettes. Six-hundred-year-old cigarettes were found at the Red Bow Cliff Dwelling in Arizona, some adorned with miniature bows. The tobacco was lit and smoked and because the tough reed exterior did not burn, it could be used again as a sort of pipe. Thirteen kinds of plants, as well as bird feathers, were smoked.
- Although not considered medically useful today, in the past, common reed was widely employed for medicinal purposes. The leaves and rhizomes were used to treat numerous illnesses in folk medicine, most commonly as a diuretic (to induce urination), but also for arthritis, rheumatism, leukemia and other cancers, breast problems, diabetes, gout, jaundice, nausea, typhoid, and many other conditions.
- In China, severe bad breath was treated with a mouthwash, prepared by combining the rhizome of reed with gypsum (calcium sulfate).
- Reedbeds as old as 1000 years are known, and it has been speculated that some may be up to 6000 years of age (although no living part of the plants is more than 8 years old).
- "Ayahuasca" (pronounced EYE-a-wasca, literally "spirit-vine" in the Andean language Quechua) refers to drug mixtures from South American plants, particularly the vine *Banisteriopsis caapi* (Spruce ex Griseb.) C.V. Morton, but also many other plants such as *Psychotria viridis* Ruíz and Pav., which have been used for millennia in the Amazon Basin by Indian tribes and shamans for healing and spiritual development. The term ayahuasca is also used generically to refer to plant preparations of other regions of the world used in the same manner. The active constituents are alkaloids, including tryptamines, which are powerful and dangerous hallucinogens that are popular in the illicit drug trade. Similarly, the skin and glands of several toads were used by ancient shamans, and in the 1960s some people licked or smoked the secretions of the rare venomous Colorado River toad (also known as the Sonoran Desert toad and the psychedelic toad), *Bufo alvarius*, which is known to contain high amounts of hallucinogenic tryptamine. Extracts of the rhizome of common reed have ayahuasca analogues (similar chemicals), particularly dimethyltryptamine, and dried extracted resin has psychoactive properties when smoked. Although some early cultures consumed tryptamines, probably with great care, use of tryptamine in any of these forms is quite risky. The presence of high amounts of tryptamines in some animals, plants, and fungi likely reflects a common protective function of these powerful drugs against insects and microbes.
- Parts of common reed yield up to 13% of the chemical furfural, so the species is a potential major source of this major industrial chemical. Furfural is currently derived by steam distillation from agricultural residues (straw, husks, etc.) treated with dilute sulfuric acid. This organic solvent is widely used in the chemical, petrochemical, and pharmaceutical industries for an amazing variety of purposes. It is used as a refining solvent in the manufacture of synthetic rubber and nylon, and in the production of resins for molded plastics and metal coatings. It is also a component of insecticides, embalming and disinfectant fluids, an additive to rocket fuel, and for manufacturing pharmaceutical products such as antibiotics, anesthetics, bacteriostatics (antibacterial compounds), and fungicides.
- "Biomass crops" are ones that produce a large amount of dry matter that can be used for fuel, pulp, or industrial purposes. The potential for fuel production is well recognized in the colder regions of Europe. An astonishing 300 metric tons/ha (135 tons/acre) has been recorded for common reed in Australia, but even in the less hospitable climate of Europe, as much as 100 metric tons/ha (45 tons/acre) seems possible.

## KEY INFORMATION SOURCES


Bellevance, M.E., and Brisson, J. 2010. Spatial dynamics and morphological plasticity of common reed (*Phragmites australis*) and cattails (*Typha* sp.) in freshwater marshes and roadside ditches. *Aquatic Bot.* 93: 129–134.

Bjoerndahl, G., and Egneus, H. 1980. *Ecology and physiology of reed.* Statens Naturvaardsverk, Solna, Sweden. (In Swedish; presents a literature review for the evaluation of reed as a source of energy)

Björk, S. 1967. Ecological investigations of *Phragmites communis*: studies in theoretical and applied limnology. *Folia Limnol. Scandinav.* 14: 1–248.

Brisson, J., Paradis, E., and Marie-Eve, B. 2008. Evidence of sexual reproduction in the invasive common reed (*Phragmites australis* subsp. *australis*; Poacaeae) in Eastern Canada: a possible consequence of global warming. *Rhodora.* 110: 225–230.

Brix, H. (Editor). 1999. Special issue on genetic diversity, ecophysiology and growth dynamics of the common reed (*Phragmites australis*). *Aquatic Bot.* 64: 179–415.

Catling, P.M., and Carbyn, S. 2006. Recent invasion, current status and invasion pathway of European common reed, *Phragmites australis* subspecies *australis*, in the southern Ottawa district. *Can. Field-Nat.* 120: 207–312.

Catling, P.M., and Mitrow, G. 2011. Major invasive alien plants of natural habitats in Canada. 1. European common reed (often just called phragmites), *Phragmites australis* (Cav.) Trin. ex Steud. subsp. *australis*. *Can. Bot. Assoc. Bull.* 44(2): 52–61.

Catling, P.M., Schueler, F.W., Robichaud, L., and Blossey, B. 2003. More on *Phragmites*—native and introduced races. *Can. Bot. Assoc. Bull.* 36(1): 4–7.

Chambers, R.M., Meyerson, L.A., and Saltontall, K. 1999. Expansion of *Phragmites australis* into tidal wetlands of North America. *Aquatic Bot.* 64: 261–273.

Clayton, W.D. 1968. The correct name of the common reed. *Taxon* 17: 168–169.

Clevering, O., and Jorgen, L. 1999. Taxonomy, chromosome numbers, clonal diversity and population dynamics of *Phragmites australis*. *Aquatic Bot.* 64: 185–208.

Coops, H. 1999. Bank protection using common reed. *Levende Natuur.* 100(2): 46–49.

Cross, D.H., and Fleming, K.L. 1989. *Waterfowl management handbook.13.4.12, control of* Phragmites *or common reed.* U.S. Department of the Interior, Fish and Wildlife Service, Washington, DC. pp. 5.

Fiala, K. 1976. Underground organs of *Phragmites communis*, their growth, biomass and net production. *Folia Geobot. Phytotax.* 11: 225–259.

Gervais, C., Trahan, R., Moreno, D., and Drolet, A.M. 1993. *Phragmites australis* in Quebec: Geographic distribution, chromosome numbers and reproduction. *Can. J. Bot.* 71: 1386–1393. (In French, English summary)

Graneli, W. 1984. Reed *Phragmites australis* (Cav.) Trin. ex Steudel as an energy source in Sweden. *Biomass* 4: 183–208.

Hara, T., Toorn, J. van der, and Mook, J.H. 1993. Growth dynamics and size structure of shoots of *Phragmites australis*, a clonal plant. *J. Ecol.* 81: 47–60.

Haslam, S.M. 1969. *The reed: a study of* Phragmites communis *Trin., in relation to its cultivation and harvesting in East Anglia for the thatching industry.* Norfolk Reed Growers Association, Norwich, UK. pp. 45.

Haslam, S.M. 1972. Biological flora of the British Isles: Phragmites communis Trin. *J. Ecol.* 60: 585–610.

Haslam, S.M. 1973. Some aspects of the life history and autecology of *Phragmites communis* Trin. A review. *Pol. Arch. Hydrobiol.* 20: 79–100.

Howard, R., Rhodes, D.G., and Simmers, J.W. 1978. *A review of the biology and potential control techniques for* Phragmites australis. U.S. Army Engineer Waterways Experiment Station, Vicksburg, MS. pp. 83.

Hocking, P.J., Finlayson, C.M., and Chick, A.J. 1983. The biology of Australian weeds. 12. *Phragmites australis* (Cav.) Trin. ex Steud. *J. Austral. Inst. Agric. Sci.* 49: 123–132.

Kiviat, E., and Hamilton, E. 2001. *Phragmites* use by Native North Americans. *Aquatic Bot.* 69: 341–357.

Kopppitz, H. 1999. Analysis of genetic diversity among selected population of *Phragmites australis* worldwide. *Aquatic Bot.* 64: 209–221.

Mal, T.K., and Narine, L. 2005. The biology of Canadian weeds. 129. *Phragmites australis* (Cav.) Trin. ex Steud. *Can. J. Plant Sci.* 84: 365–396.

Marks, M., Lapin, B., and Randall, J. 1994. *Phragmites australis* (*P. communis*): Threats, management, and monitoring. *Natural Areas J.* 114: 285–294.

Mason, C.F., and Bryant, R.J. 1975. Production, nutrient content and decomposition of *Phragmites communis* Trin. and *Typha angustifolia* L. *J. Ecol.* 63: 71–95.

Matoh, T., Matsushita, N., and Takahashi, E. 1988. Salt tolerance of the reed plant *Phragmites communis*. *Physiol. Plant.* 72: 8–14.
McKee, L., and Richards, A.J. 1996. Variation in seed production and germinability in common reed (*Phragmites australis*) in Britain and France with respect to climate. *New Phytol.* 133: 233–243.
Meyerson, L.A., Lambert, A.M., and Saltonstall, K. 2010. A tale of three lineages: expansion of common reed (*Phragmites australis*) in the U.S. Southwest and Gulf Coast. *Weed Sci. Soc. Am.* 3: 515–520.
Meyerson, L.A., Viola, D., and Brown, R. 2009. Hybridization of invasive *Phragmites australis* with a native subsp. in North America. *Biol. Invasions* 12: 103–111.
Meyerson, L.A., Saltonstall, K., Windham, L., Kiviat, E., and Findlay, S. 2000. A comparison of *Phragmites australis* in freshwater and brackish marsh environments in North America. *Wetlands Ecol. Manage.* 8: 89–103.
Paul, J., Vachon, N., Garroway, C.J., and Freeland, J.R. 2010. Molecular data provide strong evidence of natural hybridization between native and introduced lineages of *Phragmites australis* in North America. *Biol. Invasions* 12: 2967–2973.
Saltonstall, K. 2002. Cryptic invasion by non-native genotypes of the common reed, *Phragmites australis*, into North America. *Proc. Nat. Acad. Sci.* 99: 2445–2449.
Saltonstall, K., Lambert, A., and Meyerson, L.A. 2010. Genetics and reproduction of common (*Phragmites autralis*) and giant reed (*Arundo donax*). *Invasive Plant Sci. Manage.* 3: 495–505.
Saltonstall K., Peterson, P.M., and Soreng, R.J. 2004. Recognition of *Phragmites australis* (Poaceae: Arundinoideae) in North America: Evidence from morphological and genetic analysis. *Sida* 21: 683–692.
Small, E., and Catling, P.M. 2001. Poorly known economic plants of Canada—29. Common reed, *Phragmites australis* (Cav.) Trin. ex Steud. *Can. Bot. Assoc. Bull.* 34(2): 21–26.
Soukup, A., Votrubova, O., and Cizkova, H. 2000. Internal segmentation of rhizomes of *Phragmites australis*: protection of the internal aeration system against being flooded. *New Phyol.* 145: 71–75.
Tewksbury, L.T., Casagrande, R.A., Blossey, B., Schwartzlaender, M., and Haefliger, P. 2002. Potential for biological control of *Phragmites australis* in North America. *Biol. Control* 23: 191–212.
Ward, D.B. 2010. North America has two species of *Phragmites* (Gramineae). *Castanea* 75: 394–401.
Weinstein, M.P. (Editor). 2003. *Phragmites australis*: A sheep in wolf's clothing? *Estuaries* 26(2B): 397–630. [Conference workshop issued by Estuarine Research Federation, Lawrence KS]
Zeidler, A., Schneider, S., Jung, C., Melchinger, A.E., and Dittrich, P. 1994. The use of DNA fingerprinting in ecological studies of *Phragmites australis* (Cav.) Trin. ex Steudel. *Botanica Acta (Germany)* 107: 237–242.

## Specialty Cookbooks

(*Note:* the rhizomes and young shoots of *Phragmites* can be prepared in the same way as those of cattail, and Chapter 24 on it can be consulted for additional information.)

Havens, K. 2002. *Phragmites australis* (reed grass) bane or beneficial? *The Virginia Wetlands Report* 17(2): 1–2. [Has a recipe entitled Roasted *Phragmites* rhizomes and another titled *Phragmites* gruel (which uses the seeds).]

# 80 Roses

Family: Rosaceae (rose family)

## NAMES

Scientific name: *Rosa* species; see Table 80.1 for information on the nine North American species that are especially notable for producing tasty fruit.

**TABLE 80.1**
**Native North American Rose (*Rosa*) Species with Fruits That Have Been Used Notably as Food**

| Scientific Name | Common Names | Distribution | Notes |
|---|---|---|---|
| *R. acicularis* Lindl. | Prickly rose | Eurasia, Canada, northern United States | Extensive historical use of fruit by Native Americans[a]; fruits recommended for culinary usage[b] |
| *R. arkansana* Porter | Prairie rose | Western Canada, northern and southcentral United States | Minor historical use of fruit by Native Americans; fruits recommended for culinary usage |
| *R. blanda* Aiton | Smooth rose, Labrador rose | Eastern Canada, northeastern and northcentral United States | Fruits recommended for culinary usage |
| *R. californica* Cham. & Schltdl. | California rose | Oregon, California; Baja Norte, Mexico | Minor historical use of fruit by Native Americans |
| *R. carolina* L. | Carolina rose, pasture rose | Central and southeastern United States | Fruits recommended for culinary usage |
| *R. gymnocarpa* Nutt. | Dwarf rose | Western Canada and western United States | Minor historical use of fruit by Native Americans |
| *R. nutkana* C. Presl | Nootka rose, nutka rose | Alaska, western Canada, western and southcentral United States | Extensive historical use of fruit by Native Americans; fruits recommended for culinary usage |
| *R. pisocarpa* A. Gray | Cluster rose, pea rose | Western Canada, western United States | Minor historical use of fruit by Native Americans |
| *R. woodsii* Lindl. | Fendler rose, interior rose, Woods' rose | Western North America from Alaska to California; Ontario; central and southwestern United States; Mexico | Moderate historical use of fruit by Native Americans; fruits recommended for culinary usage |

[a] Remarks on historical use of fruit by Native Americans are as indicated in Moerman, D.E. 1998. *Native American ethnobotany*. Timber Press, Portland, OR. pp. 927.

[b] Remarks on culinary usage are as indicated in Facciola, S. 1998. *Cornucopia II. A source book of edible plants*. Kampong Publications, Vista, CA. pp. 713.

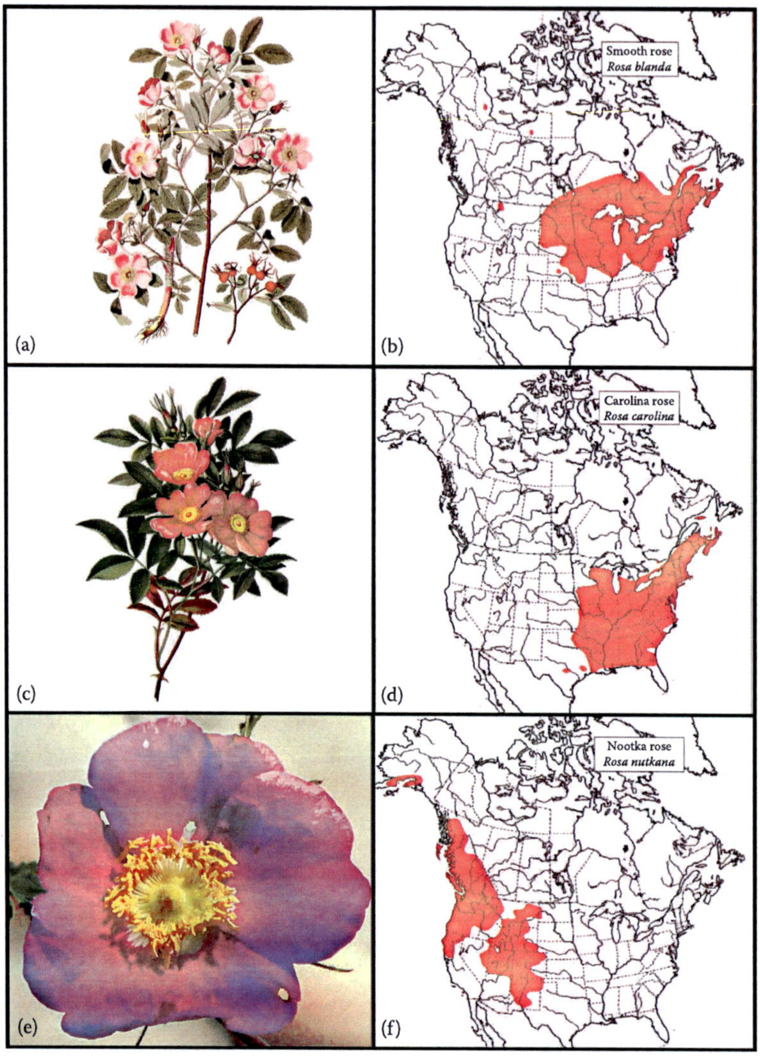


**FIGURE 80.1** Edible North American roses (*Rosa* species). (a) Branches of smooth rose (*R. blanda*) in flower and in fruit. (From Jacquin, N.J., *Fragmenta botanica, figures coloratis illustrata*, M.A. Schmidt, Vienna, Austria, 1809.) (b) Distribution map of smooth rose. (c) Carolina rose (*R. carolina*) in flower. (From Goodale, G.L., *The wild flowers of America*, Bradlee Whidden, Boston, MA, 1886.) (d) Distribution map of Carolina rose. (e) Flower of Nootka rose (*R. nutkana*). (Courtesy of Peter Stevens [Flickr/CC-attribution].) (f) Distribution map of Nootka rose.

- Rose is translated as rose in Danish, French, German, and Norwegian; and as rosa in Italian, Portuguese, Russian, and Spanish.
- Not all plants with "rose" in their name are true roses. For example, "rose-apple" is *Syzgium jambos* (L.) Alston of the Myrtaceae family. This tree is widely cultivated in tropical regions for its dry, crisp fruit that is insipid raw, but prized for confections and jellies. The "Christmas rose" is *Helleborus niger* L., a perennial herb of the Ranunculaceae family. Several ornamental cultivars of this are available, and the plant has also been used medicinally. However, all parts of the plant are extremely poisonous, and it should not be consumed. The "rose of Jericho" is any of several desert plants: *Anastatica hierochuntica* L. (mustard family) of northern Africa and southwest Asia; *Selaginella lepidophylla*

**FIGURE 80.2** More edible North American roses (*Rosa* species). (a) Prickly rose (*R. acicularis*) in fruit. (Courtesy of Steve Hillebrand, U.S. Fish and Wildlife Service.) (b) Distribution map of prickly rose. (c) Fruiting branch of Fendler rose (*R. woodsii*). (Courtesy of J. Schmidt, U.S. National Park Service.) (d) Distribution map of Fendler rose. (e) Flowers of prairie rose (*R. arkansana*). (Courtesy of Alexwcovington [CC By 3.0].) (f) Distribution map of prairie rose.

(Hook. & Grev.) Spring (club moss family) distributed from Texas and Arizona south to El Salvador; *Odontospermum pygmaeum* O. Hoffm. (sunflower family) of Asia Minor. These plants form tight balls when dry and unfold and grow under moist conditions. In the dry condition, they are tumbleweeds, scattering seeds as they roll about in the wind. Some plants called rose of Jericho, especially *Selaginella lepidophylla*, are also called "resurrection plants," because unlike most tumbleweeds which are dead in the dry condition, the plants remain alive but dormant while they are tumbling about, and when provided with water they begin to grow again.

- The genus name *Rosa* is based on the ancient Latin name for roses, *rosa*, which in turn originated from the Akkadian (the ancient language of Mesopotamia) word *russu*, "red."

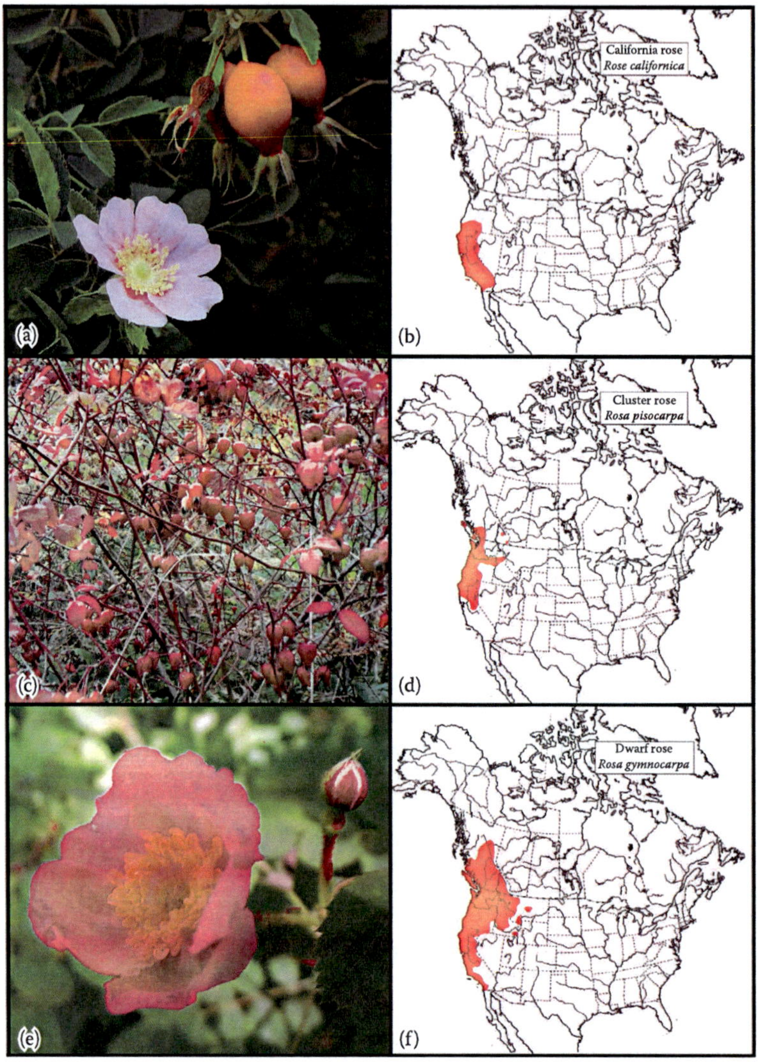


FIGURE 80.3 Still more edible North American roses (*Rosa* species). (a) Fruits of California rose (*R. californica*). (Courtesy of Paula Kate Marmor PKM [CC By 3.0].) (b) Distribution map of California rose. (c) Autumn appearance of cluster rose (*R. pisocarpa*). (Courtesy of Regina Johnson [CC By 3.0].) (d) Distribution map of cluster rose. (e) Flower of dwarf rose (*R. gymnocarpa*). (Courtesy of Stickpen [CC By 1.0].) (f) Distribution map of dwarf rose.

## GEOGRAPHY AND ECOLOGY OF WILD PLANTS

Rose species grow in a variety of habitats throughout North America, usually in open situations. Some are fairly widespread, others have narrow distributions and indeed are considered endangered. Depending on species, a range of soil types and tolerance of dry and wet soils is found.

## PLANT PORTRAIT

The genus *Rosa* consists of about 190 species of prickly shrubs (a few are rambling vines), distributed throughout the temperate and subtropical areas of the northern hemisphere. In North America there are 15–20 native species, and perhaps 12–15 introduced species. Roses are famous as ornamentals; at least 13,000 cultivars have been recorded, of which at least 5000 are commercially available.

A few roses yield important essential oils used in perfumery. To make a kilogram (2.2 pounds) of costly rose essence for use in fragrances and perfumes, 5 tons of roses or about 60,000 blossoms are required. The perfumery agent attar of roses is an oil, which rises to the surface of rose water left to stand and kept cool. *Rosa* × *damascena* Mill. (known most commonly as Damask rose, also four-seasons rose, Portland rose, and York and Lancaster rose) is the main source. It is grown commercially mainly in Bulgaria and Turkey.

Rose "hips" (fruits) have significant market value as sources of medicinal preparations. Rose hip oil, which is pressed from rose hips of some species, is used medicinally. It is high in essential fatty acids, which contribute to skin health, and so is used in skin care products. Extracts from rose hips have been shown to alleviate rheumatoid arthritis. Rose hips also have health values for pets and livestock. They are fed to pet guinea pigs—rodents which (like humans) can not synthesize their own vitamin C. Rose hips are also used as a vitamin C source for chinchillas, in the mistaken impression that they have the same need for vitamin C as guinea pigs. Rose hips are also fed to horses in the form of a dried powder, with the intention of improving coat condition and stimulating hoof growth. Most medicinal products made with rose hips are not produced from native North American species, but there have been recent attempts to develop the native species as sources of medicinal rose hips.

From a food viewpoint, the most important part of roses is their fruits, usually called rose hips (infrequently termed rose haws). Technically, rose hips are accessory or false fruits, not true fruits. The fleshy, outer portion that surrounds the "seeds" is a "hypanthium," developed from the top of the flower stalk; in contrast, the flesh of true fleshy fruits is developed from one or more ovary walls. The "seeds" of a rose are in fact the true fruits; these are categorized as "achenes," defined as single-seeded, dry, indehiscent fruits in which the seed coat is separate from the fruit wall (sunflower "seeds" are another example of achenes).

## CULINARY PORTRAIT

The use of rose fruits as food is minor, but nevertheless has a long history. Neolithic Old World archaeological sites dating back to 5000 years BP show evidence of humans intentionally collecting rose fruits for food, and they were similarly consumed in ancient times by the indigenous peoples of North America. Rose hips are often covered with hooked, hairy outgrowths which should be removed before the fruits are cooked. In most species, the hairs on the outside are less troublesome than the fuzz on the inside. De-seeding the hips and removing the fuzz and hairs are tedious tasks. Rose hips should be picked when they are plump and red, but not softly overripe. They can be left on the bush until the first light frost, and picked when slightly soft to the touch. They should be processed quickly after harvest to prevent loss of vitamin C (in relation to its small amount of edible matter, the rose fruit is extraordinarily rich in this vitamin). The stem and blossom ends should be trimmed, the hips cut in half lengthwise, the seeds and fibers scooped out with a small spoon, and the halves dried on a screen in an airy, shaded room indoors. Alternatively, they can be dried by hot air from an electric heater, care being taken to prevent overheating and loss of natural color. The hips can be screened in a rough wire container to remove their hair (unless the hairs are strained away, they can be irritating in foods prepared with the hips). Storage should be in a cool place, and as vitamin C decreases gradually, the hips and their products should not be kept longer than a year. The principal species used for rose hips are dog rose (*R. canina* L.) and Turkestan rose (*R. rugosa* Thunb.), both Old World species that are cultivated and have become naturalized in North America.

Some cautions should be kept in mind for those wishing to harvest and prepare rose hips. Rose bushes may have been sprayed with pesticides, and rose hips should not be harvested unless one is certain that the plants are safe. Very acidic foods should not be prepared in non-stainless-steel metal containers because the high content of acid may leach metal ions into the food. Because of the high ascorbic acid (vitamin C) content of rose hips, copper, aluminum, and iron containers should not be used to cook them. Stainless steel is best, glass containers and wooden spoons are also acceptable, earthenware, China bowls, and enamel are likely safe, but keep in mind that imported containers may not have been manufactured to domestic safety requirements, and toxic constituents could be set free by acid foods.

Rose flowers have limited, but interesting culinary uses. Drinks have been perfumed with roses, and rose petal wines have been made since classical times. During the Elizabethan era, rose petals were not only extensively used to scent food but even washing water. Rose water, used for flavoring sweet dishes, is the result of distilling rose flowers with a little water. The Damask rose, mentioned earlier, is the main source. It is also used to produce rose petal jam. The French rose (apothecary rose), *R. gallica* L., provides petals that are often crystallized, preserved in syrup, or dried and used to flavor teas, beverages, cakes, honey, and liqueurs. Rose water, rose petal jam, rose petal-flavored honey, rose-flavored candy, and rose-flavored vinegar are still marketed, albeit on a very limited scale. To a minor extent, the food industry uses rose extracts to flavor beverages, candy, ice cream, bakery goods, gelatin desserts, and jelly.

### Culinary Vocabulary

- *Palinka* is a traditional Hungarian fruit brandy, sometimes made with rose hips.
- *Nyponsoppa* is a rose hip soup which is very popular in Sweden, and is often served as a beverage or as a dessert with milk, cream, or vanilla ice cream.
- *Rhodomel* is a type of mead (a wine prepared at least partly from fermented honey) made with rose hips.
- "Stinking rose" is a euphemism for garlic.
- A "Russian Rose" is a cocktail prepared with vodka, grenadine, and orange bitters.
- In old lunch counter slang, "pin a rose on it" meant add onion to an order. "Burn one, take it through the garden, and pin a rose on it" meant hamburger with lettuce, tomato, and onion.
- "China rose tea" is a blend of freshly cut rose petals and Chinese black tea, the resulting brew having a strong rose scent. Mei Kwei (known as rose tea) is a similar blend, combining Chinese black tea and rosebuds.

## PROSPECTS

Roses grown for ornament and cut flowers are the basis of a huge industry. Rose fruits (hips) from wild and cultivated plants, which are harvested for food and medicinal applications, are of much smaller commercial significance, but these products have appreciable growth potential. At present, the rose species native to North America have very limited commercial value, for ornament, medicines, or food. By contrast, several rose species of Eurasia have become significant cultivated sources of rose hips, and indeed should the rose hip industry develop in North America, these foreign rose species, generally with quite large fruits, may well become the dominant crops. However, native North American roses are naturally adapted to the climates and soils of the continent, and research in progress may well show that one or more are advisedly developed as a new crop for food and associated health benefits. By comparison with most shrub species of North America, the native roses have reasonable possibilities of becoming new sources of marketable rose hip products.

## CURIOSITIES OF SCIENCE AND TECHNOLOGY

- The Roman Emperor Nero (37–68 AD) reputedly used up the equivalent in modern currency of $100,000 in cut roses to decorate a single night's feast. He is also believed to have had pipes installed under banquet plates to allow his guests to be spritzed with rose scent between dinner courses.

- Some privileged Romans slept on beds of rose petals, giving rise to the saying "It's not a bed of roses."
- "Apicius," a third or fourth century Roman, is supposedly the author of the oldest cookbook of Western nations, entitled *De Re Coquinaria* (On cooking), or alternately *De opsoniis et condimentis sive de re culinaria libri decem* (Ten books on catering and seasoning, or on cookery). This work has about 500 recipes divided into ten books. There were three Roman culinary experts named Apicius. Marcus Gavius Apicius, a famous gourmet, is thought to be most likely the original author. He poisoned himself when his fortune dwindled to the point that he could no longer support his extravagant meals. Manuscripts were copied repeatedly over the centuries, and although changes were introduced, presently available versions of Apicius' cookbook are thought to reflect well the eating habits of classical Rome. A much later writer, Coelius or Caelius, linked his own name with that of Apicius (he called himself "Caelius Apicius") and produced an especially popular version of the book which appeared sometime between 1483 and 1486. The recipe for rose wine in that version follows: "Make rose wine in this manner: Sew rose petals, the lower white part removed, into a linen bag and immerse in wine for 7 days. Then add a bag of new petals and repeat for another seven days. Again remove the old petals and replace them by fresh ones for another week; then strain the wine through a colander. Before serving, add honey, sweetening to taste. Take care that only the best petals free from dew are used."
- Western North American Indians thought wild roses repelled ghosts and evil spirits. Rose branches were scattered around the house and yard of recently deceased persons to prevent the return of the ghost, and relatives and friends drank a tea made from rose branches to protect them from it. A rose branch was used to sweep out the grave before burial to prevent the living from being drawn down.
- Since the time of the ancient Egyptians, roses have been associated with secrecy. According to Greek mythology, Cupid, shocked by the conduct of his mother Venus, took her flower, the rose, to Harpocrates, god of silence, to keep her scandal confidential. This story is said to have brought about the expression "sub rosa," referring to activities conducted confidentially. Roses were customarily suspended over the dinner table as a sign that confidences revealed under the influence of wine were to be kept secret. For many years the ceilings of European dining rooms featured formed plaster roses, with the same symbolism. The plaster ornament in the center of a ceiling is still known as "the rose" or "rosette," even if it is no longer in the form of a rose. Beginning in the Middle Ages, a wood carving of a rose was placed in the space over confessional doors in Catholic churches, once again symbolizing confidentiality.
- Rosaries are used to count prayers to Our Lady (the Virgin Mary). The first "rosaries" were made out of pounded rose petals, molded into beads and threaded (the word rosary is derived from the Latin *rosarium*, rose garden). Rosaries were later made out of rose hips, and eventually out of various materials. The scent of the beads was supposed to remind one of the roses in Mary's garden.
- Roses were important medicinal plants during the Middle Ages, used to reduce fever, inflammation, pain, and to stop any kind of excessive flow, be it tears, diarrhea, or hemorrhage. The popularity of the rose eventually resulted in its elevation to the status of a wonder drug, used even for epilepsy, tuberculosis, goiter, and gout. During plagues, rose fragrance was used to purify the air in public places. Apothecaries even used rose galls, employing them to induce urination.
- Roses are the most popular of gravestone-inscribed flowers. In Victorian cemeteries where a husband and wife were buried next to each other, the gravestone of the wife often showed a hand holding a lily (symbolic of innocence and purity), whereas that of the husband showed a hand with a rose. Children's gravestones often showed rose buds as symbols of

innocence, and sometimes the stem of the rosebud was carved as broken, suggesting their short lives.

- The Cherokee rose (*R. laevigata* Michx.) is a native of China that was imported into the United States, and now grows wild in the south. According to legend, it acquired its name when a Cherokee maiden carried one of these roses to her lover, a soldier who was wounded and convalescing in a tent. The white-flowered Cherokee rose has long been used as a peace symbol following fighting. During the American civil war, it was often planted as a memorial on the graves of the fallen.
- During World War II, rose hips were eaten for their content of vitamin C while supplies of foreign fruit were limited. The British Ministry of Health distributed "National Hip Syrup." Similarly, German submarines and ships were supplied with rose hip syrup.
- In England, it was found that the vitamin C content of rose hips increases toward the north, with four times as much in plants grown in Scotland compared to those in southern England.
- In 1939 Frances Meilland, a rose specialist in France, found a rose with magnificent pale gold blossoms growing from one seed he had nurtured. He sent cuttings to a Pennsylvania rose grower, which were transported on the last American plane that fled France in November 1940, a step ahead of the invading Nazis. The cuttings were used to propagate this rose which many experts consider the best ever developed, the Peace rose, which blooms on more than 30 million bushes throughout the world.
- "Rosarians" are devoted rose hobbyists.
- The naming of roses has become a commercial venture. Some companies offer clients the possibility of having new rose varieties named for them, for a price, sometimes ranging up to $75,000. New rose varieties are being named for corporations ('Barbie' for the doll; 'Weight Watcher's Success' rose), causes ('Arthritis' rose, 'Veterans' Honor' rose), and of course famous personalities (including Billy Graham, Céline Dion, Chris Evert, Vidal Sassoon, Rosie O'Donnell, Barbra Streisand, and George Burns).
- Roses have been very popular choices as floral symbols of political regions, although often a particular species is not identified. Alberta and Iowa chose a "wild rose" as its emblem. The rose is the New York state flower. Georgia adopted the white Cherokee rose (which was once also the state flower of North Carolina). The District of Columbia is associated with the American beauty rose. North Dakota chose the prairie rose. Roses have national significance in Bulgaria, England, Iran, Luxembourg, and the Maldive Islands in the Indian Ocean. The rose served as the national flower of Honduras from 1946 to 1969, but an orchid was adopted as the national flower in 1946. In 1986, the rose became the "national floral emblem" of the United States, when President Reagan signed the proclamation in the White House Rose Garden; according to one newspaper report, the president "rose to the occasion."

## KEY INFORMATION SOURCES

(*Note:* In addition to the articles cited below, there are numerous papers on rose hip production and utilization in European languages.)

Barry, R., Sanderson, K., and Fillmore, S. 2008. Establishment of wild roses for commercial rose hip production in Atlantic Canada. *Internat. J. Fruit Sci.* 8: 266–281.

Beales, P. 1985. *Classic roses: An illustrated encyclopaedia and grower's manual of old roses, shrub roses, and climbers*. Holt, Rinehart, and Winston, New York, NY. pp. 432.

Bruneau, A., Joly, S., Starr, J.R., and Drouin, J.M. 2005. Molecular markers indicate that that the narrow Quebec endemics *Rosa rousseauiorum* and *Rosa williamsii* are synonymous with the widespread *Rosa blanda*. *Can. J. Bot.* 84: 386–398.

Bruneau, A., Starr, J.R., and Joly, S. 2007. Phylogenetic relationship in the genus *Rosa*: New evidence from chloroplast DNA sequences and an appraisal of current knowledge. *Syst. Bot.* 32: 366–378.

Christensen, R., Bartels, E.M., Altman, R.D., Astrup, A., and Bliddal, H. 2008. Does the hip powder of *Rosa canina* (rosehip) reduce pain in osteoarthritis patients?—A meta-analysis of randomized controlled trials. *Osteoarthr. Cartil.* 16: 965–972.

Cinar, I., and Colakoglu, A.S. 2005. Potential health benefits of rose hip products. *Acta Hortic.* 690: 253–257.

Cooper, B. 1994. The rambunctious rose hip. *Tea Coffee Trade J.* 166(9): 74–76, 79.

Crockett, J.U. 1978. *Roses*. Time-Life Books, Alexandria, VA. pp. 160.

Dobson, B., and Schneider, P. 2002. *Combined rose list, 2002: Roses in commerce, rose registrations update, hard-to-find roses & where to find them.* Peter Schneider, Mantua, OH. pp. 240.

Erenturk, S., Gulaboglu, M.S., and Gultekin, S. 2005. Vitamin C degradation of rose hips during drying and effects of drying medium on vitamin C degradation. *Acta Hortic.* 690: 303–310.

Gao, X., Uggla, M., and Rumpunen, K. 2005. Antioxidant activity of dried and boiled rose hips. *Acta Hortic.* 690: 239–243.

Joublan, J.P., Berti, M., Serri, H., Wilckens, R., Hevia, F., and Figueroa, I. 1996. Wild rose germplasm evaluation in Chile. In *Progress in new crops*. Edited by J. Janick. ASHS Press, Arlington, VA. pp. 584–588.

Jovel, E. M., Zhou, X.L., Ming, D.S., Wahbe, TR., and Towers, G.H.N. 2007. Bioactivity-guided isolation of the active compounds from *Rosa nutkana* and quantitative analysis of ascorbic acid by HPLC. *Can. J. Physiol. Pharmacol.* 85: 865–871.

Kaack, K., and Kuehn, B.F. 1991. Evaluation of rose hip species for processing of jam, jelly and soup. *Tidsskrift for Planteavl (Denmark)* 95: 353–358.

Krüssman, G. 1981. *The complete book of roses*. Timber Press, Portland, OR. pp. 436.

Kwaselow, A., Rowe, M., Sears-Ewald, D., and Ownby, D. 1990. Rose hips: a new occupational allergen. *J. Allergy Clinic. Immunol.* 85: 704–708.

Lewis, W.H. 1959. A monograph of the genus *Rosa* in North America. 1. *R. acicularis*. *Brittonia* 1: 1–24.

Lewis, W.H. 2007. Subspecies of *Rosa nutkana* and *R. woodsii* (Rosaceae) in western North America. *Novon* 17: 341–353.

Lewis, W.H. 2008. *Rosa carolina* (Rosaceae) subspecies and hybrids in eastern and Midwestern United States, Canada, and Mexico. *Novon* 18: 192–198.

Macoboy, S. 1993. *The ultimate rose book: 1,500 roses—Antique, modern (including miniature), and wild—all shown in color and selected for their beauty, fragrance, and enduring popularity.* H.N. Abrams, New York pp. 472.

Mamadrizohonov, A.A., Mirzobekov, R.S., and Buribekov, Z.H. 1994. Storage of rose-hip. *Acta Hortic.* 368: 706–711.

Mapson, L.W., and Tomalin, A.W. 1958. Preservation of ascorbic acid in rose hips during storage. *Arch. J. Sci. Food Agric.* 9: 424–430.

Mazza, G. 1979. Development and consumer evaluation of a native fruit product: saskatoon, chokecherry and rose hip. *J. Can. Inst. Food Sci. Technol.* 12(4): 166–169.

McCann, S. 1993. *The rose: An encyclopedia of American roses, rosarians, and rose lore*. Stackpole Books, Harrisburg, PA. pp. 198.

McKeon, J.C. 1995. *The encyclopedia of roses: An organic guide to growing and enjoying America's favorite flower*. Rodale Press, Emmaus, PA. pp. 192.

Moody, M., and Harkness, P. 1992. *The illustrated encyclopedia of roses*. Timber Press, Portland, OR. pp. 304.

Nobbs, K.J. 1980. The rose as a fruit. *Pomona* 13(3): 123–134.

Nybom, H., and Rumpunen, K. (Editors). 2005. *Proceedings of the 1st international rose hip conference, Gümüşhane, Turkey, Sept. 7–10, 2004*. International Society for Horticultural Science, Leuven, Belgium. pp. 310.

Ochoa, M.R., Kesseler, A.G., Pirone, B.N., Marquez, C.A., and Michelis, A. de. 2002. Shrinkage during convective drying of whole rose hip (*Rosa rubiginosa* L.) fruits. *Lebensmittel Wiisenschaft + Technologie [Food science+technology]* 35: 400–406.

Olsson, M.E., Andersson, S., Werlemark, G., Uggla, M., and Gustavsson, K.E. 2005. Carotenoids and phenolics in rose hips. *Acta Hortic.* 690: 249–252.

Phillips, R., and Rix, M. 1988. *Roses*. Random House, New York. pp. 224.

Quest-Ritson, C., and Quest-Ritson, B. 2003. *American Rose Society encyclopedia of roses*. DK, New York. pp. 448.

Richardson, J.E. 1996. *Rosa acicularis*: The Aarctic or circumpolar rose. *New Plantsman* 3: 147–149.

Rossnagel, K., Roll, S., and Willich, S.N. 2007. The clinical effectiveness of rosehip powder in patients with osteoarthritis. A systematic review. *MMW Fortschritte der Medizin* 149(Suppl. 2): 51–56. (In German)

Rouhani, I., Khosh-Khui, M., and Bassiri, A. 1976. Changes in ascorbic acid content of developing rose hips. *J. Hort. Sci.* 51: 375–378.

Scanniello, S., and Bayard, T. 1990. *Roses of America: The Brooklyn Botanic Garden's guide to our national flower*. H. Holt, New York, NY. pp. 224.
Schneider, P. 1995. *Taylor's guide to roses*. Revised edition. Houghton Mifflin Co., Boston, MA. pp. 451.
Small, E. 2012. Top Canadian ornamental plants. 2. Roses. *Bull. Can. Bot. Assoc.* 45(2): 78–87.
Szentmihalyi, K., Vinkler, P., Lakatos, B., Illes, V., and Then, M. 2002. Rose hip (*Rosa canina* L.) oil obtained from waste hip seeds by different extraction methods. *Bioresource Technol.* 82: 195–201.
Uggla, M., and Martinsson, M. 2005. Cultivate the wild roses—experiences from rose hip production in Sweden. *Acta Hortic.* 690: 83–89.
Uggla, M., and Nybom, H. 1999. Domestication of a new crop in Sweden—Dog roses (*Rosa* sect. *Caninae*) for commercial rose hip production. *Acta Hortic.* 484: 147–153.
Werlemark, G. 2006. *Rosa* spp., dogroses. In *The encyclopedia of fruit & nuts*. Edited by J. Janick and R.E. Paull. CABI, Wallingford, Oxfordshire, UK. pp. 745–748.
Winther, K., Rein, E., and Kharazmi, A. 1999. The anti-inflammatory properties of rose-hips. *Inflammopharmacology* 7: 63–68.
Wright, P.H. 1976. "Orchards" for rose hips. *North Amer. Pomona* 9(1): 11–12.

## Specialty Cookbooks

Clements, J. 1958. *My roses*. Collingridge, London, UK. pp. 88.
Clements, J. 1984. *Julia Clements book of rose arrangements*. Arco Pub., New York. pp. 127.
Gordon, J. 1958. *Rose recipes: customs, facts, fancies*. Red Rose Publications, Woodstock, VT. pp. 100.
Gordon, J. 1968. *The art of cooking with roses*. Noonday Press, New York. pp. 159.
Leggatt, J. 1987. *Cooking with flowers*. Century Hutchinson Ltd., London, U.K. pp. 142.
Love, G. 1998. *Rose crafts: using fresh and dried roses in crafts, gifts and displays*. Hermes House, London, UK. pp. 96.
Mudge, P. 1976. *Cooking with roses*. Northport Historical Society, Northport, NY. pp. 8.
Ohrbach, B.M. 2000. *Roses for the scented room: Beautiful ideas for entertaining, gift-giving and the home*. C. Potter, New York, NY. pp. 143.
Puma, S. (Editor). 1973. *Rose Haven recipes: Cooking and crafting with the roses we love*. Temecula Valley Rose Society, Temecula, CA. pp. 181.
Reeder, D. 1995. *From blueberries to wild roses: A northwoods wild foods cookbook*. Explorer's Guide Publishing, Rhinelander, WI. pp. 122.
Rohde, E.S. 1973. *Rose recipes from olden times*. Dover Publications, New York, NY. pp. 94. (reprint of 1939 edition).
Smith, L.W. 1972. *Roses are for eating: Recipes, facts, fancies*. Le Bouquet, Ft. Lauderdale, FL. pp. 20.
Smith, L.W. 1973. *The forgotten art of flower cookery*. Harper & Row, New York, NY. pp. 180.

Robe-Terry (1997) has 7 rose recipes intended for material collected from the wild, Russell (1975) has 19, and Stanek and Butcher (2007) have 11. Williamson (1995) has 12 recipes formulated for Fendler rose (*R. woodsii*). (See the Appendix to this book for details of the publications cited.)

# 81 Saguaro

Family: Cactaceae (cactus family)

## NAMES

Scientific name: *Carnegiea gigantea* (Engelm.) Britton & Rose (*Cereus giganteus* Engelm.)

- The name saguaro (occasionally spelled sahuaro and suwaro) is a Mexican Spanish word for the plant, probably derived from the Opata *sahuaro* (the Opata language and peoples are indigenous to the northeastern part of the state of Sonora, Mexico).
- Pronunciation: sah-WAH-roe.
- The saguaro is also known as giant cactus and saguaro cactus.
- The Spanish conquistadores and missionaries called the saguaro "the great thistle."
- The genus name *Carnegiea* commemorates Andrew Carnegie (1835–1919), American (Scottish-born) industrialist, philanthropist, and patron of science.
- *Gigantea* in the scientific name *C. gigantea* is the Latinized form of the Greek word for giant, pointing out the huge size of the plant.

## GEOGRAPHY AND ECOLOGY OF WILD PLANTS

This "monarch of the desert," as it is sometimes known, is native to the Sonoran Desert of extreme southeastern California, southern Arizona, and adjoining northwestern Mexico. The fluted surface of the trunk allows the cactus to absorb water by expanding like an accordion during rains and shrinking during dry periods. Plants may weigh 5–12 tons, and can gain a ton of weight from water absorption during a rain. Depending on how recently it rained, the water content of a saguaro varies from 80% to 95%. The saguaro does not tolerate saline or waterlogged soils, or prolonged freezing.

## PLANT PORTRAIT

The saguaro cactus is a spectacular plant, the largest cactus in the United States, and probably the most recognized cactus in the world, commonly shown as background in Wild West movies and Roadrunner/Coyote cartoons. The plant grows to a height of 5–20 m (16–65 feet), but very slowly—often only 2.5 cm (1 inch) in a year and 1 m (about a yard) in 30 years. A young saguaro can take nearly a decade to achieve a height of 2.5 cm (1 inch). Average saguaros are 9–13 m (30–43 feet) tall. The plant remains unbranched for many years, and does not flower until the age of 50 to 75 years. Eventually the saguaro develops a tall, single, thick, fluted trunk 30–75 cm (12–30 inches) in diameter, often with several large branches ("arms") curving upward. Plants live as long as 200 years, at which point they often have more than five, sometimes as many as 50 arms. (Saguaros rarely produce a very odd "crested" or "cristate" form in which the young arms are fused into a flat, fan-like form.) The trunk and branches have stout, 5 cm (2 inches) long protective spines clustered on their ribs. Creamy-white flowers with yellow centers bloom in May and June, clustered near the ends of branches. These are about 7.5 cm (3 inches) wide, opening just once, during the cooler desert nights, and closing again by the next midday. The flowers produce a sweet nectar that attract pollinators (birds, bats, and insects), and when fertilized they

**FIGURE 81.1** Saguaro (*Carnegiea gigantea*). (a) Huge plant. (Courtesy of Rocky Mountain Research Station/Forest Pathology Archive, USDA Forest Service, Bugwood.org [CC By 3.0]). (b) Papago girl harvesting fruit. (c) Crested saguaro (see text). (Courtesy of U.S. National Park Service, Department of the Interior.) (d) Distribution map.

develop into fruits. Mature plants commonly have more than 100 fruits. The fruit is 5.5–9 cm (2.2–3.5 inches) long, green, scaly, and oval, and when ripe it splits open to reveal a bright-red, pulpy flesh relished by desert animals. The fruits somewhat resembles tiny watermelons when split open by hand, with a red interior pulp and as many as 2500 black-red seeds. With their red interior exposed, the fruits are easily misinterpreted as red flowers. Saguaro fruit was an especially important food source for Native Americans of the region, who are thought to have selected a practically thornless fruit.

Inevitably, desert travelers have tried to obtain water from the succulent saguaro, only to find that the sap is too bitter to drink. For the same reason, most animals do not use the saguaro as a source of water. The unpleasant taste is due to the present of alkaloid chemicals, which are

FIGURE 81.2 Saguaro (*Carnegiea gigantea*). (a) White winged dove on fruiting top of saguaro cactus. (Courtesy of William Heron [CC By 2.0].) (b) Fruit. (Courtesy of Steven J. Baskauf [http://bioimages.vanderbilt.edu].) (c) Flowers. (Courtesy of Ehiris at en.wikipedia [http://en.wikipedia.org] [CC By 3.0].) (d) Split fruit. (Courtesy of U.S. National Park Service.) (e) Spiny surface of trunk. (Courtesy of Mwinog2777 [CC By 3.0].) (f) Saguaro boot (see text). (Courtesy of Sharktapus aka Sharktopus [CC By 3.0].)

closely related to the alkaloids found in the hallucinogenic peyote cactus (*Lophophora williamsii* (Lem. ex Salm-Dyck) J.M. Coult.). However, there is almost no evidence of the narcotic use of the saguaro.

Habitat destruction, vandalism, and "cactus rustling" for sale of the plant as an ornamental have taken a toll on the saguaro. To protect it, in 1933 the 337 square km (130 square mile), Saguaro National Monument (upgraded to a National Park in 1994) was established in southeastern Arizona. The cactus is protected by the Arizona Native Plant Law. No part of the cactus may be harvested on public land without a permit. On private property the plants can be destroyed, although some cities impose tough requirements. Special exemption applies to the indigenous Native Americans of the region, who are permitted to carry on their tradition of utilizing the fruit.

## CULINARY PORTRAIT

The Tohono O'odham Indians (formerly known as Papago) are the major traditional inhabitants of the Sonoran Desert on both sides of the United States–Mexican border. They and the Pima, as well as several other groups of Native Americans, extensively used the saguaro's fruits as food, and this tradition has continued to the present. The fruit is eaten raw, cooked into a jelly, jam, preserves, or syrup, or dried for winter storage. The taste of the pulp has been compared to that of a fig, with a hint of strawberry. The juice is used to prepare drinks or boiled into a sweet syrup, which is fermented into a mildly alcoholic wine (often for the celebration of the summer rain ceremonies). The seeds are ground into oil-rich spreads resembling butter or peanut butter; roasted and ground, and mixed with water to make a mushy cereal; made into meal cakes; and ground along with corn into a pudding. Although few people today will have the opportunity to consume saguaro fruit, saguaro syrup is sometimes available as a specialty item, often prepared over mesquite fires, which give it a smoky flavor. This can be used as a garnish, glaze, or simply drizzled over ice cream or fresh fruit. Saguaro tea is also sold commercially, but is not often found in stores.

### Culinary Vocabulary

- An "olla" is a round earthenware pot with a wide mouth and handles, and a globular body. In Spain and South and Central America, it is used to heat or hold water and to cook stews. The Tohono O'odham (Papago) Indians often prepare saguaro wine in ollas.

## PROSPECTS

With the exception of American chestnuts, saguaro fruit is the rarest plant food delicacy discussed in this book. Some saguaro cactus grows naturally on private lands, but much of it is reserved for Native Americans, and only a small amount is sold to the general public. Given the limited natural habitat where the plant grows and its apparently low productivity, the supply will probably always be much less than the potential market. However, Native Americans historically appear to have used saguaro fruit as a major supply of nutrition. Records indicate that a Native American in the Sonoran Desert could harvest 350 kg (770 pounds) of saguaro fruit in a season. In the seventeenth century, a colonial priest reported that "Some Indians became so corpulent after eating huge quantities of the sugar-rich fruit that he was sometimes unable to recognize at first sight individuals otherwise perfectly familiar to him—and this was after feeding for about 3 weeks on the fruit" (quotation from Niethammer 1999; see the Appendix to this book). This suggests that the potential of saguaro as a fruit source may be greater than one might think. However, climate change and urbanization pose threats to the future of saguaro cactus, despite its legislated conservation. Notwithstanding the improbability of ever becoming a standard supermarket fruit, saguaro fruit is so unique and interesting that those who get the opportunity to sample it should consider themselves privileged. Saguaro seeds can be purchased online, and the plants can be grown for a personal fruit supply by those living in a suitable climate, although it may take a century before the plants flower and produce edible fruit!

## CURIOSITIES OF SCIENCE AND TECHNOLOGY

- Native Americans where the saguaro occurs typically consider cutting it down to be a crime, and in the past often curved their roads to preserve the plants.
- The tiny seeds of the saguaro fruit when ingested pass through the body almost unchanged. Peoples of the Sonoran region sometimes collected the seeds from feces, a considerable task considering that they are smaller than poppy seeds, to grind them into an edible meal. This seems like an extreme measure by today's standards, with food often so abundant it is

not respected, but was a means of surviving famine in a harsh environment. The seeds of the saguaro contain about 16% protein, and so have considerable nutritional value.

- When a hole is made in the saguaro by animals, especially by woodpeckers, the plant protects itself from losing water by walling off the damaged area, that is, by forming a hard, water-proof, cup of callus tissue (often called a "saguaro boot"; see Figure 81.2f) around the wound. In the past, when the cactus died, these water-proof structures were collected and used as cups by Native Americans.
- The saguaro trunk is not woody like a tree, but is strengthened by tough, wooden ribs arranged in a vertical circle inside the trunk. These ribs are used in numerous ways by Native Americans. Two ribs tied together can make a pole 5 m (16 feet) long for knocking the fruits off the top of the cactus. The ribs were used as splints for a broken leg, as spears and arrows, for house construction, and for fencing.
- Saguaro thorns were used by Native Americans as a tattoo tool.
- The Papago and Pima Indians began their calendar year (in June) with the harvest of the saguaro's fruit.
- The root system of the saguaro is shallow but often extensive to collect water when it falls. Roots have been observed extending as far as 30 m (98 feet) from the trunk.
- In Arizona, target shooting at saguaro cacti became so frequent that the state made this behavior a felony, punishable by a $100,000 fine and 3 years in prison. In 1982, David Grundman, on parole from a criminal sentence for robbery, opened fire on a saguaro with his 16-gauge shotgun. He died when the plant, estimated to have been 7.6 m (25 feet) tall, 125 years of age, and weighing 1.5 tons, fell on him. The event seems so bizarre that many people learning of it today consider that it is merely an "urban legend." (For a detailed account, see *Jack Ruby's Kitchen Sink*, published by National Geographic in 2000, authored by Arizona journalist Tom Miller.)
- As the symbol of Arizona, the saguaro appears on the state's car license plates.
- Of the 2000 or so species of cacti, the giant cardón or saguesa (*Pachycereus pringlei* (S. Wats.) Britt. & Rose) of northwestern Mexico is the world's largest. Its trunk can exceed a meter (over 3 feet) in diameter, and the largest plants are more massive than the saguaro, although the tallest specimens are about the height of the tallest saguaro cacti. The species is similar in appearance and related to the saguaro, and is often mistaken for it. In comparison with the saguaro, cardón has a shorter main trunk and tends to form multiple branches quite close to the base, so the plants often have a number of long, straight, vertical arms.
- "Nurse plants" assist other plants to become established from seed by providing protection from animals or by moderating the effects of climate. Saguaro seedlings are very vulnerable to damage from freezing, extreme heat, and other hazards, so they generally only grow successfully under a protective nurse plant, which provides shade and greater moisture in the summer, and warmer temperatures in the winter. Perennial shrubs, such as foothills palo verde (*Parkinsonia microphylla* Torr.), are important as nurse plants for facilitating saguaro establishment.
- "Keystone species" are species that have exceptional importance to the health of an ecosystem. When they are reduced or eliminated, their removal results in notable, often extremely harmful changes in the ecosystem, such as the extermination of other species and the introduction of less desirable organisms. The saguaro is considered to be a keystone species in the Sonoran Desert.
- The "Great Bear Constellation" (part of which, the Big Dipper, is familiar to most people), is known as the "Cactus Hook" by many of the Native Americans living where the saguaro grows. The "hook" is a reference to the hook on the end of the long pole used to knock fruits off the top of the cactus, it being imagined that the constellation resembles it.

## KEY INFORMATION SOURCES


Arizona Highways. 1969. *Saguaro, monarch of the desert*. Arizona Highways, Phoenix, AZ. pp. 39.

Betton, H.B. 1972. *Physical and biological factors affecting the maintenance and management of the giant (saguaro) cactus,* Cereus giganteus, *Engelm.* Bureau of Land Management, Dept. of the Interior, Washington, DC. pp. 38.

Bowers, J.E. 1996. Environmental determinants of flowering date in the columnar cactus *Carnegiea gigantea* in the northern Sonoran Desert. *Madroño* 43: 69–84.

Bowers, J.E. 2003. *Frequently asked questions about the saguaro*. Western National Parks Association, Tucson, AZ. pp. 18.

Bruhn, J.G. 1971. *Carnegiea gigantea*: the saguaro and its uses. *Econ. Bot.* 25: 320–329.

Ceotto, E. 2009. Cultivation of *Carnegiea gigantea* from seeds: A journey in desert ecology. *Desert Plants* 25: 10–15.

Clemensen, A.B. 1986. *Cattle, copper, and cactus: The history of Saguaro National Monument, Arizona.* National Park Service, Denver Service Center, Denver, CO. pp. 271.

Despain, D.G. 1974. The survival of saguaro (*Carnegiea gigantea*) seedlings on soils of differing albedo and cover. *J. Ariz. Acad. Sci.* 9(3): 102–107.

Drezner, T.D. 2003. Saguaro (*Carnegiea gigantea*, Cactaceae) age–height relationships and growth: the development of a general growth curve. *Am. J. Bot.* 90: 911–914.

Drezner, T.D. 2003. Revisiting Bergmann's Rule for saguaros (*Carnegiea gigantea* (Engelm.) Britt. and Rose): Stem diameter patterns over space. *J. Biogeog.* 30: 353–359.

Drezner, T.D. 2005. Saguaro (*Carnegiea gigantea,* Cactaceae) growth rate over its American range and the link to summer precipitation. *Southwest. Nat.* 50: 65–68.

Drezner, T.D. 2006. Plant facilitation in extreme environments. The non-random distribution of saguaro cacti (*Carnegiea gigantea*) under their nurse associates and the relationship to nurse architecture. *J. Arid Environ.* 65: 46–61.

Drezner, T.D. 2008. Variation in age and height of onset of reproduction in the saguaro cactus (*Carnegiea gigantea*) in the Sonoran Desert. *Plant Ecol.* 194: 223–229.

Drezner, T.D., and Balling, R.C. Jr. 2008. Regeneration cycles of the keystone species *Carnegiea gigantea* are linked to worldwide volcanism. *J. Veg. Sci.* 19: 587–596.

Evans, D. 1980. *Saguaro.* Southwest Parks and Monuments Association, Globe, AZ. pp. 56.

Gibson, A.C. 2003. *Carnegiea* In *Flora of North America north of Mexico*. Vol. 4. Edited by Flora of North America Editorial Committee. Oxford University Press, New York, NY. pp. 184–186.

Hazen-Hammond, S. 1997. *The great saguaro book.* Ten Speed Press, Berkeley, CA. pp. 144.

Humphreys, A., and Lowell, S. 2002. *Saguaro: The desert giant.* Rio Nuevo Publishers, Tucson, AZ. pp. 59.

Hutto, R.L., McAuliffe, J.R., and Hogan, L. 1986. Distributional associates of the saguaro (*Carnegiea gigantea*). *Southwest. Nat.* 31: 469–476.

Keasey, M.S. 1981. *The saguaro book.* Kendall/Hunt Pub. Co., Dubuque, IA. pp. 48.

Kolberg, K., and Lajtha, K. 1997. Population dynamics and trace metal biogeochemistry of the saguaro cactus (*Carnegiea gigantea*) in the Sonoran desert. *Ecoscience* 4: 223–231.

Lajtha, K., Kolberg, K., and Getz, J. 1997. Ecophysiology of the saguaro cactus (*Carnegiea gigantea*) in the Saguaro National Monument: Relationship to symptoms of decline. *J. Arid Environ.* 36: 579–590.

Lowe, C.H., and Steenbergh, W.F. 1981. On the Cenozoic ecology and evolution of the sahuaro (*Carnegiea gigantea*). *Desert Plants* 3: 83–86.

Mitich, L.W. 1972. The saguaro—a history. *Cactus Succulent J.* 44(3): 118–129.

Mitich, L.W., and Bruhn, J.G. 1982. The saguaro: A bibliography. II. *Cactus Succulent J.* 54(3): 101–108.

Nobel, P.S. 1980. Morphology, nurse plants, and minimum apical temperatures for young *Carnegiea gigantea*. *Bot. Gaz.* 141: 188–191.

Pierson, E.A., and Turner, R.M. 1998. An 85-year study of saguaro (*Carnegiea gigantea*) demography. *Ecology* 79: 2676–2693.

Shelton, N., and Dodge, N.N. 1972. *Saguaro National Monument, Arizona.* National Park Service, U.S. Dept. of the Interior, Washington, DC. pp. 98.

Steenbergh, W.F. 1983. *Ecology of the saguaro III: Growth and demography.* U.S. Dept. of the Interior, National Park Service, Washington, DC. pp. 228. [I & II of this series are in Steenbergh and Lowe, listed below.]

Steenbergh, W.F., and Lowe, C.H. 1977–1983. *Ecology of the saguaro.* National Park Service, Washington, DC. 2 vols.

Stone, C.P., and Bellantoni, E.S. 1992. *Proceedings of the symposium on research in Saguaro National Monument, 23–24 January 1991.* Rincon Institute, Tucson, AZ. pp. 277.

Wild, P., and Coss, H. 1986. *The saguaro forest*. Northland Press, Flagstaff, AZ. pp. 65.
Yeaton, R.I., Karban, R., and Wagner, H.B. 1980. Morphological growth patterns of saguaro (*Carnegiea gigantea*: Cactaceae) on flats and slopes in Organ Pipe Cactus National Monument, Arizona. *Southwest. Nat.* 25: 339–349.

## Specialty Cookbooks

Niethammer (1999) is an excellent source of information on culinary preparation of saguaro fruit, providing six recipes. Williamson (1989) has 18 recipes. (See the Appendix to this book for details of the publications cited.) See Chapter 20 on Cactus Pear for additional suggestions on recipes for cactus fruits. A few saguaro fruit recipes and some seed recipes can be found online.

# 82 Salal

Family: Ericaceae (heath family)

## NAMES

Scientific name: *Gaultheria shallon* Pursh

- "Salal" is based on the Chinook jargon word *sálal* for *G. shallon*, a usage dating to the early nineteenth century. (Chinook refers to a Native American people, and their language, centered formerly in the lower Columbia River valley and adjoining coastal regions of Washington and Oregon. The Chinook traded widely throughout the Pacific Northwest, and Chinook became a language of trade. For additional information, see Chapter 35 on Duck Potato.)
- Salal is also known as Oregon wintergreen, shallal, shallon, and western wintergreen.
- In England, salal is usually known as gaultheria, sometimes as shallon.
- The genus name *Gaultheria* commemorates French botanist and Quebec physician Jean François Gaultier (ca. 1708–1756).
- *Shallon* in the scientific name is derived from *Kikwu-salu*, the Chinook name for the plant (compare the derivation of "salal," above.)

## GEOGRAPHY AND ECOLOGY OF WILD PLANTS

The native distribution of salal includes the West coast of North America from southeast Alaska to southern California. The plant has become naturalized in northwestern Europe, where it is sometimes regarded as a pest. The species occurs in woods, coastal bluffs, cliffs, bogs, forest edges, clear-cuts, and roadsides, in moist or dry, rocky or sandy soils. Salal is adapted to a humid, mild climate, and is intolerant of freezing. It grows best in semi-shade, but tolerates full sun and deep shade, considerable drought, and poor soils. Salal is one of the most common shrubs on the North American Pacific Coast, and often forms continuous, almost impenetrable stands in coniferous forests. Some foresters consider salal an undesirable weed since it outcompetes many tree seedlings.

## PLANT PORTRAIT

This highly branched evergreen shrub grows creeping or erect to a height of 0.5–2 m (1.5–6.6 feet), occasionally up to 5 m (16 feet), but frequently is very short. It spreads by rhizomes (underground stems), sometimes forming dense thickets. The rhizome system can persist for hundreds of years. The leaves are leathery, thick, lustrous, and dark green, 3–10 cm (0.5–4 inches) long. The waxy, white, pink, or rose, urn-shaped flowers are 6–11 mm (0.2–0.4 inch) long and are produced in the spring and early summer. The fruits are dark blue, bluish-purple, reddish, or purple-black, round, 6–10 mm (0.2–0.4 inch) in diameter, fleshy, thick-skinned, covered with sticky hairs, and contain many (often more than 100) tiny seeds. The fruits have been referred to as "pseudoberries," since they are developed from the flower's sepals rather than from the ovaries as in most other plants. They are developed in late summer and autumn, often persist on the plant for several months, and are easily collected in quantity by stripping the fruits off from their elongated clusters.

The evergreen branches of salal are extensively harvested for use in floral arrangements. The foliage complements long-stemmed roses and other bouquets. The leaves have become a staple of

FIGURE 82.1 Salal (*Gaultheria shallon*). (a) Branch with fruits ([a] and [b] Courtesy of Peter Stevens [Flickr/CC-attribution]). (b) Shrubs. (c) Flowers. (Courtesy of Walter Siegmund [CC By 3.0].) (d) Distribution map.

florists, who refer to the attractive foliage as "lemon leaf." The value of the harvested greenery in British Columbia alone has been estimated to approach $50 million annually and to employ up to 15,000 people. The species is also used to some extent as an ornamental shrub. It is an excellent groundcover in coastal North America and is grown internationally.

## CULINARY PORTRAIT

Salal berries were the most important staple fruit of most indigenous peoples of the Pacific Northwest coast. The aromatic berries remain an important food for them. They are consumed fresh or dried. Commonly the berries were cooked and preserved as cake-like masses, subsequently rehydrated and eaten in various foods or used as a sweetener or flavorant. Immigrants to the West coast frequently adopted the Indian practice of preparing salal berry preserves. Today, the fruits are often made into jams, jellies, preserves, and pies. The taste of the berries has variously been described as bland, very sweet, acidic, unique, and with a mealy, unpleasant texture. These divergent evaluations suggest that the fruit varies considerably in quality and taste. The dried leaves have been used for tea, seasoning, and flavoring. Curiously, the fruit has been used in the past for alleged appetite-suppressing properties.

### Culinary Vocabulary

- The word "halal" can be confused with "salal" when food is under discussion. Halal is an Arabic term designating permitted objects and actions according to Islamic Law. Halal frequently designates permitted foods, which excludes pork products, animals improperly slaughtered, alcoholic drinks and other intoxicants, and carnivorous animals and birds.

## PROSPECTS

The following passage in Hobby et al. (2010) suggests the prospects for agricultural development of salal for the florist market: "Salal production on private lands could move to a 'forest farming' system that should improve production quantity and quality through the adoption of compatible management methods. Salal production has been estimated to produce an average of $1000 per hectare per year on the Pacific Northwest coast and recent agreements in the United States Pacific Northwest have provided forest landowners $25 per hectare rents from salal agreements with regional floral companies (N. Schaaf, pers. comm.)."

Although salal is a multimillion dollar crop for foliage harvest for the use of the floral greens industry, the fruit harvest has not progressed beyond the cottage industry level. However, the berries were once the leading fruit of North American West coast native peoples. Salal berries vary in taste, some quite sweet and juicy, and clearly superior pomological forms could be readily selected. The species grows extremely aggressively, to the alarm of foresters who often consider it to be a weed, but this vigorous growth is an excellent indicator of a plant that can be successful as a domesticated crop. Although it achieves best growth in semi-shade, it grows well in full sun, and indeed berry production increases, so field cultivation for fruit production is feasible. Because salal is widely harvested for foliage, and is frequently grown as an ornamental and landscape plant, a considerable body of knowledge has been accumulated for it, and this information also can be useful for developing the species for harvest of fruit. Plant species such as salal that are dominant and widespread in particular habitats can often be managed for yield, in the manner of lowbush blueberry (see Chapter 15 on Blueberries); indeed, salal is already receiving some management for harvest of foliage. However, it is very difficult to introduce new berry crops into the marketplace. The natural supply of salal berries is not sufficiently tasty to compete with the established berries, and selection of superior cultivars combined with field cultivation would be necessary to transform salal into a modern fruit crop.

## CURIOSITIES OF SCIENCE AND TECHNOLOGY

- In England, it was once common for the very rich to establish "shooting estates"—private, rural hunting grounds (thousands still exist in the United Kingdom). During the heyday of this practice in the nineteenth century, salal was widely planted as cover for pheasant and other game birds. The dense, evergreen thickets that were formed seemed desirable, but eventually the plants spread, often smothering species of native vegetation, and proving to be capable of regenerating even when seemingly cleared from an area.
- A good indication of the importance of a plant species and how well it is regarded in popular culture is the extent to which it is recognized in nonplant names. On the West coast, "salal" is used in the names of numerous businesses (notably those connected with plant sales), publications, sports teams, housing developments, and streets, all of which points out the favorable view that has developed for the species.

## KEY INFORMATION SOURCES

Bennett, J.N., Lapthorne, B.M., Blevins, L.L., and Prescott, C.E. 2004. Response of *Gaultheria shallon* and *Epilobium angustifolium* to large additions of nitrogen and phosphorus fertilizer. *Can. J. For. Res.* 34: 502–506.

Boateng, J.O., and Comeau, P. (Coordinators). 2002. *Operational summary for vegetation complexes: salal complex*. Forest Practices Branch and Forest Renewal BC (co-publishers), Victoria, BC. pp. 11.

Bunnell, F. 1990. Reproduction of salal (*Gaultheria shallon*) under forest canopy. *Can. J. For. Res.* 20: 91–100.

Chou, Y.L. 1952. Floral morphology of three species of *Gaultheria*. *Bot. Gaz.* 114: 198–221.

Cocksedge, W., and Titus, B.D. 2006. Short-term response of salal (*Gaultheria shallon* Pursh) to commercial harvesting for floral greenery. *Agrofor. Syst.* 68: 103–111.

Dimock, E.J. II., Johnston, W.F., and Stein, W.I. 1974. *Gaultheria* L. wintergreen. In *Seeds of woody plants in the United States*. Agriculture Handbook No. 450. Edited by C.S. Schopmeyer. U.S. Department of Agriculture, Forest Service. Washington, DC. pp. 422–426.

Dorworth, C., Sieber, T., and Woods, T. 2001. Early growth performance of salal (*Gaultheria shallon*) from various North American west-coast locations. *Ann. For. Sci.* 58: 597–606.

Fraser, L.H., Chanway, C.P., and Turkington, R. 1995. The competitive role of *Gaultheria shallon* on planted western hemlock and western red cedar saplings on northern Vancouver Island. *For. Ecol. Manage.* 75: 27–39.

Fraser, L., Turkington, R., and Chanway, C.P. 1993. The biology of Canadian weeds. 102. *Gaultheria shallon* Pursh. *Can. J. Plant Sci.* 73: 1233–1247.

Hobby, T., Dows, K., and MacKenzie, S. 2010. Commercial development of salal on south Vancouver Island [B.C.]. *J. Ecosystems Manage.* 11(1/2): 63–71.

Huffman, D.W., Tappeiner, J.C. II., and Zasada, J.C. 1994. Regeneration of salal (*Gaultheria shallon*) in the central Coast Range forests of Oregon. *Can. J. Bot.* 72: 39–51.

Huffman, D.W., Zasada, J.C., and Tappeiner, J.C. II. 1994. Growth and morphology of rhizome cuttings and seedlings of salal (*Gaultheria shallon*): Effects of four light intensities. *Can. J. Bot.* 72: 1702–1708.

Messier, C. 1992. Effects of neutral shade and growing media on growth, biomass allocation, and competitive ability of *Gaultheria shallon*. *Can. J. Bot.* 70: 2271–2276.

Messier, C., Honer, T.W., and Kimmins, J.P. 1989. Photosynthetic photon flux density, red:far-red ratio, and minimum light requirement for survival of *Gaultheria shallon* in western red cedar–western hemlock stands in central British Columbia. *Can. J. For. Res.* 19: 1470–1477.

Mirick, S., and Quinn, J.A. 1981. Some observations on the reproductive biology of *Gaultheria procumbens* (Ericaceae). *Am. J. Bot.* 68: 1298–1305.

Prescott, C.E., Coward, L.P., Weetman, G.F., and Gessel, S.P. 1993. Effects of repeated nitrogen fertilization on the ericaceous shrub, salal (*Gaultheria shallon*), in two coastal Douglas-fir forests. *For. Ecol. Manage.* 61: 45–60.

Ricou, L. 2007. *Salal: listening for the Northwest understory*. NeWest Press, Edmonton, AB. pp. 263.

Schmidt, M.G. 1979. Natives for your garden; salal (*Gaultheria shallon*). *Fremontia* 6(4): 20–21.

Smith, N.J. 1991. Sun and shade leaves: Clues to how salal, *Gaultheria shallon* responds to overstory stand density. *Can. J. For. Res.* 21: 300–305.

Trock, D.K. 2009. *Gaultheria*. In *Flora of North America North of Mexico*. Vol. 8. Edited by Flora of North America Editorial Committee. Oxford University Press, Oxford, U.K. pp. 512–515.

Wilkin, J.E., Shamoun, S.F., Ritland, C., Ritland, K., and El-Kassaby, Y.A. 2005. Population genetics of *Gaultheria shallon* in British Columbia and the implications for management using biocontrol. *Can. J. Bot.* 83: 501–509.

Xiao, G., and Berch, S.M. 1995. The ability of known ericoid mycorrhizal fungi to form mycorrhizae with *Gaultheria shallon*. *Mycologia* 87: 467–470.

Xiao, G., and Berch, S.M. 1996. Diversity and abundance of ericoid mycorrhizal fungi of *Gaultheria shallon* on forest clearcuts. *Can. J. Bot.* 74: 337–346.

Xiao, G., and Berch, S.M. 1999. Organic nitrogen use by salal ericoid mycorrhizal fungi from northern Vancouver Island and impacts on growth in vitro of *Gaultheria shallon*. *Mycorrhiza* 9: 145–149.

### Specialty Cookbooks

Stewart (2002) provides information on the culinary preparation of salal berries. Freitus (1980) has four salal recipes and Freitus and Haberman (2005) have eight. Marie (2008) has a recipe for salal berry wine, and Hahn (2010) has recipes entitled Wild salal–cranberry compote and Oregon grape–salal berry jelly. (See the Appendix to this book for details of the publications cited.)

# 83 Salmonberry

Family: Rosaceae (rose family)

## NAMES

Scientific name: *Rubus spectabilis* Pursh

- Salmonberries are often salmon-colored, and the raspberry-like clusters of tiny fruitlets are reminiscent of salmon roe, likely accounting for the name. The species flowers early, often when salmon runs are occurring, and native peoples commonly ate salmon with salmon berries, and these considerations perhaps contributed to the name.
- Salmonberry has also been called Russian berry (a phrase used in Kodiak, Alaska) and salmon raspberry.
- The name salmonberry is sometimes applied to the cloudberry (see Chapter 29).
- For information on the genus name *Rubus*, see Chapter 77 on Raspberries.
- *Spectabilis* in the scientific name is Latin for noteworthy or outstanding.

## GEOGRAPHY AND ECOLOGY OF WILD PLANTS

Salmonberry is native to the West coast of North America, from west central Alaska to California. It has rarely been observed in northern Idaho. The species has escaped cultivation and become established in northwestern Europe, including the British Isles. The plants occur in moist open forests and woods, in shaded swamps, in moist clearings along roads, along the margins of streams, and on mountain slopes. Salmonberry rapidly colonizes open areas and often develops large, dense thickets. In addition to reproducing by seeds, salmonberry spreads very extensively by root suckers and rhizomes (underground stems). Sometimes the canes arch over, and where they touch the ground new plants develop; occasionally broken pieces of canes take root. The thickets produced (occasionally over 100 m or 330 feet in diameter) are sometimes so impenetrable that they affect forest regeneration and so are of concern to foresters. Salmonberry is sometimes considered to be a weed and measures to control it have been carried out in some areas.

## PLANT PORTRAIT

Salmonberry is a deciduous shrub, 1–4 m (3–13 feet) in height, with prickles on the stems. The stems of most species of *Rubus* are biennial, and salmonberry appears to conform to this pattern. The pink, purple, or red flowers are 2–3 cm (0.8–1.2 inches) in diameter. The fruits are round, ovoid, or conical, resemble raspberries, and are yellow to orange-red, occasionally almost black, 1.5–2.5 cm (0.6–1 inch) long, with bushes sometimes bearing fruit of different colors. The fruits mature in late summer and early autumn. This shrub is widely grown as an ornamental. Ornamental cultivars with doubled flowers are available.

## CULINARY PORTRAIT

Coastal First Nations peoples and wild food enthusiasts often consume the juicy berries fresh. However, the fruits are seedy, often insipid, and do not keep well. Accordingly, they are often processed into jam, jelly, wine, and candy. The leaves have been used for medicinal and culinary teas.

**FIGURE 83.1** Salmonberry (*Rubus spectabilis*). (a) Fruit. (Courtesy of jfh686 [Flickr/CC-attribution].) (b) Golden berries. (Courtesy of fishermansdaughter [Flickr/CC-attribution].) (c, d) Branches bearing fruit. (Courtesy of Vernon Smith.) (e) Bowl of salmonberries. (Courtesy of Ariasnow.) (f) Flowering branch. (From Loddiges, C. & Sons, *The botanical cabinet*, J. & A. Arch, Cornhill, U.K. Vol. 17, Plate 1602, 1830.) (g) Distribution map.

### Culinary Vocabulary

- Salmonberry "sprouts" are young, tender, developing (branch) shoots. West Coast indigenous people traditionally peel and eat these raw, or sometimes steam them like asparagus before consumption.

## PROSPECTS

Salmonberry is not a cultivated food crop, although it is grown as an ornamental and ground stabilizer. However, it is a very popular wild fruit, extensively collected on the West coast of North America by indigenous people and wild food enthusiasts, and employed to some extent in cottage industry sales of preserved products. Salmonberry is naturally adapted to West coast climates and has appreciable cold tolerance. It grows aggressively, a feature that is generally desirable in crop plants. The taste of different clones varies from delicious to insipid, so it should be possible to select forms with attractive fruits. The berry market is very competitive, and it is difficult to find support for the research and development that is necessary to introduce new crops. Nevertheless, salmonberry is an interesting new crop possibility. Salmonberry has been hybridized with raspberries, and it is feasible that new fruits of hybrid origin will be created in the future.

## CURIOSITIES OF SCIENCE AND TECHNOLOGY

- Birds have been shown to prefer red over orange salmonberries (see Traveset and Willson, 1998 and Gervais et al. 1999).
- Domestic sheep have been used with some success to control the aggressive growth of salmonberry.
- The extent of underground spread of salmonberry is astonishing. Some stands develop more than 70 km (40 miles) of rhizomes (underground stems).

## KEY INFORMATION SOURCES

Barber, H.W., Jr. 1976. *An autecological study of salmonberry in western Washington*. Ph.D. thesis. University of Washington, Seattle, Washington, DC. pp. 152.

Gervais, J.A., Noon, B.R., and Willson, M.F. 1999. Avian selection of the color-dimorphic fruits of salmonberry, *Rubus spectabilis*: a field experiment. *Oikos* 84: 77–86.

Gratkowski, H. 1971. *Midsummer foliage sprays on salmonberry and thimbleberry*. USDA Forest Service Research Note PNW-171. Pacific Northwest Forest and Range Experiment Station, Portland, OR. pp. 5.

Henderson, J.A. 1978. Plant succession on the *Alnus rubra/Rubus spectabilis* habitat type in western Oregon. *Northwest Sci.* 52: 156–167.

Jensen, E.C., Anderson, D.J., Zasada, J.C., and Tappeiner, J.C. II. 1995. *The reproductive ecology of broad-leaved trees and shrubs: salmonberry,* Rubus spectabilis *Pursh*. Research Publication 9e. Forest Research Laboratory, Oregon State University, Corvallis, OR. pp. 7.

Kellogg, J., Wang, J., Flint, C., Rubnicky, D., Kuhn, P., De Mejia, E.G., Raskin, I., and Lila, M.A. 2010. Alaskan wild berry resources and human health under the cloud of climate change. *J. Agric. Food Chem.* 58: 3884–3900.

Knight, V.H. 1991. Use of the salmonberry, *Rubus spectabilis* Pursh, in red raspberry breeding. *J. Hort. Sci.* 66: 575–582.

Krygier, J.T., and Ruth, R.H. 1961. Effect of herbicides on salmonberry and on sitka spruce and western hemlock seed. *Archive Weeds* 9: 416–422.

Madison, R.W., and Freed, V.H. 1962. Basal treatments for control of salmonberry. *Archive Weeds* 10: 247–248.

Maxwell, B.D. 1990. *The population dynamics and growth of salmonberry and thimbleberry*. Ph.D. thesis. Oregon State University, Corvallis, OR. pp. 206.

Maxwell, B D., Zasada, J.C., and Radosevich, S.R. 1993. Simulation of salmonberry and thimbleberry population establishment and growth. *Can. J. For. Res.* 23: 2194–2203.

Newton, M., and White, D.E. 1983. Effect of salmonberry on growth of planted conifers. *Proc. West. Soc. Weed Sci.* 36: 59–64.

Oleskevich, C., Shamoun, S.F., and Punja, Z.K. 1996. The biology of Canadian weeds. 105. *Rubus strigosus* Michx., *Rubus parviflorus* Nutt., and *Rubus spectabilis* Pursh. *Can. J. Plant Sci.* 76: 187–201.

Runeson, L., and Segerback, L. 2008. *Rubus spectabilis* in Skane, southernmost Sweden. *Svensk Botanisk Tidskrift* 102: 263–264. (In Swedish)

Ruth, R.H. 1970. *Effect of shade on germination and growth of salmonberry*. U.S. Forest Service Research Paper PNW-96. Pacific Northwest Forest and Range Experiment Station, Portland, OR. pp. 10.

Steiner, G.C., Arnold, R.W., Roth, R.R., and Ice, J.S. 1991. Stevens-Johnson syndrome secondary to ingestion of salmon berries. *Alaska Med.* 33(2): 57–59.

Tappeiner, J.C., and Zasada, J.C. 1993. Establishment of salmonberry, salal, vine maple, and bigleaf maple seedlings in coastal forests of Oregon. *Can. J. For. Res.* 23: 1775–1780.

Tappeiner, J.C. II., Zasada, J.C., Huffman, D.W., and Ganio, L.M. 2001. Salmonberry and salal annual aerial stem production: The maintenance of shrub cover in forest stands. *Can. J. For. Res.* 31: 1629–1638.

Tappeiner, J., Zasada, J., Ryan, P., and Newton, M. 1991. Salmonberry clonal and population structure: the basis for a persistent cover. *Ecology* 72: 609–618.

Traveset, A., and Willson, M.F. 1998. Ecology of the fruit-colour polymorphism in *Rubus spectabilis*. *Evol. Ecol.* 12: 331–345.

Virdi, B., and Eaton, G.W. 1969. Interspecific hybridization of the red raspberry and salmonberry. *Can. J. Bot.* 47: 1820.

Zasada, J.C., Tappeiner, J.C., Maxwell, B.D., and Radwan, M.A. 1994. Seasonal changes in shoot and root production and in carbohydrate content of salmonberry (*Rubus spectabilis*) rhizome segments from the central Oregon Coast Ranges. *Can. J. For. Res.* 24: 272–277.

Zasada, J., Tappeiner, J., and O'Dea, M. 1992. Clone structure of salmonberry and vine maple in the Oregon Coast Range. In *Proceedings symposium on the ecology and management of riparian shrub communities*. Compiled by W.P. Clary, E.D. McArthur, D. Bedunah, and C.L. Wambolt. General Technical Report INT-289. USDA Forest Service, Intermountain Research Station, Ogden, UT. pp. 56–60.

## Specialty Cookbooks

(*Note*: Salmonberries can be substituted for raspberries (Chapter 77) and blackberries (Chapter 12). See the Chapters on these fruits for additional recipes.)

Stewart (2002) provides information on the culinary preparation of salmonberries. Freitus and Haberman (2005) have eight salmonberry recipes, and Stanek and Butcher (2007) have five. Marie (2008) has a recipe for salmonberry wine. (See the Appendix to this book for details of the publications cited.)

# 84 Saskatoon

Family: Rosaceae (rose family)

## NAMES

Scientific name: *Amelanchier alnifolia* (Nutt.) Nutt. ex M. Roem.

- The word saskatoon is said to be an anglicized version of the Cree name for the fruit, *mis-sask-qua-too-mina* or *mis-sask-a-too-mina*. It has also been suggested that the name was derived from the Cree term for the place where stems of saskatoon bushes were collected for arrow shafts: *Mane-me-sas-kwa-tan*. Saskatoon is the usual name in Canada and the western United States. The city of Saskatoon in central Saskatchewan is believed to have been named for the saskatoon (not vice versa), in recognition of the abundance of the fruit along river banks. (Saskatoon was named by John H. Lake as a temperance colony for teetotalling Methodists from Ontario.) Nevertheless, apparently under the impression that the saskatoon is named for the city of Saskatoon, the initial s in the word is frequently capitalized ("Saskatoon"). (The similar-sounding name of the province of Saskatchewan comes from a different Cree source: *kisiskatchewani sipi*, meaning rapidly flowing river.)
- The saskatoon is also called Juneberry, lancewood, mountain Juneberry, Pacific serviceberry, pigeonberry (a name also used for pokeweed; see Chapter 75), Rocky Mountain service tree, sarvis (tree), Saskatoon serviceberry, Saskatoon berry, serviceberry (and its corruption sarvisberry), shadblossom, shadblow, shadbush, shadwood, western serviceberry, and western shadbush. In addition, there are numerous local names, such as bilberry, boxwood, Canadian medlar, grape pear, Indian pear, May cherry, Rocky Mountain blueberry, snowy mespilus, swamp pear, sugar pear, and sugarplum. Most of these names could lead to confusion with the established colloquial names of other species.
- "Serviceberry" is a name of European origin, thought to be from the classical Latin *sorbus*, the service tree. The currently recognized genus *Sorbus* denotes the mountain ash, a different genus, but one close enough that it can be hybridized with *Amelanchier.* It has also been suggested that the name serviceberry was derived from the practice of cutting off branches of *Amelanchier* species in midwinter and forcing them to bloom for church services in England. An imaginative interpretation holds that the name serviceberry came into being in North America, because the species bloomed when the dirt roads had thawed and dried enough so the circuit riding preachers could visit and conduct spiritual services. Still other contrived interpretations are that the berries were used to make wine used for religious services, and that the plant's blooming was an indication that the ground was sufficiently thawed to bury and hold ceremonies for all who had died over the winter. The name serviceberry is most commonly employed in the New World in the midwestern and southern United States.
- The "pear" in several of the common names is due to the pear shape of the fruit of some of the species of *Amelanchier.*
- The name "Juneberry" is for the flowering period, especially in the north central United States.

**FIGURE 84.1** Saskatoon (*Amelanchier alnifolia*). (a) Berries. (Courtesy of NatShots Photography [Flickr/CC-attribution].) (b) Shrub. (Courtesy of E. Small.) (c, d) Branches with fruit. (Courtesy of Louis-M. Landry, reproduced with permission.) (e) Distribution map.

- The names shadbush and shadblow are based on the observation that the plants are in full bloom when a fish, the American shad (*Alosa spidissima* (Wilson)), begins to ascend streams to spawn. (This fish is "anadromous" like the salmon, returning from the sea to breed in fresh water streams. The species inhabits the waters of the Atlantic coast from Newfoundland to Florida and has been introduced on the West coast of North America.) The name shadbush is mostly used on the East coast of North America.
- The genus name *Amelanchier* is based on the French name *amélanchier*, which in turn is based on *amelanche*, the Provençal name (Provence is a section of southeastern France) of European serviceberry, *A. ovalis* Medik. It has also been suggested that *amelanche* was the name used in Savoy (France) for the closely related medlar (*Mespilus germanica* L.).
- *Alnifolia* in the scientific name *A. alnifolia* is based on the Latin *alnus*, alder + *folium*, leaf, literally "with leaves like those of the alder."

## GEOGRAPHY AND ECOLOGY OF WILD PLANTS

The Saskatoon is native from Alaska south to California, extending east to Ontario in Canada and the midwestern United States. Saskatoons occur in open prairies, coulees (deep gulches or ravines that are often periodically dry; a western North American term), open woods, and on bluffs, hillsides, and stream banks. Soils range from dry and rocky to deep, fertile, and moist. The species is often found in areas subject to fire. Although the aboveground parts are killed by flames, the plants sprout vigorously afterward. Deep rhizomes (underground stems) protect the plants from even the most intense fires. Wild saskatoon has become scarce in parts of its range due to increasing alteration of the natural landscape, because of urban, agricultural, and industrial pressures.

## PLANT PORTRAIT

The saskatoon is a shrub or a small tree 1–7 m (3–23 feet) high with a trunk up to 25 cm (10 inches) in diameter. It often has several stems and may form a large thicket. Cultivated plants are generally shrubs, and these produce suckers at the base, which can develop into new plants. The plants are very ornamental, many small white to cream flowers appearing in early spring, and the foliage becoming highly colored in autumn. The fruit is borne in clusters of 6–12, and is usually blue-purple, but ranges from cream to black. The berries vary in diameter from 0.6–1 cm (0.25–0.4 inch) in the wild, to 1.6 cm (0.6 inch) in some cultivated varieties. The fruits have a heavy surface bloom and resemble blueberries. There are usually two to five seeds contained in an apple-like core at the centre of the fruit; these are small enough to be eaten. The taste of the flesh has been described as a combination of apple and sweet cherry, with a hint of almond. The fruits are soft, juicy, and sweet in the best cultivated varieties, most of which were bred in Canada. The first notable cultivated variety of *Amelanchier* was 'Success', apparently selected in the early twentieth century, not from *A. alnifolia* but from the downy serviceberry, *A. arborea* (F. Michx.) Fernald, another native of North America. Since then, more than two dozen cultivated *Amelanchier* varieties have been derived, primarily from *A. alnifolia*, often with some parentage from the downy serviceberry and other species. Saskatoon must compete with blueberry, which it resembles, and indeed it has often been called the "blueberry of the northern plains." The crop has achieved some importance in Canada's prairie provinces, with annual value in the millions of dollars.

## CULINARY PORTRAIT

Saskatoon berries can be substituted freely for blueberries in most recipes and dishes. They are good eaten fresh, cooked in pies and other desserts, or made into wine, jams, jellies, syrups, ice cream toppings, liqueurs, and flavor concentrates for baked goods. The berries can be home-canned, frozen, and dried to yield "raisins." Barely ripe fruit jells more readily and is better for freezing and preserving than mature fruit. Fully ripe fruit has higher sugar content and is more suitable for making wine. Fresh fruit contains about 10% sugar (mostly fructose; as much as 19% has been recorded) and 1%–2% pectin. Pectin content of most varieties is too low to jell without additional pectin. Early settlers often added a little lemon or rhubarb juice to improve the mild flavor of saskatoon berries, and this practice is still useful.

### Culinary Vocabulary

- "Pemmican" (from the Cree Indian word meaning "fat") was frequently prepared by North American Indians using saskatoon berries. Pemmican was made by stirring berries into dried (or a boiling mixture of) fat and pulverized meat of deer, moose, caribou, or buffalo, and pressing the resulting material into cakes. This was the main winter food of prairie Indians and was also used by fur traders and voyageurs. Native Americans ate the sweet berries fresh from the trees and picked large quantities to dry for use throughout the year. The Shoshone added the fruits to their bread to make a sort of raisin bread.

## PROSPECTS

Saskatoon has become a significant minor fruit crop, especially in Canada, where about 2000 orchards are managed in the western provinces, particularly in Alberta. There are more than two dozen cultivars, and breeding of improved varieties is continuing. The berries have reasonable keeping qualities and ship well. Because much of the crop is processed, its value is increased considerably. Saskatoon's chief rival is blueberry, which has much greater popularity. With its resistance to low temperature and drought, and tolerance of calcareous soils, saskatoon can be grown in areas where blueberry can not, particularly in the North American plains. The cold hardiness of the saskatoon (down to −60°C or 76°F) makes it particularly appropriate for areas with cold winters. The supply of saskatoon has frequently been insufficient to meet consumer demand by the food processing industry, restaurants, caterers, and others, indicating that considerable expansion of the crop is likely.

## CURIOSITIES OF SCIENCE AND TECHNOLOGY

- Indigenous peoples used preparations of saskatoon berries, twigs, and bark as a tonic, laxative, and contraceptive, and to treat coughs, colds, fever, toothache, diarrhea, dysentery, gonorrhea, and the discomforts of pregnancy and child birth. Extracts of saskatoon have recently been shown to have antiviral activity, suggesting the validity of some of these uses.
- Indigenous peoples employed saskatoon as a source of wood for arrows, spears, and harpoons, the handles of implements, frames for baskets and canoes, and for making armor and shields. One of the most interesting uses was for making frames for "bull boats"—so named because the hide of an entire buffalo was simply stretched over the frame to make a small, round boat. In more recent times, the wood has been used for fishing poles and umbrella handles.
- Explorers of the Lewis and Clark expedition of 1804–1806 in western North American relied heavily on the fruit of saskatoon when food supplies were low.
- Thomas Nuttall (1786–1859), an American botanist and ornithologist, recognized *A. alnifolia* as a new species in 1811, based on plants he collected in North Dakota, which was then far into a dangerous wilderness with roaming war parties. He was a most interesting character who wandered about the Great Plains searching for new species, oblivious to hunger, fatigue, illness, and biting insects. Although considered an entertaining eccentric by European settlers, the Indians held him in the highest regard, because like them he had a very deep knowledge of nature. One day he collapsed on the open prairie from extreme fatigue but was saved by a passing Indian who transported him in a canoe to the nearest fort.
- Fresh saskatoon berries are about 80% water.
- Saskatoon is of considerable significance as a source of food for wildlife and is often planted to attract birds and mammals. Birds, especially robins, are extremely fond of saskatoon and will quickly strip a tree of its fruit. Attempts have been made to plant saskatoon near cherry orchards to lure birds away from the cherries, but the experiments failed (the birds simply ate the saskatoon berries and returned to the cherries). *Amelanchier* species are often planted around the home for ornament and for fruit. They are excellent for attracting wildlife, but birds frequently eat the berries as soon as they begin to ripen, often before they are mature enough for human consumption.
- The twigs of saskatoon are a preferred browse plant for rangeland livestock and wildlife. Unfortunately a toxic substance (prunasin) is present in at least some plants of the species, and may be hazardous during the period of new growth of leaves and twigs, when the highest concentrations are produced.

## KEY INFORMATION SOURCES


Alberta Agriculture, Food and Rural Development. 2002. *Economics of saskatoon berry production: A ten-acre enterprise*. Alberta Agriculture, Food and Rural Development, Edmonton, AB. pp. 6.

Campbell, C.S., and Wright, W.A. 1996. Apomixis, hybridization, and taxonomic complexity in eastern North American *Amelanchier* (Rosaceae). *Folia Geobot. Phytotax.* 31: 345–354.

Catling, P.M., and Small, E. 2003. Poorly known economic plants of Canada—37. Saskatoon, *Amelanchier alnifolia*—Canada's national fruit? *Bull. Can. Bot. Assoc.* 36(2): 20–26.

Cinq-Mars, L. 1971. The genus *Amelanchier* in Quebec. *Nat. Can.* 98: 329–346. (In French)

Davidson, C.G., and Mazza, G. 1991. Variability of fruit quality and plant height in populations of saskatoon berries (*Amelanchier alnifolia* Nutt.). *Fruit Var. J.* 45: 162–165.

Davidson, J.G.N. 1987. The principal diseases of commercial saskatoons. *Agric. For. Bull. Alberta Univ. Fac. Ext. Edmonton* 10(1): 6–8.

Dey, D.H. 1998. *Commercial saskatoon berry industry*. Revised edition. Alberta Agriculture, Food and Rural Development, Edmonton, AB. pp. 12.

Green, R.C., and Mazza, G. 1986. Relationships between anthocyanins, total phenolics, carbohydrates, acidity and colour of saskatoon berries. *Can. Inst. Food Sci. Technol. J.* 19: 107–113.

Harris, R.E. 1973. *The saskatoon*. Agriculture Canada Publication 1246. Agriculture Canada, Ottawa, ON. pp. 8.

Harris, R.E. 1976. The saskatoon—Canada's national fruit? *Can. Agric.* 21(2): 28–29.

Hilton, R.J. 1982. Registration of *Amelanchier* cultivar names. *Fruit Var. J.* 36(4): 108–110.

Jones, G.N. 1946. *American species of Amelanchier*. University of Illinois Press, Urbana, IL. pp. 126. [Illinois Biological Monographs 20(2): 1–126.]

Junttila, O., and Stushnoff, C. 1983. De-hardening in flower buds of saskatoon-berry *Amelanchier alnifolia*, in relation to temperature, moisture content, and spring bud development. *Can. J. Bot.* 61: 164–170.

Kaurin, A., Stushnoff, C., and Junttila, O. 1984. Cold acclimation and dormancy of *Amelanchier alnifolia*. *J. Am. Soc. Hortic. Sci.* 109: 160–163.

Majak, W., Udenberg, T., Clark, L.J., and McLean, A. 1980. Toxicity of saskatoon serviceberry to cattle. *Can. Vet. J. Rev. Vet. Can.* 21(3): 74–76.

Mazza, G. 1982. Chemical composition of saskatoon berries (*Amelanchier alnifolia* Nutt.). *J. Food Sci.* 47: 1730–1731.

Mazza, G. 1986. Anthocyanins and other phenolic compounds of saskatoon berries (*Amelanchier alnifolia* Nutt.). *J. Food Sci.* 51: 1260–1264.

Mazza, G., and Davidson, C.G. 1993. Saskatoon berry: A fruit crop for the prairies. In *New crops*. Edited by J. Janick and J.E. Simon. Wiley, New York, NY. pp. 516–519.

McConkey, M. 1979. What's a saskatoon? The pioneering and popularizing of a native fruit. *North American Pomona* 12(1): 2–10.

McGarry, R., Ozga, J.A., and Reinecke, D.M. 1998. Patterns of saskatoon (*Amelanchier alnifolia* Nutt.) fruit and seed growth. *J. Am. Soc. Hortic. Sci.* 123: 26–29.

McGarry, R., Ozga, J.A., and Reinecke, D.M. 2001. Differences in fruit development among large— and small-fruited cultivars of saskatoon (*Amelanchier alnifolia*). *J. Am. Soc. Hortic. Sci.* 126: 381–385.

Miller, W.S., and Stushnoff, C. 1971. A description of *Amelanchier* species in regard to cultivar development. *Fruit Var. Hortic. Digest* 25: 3–10.

Nielsen, E.L. 1939. A taxonomic study of the genus *Amelanchier* in Minnesota. *Am. Midl. Nat.* 22: 160–208.

Olson, A.R. 1984. Structural aspects of pollination in *Amelanchier alnifolia* (Maloideae). *Can. J. Bot.* 62: 858–864.

Olson, A.R., and Steeves, T.A. 1982. Structural changes in the developing fruit wall of *Amelanchier alnifolia*. *Can. J. Bot.* 60: 1880–1887.

Olson, A.R., and Steeves, T.A. 1983. Frost damage in flowers and immature fruits of *Amelanchier alnifolia* Nutt. (Maloideae). *Can. J. Plant Sci.* 63: 461–466.

Pruski, K., Mohyuddin, M., and Grainger, G. 1991. Saskatoon (*Amelanchier alnifolia* Nutt.). *Biotechnol. Agric. For.* 16: 164–179.

Rogiers, S.Y., and Knowles, N.R. 1997. Physical and chemical changes during growth, maturation, and ripening of saskatoon (*Amelanchier alnifolia*) fruit. *Can. J. Bot.* 75: 1215–1225.

Rogiers, S.Y., and Knowles, N.R. 1998. Effects of storage temperature and atmosphere on saskatoon (*Amelanchier alnifolia* Nutt) fruit quality, respiration and ethylene production. *Postharvest Biol. Technol.* 13: 183–190.

Rogiers, S.Y., and Knowles, N.R. 2000. Efficacy of low $O_2$ and high $CO_2$ atmospheres in maintaining the post-harvest quality of saskatoon fruit (*Amelanchier alnifolia* Nutt.). *Can. J. Plant Sci.* 80: 623–630.

St-Pierre, R.G. 1997. *Growing saskatoons—A manual for orchardists.* Fifth edition. Dept. of Horticultural Science, University of Saskatchewan, Saskatoon, SK. Irregularly paginated.
St-Pierre, R.G. 2008. *Amelanchier alnifolia*, saskatoon. In *The encyclopedia of fruit & nuts.* Edited by J. Janick and R.E. Paull. CABI, Wallingford, U.K. pp. 619–622.
St-Pierre, R.G., and Steeves, T.A. 1990. Observations on shoot morphology, anthesis, flower number, and seed production in the saskatoon, *Amelanchier alnifolia* (Rosaceae). *Can. Field-Nat.* 104: 379–386.
Steeves, T.A., Steeves, M.W., and Olson, A.R. 1991. Flower development in *Amelanchier alnifolia* (Maloideae). *Can. J. Bot.* 69: 844–857.
Stushnoff, C. 1990. *Amelanchier* species. In *Genetic resources of temperate fruit and nut crops.* Edited by J.N. Moore and J.R. Ballington, Jr. International Society of Horticultural Science, Wageningen, the Netherlands. pp. 549–566.
Taylor, S. 1976. *Amelanchier alnifolia* (Nuttall) Nuttall. *Davidsonia* 7: 5–12.
Wallace, J.A., and Graham, T.O. 1976. Outlook in Western Canada towards improvement of the genus *Amelanchier. Proc. Western Canad. Soc. Hortic.* 32: 81–107.
Weaver, R.E. 1974. The shadbushes. *Arnoldia* 34: 22–31.
Wiegand, K.M. 1912. *The genus Amelanchier in eastern North America. Rhodora* 14: 116–164.
Williams, S. 1994. *Commercial saskatoon berry production on the prairies: A growers' guide.* Second edition. University Extension Press, Extension Division, University of Saskatchewan, Saskatoon, SK. Variously paginated.
Zatylny, A.M., and St-Pierre, R.G. 2005. Revised international registry of cultivars and germplasm of the genus *Amelanchier. Small Fruits Rev.* 2(1): 51–80.
Zatylny, A.M., St-Pierre, R.G., and Tulloch, H.P. 2002. Comparative agronomic performance of 15 saskatoon (*Amelanchier alnifolia* Nutt.) cultivars during their first seven years of growth. *J. Am. Pomol. Soc.* 56: 118–128.

## Specialty Cookbooks

Saskatoons can be substituted in blueberry recipes (see Chapter 15 on Blueberries).

Baird (1980) and Krumm (1991) each have 12 saskatoon recipes. Freitus and Haberman (2005) have 11 recipes, Stanek and Butcher (2007) have 3, Marrone (2009) has 15, and Williamson (1995) has 13. Other recipes in cookbooks include Saskatoon pie (Borman 1969), Juneberry pancakes, Juneberry–almond cookies, and Juneberry ice cream (Brill 2002); Saskatoon berry and rhubarb pie (Stewart 2000); Juneberry–raspberry ice cream topping, and Juneberry–Raspberry pie (Hibler 2004); Saskatoon berry tea (Watts and Watts 2007); Serviceberry wine (Marie 2008); Saskatoon-berry pie (Gray 2011); and Serviceberry coffee cake (Schufer 2011). (See the Appendix to this book for details of the publications cited.)

# 85 Sassafras

Family: Lauraceae (laurel family)

## NAMES

Scientific name: *Sassafras albidum* (Nutt.) Nees (*S. variifolium* (Salisb.) Kuntze, *S. officinale* Nees & C.H. Eberm.)

- The common name "sassafras" and the genus name *Sassafras* are based on the Spanish *sasafrás*, from Late Latin s*axifragia*, referring to a kind of herb. This has also been equated with the Latin *saxum fragrans* meaning "stone breaker," referring to the use of sassafras against kidney stones, but this explanation seems contrived. (The genus name *Saxifraga*, which includes herbs unrelated to sassafras, is based on the Latin for "stone breaker," based on the idea that the roots of herbs of this species can expand in crevices, even breaking large rocks apart.)
- Sassafras has also been called ague tree, American sassafras, black ash, chewing stick, cinnamon wood, common sassafras, filé gumbo, golden elm, green stick, gumbo filé, saloop, sassafrac, saxifrax, silky sassafras, smelling-stick, tea tree, and white sassafras. Several of these names are used quite locally and are easily confused with the names of other species.
- "Red sassafras" refers to a wild variety (var. *molle* (Raf.) Fern.) with quite hairy leaves. The wood is said to be somewhat reddish. However, the merit of recognizing this variety has not been established.
- "Brazilian sassafras" (often simply called "sassafras" in the essential oil trade) is *Ocotea odorifera* (Vell.) Rohwer (*O. pretiosa* (Nees) Mez, often called "*O. cymbarum*"). This tree is native to Brazil, Paraguay, and Columbia, and is also in the laurel family. Like sassafras, the main component is safrole. The species is a major source of commercial safrole.
- "Chinese sassafras" (*Cinnamomum camphora* (L.) J. Presl) is a tree that occurs throughout much of Southeast Asia. It is the world's main commercial source of safrole.
- The name "ague tree" reflects the old usage of sassafras as a cure for ague (i.e., malaria).
- Sassafras was called "green stick" by Native Americans because of its bright green twigs.
- *Albidum* in the scientific name *S. albidum* is Latin for whitish. The "whiteness" has been interpreted as both (1) referring to the light-colored wood and (2) the sometimes whitish undersides of the leaves.

## GEOGRAPHY AND ECOLOGY OF WILD PLANTS

Sassafras is native to eastern North America, reaching northward to southern Ontario and southern Maine, and southward to Texas and Florida. The species occurs in rich soils in deciduous woods and forests, old fields, along fencerows, and in disturbed sites. It grows best in well drained, sandy loams, with good exposure to light. Root sprouts frequently arise from the base of the tree and generate new trees, this mode of vegetative reproduction often resulting in thickets.

**FIGURE 85.1** Sassafras (*Sassafras albidum*). (a) Parts of the tree. (From Trew, C.J., *Herbarium Blackwellianum*, I. Fleischmanni, Norimbergae. 10 vols. bound in one; Plate 267, 1750–1773.) (b) Branch with fruit (left) and branch with male flowers (right). (From Köhler, H.A., *Köhler's Medizinal Pflanzen*, Verlag von F.E. Köhler, Germany. Vol. 2, 1887.) (c) Tree. (From Trew [cited above], Plate 69.) (d) Root beer, traditionally flavored with sassafras. (Left photo courtesy of Jonrev; right photo courtesy of Sonic3KMaster [CC By 3.0].) (e) Distribution map.

## PLANT PORTRAIT

Sassafras is a deciduous tree, rarely growing to 30 m (100 feet) in height in the southern states, but often just a dwarf tree or shrub in the northern part of its range. Typically, sassafras trees are 6–9 m (20–30 feet) tall. The leaves are remarkably variable in shape, some unlobed, some with three lobes, and others with just one lobe (such leaves shaped like mittens). Male and female flowers are borne on different trees. Pea-sized, dark blue fruits ripens in the fall on the female trees. All parts of the tree are aromatic.

The Choctaw Indians of Louisiana were the first to use dried, ground sassafras leaves as a seasoning. This is now known as filé or gumbo filé, familiar in Creole and Cajun cooking. (Filé is French for "thread," for the long threads formed when the powder is added to hot liquids.) Filé powder made from the young, dried leaves is the most significant commercial product now prepared from sassafras. Today, gumbo is generally made with okra (*Abelmoschus esculentus* (L.) Moench) instead of filé.

In 1586, English admiral Francis Drake (1540 or 1543–1596) landed at Roanoke, Virginia, and heard tales of colonists who had survived on sassafras soup. He returned to England with a shipment of this plant. By the early sixteenth century, sassafras was in great demand in English coffeehouses. Sassafras tea had a reputation among Native Americans as being capable of curing a wide range of diseases, and the Europeans also took up the medicinal use of sassafras for treating such varied ailments as parasitic worms, syphilis, colds, fevers, measles, diarrhea, and constipation. Early American settlers fermented sassafras roots with molasses to produce beer, and during the Civil War sassafras tea also became popular in America.

True sassafras oil, from the roots of North American *S. albidum*, is no longer produced commercially. Chinese sassafras and Brazilian sassafras, rather than the American sassafras species discussed in this chapter, are now used as commercial sources of sassafras oil, which is employed as a precursor for the synthesis of compounds used in the manufacture of perfumes, flavoring agents, and insecticides.

## CULINARY PORTRAIT

The inner bark of sassafras, especially from the roots, is a source of oil of sassafras, which was once the chief flavor ingredient of the uniquely American "root beer." Root beer is an original American soft drink, first made by indigenous peoples of the New World. Rural inhabitants during the eighteenth and nineteenth centuries also made elementary root beers, using available wild herbs and roots and bark of trees. In the past, root beer was also sometimes made at home by fermenting a combination of commercially produced flavorings and extracts, sugar, and yeast. The resulting beverage was naturally effervescent and slightly alcoholic. However, Charles E. Hires (1851–1937) is generally credited with creating the root beer beverage familiar today and has been called "the Father of Root Beer." He first experimented with various recipes in his Philadelphia pharmacy in 1869, sold root beer as a cold drink at the Centennial Exposition in 1876 in Philadelphia, and organized his own company in 1890. He intended to name his new concoction "root tea," but used the name "root beer" to appeal to the large market of hard-drinking Pennsylvania miners. (The Hires Root Beer Company was precluded from patenting the name "root beer" in 1879, when Congress passed a law stating that no word in the dictionary could be registered—a law that was repealed in 1920.) Since it was not made with alcohol, the temperance movement strongly supported the drinking of root beer. Hires himself was active in the temperance and Quaker movements. Prohibition (1920–1933) in the United States, sometimes called "The Great Dry Spell," contributed to the popularity of root beer. Hires became the largest manufacturer of the beverage in the world and extremely wealthy. At present, "root beer" sales in the United States are almost 2 billion dollars annually, and the beverage has a loyal following. Although root beer is still made from roots, today it is a soft drink, not a beer (i.e., it is nonalcoholic). Modern commercial root beer includes some of the original flavoring ingredients but is mostly made with sugar, caramel coloring, and artificial flavorings, with carbonated water for sparkle.

Oil of sassafras became obsolescent in the early 1960s when it was appreciated that its main component, the chemical safrole, is carcinogenic, causing liver cancer in the small intestine of experimental mice. Oil of sassafras has even caused abortions. Safrole is also thought to be hepatotoxic (poisoning the liver) and is suspected of being hallucinogenic and causing dermatitis. In 1976, the American Food and Drug Administration prohibited the sale of sassafras tea, roots, and oil.

Filé powder, made from the safrole-free leaves, was allowed to stay on the market. Root beer is still made using the essential oil of sassafras from the root (as well as with synthetic flavorings), but the safrole is removed. The taste of oil of sasssafras from which the safrole has been removed is inferior, but the product is safe to consume. Today, safrole-free root bark extract is used in perfumery, and as flavoring for candy, beverages, and tea.

Filé powder is a thickening and flavoring agent used in soups, stew, and gumbo. If boiled, it becomes stringy and unfit to eat, so it is stirred in after soup or stew has been removed from the heat. Filé powder is usually stirred into individual plates, not into a large pot of stew or soup, because once the powder has been added, reheating would result in stringiness. One to two teaspoons of filé thicken 3 L (about 3 American quarts) of soup. Filé is an important seasoning of Creole gumbo, a dark, rich, seafood or chicken soup, sometimes with added green vegetables. Purists never use both filé and okra, both natural thickeners for food preparations, in the same dish.

### Culinary Vocabulary

- The name "saloop" was used for sassafras tea when it was sold as a morning beverage on the streets of London, England.
- A "black cow" (a name that arose in Chicago) is a float made with root beer and vanilla ice cream.
- "Safrole-free extract of sassafras" is a recognized food additive that can be used as a flavoring agent.

## PROSPECTS

Today, the American sassafras is economically important only as a minor culinary product, the thickening and flavoring agent filé powder, from which the toxic chemical safrole has been removed. Oil of safrole, once an important extract, is now obtained from foreign sources. It is unlikely that the modest market for North American sassafras products will increase.

## CURIOSITIES OF SCIENCE AND TECHNOLOGY

- Christopher Columbus (1451–1506) is reported to have sensed the nearness of the New World from the strong smell of sassafras.
- In 1610, sassafras was so highly prized in England that the king demanded sassafras oil from the colony of Virginia as a condition of its charter.
- In pioneer days, pleasant tasting toothbrushes were made out of sassafras twigs, frayed at the end.
- Early settlers used oil of sassafras to drive away bed bugs, chicken lice, and dog fleas.
- "Godfrey's Cordial" (named after Englishman Thomas Godfrey in the early eighteenth century) was a patent medicine given to children troubled with colic. It consisted of sassafras, opium in some form, brandy or rectified spirit, caraway seed, and treacle. Sassafras was present chiefly to mask the taste of the opium. Godfrey's Cordial was highly effective at quieting infants, but at the risk of severe poisoning, permanent damage, and death.
- In some rural areas of the United States, sassafras was added as a flavoring ingredient to moonshine.
- In the nineteenth century, American cowboys often ordered "sarsaparilla" (which in fact was root beer) because it was both the most widely used treatment for syphilis and was also considered a male aphrodisiac. Sarsaparilla was generally the "good guy" drink of the early American Western movies (it was like ordering a glass of milk in a saloon while everyone else was drinking whisky); most movie goers were unaware that in fact it was usually requested in the hope of curing venereal diseases.

- A & W Root Beer is the most popular root beer in North America. It began when Roy W. Allen (died 1968) opened his first root beer and hamburger stand in Lodi, California in 1919, using a recipe purchased with 10 cents from a pharmacist in Arizona. In 1922, Frank Wright joined Allen in the business, and the brand name was born—A & W Root Beer. By 1923, A & W had established the first car hop restaurant in the United States, reshaping American fast food culture. (Car hops are waiters or waitresses who serve patrons in their cars at a drive-in restaurant, a concept now basically obsolete.)
- It has been estimated that as many as 2000 different brands of root beer have entered the market since C.E. Hires marketed his brand. National brands in the United States include A & W, Hires', Barq's, Mug, and Dad's Old Fashioned. Some of the more creative names include "Witches' Brew" (from Salem, Massachusetts, where women identified as witches were once hung), and "Big Al's Sarsaparilla" (from Lansing, Michigan, honoring gangster Al Capone, who used to frequent the local restaurants). "Root 66" Root Beer (from central Virginia) commemorates Route 66, the legendary "Highway of Dreams," or "The Main Street of America," which started in Chicago and ended in Santa Monica, Los Angeles, crossing eight states and three time zones (the highway was officially decommissioned in 1985, but there are still some stretches in existence).
- The world's largest sassafras tree is in Owensboro, Kentucky. It has a trunk circumference of 6.4 m (21 feet), a height of 30 m (100 feet) and is thought to be between 250 and 300 years of age.
- Sassafras trees may live to be 800 years old.
- Pyrethrins are the most widely used natural botanical insecticides, and are extracted from the pyrethrum daisy, *Tanacetum cinerariifolium* (Trevir.) Sch. Bip. Another naturally occurring material, often used in conjunction with pyrethrins because it acts as a synergist, is piperonyl butoxide, which is derived from safrole.
- The popular but dangerous, stimulant and hallucinogenic street drug Ecstasy (methylenedioxymethamphetamine or MDMA) is a synthetic chemical that is derived from safrole. It was first patented in 1914 by the German pharmaceutical company Merck, but only became the popular choice at all-night dance parties or "raves" in recent years because it heightens the receptivity of the senses, particularly to music. This so-called "love drug" creates a powerful sense of euphoria, but according to various reports it can also kill, causing brain damage, liver and kidney failure, and internal bleeding. Safrole is controlled, but illicit laboratories nevertheless find sources of it to manufacture illegal drugs.

## KEY INFORMATION SOURCES


Bonner, F.T., and Maisenhelder, L.C. 1974. *Sassafras albidum* (Nutt.) Nees, sassafras. In *Seeds of woody plants in the United States*. Technically coordinated by C.S. Schopmeyer. Agric. Handb. 450. U.S. Department of Agriculture, Washington, DC. pp. 761–762.

Bosela, M.J., and Ewers, F.W. 1997. The mode of origin of root buds and root sprouts in the clonal tree *Sassafras albidum* (Lauraceae). *Am. J. Bot.* 84: 1466–1481.

Bronaugh, W. 1994. The biggest sassafras. *Am. Forests* 100(5/6): 43.

Carlson, M., and Thompson, R.D. 1997. Liquid chromatographic determination of safrole in sassafras-derived herbal products. *J. AOAC Int.* 80: 1023–1028.

Clepper, H. 1989. The singular sassafras. *Am. Forests* 95(3/4): 33–34, 57.

Dathe, M. 1984. Sassafras in spite of itself. *Horticulture* 63(9): 19–24. [Ornamental & landscape use of *Sassafras* species.]

De Soyza, A. G., and Kincaid, D. T. 1991. Patterns in leaf morphology and photosynthesis in shoots of *Sassafras albidum* Lauraceae. *Am. J. Bot.* 78: 89–98.

Enkema, L.A. 1952. *Root beer; how it got its name; what it is; how it developed from a home-brewed beverage to its present day popularity*. Hurty-Peck & Co., Indianapolis, IN. pp. 10.

Flanagan, M. 1998. Enigmatic sassafras. *New Plantsman* 5: 43–47.

Gant, R.E., and Clebsch, E.E.C. 1975. The allelopathic influences of *Sassafras albidum* in old-field succession in Tennessee. *Ecology* 56: 604–615.
Ghent, A.W. 1973. Gravity and the distribution of leaf shape in the trees of *Sassafras albidum*. *New Phytol.* 72: 1141–1158.
Graham, A.W. Jr. 1973. History and uses of *Sassafras albidum. Longwood Program Semin.* 5: 42–45.
Hamilton, T.S. Jr. 1974. *Sassafras albidum* (Nutt.) Nees. *USDA Forest Service Tech. Rep.* 9: 122–125. [Forest ranges.]
Hutson, D., and Cupp, M.J. 2000. Sassafras. In *Toxicology and clinical pharmacology of herbal products*. Edited by M.J. Cupp. Humana Press, Totowa, NJ. pp. 245–252.
Kamdem, D.P., and Gage, D.A. 1995. Chemical composition of essential oil from the root bark of *Sassafras albidum*. *Planta Medica* 61: 574–575.
Kozel, P.C., and Jansen, M. 1976. The exciting sassafras. *Am. Nurserym.* 144(8): 10, 111–113.
Leonard, R.G. 1977. Sassafras/*Sassafras albidum* (Nutt.) Nees. *USDA Forest Service Tech. Rep.* 16: 91–92. [Use as wildlife food.]
McDaniel, J.C. 1975. Variation in sassafras. *Am. Nurserym.* 142(8): 11, 71–73.
Poole, I., Richter, H.G., and Francis, J.E. 2000. Evidence for Gondwanan origins for *Sassafras* (Lauraceae)? Late Cretaceous fossil wood of Antarctica. *IAWA Journal* 21: 463–475.
Quarantiello, L.E. 1997. *The root beer book: A celebration of America's best-loved soft drink*. Limelight Books, Lake Geneva, WI. pp. 96.
Tucker, A.O., Maciarello, M.J., and Broderick, C.E. 1994. Filé and essential oils of the leaves, twigs, and commercial root teas of *Sassafras albidum* (Nutt.) Nees (Lauraceae). *Dev. Food Sci.* 34: 595–604.
Tyler, V.E. 1980. Sassafras: Toxic tonic. *Garden J. (Bronx)* 4(4): 24–25. [Carcinogens.]
Weaver, R.E. Jr. 1976. Sassafras: A neglected native ornamental. *Arnoldia* 36: 23–27.
Wendel, G.W. 1977. *Longevity of black cherry, wild grape, and sassafras seed in the forest floor*. USDA, Forest Service, Northeastern Forest Experiment Station, Newtown Square, PA. pp. 375.
Yates, D., and Yates, E. 2003. *Ginger beer & root beer: Heritage 1790 to 1930*. Donald Yates Publishers, Homerville, OH. pp. 356.
Zwaving, J.H., and Bos, R. 1996. Composition of the essential oil from the root of *Sassafras albidum* (Nutt.) Nees. *J. Essent. Oil Res.* 8: 193–195.

## Specialty Cookbooks

Cresswell, S.E. 1998. *Homemade root beer, soda, & pop*. Storey Books, Pownal, VT. pp. 121.
Holmes, W.C. 1984. *The story of the gumbo filé tree*. Walter C. Holmes, Natchitoches, LA. pp. 8.
McKee, G. 1986. *The little gumbo book: Twenty-seven carefully created recipes that will enable everyone to enjoy the special experience of gumbo*. Quail Ridge Press, Baton Rouge, LA. pp. 64.
Shelton, C. 2001. *Angelina's zydeco gumbo cookbook*. Cookbooks, Videos, and Seasonings, Baton Rouge, LA. pp. 57.
St. Pierre, T.-M. 2000. *Jambalaya, crawfish pie, filé gumbo: Cajun and Creole cuisine*. Writers Club Press, San Jose, CA. pp. 121.
Weber, B., and Weber, F. 1998. *The Louisiana gumbo cookbook*. Acadian House, Lafayette, LA. pp. 186.

Beatty (1997) has two recipes for wild-collected sassafras, Freitus (1980) has three, Kluger (1973) has six, Robe-Terry (1997) has five, and Tatum (1976) has five. Schufer (2011) has two recipes, one of which is based on sassafras flowers. (See the Appendix to this book for details of the publications cited.)

# 86 Saw Palmetto

Family: Arecaceae (Palmae; palm family)

## NAMES

Scientific name: *Serenoa repens* (W. Bartram) Small (*Sabal serrulata* (Michx.) Nutt. ex Schult. & Schult. f.; *Serenoa repens* (W. Bartram) Small; *Serenoa serrulata* (Michx.) G. Nicholson)

- "Saw" in the name saw palmetto is based on the dangerous, saw-like teeth on the leaf stalk. For an explanation of "palmetto" in the name, see Chapter 19 on Cabbage Palmetto.
- Saw palmetto (saw-palmetto) has also been called palmetto, palmetto berry, palmetto scrub, saw palmetto berry, and scrub palmetto. It is sometimes called sabal in Europe.
- The genus name *Serenoa* commemorates American botanist Sereno Watson (1826–1892), a curator of the Gray Herbarium at Harvard University.
- *Repens* in the scientific name is Latin for creeping, an allusion to the growth form.

## GEOGRAPHY AND ECOLOGY OF WILD PLANTS

Saw palmetto is native to the coastal plain of the southeastern United States, occurring in southern Alabama, Florida, southeastern Georgia, southeastern Louisiana, southern Mississippi, and South Carolina. The species is most abundant in Florida, where it is present in every county, not just coastal areas. The plant grows in various habitats, including coastal sand dunes, moist hammocks, scrub pine associations, and as undergrowth in pinelands and woodlands. Saw palmetto thrives in disturbed areas, including those that have been clear-cut and burned. It also grows as a weed in pastures and for centuries it was regarded as an obstacle to urbanization and farming, and vigorous attempts were made at eradication. The species forms colonies, sometimes developing into vast stands. Saw palmetto withstands severe conditions, including salt spray, drought, and flooding, but does not tolerate long-term flooding by brackish or saline water. The fruits are eagerly consumed by birds and mammals. Both black bears and feral pigs have been observed to be extremely adept at removing and consuming the "hearts" (growing tissue at the top of the stems), which are the edible part of the saw palmetto featured in this chapter as having crop potential.

## PLANT PORTRAIT

Saw palmetto is a creeping or sprawling shrub 1–4 m (3–13 feet) tall. The trunks typically lie prostrate on the soil surface, sometimes growing partly or completely underground, but sometimes growing upright. A single plant commonly has several trunks. Trunks which creep along the ground often root and produce new plants. The species is a "fan palm," the leaf stalks terminating in a round fan of sword-like leaflets. The leaves are usually 30–100 cm (1–3 feet) long, the leafstalks representing about half this length. The foliage usually lasts for 2 years before dying. The dead leaves remain on the plant for a year or more, becoming very flammable and subject to fire. The flowers are creamy white and fragrant, 4–5 mm (0.16–0.2 inch) long, and borne in dense, branching, clusters ("spadices") up to 60 cm (2 feet) long. The plumelike clusters typically have thousands of the tiny flowers. The fruits are ellipsoid or subglobose, 16–25 mm (0.6–1 inch) long, green when immature,

**FIGURE 86.1** Saw palmetto (*Serenoa repens*). (a) plants in woods. (Courtesy of Miguel Vieira [CC By 2.0].) (b) Small plant bearing fruiting stalks. (Courtesy of Scott Zona [CC By 2.0].) (c) Large plant. (Courtesy of Javier Martin.) (d) Fruit. (Courtesy of Javier Martin.) (e) Distribution map.

ripening through yellow and orange to black. The fruits contain one seed embedded in fleshy pulp. The plants are often long-lived, some estimated to be 700 years of age.

For more than a century, saw palmetto has been known as the source of extracts used to treat urinary tract infections, baldness, and especially benign prostatic hyperplasia (BPH, a noncancerous enlargement of the prostate gland, common in older men, and affecting about half of those over 60 years). More than 2 million men in the United States have used saw palmetto for BPH. The fruits are collected from the wild and from private property and sold to pharmaceutical firms. Once regarded as a nuisance, saw palmetto is now viewed as a valuable source of income, and landowners often manage the plants on their property to maximize fruit harvest. Saw palmetto products are sold both as prescription pharmaceuticals and as over-the-counter products. In the United States, saw palmetto is available as a "supplement," not as a "medication," because there is still uncertainly regarding the net benefits from its use. In 2010 in the United States, sales amounted to $19 million, and saw palmetto was second only to cranberry as a herbal supplement. In Germany, where herbal medicine is very popular, saw palmetto extracts must contain 85%–95% fatty acids and sterols, thought to be the active medicinal ingredients. Saw palmetto is thought to reduce BPH by reducing the effects of testosterone that stimulates prostate growth. A recent analysis (Barry et al. 2011) concluded that saw palmetto was no more effective than placebo in treating BPH.

Saw palmetto is often cultivated as an ornamental and landscape shrub. It is useful in these regards because of its considerable range of adaptability, tolerating wet, dry, shady, and sunny sites. Curiously, this extraordinarily vigorous palm is very difficult to transplant, and nurseries routinely grow it from seed rather than attempting to dig up the huge supply of plants that grow naturally. The species has been grows well outside of its native range, including England and the Vancouver area of British Columbia.

Seminole and other native peoples fashioned baskets and ornamental items from the leaf stalks, a craft that persists today. In past times, saw palmetto was employed on a small scale for commercial products made from tough parts of the plant. Fiber from the leaf sheaths, stems, and roots was used in scrubbing brushes and upholstery fill, and a cork substitute was once manufactured from the stem.

## CULINARY PORTRAIT

The culinary potential of saw palmetto is largely confined to the soft internal tissues (hearts) at the tips of the trunks. Extensive information on palm hearts is given in Chapter 19 on Cabbage Palmetto and is not repeated here. Native Americans and subsequently European settlers consumed the hearts, albeit not to a major extent.

Indigenous peoples in the natural geographical range of saw palmetto consumed the fruits, fresh or in processed products. However, the fruits are not to the taste of most people (they have often been described as repulsive), although they are still collected and eaten by some wild food enthusiasts. Some "saw palmetto honey" is produced by honeybees in regions where the plant is common, and sold commercially.

### Culinary Vocabulary

- "Metto" was a soft drink manufactured in the early 1900s in Miami, by mixing saw palmetto fruit juice with carbonated water.
- "Shiope sofkee" is a traditional drink prepared by Florida's Seminoles, made by adding sugar to palmetto fruit juice.

## PROSPECTS

Saw palmetto is the most common palm in North America north of Mexico, and in the southeastern United States it is a very widespread wild plant. This indicates that managing, rather than cultivating the plant suffices for crop production in its natural area of distribution, but the species could be established in other areas with suitable climates. Harvesting the hearts from the top of the trunk kills it, but often causes the plant to produce new trunks at its base, so that harvesting is far less destructive than for many other palm species. At present medicinal extracts from the fruits represent the principal, indeed the exclusive commercial product from the species. However, palm hearts are a popular food item, and the commercial supply needs to be imported into northern regions because the palm species utilized cannot be grown in cooler climates, even in the southern United States. Saw palmetto is already a highly successful crop for medicinal purposes, and considerable research has clarified its agricultural properties. The utilization of a given species for multiple purposes increases its value and flexibility in the marketplace. Accordingly, saw palmetto is a good possible candidate for additional exploitation as a food source for its tender hearts.

## CURIOSITIES OF SCIENCE AND TECHNOLOGY

- The prostate gland, which saw palmetto is alleged to treat, is a part of the male reproductive system of most mammals. It secretes a slightly alkaline fluid that prolongs the life of sperm by neutralizing the acidity of the vaginal tract. Muscles in the prostate increase the propulsion of the sperm during ejaculation. Whether or not females have or lack a prostate is a matter of semantics. The term "female prostate" (accepted by some medical organizations) refers to Skene's glands—structures homologous to the male prostate, located in the vagina near the vulva, postulated to secrete a lubricant to assist in intercourse.
- Black bears in Florida (where they are rare) have a special relationship with saw palm etto. They have been observed to abandon home ranges when they find new sources of the fruit, and when available the bears gorge on it. A road-killed bear was found to have almost 14 kg (more than 30 pounds) in her stomach. The bears serve to disperse the seeds, which probably germinate better after passing through the digestive system of the animals. The bears also are quite dependent on saw palmetto hearts for food. To obtain the succulent hearts from the trunks, they carefully rip away the young leaves at the tips with their mouths and then consume the hearts. The bears especially eat the hearts in the spring (a time of food shortage for the animals) and early summer, when the starch in the hearts is being converted to sugar.

## KEY INFORMATION SOURCES


Abrahamson, W.G. 1999. Episodic reproduction in two fire-prone palms, *Serenoa repens* and *Sabal etonia* (Palmae). *Ecology* 80: 100–115.

Agbabiaka, T.B., Pittler, M.H., Wider, B., and Ernst, E. 2009. *Serenoa repens* (saw palmetto): A systematic review of adverse events. *Drug Safety* 32: 637–647.

Barry, M.J., Meleth, S., Lee, J.Y., Kreder, K.J., Avins, A.L., Nickel, J.C., Roehrborn, C.G. et al. 2011. Effect of increasing doses of saw palmetto extract on lower urinary tract symptoms: A randomized trial. *J. Am. Med. Assoc.* 306(12): 1344–1351.

Bennett, B.C., and Hicklin, J.R. 1998. Use of saw palmetto (*Serenoa repens*, Arecaceae) in Florida. *Econ. Bot.* 52: 381–393.

Bent, S., Kane, C., Shinohara, K., Neuhaus, J., Hudes, E.S., Goldberg, H., and Avins, A.L. 2006. Saw palmetto for benign prostatic hyperplasia. *New England J. Med.* 354: 557–566.

Boyle, P., Robertson, C., Lowe, F., and Roehrborn, C. 2004. Updated meta-analysis of clinical trials of *Serenoa repens* extract in the treatment of symptomatic benign prostatic hyperplasia. *B.J.U. Int.* 93: 751–756.

Carrington, M.E., Gottfried, T.D., and Mullahey, J.J. 2003. Pollination biology of saw palmetto (*Serenoa repens*) in southwestern Florida. *Palms* 47: 95–103.

Carrington, M.E., Mullahey, J.J., Krewer, G., Boland, B., and Affoler, J. 2000. Saw palmetto (*Serenoa repens*): an emerging forest resource in the southeastern United States. *J. Appl. For.* 24: 129–134.

Carrington, M.E., Roberts, P.D., Urs, N.V.R.R., McGovern, R.J., Seijo, T.E., and Mullahey, J.J. 2001. Premature fruit drop in saw palmettos caused by *Colletotrichum gloeosporioides*. *Plant Disease* 85: 122–125.

Dedhia, R.C., and McVary, K.T. 2008. Phytotherapy for lower urinary tract symptoms secondary to benign prostatic hyperplasia. *J. Urol.* 179: 2119–2125.

Essig, F.B., Taylor, Y.R., and TeStrake, D. 2000. Florida's wax palm: The silver form of *Serenoa repens* (Arecaceae). *Florida Sci.* 63: 13–16.

Gholz, H.L., Guerin, D.N., and Cropper, W.P. 1999. Phenology and productivity of saw palmetto (*Serenoa repens*) in a north Florida slash pine plantation. *Can. J. For. Res.* 29: 1248–1253.

Habib, F.K., and Wyllie, M.G. 2004. Not all brands are created equal: A comparison of selected components of different brands of *Serenoa repens* extract. *Prostate Cancer Prostatic Dis.* 7: 195–200.

Hale, E.M. 1898. *Saw palmetto* (Sabal serrulata, Serenoa serrulata): *Its history, botany, chemistry, pharmacology, provings, clinical experience and therapeutic applications*. Boericke & Tafel, Philadelphia, PA. pp. 103.

Hawkes, A.D. 1950. Notes on the palms. 2. Saw palmetto *Serenoa repens* Small. *Nat. Hort. Mag.* 29: 93–95.

Hilmon, J.B. 1968. *Autecology of saw palmetto (*Serenoa repens *(Bartr.) Small)*. Ph.D. Diss. Duke University, Durham, NC. pp. 191.
Maehr, D.S., and Layne, J.N. 1996. Florida's all-purpose plant, the saw palmetto. *Palmetto* 6: 6–15, 22.
Makus, D.J. 2008. Seed germination methods and establishment of saw palmetto, *Serenoa repens*, in south Texas. *Acta Hort.* 782: 381–385.
Mcnab, W.H. and Edwards, M.B., Jr. 1980. Climatic factors related to the range of saw-palmetto (*Serenoa repens* (Bartr.) Small). *Am. Midl. Nat.* 103: 204–208.
Pitman, W.D. 1993. Evaluation of saw palmetto for biomass potential. *Bioresource Tech.* 43: 103–106.
Rosler, T.W., Matusch, R., Weber, B., and Schwarze, B. 2009. Analysis of the hydrodistillate from the fruits of *Serenoa repens*. *Planta Medica* 75: 184–186.
Shimada, H., Tyler, V.E., and McLaughlin, J.L. 1997. Biologically active acylglycerides from the berries of saw palmetto (*Serenoa repens*). *J. Nat. Prod.* 60: 417–418.
Smith, D. 1972. Fruiting in the saw palmetto. *Principes* 16: 30–33.
Stratman, M.R., and Pelton, M.R. 1999. Feeding ecology of black bears in northwest Florida. *Florida Field Nat.* 27(3): 95–102.
Tanner, G., Mullahey, J.J., and Maehr, D. 1996. *Saw palmetto: An ecologically and economically important native palm.* IFAS Circular WEC-109, University of Florida, Gainesville, FL. pp. 4.
Zona, S.A. 2000. *Serenoa*. In *Flora of North America North of Mexico*. Vol. 22. Edited by Flora of North America Editorial Committee. Oxford University Press, Oxford, U.K. pp. 104–105.

## Specialty Cookbooks

See Chapter 19 on Cabbage Palmetto.

# 87 Scotch Lovage

Family: Apiaceae (Umbelliferae; carrot family)

## NAMES

Scientific name: *Ligusticum scoticum* L. (including *L. hultenii* Fernald = *L. scoticum* subsp. *hultenii* (Fernald) Calder & R.L. Taylor)

- The word lovage in the name Scotch lovage is often said to be based on the use of true lovage (discussed below) in love potions. Indeed, it was once thought that wearing sachets of lovage and orris root (*Iris* ×*germanica* L. var. *florentina* (L.) Dykes) would ensure a strong attraction from one's secret love. However, the name is a corruption of the Latin name, *ligusticum*, explained below.
- The "Scotch" in the common name reflects the fact that, true to its name, the plant was (and continues to be) used in Scotland as a cooked potherb. *Scoticum* (sometimes misspelled "*scothicum*") in the scientific name is Latin for Scotch.
- Scotch lovage has also been called beach lovage, northern lovage, Scot lovage, Scotch parsley, Scots parsley, Scots lovage, Scottish licorice-root, Scottish lovage, sea lovage, wild beach celery, and wild celery.
- A Google examination of the frequency of occurrence of the phrases with adjectives referring to Scotland coupled with "lovage" follows: Scots lovage: 7180; Scotch lovage: 4630; Scottish lovage: 1500; Scot lovage: 144. The adjective "Scotch" has become somewhat obsolescent, and indeed is often considered pejorative unless related to food or drink (e.g., Scotch whiskey). The name "Scotch lovage" is used here because it is still widely used by botanists as well as in legislation, and this usage for an edible plant is consistent with the use of "Scotch" for food.
- "Black lovage" (also known as Alexanders) is *Smyrnium olusatrum* L., an herb of the Mediterranean area that has occasionally been used like celery.
- Scotch lovage and black lovage, mentioned above, are in the carrot family. Several other species of the carrot family also have "lovage" as part of their names.
- The genus name *Ligusticum* is of Greek origin, based on Liguria in western Italy where Scotch lovage once grew abundantly.

## GEOGRAPHY AND ECOLOGY OF WILD PLANTS

Scotch lovage is native to the north temperate region, including Eurasia and Greenland. In North America it is indigenous to western British Columbia, eastern Canada, the states of Alaska, Connecticut, Maine, Massachusetts, New Hampshire, and Rhode Island. In some regions where the species is rare, it is protected by legislation and should not be collected from nature. Scotch lovage is naturalized in New Zealand and other locations. The plant grows on sandy and rocky seashores and seacliffs and along the edges of salt marshes, often among beach grasses. It can tolerate some salt spray. It also grows inland, often as a weed.

**FIGURE 87.1** Scotch lovage (*Ligusticum scoticum*). (a) Flowering branch. (From Syme, J.T.B., *English botany*, George Bell & Sons, London, U.K. Vol. 4, Plate 603, 1873.) (b, c) leaf and flowering stalk. (Courtesy of Louis-M. Landry, reproduced with permission.) (d) Canadian 2006 50¢ sterling silver short-tailed swallowtail butterfly coin, showing the insect, which is dependent on Scotch lovage for its existence. (Coin image © Royal Canadian Mint—All Rights Reserved.) (e) Distribution map of this coastal species.

## PLANT PORTRAIT

True lovage (*Levisticum officinale* W.D.J. Koch), usually known simply as lovage, is similar to Scotch lovage but is much better known and provides a standard against which Scotch lovage needs to be judged. Lovage has also been called common lovage, Cornish lovage, English lovage, Italian lovage, and love parsley. It is a robust perennial herb growing as tall as 2 m (6.6 feet). The species is native to the eastern Mediterranean and Afghanistan and has escaped from cultivation in much of Europe and parts of North America. In North America, lovage grows outside of cultivation as far north as Saskatchewan, the southern parts of Ontario and Quebec, and Nova Scotia, and also grows as a weed in much of the eastern United States. The majority of cultivated lovage is grown in central Europe, largely for essential oil. In North America, this herb is most familiar to those of central and eastern European origin. The leaves, roots, and stems have a pleasant flavor that is reminiscent of celery and parsley with a spicy tang. Many consider lovage the best of culinary herbs while others find it too strong in flavor. Both the leaves and shoots are used as savory herbs.

Scotch lovage is a perennial clump-forming herb 15–90 cm (0.5–3 feet) tall, the aboveground portion dying back to ground level by winter. It has a robust, hollow, angled, and branched stem, and large, long-stalked, divided leaves with glossy dark-green toothed leaflets. The flowers are tiny, pinkish-white or greenish-yellow. The fruits are 4–8 mm (0.16–0.3 inch) long. The species is occasionally cultivated, but mostly collected from the wild. It is grown to some extent as an ornamental, particularly appreciated for forming dense patches with reddish stems that are attractive in flower beds. A variegated garden form is available. In Scotland, the plant was once used to treat gas, indigestion, nerve problems, and tuberculosis, and as a tonic and aphrodisiac.

## CULINARY PORTRAIT

The leaves, young shoots, stems, and sometimes also the flowers and roots are consumed raw in salads or cooked in soups and stews. The stems are best picked before the plants flower, and younger leaves are preferable to older foliage. Blanching (covering the shoots so that they develop without exposure to sunlight) makes the strong, pungent, sometimes unpleasant taste milder. The flavor of Scotch lovage is reminiscent of celery and parsley, and indeed the blanched shoots and leafstalks are employed as a celery substitute. Scotch lovage makes an interesting replacement for a celery stick in a Bloody Mary. The seeds have been ground into a powder and used as a peppery flavoring in soups and stew. The seeds can be employed in dishes in the same manner as celery seeds. Occasionally the young shoots and roots are candied like angelica. Some upscale restaurants in Quebec serve wild-collected Scotch lovage.

### Culinary Vocabulary

- "Lovage salt" is a mixture of crushed (true) lovage seeds and salt, used as a condiment.
- In Europe a flavoring sauce (also available in dry cube form) called "Maggi" was popular in the 1960s and 1970s. This is a concentrated, salted solution of browned and partially hydrolyzed meat proteins in water, which seems to have been either flavored by true lovage or had an aroma similar to that of lovage.
- "Shunis" is a salad of Scotch lovage, popular in the Hebrides of Scotland.

## PROSPECTS

As a vegetable, Scotch lovage competes directly with the similar celery and lovage. Historically, celery has been credited with making Scotch lovage obsolete, and it is unlikely that Scotch lovage will now compete well with celery. However, experience in Quebec restaurants has indicated that there is a demand for it for gourmet meals, and the species could well find a broader market for those seeking different taste experiences. Scotch lovage would not be the first obsolescent vegetable to be resurrected as a "new" specialty crop with snob appeal.

## CURIOSITIES OF SCIENCE AND TECHNOLOGY

- In the British Isles, Scotch lovage was used to deworm livestock.
- Scotch lovage was eaten by sailors after long voyages in the expectation that it would reduce the symptoms of scurvy.
- Alexander Point and the Alexander River in Labrador were named by Sea Captain George Cartwright (1739/1740–1819) after he saw what he thought was Alexanders (also known as black lovage, noted above) growing abundantly on the banks. In fact, what he must have observed is the similar seashore plant, Scotch lovage.

- Another species of *Ligusticum*, *L. porteri* Coult. & Rose, grows in the Rocky Mountains and is known as "osha" (a Native American word meaning "bear root"). It is often referred to as "bear medicine" since, according to Native American lore, the plants seem to be eaten by bears for medicinal purposes. Bears are said to search for stands of osha and consume the plants roots directly after emerging from winter hibernation or when wounded or sick. Unfortunately, osha root has acquired a large reputation as an herbal medicine and has been seriously overcollected, endangering the wild plants.
- In 2006, Canada's Royal Canadian Mint issued a collectors sterling silver 50 cent coin showing a short-tailed swallowtail butterfly (*Papilio brevicauda* Saunders) on a Scotch lovage flower (Figure 87.1d). The butterfly is found in Newfoundland and nearby areas. Its larvae are specialist feeders on Scotch lovage, but will consume other herbs of the carrot family. As with most collector coins, the cost of the coin was much higher ($45.00) than the face value.

## KEY INFORMATION SOURCES


Anonymous. Undated. Fact sheet, Gaspé Peninsula's Non-Timber Forest Products: *Ligusticum scoticum*, Scotch lovage. L'Union des Producteurs Agricoles. pp. 2. [www.gaspesielesiles.upa.qc.ca, URL examined Jan. 2012]

Gelting, K.C. 1993. Changes in the distribution of *Ligusticum scoticum* in Norden since 1943. *Svensk Botanisk Tidskrift* 87: 255–262. (In Swedish)

Haeggstrom, C.-A., Tornroth, H., and Stjernberg, T. 1992. Records of *Ligusticum scoticum,* Apiaceae on the Aland Islands Southwest Finland. *Memoranda Societatis Fauna Flora Fennica* 68: 11–15. (In Swedish)

Hata, K., Kozawa, M., Bab, K., and Imisui, M. 1973. Chemical components of the roots of *Ligusticum hultenii* Fernald and *Angelica shikokiana* Makino. *Yakugaku Zasshi* 93: 248–251. (In Japanese)

Hatano, K., Nishioka, I., and Iwasa, S. 1977. Cytogenetical studies of umbelliferous plants, part 6. Karyotype of *Ligusticum hultenii. Shoyakugaku Zasshi* 31: 114–116. (In Japanese)

Hebda, R.J. 1985. Pollen morphology of *Ligusticum* (Apiaceae) in Canada. *Can. J. Bot.* 63: 1880–1887.

Jean, F.I., Deslauriers, H., Collin, G.J., Gagnon, M., and Hachey, J.M. 1990. The essential oil of *Ligusticum scoticum* L. *J. Essent. Oil Res.* 2: 37–44.

Meepagala, K., Sturtz, G., Wedge, D.E., Schrader, K.K., and Duke, S.O. 2005. Phytotoxic and antifungal compounds from two Apiaceae species, *Lomatium californicum* and *Ligusticum hultenii*, rich sources of Z-ligustilide and apiol, respectively. *J. Chem. Ecol.* 31: 1567–1578.

Palin, M.A. 1988. Biological flora of the British Isles. *Ligusticum scoticum* L. (*Haloscias scoticum* (L.) Fr.). *J. Ecol.* 76: 889–902.

Rydgren, K., and Often, A. 1993. Establishment and growth characteristics of inshore populations of *Ligusticum scoticum* (Apiaceae). *Nordic J. Bot.* 13: 83–94.

Sinclair, S. 1998. Chinese herbs: a clinical review of *Astragalus*, *Ligusticum*, and *Schizandrae*. *Altern. Med. Rev.* 3: 338–344.

Smith, K.M., and Markham, R. 1944. A virus disease of lovage (*Ligusticum scoticum*). *Phytopathology* 34: 335–340.

Suk, K.-T., Nitta, A., and Konoshima, M. 1974. Studies on origin of Japanese chuanxiong, part 4, *Ligusticum scothicum* and *Tilingia* spp. *Yakugaku Zasshi.* 94: 1257–1264. (In Japanese)

Sun, N., He, X.-J., and Zhou, S.-D. 2008. Morphological cladistic analysis of *Ligusticum* (Umbelliferae) in China. *Nordic J. Bot.* 26: 118–128.

Valiejo-Roman, C.M., Shneyer, V.S., Samigullin, T.H., Terentieva, E.I., and Pimenov, M.G. 2006. An attempt to clarify taxonomic relationships in "Verwandtschaftskreis der Gattung *Ligusticum*" (Umbelliferae-Apioideae) by molecular analysis. *Pl. Syst. Evol.* 257: 25–43.

Von Numers, M., Haeggstrom, C.-A., Haeggstrom, E., Franzen, H., Franzen, J., and Carlsson, R. 2009. Scots lovage, *Ligusticum scoticum*, is spreading in the Aland Islands, SW Finland. *Memoranda Societatis Fauna Flora Fennica* 85: 1–5.


### SPECIALTY COOKBOOKS


Evans, H. 1996. *Lavender, lovage & lemongrass.* JG Press, North Dighton, MA. pp. 61.

# 88 Scurvy Grass

Family: Brassicaceae (Cruciferae; mustard family)

## NAMES

Scientific name: *Cochlearia officinalis* L.

- Scurvy grass leaves are quite rich in vitamin C, which prevents the development of scurvy, leading to the common use of the herb for this purpose in the past, and its name.
- Scurvy grass (scurvy-grass, scurvygrass) has also been called bad man's oatmeal, common scurvy grass, round leaved scurvy grass, scorbute grass, scruby grass, scurvy cress, scurvy weed, and spoonwort.
- The name "scurvy grass" is most commonly used to designate *C. officinalis* but is also often used for other members of the genus *Cochlearia*. In addition, some species of other genera may bear this name, for examples, *Rhodiola rosea* L. (*Sedum rosea* (L.) Scop.), a circumpolar wild plant occurring in Canada and the northeastern United States, sometimes used as a potherb; *Oxalis enneaphylla* Cav., a native of the Falkland Islands and Patagonia, occasionally used as a culinary herb but mostly grown as an ornamental; and *Crambe maritima* L., an edible Old World herb known as sea kale. None of these species is a "grass," that is, a member of the Poaceae (Gramineae) family.
- The genus name *Cochlearia* comes from the Greek *cochlear*, a spoon, in reference to the spoon-shaped basal leaves. The genus name has also been used as a common name, especially in England in past centuries, when scurvy grass was relatively popular, and "cochlearia" seemed like a more attractive name.
- *Officinalis* in the scientific name *C. officinalis* means "of the shops," used to indicate herbs that were employed medicinally in the past, hence sold in medicinal shops.

## GEOGRAPHY AND ECOLOGY OF WILD PLANTS

In North America, scurvy grass is distributed along the Arctic coast, from Greenland to Newfoundland to Alaska, and down the West and Atlantic coasts of Canada. Isolated colonies are found on the coasts of the states of Washington, Oregon, and California. Scurvy grass is a northern, salt-tolerant sea-beach plant that thrives near nesting sites of sea birds, where their droppings accumulate. It is found in various habitats, including sandy or rocky seashores and sea cliffs, stream banks, meadows, and tundra.

In Europe, scurvy grass occurs along the shores and some rivers of Scotland, in stony muddy and sandy soils in England and Ireland, along the seacoasts of northern and western Europe, in many salt marshes in northern and western Europe, and at elevations up to 2200 m (7200 feet) in the great European mountain chains such as the Alps. Occasionally it grows in lowland, inland situations in Britain, usually on roadsides, generally where salt has been used for the clearance of ice.

## PLANT PORTRAIT

The classification of *Cochlearia* species is problematical, particularly concerning the relationships of North American and European circumpolar and subarctic forms. For this chapter, *C. officinalis* is interpreted as a comprehensive species of both Eurasia and North America (others split off

**FIGURE 88.1** Scurvy grass (*Cochlearia officinalis*). (a) Flowering plants. (Courtesy of H. Zell [CC By 3.0].) (b) Flowering branch and vegetative plant. (From Köhler, H.A., *Köhler's Medizinal Pflanzen*, Verlag von F.E. Köhler, Germany. Vol. 1, 1887.) (c) Distribution map of this coastal species.

most plants of North America as *C. groenlandica* L., while recognizing *C. officinalis* as a strictly Eurasian species). Scurvy grass is a biennial, rarely a perennial, although normally cultivated as an annual. In its first year, it forms a low rosette (basal cluster) of leaves. In the spring of the second year, it produces a flowering stalk about 30 cm (1 foot) in height with small white flowers. The plant continues to flower into summer. Dwarfed plants may have stems that are only 5 cm (2 inches) long, whereas the stems of robust plants may be as long as 35 cm (14 inches). The plant has an unpleasant odor like that of horseradish. Scurvy, a disease marked by spongy gums, loosening of the teeth, and bleeding into the skin and mucous membranes, is caused by a lack of ascorbic acid (vitamin C). In the past, people who did not have access to fresh fruit for prolonged periods often suffered from the disease. It has been estimated that between 1500 and 1800, two million sailors died from scurvy. Historically, scurvy grass was used by mariners, explorers, prospectors, and traders to combat scurvy resulting from vitamin C deficiency. Also, the mustardy leaves were consumed as a salad in France and Ireland and, during the late eighteenth and early nineteenth centuries, scurvy grass was added to sandwich fillings in England. To a minor extent, northern people in North America and Europe still make use of it for food and health purposes. The species has no commercial importance, although it is occasionally sold through seed catalogs. Today, it is infrequently cultivated as a culinary herb and garden ornamental and is sometimes collected as a wild edible plant.

## CULINARY PORTRAIT

Scurvy grass is sometimes grown as a salad plant and pot herb with a warm flavor reminiscent of cress, although opinions differ on its palatability. Some authors have characterized the taste as bitter, unappetizing, and tasting too much like tar to be liked. Others have found it delicious, like a cross between cucumber and mustard and cress. Such differences in opinion may not simply reflect different personal tastes but the different tastes of different races of scurvy grass. It has been recommended that before serving scurvy grass in a salad, it should be soaked in fresh water for at least half an hour, and salt omitted from the salad dressing as the scurvy grass has enough of its own. The succulent, slightly salty leaves can be eaten fresh or cooked. The taste of fresh leaves is very pungent. They can be used in tossed green salads like watercress, in sandwiches along with other ingredients, and in soups, sauces, and stuffed vegetables. The flower-heads are also sometimes eaten. Danish scurvy grass (*C. danica* L.) is cultivated and considered by some to be superior to the common scurvy grass (*C. officinalis*).

### Culinary Vocabulary

- "Scurvy grass ale" was a bitter alcoholic brew made with scurvy grass (or sometimes with other herbs) and frequently used in England as an antiscorbutic (preventative of scurvy) before citrus fruits became available.
- "Dr. Butler's Ale," popularized around the beginning of the seventeenth century in London, England, was another famous early alcoholic "root beer" made with herbs, including scurvy grass. Its medicinal values were extensively advertised, giving people a good excuse to drink.

## PROSPECTS

The possibility of developing scurvy grass into a commercial crop is quite limited but not as improbable as might appear on first consideration for a plant that has generally been reported as unpleasant in taste. The poor taste can be overcome by breeding, as it is evident that palatable forms exist in nature. There are very few vegetables that can be grown in the coldest areas of the world, and the most desirable of these are harvested for their berries. Scurvy grass is a vegetable, and its tolerance to cold temperatures and saline soils suggest that it could fill a marketplace niche. Moreover, some odd-tasting salad vegetables have proven to be attractive to the "arugula crowd," and so could develop into an item with snob appeal.

## CURIOSITIES OF SCIENCE AND TECHNOLOGY

- Scurvy grass was once used to treat paralysis and rheumatism.
- Danish navigator Vitus Jonassen Bering (1681–1741), commemorated by the Bering Strait, the Bering Sea, and Bering Island, died of scurvy on the island named after him, amid thick growths of scurvy grass that could have saved him.

- In the 1700s, before the British navy issued lime juice (with limited effectiveness) to combat scurvy, mariners commonly used scurvy grass for the purpose. Unfortunately, they usually did not appreciate that the benefit was mainly in the fresh herb, and they collected and stowed large bales of dried leaves. Nevertheless, the leaves steeped in ale were highly valued by sailors, perhaps as much for the alcohol as for preventing scurvy.
- In France, the medicinal use of scurvy grass became quite popular in 1987, following favorable press reports, and the value increased from $4.00/kg ($9.00/pound) to 20.00/kg ($44.00/pound).
- Scurvy is rare among mammals, most of which can manufacture their own vitamin C. Primates (including humans) and guinea pigs are exceptions.

## KEY INFORMATION SOURCES

Alm, T. 1996. Use of scurvy grass (*Cochlearia officinalis*) in Northern Norway. *Blyttia* 54: 185–192.

Al-Shehbaz, I.A., and Koch, M. 2010. *Cochlearia*. In *Flora of North America*. Vol. 10. Edited by Flora of North America Editorial Committee. Oxford University Press, New York. pp. 514–516.

Burton, R.M. 1996. Danish scurvy-grass *Cochlearia danica* in the London area. *London Nat.* 0(75): 133–135.

De Lange, P.J., and Norton, D.A. 1996. To what New Zealand plant does the vernacular "scurvy grass" refer? *New Zealand J. Bot.* 34: 417–420.

Eriksen, A.B., and Nordal, I. 1989. Ecotypic differentiation in relation to soil nitrogen in northern Scandinavian *Cochlearia officinalis*. *Holarctic Ecol.* 12: 31–38.

Gill, J.J.B. 1971. Cytogenetic studies in *Cochlearia*. L. The chromosomal homogeneity within both the 2n = 12 diploids and the 2n = 14 diploids and the cytogenetic relationship between the two chromosome levels. *Ann. Bot.* 35: 947–956.

Gill, J.J.B. 1973. Cytogenetic studies in *Cochlearia* L. (Cruciferae). The origins of *Cochlearia officinalis* L. and *Cochlearia micacea* Marshall. *Genetica* 44: 217–233.

Gill, J.J.B. 1976. Cytogenetic studies in *Cochlearia* L. (Cruciferae). The chromosomal constitution of *Cochlearia danica* L. *Genetica* 46: 115–127.

Gill, J.J.B., McAllister, H.A., and Fearn, G.M. 1978. Cytotaxonomic studies on the *Cochlearia officinalis* L. group from inland stations in Britain. *Watsonia* 12: 15–21.

Johnston, W.R., and Proctor, J. 1979. Ecological studies on the Lime Hill serpentine, Scotland assemblages of characteristics species such as *Cochlearia officinalis*, *Minuartia verna* and *Armeria maritima*. *Trans. Bot. Soc. Edinb.* 43(pt. 2): 145–150.

Koch, M., Hurka, H., and Mummenhoff, K. 1996. Chloroplast DNA restriction site variation and RAPD-analyses in *Cochlearia* (Brassicaceae): biosystematics and speciation. *Nordic J. Bot.* 16: 585–603.

Koch, M., Huthmann, M., and Hurka, H. 1998. Isozmyes, speciation and evolution in the polyploid complex *Cochlearia* L. (Brassicaceae). *Botanica Acta* 111: 411–425.

Koch, M., Mummenhoff, K., and Hurka, H. 1999. Molecular phylogenetics of *Cochlearia* (Brassicaceae) and allied genera based on nuclear ribosomal ITS DNA sequence analysis contradict traditional concepts of their evolutionary relationship. *Plant Syst. Evol.* 216: 207–230.

Lester, G.S., Vasey, M.C., and Rodstrom, W.E. 1981. *Cochlearia officinalis* L. (Brassicaceae). *Madroño* 28: 86.

Nordal, I., and Laane, M.M. 1996. Taxonomic delimitation within *Cochlearia officinalis* s. lat. with particular discussion on the rank of *C. anglica* (Brassicaceae). *Acta Univ. Upsal. Symb. Bot. Upsal.* 31(3): 47–57.

Nordal, I., and Stabbetorp, O.E. 1990. Morphology and taxonomy of the genus *Cochlearia* (Brassicaceae) in Northern Scandinavia. *Nordic J. Bot.* 10: 249–264.

Nordal, I., Eriksen, A.B., Laane, M.M., and Solberg, Y. 1986. Biogeographic and biosystematic studies in the genus *Cochlearia* in Northern Scandinavia. *Acta Univ. Ups. Symb. Bot. Ups.* 27(2): 83–93.

Pegtel, D.M. 1999. Effect of ploidy level on fruit morphology, seed germination and juvenile growth in scurvy grass (*Cochlearia officinalis* L. s.l., Brassicaceae). *Plant Species Biol.* 14: 201–215.

Vogel, G. 1994. Vegetable biographies. 23: *Cochlearia officinalis*. *Taspo Gartenbaumagazin (Germany)*. 3(7): 37. (In German)

Welch, D., and Welch, M.J. 1998. Colonisation by *Cochlearia officinalis* L. (Brassicaceae) and other halophytes on the Aberdeen–Montrose main road in North-East Scotland. *Watsonia* 22: 190–193.
Wyse Jackson, P.S., and Akeroyd, J.R. 1993. *Cochlearia.* In *Flora Europaea.* Vol. 1. Second edition. Edited by G.T. Tutin, N.A. Burges, A.O. Chater, J.R. Edmondson, V.H. Heywood, D.M. Moore, D.H. Valentine, S.M. Walters, and D.A. Webb. Cambridge University Press, UK. pp. 378–380.

## Specialty Cookbooks

Some recipes for scurvy grass can be found on the Web and in books on edible wild foods of northern regions. It can be used as a salad plant (particularly as a substitute for watercress) or as a cooked herb, as noted above. Michael (1980) has a recipe for Scurvy grass sandwiches, and Szczawinski and Turner (1980) have recipes for Scurvy grass salad and Scurvy grass salad with rice. (See the Appendix to this book for details of the publications cited.)

# 89 Sea Grape

Family: Polygonaceae (knotweed family, buckwheat family)

## NAMES

Scientific name: *Coccoloba uvifera* (L.) L.

- Sea grape (sometimes spelled seagrape or sea-grape) is named for its grape-like fruits borne in grape-like clusters.
- Sea grape is also known as baygrape, common sea grape, grape, Jamaican kino (discussed below), platterleaf, seaside grape, and shore sea grape. Rarely used names include pigeon wood, seaside plum, and wild grape. The Spanish name *uva de playa* (literally "grape of the sea") is often encountered.
- The name sea grape is also used for: (1) seaweeds (large marine algae) of the genus *Caulerpa*, especially *C. lentillifera* J. Agardh, often eaten in Southeast Asia; (2) species of *Ephedra*, especially the Asian *E. distachya* L., used for medicinal purposes; and tunicates, which are invertebrate aquatic animals living mostly in oceans.
- The genus name *Coccoloba* is based on *cocolobis* or *cocolubis*, a Spanish name for a kind of grape, tracing to the Greek *kokkos*, berry + *lobos*, pod, referring to the grape-like fruits.
- *Uvifera* in the scientific name is Latin for grape-bearing.

## GEOGRAPHY AND ECOLOGY OF WILD PLANTS

Sea grape is native to southern Florida, the Bahamas, Bermuda, the West Indies, Mexico, Central America, northern and eastern South America to Brazil, and the Pacific coast of South America south to Peru. The species has been grown as an ornamental in most tropical areas and has become widely naturalized outside of its native range, including Hawaii.

Sea grape is adapted to living in sandy beaches near seashores, often growing very close to the ocean, like coconut palms. The plant also occurs on coastal sand dunes and rocky seashores. It cannot survive temperatures below freezing except for a brief period. The species is very tolerant of salt (both salt spray and salty soils) and is often used to stabilize beaches. Sea grape is also tolerant of wind, heat, drought, and waterlogged soil, and can withstand moderate shade (but prefers full sun). The species is often one of the first to colonize sandy and rocky seashores. It occupies several plant associations, including coastal hammocks, coastal scrub, coastal grasslands, and beach strands. In its native range, the species is often quite common. Rarely, sea grape occurs in moist, inland forests, such as in Cuba and Jamaica. Sea grape is an important natural sand stabilizer preventing shoreline erosion. The large leaves are credited with stopping wind-borne sand from blowing away. In Florida, legislation protects shoreline sea grape stands from damage by humans, to conserve the environment and oceanfront property. The plant is also ecologically valuable in that it can be grown with minimal care and irrigation, and in its native area is considered ideal for xeriscaping (cultivation of plants that need no or minimal watering).

FIGURE 89.1 Sea grape (*Coccoloba uvifera*). (a) Fruits. (Courtesy of Daniel diPalma [CC By 3.0].) (b) Branch with immature fruit. (Courtesy of Wilfredo Rodríguez [CC By 3.0].) (c) Shrub. (d) Distribution map.

## PLANT PORTRAIT

This species is an evergreen shrub or small tree growing up to 15 m (50 feet) in height. Most plants are shrubs no more than 2 m (6.6 feet) high. Plants near the sea, and exposed to wind, sand spray, and salt, usually become sprawling shrubs, but those in protected areas can become symmetrical trees. In the northern part of the species' range, tree stature is usually not attained, and plants often show asymmetrical shapes due to dieback caused by frost damage. The leaves are large (often up to 20 cm or 8 inches across, occasionally as wide as 25 cm or 10 inches), stiff, leathery, glossy, rounded or kidney-shaped in outline, with a red midrib. The young leaves are bronze, and the old ones turn red when they senesce (foliage generally remains on the plant 2 or 3 years), making the plants rather attractive. Male flowers and female flowers occur on separate individuals, and both male and female plants are necessary for fruit formation. The flowers are small, whitish or greenish-white, and fragrant. The fruits are green when young, maturing in late summer to orange, red, or typically purple. The fruits are pear-shaped or round, 1.2–2.5 cm (0.5–1 inch) in diameter, and occur in grape-like bunches, often more than 40 to a cluster. Much of the fruit is filled with a large pit, about 1 cm (0.4 inch) long. Unlike common grapes, the fruits do not ripen simultaneously.

Sea grape is widely cultivated as an ornamental in tropical areas. A form with variegated leaves is available. In coastal cities, it is often grown as a street tree, and on beaches it is employed as a wind break and a sand stabilizer. The species is often planted as a hedge, screen, or barrier for roadside medians, parking lots and shopping centers. Sea grape is also grown as a potted plant, occasionally as bonsai.

Sea grape trees are used to a small extent as sources of lumber, but mostly for firewood and charcoal. The wood is used for furniture, cabinetry, and handicrafts. The astringent (due to tannins) bark and roots are employed in traditional medicine in the native range of the species. Astringent red sap exuded or extracted from the cut bark or roots, known as West Indian and Jamaica kino, was once exported to Europe for use in tanning and dyeing.

## CULINARY PORTRAIT

Sea grape fruits are tougher than grapes but very tasty (tartly sweetish or musky and acidic) and are frequently eaten raw. They contain considerable pectin and are commonly used for jam and jelly. Sea grapes are also sometimes fermented like grapes into wine. Bees are attracted to the abundant nectar in sea grape flowers, and the resulting honey is light in color, spicy, and of good quality.

### Culinary Vocabulary

- As noted above, "sea grape" can refer to edible seaweeds of the genus *Caulerpa*, especially *C. lentillifera*, which is slightly reminiscent of a bunch of grapes in appearance. This is also called "green caviar," and is especially popular in the Philippines, where it is consumed as a snack, vegetable, and in salads. It is also well known in Japan. To avoid confusion, the phrase "seaweed sea grape" is sometimes used for *Caulerpa*.

## PROSPECTS

Sea grape is mostly consumed locally by wild food collectors and by knowledgeable individuals who appreciate the taste of plants grown on their properties for ornamental and shade purposes. A small amount of fruit is processed for production of jam, jelly, and wine that is sold commercially. The large pit leaves little flesh on the berries, but they occur extremely abundantly and are easily collected in large amounts, although uneven ripening limits the harvest of fruit from a given plant at a given time. Despite these limitations, sea grape is an attractive candidate for development of a more substantial industry than exists at this time.

## CURIOSITIES OF SCIENCE AND TECHNOLOGY

- Early Spanish colonists sometimes used the fresh thick leaves of sea grape as paper, scratching messages on them with a pin or sharply pointed object.
- American botanist Nathaniel Lord Britton (1859–1934), founder of the New York Botanical Garden, wrote that sea grape was probably the first land plant of the New World seen by Christopher Columbus, since it is very conspicuous on the seashores of West Indies islands.
- Sea turtles often lay their eggs in sand in the shelter of sea beach vegetation, especially sea grape. The amount of shade provided by the plant can be critical for determining the sex of the turtles: shaded, cooler areas favors the development of males, more open, warmer areas produces more females.

## KEY INFORMATION SOURCES


Bandou, E., Lebailly, F., Muller, F., Dulormne, M., Toribio, A., Chabrol, J., Courtecuisse, R., et al. 2006. The ectomycorrhizal fungus *Scleroderma bermudense* alleviates salt stress in seagrape (*Coccoloba uvifera* L.) seedlings. *Mycorrhiza* 16: 559–565.

Burke, J.M., Sanchez, A., Kron, K., and Luckow, M. 2010. Placing the woody tropical genera of Polygonaceae: a hypothesis of character evolution and phylogeny. *Am. J. Bot.* 97: 1377–1390.

Duarte, O., and Paull, R.E. 2008. *Coccoloba uvifera*, sea grape. In *The encyclopedia of fruit & nuts*. Edited by J. Janick and R.E. Paull. CABI, Wallingford, Oxfordshire, U.K. pp. 597–599.

Freeman, C.C. 2005. *Coccoloba*. In *Flora of North America North of Mexico*. Vol. 5. Edited by Flora of North America Editorial Committee. Oxford University Press, Oxford, U.K. pp. 483–484.

Gilman, E.F., and Watson, D.G. 1993. *Coccoloba uvifera, seagrape*. Fact Sheet ENH 334. Environmental Horticulture Department, Florida Cooperative Extension Service, Institute of Food and Agricultural Sciences, University of Florida, Gainesville, FL. 1 page.

Howard, R.A. 1958. A history of the genus *Coccoloba* in cultivation. *Baileya* 6: 204–212.

Howard, R.A. 1996. *Polygonaceae:* Coccoloba *P. Browne*. Royal Botanic Gardens, Kew, UK. pp. 42.

Madriz, R, and Ramirez, N. 1996–1997. Reproductive biology of *Coccoloba uvifera* (Polygonaceae) a polygamous-dioecious species. *Revista de Biologia Tropical* 44–45(3–1): 105–115. (In Spanish)

Malathi, S., Masilamani, P., Balasubramanian, V., Rao, R.B., and Brindha, P. 1995. Constituents of *Coccoloba uvifera* leaves. *Fitoterapia* 66: 277.

Mayo, E.M. 2002. *Coccoloba uvifera* (L.) L. In *Tropical tree manual. Agriculture Handbook 721*. Edited by J.A. Vozzo. U.S.D.A. Department of Agriculture, Forest Service, Washington, DC. pp. 390–398.

Parotta, J.A. 1994. Cocoloba uvifera *(L.) L. Sea grape, Uva de playa*. International Institute of Tropical Forestry, SO-ITF-SM-74. U.S.D.A. Forest Service, New Orleans, LA. pp. 5.

Phillips, R.L., and Joyner, G. 1994. *The seagrape*. Fruit crops facts sheet HS-37. Horticultural Sciences Department, Florida Cooperative Extension Service, University of Florida, Gainesville, FL. pp. 2.

Shaw, P.E., Moshonas, M.G., and Baldwin, E.A. 1992. Volatile constituents of *Coccoloba uvifera*. *Phytochemistry* 31: 3495–3497.

Silveira, J.E.P.S., Pereda, M. del, C.V., Eberlin, S., Dieamant, G.C., and Di Stasi, L.C. 2008. Effects of *Coccoloba uvifera* L. on UV-stimulated melanocytes. *Photoderm. Photoimmun. Photomed.* 24: 308–313.


### SPECIALTY COOKBOOKS

(*Note:* Recipes in English that use sea grape are most often encountered in English-speaking regions where the plant occurs. Recipes for preparing sea grape wine may be available in books dedicated to making wines from vegetable products other than grapes.)


De Verteuil, M. 1979. *Maurice's tropical fruit cookbook*. Great Outdoors Publishing Company, Saint Petersburg, FL. pp. 79.

Mendelson, M., and Sawyer, M. 1979. *Gourmet Bahamian cooking*. Best-way Publishing, Iowa City, IA. pp. 264. (Has a recipe for sea grape jelly.)


Freitus (1980) has four sea grape recipes, and Freitus and Haberman (2005) have six. (See the Appendix to this book for details of the publications cited.)

# 90 Spicebush

Family: Lauraceae (laurel family)

## NAMES

Scientific name: *Lindera benzoin* (L.) Blume

- Spicebush is so named for its spicy aroma.
- Spicebush (spice bush) is also known as American spicebush, benjamin bush, benjamin spice bush, common spicebush, fever bush, feverwood, northern spicebush, spicewood, wild allspice, wild forsythia (reflective of yellow flowers in spicebush and forsythia), and wild spicebush.
- "Benzoin" is a resin obtained from the bark of Asian trees of the genus *Styrax*. Because of its vanilla-like aroma, it is used in perfumery and incense. Benzoin in the scientific name *L. benzoin* is an allusion to the aromatic leaves of spicebush. (*Benzoin* was once recognized as a genus, in which spicebush was placed; the word benzoin is derived from an Arabic term meaning aromatic gum.) "Benjamin" in the common name benjamin bush (often capitalized as Benjamin in the mistaken belief that it is based on a person) is a language transformation based on bezoin.
- "Spicebush" is also used as the name of several other species of *Lindera* and for the genus *Calycanthus* (two species of North America). The use of the same common name for *Lindera* and *Calycanthus* (of the Calycanthaceae or strawberry shrub family) has led to considerable confusion.
- The genus name *Lindera* is of uncertain origin, possibly coined in honor of Johann Linder (later Lindestolpe), 1676–1723, a Swedish physician and botanist.

## GEOGRAPHY AND ECOLOGY OF WILD PLANTS

Spicebush is native to eastern North America, from Ontario and Maine in the north, south to Texas and northern Florida, and west to Kansas. It has endangered status in some states. The species usually occurs in moist to wet woods and on the borders of wetlands and streams. It tolerates moderate shade, and is often found as an understory plant beneath trees.

## PLANT PORTRAIT

The genus *Lindera* has about 80 species of deciduous and evergreen shrubs and trees, but only three of these are native to the United States (just the one featured in this chapter reaches into Canada). In addition to the widespread *L. benzoin* discussed here, two other similar species of *Lindera* are restricted to areas of the southeastern United States, also as understory plants of moist woodlands. Spicebush is a deciduous shrub or small tree up to 5 m (16 feet) in height. Male and female flowers are on different plants. The species produces showy clusters of fragrant, yellow or greenish-yellow flowers, each about 3 mm (0.12 inch) wide. The flowers appear in the spring before the foliage develops, sometimes as early as January in the southern part of the range. The berries are ovoid, shiny, and red, about 1 cm (0.4 inch) in length, and contain a large seed. The fruits mature in August and September, of course only on the female plants. Spicebush is commonly cultivated as an ornamental, especially suited to damp soils. (Several foreign species of *Lindera* are also often grown as ornamentals.) Spicebush is frequently established to attract wildlife, especially birds and butterflies (it is the larval host for the eastern tiger

**FIGURE 90.1** Spicebush (*Lindera benzoin*). (a) Branch with fruits. (Courtesy of Steven J. Baskauf [http://bioimages.vanderbilt.edu].) (b) Painting of flowering (above) and fruiting (below) branches. (From Eaton, M.E., *Addisonia*, Vol. 5, Plate 168, New York Botanical Garden, Brooklyn, NY, 1920.) (c) Seeds. (Courtesy of L. Pfiffner, U.S. Department of Agriculture.) (d) Distribution map.

swallowtail and the spicebush swallowtail), but has the advantage of being unpalatable to grazing deer. In order for the attractive berries to be produced, at least one male plant is necessary for every few female plants. The leaves, bark, and fruit have been used by Native Americans and in herbal medicine for treatment of fever, dysentery, coughs, colds, and other illnesses.

## WARNING

Like its relative sassafras (see Chapter 85), spicebush contains safrole, a flavorant chemical that has been banned as a food additive in the United States and Canada because of its toxicity. The presence of safrole in sassafras led to a great decrease in its culinary usage, except when the safrole has been removed. The degree of concern that should be exercised in this regard with respect to consumption of spicebush is unclear, and it has been suggested that the problem is not as serious as it is for sassafras.

## CULINARY PORTRAIT

The dried, powdered fruit was employed in the past as a wild-collected spice in areas where spicebush grows naturally. The lemony-pepper-tasting berries were often used as a substitute for allspice (the dried, unripe fruit of *Pimenta dioica* (L.) Merrill). A tea was also made from the leaves, especially in Appalachia. All of these usages are obsolete, and spicebush has never attained commercial importance. Spicebush is one of the more recommended plants discussed in wild food collection and preparation guides. The berries are a good flavoring for chutney, marinades, jams, and sauces. The dried berries keep well, although the leaves do not produce as tasty material when dried. According to Marrone (2009, cited in the Appendix to this book), the fruits tend to develop a musty flavor when stored at room temperature for more than a few weeks, and are best stored in the refrigerator or freezer.

As a tea, spicebush leaves, twigs, and bark are simply boiled and the material filtered off (for example, through a coffee filter). As a spice, the leaves, twigs, and bark can be used for seasoning; or the dried, de-seeded fruits can be ground up using a coffee grinder or mortar and pestle. The ground spice loses its flavor quickly and grinding should be done just before use.

### Culinary Vocabulary

- Many writers refer to "spicebush tea," but there are several technical distinctions that have been made with regard to preparations made by soaking plant materials in water. "Tea" has acquired the general meaning of a hot beverage prepared by soaking an herb in warm or hot water. However, many purists reserve the word tea for beverages made with common tea (*Camellia sinensis* (L.) Kuntze). Often the word "tisane" is used to designate an "herbal tea," indicating one that is produced from the leaves or other parts of plants other than the tea plant. Another distinction concerns how long the plant material is exposed to water that is maintained at boiling temperatures. There are two kinds of preparations: "infusions" (made by pouring hot water over herbal material) and "decoctions" (prepared by boiling herbal material in water). In the case of spicebush, both infusions and decoctions are made, depending on recipe.

## PROSPECTS

Excellent material for a spicy condiment and tea base is produced by spicebush, which nevertheless is of no market significance, and indeed no serious attempts have been made to develop it as a crop. The problem spicebush faces is simply that both the spice and tea markets are dominated by tropical crops, which have become established, indeed entrenched in commerce. In general, tropical crops can be produced very cheaply by comparison with temperate zone crops. Moreover, spices and tea are very compact commodities (i.e., a small amount of material has a high value), and so shipping to distant markets is quite economical. Nevertheless, spicebush is a genuinely attractive culinary product, and it remains a very interesting native North American plant with potential for development.

## CURIOSITIES OF SCIENCE AND TECHNOLOGY

- Spicebush was one of many plants used in the past to destroy intestinal parasites in humans and livestock. The difficulty with most of these preparations is that the effective dosage had severe side effects, indeed sometimes was close to the lethal dosage, so that usage was very hazardous.

- Spicebush generally grows on rich, damp soil. In the past, surveyors, settlers, and farmers used the presence of spicebush as a clue to how good land was for agricultural purposes.
- During the American Civil War, when common (Chinese tea) and other beverage herbs were in very short supply, spicebush was one of several plants used to prepare tea in the blockaded South.

## KEY INFORMATION SOURCES

Anderson, J.E., Ma, W., Smith, D.L., Chang, C.J., and McLaughlin, J.L. 1992. Biologically active gamma-lactones and methylketoalkenes from *Lindera benzoin*. *J. Nat. Prod.* 55: 71–83.
Champlin, R.L. 1977. Red form of *Lindera benzoin*. *Rhodora* 79: 166.
Cipollini, M.L., Wallace-Senft, D.A., and Whigham D.F. 1994. A model of patch dynamics, seed dispersal, and sex ratio in the dioecious shrub *Lindera benzoin* (Lauraceae). *J. Ecol.* 82: 621–633.
Cipollini, M.L., and Whigham, D.F. 1994. Sexual dimorphism and cost of reproduction in the dioecious shrub *Lindera benzoin* (Lauraceae). *Am. J. Bot.* 81: 65–75.
Edwards, M.J., and Niesenbaum, R.A. 2007. Eleven polymorphic microsatellite loci in *Lindera benzoin*, Lauraceae. *Molec. Ecol. Notes* 7: 1302–1304.
Godt, M.J.W., and Hamrick, J.L. 1996. Allozyme diversity in the endangered shrub *Lindera melissifolia* (Lauraceae) and its widespread congener *Lindera benzoin*. *Can. J. For. Res.* 26: 2080–2087.
Hoss, G. 2006. Propagation protocol for spicebush, *Lindera benzoin*. *Native Plants J.* 7: 134–136.
Ingersoll, C.M., Niesenbaum, R.A., Weigle, C.E., and Lehman, J.H. 2010. Total phenolics and individual phenolic acids vary with light environment in *Lindera benzoin*. *Botany* 88: 1007–1010.
Luken, J.O., Kuddes, L.M., Tholemeier, T.C., and Haller, D.M. 1997. Comparative responses of *Lonicera maackii* (Amur honeysuckle) and *Lindera benzoin* (spicebush) to increased light. *Am. Midl. Nat.* 138: 331–343.
Medley, K.E. 1996. Dieback in the native shrub, *Lindera benzoin*: A subtle effect of forest fragmentation. *Ohio J. Sci.* 96: 76–80.
Mooney, E., Edwards, M., and Niesenbaum, R. 2010. Genetic differentiation between sun and shade habitats in populations of *Lindera benzoin*. *Pop. Ecol.* 52: 417–425.
Mooney, E.H., Tiedeken, E.J., Muth, N.Z., and Niesenbaum, R.A. 2009. Differential induced response to generalist and specialist herbivores by *Lindera benzoin* (Lauraceae) in sun and shade. *Oikos* 118: 1181–1189.
Moore, J.A., and Wilson, M.F. 1982. The effect of microhabitat, spatial distribution, and display size on dispersal of *Lindera benzoin* by avian frugivores. *Can. J. Bot.* 60: 557–560.
Niesenbaum, R.A. 1992a. Sex ratio, components of reproduction, and pollen deposition in *Lindera benzoin* (Lauraceae). *Am. J. Bot.* 79: 495–500.
Niesenbaum, R.A. 1992b. The effect of light environment on herbivory and growth in the dioecious shrub *Lindera benzoin* (Lauraceae). *Am. Midl. Nat.* 128: 270–275.
Niesenbaum, R.A. 1993. Light or pollen—Seasonal limitations on female reproductive success in the understory shrub *Lindera benzoin*. *J. Ecol.* 81: 315–323.
Niesenbaum, R.A. 1996. Linking herbivory and pollination: Defoliation and selective fruit abortion in *Lindera benzoin*. *Ecology* 77: 2324–2331.
Primack, R.B. 1985. Sex ratio and sexual constancy in spicebush (*Lindera benzoin*). *Rhodora* 87: 305–308.
Skoczylas, D.R., Muth, N.Z., and Niesenbaum, R.A. 2007. Contribution of insectivorous avifauna to top down control of *Lindera benzoin* herbivores at forest edge and interior habitats. *Acta Oecol.* 32: 337–342.
Tucker, A.O., Maciarello, M.J., Burbage, P.W., and Sturtz, G. 1994. Spicebush. *Lindera benzoin* (L.) Blume var. *benzoin*, Lauraceae: A tea, spice, and medicine. *Econ. Bot.* 48: 333–336.
Veres, J.S., and Pickett, S.T.A. 1982. Branching pattern of *Lindera benzoin* beneath gaps and closed canopies. *New Phytol.* 91: 767–772.

### SPECIALTY COOKBOOKS

Freitus (1980) has two spicebush recipes, Hall and Hall (1980) have one, Marrone (2009) has three, and Medve and Medve (1990) have two. (See the Appendix to this book for details of the publications cited.)

# 91 Squash (*Cucurbita pepo* squash)

Family: Cucurbitaceae (gourd family)

## NAMES

Scientific name: *Cucurbita pepo* L.

- *Cucurbita pepo* includes vegetables called summer squash and pumpkin, less frequently marrow and winter squash.
- "Squash" is derived from the Massachuset Indian (Narraganset and Natick) word *askutasquash*, meaning "eaten raw or uncooked," probably indicating that some forms of squash were eaten uncooked by Native Americans.
- The terms "squash" and "pumpkins" have been applied so inconsistently to different varieties of several species of *Cucurbita* that they cannot be reliably distinguished. This is illustrated by the claim, often made, that "canned pumpkin" is really "canned squash." In reality, both are the same thing. Confusingly, some *Cucurbita* species contain some varieties that are called pumpkins, and other varieties that are called squashes. In Britain, "pumpkin" is often interpreted in a broader sense than in North America, including many fruits that North Americans would call "squash." Sometimes edible *Cucurbita* fruits that are round (i.e., spherical) or nearly so and consumed when fully ripe are distinguished as "pumpkins," whereas edible *Cucurbita* fruits that are not round and consumed for the most part when immature are distinguished as "squash."
- For information on the genus name *Cucurbita*, see Chapter 17 on Buffalo Gourd.
- *Pepo* in the scientific name *C. pepo* is Latin for a kind of melon; the word is related to the Greek *pepon*, ripe, mellow.

## GEOGRAPHY AND ECOLOGY OF WILD PLANTS

As detailed below, *C. pepo* is a complex species made up of (1) wild populations, some of which are considered to be ancestral to modern cultivated forms; (2) cultivated forms (both "land races", i.e., primitive, usually variable kinds selected before the era of modern plant breeding, and "cultivars", i.e., named, usually uniform kinds selected by modern plant breeders); and (3) feral forms (weedy kinds that are self-perpetuating in nature and usually have arisen by escape from cultivated forms). The wild forms, as noted below, belong to *C. pepo* var. *pepo* (in Mexico), var. *texana* (in Texas and some adjacent states), and var. *ozarkana* (in south central United States). Some of these populations have proven to be serious weeds of crops, and herbicides and mycoherbicides (fungi used to kill weeds) have been employed to control them. Feral escapes of *C. pepo* occur in much of the southern United States and eastern North America.

Plants of *C. pepo* growing without deliberate human cultivation may be (1) wild forms that have probably persisted in North America since times when humans had not yet arrived in North America (these are placed in *C. pepo* vars. *texana*, *ozarkana*, and *fraterna*); (2) escapes from cultivated forms (mostly from *C. pepo* var. *pepo*); and (3) hybrids or at least forms that have been genetically changed by past hybridization, involving any two of the four recognized varieties of *C. pepo*. In the United States, typical habitats include cultivated and abandoned fields, roadsides, riverbanks, and moist thickets. Populations may be very extensive in disturbed habitats, but in places not influenced by human activities there are typically only one or just a few plants present.

**FIGURE 91.1** Squash (*Cucurbita pepo*). (a) Ornamental gourds (*C. pepo* var. *ovifera*). (Courtesy of Prima Civitas Foundation [Flickr/CC-attribution].) (b) Flower and fruit. (Courtesy of Kirybabe [Flickr/CC-attribution].) (c) Ornamental patty pan squash (*C. pepo* var. *ovifera*); photo by Vasenka (Flickr/CC-attribution). (d) Fruits of Texas gourd (*C. pepo* var. *texana*). (Courtesy of John D. Byrd, Mississippi State University, online at Bugwood.org [CC By 3.0].) (e) Distributions of the native American varieties, Ozark melon and Texas gourd.

## PLANT PORTRAIT

The genus *Cucurbita* is native mostly to southern Mexico and northern Central America. The plants are yellow-flowered, herbaceous, annual or perennial, trailing, frost-sensitive vines or bushes, which climb by tendrils ("summer squashes" are usually of bush form and lack tendrils; see "Culinary Vocabulary" below for more information). The plants have large, palmate leaves. The large orange-yellow, nectar-producing, unisexual flowers are pollinated by bees. The fruits are remarkably varied in shape, size, and color and include some of the largest fruits in the plant kingdom. Squashes and pumpkins were unknown in the Old World prior to Columbus' voyage of 1492. The several species now cultivated may have been the earliest domesticated plants of the New World, perhaps dating

**FIGURE 91.2** Squash cultivars of Native American origin (*Cucurbita pepo* var. *ovifera*). (a) Ornamental form of acorn squash. (Courtesy of Shaferlens/Dave Shafer [Flickr/CC-attribution].) (b) Long section of fruit of acorn squash. (Courtesy of Watashiwani [Flickr/CC-attribution].) (c) Ornamental form of scallop squash. (Courtesy of ilovebutter [Flickr/CC-attribution].) (d) Patty pan squash. (Courtesy of thebittenword.com [Flickr/CC-attribution].) (e) Crookneck squash. (Courtesy of E. Small and B. Brookes.) (f) Sliced crookneck squash. (Courtesy of cogdogblog/Alan Levine [Flickr/CC-attribution].)

back 10,000 years, and certainly well known by 5000 BC. Most varieties of *Cucurbita* are extraordinarily productive plants, and home gardeners often find themselves with an embarrassingly large harvest, especially of zucchini ("zucchini fairies" are gardeners who dispose off their excess zucchinis by surreptitiously leaving them at neighbors' doorsteps). Summer squash develop with startling rapidity and may be harvested in as little as 2 days after the flowers are fertilized. Most gardeners make the error of allowing summer squash to mature excessively. Winter squash can be stored for up to 2 months without refrigeration. Some forms of *Cucurbita* are grown for their production of ornamental, nonedible gourds. In addition to *C. pepo* highlighted in this chapter, the most important cultivated species and their names are

- *C. maxima* Duchesne—Winter squash (less commonly marrow, pumpkin, and turban squash)
- *C. moschata* Duchesne—Winter squash (less commonly pumpkin)
- *C. argyrosperma* C. Huber (*C/mixta* Pangalo)—Winter squash, cushaw, cymlin (cymling, cymbling), and pumpkin
- *C. ficifolia* Bouché—Malabar gourd (also Angora gourd, black-seed squash, black-seeded squash, chilacayote, fig-leaf pumpkin, and Siamese gourd)

*Cucurbita pepo,* highlighted in this chapter, is the most variable of the species of *Cucurbita,* and includes a very widely grown group of varieties, the fruits differing in size, shape, skin color, flesh color, flesh texture, and other characteristics. Illustrative of the range of variation are small inedible ornamental gourds, and huge jack-o'-lantern pumpkins. The best known kind is the zucchini squash.

Much more so than any other species included in this book, the geographical origin of *C. pepo* has been disputed. The following summary is intended to simplify the complex situation.

A recent classification of *C. pepo* recognizes three subspecies, one of which is subdivided into three taxonomic varieties (technically this is an "artificial classification" since it places some ancestors and their derivatives in different groups):

1. Subspecies *pepo*—cultivated forms only: all "pumpkins," "marrows" (including zucchini), and some ornamental gourds (apparently domesticated in Mexico from unidentified wild plants)
2. Subspecies *texana* (Scheele) Filov (subsp. *ovifera* (L.) D.S. Decker)
   a. var. *ovifera*—domesticated forms (including cultivars, cultigens, and land races)
   b. var. *ozarkana* D.S. Decker (Ozark melon)—wild populations in the Greater Mississippi Valley and the Ozark Plateau (Arkansas, Illinois, Louisiana, Missouri, Oklahoma)
   c. var. *texana* (Scheele) Filov (Texas gourd; also field pumpkin, Texan wild pumpkin, Texan wild marrow, and wild marrow)—wild populations mostly in Texas, but also in Louisiana and New Mexico
3. Subpecies *fraterna* (L.H. Bailey) Lira, Andres & Nee—wild populations in north-eastern Mexico

The following conclusions regarding the origin of squashes (and pumpkins) provide the context for inclusion of squash in this book dedicated exclusively to crop plants that are native to North America north of Mexico.

1. In the past, it has not been clear that any *Cucurbita* crop arose in the United States, but recent DNA studies have strongly supported the conclusion that some cultivated forms are U.S. native crops.
2. All cultivated squashes, pumpkins, and ornamental gourds classified in species other than *C. pepo* arose by domestication south of the United States–Mexican border from wild populations that were also south of the border.
3. Most cultivated squashes, pumpkins, and ornamental gourds derived from *C. pepo* also arose by domestication south of the United States–Mexican border from wild populations that were also south of the border. *Cucurbita. pepo* var. *fraterna*, made up of wild populations south of the border, may have been one of the wild sources that provided plants for domestication. It is thought to have been the basis of some domesticated ornamental gourds (placed in subsp. *pepo*).

4. A few cultivars of *C. pepo* var. *ozarkana* currently recognized as "squashes" appear to have been derived in the eastern United States from wild populations of this group. This conclusion that some squashes of *C. pepo* arose in the United States from wild squashes in the United States is the basis upon which squash is included in this book as a North American (i.e., U.S.) native crop.
5. *Cucurbita. pepo* var. *texana*, a variety native to the eastern United States, once proposed as an ancestor of some cultivars, does not appear to have given rise to any cultivars. However, additional research is required to clarify this.
6. Feral (i.e., weedy) plants growing in North America may belong to varieties *ozarkana* or *texana*, but may also be escapes from cultivation of cultivars of foreign origin, and therefore could belong to *C. pepo* var. *pepo*, or indeed to any of several other species of *Cucurbita*. Feral forms of *C. pepo* have been collected from the southern half and the eastern half of the United States and Canada. As was noted in the introduction (Chapter 1), this book excludes crop species that are cultivated in the United States and/or Canada but are entirely of foreign origin. Squash is a species that is partly of American origin.
7. All forms of *C. pepo* can interbreed freely, and therefore past exchange of genes may have complicated the genetic heritage of both cultivated and wild forms. (The recent interbreeding of genetically modified cultivars of *C. pepo* and feral and wild forms of the species has been a contentious subject, since some conservationists object to the release of engineered genes into wild plants.)

The following cultivated edible forms have been interpreted as belonging to subsp. *ovifera* var. *ovifera* (and therefore are of American origin): acorn squash, crookneck squash, fordhook squash, patty pan squash, scallop squash, straightneck squash, summer crookneck squash, and table queen squash. Most ornamental *C. pepo* gourds have also been interpreted as belonging to var. *ovifera*.

## CULINARY PORTRAIT

The flesh of pumpkins and squashes is consumed immature as a fresh vegetable, or used in cooking, and the mature flesh is used in cooking and in canning, especially for pie-stock. Pumpkin pie was served at the Pilgrims' second Thanksgiving in 1623 and has remained a traditional Thanksgiving dessert in the United States ever since. In North America, the top Thanksgiving Day pie preference is still pumpkin, followed by apple. Traditional pumpkin dishes in New England include pumpkin pudding, pancakes, bread, butter, dried chips, and even beer (pumpkin beer was popular among the Pilgrims). Mature pumpkins and winter squashes are roasted (especially with butter and brown sugar), stewed, steamed, boiled as vegetables, added to soups, and candied.

Pumpkin seed is the source of a high-quality vegetable oil, mostly produced in Europe. Mutant varieties of pumpkin with hull-less (hulless) seeds arose in Europe centuries ago. The most popular is the Steiermark mutation, which has been recognized as *C. pepo* var. *styriaca* Greb., named for the Styria region of Austria, where it is the basis of the popular Steiermark pumpkin seed oil. This oil is used primarily for salads and is not suitable as cooking oil. When toasted (the usual form in which it is available), Steiermark pumpkin seed oil has a strong flavor (which takes getting used to), which is useful for increasing the pumpkin taste of pumpkin dishes. The Steiermark mutant arose as a trailing plant (like most pumpkin plants), but is also available as a bush form (*C. pepo* var. *oleifera* Pietsch). This latter name is used more broadly to include var. *styriaca*, that is all hull-less central European oilseed cultivars, and also sometimes to include edible-seed varieties developed in the Americas. Some soft-seeded types of pumpkin or squash seeds are sold as a confection. Roasted, salted seeds are a popular snack in Latin America. The seeds have a nutty flavor and are consumed worldwide raw, boiled, or roasted, usually with the coat removed. 'Lady Godiva', a variety of *C. pepo* with seeds lacking a seed coat, is favored for snacks in the United States.

Aboriginal Americans dating back to Aztec times have eaten the relatively large male flowers of squash or pumpkin species. The big blossoms can be used to add flavor and color to stews, soups, and salads, and may be stuffed, fried, or battered. In Mexico and Asia, young leaves and shoots are used as potherbs.

Summer squashes are picked while very young, typically 2 to 7 days after flowering. Their skin and seeds are tender and edible. These squashes are quite perishable, and do not store well. The best known is the zucchini or courgette, which is at its most flavorful when measuring between 15 and 20 cm (6–8 inches) in length, the flavor and quality declining as the fruit becomes larger. Zucchini flowers are often stuffed or deep fried in batter.

Winter squashes are harvested when completely ripe. Like melons, they have a hollow central cavity containing hard, fully developed seeds. The seeds are edible, but the thick, hard shells of winter squashes are not. The rind of some varieties is so hard that a mallet is required to strike a knife hard enough to penetrate the gourd. Depending on variety, winter squashes can be stored for 1 to 6 months.

Pumpkin and squash can be boiled, steamed, baked, microwaved, and cooked in a pressure cooker. Boiling tends to make squash watery and less tasty. Steaming is highly recommended.

### Culinary Vocabulary

- Generally "summer squashes" are the fruits of certain varieties of *C. pepo*, used in the immature state, while "winter squashes" are the fruits of certain varieties of several species of *Cucurbita*, including *C. mixta*, *C. moschata*, and *C. maxima*.
- "Marrow," in Great Britain, denotes fruits of varieties of *C. pepo* and *C. maxima* used in the mature state. The word marrow arose because it appeared to some that the soft flesh of these squashes was like relatively soft bone marrow. In Great Britain, "marrow squash" is often used to mean summer squash in general.
- "Squash" has an unrelated meaning in England: a drink prepared by squeezing the juice from the pulp of citrus fruit. Usually soda water is added.
- "Gourd" refers to fruits of the gourd family, irrespective of whether they are edible or merely grown for ornament. Gourds of the calabash or gourd tree, *Crescentia cujete* L., belong to another family (the Bignoniaceae).
- *Pepitas* (pronounced peh-PEE-tahs or puh-PEE-tahs) are hulled, roasted pumpkin seeds, consumed in Mexican cuisine as snacks. When the seeds are ground to powder and used as a thickener, the preparation is known as *pepitoria* (pronounced peh-PEE-toh-ree-ah).

## PROSPECTS

Squash is the most important vegetable native to North America north of Mexico, and one of the major vegetables of the world. The cultivated forms that trace their ancestry directly to wild plants indigenous to the United States constitute only a fraction of squash cultivars, most of which belong to species other than *C. pepo* and to cultivars of *C. pepo* that are derived from wild plants native to Mexico. Nevertheless, the cultivated plants derived from U.S. wild plants are very significant in the squash market, and very valuable for continued breeding of new forms of squash and pumpkin that can be expected to be important in the future.

## CURIOSITIES OF SCIENCE AND TECHNOLOGY

- The advanced pre-Columbian civilizations of the Americas were dependent on a combination of corn, beans, and squash, a mixture which produces excellent nutrition. Beans are rich sources of protein, particularly the essential amino acid lysine, but are usually deficient in the sulfur-containing amino acids methionine and cystine. By contrast, cereal grains like corn contain lower amount of proteins, and these are deficient in lysine but adequate in sulfur-containing amino acids. Together, beans and corn provide a balance of essential protein components, which is especially necessary for survival in diets deficient in meat. The combination of corn, beans, and squash was referred to by several Indian groups as "the three sisters." The three crops benefit by being grown together. The corn provides a framework for the beans to climb, eliminating the need for poles. Nitrogen-fixing bacteria in nodules on the roots of the bean plants provide nitrogen that fertilizes the soil. The squash spreads on the ground, shading out weeds and creating a microclimate to retain moisture in the soil. The prickly hairs of the squash vines also deter pests.

- In ancient times in North America, Native American women, as the gatherers and preparers of plant foods, were very likely responsible for selecting superior forms of plants for planting. In the case of *C. pepo*, it is probable that selecting large fruits was valuable not just to improve the quantity of food harvested, but also because the gourds could be used as lightweight containers, of great value to people who were semi-nomadic.
- Squash seeds were used as a cure for tapeworms and roundworms by Native American medicine men, and European settlers adopted the practice.
- A widespread fallacy is that squashes and pumpkins will cross-pollinate with other genera of the gourd family, including watermelon and muskmelon, causing the fruit of one to take on the taste attributes of the other.
- The flowers of pumpkin and squash species close about dusk and open about sunrise. Squash and gourd bees, the natural pollinators, commonly enter the closing flowers at dusk to pass the night protected, and leave at dawn when the flowers open.
- Botanically speaking, pumpkins and squashes are a type of fruit technically classified as a "berry," but are sold as "vegetables."
- The racket ball game of squash, which has nothing to do with the vegetable squash, was invented in Harrow school in Britain around 1830, at which time the similar game of rackets ball was being played. The pupils discovered that a punctured rackets ball, which "squashed" on impact with the wall, produced a game with a greater variety of shots and a more interesting game. In 1864, the first squash courts were constructed at the school and squash was officially founded as a sport in its own right.
- In 1988, Oklahoma declared squash to be one of a dozen or so foods constituting its official "state meal."

## KEY INFORMATION SOURCES

Bailey, L.H. 1943. The domesticated cucurbitas. *Gentes Herb.* 2: 62–115.

Bird, R. 2003. *Growing squashes and pumpkins: A directory of varieties and how to cultivate them successfully.* Lorenz, London, U.K. pp. 64.

Boyette, G., Templeton, E., and Oliver, L.R. 1984. Texas gourd (*Cucurbita texana*) control. *Weed Sci.* 32: 649–655.

Cutler, H.C., and Whitaker, T.W. 1961. History and distribution of the cultivated cucurbits in the Americas. *Am. Antiquity* 26: 469–485.

Decker, D.S. 1988. Origin(s), evolution, and systematics of *Cucurbita pepo* (Cucurbitaceae). *Econ. Bot.* 42: 4–15.

Decker D.S., and Wilson H.D. 1986. Numerical analysis of seed morphology in *Cucurbita pepo*. *Syst. Bot.* 11: 595–607.

Decker, D.S., and Wilson, H.D. 1987. Allozyme variation in the *Cucurbita pepo* complex: *Cucurbita pepo* var. *ovifera* vs. *Cucurbita texana*. *Syst. Bot.* 12: 263–273.

Decker-Walters, D.S. 1990. Evidence for multiple domestications of *Cucurbita pepo*. In *Biology and utilization of the Cucurbitaceae*. Edited by D.M. Bates, R.W. Robinson, and C. Jeffrey. Cornell University Press, Ithaca, NY. pp. 96–101.

Decker-Walters, D.S., and Walters, T.W. 2000. Squash. In *The Cambridge world history of food*. Edited by K.F. Kiple and K.C. Ornelas. Cambridge University Press, Cambridge, U.K. pp. 335–351.

Decker-Walters, D.S., Staub, J.E., Chung, S.-M., Nakata, E., and Quemada, H.D. 2002. Diversity in free-living populations of *Cucurbita pepo* (Cucurbitaceae) as assessed by random amplified polymorphic DNA. *Syst. Bot.* 27: 19–28.

Decker-Walters, D.S., Walters, T.W., Posluszny, U., and Kevan, P.G. 1990. Genealogy and gene flow among annual domesticated species of *Cucurbita*. *Can. J. Bot.* 68: 782–789.

Erwin, A.T. 1938. An interesting Texas cucurbit *Cucurbita pepo* L. var. *ovifera* Alef. (*C. texana* Gray). *Archive Iowa State College J. Sci.* 12: 253–261.

Fuchs, M., Chirco, E.M., and Gonsalves, D. 2004. Movement of coat protein genes from a commercial virus-resistant transgenic squash into a wild relative. *Environ. Biosafety Res.* 3: 5–16.

Fuchs, M., Chirco, E.M., Mcferson, J.R., and Gonsalves, D. 2004. Comparative fitness of a wild squash species and three generations of hybrids between wild × virus-resistant transgenic squash. *Environ. Biosafety Res.* 3: 17–28.

Goldman, A. 2004. *The compleat squash: A passionate grower's guide to pumpkins, squash, and gourds.* Artisan, New York, NY. pp. 224.

Gong, L., Paris, H.S., Nee, M.H., Stift, G., Pachner, M., Volmann, J., and Lelley, T. 2012. Genetic relationships and evolution in *Cucurbita pepo* (pumpkin, squash, gourd) as revealed by simple sequence repeat polymorphisms. *Theor. Appl. Genet.* 124: 875–891.

Heiser, C.B., Jr. 1989. Domestication of Cucurbitaceae: *Cucurbita* and *Lagenaria*. In *Foraging and farming: the evolution of plant exploitation.* Edited by D.R. Harris and G.C. Hillman. Unwin Hyman Limited, London, UK. pp. 471–480.

Kirkpatrick, K.J., and Wilson, H.D. 1988. Interspecific gene flow in *Cucurbita*: *Cucurbita texana* vs. *Cucurbita pepo*. *Am. J. Bot.* 75: 519–527.

Kirkpatrick, K.J., Decker, D.S., and Wilson, H.D. 1985. Allozyme differentiation in the *Cucurbita pepo* complex: *Cucurbita pepo* var. *medullosa* vs. *Cucurbita texana*. *Econ. Bot.* 39: 289–299.

Loy, J.B. 1990. Hull-less seeded pumpkins: a new edible snackseed crop. In *Advances in new crops.* Edited by J. Janick and J.E. Simon. Timber Press, Portland, OR. pp. 403–407.

Merrick, L.C. 1995. Squashes, pumpkins and gourds. In *Evolution of crop plants.* 2nd edition. Edited by J. Smartt and N.W. Simmonds. Longman Scientific & Technical, Burnt Mill, Harlow, Essex, UK. pp. 97–105.

Nee, M. 1990. The domestication of *Cucurbita* (Cucurbitaceae). *Econ. Bot.* 44(3 Supp.): 56–68.

Paris, H.S. 1989. Historical records, origins, and development of the edible cultivar groups of *Cucurbita pepo* (Cucurbitaceae). *Econ. Bot.* 43: 423–443.

Paris H.S. 2001. History of the cultivar-groups of *Cucurbita pepo*. *Hort. Rev.* 25: 71–170.

Paris, H.S., Lebeda, A., Křistkova, E., Andres, T.C., and Nee, M.H. 2012. Parallel evolution under domestication and phenotypic differentiation of the cultivated subspecies of *Cucurbita pepo* (Cucurbitaceae). *Econ. Bot.* 66: 71–90.

Paris, H.S., Yonash, N., Portnoy, V., Mozes-Daube, N., Tzuri, G., and Katzir, N. 2003. Assessment of genetic relationships in *Cucurbita pepo* (Cucurbitaceae) using DNA markers. *Theor. Appl. Genet.* 106: 971–978.

Sanjur, O.L., Piperno, D.R., Andres, T.C., and Wessel-Beaver, L. 2002. Phylogenetic relationships among domesticated and wild species of *Cucurbita* (Cucurbitaceae) inferred from a mitochondrial gene: Implications for crop plant evolution and areas of origin. *Proc. Natl. Acad. Sci. USA* 99: 535–540.

Smith, B.D. 2006. Eastern North America as an independent center of plant domestication. *Proc. Natl. Acad. Sci. USA*. 103: 12223–12228.

Spencer, L.J., and Snow, A.A. 2001. Fecundity of transgenic wild-crop hybrids of *Cucurbita pepo* (Cucurbitacaeae): Implications for crop-to-wild gene flow. *Heredity* 86: 694–702.

U.S. Agricultural Research Service. 1969. *Growing pumpkins and squashes.* Revised edition. U.S. Agricultural Research Service, Washington, DC. pp. 21.

Whitaker, T.W. 1971. Squash and gourd bees (*Peponapis xenoglossa*) and the origin of the cultivated *Cucurbita*. *Evolution* 25: 218–234.

Whitaker, T.W., and Bemis, W.P. 1975. Origin and evolution of the cultivated *Cucurbita*. *Bull. Torrey Bot. Club* 102: 362–368.

Whitaker, T.W., and Bohn, G.W. 1950. The taxonomy, genetics, production and uses of the cultivated species of *Cucurbita*. *Econ. Bot.* 4: 52–81.

Whitaker, T.W., and Robinson, R.W. 1986. Squash breeding. In *Breeding vegetable crops.* Edited by M.J. Bassett. AVI Publishing Company, Westport, CT. pp. 209–242.

Wien, H.C. 1997. The cucurbits: Cucumber, melon, squash and pumpkin. In *The physiology of vegetable crops.* Edited by H.C. Wien. CABI International, Wallingford, Oxon, U.K. pp. 345–386.

Wilson, H.D. 1990. Gene flow in squash species. *Bioscience* 40: 449–455.

Wilson, H.D., Doebley, J., and Duvall, M. 1992. Chloroplast DNA diversity among wild and cultivated members of *Cucurbita* (Cucurbitaceae). *Theor. Appl. Genet.* 84: 859–865.

## Specialty Cookbooks

(*Note:* there are dozens of cookbooks dedicated to squash and/or pumpkin. The following are selected examples.)

Bard, R., and Kellogg, C. 1985. *Zucchini and all that squash.* W.H. Publishing, Tacoma, WA. pp. 103.
Boisset, C. 1997. *Pumpkins & squashes.* The Reader's Digest Association, New York, NY. pp. 120.
Brabb, E., Krupski, H., and Paddock, B.T. 1996. *Pumpkin companion.* Brick Tower Press, New York, NY. pp. 94.
Damerow, G. 1997. *The perfect pumpkin.* Storey Publishing, Pownal, VT. pp. 219.
Dandar, H.O., and Dandar, E.B. 1977. *The complete pumpkin cook book.* Sterling Specialities Cook Books, Penndel, PA. pp. 108.
Gabbert, B.B. 1995. *Pumpkin lovers cook book.* Golden West Publishers, Phoenix, AZ. pp. 126.
Gilberg, R.L. 1982. *Pumpkin cookbook.* Gilmar Press, Newcastle, CA. pp. 120.
Hill, N. 1996. *The pumpkin cookbook.* Hamlyn, London, UK. pp. 64.
Kowbel, A.K. 1987. *Fabulous cooking with zucchini, pumpkin and squash.* Fabulous Cooking, Saskatoon, SK. pp. 112.
Libby, McNeill & Libby. (Corporate author). 1984. *The great pumpkin cookbook: A harvest of Libby's favorite recipes.* Rutledge Books, New York, NY. pp. 128.
MacCallum, A.C. 1986. *Pumpkin, pumpkin!: Lore, history, outlandish facts, and good eating.* Heather Foundation, San Pedro, CA. pp. 203.
McMonigal, V. 1994. *The perfect pumpkin and other squash.* Bay Books, Pymble, N.S.W., Australia. pp. 96.
Moon, R. 1998. *The American harvest cookbook: Cooking with squash, zucchini, pumpkins, and more.* Chartwell Books, Edison, NJ. pp. 128.
Nizzoli, A. 1998. *The squash: History, folklore, ancient recipes, celebrated in recipes from ancient times to the present.* Könemann, Köln, Germany. pp. 141.
Petersen-Schepelern, E. 2000. *Pumpkin, butternut & squash: 30 sweet and savory recipes.* Ryland Peters & Small, London, UK. pp. 62.
Ralston, N.C., Jordan, M., and Chesman, A. 2002. *The classic zucchini cookbook: 225 recipes for all kinds of squash.* Storey Books, North Adams, MA. pp. 311.
Tarr, Y.Y. 1978. *The squash cookbook.* Random House, New York, NY. pp. 223.

Niethammer (1999; see the Appendix to this book) presents five recipes for squash that were employed by Native Americans in the southwestern United States.

# 92 Strawberries
## *North American Species*

Family: Rosaceae (rose family)

### NAMES

Scientific names: *Fragaria* species

- Virginia strawberry—*F. virginiana* Mill. (commonly "*F. virginiana* Duchesne")
- Woodland strawberry—*F. vesca* L.
- The word "strawberry," derived from the Anglo-Saxon *streoberie*, is peculiar to the English language and does not seem to have originated from other languages. It was used as early as 1000 AD in England. The name has a variety of possible origins. Straw was commonly used to mulch the plants during the winter, and as weed and soil control to keep the berries cleaner. In London, England, children used to collect the berries, string them on pieces of straw, and sell them at the markets as "straws of berries." The runners which the plants produce are said to be strewn or dispersed around the plant, and indeed in some literature, the fruit is called strewberry, so the "straw" in strawberry may be a corruption of "strew."
- The name Virginia strawberry reflects presence in the old territory of "Virginia," which occupied a substantial portion of the eastern United States (see Chapter 5 on American Persimmon for additional information).
- The Virginia strawberry is also known as scarlet strawberry and wild strawberry.
- The name woodland strawberry is based on the frequent presence in open woodlands and on the borders of wooded areas.
- The woodland strawberry is also known as alpine strawberry, European strawberry, wild strawberry, wood strawberry, and woods strawberry.
- The name "wild strawberry" for the two species noted in this chapter is somewhat problematical, since cultivated varieties have been bred, and these domesticated forms are not "wild" (for a fuller discussion of such contradictory meanings of "wild," see Chapter 98 on Wild Rice).
- *Vesca* in the scientific name of the wood strawberry is Latin for small or slim, descriptive of the plant.
- The ancient Romans were very fond of *F. vesca*, the wild strawberry native to the Alps. The Romans called it *fragaria* (which was adopted as the genus name), based on the Latin word *fragrans*, fragrant, in tribute to its intense aroma.
- The "strawberry tree" is *Arbutus unedo* L., a shrub or small tree native to Eurasia and Africa, and cultivated in warmer regions of the world, including the southern United States. Its strawberry-like fruit is used in preserves and alcoholic beverages.

### GEOGRAPHY AND ECOLOGY OF WILD PLANTS

The Virginia strawberry is a widely distributed plant of North America. It typically occupies moist habitats, usually in open areas, often in disturbed sites. Habitats include meadows, forest openings, slopes, abandoned fields, and roadsides.

The woodland strawberry occurs throughout the Northern Hemisphere, including most of the United States and Canada, as well as in Baja California of Mexico. Although the species is native to

FIGURE 92.1 Woodland strawberry (*Fragaria vesca*). (a) Plant with ripe fruits. (Courtesy of H. Zell [CC By 3.0].) (b) Fruits (notice fingers for scale). (Courtesy of tasselflower/Mervi Emilia Eskelinen [Flickr/CC-attribution].) (c) Flowers. (Courtesy of nordique/Peter Stevens [Flickr/CC-attribution].) (d) Distribution map.

North America, some American populations are believed to be introductions from Europe. Habitats include open woods and forests, the edges of woodlands, meadows, and roadsides. The plants occupy both dry and moist soils. They tolerate some shade, but not deep shade.

Wild strawberry species are very widely distributed in the temperate world, in part because it is impossible for animals (or humans) to eat the fruit without eating the seeds. The seeds pass through the digestive system to be deposited unharmed with a useful application of natural fertilizer.

## PLANT PORTRAIT

The two North American strawberry species highlighted in this chapter are of quite limited importance at present, their significance dwarfed by the commercial strawberry. The modern cultivated strawberry of commerce (*F.* × *ananassa* Duchesne ex Rozier) is a hybrid of two American species, the Virginia

**FIGURE 92.2** Strawberries (*Fragaria* species). (a) Large commercial strawberry (*F.* × *ananassa*). (Courtesy of M1sslontomars2k4.) (b–f) Virginia strawberry (*Fragaria virginiana*). (b) Handful of berries. (Courtesy of Thomas & Dianne Jones [Flickr/CC-attribution].) (c) Fruits. (Courtesy of Jason Hollinger [Flickr/CC-attribution].) (d) Fruits. (Courtesy of frankenstoen [Flickr/CC-attribution].) (e) Flowers. (Courtesy of Walter Siegmund [CC By 3.0].) (f) Distribution map.

strawberry of eastern North America and the Chilean strawberry (*F. chiloensis* (L.) Mill.), which grows from Chile to the mountains of western North America. The Chilean strawberry was probably cultivated in South America long before the Spanish arrived, while North American Indians regularly ate the wild Virginia strawberry. The cross of the two strawberry species to produce the modern strawberry was first made in France about 1750. Captain François Amédée Frézier, a French explorer-spy, after observing Spanish fortifications on the west coast of South America, in 1712 brought back to France five plants of the Chilean strawberry. The French word for strawberry, *fraise*, commemorates the critical role played by the captain in making the Chilean strawberry available in France. Most of today's hundreds of strawberry varieties were selected from the original hybrid. Major producers include the United States, Poland, Japan, Spain, Italy, Russia, and Korea, but many other countries grow strawberries.

Strawberry species are low-growing, deciduous, perennial herbs, with prostrate stems ("runners") that spread out from the base of the plant, take root, and grow into new plants. The fruit is a fleshy body with numerous seeds sunken in pits on the surface. Botanically speaking, a strawberry is not a true berry (berries have seeds on the inside, not the outside), nor is the fleshy "fruit" a true fruit (true fruits arise from flowers; most of the strawberry is actually the enlarged swollen top of the stalk that held up many flowers), nor are the little "seeds" on the surface true "seeds" (they are actually true fruits, which admittedly do contain seeds). The fruits of the species discussed in this chapter are quite small, 0.5–2 cm (0.2–0.8 inch) in diameter in wild populations, larger in some cultivars. They are miniature versions of commercial strawberries.

Wild woodland strawberries were no doubt collected in prehistoric times, and the classical Persians, Greeks, and Romans ate them. In fact, the Romans cultivated the species. The woodland strawberry is now raised to a minor extent in home gardens, mostly as the "alpine strawberry." Wild woodland strawberries are quite small, but often have a delightful taste, and to this day continue to be collected, and sometimes offered in elite restaurants at very expensive prices. In Turkey, hundreds of tons of wild fruit are harvested every year, mostly for export. Some cultivated varieties have been selected, many of these with notably larger fruit than the related wild plants (some with fruits the size of the smallest fingernail). Several of the cultivated varieties are propagated vegetatively by runners, rather than by seed, in order to maintain their characteristics.

The Virginia strawberry is most significant as a source of breeding material for the cultivated commercial strawberry, of which it is one of the two ancestors. Before the creation of the modern cultivated strawberry, the Virginia strawberry was often collected from the wild by indigenous peoples and European settlers. Its fruits are still commonly harvested from wild plants. Several cultivars have been selected, including the well-known heirloom variety 'Little Scarlet' and a white-fruited form ('Christina White'), with fruits that are larger and pleasanter than those of the wild plants.

> If you have never eaten a wild strawberry, you have missed a summer delight unparalleled in the grocery store. Though much smaller in size than the cultivated fruit, wild strawberries furnish a sweeter and more flavorful treat.
>
> **—Tull, D. 1999. *Edible and Useful Plants of Texas and the Southwest: A Practical Guide.***

## CULINARY PORTRAIT

Commercial strawberries are often consumed fresh—whole, sliced, or crushed—often after being preserved frozen. Fresh strawberries are excellent with yogurt, ice cream, cream, or sprinkled with liqueur, and they may be added to fruit salads, omelets, and other dishes. The fruit is commonly placed on appetizer and cheese platters. Different traditions of consuming the fruit have developed in various countries: in Venice—with a squeeze of lemon and sugar; in France—splashed with red wine vinegar; in Greece—half dipped in chocolate fondant; in Belgium—with a lemon wedge and fresh ground pepper. Strawberries are also widely used in cooked preparations, including jams, preserves, confectionery, flavoring syrups, sorbets, custards, and pastries.

When consumed in large quantities, strawberries may act as a laxative. Curiously, strawberry leaf tea is sometimes used to relieve diarrhea, while strawberry plant root is used to increase urination. Some people develop a temporary skin rash after eating strawberries.

The woodland strawberry has well-flavored fruit. The berries from both wild and cultivated plants are employed commercially on a small scale for high-end consumption. The berries are also used commercially to prepare jam, sauces, liqueurs, cosmetics, and alternative medicines.

The small fruits of the Virginia strawberry are juicy and delicious, with a flavor much richer than found in the commercial strawberry varieties found in stores, which have considerable water (of the order of 90% by weight; strawberries with more water tend to have less-intense flavor).

### Culinary Vocabulary

- "Arizona strawberries," "prairie strawberries," and "Mexican strawberries" are not strawberries. They are joke names created by American cowboys for beans (which were often red, like strawberries). Similarly, "Boston strawberries" is an obsolete nineteenth-century term for "Boston baked beans."
- "In the hay" is old lunch counter jargon for a strawberry milkshake (a pun on hay as straw).
- "Bagatelle" (also known by the French *le fraisier*) is a French strawberry cake.
- *Schweinerei* (pronounced SHVINE-eh-ree) is an American Jewish dish prepared with cottage cheese, strawberry preserves, and chopped cold vegetables.
- "Chambraise" is a light French vermouth flavored with wild alpine strawberry juice.
- "Strawberries Romanoff" (sometimes rendered Romanov) was created by renowned French cook Marie Antoine Carême (1784–1833) for Czar Alexander I (1777–1825) of Russia (the dish has no connection to the late Mike Romanoff of Hollywood). The original recipe called for crushing very ripe strawberries, adding a good red Port wine, stirring, and allowing the concoction to stay in a cool spot overnight, and the next day sieving the mixture and pouring it over fresh strawberries. More recently, the dish is often prepared by soaking strawberries in orange-flavored liqueur (or orange juice and a liqueur) and topping it with whipped cream.

## PROSPECTS

The two American species of strawberry discussed in this chapter are interesting specialty fruits that have a small, connoisseur market. They cannot compete with the modern commercial strawberry, which is a much larger fruit that can be produced much more easily. However, the American species produce fruit that is at least the equal of any of the many other wild strawberry species, and it is probable that there will be continued interest in developing improved cultivars for the niche fruit market.

## CURIOSITIES OF SCIENCE AND TECHNOLOGY

- Medieval stone masons carved strawberry designs, symbolizing perfection and righteousness, on altars and around the tops of pillars in churches and cathedrals. Consistent with this symbolism, strawberries were served at important occasions of state and festivals to ensure peace and prosperity.
- Cardinal Wolsey (1475?–1530) introduced the English to wild strawberries and cream, before modern strawberries were invented.
- Madame Tallien, a prominent figure at the French court of Emperor Napoleon, was famous for bathing in the juice of 10 kg (22 pounds) of fresh strawberries.
- Several hundred years ago, strawberry leaves symbolized the rank of a Duke in England, which indicates how high the berries were regarded.
- Strawberries were once considered to be a medicinal cure for almost everything; it was even thought that a lotion made of the roots could fasten loose teeth by strengthening the gums. Strawberry was used for the treatment of gout by the eighteenth-century botanist Linnaeus.
- Native Americans used to crush strawberries in a mortar and mix them with meal to make a strawberry bread. After trying this bread, the colonists developed their own version: strawberry shortcake.
- The most popular ice cream toppings in the United States, in decreasing order, are hot fudge, chocolate fudge, caramel, butterscotch, and strawberry.
- Juan Metzger, a former executive of the Dannon Company located in the Bronx, is credited with first adding fruit to the bottom of containers of yogurt. The first flavor was strawberry.

- Louisiana adopted the strawberry as its state fruit in 1980, and Oklahoma did so in 1988 (to satisfy the watermelon lobby, in 2007 Oklahoma declared the watermelon to be its "state vegetable," since the state fruit honor had been given to the strawberry). North Carolina adopted the strawberry as its "state red berry" in 2001.
- A "strawberry moon" is a traditional American name for a full moon in June.
- In baseball sports medicine, a "strawberry" refers to the bruised thigh suffered by sliding base runners.
- Worldwide sales of Jell-O are over 1,000,000 packages a day, or 13 boxes per second. The most popular flavor is strawberry.

## KEY INFORMATION SOURCES

Almenar, E., Hernandez-Munoz, P., Lagaron, J.M., Catala, R., and Gavara, R. 2006. Controlled atmospheric storage of wild strawberry fruit (*Fragaria vesca* L.). *J. Agric. Food. Chem.* 54: 86–91.

Angevine, M.W. 1983. Variations in the demography of natural populations of the wild strawberries *Fragaria vesca* and *Fragaria virginiana*. *J. Ecol.* 71: 959–974.

Avigdori-Avidov, H. 1986. Strawberry. In *CRC handbook of fruit set and development*. Edited by S.P. Monselise. CRC Press, Boca Raton, FL. pp. 419–448.

Bertelsen, D.R. 1995. *The U.S. strawberry industry*. U.S. Department of Agriculture, Economic Research Service, Washington, DC. pp. 34.

Bringhurst, R.S. 1990. Cytogenetics and evolution in American *Fragaria*. *Hortic. Sci.* 25: 879–881.

Darrow, G.M. 1966. *The strawberry: History, breeding, and physiology*. Holt, Rinehart and Winston, New York, NY. pp. 447.

Fletcher, S.W. 1917. *The strawberry in North America; history, origin, botany, and breeding*. The Macmillan Company, New York, NY. pp. 234.

Hancock, J. 1999. *Strawberries*. CABI Publishing, Oxon, U.K. pp. 237.

Hancock, J.F. Jr., and Bringhurst, R.S. 1978. Inter-populational differentiation and adaptation in the perennial diploid species *Fragaria vesca*. *Am. J. Bot.* 65: 795–803.

Hancock, J.F., and Luby, J.J. 1993. Genetic resources at our doorstep: the wild strawberries. *BioScience* 43: 141–147.

Hancock, J.F., Callow, P.W., Serce, S., and Son, P.Q. 2003. Variation in the horticultural characteristics of native *Fragaria virginiana* and *F. chiloensis* from North and South America. *J. Am. Soc. Hort. Sci.* 128: 201–208.

Hancock, J.F., Serce, S., Portman, C.M., Callow, P.W., and Luby, J.J. 2004. Taxonomic variation among North and South American subspecies of *Fragaria virginiana* Miller and *Fragaria chiloensis* (L.) Miller. *Can. J. Bot.* 82: 1632–1644.

Hancock, J.F., Maas, J.L., Shanks, C.H., Breen, P.J., and Luby, J.J. 1990. Strawberries (*Fragaria*). In *Genetic resources of temperate fruit and nut crops*. Edited by J.N. Moore and J.R. Ballington, Jr. International Society for Horticultural Science, Wageningen, the Netherlands. pp. 491–546.

Hancock, J.F., Luby, J.J., Dale, A., Callow, P.A., Serce, S., and El-Shiek, A. 2002. Utilizing wild *Fragaria virginiana* in strawberry cultivar development: Inheritance of photoperiod sensitivity, fruit size, gender, female fertility and disease resistance. *Euphytica* 126: 177–184.

Harrison, R.E., Luby, J.J., Furnier, G.R., and Hancock, J.F. 1997. Morphological and molecular variation among populations of octoploid *Fragaria virginiana* and *F. chiloensis* (Rosaceae) from North America. *Am. J. Bot.* 84: 612–620.

Hokanson, K.E., Smith, M.J., Connor, A.M., Luby, J.J., and Hancock, J.F. 2006. Relationships among subspecies of New World octoploid strawberry species, *Fragaria virginiana* and *Fragaria chiloensis*, based on simple sequence repeat marker analysis. *Can. J. Bot.* 84: 1829–1841.

Jones, J.K. 1966. Evolution and breeding potential in strawberries. *Sci. Hortic.* 18: 121–130.

Jones, J.K. 1995. Strawberry. In *Evolution of crop plants*. Second edition. Edited by J. Smartt and N.W. Simmonds. Longman Scientific & Technical, Burnt Mill, Harlow, Essex, U.K. pp. 412–418.

Luby, J.J., and Stahler, M.M. 1993. Collection and evaluation of *Fragaria virginiana* in North America. *Acta Hortic.* 345: 49–53.

Preeda, N., Yanagi, T., and Okuda, N. 2009. Evaluation of wood strawberry (*Fragaria vesca*) productivity. *Acta Hortic.* 842: 87–90.

Scott, D.H., and Lawrence, F.J. 1975. Strawberries. In *Advances in fruit breeding*. Edited by J. Janick and J.N. Moore. Purdue University Press, West Lafayette, IN. pp. 71–97.

Scott, D.H., Lawrence, F.J., and Draper, A.D. 1979. *Strawberry varieties in the United States*. Revised edition. U.S. Department of Agriculture, Washington, DC. pp. 26.

Smith, C.R., and Childers, N.F. 1963. *The strawberry: Varieties, culture, marketing, pest control*. Rutgers–The State University, New Brunswick, NJ. pp. 184.

Staudt, G. 1989. The species of *Fragaria*, their taxonomy and geographical distribution. *Acta Hortic.* 265: 23–33.

Staudt, G. 1999. *Systematics and geographic distribution of the American strawberry species: Taxonomic studies in the genus* Fragaria *(Rosaceae: Potentilleae)*. University of California Press, Berkeley, CA. pp. 162.

Staudt, G. 2009. Strawberry biogeography, genetics and systematics. *Acta Hortic.* 842: 71–84.

Wilhelm, S. 1974. The garden strawberry: a study of its origin. *Am. Sci.* 62: 264–271.

Wilhelm, S., and Sagen, J.E. 1974. *A history of the strawberry, from ancient gardens to modern markets*. University of California, Division of Agricultural Sciences, Berkeley, CA. pp. 298.

## Specialty Cookbooks

(*Note:* there are over 100 books dedicated to strawberry cuisine. The following are examples.)

Allardice, P. 1993. *Strawberries*. Chronicle Books, San Francisco, CA. pp. 71.

Anonymous. 1974. *National strawberry festival cookbook*. West Graf, Manistee, MI. pp. 125.

Buszek, B.R. 1984. *The strawberry connection: Strawberry cookery with flavour, fact, and folklore, from memories, libraries, and kitchens of old and new friends and strangers*. Nimbus Publishing, Halifax, NS. pp. 212.

Clark, M., and Clark, G. 1994. *Cooking with strawberries*. Bearly Cooking, La Crescenta, CA. pp. 112.

Elliot, E., and Beveridge, J. 1998. *Strawberries*. Formac Publishing, Halifax, NS. pp. 64.

North American Strawberry Growers Association. 1997. *Strawberry eats & treats: The guide to enjoying strawberries*. Amherst Press, Amherst, WI. pp. 108.

Pitzer, S. 1985. *Simply strawberries*. Garden Way Publishing, Pownal, VT. pp. 123.

Shirkey, C. 1992. *Strawberries: Capture the essence*. New Win Publishing, Clinton, NJ. pp. 174.

Siegel, H., and Gillingham, K. 1999. *The totally strawberries cookbook*. Celestial Arts, Berkeley, CA. pp. 95.

Waldron, M. 1977. *Strawberries*. 101 Productions, San Francisco, CA. pp. 96.

Whiteaker. S. 1985. *The compleat strawberry*. Crown Publishers, New York, NY. pp. 128.

York, L. 2001. *Simple strawberry sensations*! Simple Sensations Marketing, Tampa, FL. pp. 76.

Boorman (1969) and Freitus and Haberman (2005) each have 13 wild strawberry recipes, Kluger (1973) has 16, Krumm (1991) has 6, Marie (2008) has 2, Mogelon (2001) has 4, Robe-Terry (1997) has 1, and Stanek and Butcher (2007) have 10. Stewart (2002) provides information on the culinary preparation of leaves as well as of berries of wild strawberries. (See the Appendix to this book for details of the publications cited.)

# 93 Sugar Maple

Family: Aceraceae (maple family)

## NAMES

Scientific name: *Acer saccharum* Marsh.

- The name "maple" apparently originated in England, based on the Old English *mapul*, which may be related to *möpurr*, the Old Norse name for the maple.
- The name "sugar maple" is based on the sweet, sugary sap. No other tree in the New World produces sap with as high a concentration of sugar.
- The sugar maple (also spelled sugar-maple) is also known as: bird's-eye maple and curly maple (allusions to the wood grain of some kinds of sugar maple, as explained below); and hard maple and rock maple (allusions to the hard wood).
- In England, maples were once called "maser trees," based on the word "maser," which was a bowl or drinking cup made from the knotty part of the wood (*maser* in Old High German was the word for hard, knotty wood). In the past, bowls made of silver and gold were also sometimes called masers.
- The names "flowering maple" and "parlor maple" refer to species of the genus *Abutilon*, quite unrelated to *Acer*.
- The genus name *Acer* is derived from *acer*, the classical Latin word for maple. The Latin term has been variously interpreted as meaning: hard, referring to the wood; or tracing to a Proto-Indo-European word meaning "sharp," referring to the sharp leaf edges; or tracing to the ancient Akkadian *arku*, meaning long or tall, perhaps a reference to the height of the trees.
- *Saccharum* in the scientific name *A. saccharum* is Latin for sweet or sugary, referring to the sap; the word traces to the Sanskrit *sarkarà*, meaning sugar, through the Greek *sakcharon*, a word used to denote the sweet liquid obtained from bamboo joints.

## GEOGRAPHY AND ECOLOGY OF WILD PLANTS

The sugar maple is one of the most common trees of eastern North America. Several allied species are sometimes treated as varieties of the sugar maple. These include: *A. grandidentatum* Nutt. = *A. saccharum* subsp. *grandidentatum* (Nutt.) Desmarais, Rocky Mountain sugar maple or big-tooth sugar maple, which grows from Wyoming to Utah and the northern part of Mexico; *A. leucoderme* Small = *A. saccharum* subsp. *leucoderme* (Small) Desmarais, chalk maple, which ranges from southern Virginia to northern Alabama; and *A. barbatum* Michx. = *A. saccharum* subsp. *floridanum* (Chapm.) Desmarais, southern sugar maple, found on the Coastal Plain from Texas to East Virginia, and in the Ozark Plateau of Missouri and Arkansas. A few remnant populations in Mexico and Guatemala have been assigned to *A. saccharum* subsp. *skutchii* (Rehder) Murray. Excluding the above, the northern distribution of *A. saccharum* ranges from the Gaspé Peninsula in Quebec to the southeastern corner of Manitoba. The southern limit extends from Oklahoma to New Jersey. The tree grows best on moderately coarse-textured, moist, well-drained, deep soils that are somewhat acidic to slightly basic. The species is quite cold tolerant and shade tolerant.

FIGURE 93.1 Sugar maple (*Acer saccharum*). (a, b) Children enjoying a maple sugar excursion. (Courtesy of the Agriculture and Agri-Food Canada photo library.) (c) American red squirrel eating maple seeds. (Courtesy of Gilles Gonthier [Flickr/CC-attribution].) (d) Distribution map.

## PLANT PORTRAIT

Although many of the species of *Acer* can be used to produce sugar, commercial production of maple sugar syrup is virtually all from the sugar maple. When growing in a forest, sugar maple trees produce a straight, branch-free trunk as high as 40 m (131 feet). Open-grown trees develop a rounded crown. The trunk may exceed 1.5 m (5 feet) in diameter at maturity. There are a number of different subspecies of the sugar maple. The most prominent of these is the black maple (subsp. *nigrum* (F. Michx.) Desmarais), which ranges from southern Quebec to southern Minnesota, south to west Virginia and Arkansas, but is abundant only in the eastern portion of its range. It is known as black maple and black sugar maple, because the bark is relatively black.

The native peoples of North America collected sap in crude wooden, birchbark, or pottery vessels in the early spring after simply hacking wounds on the trunks of the trees. Since metal containers

**FIGURE 93.2** Nineteenth century lithographic prints by Currier & Ives of paintings of maple sugar harvests (from *Acer saccharum*). (a) "Maple Sugaring" published in 1872. (b) "American Forest Scene—Maple Sugaring" published in 1856.

were not available prior to Columbus' discovery of the New World, present-day technology of lighting fires under containers of sap could not be used. The sap was concentrated by boiling, achieved by dropping hot rocks into the filled vessels that were available. Alternatively, the containers of sap were left until a layer of ice formed at the top; this (mostly sugar-free) ice was removed, and the process repeated until the sap was sufficiently concentrated. Maple syrup was the principal confection of the Eastern Woodlands Indians, who used the syrup and its sugar to flavor and season vegetables, fruits, stews, cereals, and fish. A variety of beverages, including fermented drinks, were also prepared from sugar maple. By the seventeenth century, European colonists improved on the Indian techniques of collection. Until cane sugar became available, about 1860 in New England, maple syrup was an important component of the colonists' limited culinary fare. North American settlers often apologized for serving maple sugar instead of the rather inferior, costly, scarce manufactured brown sugar of the time. Today, the mantle of inferiority has passed to sugared, maple-flavored imitations of real maple syrup.

Sugar maple syrup is generally considered a luxury item, since cane and beet sugar, and corn syrup are produced very much more cheaply. The quantity of sugar maple syrup is only commercially important in primarily northern or alpine areas, where alternate spring freezing and thawing produces an exceptionally high sap yield. Sugar maple production is a uniquely North American industry, with about 40 million L (over 10 million U.S. gallons) often produced annually. Canada accounts for three quarters of world production, with the province of Quebec yielding three quarters of Canada's output. In the remainder of Canada, the industry is located in eastern Ontario, Nova Scotia, and New Brunswick. The Canadian maple syrup industry was valued at $291 million in 2010, and $349.5 million in 2011. In the United States, maple sugar syrup is produced commercially in 13 northern states from Maine to Minnesota. The leading states are Vermont, New York, Pennsylvania, Michigan, and Ohio. In 2012, Vermont harvested 39% of American production. Sugar maple syrup production is a seasonal industry, conveniently occurring during a farm's slack season. Although a multi-million dollar industry, few producers rely on it as the chief source of income. Harvesting syrup is labor intensive, and in North America approximately 60,000 part-time workers are employed each year. There are more than 10,000 maple syrup producers in Canada, mostly in Quebec, with the rest in Ontario, New Brunswick, and Nova Scotia. Preparing maple syrup is generally a task for professionals, although the syrup has been made in home kitchens, boiling sap down on the kitchen stove. A frequent result is that the huge volume of water vapor produced has condensed on walls, sometimes peeling wallpaper away.

Sugar concentration averages less than 3% sucrose in the sap (some trees have as much as 8% sugar content), so that 30–40 units of sap often must be boiled down to produce one unit of syrup. Production varies considerably from year to year, and the factors responsible for this are largely unknown. The running of sap in the spring is poorly understood. In the previous growing season, sugars formed in the leaves by photosynthesis and moved down to the trunk and roots, where they were converted into starch and stored in living ray cells of the wood. When the trunk warms up sufficiently in the spring, the starch is first converted into glucose sugar, then into invert sugar (a mixture of glucose and fructose), and finally into sucrose. Ideal temperatures for the sap to run are 7°C (45°F) in the daytime and −4°C (25°F) at night. Warm days and freezing nights create a pumping action. Cold nights induce a negative pressure within the tree, with moisture being absorbed through the roots. Warm days provide positive pressure, forcing the sap out of the tree. However, the actual physiological mechanisms causing the development of positive and negative pressure in maple trees are poorly appreciated. Only a small percentage of the sap of the tree is removed each year, and trees can be used repeatedly for a century or more. An excessive number of holes in the trees or extra-large holes may damage trees by encouraging decay.

The early runs of sap produce light-colored, aromatic "fancy" grades of syrup, which are slightly less concentrated. However, sugar concentration remains more or less constant during the season. As the season progresses the syrup darkens, and eventually the maple taste becomes strong and almost unpleasant, so that the product is used mainly to flavor breakfast cereals and junk food. Although high in sugar content, maple syrup spoils at room temperature; once opened, containers should be stored in the refrigerator. Commercially, potassium sorbate and sodium citrate are widely used as preservatives. Maple sugar is prepared by concentrating the sucrose to the point that it will crystallize out of solution when the syrup cools; this occurs at −114°C (−173°F) at sea level.

As well as being a source of syrup, the sugar maple is one of the most important commercial hardwoods in Canada and the northern United States. It is the chief source of maple wood in North America and is considered the most valuable of all maples for commercial timber production. Sugar maple is used for the production of furniture, flooring, farm tools, turnery, veneer, plywood, dies, and cutting blocks. It is the preferred wood for bowling pins, piano frames, shoe lasts, billiard cues, and many other products. Its strength makes it an outstanding choice for factory, dance-hall, and bowling-alley floors.

Maples are grown as ornamental trees in lawns and parks, and along streets. Almost all maples produce good fall coloration, particularly the sugar maple. Although many of the species are tall trees, *Acer* is the most important deciduous genus of plants employed in bonsai, several Asian species used especially for this purpose.

## CULINARY PORTRAIT

Most popularly, sugar maple syrup is served "as is" as a topping on pancakes, waffles, ice cream, and other desserts. It is a valued ingredient poured over or used as a flavoring constituent of breakfast cereals and various confectionery items. Maple syrup is essential for Vermont Baked Beans, and it is often used to flavor vegetables such as carrots and sweet potatoes. The syrup is sometimes used as a replacement for sugar in coffee, tea, and a variety of dishes. It is occasionally employed for adding sweetness to meats such as ham and sausages. Maple syrup can be processed into taffy, soft and hard sugar, and maple butter. The ritual of the sugar-on-snow party is a familiar one to many northern North Americans. Hot, thick syrup is first produced by boiling the sap beyond the syrup point until it reaches the consistency of molasses, and this is ladled onto fresh snow, where it quickly hardens into a taffy-like mass. Maple sugar pie is a New England specialty. Maple butter, maple honey, and maple cream are other forms of maple sugar. The tobacco industry purchases considerable maple syrup for flavoring its products.

Closed containers of maple syrup can be stored in a cool place, but should be refrigerated once opened. If mold appears on the surface, the container should be discarded.

Because of its attractive flavor and relative expense, maple syrup is often imitated. Synthetic chemicals like cyclotene, maraniol, and butylidene phthalides are often used to obtain maple taste. So are natural materials such as fenugreek, celery seed, and lovage. Artificial maple syrup is based on sugar syrup with the addition of natural and synthetic flavorings. Possibly 10 times as much artificial maple syrup as the real thing is sold.

Maple is one of the most popular woods used to make liquid wood smoke for flavoring meat and other foods. Several major toxicity studies have shown that liquid smoke appears safe, but a 1988 Massachusetts Institute of Technology study indicated that liquid smoke was mutagenic (producing genetic changes) when applied to human white blood cells.

Apart from the syrup, other products of maple species have been used occasionally as food by indigenous peoples. In Canada, big-leaf maple (*Acer macrophyllum* Pursh) was used by various tribes: the Lower Nlaka'pamux (Thompson) of British Columbia ate raw young shoots and boiled, sprouted seeds; the Sechelt also consumed the seeds of this species and the Saanich ate the inner bark. The Micmac of the Maritimes made tea from the bark of striped maple (*Acer pensylvanicum* L.). The Blackfoot used the dry crushed leaves of several species of maples to spice stored meat. The Iroquois of the Lake Ontario region pounded the dried bark of red maple (*Acer rubrum* L.) into flour for bread, as did the Iroquois who employed the bark of silver maple (*Acer saccharinum* L.) for this purpose. In the United States, Native Americans also once prepared bread from the inner bark of maples. Outside of North America, maple seeds have also been used as food. De-winged maple seeds were cooked in butter and milk by the Calmucks of Mongolia. However, some maple seeds are toxic.

### Culinary Vocabulary

- "Vermont" is old U.S. lunch counter jargon for maple syrup, because most of the American supply came from the state.
- "Vermont Vigor" is a cocktail made with gin, lemon juice, and maple syrup.
- A "Lumberjack" is a cocktail prepared with gin, applejack, Southern Comfort, and maple syrup.
- A "Mule's Hind Leg" is a cocktail prepared with gin, apple brandy, Benedictine, apricot brandy, and maple syrup.
- "Vermont Madness" is a New England dish made with a dozen plain doughnuts, a quart of maple syrup, and a jar of dill pickles. The syrup is boiled down by a third to the consistency of a sugar glaze and is sopped up with the doughnuts, which are munched alternately with the pickles.

- *Tarte au suif* (pronounced tahrt-ta oh soo-if) is a French-Canadian pie with a suet crust filled with chopped nuts, beaten eggs, and maple syrup, and served cold after baking.
- "Maple cream" is a malleable mixture of very fine crystals in a small amount of dispersed syrup, prepared by cooling the syrup very rapidly to about 21°C (70°F) by immersing a pan of syrup in a bath of iced water, and then beating it continuously until it become very stiff. The mass of material is then rewarmed in a double boiler, becoming smooth and semisoft.

## PROSPECTS

Maple syrup products are obtained from wild trees, albeit these are often managed to promote good productivity. While there are cultivated varieties with superior sugar production, these are generally not grown for the purpose. There is very limited likelihood that the species would ever be cultivated for syrup outside of its indigenous area, since it is perfectly adapted to its native range and has difficulty growing in many other locations. Maple syrup products are unique, very desirable compared to competitive sweeteners, and sold at premium prices. The availability of artificial maple syrup flavoring limits the expansion of the market somewhat. The demand for maple syrup is stable and is likely to remain so. The chief threat to sugar maple trees is diseases, particularly "maple decline," a condition that has reduced yield for several decades. Climate change and continuing urbanization are additional long-term threats to the industry.

## CURIOSITIES OF SCIENCE AND TECHNOLOGY

- On journeys, the Iroquois carried maple syrup in empty quail and duck eggs, the first "no-deposit-no-return" biodegradable containers.
- North American Indians used sugar maple for medicinal purposes. The Iroquois employed the sap for shortness of breath. The inner bark was utilized by the Mohegan for cough, by the Ojibwa for diarrhea and to promote urination, and by the Potawatomi to clear mucus from the respiratory tract. The Walpole Island Ojibwa employed an infusion of the bark for cataracts and as an eyewash. During the maple syrup season, New Englanders used to drink collected sap as a spring tonic.
- Squirrels are fond of maple syrup, and commonly lap it up from natural wounds on maple trees. The eastern grey squirrel (*Sciurus carolinensis*) is said to deliberately bite into maple trees to stimulate the flow of sap, upon which it can later feed. Legends relate how observing such squirrel behavior taught North American Indians how to collect maple syrup.
- In pioneering days of the United States, apples and such root crops as carrots and potatoes were stored separately between layers of maple leaves to help in their preservation.
- Log Cabin syrup, which combines maple syrup and sugar, was invented in 1887 by P.J. Towle, a Minnesota grocer. The name "Log Cabin" was based on Towle's boyhood hero, Abraham Lincoln, who was born in a log cabin.
- Sugar concentration averages less than 3% sucrose in the sap of the sugar maple, although some trees have been reported to have as much as 11%. By comparison, the sucrose in sugar cane is between 16% and 17%, and in sugar beets, it is about 18%. The date palm (*Phoenix dactylifera* L.) yields the strongest sugar solution of any known tree used for sugar extraction—15% to 20%.
- All maples are "hardwoods," a term used to designate most deciduous trees, as opposed to the "softwoods," the contrasting term referring in general to coniferous, evergreen trees, which generally have softer wood than the hardwoods. Most of the dozen or so North American species of *Acer*, including sugar maple, are "hard maples," so-called because of the very hard texture of the wood. Some maples, like silver maple and red maple are "soft maples," with softer, although still quite hard, wood. Maple wood can be identified as hard

or soft by applying a solution of ferric salt to the sapwood; blue stain indicates a soft maple and green stain identifies the tree as a hard maple. In Maine, it was once thought that too much fertilizer and rich earth could change a hard maple into a soft maple.

- "Birdseye" is a distortion of the grain in sugar maple wood, associated with a pattern of conical indentations in the growth rings. Its causes are poorly understood, but the attractive pattern of "eyes" in the wood grain make the lumber from affected trees as much as 40 times as valuable as normal sugar maple lumber.
- Maple sap that flows in the spring appears to come from all directions (not just from the roots): from above, below, and sideways within the tree, apparently to prepare critical areas of the plant for growth. The lateral flow provides sugar to the growing part of the tree trunk (cambium), located near the bark, for active growth in the spring.
- The biggest known sugar maple, located in Norwich Connecticut, in 1984 possessed the following dimensions: circumference at "breast height" (1.37 m or 4.5 feet): 683 cm (22 feet 5 inches); height: 28.3 m (93 feet); crown spread: 24.4 m (80 feet).
- Maple keys (the winged seeds, technically called samaras) have an exquisite aerodynamic design. With a moderate wind blowing, these can travel as much as 160 m (525 feet) away from the tree. The key design features that seem to provide better flight performance are the thickened leading edge of the wing part and the roughened surface, characteristics similar to bird wings. The samaras are often compared to helicopters but, in flight, they cut into the air at very sharp angles that would cause stalling in a helicopter.
- Decline of certain forest tree species in parts of their distribution range has been documented in various regions of the world for hundreds of years, but especially during the last century. Maple dieback or decline in recent years has caused great concern in the sugar maple industry. It seems clear that a primary cause is insect defoliation followed by fungus attack. Deep soil freezing at times of low snow cover can result in dieback. Atmospheric deposition, notably acid rain, has been hypothesized as an important additional factor stressing trees and resulting in dieback. Acidic substances, perhaps ozone and heavy metals, might be causal agents. However, there is disagreement regarding the role, if any, of atmospheric deposition in maple decline.
- "Big Yo," the world's largest yo-yo, displayed in the National Yo-yo Museum in Chico, California, weighs 116 kg (256 pounds) and is partly made of maple.

## KEY INFORMATION SOURCES


Blum, B.M. 1973. Relation of sap and sugar yields to physical characteristics of sugar maple trees. *For. Sci.* 19: 175–179.

Bragg, D.C., and Stokke, D. 1994. *Field identification of birdseye in sugar maple* (Acer saccharum *Marsh.)* Research paper NC 317. U.S. Department of Agriculture, Forest Service, North Central Forest Experiment Station, Saint Paul, MN. pp. 16.

Bronaugh, W. 1992. The biggest sugar maple. *Am. For.* 98(11/12): 43.

Dansereau, P., and Desmarais, Y. 1947. Introgression in sugar maples—II. *Am. Midl. Nat.* 37: 146–161.

Dent, T.C. 1992. Some preliminary implications from a limited statistical analysis concerning the distribution of sugar maple in Central America. In *Biogeography of MesoAmerica, Proceedings of a Symposium, Oct. 1984*. Edited by P. Steven and A.L. Darwin. *Tulane Stud. Zool. Bot.* Suppl. Publ. No. 1. pp. 73–77.

Deschênes, J.-M. 1970. The history of the genus *Acer*, a review. *Nat. Can.* 97: 51–59.

Dore, W.G. 1967. A mingling of maples. *Trail & Landscape* (Ottawa Field-Naturalist Club) 1: 64–67.

Gunter, L.E., Tuskan, G.A., Gunderson, C.A., and Norby, R.J. 2000. Genetic variation and spatial structure in sugar maple (*Acer saccharum* Marsh.) and implications for predicted global-scale environmental change. *Glob. Chang. Biol.* 6: 335–344.

Harris, J. 2000. *The gardener's guide to growing maples.* David & Charles, Newton Abbot, Devon, U.K. pp. 160.

Horsley, S.B., and Long, R.P. (Editors). 1999. *Sugar maple ecology and health: Proceedings of an international symposium, June 2–4, 1998, Warren, Pennsylvania.* USDA Forest Service, Northeastern Research Station, Radnor, PA. pp. 120.

Horsley, S.B., Long, R.P., Bailey, S.W., Hallett, R.A., and Wargo, P.M. 2002. Health of eastern North American sugar maple forests and factors affecting decline. *Northern J. Appl. For.* 19: 34–44.
Janerette, C.A. 1979. Cold soaking reduces the stratification requirement of sugar maple seeds. *Tree Planters' Notes* (U.S. Dept. Agric. Forest Service) 30(2): 3.
Klein, R.M. 1979. Maple syrup: Confection from the forest. *Garden* 3(2): 9–11.
Kriebel, H.B. 1957. Patterns of genetic variation in sugar maple. *Ohio Agric. Expt. Stat. Res. Bull.* 791: 56.
Kriebel, H.B. 1989. Genetic improvement of sugar maple for high sap sugar content. I. Clone selection and seed orchard development. *Can. J. For. Res.* 19: 917–923.
Kriebel, H.B. 1990. Genetic improvement of sugar maple for high sap sugar content. II. Relative effectiveness of maternal and biparental selection. *Can. J. For. Res.* 20: 837–844.
Le Hardÿ de Beaulieu, A., and Mechelynck, A.L. 2003. *An illustrated guide to maples*. Timber Press, Portland, OR. pp. 464.
Millers, I., Lachance, D., Burkman, W.G., and Allen, D.C. 1991. *North American sugar maple decline project: organization and field methods*. U.S. Department of Agriculture Forest Service General Technical Report. NE 154. U.S. Department of Agriculture, Forest Service, Northeastern Forest Experiment Station, Radnor, PA. pp. 26.
Munson, P.J. 1989. Still more on the antiquity of maple sugar and syrup in aboriginal eastern North America. *J. Ethnobiol.* 9: 159–170.
Murray, A.E. Jr. 1975. North American maples. *Kalmia* 7: 1–19.
Nearing, H., and Nearing, S. 1950. *The maple sugar book.* The John Day Company, New York, NY. pp. 271.
Northeastern Forest Experiment Station. 1982. *Sugar maple research: Sap production, processing, and marketing of maple syrup*. U.S. Department of Agriculture Forest Service NE Forest Experiment Station General Technical Report NE–72. U.S. Department of Agriculture, Forest Service, Northeastern Forest Experiment Station, Radnor, PA. pp. 109.
Pendergast, J.F. 1982. *The origin of maple sugar*. National Museums of Canada, National Museum of Natural Sciences, Ottawa, ON. pp. 79.
Raulier, F., and Bernier, P.Y. 2000. Predicting the date of leaf emergence for sugar maple across its native range. *Can J. For. Res.* 30: 1429–1435.
Robitaille, G., Boutin, R., and Lachance, D. 1995. Effects of soil freezing stress on sap flow and sugar content of mature sugar maples (*Acer saccharum*). *Can. J. For. Res.* 25: 577–587.
Schaberg, P.G., Van den Berg, A.K., Murakami, P.F., Shane, J.B., and Donnelly, J.R. 2003. Factors influencing red expression in autumn foliage of sugar maple trees. *Tree Physiol.* 23: 325–333.
van Gelderen, D.M., de Jong, P.C., Oterdoom, H.J., and van Hoey Smith, J.R.P. 1994. *Maples of the world.* Timber Press, Inc., Portland, OR. pp. 512.
van Gelderon, C.J., and van. Gelderon, D.M. 1999. *Maples for gardens: A color encyclopedia.* Timber Press, Portland, OR. pp. 294.
Yawney, H.W. 1984. How to root and overwinter sugar maple cuttings. *Am. Nurseryman* 160(8): 95–96, 98–102.
Young, A.G., Merriam, H.G., and Warwick, S.I. 1993. The effects of forest fragmentation on genetic variation in *Acer saccharum* Marsh. (sugar maple) populations. *Heredity* 71: 277–289.

## Specialty Cookbooks

Buszek, B.R. 1984. *The sugar bush connection: maple cookery—"Sweet water" folklore—sugaring off facts, sucré d'érable recettes délicieuses*. Second edition. Numbus Pub. Ltd., Halifax, NS. pp. 202.
Davis, C. 1997. *The naturally sweet baker: 150 decadent desserts made with honey, maple syrup, and other delicious alternatives to refined sugar*. Macmillan, New York, NY. pp. 241.
Davy, P., and Deugo, S. 1987. *Microwaved maple*. SAP, Almonte, ON. pp. 202.
Haedrich, K. 2001. *Maple syrup cookbook.* Revised edition. Storey Books, North Adams, MA. pp. 137.
Lawrence, J., Martin, R., and Boisvert, P. 1993. *Sweet maple: Life, lore and recipes from the sugarbush.* Vermont Life, Montpelier, VT. pp. 223.
Lewis, D. 1977. *The maple harvest cookbook.* Stein and Day, New York, NY. pp. 94.
Messer, B. (Editor). 2001. *A collection of maple recipes: Pancakes—and more!!!*. Revised edition. New Hampshire Maple Producers' Association/Hanover Printing Co., Lebanon, NH. pp. 97.
Mongrain-Dontigny, M. 2003. *A taste of maple: History and recipes*. Éditions La Bonne recette, St-Irénée, QC. pp. 127.
Muir, R.L. 1974. *The Vermont maple syrup cook book*. Second edition. Phoenix Pub., Woodstock, VT. pp. 180.
Myott, L. 1998. *Vermont maple cookbook.* Vermont Maple Festival, St. Albans, VT. pp. 98.

Thompson, B. 1978. *Syrup trees*. Walnut Press, Fountain Hills, AZ. pp. 164.
Thompson, J.T. 2003. *Very maple syrup*. Celestial Arts, Berkeley, CA. pp. 80.
van der Zeijst Bates, J. 2001. *Maple moon: A maple syrup cookbook*. Maple Moon Pub., Calgary, AB. pp. 172.
Vargas-Rodriguez, Y.L., and Platt, W.J. 2012. Remnant sugar maple (*Acer saccharum* subsp. *skutchii*) populations at their range edge: Characteristics, environmental constraints and conservation implications in tropical America. *Biol. Conserv.* 150: 111–120.
Vermont Maple Promotion Board, 1999. *The official Vermont maple cookbook.* Second edition. Vermont Maple Promotion Board and the Vermont Department of Agriculture, Burlington, VT. pp. 64.
Vidler, V. 1979. *Sugar-bush antiques*. A.S. Barnes, South Brunswick, NJ. pp. 119.

# 94 Sunflower

Family: Asteraceae (Compositae; sunflower family)

## NAMES

Scientific name: *Helianthus annuus* L.

- The name "sunflower" is an allusion to the similarity of the flower to the sun's shape, having a circular head around which prominent petals look like emanating rays.
- The Spanish name for the sunflower, *girasol*, and the French name, *tournesol*, mean "turn with the sun." The flowers in a field of sunflowers are generally aligned to the east, facing the rising sun in the morning. While many green plants, particularly the sunflower, grow toward light sources, and during early growth bend toward the east in the morning and toward the west in the evening, once the sunflower opens its flowers this "phototropism" no longer occurs; the stem at this stage has become quite stiff and does not bend. Nevertheless the belief is extremely widespread that the sunflower head actually follows the sun. However, some phototropism may be exhibited by the leaves.
- The sunflower is also known as the common sunflower and has a number of other much less frequently encountered names.
- The genus name *Helianthus* is from the Greek *helios*, the sun, and *anthos*, a flower, named for the same reason as explained above.
- *Annuus* in the scientific name is Latin for annual.

## GEOGRAPHY AND ECOLOGY OF WILD PLANTS

The sunflower is native to western North America, but weedy forms and escapes from cultivated plants have been found in every state of the United States, every province of Canada (except Newfoundland and Labrador), and southern Mexico. The species is grown in many areas of the world and has escaped and become established as a wild plant in many countries. The wild plants grow in fields, pastures, and along roadsides and riverbanks. They constitute a serious weed problem in some areas. The wild sunflower is adapted to prairies and open, dry habitats. It grows best in sunny, moist locations, especially in disturbed soils. Soils occupied vary in texture, range in pH from somewhat acidic to slightly basic, and are usually well drained.

## PLANT PORTRAIT

The apparent "flower" of species in the sunflower family is actually a head of miniature flowers. In many of the "radiate" species, including the sunflower, there are two kinds of flowers, the inner ones with quite inconspicuous petals, and the outer ones with a large strap-shaped petal facing outward. Reminiscent of the rays of the sun, the large petals facing outward are called "ray petals." In *Helianthus,* the ray flowers are always sterile, whereas the inner flowers produce the seeds. The sunflower "flower" (i.e., head of flowers) has more flowers than any other genus in the family. There are often between 1000 and 8000 individual flowers present in the flowering head. Each inner flower of the sunflower head produces a small, one-seeded fruit, botanically called an achene. In common language, these are termed "seeds," and this practice is followed in the remainder of this chapter.

**FIGURE 94.1** Sunflower (*Helianthus annuus*). (a) Field of plants. (Courtesy of Bruce Britz, U.S. Department of Agriculture.) (b) A handful of seeds. (Courtesy of wetwebwork [Flickr/CC-attribution].) (c) Hulled seed at left (fruit wall has been removed), entire confectionary-type striped "seed" (technically a fruit) in center. (Courtesy of Kaldari.) Oilseed type, black seed at right. (Courtesy of Amada44 [CC By 3.0].) (d) Fruiting head with mature seeds. (Courtesy of Amada44 [CC By 3.0].) (e) Distribution map.

The sunflower is a fast-growing, annual herb, generally ranging from 0.6 to 4.2 m (2–14 feet) in height. "Dwarf" types are considered to be no higher than 1.4 m (4.6 feet) tall. The taproot sometimes penetrates to a depth of 3 m (10 feet). Although there have been claims that sunflower was domesticated as early as 5000 BC in Mexico, there is well-supported evidence that there was a single domestication center, located in the eastern United States, from where the domesticated sunflower was taken to other areas of North America. The first Europeans in the New World observed sunflower cultivated in many places from southern Canada to Mexico. Spaniards brought the sunflower to Europe in 1510, but it did not develop into a significant European crop until the late 1800s, when Russians selected varieties with very high oil content. In the 1940s, Americans brought back to

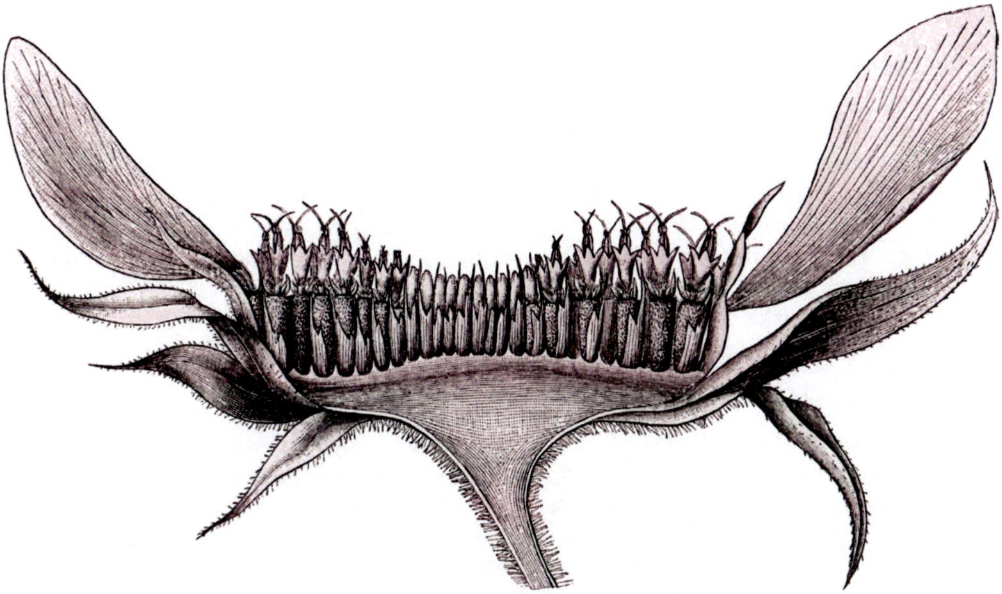

**FIGURE 94.2** Sunflower (*Helianthus annuus*). Sectioned "flower" (flower head). (From Engler, H.G.A., and Prantl, K.A.E. (Editors), *Die natürlichen Pflanzenfamilien*, First edition, Wilhelm Engelmann, Leipzig, Germany. Vol. 4(5): 88, 1889.)

North America Russian sunflower varieties with seed oil contents of nearly 50%—twice as high as existing American varieties. Over 2000 cultivated varieties of sunflowers have been recognized. Confectionery type seeds are large, with striped hulls that can be separated relatively easily. The hulls represent about 15% of the weight of the seeds, and they are used for livestock bedding, and occasionally as a source of fiber for cattle feed. The main areas of sunflower production are Russia, Argentina, eastern and western Europe, China, and the United States. Most North American sunflower seed is produced in North and South Dakota and Minnesota. In Canada, sunflower is a minor crop in southern Manitoba and southeastern Saskatchewan. Sunflower is cultivated mostly for its seeds. About 95% of world production is the oilseed type of plant, and only about 5% is the confectionery type. In addition to its culinary uses noted below, sunflower oil is used for producing lubricants, soaps, paints, and varnishes. As with most other oilseeds, the press-cake left after expressing the oil—is used as food for livestock. Both wild and domesticated birds love sunflower seeds.

## CULINARY PORTRAIT

One of the first uses of sunflowers when brought to Europe from the New World in the sixteenth century seems to have been consumption of the leaf stalks and young flowers as vegetable delicacies. Indeed, the flower buds can be eaten like artichokes. However, the seeds and the expressed oil are the principal sunflower foods. Sunflower oil is used for cooking and in margarine, shortening, and salad dressings. It is generally considered to be a premium oil because of its light color, high level of unsaturated fatty acids, bland flavor, and high smoke point (a measure of ability to withstand high cooking temperatures). Sunflower kernels are eaten by humans as a snack food, raw or roasted and salted, or made into flour for bread, pancakes, cookies, and cakes. The seeds can be added to many dishes, including salads, stuffings, sauces, and yogurt, to add crunchiness and a nutty taste. Although sunflower seeds are nutritionally excellent, commercially sold snack seeds are often roasted in saturated oil and bathed in salt and additives. Sunflowers are easily grown around the home, and the seeds harvested. Shelling the seeds, however, is more difficult. A seed mill is best, but an electric mixer can be used (pour a small amount of seeds into the bowl, turn the mixer on for a few seconds, then separate the seeds from the shell by putting them in water—the light shells will float to the surface and can be skimmed away; strain the seeds and dry them). Fresh sunflower seeds can be purchased (shelled or unshelled) and cooked with little effort. They can be roasted in a frying pan, without oil, over medium heat, stirring constantly; or baked for 10 minutes in an oven at 93°C (200°F), stirring occasionally. A small amount of oil may be added afterwards so that salt will stick to the seeds. Sunflower seeds often turn an olive-green when cooked, because of a reaction that occurs under alkaline pH conditions, between the protein and chlorogenic acid that is present (this is harmless). Sunflower seeds can be stored in a cool, dry place, but once shelled they should be stored in a refrigerator or frozen to prevent them from becoming rancid.

### Culinary Vocabulary

- "Iron rations" refers to rationed supplies of emergency food, especially when issued to the military at times of extreme food shortages. During the reign of the Czars in Russia, sunflower seeds were a staple food for many. Solders in the field were sometimes issued 1 kg (2.2 pounds) of sunflower seeds as an iron ration, and at times had little else to eat.
- "In-shell" is an industry term used for confection-type sunflower seeds (which are normally black with white stripes). The "shell" is the seed covering (technically a fruit wall), and in-shell means that the kernel (technically the seed embryo) is left intact within the shell.
- "Chew and spit" is an expression referring to the typical manner of consumption of the roasted, seasoned confection-type sunflower seeds eaten as a snack. The shell is cracked (chewed) using one's teeth, and discarded (spit) away, leaving the edible interior morsel to be enjoyed.

## PROSPECTS

The sunflower is one of the world's most important sources of edible vegetable oil (after soybean, palm, and canola/rapeseed). The oil is high in polyunsaturated fatty acids, and so is considered to be very healthy. The sunflower is the most important of all the food crops native to the United States and canada, and appears destined to maintain this status in the future.

## CURIOSITIES OF SCIENCE AND TECHNOLOGY

- American Indians used sunflower as a hunting calendar. When the plants were tall and in bloom, the buffalo were fat and the meat good.
- A number of Native American tribes of the southwestern United States, including the Hopi, extracted yellow and purple dyes from sunflowers, which were used to color basketry and decorate bodies.
- Stems of sunflowers were used by Native Americans to construct hoods over baking stoves.
- The Holy Orthodox Church of Russia imposed very strict dietary regulations during the 40 days of Lent and during Advent, and in the early nineteenth century nearly all foods rich in oil were prohibited. But the sunflower was not on the list, perhaps because the church authorities had not yet become familiar with it. As a result, the Russian people eagerly took to consuming sunflower seeds and oil, and by midcentury, Russia had become the world's leader in sunflower production.
- In Victorian times, the sunflower became a symbol for the Aesthetic movement, a reaction against the Industrial Age. The flower's image was carved into chair backs, glazed onto vases, and emblazoned on iron railings.

- Sunflower stems were used to fill lifejackets before the advent of modern flotation materials.
- Sunflower seeds in a bird feeder attract a greater variety of wild birds than any other bird food.
- Species of the sunflower family are popular with butterflies because the wide flower head makes a good landing platform and the numerous individual flowers make for a high probability of finding nectar. Monarch butterflies are commonly seen sucking nectar from sunflowers during their fall migration.
- During their peak growing period, sunflowers can grow as much as 30 cm (12 inches) per day.
- According to the *2001 Guinness Book of World Records*, in 1986 M. Heijms of Oirschot, the Netherlands, grew a sunflower that was 7.7 m (25 feet 5 inches) tall.
- According to the *2002 Guinness Book of World Records*, the most heads on one sunflower was produced in 2000 by Rose Marie Roberts of Waterford, Michigan, who grew a sunflower with 129 heads.
- The shortest recorded mature sunflower was just over 5 cm (2 inches) tall and was grown in Oregon using a bonsai technique.
- A record size sunflower head was grown in Maple Ridge, British Columbia, measuring 82 cm (32.3 inches) in diameter.
- "Sunflowers instead of missiles in the soil would ensure peace for future generations." This was a statement by U.S. Secretary of Defense, William J. Perry, on June 4, 1996, while celebrating Ukraine officially giving up its nuclear weapons. Perry joined Russian and Ukrainian defense ministers in a ceremony planting sunflowers on a former missile base in the Ukraine. The symbolism of using sunflowers arose from their use to soak up irradiation-contaminated soil resulting from the Chernobyl nuclear plant explosion.
- Some specialists contend that the largest family of flowering plants is the sunflower family, with about 24,000 species. The sunflower family is rivaled in size only by the orchid family (Orchidaceae) with at least 20,000 species (some specialists think this is the largest family), and the legume family (Leguminosae or Fabaceae) with about 18,000 species.
- Living things often grow in mathematical patterns, and the best known of such patterns are Fibonacci series. Fibonacci numbers are a sequence of numbers with several fascinating properties. The most evident property is that each term is the sum of the two previous terms: $0 + 1 = 1$, $1 + 1 = 2$, $1 + 2 = 3$, $2 + 3 = 5$, $3 + 5 = 8$, and so on, leading to: 0, 1, 1, 2, 3, 5, 8, 13, 21, 34, 55, 89, 144, 233, 377, 610, 987, 1597... Another property has to do with the "golden ratio" or "divine ratio" which is approximately 0.62 (e.g., a picture is said to look best if its height is 0.62 times its width). The sequence of ratios between every consecutive pair of Fibonacci numbers is alternatively greater than, or less than the golden ratio, and gets closer and closer to it (e.g., $1/1 = 1.00$, $1/2 = 0.500$, $2/3 = 0.667$, $3/5 = 0.6000$, $5/8 = 0.625$, $8/13 = 0.615$). Fibonacci numbers were first discussed by an Italian mathematician, Leonardo of Pisa (1175–1250), whose nickname was Fibonacci. Fibonacci numbers are most apparent in plants in *phyllotaxis*, the spiral patterned arrangement of organs on plants, such as the leaves on stems. Demonstrate this for yourself with the stem of any herbaceous plant. Start at the bottom with a given leaf, move up the stalk, counting the leaves until you reach a leaf that is directly above the first one (do not count the first one), and you will have a Fibonacci number. In addition, the number of times you have circled the stalk will be another Fibonacci number. Leaf arrangements in most plant species show a ratio of Fibonacci numbers, the one representing the number of units (leaves in this case) before one gets to the same position, and the other the number of spirals to get to the same position. The arrangements of seeds in sunflower heads spectacularly show Fibonacci ratios, but these are more difficult to count than leaves on a stem. A sunflower head may have Fibonacci ratios of 21/34, 34/55, 55/89, or 89/144. In a good specimen of a sunflower, observe two sets of spirals superimposed or intertwined, one spiral turning

clockwise and the other counterclockwise, with each seed filling a dual role by belonging to both spirals.

- Most ornamental sunflower varieties available to home gardeners today are hybrids. Those who find that they have grown a particularly huge plant or one with a huge seed head are often tempted to collect the seeds to grow next year. However, this is generally unwarranted for hybrid plants as the seeds will not produce plants with the same characteristics as the parental plant.
- Mature sunflower heads in the Northern Hemisphere generally face east, probably (it has been claimed) because this prevents overheating of the seeds by excessive exposure to the hot sun. (As noted earlier, only young plants bend toward the sun as it progresses across the sky.) Breeders have created some plants with heads that face downward so that birds find it harder to reach the seeds.

## KEY INFORMATION SOURCES

Berglund, D.R. 1994. *Sunflower production*. North Dakota State University Extension Service, Fargo, ND. pp. 98.

Blackman, B.K., Scascitelli, M., Kane, N.C., Luton, H.H., Rasmussen, D.A., Bye, R.A., Lentz, D.L., and Rieseberg, L.H. 2011. Sunflower domestication alleles support single domestication center in eastern North America. *PNAS* 108(34): 14360–14365.

Blamey, F.P.C., Edwards, D.G., and Asher, C.J. 1987. *Nutritional disorders of sunflower*. Department of Agriculture, University of Queensland, St. Lucia, Qld., Australia. pp. 72.

Carter, J.F. (Editor). 1978. *Sunflower science and technology*. American Society of Agronomy, Madison, WI. pp. 505.

Cobia, D.W., and Zimmer, D.E. 1978. *Sunflower: Production and marketing*. North Dakota State University of Agriculture and Applied Science, Fargo, ND. pp. 73.

Dedio, W. 1980. *Sunflower seed crops*. Agriculture Canada, Ottawa, ON. pp. 31.

Fick, G.N. 1989. Sunflower. In *Oil crops of the world*. Edited by G. Röbbelen, R.K. Downey, and A. Ashri. McGraw-Hill, New York. pp. 301–318.

Heiser, C.B. 1951. The sunflower among the North American Indians. *Proc. Am. Philosoph. Soc.* 98: 432–448.

Heiser, C.B. 1976. *The sunflower*. University of Oklahoma Press, Norman, OK. pp. 198.

Heiser, C.B. Jr. 1995. Sunflowers. In *Evolution of crop plants*. Second edition. Edited by J. Smartt and N.W. Simmonds. Longman Scientific & Technical, Burnt Mill, Harlow, Essex, U.K. pp. 51–53.

Heiser, C.B. Jr. 2000. Sunflower. In *The Cambridge world history of food*. Edited by K.F. Kiple and K.C. Ornelas. Cambridge University Press, Cambridge, U.K. pp. 427–430.

Heiser, C.B. Jr. 2008. Comment: sunflowers among Aztecs? *Int. J. Plant Sci.* 169: 980.

Heiser, C.B., Smith, D.M., Clevenger, S.B., and Martin, W.C. Jr. 1969. The North American sunflowers (*Helianthus*). *Mem. Torrey Bot. Club* 22: 1–218.

Kansas State University. 1993. *Sunflower production handbook*. Kansas State University, Cooperative Extension Service, Manhattan, KS. pp. 24.

Kolte, S.J. 1985. *Diseases of annual edible oilseed crops. Sunflower, safflower and nigerseed diseases*. CRC Press, Boca Raton, FL. 3: 97–136.

Lentz, D.L., Pohl, M.D., and Bye, R. 2008. Reply to Rieseberg and Burke, Heiser, Brown, and Smith: molecular, linguistic, and archaeological evidence for domesticated sunflower in pre-Columbian Mesoamerica. *PNAS* 105: E49–E50.

Lentz, D.L., Pohl, M.L., Alvarado, J.L., Tarighat, S., and Bye, R. 2008. Sunflower (*Helianthus annuus* L.) as a pre-Columbian domesticate in Mexico. *PNAS* 105: 6232–6237.

McCormick, I., Davison, C.W., and Hoskin, R.L. 1992. *The U.S. sunflower industry*. U.S. Department of Agriculture, Economic Research Service, Washington, DC. pp. 58.

McMullen, M.P. 1985. *Sunflower: Production and pest management*. North Dakota State University, Fargo, ND. pp. 76.

Nwokolo, E. 1996. Sunflower (*Helianthus annuus* L.). In *Food and feed from legumes and oilseeds*. Edited by J. Smartt and E. Nwokolo. Chapman and Hall. London, U.K. pp. 259–269.

Poncavage, J. 1996. *Totally sunflowers*. Celestial Arts, Berkeley, CA. pp. 94.

Pustovoit, V.S. 1976. *Selection, seed culture and some agrotechnical problems of sunflower.* Indian National Scientific Documentation Centre, New Delhi, India. pp. 461.
Riesberg, L., and Seiler, G. 1990. Molecular evidence and the origin and development of the domesticated sunflower (*Helianthus annuus*, Asteraceae). *Econ. Bot.* 44(3 Suppl.): 79–91.
Rogers, C.E. 1980. *Natural enemies of insect pests of sunflower: A world view.* Texas Agricultural Experiment Station, Texas A & M University System, College Station, TX. pp. 30.
Rogers, C.E., Thompson, T.E., and Seiler, G.J. 1982. *Sunflower species of the United States.* National Sunflower Association, Bismarck, ND. pp. 75.
Schilling, E.E., and Heiser, C.B. 1981. Infrageneric classification of *Helianthus. Taxon* 30: 293–403.
Schneiter, A.A., Seiler, G.J., Miller, J.F., Charlet, L.D., and Bartels, J.M. (Editors). 1997. *Sunflower technology and production. Agronomy Series No. 35.* American Society of Agronomy, Madison, WI. pp. 834.
Seiler, G. 1992. Utilization of wild sunflower species as sources of genetic diversity for improvement of cultivated sunflower. *Field Crops Res.* 30: 195–230.
Skoric, D. 1992. Achievements and future direction of sunflower breeding. *Field Crops Res.* 30: 231–270.
van der Vossen, H.A.M., and Duriyaprapan, S. 2001. *Helianthus annuus* L. In *Plant resources of South-East Asia, 14, vegetable oils and fats.* Edited by H.A.M. van der Vossen and B.E. Umali. Backhuys Publishers, Leiden, the Netherlands. pp. 101–107.

## Specialty Cookbooks

Flores, B. 1997. *The great sunflower book: A guidebook with recipes.* Ten Speed Press, Berkeley, CA. pp. 130.
Halperin, A. 1970. *Cooking and baking for low fat diets: A selection of recipes compiled particularly for sufferers from coronary thrombosis and kindred ailments, using ingredients free from animal fats and the whites of eggs only and sunflower or corn oil.* Bailey Bros. & Swinfen, Folkestone, U.K. pp. 113.
Millette, R.A. 1973. *Seeds from the sunflower.* Cooperative Extension Service, North Dakota State University, Fargo, ND. pp. 4.
Peel, L. 1997. *The ultimate sunflower book.* HarperCollins Publishers, New York. pp. 110.
Sitton, D.M. 1995. *Sunflowers: Growing, crafting, and cooking with the sunniest of plants.* Gibbs Smith, Salt Lake City, UT. pp. 54.

# 95 Sweet Gale

Family: Myricaceae (bayberry family, wax myrtle family, less frequently called sweet gale family)

## NAMES

Scientific name: *Myrica gale* L.

- The genus name *Myrica* is based on the Greek *myrike*, tamarisk (an unrelated plant species), which is based on *myron*, a perfume or scent.
- "Sweet" in the name sweet gale is an allusion to the sweet, resinous odor of the plant (which has bitter leaves). "Gale" is an old name for a group of plants, the name seemingly derived from older European languages in which similar names occur.
- Sweet gale has many common names, including bog myrtle (one of the most popular names), gale, meadow fern, and sweet willow. There are also many local names, such as English bog myrtle, English gale, Dutch gale, and Dutch myrtle.
- *Gale* in the scientific name is explained above. A traditional pronunciation is gah-lee.

## GEOGRAPHY AND ECOLOGY OF WILD PLANTS

The genus *Myrica* has about 50 species of shrubs and small trees distributed throughout the temperate and subtropical areas of both hemispheres. *Myrica gale* is the most widespread species. In North America, sweet gale is native from Alaska to the Maritime Provinces, and occurs in the northeastern and north central United States, with scattered locations elsewhere. The species also is found in western Europe and northeastern Asia (including eastern Siberia and northern Japan).

*Myrica gale* grows on a variety of soils, including sands and peats. The plants are tolerant of both full sun and moderate shade. The species is adapted to wet areas, preferring sites that are waterlogged. It is usually found in shallow waters of swamps and bogs, and along shores of lakes and streams. Two spongy wing-like bracts surrounding the fruit assist in dispersal of the seed by flotation, but seeds may also be distributed by adhesion to waterfowl and mammals. However, seedlings are rarely encountered, and vegetative reproduction by suckers is widespread.

Nitrogen is the most critical element limiting growth of plants in nature. Hundreds of genera in the pea family (Fabaceae) have symbiotic relationships with advanced bacteria that are able to take nitrogen gas out of the air and make it available to the plants. By contrast, only about two dozen other plant genera have developed nitrogen-fixing relationships, mostly with a primitive class of bacteria called the Actinomycetes (such plants are termed actinorhizal). Because these bacteria form long, thread-like branched filaments, they were considered to be fungi until fairly recently. The roots of sweet gale are associated with the actinomycete *Frankia* that is housed in root nodules. The bacterium enables the plant to grow in soils that are so nitrogen-deficient the majority of plants could not survive. Most actinorhizal plants are pioneers on nitrogen-poor, open sites.

Sweet gale produces "cluster roots." These are regions of root systems where many short rootlets are produced in a compact grouping giving the appearance of a bottle brush. Cluster roots are characteristic of a number of actinorhizal plants; they are known to occur in only a few plant families and particularly in certain ecosystems. This suggests that they play an important ecological role, likely related to nutrient uptake. There is evidence that cluster roots increase phosphorus absorption.

FIGURE 95.1 Sweet gale (*Myrica gale*). (a) Branch with male catkins (above) and branch with fruits (below). (From Thomé, O.W., *Prof. Dr. Thomé's Flora von Deutschland, Österreich und der Schweiz*, H.V. Verlag, Berlin-Lichterfelde, Germany. Vol. 2, Plate 170, 1904.) (b) Male catkins. (Courtesy of Sten Porse [CC By 3.0].) (c) Clusters of fruits. (Courtesy of Hajotthu [CC By 3.0].) (d) Distribution map.

## PLANT PORTRAIT

This low, bushy, deciduous shrub is usually about a meter (about 3 feet) high, occasionally twice this height. Numerous branching stems often arise from the base of the plant. The twigs are slender, dark or reddish brown, and the bark has prominent lighter lenticels (breathing pores). The leaves are alternately arranged on the stems. The leaf blades are 1.5–6.5 cm (0.5–2.5 inches) long, and densely covered with tiny oil glands that look like dots (these are more numerous on the lower surfaces). The degree of pubescence of the plants varies greatly, and varieties have been recognized depending on whether the plants are very hairy (var. *tomentosa*) or have few hairs (var. *subglabra*). Flowering occurs in the spring, both male and female flowers lacking petals and sepals. Although male and female flowers sometimes occur on the same plants, most shrubs bear either male flowers only or

**FIGURE 95.2** Sweet gale (*Myrica gale*). (a) Sweet gale beer being dispensed at a craft beer festival in 2012 in Norway. (Courtesy of Bernt Rostad [Flickr/CC-attribution].) (b) Coat of Arms of Wirral Metropolitan Borough Council, England; sprigs of sweet gale are shown on either side of the bird at the top. (Courtesy of Jza84 [CC By 3.0].) (c) Lamb loin prepared with jus de sweet gale. (Courtesy of Theresa Power [Flickr/CC-attribution].)

female flowers only, and these are grouped into clusters ("catkins"). Male plants are generally much more numerous than females (one study found males were 20 times as frequent). Plants occasionally change sex from one year to the next. The male catkins are unbranched, about 10 mm (0.4 inch) long, with red-brown bracts (tiny leaves), and are borne on leafless branches of the previous year's growth. The female catkins are slightly smaller but thicker and closely set with green bracts. The plants are wind-pollinated (some people are allergic to the pollen). The female catkins mature in the fall, and may remain on the plants overwinter until the spring. The fruit is dry, compressed, resinous, and two-winged.

Modern commercial interest in sweet gale has focused on new skincare products from European pharmaceutical firms (particularly Boots and Highland Natural Products). Sweet gale essential oil is said to be useful for skin care because it is antibacterial and has a very pleasant odor, and the plant has been used historically to treat skin conditions. It has also been claimed that the essential oil has antioxidant properties, which can help protect the skin against damage from pollution, the atmosphere, and the sun. Among the products under development are face wipes, skin cleansers, moisturizers, wrinkle cream, and acne ointment. The name sweet gale is attractive, and the appearance and natural habitat of the species contribute to a healthful image, which is important for a plant used to produce cosmetics.

In the past, sweet gale has been used to treat a variety of medical problems, including ulcers, intestinal worms, cardiac disorders, and aching muscles. The Bella Coola Indians of North America prepared teas from branches and used these as a diuretic (to stimulate urination) and to treat gonorrhea. Today, medical researchers are examining the potential of sweet gale chemical compounds for use in pharmaceuticals. Although there are indications that there might be medically important constituents, most of the essential oil chemicals are found in numerous other plants, and it remains to be seen whether or not sweet gale is a significant source of medicinals.

Aside from the development of cosmetics and pharmaceuticals, the greatest potential of sweet gale lies in its possible anti-insect uses. Leaf and fruit infusions (material left to soak in water for a period) of sweet gale have been used as an insect repellent for centuries. The odor, said to resemble sage, remains intense in dried leaves. Europeans used dried leaves much like moth balls to protect linens, and even in beds to get rid of biting insects. North American Indians used the foliage to repel insects from tents, especially mosquitoes, and sweet gale has also been used as a mosquito repellent in Europe. Modern campers still use the foliage of sweet gale to keep biting insects out of tents. Many of the volatile constituents of sweet gale are in fact well known to have insect-repellent and insecticidal properties. A commercial midge repellent based on sweet gale appeared on the European market under the name "Myrica," and similar products are now available.

Sweet gale is a useful ornamental shrub for cold climate areas with wet soils, and it is often grown in damp ground that could not support most other plants. It has potential for use in reclamation and stabilization of wet soils because of its vigorous growth.

In Europe, large numbers of branches have been collected from wild plants for the cut-flower trade. The scented catkins remain on the stems, giving a long-lasting menthol-like aroma to bouquets.

The following are antiquated uses. An aromatic wax suitable for candles can be obtained from the catkins and fruits, and this gave rise to the name "candle berry." However, the fruits of several other species of *Myrica* have a much higher content of wax and are more suitable for the purpose. From very early times, aromatic herbs have been used to mask bad odors in households, and it appears that sweet gale was among the plants used. North American Indians used the catkins as fish lures. The boiled seeds, roots, and bark have been used to produce a yellow dye, sometimes applied to calfskin and wool.

## CULINARY PORTRAIT

The leaves of sweet gale have a bittersweet taste because of the essential oils produced in the tiny leaf glands, and are astringent because of the presence of bitter tannins. Sweet gale appears to have been used thousands of years ago in Denmark to spice fermented drinks, as evidenced by residues from an ancient birch-bark bucket. The Celts (who were dominant in parts of Europe over 2000 years ago) used sweet gale in their beer. Until modern refrigeration, herbs were used in foods not simply for flavor but also as a preservative, since the chemicals present are protective against microorganisms. Hop (see Chapter 51) was not used in beer until the ninth century, and sweet gale was the principal beer herb until the early Middle Ages. Beer was in fact the main beverage of much of Europe in the past (water was often unsafe to drink), and so sweet gale was an extremely important food plant. In modern times, however, the herb is merely one of many minor plants used to prepare specialty beers and liquors. It has been claimed that sweet gale increases foam in beer, and in Yorkshire, England "gale beer" is especially popular.

Sweet gale has also been widely used to flavor tea but, on the whole, the contemporary culinary usage of the species is very limited. Nevertheless, sweet gale fruits and leaves are still used to some extent to flavor broths, soups, stews, and meats. It is recommended that when the herb is added, it should be used like bay leaves: used whole and steeped in soups and sauces, but removed before serving.

Sweet gale has been used to promote abortion, and the essential oil contains toxic compounds (flavonoid glycosides or sesquiterpenes). As a result, some authorities have recommended that consumption should be avoided during pregnancy and lactation. It has also been suggested that constituents may be present that are neurotoxic (adversely affecting nerve tissue), and accordingly consumption should generally be limited.

### Culinary Vocabulary

- "Bog Water" is the colorful name of a seasonal strong ale made with sweet gale collected from northern Quebec by Beau's All Natural Brewing Company, of Vankleek Hill, Ontario, Canada.
- "Gruit" was a mixture of herbs and spices used to flavor and preserve beer before hops was used for the purpose. Today, brewers occasionally produce a "gruit ale," that is, one lacking hops and flavored with other materials (pasteurization makes the preservation function of herbs obsolete).

## PROSPECTS

As a food plant, sweet gale is insignificant, and is likely to remain minor. However, the food use may increase as a result of the considerable potential of the species as a source of essential oils for use as skin care products, flavorants, pharmaceuticals, aromatherapeutics, and insect repellents. On the negative side, these markets are very competitive. Considerable variation has been observed in essential oil constituents and overall yield, but it is not clear whether this is due to genetic differences, environmental effects, or stage of maturity, and there is thus a need to clarify the chemistry of the plant. A chief advantage of sweet gale is its adaptation to wetlands, which are usually considered to have little value for agricultural and urban purposes. Further development of the sweet gale industry may emphasize either continued harvest from the wild or cultivation. Cultivation would likely require conversion of natural wetlands. Due to the widespread recognition of the need for wetland protection, successful development of sweet gale may depend on ample consideration being given to prevent loss of essential wetland habitat and biodiversity.

## CURIOSITIES OF SCIENCE AND TECHNOLOGY

- According to Amerindian belief, consuming sweet gale produces vivid dreams at night and stimulates meditation during the day. In modern times, some herbalists recommend adding a few drops to a beverage before meditation or before going to bed. The plant has been used since the Dark Ages to treat depression, and there is modern evidence that sweet gale has a sedative and calming effect.
- The Vikings are reputed to have used sweet gale as a stimulant before going into battle. This idea probably arose because sweet gale was widely used in preparation of alcoholic beverages, which indeed were consumed in large amounts by the Vikings. However, sweet gale does not seem to result in aggression.
- In Scotland, plants were often used to distinguish members of different clans when at war. During battles, sweet gale was worn in bonnets as a distinguishing plant badge by the Scottish clan Campbell ("badge of the Campbells" is an occasional name of sweet gale).
- Hemoglobin has been discovered in the root nodules of *Myrica gale*. The function of this is unclear, but since hemoglobin in the blood of animals transports oxygen throughout the

body, it has been suggested that in the plant it functions to provide oxygen to the roots that grow in saturated soil that is very low in oxygen.

- Sweet gale branches were once attached to the eaves of houses in Scandinavia to repel witches. In modern times, a bouquet may be kept indoors for good luck.
- In the Royal Botanic Gardens at Edinburgh, the central feature of the Queen Mother Memorial Garden (dedicated in 2006 to the late Queen Elizabeth) is a maze constructed of sweet gale shrubs.

## KEY INFORMATION SOURCES

Behre, K.-E. 1999. The history of beer additives in Europe—A review. *Veg. Hist. Archaeobot.* 8: 35–48.

Bornstein, A.J. 1997. *Myrica*. In *Flora of North America*. Vol. 3. Edited by Flora of North America Editorial Committee. Oxford University Press, Oxford, U.K. pp. 430–434.

Carlton, R.R., Waterman, P.G., and Gray, A.I. 1992. Variation of leaf gland volatile oil within a population of sweet gale (*Myrica gale*) (Myricaceae). *Chemoecology* 3: 45–54.

Crocker, L.J., and Schwintzer, C.R. 1994. Soil conditions affect the occurrence of cluster roots in *Myrica gale* L. in the field. *Soil Biol. Biochem.* 26: 615–622.

Fordham, A.J. 1983. Of birds and bayberries: Seed dispersal and propagation of three *Myrica* species. *Arnoldia* 43: 20–23.

Hausen, B.M., Kroidl, R.F., and Hesselkamp, J. 1984. Pollen allergy due to gale *Myrica gale*. *Allergologie* 7: 457–460.

Heckmann, A.B., Hebelstrup, K.H., Larsen, K., Micaelo, N.M., and Jensen, E.O. 2006. A single hemoglobin gene in *Myrica gale* retains both symbiotic and non-symbiotic specificity. *Plant Molec. Biol.* 61: 769–779.

Lawrence, B.M., and Weaver, KM. 1974. Essential oils and their constituents. XII. A comparative chemical composition of the essential oils of *Myrica gale* and *Comptonia peregrina*. *Planta Medica* 25: 385–388.

Lloyd, D.G. 1981. The distribution of sex in *Myrica gale*. *Plant Syst. Evol.* 138: 29–45.

Pathirana, S.M., and Tjepkema, J.D. 1995. Purification of hemoglobin from the actinorhizal root nodules of *Myrica gale* L. *Plant Physiol.* 107: 827–831.

Simpson, M.J.A., MacIntosh, D.F., Cloughley, J.B., and Stuart, A.E. 1996. Past, present and future utilisation of *Myrica gale* (Myricaceae). *Econ. Bot.* 50: 122–129.

Skene, K.R., Sprent, J.I., Raven, J.A., and Herdman, L. 2000. Biological flora of the British Isles: *Myrica gale* L. *J. Ecol.* 88: 1079–1094.

Small, E., and Catling, P.M. 2007. Blossoming treasures of biodiversity: 25. Sweet gale—new products from an ancient beer plant. *Biodiversity* 8(3): 27–32.

Sprent, J.I., Scott, R., Perry, K.M. 1978. The nitrogen economy of *Myrica gale* in the field. *J. Ecol.* 66: 657–668.

Svoboda, K.P., Inglis, A., Hampson, J., Galambosi, B., and Asakawa, Y. 1998. Biomass production, essential oil yield and composition of *Myrica gale* L. harvested from wild populations in Scotland and Finland. *Flavour Fragrance J.* 13: 367–372.

Sylvestre, M., Legault, J., Dufour, D., and Pichette, A. 2005. Chemical composition and anticancer activity of leaf essential oil of *Myrica gale* L. *Phytomedicine (Jena)* 12: 299–304.

Sylvestre, M., Legault, J., Lavoie, S., and Pichette, A. 2006. Investigation of leaf essential oil of *Myrica gale* L. from Quebec: purification and analysis of oxygenated fractions. *J. Essent. Oil Res.* 18: 38–41.

Waterhouse, E.C., Sporring, S., and Bayfield, N.G. 1998. Sustainable harvesting of bog myrtle (*Myrica gale* L.) in Scotland. *Bot. J. Scotland* 50: 11–20.

Wilbur, R.L. 1994. The Myricaceae of the United States and Canada: genera, subgenera, and series. *SIDA* 16: 93–107.

### SPECIALTY COOKBOOKS

Stewart (2002) (see the Appendix to this book) provides information on the culinary use of sweet gale.

# 96 Tepary Bean

Family: Fabaceae (Leguminosae; bean family)

## NAMES

Scientific name: *Phaseolus acutifolius* A. Gray

- "Tepary" (occasionally "tepari") in tepary bean is based on the Spanish tépari, which in turn is based on a name of Opatan origin (Opata is a language of some indigenous people of northwestern Mexico). Another interpretation is that the word is based on *pawi*, the name for bean in another Indian language, Tohono O'odham (formerly Papago), and the name came into Spanish when the conquering Spanish asked the Papago for the name of the bean and received the reply "t'pawi," meaning "it's a bean." (The Tohono O'odham Indians, known as "the bean people," are the major traditional inhabitants of the Sonoran Desert on both sides of the United States–Mexican border.)
- Pronunciation: TEP-a-ree.
- The tepary bean is also sometimes called Texas bean. There are numerous domesticated species of the genus *Phaseolus*, but of these the tepary is the only one native to west Texas.
- Escomite or escumite is a name used in Chiapas, Mexico, for the tepary bean.
- The genus name *Phaseolus* is derived from the Greek *phaseolos*, a bean. A more imaginative interpretation translates the Greek word as a small boat, supposedly the shape of the bean pod.
- *Acutifolius* in the scientific name *P. acutifolius* is based on the Latin *acutus*, narrowly pointed (i.e., acute-angled) + *folium*, leaf.

## GEOGRAPHY AND ECOLOGY OF WILD PLANTS

The tepary bean grows as a wild plant in arid and semiarid regions, from west Texas to Arizona and New Mexico in the United States, and in Mexico it occurs in Baja California, Sonora, Chihuahua, Sinaloa, and Durango. Tepary grows wild as far south as Guatemala. The species is very well adapted to drought, heat, and a dry atmosphere. A deep, extensive root system contributes to its drought tolerance. The plant is also quite resistant to soil salinity (perhaps, it has been suggested, not because of its physiology but because of the deep root system which can evade salt). Minimum night temperatures below 8°C (46°F) are harmful. When annual rainfall exceeds 100 cm (39 inches), the plants tend to grow vegetatively and produce less seed. Light, well-drained sandy soils are preferred, clay soils are unsuitable, and waterlogged soils are not tolerated.

## PLANT PORTRAIT

The wild tepary plant is a herbaceous vine growing to a length of up to 4 m (13 feet), although usually not much longer than 2 m (6.6 feet). It often climbs up desert shrubs. Domesticated tepary plants are either semivining, or self-standing bushes often averaging 75 cm (30 inches) in height. The plant has small pink, white, or lilac flowers. This annual is fast-maturing, producing seeds in 60–90 days. The pods are 5–9 cm (2–3.5 inches) long, slightly hairy, and green when young, ripening to a light straw color. The seeds are relatively small, about 8 mm (0.3 inch) long and 6 mm (0.23 inch) wide. There are

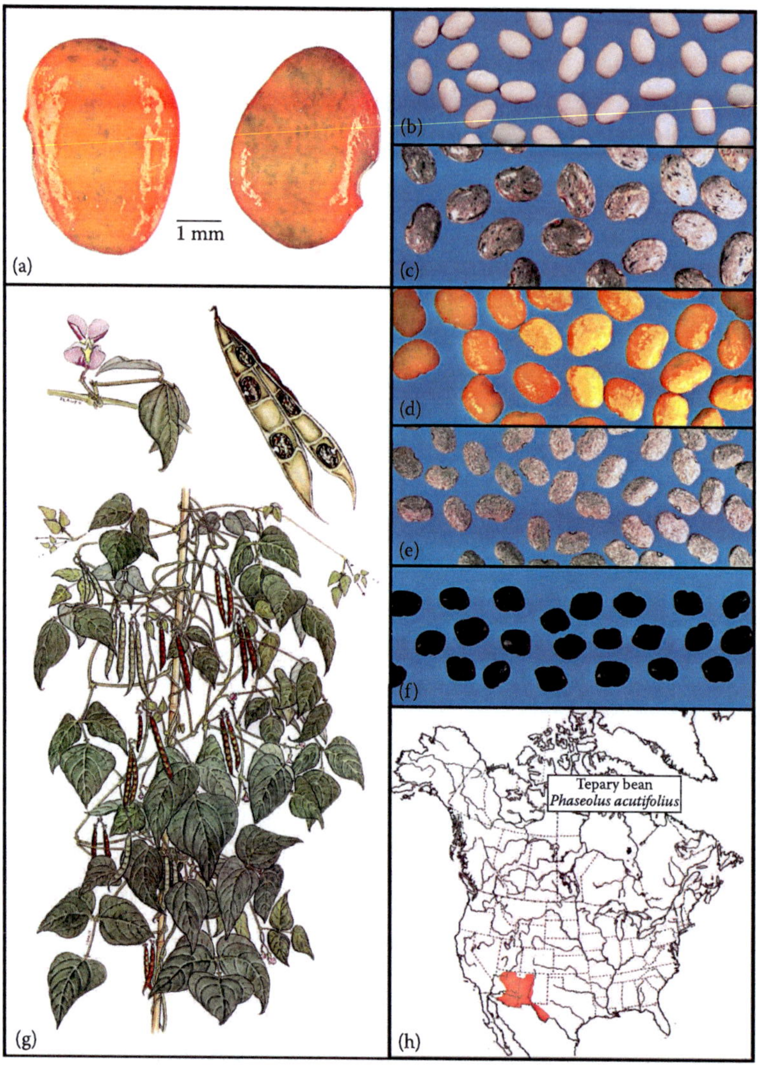


**FIGURE 96.1** Tepary bean (*Phaseolus acutifolius*). (a) Orange seeds. (Courtesy of Tracey Slotta, U.S. Department of Agriculture.) (b–e) Seeds of various colors. (Courtesy of Howard F. Schwartz, Colorado State University, online at Bugwood.org [CC By 3.0].) (f) Black seeds. (Courtesy of Lana Wheeler, digitally prepared by Phillip Stanwood, U.S. Department of Agriculture.) (g) Painting of plant (flowering twig and opened pod at top). (Painting courtesy of B. Flahey.) (h) Distribution map.

two to nine, typically five or six seeds per pod, and these are flat and resemble small butterbeans or navy beans. Seed colors in wild plants include white, black, greenish white, grey bay (reddish-brown), dark yellow, mahogany, purple-mottled, and coffee. Cultivars have been selected with white, dark yellow, and other colors of seeds. It has been hypothesized that the species originated in Central America. The plant was cultivated 5000 years ago in Mexico, but other types of beans have always been more commonly grown. The main advantage of tepary is its adaptation to a climate that is too arid for other types of beans. This drought- and heat-tolerant legume is cultivated mainly in Arizona, New Mexico, and the Mexican states of Sonora and Sinaloa. The species is raised from tropical central Mexico south to El Salvador and has been introduced into semiarid African countries. The species is raised to a limited extent in India and Burma. The tepary was a very important food plant in prehistoric times in the American southwest, but is of limited commercial importance at present.

## CULINARY PORTRAIT

Tepary beans when appropriately prepared are as tasty as other beans. The flavor is often stronger than that of common beans (*Phaseolus vulgaris* L.) but is rich and nutty. The beans are shelled and dried before use. The skin of the seeds is usually tougher than that found in common beans, so they are best suited for use as dry beans, first soaked in water (at least 12 hours), then boiled or baked (with at least three cups of water for each cup of beans). Cooking or simmering for 3 hours or more is often necessary, and even after soaking and cooking they may still be tougher than other beans. Cultivars with white seeds have a relatively permeable seed coat compared to those with black seeds and cook more quickly. Well-cooked tepary beans are light and mealy with a rich bean aroma. In northern Mexico, the beans are popular as a base for many traditional southwestern stews and casserole-like dishes, soups, burritos, and tacos. Historically, dry tepary beans were also roasted, ground into a meal, and rehydrated into a gruel simply by adding boiling water. Tepary beans are high in protein (about 25%). Antinutritional substances are present but are apparently no more of a concern than in common beans. In areas where it is grown, tepary is also used as a young tender string bean.

Eating beans is well known to cause flatulence, and tepary bean is no exception. The problem is lessened by thorough cooking and by soaking the beans in several changes of water. However, changing water results in loss of nutrients and even flavor. In the case of tepary bean, one change of water after about 1 hour of cooking has been recommended.

The Pima and Tohono O'odham tribes of Arizona are attempting to rediscover the desert foods their people traditionally consumed until as recently as the 1940s. Studies strongly indicate that people who evolved in these arid lands are metabolically best suited to the feast-and-famine cycles of their ancestors, who survived on the desert's unpredictable bounty. By contrast, the modern North American diet is making them sick. With rich food always available, obesity has become common among these once-lean people. As many as half the Pima and Tohono O'odham Native Americans now develop diabetes by the age of 35, an incidence 15 times higher than for Americans in general, and a disease that used to be rare. Similar problems have been found among other North American indigenous peoples, Australian aborigines, Pacific Islanders, and other peoples whose survival historically depended on their ability to absorb calories in times of plenty to sustain them during droughts and crop failures. Studies have indicated that a change in the Indian diet back to the beans, corn, grains, greens, and other low-fat, high-fiber plant foods that their ancestors depended on can normalize blood sugar, suppress between-meal hunger, and promote weight loss. Tepary beans are considered to be an ideal replacement food to repair the damage caused by a diet of hamburgers, fries, soft drinks, and other fatty, sugary, overly refined, fast, and "prepared" foods.

### Culinary Vocabulary

- In nineteenth century America, dried beans were not considered to be *haute cuisine* and were sometimes known by the euphemism "Alaska strawberries."
- "Refried beans" is a Mexican–American dish of mashed, cooked beans, typically served as a side dish or used as a filling for various tortillas. The English name is based on a mistranslation of the Spanish name *frijoles refritos*. The Spanish means well fried, not refried.
- "Pulses" or "grain legumes" are crops of the pea family, the seeds of which are collected, dried, and used as human food. Some enlarge the term to include crops like soybean and peanut that are used mainly for oil extraction but are also dried and subsequently eaten. Still others use the term to denote any edible seed or plant of the legume family.

## PROSPECTS

Tepary beans require considerable breeding to make them more competitive with other bean species. Problems include small seed size (compared to most beans), long cooking time, and tendency to cause flatulence. The seeds shatter (are released from the pods) easily at maturity while still on the plant, making harvest difficult, although selection of nonshattering varieties is overcoming this problem. The taste, which is too strong for some, and the odor, which is objectionable to others, also require improvement (perhaps by development of appropriate processed products). The great advantage of tepary bean is its very quick maturation in very hot, dry environments, providing extraordinary yields in relation to the water required. Still another advantage is their natural resistance to disease. With increasing interest in conserving water resources, this drought-resistant crop has considerable potential, but mostly for hot, arid regions of the world.

## CURIOSITIES OF SCIENCE AND TECHNOLOGY

- Papago Indians bit down on the tepary bean plant as a treatment for toothache.
- "Dehiscence" of dry (non-fleshy) fruits refers to splitting of the fruit wall along one or more lines of weak tissue in order to scatter the seeds at maturity. Pods of wild plants of the pea family often dehisce suddenly because of the release of tensions built up in the parchment layer lining the inner pod. This layer is made up of oriented fibers, which twist in response to humidity, building up tension, and twisting the pod valves in opposite directions. At maturity, after low humidity has resulted in great tension developing, the sutures break, the valves of the pod twist quickly in opposite direction, and the seeds are ejaculated. In cultivated varieties, pods usually have a reduction of the parchment layer, and so reduced dehiscence. Some "stringless" types lack parchment completely and are quite indehiscent. In wild forms of tepary beans, the seeds are dispersed up to 3 m (10 feet) from the mother plant by explosive dehiscence of the pods.
- The seeds of wild tepary plants are adapted to germinate only after it rains heavily in their desert habitat. The wild plants have germination-inhibiting chemicals that need to be washed out of the seeds before they will sprout. Although most of these chemicals have been bred out of the cultivated varieties, in its native arid area the tepary is normally planted at the start of the rainy period, often after the first heavy rain, to take advantage of the natural irrigation.
- On cool and overcast days, tepary leaves bend to track the sun to obtain as much light and heat as possible. Very hot days have the opposite effect, the leaves remaining motionless or curling under, reducing the surface area exposed to heat and conserving moisture.
- Beans are one of the triumvirate of important crops—maize, beans, and squash—used by New World Indians for millennia to achieve a balanced diet. Beans are rich sources of protein, particularly the essential amino acid lysine but are usually deficient in the sulfur-containing amino acids—methionine and cystine. By contrast, cereal grains like corn contain lower amount of proteins, and these are deficient in lysine but adequate in sulfur-containing amino acids. Together, beans and corn provide a balance of essential protein components, which is necessary for survival, especially when adequate meat supplies are unavailable. Vinelike climbing forms were commonly planted among maize plants so that the beans could use the corn stalks as supports on which to climb. Bean species are still grown on corn in this fashion today in Central America.

## KEY INFORMATION SOURCES

Albala, K. 2007. *Beans: A history*. Berg, New York. pp. 247.

Bhardwaj, H.L., and Hamama, A.A. 2004. Protein and mineral composition of tepary bean seed. *HortScience* 39: 1363–1365.

Bhardwaj, H.L., and Hamama, A.A. 2005. Oil and fatty acid composition of tepary bean seed. *HortScience* 40: 1436–1438.

DeMejia, E.G., Del Carmen Valadez-Vega, M., Reynoso-Camacho, R., and Loarca-Pina, G. 2005. Tannins, trypsin inhibitors and lectin cytotoxicity in tepary (*Phaseolus acutifolius*) and common (*Phaseolus vulgaris*) beans. *Plant Foods Human Nutr.* 60: 137–145.

Frederici, C.T., Ehdaie, B., and Waines, J.G. 1990. Domesticated and wild tepary bean: Field performance with and without drought-stress. *Agron. J.* 82: 896–900.

Freytag, G., and Debouck, D.G. 2002. *Taxonomy, distribution and ecology of the genus* Phaseolus *(Leguminosae—Papilionoideae) in North America, Mexico and Central America.* Botanical Institute of Texas, Fort Worth, TX. pp. 300.

Garvin, D.F., and Weeden, N.F. 1994. Isozyme evidence supporting a single geographic origin for domesticated tepary bean. *Crop Sci.* 34: 1390–1395.

Gepts, P. (Editor). 1988. *Genetic resources of* Phaseolus *beans: Their maintenance, domestication, evolution, and utilization.* Kluwer Academic Publishers, Norwell, MA. pp. 613.

Hamama, A.A., and Bhardwaj, H.L. 2002. Tepary bean: a short duration summer crop in Virginia. In *Trends in new crops and new uses.* Edited by J. Janick and A. Whipkey. ASHS Press, Alexandria, VA. pp. 429–431.

Idouraine, A., Tinsley, A.M., and Weber, C.W. 1989. Nutritional quality and sensory acceptability of akara prepared from germinated tepary beans. *J. Food Sci.* 54: 114–117.

Idouraine, A., Weber, C.W., and Kohlhepp, E.A. 1995. Composition of tepary bean (*Phaseolus acutifolius*) of the southwestern U.S. and northern Mexico. *Ecol. Food Nutrition (U.K.)* 33(3): 139–147.

Jansen, P.C.M. 1989. *Phaseolus acutifolius* A. Gray. In *Plant resources of South-East Asia, 1, pulses.* Edited by L.J.G. van der Maesen and S. Somaatmadja. Pudoc, Leiden, the Netherlands. pp. 54–55.

Kabbara, S.A.R., Abbas, I.R., Scheerens, J.C., Tinsley, A.M., and Berry, J.W. 1987. Soaking and cooking parameters of tepary beans: Effects of cooking time and cooking temperature on hardness and activity of nutritional antagonists. *Plant Foods Human Nutr. (the Netherlands)* 36(4): 295–307.

Kuruvadi, S., and Valdez, I.S. 1993. Range of yield components and phenotypic correlations in tepary beans (*Phaseolus acutifolius*) under dryland conditions. In *New crops.* Edited by J. Janick and J.E. Simon. Wiley, New York. pp. 594–496.

Lin, T.Y., and Markhart III, A.H. 1996. *Phaseolus acutifolius* A. Gray is more heat tolerant than *P. vulgaris* L. in the absence of water stress. *Crop Sci.* 36: 110–114.

Markhart, A.H. III. 1985. Comparative water relations of *Phaseolus vulgaris* L. and *Phaseolus acutifolius* Gray. *Plant Physiol.* 77: 113–117.

Mogotsi, K.K. 2006. *Phaseolus acutifolius* A. Gray. In *Cereals and pulses.* Edited by M. Brink and G. Belay. Prota Foundation, Wageningen, the Netherlands. pp. 133–137.

Muñoz, L.C., Duque, M.C., Debouck, D.G., and Blair, M.W. 2006. Taxonomy of tepary bean and wild relatives as determined by amplified fragment length polymorphism (AFLP) markers. *Crop Sci.* 46: 1744–1754.

Nabhan, G., Berry, J., Anson, C., and Weber, C. 1980. Papago Indian floodwater fields and tepary bean protein yields. *Ecol. Food Nutr.* 10(2): 71–77.

Nabhan, G.P., and Felger, R.S. 1978. Teparies in southwestern north America. A biogeographical and ethnohistorical study of *Phaseolus acutifolius. Econ. Bot.* 32: 3–19.

Nabhan, G.P., and Teiwes, H. 1983. Tepary beans, O'odham farmers, and desert fields. *Desert Plants* 5(1): 15–37.

Pratt, R.C., and Nabhan, G.P. 1988. Evolution and diversity of *Phaseolus acutifolius* genetic resources. In *Genetic resources of* Phaseolus *beans: Their maintenance, domestication, evolution, and utilization.* Edited by P. Gepts. Kluwer Academic Publishers, Dordrecht, the Netherlands. pp. 409–440.

Scheerens, J.C., Tinsley, A.M., Abbas, I.R., Weber, C.W., and Berry, J.W. 1983. The nutritional significance of tepary bean consumption. *Desert Plants* 5(1): 11–14, 50–56.

Schinkel, C. and Gepts, P., 1988. Phaseolin diversity in the tepary bean, *Phaseolus acutifolius* A. Gray. *Plant Breed.* 101: 292–301.

Sotelo, A. 1996. Escumite bean (*Phaseolus acutifolius* A. Gray). In *Food and feed from legumes and oilseeds.* Edited by J. Smartt and E. Nwokolo. Chapman and Hall, London, U.K. pp. 140–143.

Taggart, R., Storey, R., and Dower, N. 1983. Nutritional evaluation of tepary bean: Elemental analysis of seed. *J. Plant Nutr.* 6: 983–988.

Tinsley, A.M., Scheerens, J.C., Alegbejo, J.O., Adan, F.H., Krumhar, K.C., Butler, L.E., and Kopplin, M.J. 1985. Tepary beans (*Phaseolus acutifolius* var. *latifolius*): A potential food source for African and Middle Eastern cultures. *Qualitas Plant. (the Netherlands)* 35(2): 87–101.

Thomas, C.V., Manshardt, R.M., and Waines, J.G. 1983. Teparies as a source of useful traits for improving common beans. *Desert Plants* 5: 43–48.

Weis, K.G., and Webster, B.D. 1990. Flower and fruit development in tepary bean. *HortScience* 25: 119–120.

### Specialty Cookbooks

Tepary beans can be substituted for common beans in most recipes, but are harder to prepare (see the information in the section "Culinary Portrait").

Niethammer, C. 1983. Tepary cuisine beans, cooking, recipes, *Phaseolus acutifolius*. *Desert Plants* 5(1): 8–10.

Tull (1987) (see the Appendix to this book) has a recipe entitled "Vegetarian tepary bean chili."

# 97 Wild Leek (Ramp)

Family: Liliaceae (lily family; sometimes placed in Alliaceae, the onion family)

## NAMES

Scientific name: *Allium tricoccum* Ait.

- The English word "leek" traces to the Old Norse word *leek* for the plant, and ultimately to the German *lauch* (leek), and the Russian *luk* (onion).
- Wild leek is also known as ramp(s), ramson, small white leek, spring onion, wild garlic, and wood leek.
- The name "ramp" or "ramps" has been applied to several wild species of *Allium*, and probably derives from the Old English word for wild garlic, *hramsa*. The word originates from European languages (Danish, Swedish, Icelandic), where it meant "rank," for the strong odor and flavor. A more colorful (and unlikely) explanation holds that ramp comes from "ramson," meaning son of Ram, an allusion to the arrival of the plant during the appearance of the constellation Aries, the ram, the first sign of the zodiac. The European ransom (or bear leek, *A. ursinum* L.) is very similar to *A. tricoccum*.
- The quite unrelated vegetable rampion (*Campanula rapunculus* L.) is also called "ramps."
- Wild leek is sometimes called wild garlic, a name better applied to Canada garlic, *Allium canadense* (see Chapter 23).
- The genus name *Allium* is the classical Latin name for garlic.
- *Tricoccum* in the scientific name of the wild leek is Latin for three chambered, a description of the three-lobed fruit.

## GEOGRAPHY AND ECOLOGY OF WILD PLANTS

In Canada, wild leek occurs in Nova Scotia, New Brunswick, Quebec, and Ontario. In the United States, it is found from the Atlantic Coast west to North Dakota in the north and Tennessee in the south, and south as far as northern Georgia. In the area of the Great Lakes and northeastward, wild leek is a northern hardwoods forest herb, occurring on streamside bluffs, in depressions, and in upland maple woods. It prefers rich, moist soil, and is usually found in deciduous woods. In more southern areas in the United States, wild leek may grow in rocky soils on mountain slopes at higher elevations (1000–3000 m or 3300–9800 feet).

## PLANT PORTRAIT

Wild leek is typically a long-lived spring ephemeral (a plant with a brief period of above-ground appearance) of deciduous forests. Such species take advantage of the limited competition in the early spring (before full-tree canopy development) for light, and possibly also nutrients and space. The plants resume growth and emerge shortly after snowmelt, the shoots sometimes even elongating in the snow. They have one to three basal fleshy, flat leaves. Five to six weeks into the season, when the developing forest canopy has substantially reduced available light, the leaves wither rapidly and disappear. With leaf senescence, a flowering stem 10–40 cm (4–16 inches) high develops, topped by 3–50 or more white or whitish flowers in a globular cluster. This is followed by development of a

**FIGURE 97.1** Wild leek (*Allium tricoccum*). (a) Harvested plants. (Courtesy of Blueberry Files/Kate [Flickr/CC-attribution].) (b) Trimmed plants for sale. (Courtesy of Timothy Vollmer [Flickr/CC-attribution].) (c) Painting. (Painting courtesy of by B. Flahey.) (d) Distribution map.

tri-lobed capsule with one to three glossy black seeds (no more than one in each lobe), which ripen in 4–6 weeks. The seeds are shed by the end of summer. By autumn a bulb has been produced, which will overwinter and sprout in the spring. Asexual reproduction occurs by lateral buds developing into new bulbs. The brownish bulbs are slenderly ovoid, 1.5–6 cm (0.5–2.4 inches) long, with fibrous roots below. The lateral buds are smaller than the mother bulb and remain attached for several years by a short rhizome (lateral underground stem). Eventually, decomposition of the rhizome results in independent plants. Left undisturbed, vegetative reproduction may result in large, dense patches. Some clones are sterile, reproducing only vegetatively. Some populations can have millions of "plants" (i.e. bulbs), although harvesting pressures are making this less common.

Two kinds of *A. tricoccum*, alternatively recognized as varieties of this species, or as separate species, have been recognized: var. *tricoccum* and var. *burdickii* Hanes (=*Allium burdickii* (Hanes) A.G. Jones). The latter is said to have smaller bulbs, leaves, stems, and fewer flowers, and to be earlier-flowering and tending to occupation of drier habitats. However, the characteristics and geographical distributions of these groups require clarification, and some specialists have questioned whether they deserve to be recognized.

### Warning

See the warnings in Chapter 23 on Canada Garlic regarding hazards associated with *Allium* species.

## CULINARY PORTRAIT

The white bulbs and green leaves of wild leek are edible, and indeed many consider *A. tricoccum* to be the best tasting of the 90 or so native species of *Allium* of North America. On the other hand, wild leek has a pungent garlic–onion odor that seems noxious to some and has been described as "garlic with an attitude." The taste has also been described as an assertive combination of onion and strong garlic. By comparison, the much more popular commercial leek (*A. ampeloprasum* L. var. *porrum* (L.) J. Gay (*A. porrum* L.)) is a much milder vegetable. Nevertheless, wild leek was characterized by Euell Gibbons, in his 1962 classic *Stalking the Wild Asparagus*, as the sweetest and best of the North American onion bulbs, and it is considered to be a gastronomic delight by those who have acquired a taste for it. High-end restaurants often incorporate wild leek into expensive creations, and this gourmet vegetable has considerable snob appeal. However, it has traditionally been harvested by people of humble origin. Wild leek bulbs were eaten raw or dried for winter use by Iroquois and other Native American tribes in Minnesota and Wisconsin. North American Indians also commonly baked wild leek in ashes. Tons of bulbs harvested from the Appalachians are sold in the northeast United States. Many southern Appalachian communities have "ramp festivals" from mid-March to May, including one in Bradford, Pennsylvania called "Stinkfest." In Appalachia, it is widely believed that the plants are a spring tonic, and since it does provide vitamins and minerals in early spring, after an absence of fresh vegetables during the winter, it may well have had a tonic effect for many in the region. In Appalachia, wild leek is commonly fried in bacon grease along with potatoes, or scrambled with eggs and served with beans, cornbread, and bacon. Wild leeks are also incorporated into soups, stews, puddings, and other foods, and often substituted for garlic and onion. They can be chopped and used like chives, and they are excellent served as scallions in salads and savory dishes. Wild leeks can be refrigerated for at least a week, pickled, or dried for later use, or pressure canned.

### Culinary Vocabulary

- Louis Diat, chef de cuisine at the New York City Ritz-Carlton, named a famous thick, creamy leek and potato soup he created "Vichyssoise" after his home town of Vichy in France. (Although the date 1910 is commonly cited, there is debate about just when Diat created the soup. Also, Diat noted that his mother had made a leek and potato soup, so his invention may not have been original.) When Germany occupied roughly two thirds of France in 1940, the remaining portion of the country became a German puppet state with its capital at Vichy. Vichy France was hated by French Americans, many of whom were employed in the food industry. Vichy water was banned in their restaurants. (Vichy water originally referred to a naturally effervescent mineral water from the springs at Vichy, but came to refer to any sparkling mineral water resembling it.) In 1941, a conference of chefs met and voted to rename the chilled soup "crème Gauloise," or "Gallic cream," to rid it of the Vichy name. However, the name Vichyssoise survived. Vichysoisse is usually eaten cold, in accord with its full name, *crème vichyssoise glacée* ("iced cream [soup] of Vichy").

## PROSPECTS

The edible bulbs of wild leek are widely harvested by wildcrafters and there is some commercial trade. Tons of bulbs harvested from the Appalachians are sold in the northeastern United States. In Canada, wild leek is concentrated in heavily populated areas of southern Ontario and Quebec, where habitat destruction has occurred, endangering survival of the plant. Further, road building has increased accessibility. Some harvesters have been known to virtually wipe out populations by collecting everything in sight; others doom populations to extinction by harvesting all of the large plants. In Quebec, millions of newly harvested bulbs were widely available at roadside stands until the mid-1990s when legislation was enacted to prevent any form of trade in the wild plant. Even commercial cultivation is discouraged in Quebec because of fear that in the marketplace one could not distinguish legally cultivated from illegally collected wild material. In parts of the United States, such as North Carolina and Tennessee, overharvesting is also a concern. Sustainable recommendations for harvesting wild plants include the following: Collect after the period of spring photosynthesis is completed (not until early summer), to allow the bulbs to mature. If leaves are individually harvested, take only one from a plant and leave at least one. Do not collect the largest plants, since they are critical to population survival. Do not collect more than one plant in 10. Do not collect if the population has less than 1000 plants.

The possibility of cultivating wild leek is a good idea, because the plant is popular, and the wild supply is limited. Wild leek is an extremely attractive culinary novelty with such obviously high sales potential and profit margin that it should be developed agriculturally.

## CURIOSITIES OF SCIENCE AND TECHNOLOGY

- Wild leek was commonly used medicinally by indigenous North American people. The Chippewa employed boiled wild leek as an emetic, the Iroquois ate it as a spring tonic and dewormer for children, and the Cherokee consumed it for colds and used it as a mild laxative.
- The city of Chicago derives its name from the Menomini Indian word *shika'ko*, meaning "skunk place," a reference to the odor of the wild leek that was prevalent at the location.
- Most seeds of wild leek are collected by mice and are often stored in caches. Some of these seeds germinate to produce new plants, so that while the mice consume most of the seeds, they perform a service by distributing the plant.
- Morel harvesters in the United States often employ the height of flowering wild leeks as an indication of when to begin searching for the valuable mushrooms that also appear in late spring. Not surprisingly, in regions where both occur, morels are often served with wild leeks.

## KEY INFORMATION SOURCES

Anonymous. 2005. Are ramp festivals sustainable? *Non-Wood News (Food and Agriculture Organization, U.N.)* 12: 13.

Calvey, E.M., White, K.D., Matusik, J.E., Sha, D., and Block, E. 1998. *Allium* chemistry: Identification of organosulfur compounds in ramp (*Allium tricoccum*) homogenates. *Phytochemistry* 49: 359–364.

Core, E.L. 1945. Ramps. *Castanea* 10: 110–113.

Davis, J.M., and Greenfield, J. 2002. Cultivating ramps: wild leeks of Appalachia. In *Trends in new crops and new uses*. Edited by J. Janick and A. Whipkey. ASHS Press, Alexandria, VA. pp. 449–452.

Jones, A.G. 1979. A study of wild leek, and the recognition of *Allium burdickii* (Liliaceae). *Syst. Bot.* 4: 29–43.

Jones, A.G., and Shildneck, P. 1980. Note on the distribution of wild leek in Illinois *Allium burdickii*, *Allium tricoccum*. *Trans. Ill. State Acad. Sci.* 72(3): 56–59.

McNeal, D.W., Jr., and Jacobsen, T.D. 2002. *Allium*. In *Flora of North America north of Mexico*. Vol. 26. Edited by Flora of North America Editorial Committee. Oxford University Press, New York, NY. pp. 224–276.

Nantel, P., Gagnon, D., and Nault, A. 1996. Population viability analysis of American ginseng and wild leek harvested in stochastic environments. *Conserv. Biol.* 10: 608–621.

Nault, A., and Gagnon, D. 1987. Some aspects of the pollination ecology of wild leek, *Allium tricoccum* Ait. *Plant Species Biol.* 2: 127–132.

Nault, A., and Gagnon, D. 1988. Seasonal biomass and nutrient allocation patterns in wild leek (*Allium tricoccum* Ait.), a spring geophyte. *Bull. Torr. Bot. Club* 115: 45–54.

Nault, A., and Gagnon, D. 1993. Ramet demography of *Allium tricoccum*, a spring ephemeral, perennial forest herb. *J. Ecol.* 81: 101–119.

Platt, E.S. 2003. *Garlic, onion & other alliums*. Stackpole Books, Mechanicsburg, PA. pp. 150.

Rock, J.H., Beckage, B., and Gross, L.J. 2004. Population recovery following differential harvesting of *Allium tricoccum* in the southern Appalachians. *Biol. Conserv.* 116: 227–234.

Small, E., Cayouette, J., and Catling, P.M. 1999. Poorly known economic plants of Canada—22. Wild leek, *Allium tricoccum* Solander var. *tricoccum* and var. *burdickii* Hanes. *Bull. Can. Bot. Assoc.* 32(3): 39–42.

Vasseur, L. 2001. Allozymic diversity of *Allium tricoccum* (Ait.) Solander var. *burdickii* Hanes in isolated populations of Nova Scotia (Canada). *Plant Syst. Evol.* 228: 71–79.

Vasseur, L., and Gagnon, D. 1994. Survival and growth of *Allium tricoccum* Ait. transplants in different habitats. *Biol. Conserv.* 68: 107–114.

Vasseur, L., Gagnon, D., and Simon, J.P. 1990. Isoenzymatic variability among populations and varieties of wild leek (*Allium tricoccum*). *Biochem. Syst. Ecol.* 18: 321–324.

Whanger, P. D., Ip, C., Polan, C.E., Uden, P.C., and Welbaum, G. 2000. Tumorigenesis, metabolism, speciation, bioavailability, and tissue deposition of selenium in selenium-enriched ramps (*Allium tricoccum*). *J. Agric. Food Chem.* 48: 5723–5730.

## Specialty Cookbooks

Numerous recipes are available on the Web. The following book is perhaps the largest source of wild leek recipes.

Facemire, G., Jr. 2009. *Having your ramps and eating them too.* McClain Printing Co., Parsons, WV. pp. 148.

Beatty (1987) presents 7 wild leek recipes, Brill (2002) has 32, Freitus (1980) has 5, Mogelon (2001) has 3, and Tatum (1976) has 6. Berglund and Bolsby (1971), Boorman (1969), Russell (1975), Robe-Terry (1997), and Stewart (2000) each have one recipe. (See the Appendix to this book for details of the publications cited.)

# 98 Wild Rice

Family: Poaceae (Gramineae, grass family)

## NAMES

Scientific names: *Zizania* species

- Northern wild rice (northern zizania)—*Z. palustris* L. var. *palustris* L.
- Interior wild rice (interior zizania)—*Z. palustris* var. *interior* (Fassett) Dore
- Southern wild rice (southern zizania)—*Z. aquatica* L. var. *aquatica*
- The name "wild rice" was coined by analogy with the rather similar cultivated rice, *Oryza sativa* L., when European explorers observed that wild rice like rice was a cereal growing in water. The "wild" part of the name is problematical, given that this word can mean (1) undomesticated (not genetically altered by humans) and (2) growing outside of cultivation (either (a) never altered by humans or (b) altered by humans but escaped from cultivation, and perhaps having re-evolved some adaptations for living outside of cultivation). In the genus *Oryza,* there are species or varieties that are "wild" in these senses, and in fact they are called "wild rice." Indeed, use of the name "wild rice" in scientific articles is often only clear when the genus name is specified. Canadian botanist W.G. Dore (1912–1996), who specialized on wild rice, attempted to use a hyphenated phrase ("wild-rice") to designate *Zizania*, but this practice has generally not been followed. Occasionally "wildrice" is used for *Zizania*. In the past, "wild rice" was an entirely appropriate name for *Zizania* species, since the plants were entirely undomesticated and uncultivated. Today there are strains that are both cultivated and domesticated, so that the name wild rice is no longer as appropriate as in the past.
- Obsolete names for wild rice include American rice, blackbird oat(s), Canada rice, Canadian rice, Indian rice (a name also applied to *Oryza sativa* grown in India), false oat(s), squaw rice, Tuscarora, water oat(s), and zizania.
- Another obsolete name used in English for wild rice is manomin, the Ojibwe (Ojibway, Ojibwa) word for wild rice. The term has been interpreted as "good berry." It has been said that it is derived both from *Manitou*, the name of the Great Spirit, and *Meenum*, which means delicacy. Ojibwe elders called wild rice *Manitou gi ti gahn*, or "food from God's garden." Wild rice is a sacred food for the Ojibwe and is deeply imbedded in their mythology and ceremonies. The Menominee Indians of Wisconsin take their tribal name from their word for wild rice, *menomin*.
- Early French explorers called wild rice *folles avoines*, meaning "crazy oats."
- "Wild pecan rice" (also known as pecan rice) is a long-grained, nutty-tasting variety of *Oryza sativa* grown in southern Louisiana. It should not be confused with the wild rice discussed here.
- The genus name *Zizania* is derived from a Greek name for a wild grain but not the plants currently designated.
- *Palustris* in the scientific name *Z. palustris* is Latin for swampy or marshy.
- *Aquatica* in the scientific name *Z. aquatica* is Latin for aquatic.
- *Texaxa* in the scientific name *Z. texana* (discussed below) is Latin for Texan.

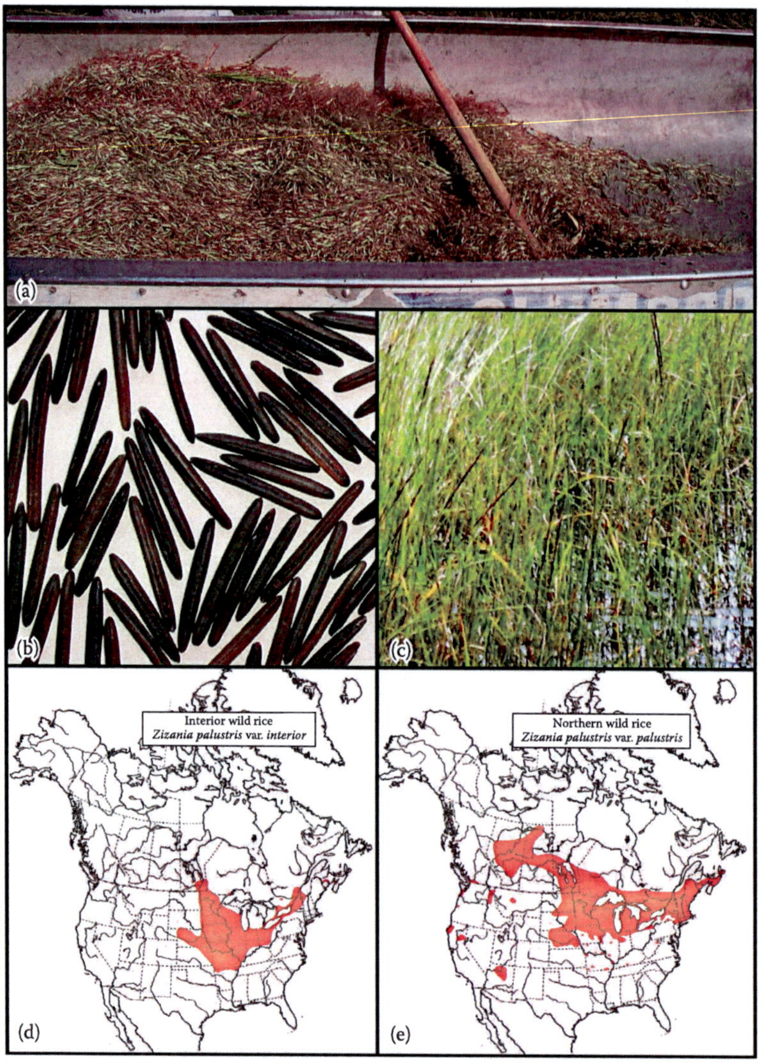


**FIGURE 98.1** Wild rice (*Zizania palustris*). (a) A canoe with collected raw wild rice. (Courtesy of U.S. Fish and Wildlife Service.) (b) "Seeds" (caryopses). (Courtesy of Stephen Darbyshire.) (c) Dense stand in northern Wisconsin. (Courtesy of Chemoqua.) (d) Distribution map of interior wild rice (*Z. palustris* var. *interior*). (e) Distribution map of northern wild rice (*Z. palustris* var. *palustris*), the predominant harvested wild rice.

## GEOGRAPHY AND ECOLOGY OF WILD PLANTS

Northern wild rice (*Z. palustris* var. *palustris*) is widespread: in the north, along shores from western Nova Scotia to Minnesota; and in the south, from northern New England to northern Indiana. Northern wild rice is widely planted, and therefore its range has been extended.

Interior wild rice (*Z. palustris* var. *interior*) is found in southeastern Manitoba and adjacent Ontario in Canada, and in the United States it is distributed from Lake Michigan to North Dakota, south to Missouri and Texas. This variety generally grows on stream banks and shallows, from which the water retreats in the summer, so that canoe harvesting is very difficult, and collecting is generally done on foot.

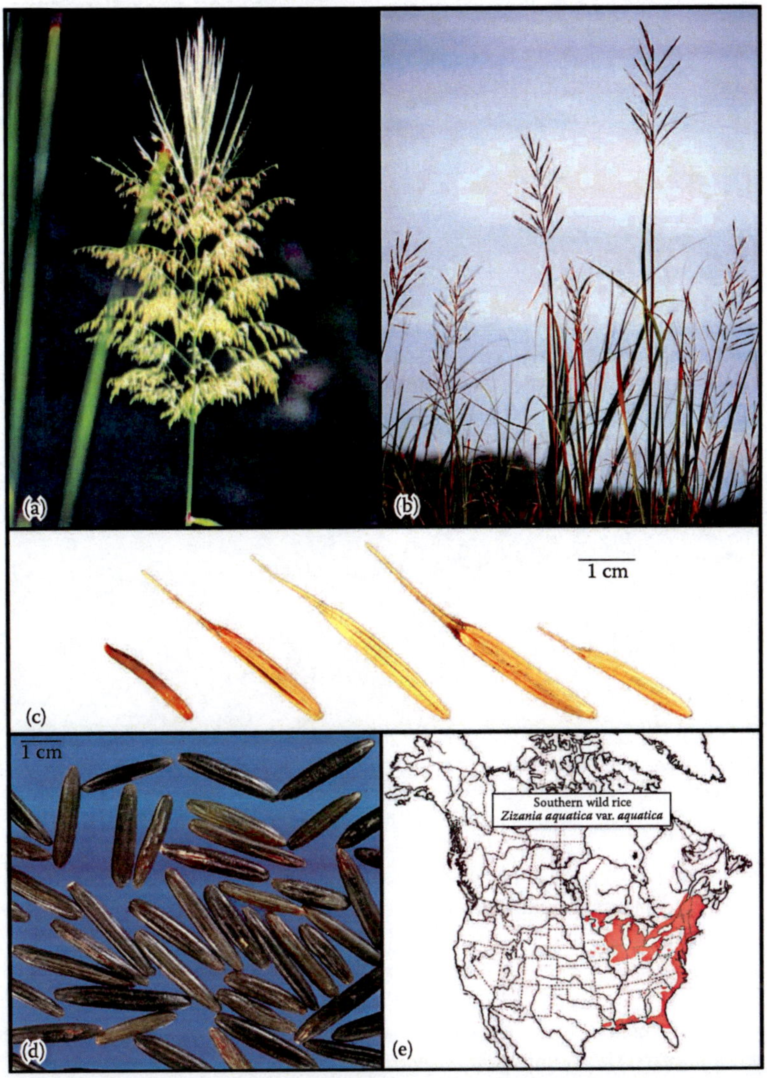


**FIGURE 98.2** Southern wild rice (*Zizania aquatica* var. *aquatica*). ([a] and [b] Courtesy of Mary Hollinger, U.S. National Oceanic and Atmospheric Administration, on coast of Maryland.) (c) Fruits ("seeds," caryopses) enclosed by bracts (palea, lemma), except for first from left, from which the bracts have been removed. (Courtesy of Steve Hurst, U.S. Department of Agriculture.) (d) "Seeds" (caryopses). (Courtesy of Howard F. Schwartz, Colorado State University, online at Bugwood.org [CC By 3.0].) (e) Distribution map.

Southern wild rice (*Z. aquatica* var. *aquatica*) occurs on the shores of the Great Lakes and on muddy shores of streams in southern Ontario and Quebec, ranging southwards to western Florida and Louisiana. It seems to be most successful in very shallow water, where the seeds are more likely to encounter the warm conditions that this variety requires for germination. It is widely distributed in the eastern United States, where it is often collected from the wild, but is present in Canada only along the Great Lakes in southern Ontario, and in the low lands of southwestern Quebec.

Wild rice species are aquatic plants that grow mostly in shallow water along the margins of rivers, ponds, and lakes. Dense stands are often produced in shallow water, with a preference for depths of 30–60 cm (1–2 feet). The plants are very intolerant of stagnant water, although the rate of water flow need not be large. Ideally a muddy bottom is present, although the species are occasionally

FIGURE 98.3 "Ojibwa Indian women gathering wild rice," an oil painting by American painter Seth Eastman (1808–1875), in the House Committee on Military Affairs, Washington, DC. (From Thwaites, R.G. (Editor), *Collections of the State Historical Society of Wisconsin*. Vol. 20, Plate 4, 1911.)

found on sand. For the most part, wild rice occurs in fresh water and coastal estuaries but not in the more saline coastal salt marshes and interior salt lakes.

## PLANT PORTRAIT

Wild rice species are generally tall, usually 1–2 m (3–6.6 feet) in height, sometimes reaching 5 m (16 feet). *Zizania palustris* is an annual, *Z. aquatica* is a perennial. Northern wild rice (*Z. palustris* var. *palustris*) is the most important commercial variety of wild rice. It produces the largest grain of all of the varieties (often 20 mm or 0.8 inch in length) and is sometimes referred to as "giant wild rice." In this case, "giant" refers to the grains, as the plants are often of smaller stature than the other two varieties of commercial importance, interior wild rice (*Z. palustris* var. *interior*) and southern wild rice (*Z. aquatica* var. *aquatica*). Interior wild rice produces grains that are not quite as long as those of northern wild rice, and are not considered quite as desirable because of the market preference for larger, long-grained wild rice. The narrowness and relative shortness of the grain of southern wild rice makes it less attractive as food.

Wild rice grains are two or more times the length of long-grained common rice, and are gray or brown. The seeds were one of the chief foods of Native American Indian tribes in the Great Lakes region, where wild rice has been gathered since prehistory. Native Americans of the Algonquian linguistic family, especially the Ojibwe and Menominee (Menomini), and certain Sioux, used wild rice as one of their staple foods, and warred for centuries for control of the fields. Native Americans gathered the seeds by pulling the grain heads over their canoes and flailing them with paddles. The harvested seeds were sun-dried or parched over a slow fire to crack the hulls, then the grain was threshed by trampling (to loosen the chaff surrounding the grain), and winnowed (separated from the chaff).

Until about 1950, natural stands were the only source of wild rice. Beginning then, experiments were conducted in Minnesota to grow wild rice as a field crop. By the 1970s, Minnesota took the lead in wild rice production, based on prepared wetland (paddy or aquacultural) cultivation. In this method, *Zizania* is raised in dyked, shallowly flooded fields, like true rice (*Oryza*). Common rice is established in paddies by transplanting shoots, not seeds. Wild rice seed, however, is sown. Yields from wild rice paddies are often 10 times as much as obtained by collecting from natural stands. California, which began paddy cultivation of wild rice in 1978, can generate two crops in a year, and by 1986 the state took the lead in production. World harvest of wild rice has grown by a factor of 10 since 1976, with the value of the crop increasing from about 1 million to 20 million dollars. California now produces 65% of the world crop, Minnesota 30%, and Canada 5%. Wild rice is a premium food, commanding a high price, and sold to the epicurean market. Freshly harvested ("green") wild rice has a moisture content of about 40% and will deteriorate quickly unless processed. To prevent decay, green rice is "cured," generally by drying and aerating the mass of grain. Drying usually is carried out over a period of days, but can be hastened by artificial heat. Many processors deliberately carry out a controlled slow drying, during which some fermentation takes place, in the belief that this improves the flavor. Subsequently, the grains are parched in ovens, a process that drives off residual water and results in the kernels becoming hard, and the grip of the hulls on the grains loosening. Hulling is carried out by mechanical agitation, coupled with air jets to drive off the lighter empty hulls. Unlike "polished" common rice, the pericarp (fruit wall) is not removed or only partially removed (partial removal is preferred when the wild rice is mixed with true rice, in order that the cooking times be compatible). Only one-third to one-half of the weight of green rice is recovered as processed rice. The rice is usually sold without further processing, although precooking, canning, or incorporation into prepared products may also be carried out.

## CULINARY PORTRAIT

Wild rice has a chewy texture and a strong, nutty flavor, often compared to hazelnut. Although expensive, it goes farther than regular rice, 1 cup of dry rice making about 3.5 cups of prepared rice (Asian, i.e., common rice, typically combines 1 cup of rice and 1.7 to 2 cups of water). Wild rice is a highly nutritious grain, richer in protein and with a better balance of protein components (amino acids) than conventional rice. The distinctive flavor of this gourmet item is admired by many. Wild rice is frequently served with game, particularly with ducks and other game birds, as it is said to enhance their flavor (indeed, wild rice goes well with all poultry). It is also often served with sautéed mushrooms, onions, or slivered almonds (or other nuts), all of which seem to complement wild rice. Wild rice also is compatible with seafood, and is used in stuffings and crêpes. To extend it, wild rice is often combined with other grains, such as brown rice or bulghur wheat (about one part wild rice to two parts of the other grain is typical). The dark color of wild rice grains contrasts attractively when combined with white rice and other light-colored foods.

Marketed wild rice tends to be dirty and needs to be thoroughly rinsed. It is cooked and eaten like common rice but requires a longer cooking time (overcooking diminishes the flavor and texture). It can be boiled in salted water, steamed like rice, or baked. Wild rice can even be popped, but this is rarely done.

Wild rice can be presoaked for several hours before boiling to reduce cooking time. A shorter method of soaking is to boil the rice for 5 minutes in four times its volume of water, cool it and allow it to soak for 1 hour in a covered container, then drain and reheat with fresh water or broth (using three times the volume of rice), along with a half teaspoon of salt. The water is brought to a boil, the rice added, the heat then reduced and the mixture simmered for about 20 minutes or until tender. Alternatively, leave out the initial soaking step and simply cook the rice for about 40 minutes in three times its volume of water. The rice will expand to about four times its initial volume. Overcooking causes a loss of flavor and firmness, and the rice becomes sticky.

### Culinary Vocabulary

- One of the most famous Indian dishes was tassimanonny: wild rice, corn, and fish boiled together.

## PROSPECTS

Minnesota and California dominate the wild rice industry, their success in recent decades the result of growing domesticated selections in artificial (paddy) aquatic systems. The collection of wild rice from nature, largely in Canada, is now responsible for less than 5% of market share. Although wild-collected wild rice has a special niche market, it is clear that the wild rice industry will remain based on cultivation of domesticated varieties. Attempts have been made to grow wild rice in Europe and Asia, but there is no indication that the American supply will encounter significant competition in the near future. Wild rice breeding is only several decades old, and cultivars produced to date are little-changed from the wild. Accordingly, future success depends on breeding superior varieties, and improving production systems. A special problem for wild rice as a crop is the need for wet environments and the associated requirement to avoid pollution. There has been some success in cultivating wild rice on (relatively) dry land, and this may become more important in the future. Wild rice is a very special, high-value, gourmet crop, in high demand, and it seems very likely to become more important in the future.

## CURIOSITIES OF SCIENCE AND TECHNOLOGY

- American Indians popped wild rice, like popcorn, and sometimes served it with maple sugar as a special treat. The popped rice was sometimes carried by hunters and fishermen, for use as an energy snack.
- In wild rice areas of North America, an elaborate burial and bereavement procedure was sometimes carried out by relatives of the deceased. On top of the grave a house was built, with a window on one side into which food was symbolically inserted for the departed soul. Those who participated in a year of bereavement were restricted from collecting rice without a taboo release. This involved their being spoon-fed some of the first wild rice gathered.
- Many wind-pollinated plants (e.g., corn) have separate male and female flowers, and in these cases the male flowers are usually located above the female flowers. This is because the wind carries pollen to other plants of the same species, so the higher up the male flowers are the more easily they will be transported by wind. In wild rice, however, the reverse pattern occurs, with the female flowers above the males. The female flowers mature first, promoting cross-fertilization.
- A critical event in the domestication of all cereals (including wheat, rice, corn (maize), barley, sorghum, millet, oat, and rye) was the selection of "nonshattering" strains—variants with grains that do not drop off the plant at maturity. The principal difficulty with wild rice as a crop has been the quick shedding of the grains at maturity. Further, because the grains mature over a period, they are shed gradually. These two factors mean that it is necessary to harvest the same plants repeatedly to obtain even half of the grains. The selection of nonshattering varieties of wild rice in the 1960s and 1970s in Minnesota has enormously benefited the wild rice industry. These nonshattering varieties retain much of their seed for long periods, making harvesting much more efficient.
- Estuarine wild rice (*Z. aquatica* var. *brevis* Fassett) is a remarkable plant that occurs on freshwater tidal flats (tides occurring twice daily) and tributary streams of the St. Lawrence River estuary in southern Quebec and eastern Ontario. The grains are rather small, and estuarine wild rice has never been harvested commercially. It is the only wild rice that

tolerates much fluctuation in water levels, and indeed the plants are often completely submerged for part of each day.

- Texas wild rice (*Z. texana* Hitchcock) is a very rare endemic species known only from Hays County, Texas. It is not currently used as a source of food, but because of its rarity and its potential usefulness as a source of genes for improving wild rice as a crop, a great deal of research has been carried out on it.
- In 1977, the Minnesota state legislature made wild rice the official state grain.
- It has been calculated that human-made wild rice paddies in Minnesota provide 3200 duck-use days/ha (a duck-use day means the equivalent of one duck using 1 hectare for 1 day—e.g., 24 ducks using 1 hectare for 1 hour each).
- Manchurian wild rice (also known as Manchurian water rice and water bamboo), *Z. latifolia* (Griseb.) Turcz. ex Stapf., is native to Asia. Its grains were used as a cereal in ancient times and are occasionally used for this purpose today. The base of the stems is often infected by a fungus that causes the shoots to swell and become tender. These swollen stems are prized as a vegetable in Asia. However, because the fungus causes the plants to become sterile, there is great concern that if it is imported to North America, there could be disastrous consequences for the American species. Fungus-infected swollen stems of Manchurian wild rice are occasionally available in cans in Chinese markets under the Cantonese name *gau sun*.

## KEY INFORMATION SOURCES


Aiken, S.G., Lee, P.F., Punter, D., and Stewart, J.M. 1988. *Wild rice in Canada*. NC Press, Toronto, ON. pp. 130.

Anderson, R.A. 1976. Wild rice: nutritional review. *Cereal Chem.* 53: 945–955.

Anderson, R.A., Navickis, L.L., Warner, K.A., Vojnovich, C., and Bagley, E.B. 1979. Quality characteristics of processed wild rice *Zizania aquatica*. *Cereal Chem.* 56: 375–379.

Anderson, R.A., Vojnovich, C., Navickis, L.L., and Bagley, E.B. 1979. Parching studies on wild rice *Zizania aquatica*. *Cereal Chem.* 56: 371–374.

Archibold, O.W. 1995. *Wild rice in Saskatchewan: Agricultural development in harmony with nature: a reference manual*. Saskatchewan Education, Training and Employment, Saskatchewan Agriculture and Food, La Ronge, SK. pp.99.

Archibold, O.W. 1999. *Wild rice research in Saskatchewan: Final report*. Department of Crop Science and Plant Ecology, University of Saskatchewan, Saskatoon, SK. pp. 269.

Archibold, O.W., and Weichel, B.J. 1986. Variation in wild rice *Zizania palustris* stands across northern Saskatchewan Canada. *Can. J. Bot.* 64: 1204–1211.

Archibold, O.W., Good, A.G., and Sutherland, J.M. 1989. Annual variation in wild rice *Zizania palustris* L. growth and potential yield in Saskatchewan Canada. *Can. J. Plant Sci.* 69: 653–666.

Burnson, B. 1972. The taming of wild rice [*Zizania aquatica*]. *Minn. Volunteer* 35(200): 44–51.

Counts, R.L. 1993. Phenotypic plasticity and genetic variability in annual *Zizania* spp. along a latitudinal gradient. *Can. J. Bot.* 71: 145–154.

Counts, R.L., and Lee, P.F. 1991. Germination and early seedling growth in some northern wild rice *Zizania palustris* populations differing in seed size. *Can. J. Bot.* 69: 689–696.

De Wet, J.M.J., and Oelke, E.A. 1978. Domestication of American wild rice (*Zizania aquatica* L., Gramineae). *J.A.I.B.A.J. Agric. Tradit. Bot. Appl. (Paris)* 25(2): 67–84.

Dore, W.G. 1969. *Wild-rice*. Canada Department of Agriculture Research Branch Publication 1393, Ottawa, ON. pp. 84.

Duvall, M.R. 1995. Wild rice (*Zizania palustris*). In *Cereals and pseudocereals*. Edited by J.T. Williams. Chapman & Hall, London, U.K. pp. 261–271.

Elliott, W.A. 1980. Wild rice, *Zizania aquatica*. In *Hybridization of crop plants*. Edited by W.R. Fehr and H.H. Hadley. American Society of Agronomy, Madison, WI. pp. 721–731.

Everett, L.A., and Stucker, R.E. 1983. A comparison of selection methods for reduced shattering in wild rice. *Crop Sci.* 23: 956–960.

Grant, R.S. 1981. Protein-rich wild rice, our only native cereal. *Can. Geogr.* 101(4): 56–61.

Hanten, H.B., Ahlgren, G.E., and Carlson, J.B. 1980. The morphology of grain abscission in *Zizania aquatica*. *Can. J. Bot.* 58: 2269–2273.

Hass, B.L., Pires, J.C., Porter, R., Phillips, R.L., and Jackson, S.A. 2003. Comparative genetics at the gene and chromosome levels between rice (*Oryza sativa*) and wildrice (*Zizania palustris*). *Theor. Appl. Genet.* 107: 773–782.
Hayes, P.M., Stucker, R.E., and Wandrey, G.G. 1989. The domestication of American wildrice (*Zizania palustris*, Poaceae). *Econ. Bot.* 43: 203–214.
Hiep, N.T. 1993. *Zizania latifolia* (Griseb.) Turcz. ex Stapf. In *Plant resources of South-East Asia, 8, vegetables*. Edited by J.E. Siemonsma and K. Piluek. Pudoc Scientific Publishers, Wageningen, the Netherlands. pp. 278–279.
Kennard, W.C., Phillips, R.L., and Porter, R.A. 2002. Genetic dissection of seed shattering, agronomic, and color traits in American wildrice (*Zizania palustris* var. *interior* L.) with a comparative map. *Theor. Appl. Genet.* 105: 1075–1086.
Lee, G.A., Davis, A.M., Smith, D.G., and McAndrews, J.H. 2004. Identifying fossil wild rice (*Zizania*) pollen from Cootes Paradise, Ontario: A new approach using scanning electron microscopy. *J. Archaeol. Sci.* 31: 411–421.
Lee, P.F., and Stewart, J.M. 1984. Ecological relationships of wild rice, *Zizania aquatica*. 3. Factors affecting seeding success. *Can. J. Bot.* 62: 1608–1615.
Lorenz, K. 1981. The starch of wild rice (*Zizania aquatica*). *Starch-Staerke* 33: 73–76.
Manitoba Agriculture, 1984. *A guide to wild rice production*. Manitoba Agriculture Agdex 116. Manitoba Agriculture, Winnipeg, MB. pp. 29.
Morton, J.F., Snyder, G.H., and Oelke, E.A. 1980. Wild rice: A promising gourmet crop for the Everglades. *Proc. Annu. Meet. Florida State Hortic. Soc.* 93: 273–278.
Oelke, E.A. 1976. Amino acid content in wild rice (*Zizania aquatica* L.) grain. *Agron. J.* 68: 146–148.
Oekle, E.A. 1993. Wild rice: Domestication of a native North American genus. In *New crops*. Edited by J. Janick and J.E. Simon. John Wiley & Sons, New York, NY. pp. 235–243.
Oelke, E.A., and Boedicker, J.L. 2000. Wild rice: Processing and utilization. In *Handbook of cereal science and technology*. Second edition. Edited by K. Kulp and J.G. Ponte, Jr. Marcel Dekker, New York, NY. pp. 275–295.
Oelke, E.A., Porter, R.A., Grombacher, A.W., and Addis, P.B. 1997. Wild rice—new interest in an old crop. *Cereal Foods World* 42: 234–247.
Pip, E., and Stepaniuk, J. 1988. The effect of flooding on wild rice, *Zizania aquatica* L. *Aquatic Bot.* 32: 283–290.
Poole, J., and Bowles, D.E. 1999. Habitat characterization of Texas wild-rice (*Zizania texana* Hitchcock), an endangered aquatic macrophyte from the San Marcos River, TX, USA. *Aquatic Conserv.* 9: 291–302.
Power, P. 2002. Resource allocation patterns and phenotypic variation in the endangered Texas wildrice (*Zizania texana*, Poaceae). *SIDA* 20: 571–582.
Schultz, B. (Editor). 1979. *The wild ricer's guide*. Appleseed Press, Berkeley, CA. pp. 171.
Small, E., and Catling, P.M. 2004. Poorly known economic plants of Canada—42. Wild rice (*Zizania aquatica and Z. palustris*). *Bull. Can. Bot. Assoc.* 37(3): 45–52.
So, M.L. 1992. Effect of storage on the changes in the nutritional content of wildrice (*Zizania latifolia*). *Int. J. Trop. Plant Diseases* 10: 189–194.
Steeves, T.A. 1952. Wild rice—Indian food and a modern delicacy. *Econ. Bot.* 6: 107–142.
Swain, E.W., Wang, H.L., and Hesseltine, C.W. 1978. *Note* on vitamins and minerals of wild rice [*Zizania aquatica*]. *Cereal Chem.* 55: 412–414.
Terrell, E.E., and Batra, L.R. 1982. *Zizania latifolia* and *Ustilago esculenta*, a grass-fungus association. *Econ. Bot.* 36: 20–26.
Terrell, E.E., Emery, W.H.P., and Beaty, H.E. 1978. Observations on *Zizania texana* (Texas wildrice), an endangered species. *Bull. Torrey Bot. Club* 105: 50–57.
Terrell, E.E., Peterson, P.M., Reveal, J.L., and Duvall, M.R. 1997. Taxonomy of North American species of *Zizania* (Poaceae). *SIDA* 17: 533–549.
Terrell, E.E., and Wiser, W.J. 1975. Protein and lysine contents in grains of three species of wild-rice (*Zizania*: Gramineae). *Bot. Gaz.* 136: 312–316.
Vennum, T., Jr. 1988. *Wild rice and the Ojibway people*. Historical Society Press, St. Paul, MN. pp. 357.
Wang, H.L., Swain, E.W., Hesseltine, C.W., and Gumbmann, M.R. 1978. Protein quality of wild rice [*Zizania aquatica*]. *J. Agric. Food Chem.* 26: 309–312.
Warwick, S.I., and Aiken, S.G. 1986. Electrophoretic evidence for the recognition of two species in annual wild rice (*Zizania*, Poaceae). *Syst. Bot.* 11: 464–473.
Watts, B.M., and Dronzek, B.L. 1981. Chemical composition of wild rice *Zizania aquatica* grain. *Can. J. Plant Sci.* 61: 437–446.

## Specialty Cookbooks

(*Note:* numerous imaginative recipes for preparing and serving wild rice are also available on the Web.)

Anderson, B. 1977. *Wild rice for all seasons cookbook.* Minnehaha Pub., Minneapolis, MN. pp. 173.

Hauser, S.C. 2000. *Wild rice cooking: Harvesting, history, natural history, and lore with 80 recipes.* Lyons Press, New York, NY. pp. 173.

Kelliher-Waskish Wild Rice Festival Committee, 1982. *Wild rice collections (Native American Cookbook).* Circulation Service, Shawnee Mission, KS. pp. 104.

Lund, D.R. 2003. *One hundred and one favorite wild rice recipes.* Adventure Publications, Cambridge, MN. pp. 76.

Ojakangas, B.A. 1992. *The best of wild rice recipes.* Adventure Publications, South Cambridge, MN. pp. 108.

The 1006 Summit Avenue Society, 1986. *Wild rice, star of the north: 150 Minnesota recipes for a gourmet grain.* McGraw-Hill, New York, NY. pp. 188.

Webber, H., and Woolsey, M. 2002. *Wild & wonderful wild rice cookbook.* Blueberries & Polar Bears Publishing, Churchill, MB. pp. 48.

# 99 Winter Purslane

Family: Portulacaceae (purslane family)

## NAMES

Scientific name: *Claytonia perfoliata* Donn ex Willd. (*Montia perfoliata* (Donn ex Willd.) Howell)

- "Summer purslane" is *Portulaca oleracea* L., which is also in the purslane family. This better-known vegetable is a native of Eurasia but has become a widespread introduced weed in North America. Winter purslane, by contrast with summer purslane, is adapted to a cooler climate, hence the "winter" in the name. Winter purslane often is available (depending on location) in late winter.
- The word "purslane" is based on the Latin word *purtulaca*, which was the basis of the genus *Portulaca*, of which summer purslane is one of the species.
- Winter purslane is also known as Cuban lettuce, Cuban spinach, Indian lettuce (a name applied to other species, particularly *Lactuca indica* L.), miner's lettuce, Naiad spring beauty, Spanish lettuce, and spring beauty.
- The common name "miner's lettuce" refers to the use of the leaves by prospectors and miners as a salad green and scurvy preventative (because of its high vitamin C content) during the time of the California gold rush in 1849.
- The genus name *Claytonia* commemorates John Clayton (1686–1773), a botanist in North America.
- *Perfoliata* in the scientific name *C. perfoliata* is a technical botanical term describing a leaf that seems to lack a leaf stalk and the stem on which it is held seems to go right through the leaf's center (see Plate 99). As noted below, a pair of leaves fused around the flowering stems of *C. perfoliata* gives the appearance of a single, perfoliate leaf.
- "Sea purslane," *Sesuvium portulacastrum* (L.) L., is an unrelated plant of coastal areas, sometimes grown as an ornamental.
- "Florida purslane," *Richardia scabra* L., a native of tropical America, has become a common weed of the southeastern United States.

## GEOGRAPHY AND ECOLOGY OF WILD PLANTS

Winter purslane is native to mountain and coastal regions of western North America. It is found from southern Alaska to British Columbia, in the Western United States, and from Mexico to Guatemala. The plants grow best in cool, damp, semi-shaded locations. They are found on moist banks and slopes, disturbed and waste ground, and dunes, often in sandy soil. In some areas winter purslane is considered to be a weed.

## PLANT PORTRAIT

Winter purslane is a small, hardy, annual, herbaceous plant, usually not taller than 30 cm (1 foot), but sometimes as high as 50 cm (20 inches). The root system is fibrous. The plants develop two forms of leaves, as noted in the following. At the base of the plant, there is a rosette of spade-shaped

**FIGURE 99.1** Winter purslane (*Claytonia perfoliata*). (a) Vegetative plant (the ideal stage for consumption). (Courtesy of Rasbak [CC By 3.0].) (b) Plants in flower. (Courtesy of Silversyrpher [Flickr/CC-attribution].) (c) Flowering plant. (From Curtis, W., *Botanical Magazine*, Vol. 33, Plate 1336, 1811.) (d) Distribution map.

leaves, 6–20 cm (2.4–8 inches) long, including the long leafstalks. Mature plants develop several flowering stalks, which arise from the base. Immediately below a cluster of tiny flowers and near the top of the flowering branches, two leaves are present, these fused around the stem; the pair of leaves appears to be a single leaf, resembling a saucer, up to 5 cm (2 inches) wide. The foliage tends to turn reddish when exposed to full sun. The flowering stems bear tiny white to pale pink flowers, and these develop into tiny fruit capsules with one to three shiny black seeds. Cultivated plants seem to be essentially unchanged from those found in the wild. Native Americans, Spanish colonists, and settlers in California ate the leaves of winter purslane, both raw and cooked, and even used it for tea. Native Americans also used tea brewed from dried winter purslane plant as a laxative. Winter purslane now grows wild in western Europe, where it has been introduced, and it is cultivated as a minor crop in several western European countries, notably Britain, France, Belgium, Germany, and the Netherlands. As is often the case, Europeans learned to appreciate this crop more than have

North Americans, despite the fact that North America is the homeland of the species. This led to confusion for some time regarding what region of the world is actually the native area.

## CULINARY PORTRAIT

The tender, juicy, succulent stems and leaves (and even the roots) of winter purslane can be used raw in salads, and indeed this species has been described as the very best wild salad plant to be found. The plant is also boiled for 2–3 minutes, or briefly steamed or stir-fried, and served like spinach. It can be puréed to make soup, or added to broths, sauces, sandwiches, and a variety of vegetable dishes. It is delicious in omelettes and can be sautéed in butter. Some consider the flavor of winter purslane to be reminiscent of water chestnut, others compare it to spinach, and still others consider it bland. Winter purslane is mucilaginous, reminiscent of okra when cooked, so to avoid the slimy texture it is best combined fresh in salads or added to soups or stews, and not served as a separate side dish. Older leaves may become bitter, and shoots should be picked before the plant starts to flower. Winter purslane has been found to be equal or better nutritionally compared to most common leafy vegetables. It is best consumed soon after picking but can be stored for 2 or 3 days in a refrigerator. For the home gardener, winter purslane is a good cool-season crop that can be used as a self-seeding, edible, mat-forming, hardy landscaping (in full sun, the leaves turn an attractive purple that can make for a decorative border).

### Culinary Vocabulary

- As noted above, winter purslane has been described as "the best wild salad plant." "Salad" is from *sel*, the Latin word for salt, reflecting the use of little more than salt to dress the first salads made in Rome. "Green salad" is more or less the same as "salad," although some use the former to mean a salad composed only of leafy vegetables, in contrast to a salad with dressing and (or) other additions such as croutons and bacon bits.

## PROSPECTS

Winter purslane is a minor vegetable, more familiar in Europe than in North America, but it probably has potential to become more popular in North America. Arugula (*Eruca sativa* Mill.) is an example of a salad and potherb leafy vegetable that came out of obscurity to become quite popular in North America in recent times; winter purslane has a decidedly superior taste compared to arugula. It does not have the keeping qualities of many leafy greens but probably storage research could overcome this problem. Winter purslane is essentially an undomesticated plant, so there is considerable scope for selection of improved cultivars.

## CURIOSITIES OF SCIENCE AND TECHNOLOGY

- California Native Americans reputedly placed winter purslane on red ant hills to pick up formic acid (secreted by ants when they defend themselves) as a piquant dressing, the formic acid acting like vinegar (acetic acid) to provide a tangy taste.
- Although winter purslane is naturally a small plant, under some stressful conditions it can be remarkably small when mature (with fruit): no taller than 1 cm (0.4 inch).
- When the fruit capsule of winter purslane becomes mature and dry, it explodes, scattering the seeds.
- The seeds of winter purslane have an edible fatty attachment (technically an "elaiosome") that is attractive to ants, which often collect them. The ants benefit by obtaining food and the plant benefits by having its seeds dispersed. This widespread relationship between ants and plants is known as myrmecochory.

## KEY INFORMATION SOURCES

Adolphi, K., and Dickore, W.B. 1977. *Claytonia perfoliata* Donn ex Willd. in MTB 4907 Leverkusen. *Gottinger Floristische Rundbriefe* 11(2): 31–33. (In German)

Bernhardt, K.G. 1994. Sociology and dynamics of *Claytonia perfoliata* associations on the East Frisian island Baltrium. *Floristische Rundbriefe* 28: 62–67. (In German)

Borkowsky, O. 1998. Occurrence, ecological niche and phytosociological behavior of *Claytonia perfoliata* Donn ex Willd. in the area of Gifhorn/Braunschweig–SE–Lower Saxony, Germany. *Braunschweiger Naturkundliche Schriften* 5: 607–616. (In German)

Gorske, S.F., and Hopen, H.J. 1978. Case of the purslane sawfly [for biological control of *Portulaca oleracea* and *Montia perfoliata*]. *Am. Veg. Grow.* 26(8): 14–15.

Hirschfeld, D. 1995. *Montia perfoliata* enriches the multiplicity of salads in winter. *Gemuese (Germany)* 31: 612–614. (In German)

Kathan, J.G. 1986. *Montia perfoliata*, an interesting winter culture. *Deutscher Gartenbau (Germany)* 40: 1104–1105. (In German)

Lindner, U. 1987. A weed species as a new vegetable crop. *Gemuese (Germany)* 23: 258–259. (In German)

Miller, J.M. 1976. Variation in populations of *Claytonia perfoliata* (Portulacaceae). *Syst. Bot.* 1: 20–34.

Miller, J.M. 1978. Diploid *Claytonia perfoliata* from southern Mexico. *Madroño* 25: 57–58.

Miller, J.M. 1979. Phenotypic variation distribution and relationships of diploid and tetraploid populations of the *Claytonia perfoliata* complex, Portulacaceae. *Syst. Bot.* 3: 322–341.

Miller, J.M. 1981. Flavonol glycoside variation in diploids of the *Claytonia perfoliata* complex (Portulacaceae). *Syst. Bot.* 6: 27–30.

Miller, J.M. 1984. Chromosome numbers and cytogeographic patterns in *Claytonia perfoliata* and *Claytonia parviflora* (Portulacaeae). *Madroño* 31: 60–61.

Miller, J. M. 2003. *Claytonia*. In *Flora of North America North of Mexico*. Vol. 4. Edited by Flora of North America Editorial Committee. Oxford University press, Oxford, U.K. pp. 465–475.

Miller, J.M., and Chambers, K.L. 2006. Systematics of *Claytonia* (Portulacaceae). *Syst. Bot. Monogr.* 78: 1–236.

Petrikova, K. 1999. *Montia perfoliata* (*Claytonia perfoliata*)—possibilities of cultivation. In *Eucarpia Leafy Vegetables '99': Proceedings of the Eucarpia meeting on leafy vegetables genetics and breeding, Olomouc, Czech Republic, Palacky University, 8–11 June, 1999*. Edited by A. Lebeda and E. Krístková. Palacký University, Olomouc, Czech Republic. pp. 53–55.

Pohl, K. 1979. *Claytonia perfoliata* Donn ex Willd., the proliferation *Claytonia*, also in Wetzlar West Germany. *Hessische Floristische Briefe* 28(3): 48–49. (In German)

Schelstraete, M, and Kennedy, B.M. 1980. Composition of miner's lettuce (*Montia perfoliata*). *J. Am. Diet. Assoc.* 77: 21–25.

Schnedler, W. 1977. *Claytonia perfoliata* Donn ex Willd. in the Giessen area. *Gottinger Floristische Rundbriefe* 11(2): 29–30. (In German)

Stieglitz, W. 1977. Additional remarks on the report of discovery of *Claytonia perfoliata* by Schnedler, Adolphi and Dickore (*Gottinger Floristische Rundbriefe* 11(2): 29–33). *Gottinger Floristische Rundbriefe* 11(3): 80. (In German)

### Specialty Cookbooks

Recipes can be found on the Web and in wild food cookbooks (e.g., Szcawinski and Turner 1978, Buishand et al. 1986). The most extensive source is perhaps Williamson (1995), which presents 10 recipes. (See the Appendix to this book for details of the publications cited).

# 100 Yampah

Family: Apiaceae (Umbelliferae; carrot family)

## NAMES

Scientific name: *Perideridia gairdneri* (Hook. & Arn.) Mathias

- The name yampah is derived from the Ute Native American tribe Yamparicas, which inhabited the Yampa River Valley in Colorado.
- Yampah is also known as common yampah, double turnip (the roots are often forked in two), false caraway, Gairdner's yampah, Gardner's yampah ("Gardener's" is a mutation of the correct name, Gairdner's), Indian caraway, Indian carrot, Indian potato (an ambiguous phrase used for several unrelated species), ipo, squawroot (considered derogatory, this term has been applied to other plants, notably black cohosh, *Actaea racemosa* L. (*Cimicifuga racemosa* (L.) Nutt.), trail potato, western false caraway, wild caraway, wild carrot (better reserved for wild forms of the true carrot, *Daucus carota* L.), yamp, and yampa. Snell (2006, see the Appendix to this book) noted the Native American name "split root."
- The genus *Perideridia* has been claimed to be based on the Greek *peri*, around + *derris*, a leather coat, perhaps an allusion to the use of yampah by Native Americans to waterproof leather. (The fruits of species of the carrot family are rich in oils, and such oil could have preservative value.)
- *Gairdneri* in the scientific name commemorates Meredith Gairdner (1809–1837), a British (Edinburgh) physician with the Hudson's Bay Company at Fort Vancouver on the Columbia River, in the early 1830s, who collected biological specimens. The steelhead trout (rainbow trout) was originally named *Salmo gairdneri* after Gairdner, but has been reclassified to *Oncorhynchus mykiss*.

## GEOGRAPHY AND ECOLOGY OF WILD PLANTS

Yampah is native to southern Canada (including British Columbia, Alberta, and Saskatchewan), and the states of Arizona, California, Colorado, Idaho, Montana, Nevada, New Mexico, Oregon, South Dakota, Utah, Washington, and Wyoming. The species occurs in a variety of moist and dry habitats, including chaparral (dry land shrub), grasslands, woodlands, hillsides, ravines, mixed evergreen forests, and meadows. The plants grow from low plains to moderate elevations in mountains.

## PLANT PORTRAIT

Yampah is a perennial herb, with a slender, erect, flowering stem growing 40–120 cm (16 inches–4 ft), or rarely up to 150 cm (5 ft) in height. The leaves, which are highly divided like carrot foliage, wither in early summer before flowering occurs. The species has numerous, tiny, white flowers, which mature into small round fruits 2–3 mm (0.8–1.2 inches) long. The plants grow from a spindle-shaped, tuberous root, the expanded portion about 5–10 cm (2–4 inches) in length. Sometimes a group of small tubers is found at the base of the plant instead of one large tuberous root. The roots were greatly valued as a food source by indigenous peoples. They were bartered in trade among tribes, and occasionally became the cause of territorial conflicts, much like wars fought over prime hunting territory.

FIGURE 100.1 Yampah (*Perideridia gairdneri*). (a) Cooked yampah. (b) Recently harvested yampah. ([a] and [b] Courtesy of T. Abe Lloyd.) (c) Tuberous roots of herbarium specimens from Saskatchewan. (Courtesy of E. Small and B. Brookes.) (d) Distribution map.

## CULINARY PORTRAIT

The roots were typically collected before flowers appeared (flowering would reduce their nutritional reserves). The tuberous roots, often the size of small, new potatoes, are consumed raw or cooked. The raw roots are laxative and were once used medicinally for this purpose. Indigenous peoples generally ate the roots like potatoes—baked, roasted, or steamed, or incorporated into stews and soups. The roots were also dried and ground into flour for porridge, flatbreads, and griddle cakes, and used as a thickener for pemmican (the roots do not store well fresh, but keep indefinitely dried). The flour could be used as a flavoring. The roots contain about 70% starch and are pleasantly sweet and nutty. The taste of the root has been characterized as resembling a resinous but sweet carrot, or a sweet, nutty parsnip. The seeds were employed as a seasoning like caraway seeds, either used in cooking or simply sprinkled on salads, porridge, and other foods.

## WARNING

Plant species in the carrot family are difficult to identify, and wild food collectors need to be certain of their identity. Some edible species can be mistaken for quite poisonous ones, such as water hemlock (*Cicuta* species) and poison hemlock (*Conium maculatum* L.).

### Culinary Vocabulary

- "Fusion food" is the combination of different kinds of ethnic foods. The phrase has been claimed to originate from the cosmopolitan mixing of foods in New York City, but blending of cuisines has occurred for millennia. A particularly notable marriage of cuisines took place during the Lewis and Clark expedition of 1803–1806 (see Chapter 85 on Sassafras for details), which began in St. Louis and traveled over the Rocky Mountains to the Pacific Ocean. In 1805 while wintering on the West coast, the explorers combined familiar foods which they had brought with them with the berries and root crops, including yampah, which they observed Native Americans eating.

## PROSPECTS

Yampah is one of the best of the dozens of "root crops" that were regularly harvested by indigenous peoples of North America. It was never domesticated and, with the introduction of potatoes and similar European vegetables, yampah became obsolete. To North Americans of native heritage, the species remains one of the most valued of the native food plants of North America. In the marketplace, yampah would have to compete directly with carrot, potato, turnip, and rutabaga, and there is little chance of it becoming a new major crop. However, there are market niches for minor root crops—parsnip is an example of a minor carrot family root crop—and a program of breeding, management, and marketing could develop yampah into a significant commercial crop.

## CURIOSITIES OF SCIENCE AND TECHNOLOGY

- Yampah tuberous roots are rich in carbohydrates, which are rapidly digested, providing quick energy. Indigenous people consumed it in the way that energy bars are used today, to provide energy while working vigorously or for sustaining endurance. Sometimes horses were fed yampah for the same reason.
- John Charles Frémont (1813–1890) was an American military officer, explorer, and politician (the first anti-slavery Republican presidential candidate). He declared yampah to be the finest of the Native American foods.
- Yampah was used medicinally by Native Americans. Mothers chewed the seeds to promote milk secretion, and fed tea prepared with the crushed seeds to their infants to treat colic.
- Yampa is a popular geographical name in Colorado (village, county, river, etc.), a heritage of the popularity of yampah with Native Americans who resided in the area.
- Potato is the chief market competitor of Yampah. Curiously, there is a cultivar of potato called 'Yampa'. (Reference: Schalal, L.A., Edmundson, W.C., and Kunkel, R. 1949. Yampa, a new scab-resistant potato. *Am. Potato J.* 26: 335–342.)
- A study of grizzly bear predation on voles found that the bears were most successful in catching the rodents in areas where yampah was most common. (Reference: Mattson, D.J. 2004. Consumption of voles and vole food caches by Yellowstone grizzly bears: exploratory analysis. *Ursus* 15(2): 218–226.)

## KEY INFORMATION SOURCES

Chuang, T.-I. 1970. A systematic anatomical study of the genus *Perideridia* (Umbelliferae–Apiodeae). *Am. J. Bot.* 57: 495–503.

Chuang, T.I., and Constance, L. 1969. A systematic study of *Perideridia* (Umbelliferae-Apioideae). *Univ. Calif. Publ. Bot.* 55: 1–74.

Downie, S.R., and Katz-Downie, D.S. 1996. A molecular phylogeny of Apiaceae subfamily Apioideae: evidence from nuclear ribosomal DNA internal transcribed spacer sequences. *Am. J. Bot.* 83: 234–251.

Downie, S.R. Sun, F.J., Katz-Downie, D.S., and Colletti, G.J. 2004. A phylogenetic study of *Perideridia* (Apiaceae) based on nuclear ribosomal DNA ITS sequences. *Syst. Bot.* 29: 737–751.

Giannasi, D.E., and Chuang, T.I. 1976. Flavonoid systematics of the genus *Perideridia* (Umbelliferae). *Brittonia* 28: 177–194.

Kaldy, M.S., Johnston, A., and Wilson, D.B. 1980. Nutritive value of Indian bread-root, squaw-root, and Jerusalem artichoke. *Econ. Bot.* 34: 352–357.

Mathias, M.E. 1936. Studies in the Umbelliferae. V. *Brittonia* 2: 239–245.

Phillips, N., Drost, D., and Varga, V. 2003. Chemical treatments enhance seed germination in *Perideridia gairdneri*. *Acta Hort.* 618: 477–482.

### SPECIALTY COOKBOOKS

Farnsworth (1999) has 1 yampah recipe, Hall and Hall (1980) have 2, Snell (2006) has 1, and Williamson (1995) has 10 yampah recipes. (See the Appendix to this book for details of the publications cited.)

# 101 Yerba Buena

Family: Lamiaceae (Labiatae; mint family)

## NAMES

Scientific name *Clinopodium douglasii* (Benth.) Kuntze (*Micromeria chamissonis* (Benth.) Greene, *Satureja douglasii* (Benth.) Briq., *Thymus douglasii* Benth.)

- *Satureja* has traditionally been considered to be a large, widespread genus of about 235 species, and in most of the pre-twenty-first-century literature the most common name given to the species discussed here has been *S. douglasii*. Many taxonomists in recent years have split off species groups from the genus *Satureja* as separate genera. The New World species are frequently placed in *Clinopodium*, not *Satureja*.
- "Yerba buena" is based on an alternate form of the Spanish name, *hierba buena*, for "good herb." Native Americans and early settlers on the Pacific coast of North America used a tea of the species to cure a variety of ailments, resulting in its common name. The region was controlled in early times by Spain, preceding the Spanish–American War, hence the common use of Spanish at the time (yerba buena is the name of the plant both in Spanish and English).
- Yerba buena is also known as Oregon tea. The species has also been called Douglas' savory, Indian mint, mountain tea, and western yerba buena.
- The name yerba buena is applied to several herbs. In most Spanish-speaking countries, the phrase "yerba buena" refers to a particular species, usually in the mint family (especially species in the genera *Satureja*, *Mentha*, and *Clinopodium*), that is important locally for its medicinal, tea, or seasoning properties.
- The genus name *Clinopodium* is based on the Latin *clinopodion*, which in turn comes from the Greek *klinopodion*, names for an unidentified mint-like plant. The Greek name is based on *kline*, a bed + *podion*, a little foot, referring to the flowers and stems.
- *Douglasii* in the scientific name commemorates David Douglas (1799–1834) of Scotland, who collected many plants in the Pacific Northwest. Over 80 species of plants and animals are named after Douglas.

## GEOGRAPHY AND ECOLOGY OF WILD PLANTS

Yerba buena is a native of the Pacific coast of western North America, from maritime islands of Alaska and southern British Columbia through the states of Washington, Idaho, Oregon, and California to the state of Baja California in Mexico. It has become established in some localities outside of its native range. The species is most common near the Pacific coast and is adapted to obtaining moisture from fog drip. It is also found in coastal shrubland and as an understory plant in woodlands. Yerba buena thrives in moist, shady places, and also occurs in disturbed places such as along roadcuts.

**FIGURE 101.1** Yerba buena (*Clinopodium douglasii*). (a) Carpet of plants. (Courtesy of docentjoyce [CC By 2.0].) (b) Flowering twig. (Courtesy of Scott Loarie [Flickr/CC-attribution].) (c) Distribution map.

## PLANT PORTRAIT

Yerba buena is a rambling, sprawling, low-growing (usually less than 15 cm or 6 inches in height), mat-forming perennial herb with a minty fragrance. The slightly woody stems sometimes develop root at places where they contact the ground. The evergreen leaves are 10–35 mm (0.4–1.4 inches) long. The flowers are tiny, white to lavender, becoming purple with age. The fruits are also tiny.

The species is cultivated as a trailing groundcover for small areas, and as a hanging potted or windowbox foliage plant, attractive for its cascading herbage rather than for flowers. A few ornamental cultivars have been selected ('Indian Mint' is the best known).

Yerba buena (often under the inappropriate name "peppermint") is internationally employed in herbal or folk medicine, mostly as an analgesic to relieve body aches and pain such as arthritis, toothache, and headache. The pungent oils from species of *Satureja* are sometimes used in toothpaste, soaps, and perfumes.

## CULINARY PORTRAIT

The leafy stems of yerba buena were dried and rolled into a ball for use as a tea by Native Americans. The flavor is minty, but more subtle than fresh spearmint or peppermint. Today, the species is still frequently used as an herbal tea, often available in specialty food outlets in dried form. Growing a plant is a simple way of obtaining a personal supply. Yerba buena is often available from Latino markets. In the marketplace, *C. douglasii* must compete with several species of *Satureja* that are used as culinary herbs, most notably summer savory (*Satureja hortensis* L.) and winter savory (*Satureja montana* L.), both of which are very popular.

### Culinary Vocabulary

- A mojito is a cocktail of light rum, lime juice, sugar and club soda, garnished with a sprig of mint. In Latino countries, yerba buena is frequently the "mint" used as a garnish.
- Albondigas soup (*sopa de albóndigas* in Spanish, literally soup of meatballs) is a Mexican and Spanish beef broth soup with meatballs and vegetables, often traditionally flavored with "mint" (yerba buena).

## PROSPECTS

There are many popular culinary herbs in the marketplace, most of these native to the Old World. Yerba buena is a very minor culinary herb, in fact considerably more popular as a folk medicine plant. It is an easily cultivated, pleasantly mild, flavoring herb, and has the capacity of achieving minor commercial success. Because it grows best in shady places, it is very low-growing and is not very vigorous, it is very unlikely to be cultivated on a large field scale, as are the major culinary herbs.

## CURIOSITIES OF SCIENCE AND TECHNOLOGY

- West coast indigenous peoples covered up human scent by rubbing their body and weapons with crushed yerba buena leaves, while waiting for game at salt licks where animals congregated.
- Yerba Buena was the original name of the city of San Francisco (or more correctly, the village that was transformed into the city). Yerba Buena is also the name of a city in Argentina, with a population of over 60,000.
- The transportation tunnel that runs between the bridges connecting San Francisco and Oakland is on Yerba Buena Island, named for the plant which once covered its shores.

## KEY INFORMATION SOURCES

Bräuchler, C., Meimberg, H., and Heubl, G. 2006. New names in Old World *Clinopodium*—the transfer of the species of *Micromeria* sect. *Pseudomelissa* to *Clinopodium*. *Taxon* 55: 977–981.

Cantino, P.D., and Wagstaff, S.J. 1998. A reexamination of North American *Satureja* s.l. (Lamiaceae) in light of molecular evidence. *Brittonia* 50: 63–70.

Gershenzon, J., Lincoln, D.E., and Langenheim, J.H. 1978. The effect of moisture stress on monoterpenoid yield and composition in *Satureja douglasii*. *Biochem. Syst. Ecol.* 6: 33–43.

Hensolt, E.A. 1966. *How Oregon Indians used the native flora*. Benton County Historical Society & Museum, Philomath, OR. pp. 22.

Lawrence, B.M., Bromstein, A.C., and Langeheim, J.H. 1974. Terpenoids in *Satureja douglasii*. *Phytochemistry* 13: 1014.

Lincoln, D.E., and Langenheim, J.H. 1976. Geographic patterns of monoterpenoid composition in *Satureja douglasii*. *Biochem. Syst. Ecol.* 4: 237–248.
Lincoln, D.E., and Langenheim, J.H. 1978. Effect of light and temperature on monoterpenoid yield and composition in *Satureja douglasii*. *Biochem. Syst. Ecol.* 6: 21–32.
Lincoln, D.E., and Langenheim, J.H. 1979. Variation of *Satureja douglasii* monoterpenoids in relation to light intensity and herbivory. *Biochem. Syst. Ecol.* 7: 289–298.
Lincoln, D.E., and Langenheim, J.H. 1981. A genetic approach to monoterpenoid compositional variation in *Satureja douglasii*. *Biochem. Syst. Ecol.* 9: 153–160.
Peer, W.A., and Langenheim, J.H. 1998. Influence of phytochrome on leaf monoterpene variation in *Satureja douglasii*. *Biochem. Syst. Ecol.* 26: 25–34.
Peer, W.A., Briggs, W.R., and Langenheim, J.H. 1999. Shade-avoidance responses in two common coastal redwood forest species, *Sequoia sempervirens* (Taxodiaceae) and *Satureja douglasii* (Lamiaceae), occurring in various light quality environments. *Am. J. Bot.* 86: 640.
Rhoades, D.G., Lincoln, D.E., and Langenheim, J.H. 1976. Preliminary studies of monoterpenoid variability in *Satureja douglasii*. *Biochem. Syst. Ecol.* 4: 5–12.
Rice, R.L. Lincoln, D.E., and Langenheim, J.H. 1978. Palatability of monoterpenoid compositional types of *Satureja douglasii* to a generalist molluscan herbivore, *Ariolimax dolichophallus*. *Biochem. Syst. Ecol.* 6: 45–53.

## Specialty Cookbooks

Evans, H. 1996. *Sage, sorrel & savory*. JG Press, North Dighton, MA. pp. 61.

Stewart (2002) (see the Appendix to this book) provides information on the culinary use of yerba buena.

# Appendix: Key Literature and Recipe Sources for Indigenous Food Plants of North America*


Adams, C. 1984. *Edible wild foods and wild game cookbook.* Friends of Deep-Portage, Hackensack, MN. pp. 214.

Alaska Magazine Editors. 1982. *Alaska wild berry guide and cookbook.* Alaska Northwest Publishing Company, Anchorage, AK. 201 pp. (Republished as: Alaska Northwest Books. 2012. The Alaska wild berry cookbook: 275 recipes from the Far North. Alaska Northwest Books, Portland, OR. pp. 135.)

Alfs, M. 2001. *Edible & medicinal wild plants of Minnesota & Wisconsin.* Old Theology Book House, New Brighton, MN. pp. 426.

Allen, C.M., Allen, A.W., and Winters, H.H. 2005. *Edible plants of the Gulf South.* Allen's Native Ventures, Pitkin, LA. pp. 291.

Angier, B. 1966. *Free for the eating.* Stackpole Books, Harrisburg, PA. pp. 191.

Angier, B. 1969. *More free-for-the-eating wild foods.* Stackpole Books, Harrisburg, PA. pp. 192.

Angier, B., and Foster, D. 2008. *Field guide to edible wild plants.* Second edition. Stackpole Books, Harrisburg, PA. pp. 285.

Anonymous. 1977. *Native cookery and edible wild plants of Newfoundland and Labrador.* Department of Rural, Agricultural and Northern Development, St. John's, NF. pp. 91.

Arnott, M.L. (Editor). 1975. *Gastronomy: The anthropology of food and food habits.* Mouton, The Hague. pp. 354.

Ayotte, E., and Stetson, M. 1980. *Collecting and using Alaska's wild berries and other wild products.* Cooperative Extension Service, University of Alaska, Fairbanks, AK. pp. 34.

Bacon, J. 1988. *Exotic vegetables A–Z.* Salem House, Topsfield, MA. pp. 128.

Bailey, F.L. 1940. Navaho foods and cooking methods. *American Anthropologist* (n.s.) 42: 270–290.

Baird, E. 1980. *Summer berries (great Canadian recipes).* James Lorimer & Company, Toronto, ON. pp. 128.

Baker, H.G. 1970. *Plants and civilization.* Second edition. Wadsworth, Belmont, CA. pp. 194.

Banks, W.H., Jr., and Kemp, S. 2004. *Plants of the Cherokee: Medicinal, edible, and useful plants of the Eastern Cherokee Indians.* Great Smoky Mountains Association, Gatlinburg, TN. pp. 149.

Bar Harbor Jam Company. 2002. *Cooking with wild berries of Down East Maine.* Bar Harbor Jam Company, Bar Harbor, ME. pp. 224.

Barash, C.W. 1993. *Edible flowers: From garden to palate.* Fulcrum Publishing, Golden, CO. pp. 250.

Beatty, B. 1987. *Bill & Bev Beatty's wild plant cookbook.* Naturegraph Publishers, Happy Camp, CA. pp. 174.

Belsinger, S. 1991. *Flowers in the kitchen.* Interweave Press, Loveland, CO. pp. 128.

Benoliel, D. 2011. *Northwest foraging: The classic guide to edible plants of the Pacific Northwest.* Revised edition. Skipstone, Seattle, WA. pp. 220.

Berglund, B., and Bolsby, C.E. 1971. *The edible wild. A complete cookbook and guide to edible wild plants in Canada and eastern United States.* Charles Scribner's Sons, New York, NY. pp. 188.

Beshoar, D. 1982. *Violet soup: Common edible plants of the Rocky Mountains.* Filter Press, Palmer Lake, CO. pp. 62.

Boorman, S. 1969. *Wild plums in brandy: A cookery book of wild foods.* Revised edition. McGraw-Hill, New York, NY. pp. 194.

Boxer, A., and Back, P. 1980. *The herb book.* Octopus Books Limited, London, U.K. pp. 224.

Brackett, B., and Lash, M. 1975. *The wild gourmet: A forager's cookbook.* D.R. Godine, Boston, MA. 160 pp.

Breeden, P. 1978. *Wild foods: A beginner's guide to identifying, harvesting, and cooking safe and tasty plants from the outdoors.* Four Winds Press, New York, NY. pp. 182.


* *Note:* Also see literature cited in Chapter 1.

Bremness, L. 1988. *The complete book of herbs*. Viking Studio Books, New York, NY. pp. 288.
Brill, S. 2002. *The wild vegetarian cookbook. A forager's culinary guide (in the field or in the supermarket) to preparing and savoring wild (and not so wild) natural foods*. Harvard Common Press, Boston, MA. pp. 500.
Brill, S., and Dean, E. 2002. *Identifying and harvesting edible and medicinal plants in wild (and not so wild) places*. HarperCollins, New York, NY. pp. 317.
Brown, T. 1985. *Tom Brown's guide to wild edible and medicinal plants*. Berkley Books, New York, NY. pp. 241.
Bryan, J.E., and Castle, C. 1974. *The edible ornamental garden*. 101 Productions, San Francisco, CA. pp. 192.
Buishand, T., Houwing, H.P., and Jansen, K. 1986. *The complete book of vegetables*. Gallery Books, W.H. Smith Publishers, New York, NY. pp. 180.
Callegari, J., and Durand, K. 1977. *Wild edible and medicinal plants of California*. J. Callegari & K. Durand, El Cerrito, CA. pp. 96.
Chase, M.C. 1965. *Field guide to edible and useful wild plants of North America: A guide to help you in using, locating and enjoying the plant life that Mother Nature has so generously supplied*. Nasco, Fort Atkinson, WI. pp. 140.
Cheatham, S., Johnston, M.C., and Marshall, L. 1995–2009. *The useful wild plants of Texas, the southeastern and southwestern United States, the Southern Plains, and Northern Mexico*. Vol. 1, *Abronia–Arundo* (1995); Vol. 2, *Asclepias–Canavalia* (2000). Vol. 3, *Canna–Celtis* (2009). Useful Wild Plants, Inc., Austin, TX.
Choyce, L., and Shaler, K. 1976. *Edible wild plants of Nova Scotia*. Eastern Shore Pub. Collective, West Chezzetcook, NS. pp. 47.
Choyce, L., and Shaler, K. 1977. *Edible wild plants of the Maritimes*. Wooden Anchor Press, West Chezzetcook, NS. pp. 48.
Clarke, C.B. 1977. *Edible and useful plants of California*. University of California Press, Berkeley, CA. pp. 280.
Couplan, F. 1998. *The encyclopedia of edible plants of North America*. Keats Publishing, New Canaan, CT. pp. 584.
Creasy, R. 1988. *Cooking from the garden*. Sierra Club Books, San Francisco, CA. pp. 547.
Crowhurst, A. 1972. *The weed cookbook*. Lancer Books, New York, NY. pp. 190.
Crowhurst, A. 1973. *The flower cookbook*. Lancer Books, New York, NY. pp. 198.
Cvancara, A.M. 2001. *Edible wild plants and herbs*. Ragged Mountain Press, Camden, ME. pp. 109.
Dahl, K. 1995. *Wild foods of the Sonoran Desert*. Arizona-Sonora Desert Museum, Tucson, AZ. pp. 22.
Darrow, G.M. 1975. Minor temperate fruits. In *Advances in fruit breeding*. Edited by J. Janick and J.N. Moore. Purdue University Press, West Lafayette, IN. pp. 269–284.
Densmore, F. 1974. *How Indian use wild plants for food, medicine, and crafts*. Dover, New York, NY. pp. 119. (Reprint of Uses of plants by the Chippewa Indians from the 44th annual report (1926/27) of the U.S. Bureau of American Ethnology.)
Derig, B.D., and Fuller, M. 2001. *Wild berries of the West*. Mountain Press Publishing Company, Missoula, MT. pp. 235.
Domico, T. 1979. *Wild harvest: Edible plants of the Pacific Northwest*. Hancock House, Seattle, WA. pp. 88.
Duke, J.A. 1992. *Handbook of edible weeds*. CRC Press, Boca Raton, FL. pp. 246.
Eaton, J.S. 1993. *Alaska's wild plants: A guide to Alaska's edible harvest*. Alaska Northwest Books, Anchorage, AK. pp. 95.
Ebeling, W. 1986. *Handbook of Indian foods and fibers of arid America*. University of California Press, Berkeley, CA. pp. 924.
Elias, T.S., and Dykeman, P.A. 2009. *Edible wild plants: A North American field guide to over 200 natural foods*. Sterling, New York, NY. pp. 286.
Elliott, B.A. 2009. *Handbook of edible and poisonous plants of western North America*. Elliott Environmental Consulting, Laramie, WY. pp. 364.
Elliott, D.B. 1976. *Roots: An underground botany and forager's guide*. Chatham Press, Old Greenwich, CT. pp. 128.
Elliott, D.B. 1995. *Wild roots: A forager's guide to the edible and medicinal roots, tubers, corms, and rhizomes of North America*. Healing Arts Press, Rochester, VT. pp. 128.
Facciola, S. 1998. *Cornucopia II. A source book of edible plants*. Kampong Publications, Vista, CA. pp. 713.
Farnsworth, K. 1999. *Going to seed: Finding, identifying, and preparing edible plants of the southwest*. Ancient City Press, Santa Fe, NM. pp. 236.
Farnsworth, K. 2005. *A taste of nature: Edible plants of the southwest and how to prepare them*. Third edition. K. Farnsworth, U.S. pp. 207.
Fernald, M.L., and Kinsey, A.C. (revised by R.C. Rollins). 1958. *Edible wild plants of eastern North America*. Harper & Row, New York, NY. pp. 452.

Fielder, M. 1983. *Wild fruits: An illustrated field guide & cookbook*. Contemporary Books, Chicago, IL. pp. 271.
Fleurbec (group: G. Lamoureux et al.). 1981. *Plantes sauvages comestibles*. Fleurbec, Saint-Cuthbert, QC. pp. 167. (In French.)
Fleurbec (group: G. Lamoureux et al.). 2005. *Plantes sauvages au menu: Cuisine raisonnée*. Second edition. Fleurbec, Saint-Henri-de-Lévis, QC. pp. 192. (In French.)
Forey, P., and Fitzsimons, C. 1989. *An instant guide to edible plants: The most familiar edible wild plants of North America described and illustrated in full color*. Crescent Books, Avenel, NJ. pp. 123.
Freitus, J. 1975. *160 Edible plants commonly found in the eastern USA*. Stone Wall Press, Lexington, MA. pp. 85.
Freitus, J. 1980. *The natural world cookbook: Complete gourmet meals from wild edibles*. Stone Wall Press, Washington, D.C. pp. 283.
Freitus, J., and Haberman, S. 2005. *Wild jams and jellies. Delicious recipes using 75 wild edibles*. Stackpole Books, Mechanicsburg, PA. pp. 322.
Furlong, M., and Pill, V.B. 1974. *Wild edible fruits & berries*. Naturegraph, Healdsburg, CA. pp. 64.
Gaertner, E.E. 1995. *Reap without sowing. Wild food from nature's cornucopia*. The General Store Publishing House, Burnstown, ON. pp. 182.
Gail, P.A. 2001. *Identifying edible wild plants, the process and the players: A directory of wild food educators in North America and elsewhere in the world*. Goosefoot Acres Press, Cleveland, OH. pp. 52.
Gardon, A. 1998. *The wild food gourmet: Fresh and savory food from nature*. Firefly Books, Willowdale, ON. pp. 174.
Garland, S. 1979. *The complete book of herbs & spices*. Viking Press, New York, NY. pp. 288.
Genders, R. 1988. *Edible wild plants: A guide to natural foods*. Van der Marck Editions, New York, NY. pp. 208.
Genest, M. 2010. *The boreal gourmet: Adventures in northern cooking*. Lost Moose, Madeira Park, BC. pp. 256.
Gibbons, E. 1962. *Stalking the wild asparagus*. David McKay, New York, NY. pp. 303.
Gibbons, E. 1966. *Stalking the healthful herbs*. David McKay, New York, N.Y. pp. 303.
Gibbons, E., and Tucker, G. 1979. *Euell Gibbons' handbook of edible wild plants*. Donning, Virginia Beach, VA. pp. 319.
Gillespie, W.H. 1959. *A compilation of the edible wild plants of West Virginia*. Scholar's Library, New York, NY. pp. 118.
Gillespie, W.H. 1986. *Wild foods of Appalachia*. Seneca Books, Morgantown, WV. pp. 159.
Gray, B. 2011. *The boreal herbal: Wild food and medicine plants of the North: A guide to harvesting, preserving, and preparing*. Aroma Borealis Press, Whitehorse, YK. pp. 440.
Hagener, L.W., and Hagener, A.R. 1977. *Free for all: Edible and useful wild plants of north central Montana*. Hill County Printing Company, Havre, MT. pp. 76.
Hahn, J. 2010. *Pacific feast. A cook's guide to West coast foraging and cuisine*. Skipstone, Seattle, WA. pp. 224.
Hall, A. 1976. *The wild food trailguide*. Revised edition. Holt, Rinehart and Winston, New York, NY. pp. 230.
Hall, W., and Hall, N. 1980. *The wild palate, a serious wild foods cookbook*. Rodale Press, Emmaus, PA. pp. 374.
Hamersley Chambers, F. 2011. *Wild berries of British Columbia*. Lone Pine Publishing, Edmonton, AB. pp. 192.
Hamersley Chambers, F., and Harris, C. 2012. *Wild berries of Ontario*. Lone Pine Publishing, Edmonton, AB. pp. 232.
Hamersley Chambers, F., and Karst, A. 2012. *Wild berries of Alberta, Saskatchewan and Manitoba*. Lone Pine Publishing, Edmonton, AB. pp. 184.
Harlan, J.R. 1992. *Crops and man*. Second edition. American Society of Agronomy, Crop Science Society of America, Madison, WI. pp. 284.
Harrington, H.D. 1972. *Western edible wild plants*. University of New Mexico Press, Albuquerque, NM. pp. 156.
Harrington, H.D. 1998. *Edible native plants of the Rocky Mountains*. University of New Mexico Press, Albuquerque, NM. pp. 392.
Heller, C.A. 1981. *Wild, edible and poisonous plants of Alaska*. Revised edition. Cooperative Extension Service, University of Alaska, Anchorage, AK. pp. 90.
Hemphill, J., and Hemphill, R. 1984. *Herbs: Their cultivation and usage*. Sterling Publisher, New York, NY. pp. 128.
Henderson, R.K. 2000. *The neighborhood forager: A guide for the wild food gourmet*. Chelsea Green Publishing Company, White River Junction, VT. pp. 226.
Hibler, J. 2004. *The berry bible: With 175 recipes using cultivated and wild, fresh and frozen berries*. William Morrow, New York, NY. pp. 333.
Hidalgo County Historical Museum. 1983. *Heritage cookbook: A round-up of wild and regional foods*. Hidalgo County Historical Museum, Edinburg, TX. pp. 90.

Hodgson, W.C. 2001. *Food plants of the Sonoran Desert*. University of Arizona Press, Tucson, AZ. pp. 313.
Holt, G. 1989. *Recipes from a French herb garden*. Stoddard, Toronto, ON. pp. 160.
Holt, G. 1991. *Geraldine Holt's complete book of herbs*. Octopus Publishing Group, London, U.K. pp. 224.
Hudson, I.B. 1950. *Medicinal and food plants of British Columbia*. I.B. Hudson, Victoria, BC. pp. 70.
Hughes, P. 1982. *Pueblo Indian cookbook*. Museum of New Mexico Press, Sante Fe, NM. pp. 64.
Hunt, D. (Editor). 1977. *Native Indian wild game, fish & wild foods cookbook. Recipes from North American native cooks*. Fox Chapel Publishing, Lancaster, PA. pp. 222.
Irvine, F.R. 1959. *Bibliography of wild food plants of United States Indians*. Smithsonian Institution, Washington, D.C. pp. 26.
Jackson, S., and Prine, L. 1978. *Wild plants of central North America for food and medicine*. Peguis, Winnipeg, MB. pp. 77.
Johns, T., and Kubo, I. 1988. A survey of traditional methods employed for the detoxification of plant foods. *Journal of Ethnobiology* 8(1): 81–129.
Johnson, C. 1989. *The wild foods cookbook*. Stephen Greene Press, Lexington, MA. pp. 236.
Kallas, J. 2010. *Edible wild plants: Wild foods from dirt to plate*. Gibbs Smith, Layton, UT. pp. 416.
Kane, C.W. 2011. *Sonoran desert food plants: Edible uses for the desert's wild bounty*. Lincoln Town Press, Oracle, AZ. pp. 52.
Kane, C.W., and Rose, F. 2006. *Herbal medicine of the American southwest: A guide to the medicinal and edible plants of the southwestern United States*. Lincoln Town Press, Tucson, AZ. pp. 307.
Kavasch, B. 1979. *Native harvests: Recipes and botanicals of the American Indian*. Random House, New York, NY. pp. 202.
Kavena, J.T. 1980. *Hopi cookery*. The University of Arizona Press, Tucson, AZ. pp. 115.
Kershaw, L.J. 2000. *Edible and medicinal plants of the Rockies*. Lone Pine, Edmonton, AB. pp. 272.
Kindscher, K. 1987. *Edible wild plants of the prairie. An ethnobotanical guide*. University Press of Kansas, Lawrence, KS. 276 pp. [1992 edition: pp. 340.]
Kirk, D.R. 1975. *Wild edible plants of the western United States*. Naturegraph, Healdsburg, CA. 307 pp. + unpaginated pages.
Kluger, M. 1973. *The wild flavor: Delectable wild foods to be found in field & forest & cooked in country kitchens*. J.P. Tarcher, Los Angeles, CA. pp. 285.
Knap, A. 1975. *Wild harvest: An outdoorsman's guide to edible wild plants in North America*. Pagurian Press, Toronto, ON. pp. 192.
Knutsen, K. 1975. *Wild plants you can eat: A guide to identification and preparation*. Doubleday and Company, Garden City, NY. 89 pp. + plates.
Kraft, K., and Kraft, P. 1977. *Exotic vegetables: How to grow and cook them*. Walker and Company, New York, NY. pp. 116.
Krochmal, A., and Krochmal, C. 1974. *A naturalist's guide to cooking with wild plants*. Quadrangle/New York Times Book Co., New York, NY. pp. 336.
Krum, B. 1991. *The Rocky Mountain berry book: A complete guide to finding, harvesting, and preparing wild berries and fruits in the Rocky Mountain West*. Falcon, Helena, MT. 162 pp. (2nd edition, 2013, pp. 192.)
Kublick, L. 1990. *The prairie herb garden*. Western Producer Prairie Books, Saskatoon, SK. pp. 141.
Kuhnlein, H.V., and Turner, N.J. 1991. *Traditional plant foods of Canadian indigenous peoples: Nutrition, botany, and use*. Gordon and Breach Science, Philadelphia, PA. pp. 633.
Latorre, D.L. 1977. *Cooking and curing with Mexican herbs. Recipes and remedies gathered in Múzquiz, Coahuila*. Encino Press, Austin, TX. pp. 178.
Lee, L. 1975. *Edible wild plants of Martha's Vineyard*. Tashmoo Press, Martha's Vineyard, MA. pp. 88.
Leggatt, J. 1987. *Cooking with flowers*. Century Hutchinson Limited, London, U.K. pp. 142.
Leyel, C.F. 1938. *Herbal delights*. Faber and Faber, London, U.K. pp. 429.
Livingston, A.D. 2012. *Forager's guide to wild foods: A fool-proof guide to identifying, gathering, and cooking the bounty*. W.W. Norton & Company Inc., New York, NY. pp. 288.
Lucas, J.M., and Carter, H. 1945. *Indian harvest: Wild food plants of America*. J.B. Lippincott, New York, NY. pp. 118.
Lund, D.R. 2005. *Fruit & nut recipes: 300 easy to prepare recipes using fruit and nuts*. Lund S&R Publications, Staples, MN. pp. 196.
Lyle, K.L. 2004. *The complete guide to edible wild plants, mushrooms, fruits, and nuts*. Lyons Press, Guilford, CT. pp. 195.
MacKinnon, A., Kershaw, L., Arnason, J.T., Owen, P., Karst, A., and Hamersley-Chambers, F. 2009. *Edible & medicinal plants of Canada*. Lone Pine, Edmonton, AB. pp. 448.

Macklin, H., and Sanders, E.B. 1978. *The backpacker's cookbook: A complete manual and handbook for cooking freeze-dried and wild foods on the trail and in the wilderness.* Pagurian Press, Toronto, ON. pp. 190.

MacLauchlan, A. 1997. *Tropical desserts: Recipes for exotic fruits, nuts, and spices.* Macmillan, New York, NY. pp. 164.

MacLeod, H., Brousseau, D., and MacDonald, B. 1976 & 1988. *Edible wild plants of Nova Scotia.* Nimbus, Halifax, NS. pp. 135.

March, K.G., and March, A.L. 1979. *Common edible and medicinal plants of Colorado: With recipes and prescriptions.* Meridian Hills Publications, Meridian Hills IN. pp. 89.

Marcin, M.M. 1983. *The complete book of herbal teas.* Congdon & Weed, New York, NY. pp. 224.

Marcus, G., and Marcus, N. 1982. *Forbidden fruits & forgotten vegetables. A guide to cooking with ethnic, exotic, and neglected produce.* St. Martin's Press, New York, NY. pp. 146.

Marie, D. 2008. *Wild wines: Creating organic wines from nature's garden. A complete guide to making wild wines using fruit, flowers, roots, and leaves.* SquareOne Publishers, Garden City Park, NY. pp. 210.

Marles, R.J. 2000. *Aboriginal plant use in Canada's northwest boreal forest.* UBC Press, Vancouver, B.C. pp. 368.

Marrone, T. 2009. *Cooking with wild berries & fruits of Minnesota, Wisconsin and Michigan.* Adventure Publications, Cambridge, MN. pp. 175.

Marrone, T. 2009. *Wild berries & fruits: Field guide, Minnesota, Wisconsin and Michigan.* Adventure Publications, Cambridge, MN. pp. 280.

McPherson, A., and McPherson, S. 1977. *Wild food plants of Indiana and adjacent states.* Indiana University Press, Bloomington, IA. pp. 215.

Medsger, O.P. 1974. *Edible wild plants.* Collier MacMillan, New York, NY. 323 pp. (Reprint of 1939 edition.)

Medve, R.J., and Medve, M.L. 1990. *Edible wild plants of Pennsylvania and neighboring states.* Pennsylvania State University Press, University Park, PA. pp. 242.

Meuninck, J. 2007. *Basic essentials: Edible wild plants and useful herbs.* Third edition. FalconGuide, Guilford, CT. pp. 87.

Michael, P. 1980. *All good things around us. A cookbook and guide to wild plants and herbs.* Holt, Rinehart and Winston, New York, NY. pp. 240.

Michalowski, A.M. 1963. *Handbook of Florida edible wild plants and those that are poisonous.* Great Outdoors Publishing Company, St. Petersburg, FL. pp. 63.

Minnis, P.E. (Editor). 2003. *People and plants in ancient North America.* Smithsonian Institution Press, Washington, D.C. pp. 423.

Moerman, D.E. 2010. *Native American food plants: An ethnobotanical dictionary.* Timber Press, Portland, OR. pp. 455.

Mogelon, R. 2001. *Wild in the kitchen: Recipes for wild fruits, weeds and seeds.* M. Evans, New York, NY. pp. 143.

Mohney, R. 1975. *Why wild edibles? The joys of finding, fixing, and tasting.* Pacific Search Press, Seattle, WA. pp. 317.

Monthey, L.G. 1973. *Some common wild food plants: Including our most edible weeds.* Recreation Resources Center, Madison, WI. pp. 54.

Morash, M. 1982. *The Victory Garden cookbook.* Knopf, New York, NY. pp. 374.

Nabhan, G.P. 1985. *Gathering the desert.* University of Arizona Press, Tucson, AZ. pp. 209.

Naegele, T.A. 1996. *Edible and medicinal plants of the Great Lakes region.* Revised edition. Wilderness Adventure Books, Davisburg, MI. pp. 200.

Newfoundland and Labrador Department of Rural Development. 1977. *Native cookery and edible wild plants of Newfoundland and Labrador.* Newfoundland. Department of Rural Development, Rural Extension Division, Saint John's, NL. pp. 87.

Niethammer, C. 1974. *American Indian food and lore.* Macmillan, New York, NY. pp. 191.

Niethammer, C. 1999. *American Indian cooking: Recipes from the Southwest.* University of Nebraska Press, Lincoln, NE. pp. 191.

Niethammer, C.J. 2011. *Cooking the wild southwest: Delicious recipes for desert plants.* University of Arizona Press, Tucson, AZ. pp. 194.

Norman, J. 1990. *The complete book of spices.* Viking Penguin, New York, NY. pp. 160.

Nyerges, C. 1982. *Guide to wild foods.* Third edition. Survival News Service, Los Angeles, CA. pp. 239.

Peterson, L., and Peterson, R.T. 1999. *A field guide to edible wild plants of eastern and central North America* (Peterson Field Guides). Houghton Mifflin Harcourt, Boston, MA. pp. 352.

Phillips, R. 1986. *Wild food.* Little, Brown and Company, Toronto, ON. pp. 192.

Porsild, A.E. 1937. *Edible roots and berries of northern Canada*. Department of Mines and Resources, National Museum of Canada, Ottawa, ON. pp. 17.

Pratt, V.E. 1995. *Alaska's wild berries and berry-like fruit*. Alaskakrafts, Anchorage, AK. pp. 127.

Prescott-Allen, C., and Prescott-Allen, R. 1986. *The first resource*. Yale University Press, New Haven, CT. pp. 529.

Reeder, D. 1995. *From blueberries to wild roses: A Northwoods wild foods cookbook*. Explorer's Guide Publishing, Rhinelander, WI. pp. 122.

Richardson, J. 1981. *Wilde edible plants of New England: A field guide: Including poisonous plants often encountered*. DeLorme Publishing, Yarmouth, ME. pp. 217.

Richmond, G.M., and Jungwirth, J. 1977. *Edible wild plants of Michigan*. Richmond, Lincoln Park, MI. pp. 67.

Robe-Terry, A.L. 1997. *Bootstraps and biscuits: 300 wonderful wild food recipes from the hills of West Virginia*. A.L. Robe-Terry, Fairmont, WV. pp. 189.

Rogers, D.J. 1980. *Edible, medicinal, useful, and poisonous wild plants of the Northern Great Plains–South Dakota region*. Buechel Memorial Lakota Museum, Saint Francis, SD. pp. 129.

Roos-Collins, M. 1990. *The flavors of home: A guide to wild edible plants of the San Francisco Bay area*. Heyday Books, Berkeley, CA. pp. 224.

Rose, J. 1979. *Jeanne Rose's herbal guide to inner health*. Grosset & Dunlap, New York, NY. pp. 239.

Ross, R. 1991. *Wild edible plants of mother nature: A guide to eating hardy on plants common to 48 states*. Plans Plus, Simi Valley, CA. pp. 92.

Russell, H.R. 1975. *Foraging for dinner: Collecting and cooking wild foods*. T. Nelson, Nashville, TN. pp. 255.

Saunders, C.F. 1976. *Edible and useful wild plants of the United States and Canada*. Dover, New York, NY. 276 pp. (A reprint of a 1934 publication.)

Schofield, J.J. 1989. *Discovering wild plants: Alaska, western Canada, the Northwest*. Alaska Northwest Books, Bothell, WA. pp. 354.

Scott, P.J. 2010. *Edible plants of Newfoundland and Labrador*. Boulder Publications, Portugal Cove-St. Philips, NL. 180 pp. [cf. Scott, P.J. 2010. Edible plants of Atlantic Canada. Boulder Publications, Portugal Cove-St. Philip's, NL. pp. 170.]

Sept, J.D. 2005. *Wild berries of the Northwest: Alaska, western Canada and the northwestern states*. Calypso Publishing, Sechelt, B.C. pp. 95.

Seymour, T. 2002. *Foraging New England: Finding, identifying, and preparing edible wild foods and medicinal plants from Maine to Connecticut*. Globe Pequot Press, Guilford, CT. pp. 183.

Shosteck, R. November/December 1979. How good are wild foods? *Mother Earth News* 60: 110–113.

Shufer, V. 2011.*The everything guide to foraging: Identifying, harvesting, and cooking nature's wild fruits and vegetables*. Adams Media, Avon, MA. pp. 291.

Smith, A.J. 2010. *Exploring the edible landscape: Wild foods cookbook*. Publish America, Baltimore, MD. pp. 196.

Snell, A.H. 2006. *A taste of heritage. Crow Indian recipes & herbal medicines*. University of Nebraska Press, Lincoln, Nebraska. pp. 191.

Springmeyer, F., and Montez, M. 1996. *Willow bark & rosehips: An introduction to common edible and useful wild plants of North America*. Falcon, Helena, MT. pp. 80.

Stahl, C.A. 1974. *Papa Stahl's wild stuff cookbook*. Grass Root Enterprises, Houston, TX. pp. 40.

Standley, P.C. 1943. *Edible plants of the Arctic Region*. U.S. Government. Printing Office, Washington, D.C. pp. 49.

Stanek, S., and Butcher. B. 2007. *Collecting and using Alaska's wild berries and other wild products*. Revised edition. University of Alaska Fairbanks, Cooperative Extension Service, Fairbanks, AK. pp. 146.

Stewart, A. 2000. *The flavours of Canada. A celebration of the finest regional foods*. Raincoast Books, Vancouver, BC. pp. 224.

Stewart, H. 2002. *Drink in the wild: Teas, cordials, jams, and more*. Second edition. Douglas & McIntyre, Toronto, ON. pp. 135.

Sweet, M. 1962. *Common edible and useful plants of the West*. Naturegraph Company, Healdsburg, CA. pp. 64.

Szczawinski, A.F. and Hardy, G.A. 1962. *Guide to common edible plants of British Columbia*. A. Sutton (Printer), Victoria, BC. pp. 90.

Szczawinski, A.F., and Turner, N.J. 1978. *Edible garden weeds of Canada*. National Museums of Canada, Ottawa, ON, Canada. pp. 184.

Szczawinski, A.F., and Turner, N.J. 1980. *Wild green vegetables of Canada*. National Museums of Canada, Ottawa, ON, Canada. pp. 179.

Tatum, B.J. 1976. *Billy Joe Tatum's wild foods cookbook and field guide*. Workman Publishing Company, New York, NY. pp. 268.

Thayer, S. 2006. *The forager's harvest: A guide to identifying, harvesting, and preparing edible wild plants.* Forager's Harvest, Ogema, WI. pp. 360.

Thayer, S. 2010. *Nature's garden: A guide to identifying, harvesting, and preparing edible wild plants.* Forager's Harvest, Birchwood, WI. pp. 512.

Thompson, B. 1978. *Syrup trees.* Walnut Press, Fountain Hills, AZ. pp. 164.

Thompson, M., and Thompson, S. 1977. *Huckleberry country: Wild food plants of the Pacific Northwest.* Wilderness Press, Berkeley, CA. pp. 183.

Thompson, S., and Thompson, M. 1972. *Wild food plants of the Sierra.* Dragtooth Press, Berkeley, CA. pp. 186.

Tilford, G.L. 1997. *Edible and medicinal plants of the West.* Mountain Press Publishing, Missoula, MT. pp. 239.

Tomikel, J. 1976. *Edible wild plants of eastern United States and Canada.* Allegheny Press, California, PA. pp. 99.

Tomikel, J. 1986. *Edible wild plants and useful herbs.* Allegheny Press, Elgin, PA. pp. 176.

Tozer, F. 2006. *The uses of wild plants: Using and growing the wild plants of the United States and Canada.* Green Man Publishing, Santa Cruz, CA. pp. 263.

Tull, D. 1987. *Edible and useful plants of Texas and the southwest.* A practical guide. University of Texas Press, Austin, TX. pp. 517.

Turner, N.J. 1995. *Food plants of coastal First Peoples.* Royal British Columbia Provincial Museum, Victoria, BC. pp. 164.

Turner, N.J. 1997. *Food plants of interior First Peoples.* Royal British Columbia Provincial Museum, Victoria, BC. pp. 215.

Turner, N.J., and von Aderkas, P. 2012. Sustained by First Nations: European newcomers' use of indigenous plant foods in temperate North America. *Acta Soc. Bot. Pol.* 81: 295–315.

Turner, N.J., Łuczaj, Ł.J., Migliornini, P., Pieroni, A., Dreon, A.L. Sacchetti, L.E., and Paoletti, M.G. 2011. Edible and tended wild plants, traditional ecological knowledge and agroecology. *Crit. Rev. Plant Sci.* 30: 198–225.

Turner, N.J., and Szczawinski, A.F. 1978. *Wild coffee and tea substitutes of Canada.* National Museums of Canada, Ottawa, ON. pp. 111.

Turner, N.J., and Szczawinski, A.F. 1979. *Edible wild fruits and nuts of Canada.* National Museum of Natural Sciences, Ottawa, ON. pp. 212.

Underhill, J.E. 1974. *Wild berries of the Pacific Northwest: On the bush, on the table, in the glass.* Superior Publishing Company, Seattle, WA. pp. 128.

United States Department of the Army. 2009. *The complete guide to edible wild plants.* Skyhorse Publishing, New York, NY. pp. 143.

Van Atta, M. 1991. *Growing & using exotic food.* Pineapple Press, Sarasota, FL. pp. 180.

Walker, M. 1984. *Harvesting the northern wild: A guide to traditional and contemporary uses of edible forest plants of the Northwest Territories.* Outcrop, Yellowknife, NWT. pp. 224.

Walker, M. 2008. *Wild plants of eastern Canada: Identifying, harvesting and using.* Nimbus Publishing, Halifax, NS. pp. 203.

Watts, D., and Watts, A. 2007. *Where people feast. An indigenous people's cookbook.* Arsenal Pulp Press, Vancouver, BC. pp. 192.

Weatherbee, E.E., and Bruce, J.G. 1982. *Edible wild plants: A guide to collecting and cooking.* Second edition. E.E. Weatherbee & J.G. Bruce, Ann Arbor, MI. pp. 127.

Welsch, R.L. 2006. *Weed 'em and reap: A weed eater reader.* Falcon Guide, Guilford, CT. pp. 47.

Willard, T. 1992. *Edible and medicinal plants of the Rocky Mountains and neighbouring territories.* Wild Rose College of Natural Healing, Limited, Calgary, AB. pp. 277.

Williamson, D. 1986. *Wild foods of the desert cookbook.* Maverick Publications, Bend, OR. pp. 206.

Williamson, D. 1995. *The Rocky Mountain wild foods cookbook.* Caxton Printers, Caldwell, ID. pp. 251.

Wiltens, J.S. 1986. *Edible and poisonous plants of northern California.* Wilderness Press, Berkeley, CA. pp. 160.

Young, K. 1993. *Wild seasons: Gathering and cooking wild plants of the Great Plains.* University of Nebraska Press, Lincoln, NB. pp. 318.